Electronic Systems Technician

Level Four

Trainee Guide
Third Edition

 **With modules endorsed by SBCA and aligned to NICET.**

PEARSON

Boston   Columbus   Indianapolis   New York   San Francisco   Upper Saddle River
Amsterdam   Cape Town   Dubai   London   Madrid   Milan   Munich   Paris   Montreal   Toronto
Delhi   Mexico City   São Paulo   Sydney   Hong Kong   Seoul   Singapore   Taipei   Tokyo

National Center for Construction Education and Research
President: Don Whyte
Director of Product Development: Daniele Stacey
Electronic Systems Technician Project Manager: Matt Tischler
Production Manager: Tim Davis

Quality Assurance Coordinator: Debie Ness
Desktop Publishing Coordinator: James McKay
Production Specialist: Heather Griffith-Gatson
Editors: Chris Wilson, Debie Ness

Writing and development services provided by Topaz Publications, Liverpool, NY
Lead Writer/Project Manager: Thomas Burke
Desktop Publisher: Joanne Hart
Art Director: Megan Paye

Permissions Editors: Andrea LaBarge, Alison Richmond
Writers: Thomas Burke, Darrell Wilkerson,
  Gerald Shannon, Wayne Adair, John Tianen

Pearson Education, Inc.
Editorial Director: Vernon R. Anthony
Executive Editor: Alli Gentile
Senior Product Manager: Lori Cowen
Operations Supervisor: Deidra M. Skahill
Art Director: Jayne Conte
Director of Marketing: David Gesell
Executive Marketing Manager: Derril Trakalo
Marketing Manager: Brian Hoehl
Marketing Coordinator: Crystal Gonzalez

Composition: NCCER
Printer/Binder: LSC Communications
Cover Printer: LSC Communications
Cover Photo: © iStockphoto.com/Chris Fertnig
Text Fonts: Palatino and Univers

Credits and acknowledgments for content borrowed from other sources and reproduced, with permission, in this textbook appear at the end of each module.

26 2021

PEARSON

Perfect bound   ISBN-13: 978-0-13-257821-9
                 ISBN-10:   0-13-257821-2

Preface

To the Trainee

NCCER is pleased to present this revision to the *Electronic Systems Technician Level Four* curriculum. In addition to a new module, *Residential and Commercial Building Networks*, this revision has been updated in full-color and covers the most recent technologies and products available to ESTs. This edition also includes two separate training pathways for NCCER credentialing: one for voice and data and the other for life safety and security. Follow NCCER's two distinct course maps in the upcoming pages to determine which training path is best for you.

New with *Electronic Systems Technician Level Four*

In recognition of the value NICET (National Institute for Certification in Engineering Technologies) certification can bring, NCCER has ensured that the competencies covered in Module 33408-12, *Fire Alarm Systems*, support the skill and knowledge statements used as the basis for NICET's Fire Alarm Installer Certification Tests. This module addresses the core learning competencies as outlined by NICET, including the proper installation of system components, wiring methods, safety, and maintenance. NICET certification is a benchmark used in the fire alarm systems industry to measure technician qualifications in a growing number of states and jurisdictions.

Additionally, NCCER is proud to carry SBCA's (Satellite Broadcasting and Communications Association) endorsement of training in support of their Satellite Fundamentals, Home Theater Fundamentals, and MDU/SMATV certifications. Module 33403-12, *Broadband Systems* prepares candidates for the SBCA Satellite Fundamentals exam; Modules 33401-12, *Audio Systems*; 33402-12, *Video Systems* ; and 33404-12, *Media Management Systems*, prepare candidates for SBCA's Home Theater Fundamentals exam. NCCER's *Media Management Systems* module also prepares candidates for SBCA's MDU/SMATV certification exam. For more information on SBCA certifications, contact **sbca.com**.

We invite you to visit the NCCER website at **www.nccer.org** for information on the latest product releases and training, as well as online versions of the *Cornerstone* newsletter and Pearson's NCCER Curricula product catalog.

Your feedback is welcome. You may email your comments to **curriculum@nccer.org** or send general comments and inquiries to **info@nccer.org**.

NCCER Curricula

NCCER is a not-for-profit 501(c)(3) education foundation established in 1996 by the world's largest and most progressive construction companies and national construction associations. It was founded to address the severe workforce shortage facing the industry and to develop a standardized training process and curricula. Today, NCCER is supported by hundreds of leading construction and maintenance companies, manufacturers, and national associations. The NCCER Curricula was developed by NCCER in partnership with Pearson, the world's largest educational publisher.

Some features of the NCCER Curricula are as follows:

- An industry-proven record of success
- Curricula developed by the industry for the industry
- National standardization providing portability of learned job skills and educational credits
- Compliance with the Office of Apprenticeship requirements for related classroom training (*CFR 29:29*)
- Well-illustrated, up-to-date, and practical information

NCCER also maintains a National Registry that provides transcripts, certificates, and wallet cards to individuals who have successfully completed modules of NCCER's Curricula. *Training programs must be delivered by an NCCER Accredited Training Sponsor in order to receive these credentials.*

Special Features

In an effort to provide a comprehensive, user-friendly training resource, we have incorporated many different features for your use. Whether you are a visual or hands-on learner, this book will provide you with the proper tools to get started as an Electronic Systems Technician.

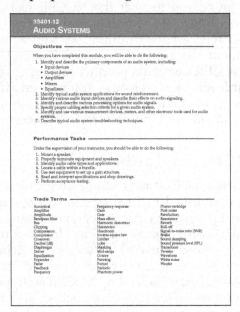

Introduction

This page is found at the beginning of each module and lists the Objectives, Performance Tasks, Trade Terms, and Required Trainee Materials for that module. The Objectives list the skills and knowledge you will need in order to complete the module successfully. The Performance Tasks give you an opportunity to apply your knowledge to the real-world duties that ESTs perform. The list of Trade Terms identifies important terms you will need to know by the end of the module. Required Trainee Materials list the materials and supplies needed for the module.

On Site

On Site features provide a head start for those entering the electronic systems field by presenting technical tips and professional practices from master technicians in a variety of disciplines. The On Site features often include real-life scenarios similar to those you might encounter on the job site.

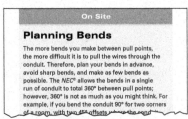

Color Illustrations and Photographs

Full-color illustrations and photographs are used throughout each module to provide vivid detail. These figures highlight important concepts from the text and provide clarity for complex instructions. Each figure reference is denoted in the text in *italic type* for easy reference.

Figure 2 Pushing down on the bender to complete the bend.

Notes, Cautions, and Warnings

Safety features are set off from the main text in highlighted boxes and are organized into three categories based on the potential danger of the issue being addressed. Notes simply provide additional information on the topic area. Cautions alert you of a danger that does not present potential injury but may cause damage to equipment. Warnings stress a potentially dangerous situation that may cause injury to you or a co-worker.

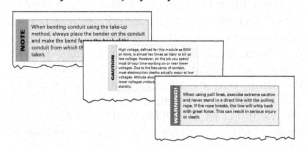

Going Green

Going Green looks at ways to preserve the environment, save energy, and make good choices regarding the health of the planet. Through the introduction of new construction practices and products, you will see how the "greening of America" has already taken root.

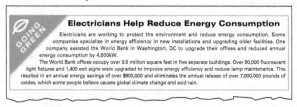

Case History

Case History features emphasize the importance of safety by citing examples of the costly (and often devastating) consequences of ignoring *National Electrical Code*® or OSHA regulations.

What's wrong with this picture?

What's wrong with this picture? features include photos of actual code violations for identification and encourage you to approach each installation with a critical eye.

What's wrong with this picture?

108SA14.EPS

Think About It

Think About It features use "What if?" questions to help you apply theory to real-world experiences and put your ideas into action.

Step-by-Step Instructions

Step-by-step instructions are used throughout to guide you through technical procedures and tasks from start to finish. These steps show you not only how to perform a task but also how to do it safely and efficiently.

Trade Terms

Each module presents a list of Trade Terms that are discussed within the text and defined in the Glossary at the end of the module. These terms are denoted in the text with blue bold type upon their first occurrence. To make searches for key information easier, a comprehensive Glossary of Trade Terms from all modules is located at the back of this book.

Review Questions

Review Questions are provided to reinforce the knowledge you have gained. This makes them a useful tool for measuring what you have learned.

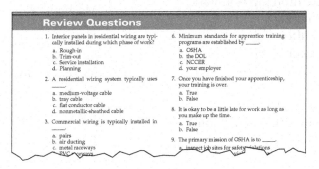

NCCER Curricula

NCCER's training programs comprise more than 80 construction, maintenance, pipeline, and utility areas and include skills assessments, safety training, and management education.

Boilermaking
Cabinetmaking
Carpentry
Concrete Finishing
Construction Craft Laborer
Construction Technology
Core Curriculum:
    Introductory Craft Skills
Drywall
Electrical
Electronic Systems Technician
Heating, Ventilating, and
    Air Conditioning
Heavy Equipment Operations
Highway/Heavy Construction
Hydroblasting
Industrial Coating and Lining
    Application Specialist
Industrial Maintenance
    Electrical and Instrumentation
    Technician
Industrial Maintenance
    Mechanic
Instrumentation
Insulating
Ironworking
Masonry
Millwright
Mobile Crane Operations
Painting
Painting, Industrial
Pipefitting
Pipelayer
Plumbing
Reinforcing Ironwork
Rigging
Scaffolding
Sheet Metal
Signal Person
Site Layout
Sprinkler Fitting
Tower Crane Operator
Welding

Green/Sustainable Construction

Building Auditor
Fundamentals of Weatherization
Introduction to Weatherization
Sustainable Construction
    Supervisor
Weatherization Crew Chief
Weatherization Technician
Your Role in the Green
    Environment

Energy

Alternative Energy
Introduction to the Power
    Industry
Introduction to Solar
    Photovoltaics
Introduction to Wind Energy
Power Industry Fundamentals
Power Generation Maintenance
    Electrician
Power Generation I&C
    Maintenance Technician
Power Generation Maintenance
    Mechanic
Power Line Worker: Distribution
Power Line Worker:
    Transmission
Solar Photovoltaic Systems
    Installer
Wind Turbine Maintenance
    Technician

Pipeline

Control Center Operations, Liquid
Corrosion Control
Electrical and Instrumentation
Field Operations, Liquid
Field Operations, Gas
Maintenance
Mechanical

Safety

Field Safety
Safety Orientation
Safety Technology

Management

Fundamentals of Crew Leadership
Project Management
Project Supervision

Supplemental Titles

Applied Construction Math
Careers in Construction
Tools for Success

Spanish Translations

Basic Rigging
    (Principios Básicos de
    Maniobras)
Carpentry Fundamentals
    (Introducción a la
    Carpintería, Nivel Uno)
Carpentry Forms
    (Formas para Carpintería,
    Nivel Trés)
Concrete Finishing, Level One
    (Acabado de Concreto,
    Nivel Uno)
Core Curriculum:
    Introductory Craft Skills
    (Currículo Básico:
    Habilidades Introductorias del
    Oficio)
Drywall, Level One
    (Paneles de Yeso, Nivel Uno)
Electrical, Level One
    (Electricidad, Nivel Uno)
Field Safety
    (Seguridad de Campo)
Insulating, Level One
    (Aislamiento, Nivel Uno)
Ironworking, Level One
    (Herrería, Nivel Uno)
Masonry, Level One
    (Albañilería, Nivel Uno)
Pipefitting, Level One
    (Instalación de Tubería
    Industrial, Nivel Uno)
Reinforcing Ironwork, Level One
    (Herreria de Refuerzo,
    Nivel Uno)
Safety Orientation
    (Orientación de Seguridad)
Scaffolding
    (Andamios)
Sprinkler Fitting, Level One
(Instalación de Rociadores,
Nivel Uno)

Acknowledgments

This curriculum was revised as a result of the farsightedness and leadership of the following sponsors:

ATI Career Training Center
B&D Industries, Inc.
Baltimore City Community College
Blonder Tongue Laboratories Inc.
DirecTV
Dish Network L.L.C.

Independent Electrical Contractors
Lincoln College of Technology
M.C. Dean, Inc.
Satellite Broadcasting & Communications
  Association

This curriculum would not exist were it not for the dedication and unselfish energy of those volunteers who served on the Authoring Team. A sincere thanks is extended to the following:

Wayne Adair
George Bish
Stephen Clare
Raymond Edwards
Dave Gilson

David Lettkeman
Clint Mayrant
Don Owens
Alton Smith

NCCER Partners

American Fire Sprinkler Association
Associated Builders and Contractors, Inc.
Associated General Contractors of America
Association for Career and Technical Education
Association for Skilled and Technical Sciences
Carolinas AGC, Inc.
Carolinas Electrical Contractors Association
Center for the Improvement of Construction
  Management and Processes
Construction Industry Institute
Construction Users Roundtable
Construction Workforce Development Center
Design Build Institute of America
Merit Contractors Association of Canada
Metal Building Manufacturers Association
NACE International
National Association of Minority Contractors
National Association of Women in Construction
National Insulation Association
National Ready Mixed Concrete Association
National Technical Honor Society
National Utility Contractors Association
NAWIC Education Foundation
North American Technician Excellence

Painting & Decorating Contractors of America
Portland Cement Association
Skills USA
Steel Erectors Association of America
The Manufacturers Institute
U.S. Army Corps of Engineers
University of Florida, M. E. Rinker School of
  Building Construction
Women Construction Owners & Executives, USA

Contents

Note: *NFPA 70*®, *National Electrical Code*®, and *NEC*® are registered trademarks of the National Fire Protection Association, Inc., Quincy, MA 02269. All *National Electrical Code*® and *NEC*® references in this textbook refer to the 2011 edition of the *National Electrical Code*®.

panel programming, testing, and troubleshooting. Explores integration of fire alarms with other systems. Examines both residential and commercial fire alarm applications, emphasizing *NEC®* requirements. (Module ID 33408-12; 40 Hours)

Module Nine

Overview of Nurse Call and Signaling Systems

Presents an overview of nurse call and signaling systems as found in hospitals and other healthcare facilities. Covers basic emergency call and duress system requirements based on facility type. Identifies installation requirements based on UL and other building code specifications. (Module ID 33409-12; 15 Hours)

Module Ten

CCTV Systems

Describes the installation and configuration of closed circuit TV systems for small, medium, and large facilities. Explains various equipment, including cameras, lenses, remote-positioning, video recording, and transmission. Covers the roles of the internet and digital technologies. Introduces test and troubleshooting equipment. (Module ID 33410-12; 30 Hours)

Module Eleven

Access Control Systems

Introduces access control systems, including applications, door locking systems, readers, biometrics, and controllers. Emphasizes installation practices as well as building and electrical codes. (Module ID 33411-12; 35 Hours)

Glossary

Index

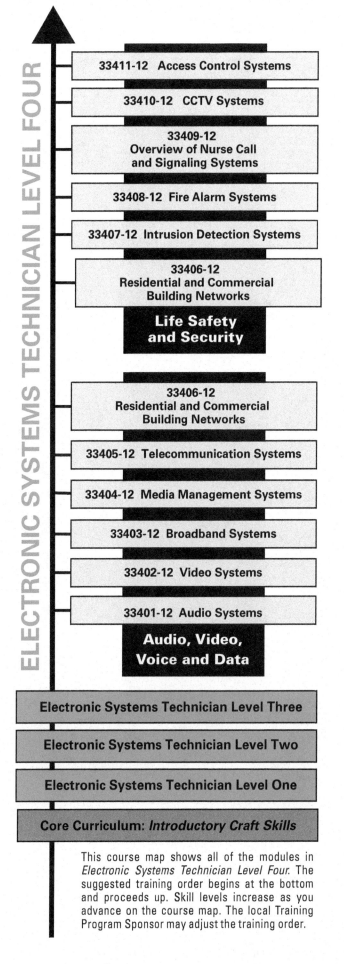

ELECTRONIC SYSTEMS TECHNICIAN LEVEL FOUR

33411-12   Access Control Systems

33410-12   CCTV Systems

33409-12
Overview of Nurse Call
and Signaling Systems

33408-12   Fire Alarm Systems

33407-12   Intrusion Detection Systems

33406-12
Residential and Commercial
Building Networks

Life Safety
and Security

33406-12
Residential and Commercial
Building Networks

33405-12   Telecommunication Systems

33404-12   Media Management Systems

33403-12   Broadband Systems

33402-12   Video Systems

33401-12   Audio Systems

Audio, Video,
Voice and Data

Electronic Systems Technician Level Three

Electronic Systems Technician Level Two

Electronic Systems Technician Level One

Core Curriculum: *Introductory Craft Skills*

This course map shows all of the modules in *Electronic Systems Technician Level Four*. The suggested training order begins at the bottom and proceeds up. Skill levels increase as you advance on the course map. The local Training Program Sponsor may adjust the training order.

33401-12

Audio Systems

Endorsed by the Satellite Broadcasting &
Communications Association (SBCA)

Module One

Trainees with successful module completions may be eligible for credentialing through NCCER's National Registry. To learn more, go to **www.nccer.org** or contact us at **1.888.622.3720**. Our website has information on the latest product releases and training, as well as online versions of our *Cornerstone* newsletter and Pearson's product catalog.

Your feedback is welcome. You may email your comments to **curriculum@nccer.org**, send general comments and inquiries to **info@nccer.org**, or fill in the User Update form at the back of this module.

Objectives

When you have completed this module, you will be able to do the following:

1. Identify and describe the primary components of an audio system, including:
   - Input devices
   - Output devices
   - Amplifiers
   - Mixers
   - Equalizers
2. Identify typical audio system applications for sound reinforcement.
3. Identify various audio input devices and describe their effects on audio signaling.
4. Identify and describe various processing options for audio signals.
5. Specify proper cabling selection criteria for a given audio system.
6. Identify and use various measurement devices, meters, and other electronic tools used for audio systems.
7. Describe typical audio system troubleshooting techniques.

Performance Tasks

Under the supervision of your instructor, you should be able to do the following:

1. Mount a speaker.
2. Properly terminate equipment and speakers.
3. Identify audio cable types and applications.
4. Locate a cable within a bundle.
5. Use test equipment to set up a gain structure.
6. Read and interpret specifications and shop drawings.
7. Perform acceptance testing.

Trade Terms

Acoustical
Amplifier
Amplitude
Bandpass filter
Bus
Clipping
Compression
Compressor
Crossover
Decibel (dB)
Diaphragm
Driver
Equalization
Expander
Fader
Feedback
Frequency

Frequency response
Gain
Gate
Haas effect
Harmonic distortion
Harmonics
Headroom
Inverse square law
Limiter
Lobe
Masking
Mid-range
Octave
Panning
Period
Periodic
Phantom power

Phono-cartridge
Pink noise
Rarefaction
Resonance
Reverb
Roll-off
Signal-to-noise ratio (SNR)
Snake
Sound damping
Sound pressure level (SPL)
Transducer
Tweeter
Waveform
White noise
Woofer

Industry Recognized Credentials

If you're training through an NCCER-accredited sponsor you may be eligible for credentials from NCCER's Registry. The ID number for this module is 33401-12. Note that this module may have been used in other NCCER curricula and may apply to other level completions. Contact NCCER's Registry at 888.622.3720 or go to nccer.org for more information.

Contents

Topics to be presented in this module include:

Contents (*continued*)

Figures and Tables

Figures and Tables (*continued*)

1.0.0 INTRODUCTION

This module will provide the electronic systems technician with the knowledge and skills needed to support the installation and troubleshooting of audio systems. It covers the components that make up an audio system. Basic audio theory will be presented, and the effects of these concepts on installed audio systems will be explained. This section begins by introducing some aspects of human hearing and then moves on to the types of audio systems that will be the focus of this module.

1.1.0 What is Sound?

At its most basic, sound is vibration. When something vibrates, it moves in two directions, forward and backward, up or down, or side to side. This back-and-forth movement is represented as a sine wave. When we speak, we cause our vocal chords to vibrate. The energy of this vibration travels through the air, moving air molecules in the process. When a wave of vibrating energy enters the ear, it causes the membrane of the eardrum to move (*Figure 1*). This energy then moves through a series of sensory organs in the inner ear. The signal is carried through the auditory nerve to the brain, and the vibration is perceived as sound.

When sound waves move through the air, they cause a small change in the air pressure along the path of the wave. These changes cause air mole-cules to move forward or backward, which in turn cause other air molecules to do the same. When a molecule moves forward, there is an increase in pressure, which is called compression. When the air molecule moves backward, there is a decrease in pressure, which is called rarefaction. Compression is caused by higher pressure, rarefaction by lower pressure.

In the sound wave shown in *Figure 2*, the portion of the wave above the zero point is the high pressure. Below the zero point is low pressure. The height of the wave above and below the zero point is called its amplitude. The amplitude of a sound wave is measured to determine the energy it carries. This is called the loudness or volume. The frequency of the wave is related to the concept of pitch; the higher the frequency, the higher the pitch.

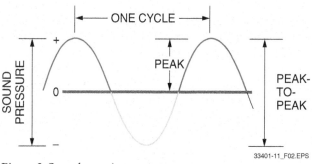

Figure 2 Sound as a sine wave.

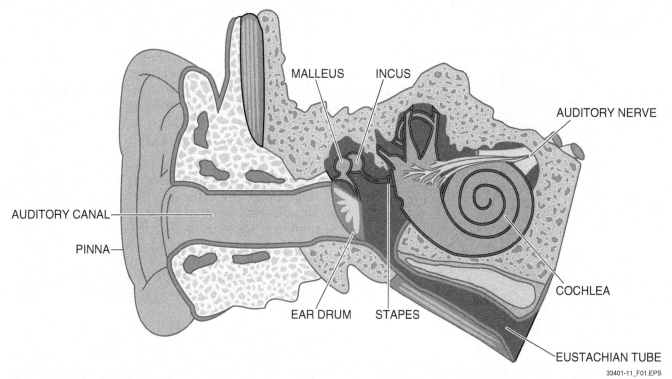

Figure 1 Cross-section of an ear.

1.2.0 Volume, Pressure, and Speed

The loudness of a sound wave is directly related to the pressure of the wave, or the sound pressure as it reaches the ear. The loudness of a sound wave is measured in decibels (dB). A decibel is one tenth of a bel, the unit of measurement used for sound intensity. The bel is named after Alexander Graham Bell. Human hearing is sensitive to sounds from zero decibels (the threshold of hearing) up to 120dB and greater (*Figure 3*). However, sounds reaching 120dB or greater can injure the ear, causing permanent damage. When the loudness of a sound wave is measured in decibels, the result is referred to as the sound pressure level (SPL).

As a general rule, the speed of sound is approximately 770 miles per hour, 12.8 miles per minute, or 1,130 feet per second. A number of factors affect the speed of sound. As seen earlier, sound involves the interaction of air molecules as they transfer energy from one to another. The temperature of the air or the amount of humidity can cause minute changes in the overall speed at which a sound wave travels. Wind or other sound waves can also affect the speed. The various speeds of sound measurements quoted here are based on a temperature of 70°F, with low relative humidity.

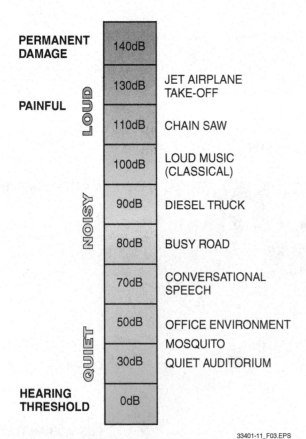

Figure 3 Sound pressure levels for common sounds.

PERMANENT DAMAGE — 140dB

PAINFUL — LOUD

130dB — JET AIRPLANE TAKE-OFF

110dB — CHAIN SAW

100dB — LOUD MUSIC (CLASSICAL)

NOISY

90dB — DIESEL TRUCK

80dB — BUSY ROAD

70dB — CONVERSATIONAL SPEECH

QUIET

50dB — OFFICE ENVIRONMENT

MOSQUITO

30dB — QUIET AUDITORIUM

HEARING THRESHOLD — 0dB

33401-11_F03.EPS

Most high-rise buildings of more than seven stories have voice notification integrated with their fire alarm system in order to deliver instructions to occupants in the event of a fire. Some use recorded messages, while others use live audio to direct evacuation. When such systems are used, they must meet the requirements of *NFPA 72*®, *National Fire Alarm and Signaling Code*. These requirements are specified in *NFPA 72, Chapters 18 and 24*.

1.2.1 Frequency and Waveforms

The frequency of a sound, or the intervals between the peaks of a sound wave, will not affect the speed of the wave. It can affect the distance that a sound wave travels, however. Frequency determines the rate with which the sound wave propagates or travels through the air. For example, a low-frequency sound has a greater distance between the peaks and will travel farther than a high-frequency wave that has smaller distances between the peaks. The distance between peaks is called the period of the wave. Periodic waves are repeating waves.

The frequency of a sound wave is measured in hertz (Hz) and represents the number of cycles per second. Human hearing is sensitive to sounds generally between 20Hz and 20,000Hz (20kHz). Sounds below the threshold are called subsonic, while those above the top threshold are called ultrasonic.

Although sound is described as a sine wave, it is really a composite of multiple sine waves. A pure sine wave is a single frequency, and sound is comprised of multiple frequencies. Usually, a single frequency forms the basis of the sound, with additional frequencies forming the harmonics. These component frequencies of a complex sound wave are called the waveform.

1.2.2 Loudness Contours

A person has a general sensitivity to sounds with frequencies between 80Hz and 10kHz. It is known that sound pressure determines the perceived and physical loudness of a sound wave. The human ear, however, does not perceive all sounds at the same sound pressure level to be of equal loudness. It actually takes more sound pressure for low frequencies or high frequencies to be perceived as being equal to the loudness of a 1kHz signal at a given SPL.

This phenomena was first researched in depth back in the 1930s by two researchers at Bell Labs and is named after them: the Fletcher-Munson equal loudness contours. Other researchers have extended their work, and the diagram in *Figure 4* shows similar results.

Notice that the bottom contour shows that almost 80dB of sound pressure is required for a 20Hz sound to be considered equally loud to a 1kHz sound at 10dB. For all of the contours, there is a reduction in required SPL at around 4.5kHz. This means that the frequency response of the human ear is not flat. It changes at various frequencies. Numerous implications of this effect will be explored throughout this module.

1.2.3 Speech Intelligibility

Speech intelligibility is used to describe several important aspects of sound reinforcement. A speaking voice must have sufficient clarity to be understood. Human voice produces a continuous waveform with primary frequencies in the range of about 100Hz to 400Hz. Male voices are typi-

cally in the 100Hz range, and female voices in the 200Hz range.

Typically, the biggest problem is the obstruction of speech clarity that results from unwanted sounds. These sounds are called masking sounds, and they interfere with the ability of the ear to discriminate between the elements of a vocalized signal. Imagine that you are listening to the radio in your car, and after a song finishes playing, you are unable to make out the name of the artist. The announcer's voice is overwhelmed by road noise, air conditioning, and even other people who are speaking in the vehicle. This is masking.

The power of a vocalization is carried by the vowels. They have a relatively long duration, on the average of 30ms to 300ms. However, it is the consonants with durations of around 10ms to 100ms that truly enable intelligibility of the speech. The amplitude of consonants, however, can be as much as 27dB lower than the vowels.

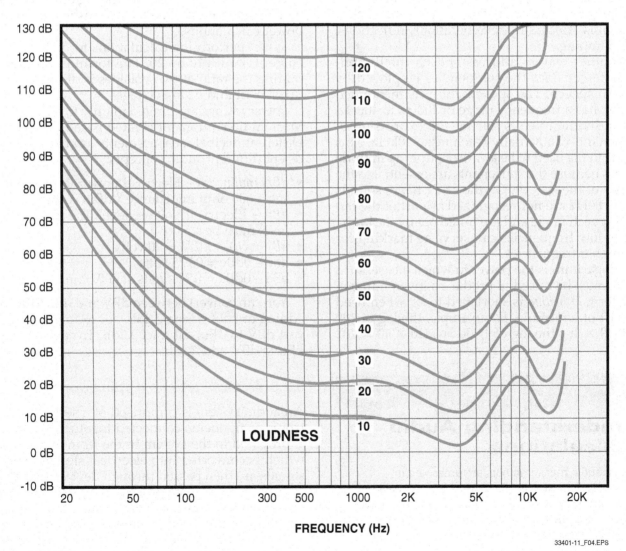

Figure 4 Equal loudness contours.

33401-11_F04.EPS

 33401-12 Audio Systems

Masking has a greater effect on speech as it overwhelms the sound of consonants.

One way to deal with the need for speech intelligibility is to address the signal-to-noise ratio (SNR). Obviously, it should be higher than 0, but how much higher is determined by the type of noise and its frequency spectrum. Typically, the speech signal should be at least 12dB above the noise. Even with this rule, recognition of only eighty percent of the individual words is considered acceptable

Experts have developed a number of tests for determining the intelligibility of speech. Many of them are empirical; that is, they are based on an applied methodology, but subject to the individual space of a given room. These tests involve humans actually speaking and listening within the environment. Word lists and instructions are spoken, and human subjects placed within the environment make subjective evaluations based on the clarity. These kinds of test are crucial, particularly because each room or position within a room is affected by reverberations, reflections, and damping.

Another way to address speech intelligibility is to consider the frequencies employed by human speech. The spoken voice, as mentioned earlier, has a frequency range of 100Hz to 400Hz, but individual peaks during the expression of consonants can actually reach up to 10kHz. This points to the need to be aware of clipping in those ranges because the consonants are essential to the clarity of the speech. On the other hand, sounds below 80Hz are not recognized as part of the human speech range and are tuned out by the brain. These low frequencies can serve as masking for the higher frequencies.

A logarithm is the power to which a base, such as 10, must be raised to produce a given number. The term *logarithm* is shortened to *log* when used as part of a formula. For example, the log of 100 is 2 (10 × 10); the log of 1,000 is 3 (10 × 10 × 10).

When working with audio calculations, always assume the base is 10.

It is important to recognize that decibels represent a specific ratio: the ratio of two different wattages, the ratio of two different voltages, or the ratio of two different sound pressure levels. It is also important to recognize that the ratio is expressed as a relative difference; the ratio is relative to the specific thing being measured—voltage, wattage, or sound pressure.

Decibels for sound pressure levels are mathematically expressed with the equation:

$$dB = 10 \times \log (P1/P2)$$

P1 and P2 are power values represented in watts. A watt is a measurement of power. It is defined as 1 joule per second. A joule is a measurement of energy. A watt, therefore, is the rate of energy transfer. This is the same as saying a watt is the rate of doing work.

Notice in the formula that the log of P1 divided by P2 is then multiplied by 10. This is true for all power calculations.

When performing calculations for voltage and current, multiply by 20. The formula for determining the ratio of two voltages is dB = 20 × Log (E1/E2). E1 and E2 are voltages.

Power ratings are limited mainly to amplifiers. Most audio equipment is rated based on voltage. Different decibel measurements are used to represent different audio concepts:

- *dBm (milliwatt)* – Electrical power level, assuming a 600 ohm impedance, 0dBm = 1mw
- *dBu* – Electrical voltage level, the impedance level must be stated
- *dBV (volt)* – 0dBV = 1.0 volts
- *dBv* – 0dBv = 0.775 volts
- *dBu* – 0dBu = 0.775 volts at 600 ohms

You can convert between dBV and dBu (dBv) by adding or subtracting 2.2dB. For example, to convert dBV to dBu, subtract 2.2dB. To convert from dBu to dBV, add 2.2dB.

1.3.0 Audio Systems Applications

An audio system is comprised of a series of electronic components connected together. An input is provided to the system in the form of a sound, which is converted into electrical signals. These signals are then passed through devices that alter or process them. The processed signals are then passed on to an output device, which converts the electrical energy back into mechanical energy.

Audio inputs are easily placed into three categories: magnetic, optical, and electrical. Magnetic inputs include phono-cartridges and tape recorders. Optical inputs include CD players, and electrical inputs include microphones. Microphones are used for live inputs, while the others are used for pre-recorded material. Likewise, outputs fall into similar categories.

Processing involves a number of signal enhancements and transformations. Audio processing includes amplification and pre-amplification, equalization, sound damping, and feedback elimination.

In this module, applications for sound systems as are viewed as sound reinforcement systems. These can be further divided into public address systems and musical systems. The focus is primarily on public address systems, but aspects of musical reinforcement are also addressed.

1.3.1 Public Address Systems

Public address (PA) systems are sound systems designed to support the communication process between speaker and listener. Public address systems are found in a variety of environments such as stadiums and sports arenas; theaters; and transportation malls, such as train stations, subway platforms, and airports, to name a few.

As a sound reinforcement system, public address applications are designed around the elements of the environment. These include physical structures such as walls, ceilings, and building supports. They also take into account sound levels from reflected sounds and ambient noise levels from people and machinery, and even consider where sound should not be heard.

Public address systems can be installed in worship halls, hospitals, and schools. Often the location determines the use or application of the system. For example, in a worship hall, both speech and music are reinforced by the system. In a hospital, the public address system may be limited to providing announcements and paging. In a school, a PA system may provide two-way communication.

1.3.2 Music Systems

Public address systems often serve double duty. They are used to carry voice and speech, but may

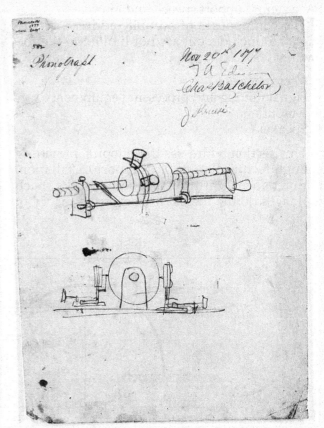

On Site

Early Sound Reinforcement

In the early 1870s, Thomas Edison began working on the recording and replay of telegraph signals. Messages were normally keyed using the Morse code of dots and dashes. Edison theorized that if he could record a message, he could then replay it very quickly over the wire. Incoming messages could also be recorded at faster rates and then transposed after the fact. Early designs of the phonograph never showed this purpose, however.

33401-11_SA01.EPS

also be used to provide for the reinforcement of music as well. PA systems can be used for musical performances or simply to provide background music. Musical systems are sound systems designed specifically for the recording and presentation of music. Music has a complex series of waveforms generated by individual instruments and voices.

For example, a recording studio must have a much greater control over environmental factors. A studio will generally have some degree of acoustical treatment. These could be anything from non-reflective materials to baffled walls that dampen specific frequencies.

A musical system is a highly refined sound reinforcement system. Modern studios have a greater reliance on digital effects processing. They use more sensitive equipment. Musical systems are often used where there is a need to separate sounds, such as with stereo, quadraphonic, or surround-sound applications.

2.0.0 ELEMENTS OF A SOUND REINFORCEMENT SYSTEM

This section identifies the individual components that make up a sound reinforcement system. Remember that the focus is on PA systems and that PA systems support speech and music.

The components of a public address system (*Figure 5*) include, but are not limited to the following:

* Microphones
* Signal mixing and processing equipment
* Amplifiers
* Speakers

This section addresses these topics, providing enough technical detail to support installation, maintenance, and troubleshooting tasks. Each topic will define the device, cover the theory of operation, provide distinctions between various technologies available, and provide an overview of the types of problems introduced into the system from their use.

2.1.0 Microphone Basics

Microphones are the most basic input for a sound reinforcement system. Public address systems rely primarily on microphones for input. Microphones are complex devices, and their characteristics greatly influence the quality of the sound being reinforced.

The purpose of a microphone is to take incoming sound and convert it from mechanical energy into an electrical signal. This section presents basic microphone theory and operation. You will learn how a microphone works, how to recognize various types of microphones, and how to select the proper microphone for a sound reinforcement system.

At its most basic, a microphone converts incoming sound into an electrical current. There are several ways of accomplishing this. Each method is often associated with a particular type of microphone, as described later in this module.

Incoming sound is a form of energy manifest as air pressure. Every microphone uses a means for converting the energy of fluctuating air pressures to create an electrical signal equivalent in amplitude and frequency to the incoming sound wave. The most basic microphone consists of a membrane and a positively charged field. As the membrane reacts to air pressure changes, it is stretched and contracted using the principles of compression and rarefaction. This in turn causes fluctuations in the charged field, which is transferred as voltage.

A microphone is a transducer. Transducers convert energy from one form to another. The accuracy with which sound energy is converted

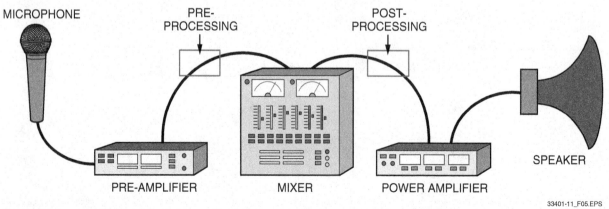

Figure 5 Elements of a sound system.

The First Public Address System

to electrical energy is determined by the method used by the microphone.

All microphones have a diaphragm and a generating element. The diaphragm reacts to sound pressure by producing vibrations. The generating element converts those vibrations into electrical voltage. Two common methods are used. A dynamic microphone uses a wire coil and a magnet. The other method operates like a capacitor and is used with a condenser microphone.

Dynamic microphones have a small magnet that surrounds a coil of wire, as shown in *Figure 6*. A thin diaphragm or membrane overlay is placed on top of the wire coil with a small air space in between. As acoustical energy strikes the diaphragm, it vibrates. These vibrations result in minute changes of air pressure between the membrane and the wire coil, causing the coil to move. As the coil moves back and forth within the magnetic field, a small electrical current is induced in the wire.

Condenser microphones use a small electrical current to charge a gold-plated ceramic backplate (*Figure 7*). A membrane or diaphragm, also gold-coated, is placed above the backplate with a small air space in between. As vibrations strike the membrane, causing it to move, minute changes in capacitance are created. These changes cause small fluctuations in the electrical current.

Another type of condenser microphone, the electret, operates basically the same way. The difference is that the backplate of an electret is positively charged at the factory. The electret microphone retains this charge throughout its life. Some electret microphones apply the positive charge to the diaphragm itself, but this requires a thicker membrane, cutting down on the responsiveness of the microphone.

A condenser microphone requires voltage to operate, whereas a dynamic microphone generates voltage from its mechanical operation. Both types of microphones require preamplifiers to boost their signals. Condenser microphones typically have a low-voltage power source of 9V to 48VDC. This is often supplied from a small battery enclosed in the microphone housing, though some draw what is called phantom power. Phan-

tom power is provided to a microphone from an audio mixer or other source, using the same wires that carry the electrical signals from the microphone to the mixer.

There are two additional methods for describing microphone operation: the ribbon microphone and the piezoelectric. Ribbon microphones operate similar to dynamic microphones. A very thin metal ribbon is stretched between two poles of a magnet. As the ribbon vibrates, it induces voltages. These microphones are fragile, though, and are rarely used outside of studio environments,

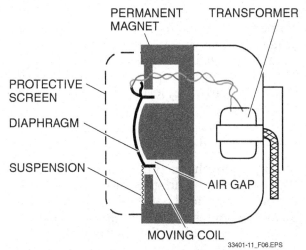

Figure 6 Cross-section of a dynamic microphone.

Invention of Audio Tape Recorders

In 1898, a Swedish telephone worker, Vlademar Poulsen, invented a process for storing the human voice using magnetic fields. Called the telegraphone, the device moved like a trolley along a stretched piano wire and applied magnetic pulses. Another device could then pick up the pulses and replay them. It was the first magnetic recording machine. Later, flat ribbon wound on spools would replace the wire.

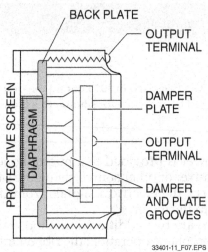

Figure 7 Cross-section of a condenser microphone.

- *Handheld microphones* – Handheld microphones (*Figure 9*) are mounted in slender tubes for easy handling. They can be affixed to a lectern or pole-stand. The tube also acts to isolate the sound element from extraneous vibrations caused by handling and movement. They can be either dynamic or condenser types.
- *Shotgun microphones* – Shotgun microphones (*Figure 10*) are long slender tubes that allow the microphone to be directed at a specific source. Sometimes called line gradient microphones, they are used frequently in theater and broadcasting applications.
- *Parabolic microphones* – Parabolic microphones (*Figure 11*) are often seen at sporting events. They consist of a microphone with a curved (parabolic) hood. They are directed at specific locations to gather sound from a distance.

where their rich enhancement of low frequencies makes them highly desirable. They are sometimes referred to as velocity microphones.

A piezoelectric microphone (*Figure 8*) is sometimes called a ceramic mic. A small crystal is connected to a diaphragm. As the diaphragm vibrates, it causes the crystal to deform, resulting in a small voltage being generated from the crystal. These types of microphones are usually considered to be fairly fragile as well, and are generally used for contact-type pickup.

2.2.0 Microphone Classifications

So far, microphone types have been based on their diaphragm and generating elements. There are other ways to classify microphones, however, including how they are mounted and how the microphone responds in the environment.

2.2.1 Mounting Styles of Microphones

The easiest way to describe microphones is by the way they are mounted for use in the sound reinforcement or public address system. The following are typical mounting classifications:

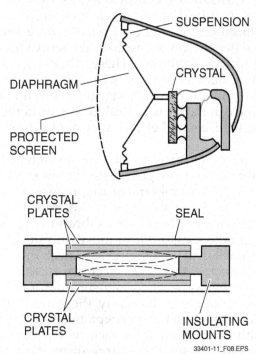

Figure 8 Cross-section of a piezoelectric microphone.

©2003 SHURE INC. USED BY PERMISSION.

33401-11_F09.EPS

Figure 9 Handheld microphones.

Most of the microphones shown in this section use cables to connect them to the audio system. Keep in mind that, like the lavalier microphone, each of them can be used with wireless transceivers such as belt-packs or other wireless connections.

Figure 11 Parabolic microphone.

33401-11_F11.EPS

- *Lavalier microphones* – Lavalier microphones (*Figure 12*) are of the electret family. They are designed to be attached to clothing or suspended from a lanyard.
- *Contact-pickup microphones* – Contact-pickup microphones, like the Korg CM100L shown in *Figure 13*, are a special type of microphone that is attached directly to a musical instrument or some other analog sound source. They are designed to pick up vibrations from solid materials, whereas standard microphones receive their acoustic input through the air. They can be used effectively in live musical performances.
- *Pressure-response microphones* – Pressure-response microphones (*Figure 14*), sometimes called boundary effect mics, are flat and are attached to a glass plate, table top, or other flat surface.

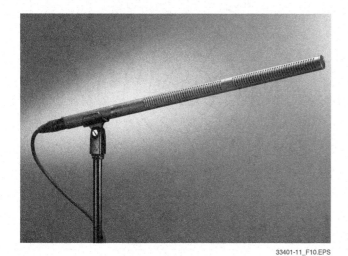

33401-11_F10.EPS

Figure 10 Shotgun microphone.

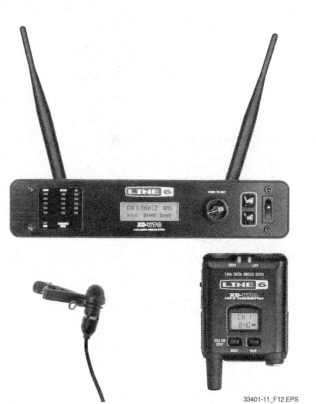

33401-11_F12.EPS

Figure 12 Wireless lavalier microphone and receiver.

2.2.2 Microphone Pickup Patterns

Another important means for classifying microphones is the way a transducer responds to sound coming from different directions. Every type of microphone that has been examined to this point has a particular pickup pattern. Though not directly related to the transducer element, certain pickup patterns can be attributed to various microphone types.

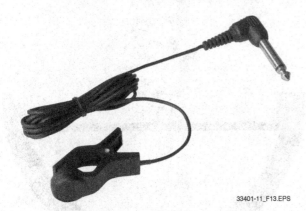

Figure 13 Contact-pickup microphone.

33401-11_F14.EPS

Figure 14 Pressure-response microphone.

The sensitivity of a microphone to sounds coming from different angles determines its directionality. This directionality is depicted in special plots that reflect the microphone's pickup pattern. The pickup pattern of a microphone has a three-dimensional character.

Microphones are usually classified as being either omni-directional or uni-directional. Omni-directional microphones receive their acoustic energy from all directions. Uni-directional microphones receive their sound from a single direction.

An omni-directional microphone will pick up sounds equally from anywhere within a 360-degree circle. *Figure 15* shows an omni-directional pickup pattern. These microphones are used where all-around sound pickup is needed, such as symphony and stage productions.

Shotgun microphones are uni-directional. A sound dampening baffle eliminates sounds from all directions except the one in which the microphone is pointing. A pressure-response mic, on the other hand, is omni-directional. It picks up sound in all directions above the mounting plane.

There are varying degrees of directionality. For example, a bi-directional microphone has sensitivity zones, or lobes, in the front and the back. It does not pick up sounds coming from the sides. This makes them useful for spot applications where only two people or instruments need to be heard.

The most prevalent of all microphone pickup patterns is the cardioid. As the name suggests, it is heart-shaped (*Figure 16*). A cardioid is a semi-directional microphone. It picks up sound primarily from the front, but will pick up sounds

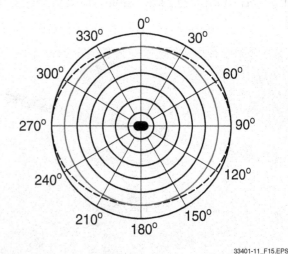

33401-11_F15.EPS

Figure 15 Omni-directional pickup pattern.

from the rear and sides with decreasing sensitivity. Cardioid microphones are relatively directional, making them useful for stage and theater applications where specific areas need additional sound reinforcement.

There are other patterns, such as the supercardioid and the hypercardioid. These microphone patterns are cardioids in shape, but with some back lobe projection. That is, they are very directional, eliminating sounds from the side, but picking up sound from behind.

When viewing typical pickup patterns, you will notice several things. First, the pattern plots are represented in two dimensions. Keep in mind that actual pickup patterns are three-dimensional. When reviewing manufacturer plots, scales are sometimes calibrated using decibels.

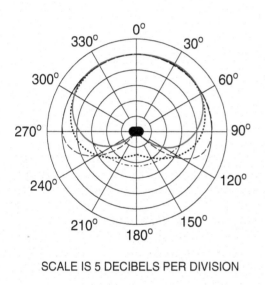

SCALE IS 5 DECIBELS PER DIVISION

LEGEND

200 Hs –··–··–··–·
1KHs ——————
6KHs ················
8KHs – – – – – –

33401-11_F16.EPS

Figure 16 A cardioid-shaped directional pickup pattern.

Decibels are a measurement for the loudness of an audio signal. This is important to keep in mind as microphone pickup patterns are examined. Decibels represent a measurement of the ratio between the loudness of the input sound and distance from the microphone. One of the problems, particularly with vocal speech, is that the distance between the speaker and the microphone can constantly change. Pickup patterns show these changes relative to the microphone as the sound source rotates in front of it.

Some pickup patterns will show microphone response at various frequencies. The human voice constantly changes in frequency. Vowels have a lower frequency than many consonants. Studies have shown that aspects of speech are affected by numerous factors. For example, emotional effects on speech can cause higher pitch or frequencies when expressing joy and lower pitches when expressing sorrow.

2.2.3 Selecting Microphones for Public Address

Microphones are available with either high- or low-output impedance. High-impedance devices are most commonly found where short cable lengths and sound quality are not critical. To use a high-impedance microphone with longer cable lengths may require a matching transformer to maintain quality. Low-impedance microphones can maintain superior sound quality over relatively long cable runs.

2.3.0 Speakers

Speakers, like microphones, operate using a simple set of concepts and principles. This section presents basic speaker theory and operation. You will learn how a speaker works, how to recognize various types of speakers, and how to select the proper speaker for a sound reinforcement system.

At its most basic, a speaker converts incoming electrical current into sound. A speaker is a transducer that converts electrical energy into acoustic energy. Speakers are sometimes called loudspeakers or, more precisely, radiators. A speaker consists of a driver, an electromagnetic device that converts electrical signals into mechanical energy. Speakers can also use a piezoelectric transducer. For sound reinforcement of public address, the electromagnetic approach is the most common.

A driver for a speaker consists of a voice coil placed in the gap of a permanent magnet (*Figure 17*). Current is applied to the coil, creating an electromagnetic field. This field interacts with the field of the permanent magnet, causing the coil to move. When applying direct current, the polarity or direction of current flow will determine the direction the coil travels. If the applied current is AC, then the coil will move in two directions as the current phase shifts from positive to negative. In either case, the coil is attached to a diaphragm. As the coils moves, it causes the diaphragm to vibrate, generating sound.

As noted in the discussion of piezoelectric microphones, certain crystals generate electricity when force is applied. This effect is reversed for piezoelectric speakers. Applying an electrical current to these crystals causes them to bend. When connected to a diaphragm, this bending causes vibrations to radiate across the surface. *Figure 18* provides a cross-sectional view of a piezoelectric speaker driver. Piezoelectric drivers have a fairly narrow range of efficiency and are generally used for high frequencies only.

2.3.1 Understanding Frequency Response

An important measurement for understanding speaker efficiency is the frequency response. The frequency response of a speaker is the ratio

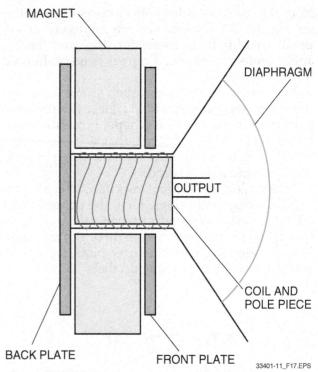

Figure 17 Cross-section of a speaker driver.

between the input power level and the output sound pressure. This measurement is used to determine how accurately and efficiently a speaker reproduces a waveform. It includes the amplitude response, or how much the frequency actually varies within the range established by the ratio. *Figure 19* shows the frequency response curve for a speaker driver. It is not an actual plot, but rather a stylized representation of a specific type of graph (a log graph) that is typical of frequency response graphs for speaker drivers.

Frequency response describes the range of frequencies that are accurately reproduced by a driver. When viewing plots or graphs, it can be

On Site

The First Phonograph Speaker

The phonograph was originally invented by Thomas Edison to improve telegraph transmission rates. Edison came to believe that it could also be used to send telegraph messages using human voice. This would eliminate the need for Morse code. His first phonograph used a single transducer, where the speaker and the microphone were the same device.

33401-11_SA02.EPS

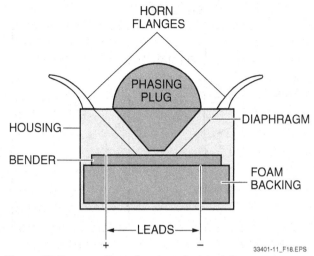

Figure 18 Cross-section of a piezoelectric driver.

seen that there is a baseline or reference level for the driver's frequency response. Though not always explicitly labeled, it is the fairly flat line between the two key frequencies (low and high). This establishes the middle area where most frequencies are reproduced.

If the sound pressure changes to a degree above or below the deviation, then the sound is clipped, or flattened. This results in a square wave instead of the usual sine wave. It also signifies that the sound is not accurately reproduced. For certain frequencies, this is not a problem. For others, significant and noticeable distortion occurs.

At either end of the plot, there is a point where the audible frequencies quickly fade away. This is called roll-off and becomes important when considering the range of sound frequencies that are produced by various types of drivers. For example, with a roll-off at 30Hz, any sound frequency below that threshold would be lost. Also, most sounds are not a single frequency. For example, if a bass sound radiated by a woofer (which has a

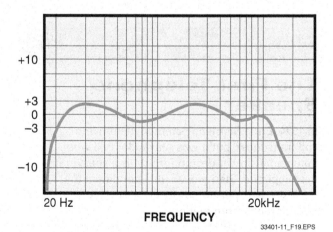

Figure 19 Plot of a frequency response curve.

frequency response in the 20Hz to 200Hz range) has a high-frequency element, it may be lost at the high-frequency roll-off.

2.3.2 Speaker Systems

A driver is the speaker component that translates electrical signals into sound signals. Unfortunately, drivers do not always reproduce all of the sounds in the frequency spectrum of human hearing. Various drivers produce different frequency response curves. Frequency roll-off occurs at different frequencies based on the driver being used. For this reason, loudspeakers are often systems consisting of multiple drivers. Each driver has its own diaphragm or cone, and is tuned to a specific set of frequencies.

A crossover is a frequency-dividing network. It separates the frequencies at their roll-off points and directs them to the appropriate driver in a speaker system. The roll-off portion of the frequency response curve is called the slope. The slope determines the rate of the drop-off. Drop rates usually occur at octave boundaries—6, 12, 18, and 24dB. A 6dB slope may allow too much overlap between drivers, so the most common crossover slopes are 18dB and 24dB.

Two common crossover types are used in sound reinforcement. The passive network crossover is designed to pass high signal levels. These crossovers are placed between the power amplifier output and the driver. Active network crossovers are inserted into the signal before it reaches the power amplifier.

Speaker systems can consist of two or more drivers. In some cases they are packaged in the same enclosure. They can also be housed separately and cabled together. Either way, they are called full-range speakers. Some full-range speakers consist of a single driver. These typically have poor performance at the extremes of the spectrum and don't generate the high-level sounds needed by a sound reinforcement system.

2.4.0 Classification of Speaker Types

Speakers are classified according to their frequency response: low-range, mid-range, and high-range drivers (20Hz to 500Hz, 500Hz to 2kHz, and 2kHz to 16kHz, respectively). These are generalizations, and the specifications for any actual speaker system vary across the frequency spectrum. Manufacturers and designers strive to enrich sounds within these spectral limits. With this in mind, speaker systems can be classified by popular names: woofer, mid-range, compression driver (tweeter), and horn.

2.4.1 Woofers and Subwoofers

A woofer is a speaker system with a low-frequency driver. It uses a cone-shaped diaphragm and is almost always enclosed in a box. Several key designs enhance its low-frequency output.

Low-frequency sound waves have a greater tendency to cancel themselves out than higher frequencies. During the first half of the sine wave, the cone is moved forward, causing rarefaction behind it. During the second half of the sine wave, the opposite occurs. Because the frequency waves are large with little directionality, the reverse compression cycle can cancel the first. Enclosures are designed to negate this effect.

Woofers are often packaged in one of three types of enclosures: vented, ducted, or sealed. A sealed cabinet is air-tight; therefore, during the compression-rarefaction cycle, no air within the enclosure moves. The sound waves are radiated outward. Sealed enclosures usually contain some form of batting, such as fiberglass, which absorbs high frequencies. In some cases, a single driver and an additional cone are mounted in a sealed enclosure. The driver moves air in the cabinet, and that in turn moves the second cone, which is mounted facing out of the enclosure.

Vented and ducted enclosures use the properties of air movement to enhance their performance. *Figure 20* shows an example of a vented speaker enclosure. With a vented enclosure, the driver is mounted on the surface of the cabinet. A vent or port provides for air movement within the enclosure, reinforcing the resonance of the front part of the wave.

A ducted enclosure has a type of baffle at the port and is called a ducted port. This serves to lower the resonant frequency of the sound. This extends the low frequency response of the enclosure, particularly smaller cabinets.

Woofers are generally omni-directional. By using a horn design, these low-frequency radiators become more directional. As frequencies move up the spectrum, their associated waveforms become shorter, reducing the wavelength of the sound. For a direct radiator, the driver size determines the low frequencies and wave size. With a horn enclosure, the horn itself determines the directional qualities. Unfortunately, the size of the horn also limits the low-frequency response. Several horns can be coupled together as an array, however, to extend the low-frequency response of the system.

A woofer reproduces sounds generally below 500Hz. A subwoofer operates in the 20Hz to 80Hz range. Sounds below 20Hz typically are not heard, but can be felt as vibration.

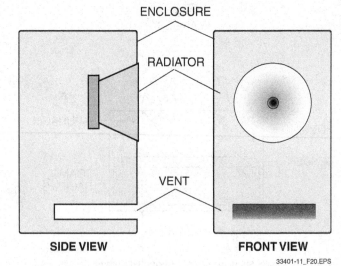

Figure 20 Vented enclosure for a woofer.

2.4.2 Mid-Range Drivers

Mid-range drivers operate using the simple fact that the larger the magnet, the more it stops motion of the cone. Mid-range frequencies have shorter wavelengths. By using a larger magnet, the pole piece can be prevented from striking the diaphragm. There is greater roll-off on the low frequencies, the bass output actually remains constant, but the output at mid-frequencies rises quickly. Mid-level frequencies have greater directionality than low frequencies.

A mid-range driver uses a dome-shaped diaphragm rather than the cone shape used in low-frequency drivers. This serves to provide a better pulse response, and domes are lighter than cones. Mid-range frequencies are shorter and therefore do not move the diaphragm as readily. Mid-range speakers are enclosed in cabinets similar to woofers, though they will generally use baffling to a greater extent.

2.4.3 Compression Drivers (Tweeters)

Compression drivers, or tweeters, are high-frequency drivers. They are usually small compared to their counterparts, the woofer and mid-range. They use a domed diaphragm and are often enclosed in horns. A tweeter may use a piezoelectric driver. The frequency response of a tweeter is in the neighborhood of 3kHz to 12kHz.

2.4.4 Horns

Horn drivers are a type of speaker system consisting of a driver and a horn-shaped or fluted enclosure. *Figure 21* shows that a typical horn speaker has a driver and an enclosure with a narrow channel that flares out. This allows the driver to be much smaller as the sound waves achieve a higher level of sound pressure before leaving the mouth of the horn. Horn drivers are highly directional.

Not all horns have the distinctively visible flute. Enclosures for horn-type speakers can be used to fold the horn portion in such a way that they will fit into traditional rectangular cabinets. This is particularly true of low-frequency horns. In fact, horn-type speakers are available for all frequencies: low, mid-level, and high.

2.5.0 Processing Equipment

A typical sound system consists of three key elements: microphones and other input devices, speakers and other outputs, and processing equipment. Processing equipment allows the various aspects of sound reinforcement to be controlled and manipulated before being reproduced at the various outputs.

2.5.1 Microphone Mixer

A microphone mixer is a simple mixer. It takes input from one or more microphones and allows the operator to control or mute the inputs. Muting is when the fader control for an input channel is set to zero, or all the way down. Some microphone mixers provide an automatic control that lowers the inputs of microphones not in use. These are referred to as automatic mixers (*Figure 22*).

2.5.2 Mixers

The mixing console, or mixer (*Figure 23*), allows audio inputs to be combined and rerouted to output devices. A simple mixer will provide for multiple inputs and outputs, with a basic degree of control over signal characteristics. A basic mixer should have controls for gain, volume, fading, and panning.

Gain is the ratio between the power of the input signal and the output signal. Volume is the loudness of the output signal. Fading is the ability to increase the strength of one signal while

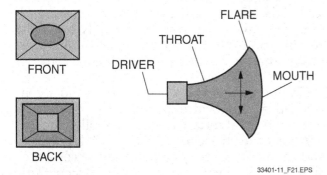

Figure 21 Horn driver.

33401-11_F21.EPS

33401-11_F22.EPS

Figure 22 Automatic microphone mixer.

On Site

Stage Monitors

There is often a need to direct sound back to the sound source. For example, a musician needs to be informed of what other musicians are doing, or a choir needs to hear an accompanying musical track. Stage monitors are speakers designed specifically for these types of applications. A monitor reproduces a full range of frequencies. Output to a monitor is usually highly equalized to prevent monitor output from re-entering the sound system and causing feedback. The use of a monitor is sometimes called foldback.

decreasing the strength of another. Panning adds directionality to fading, such as by blending stereo inputs.

Gain and fading should not be confused. Gain is used to describe the strength of a signal before it is enhanced by an amplifier or preamplifier. It relates directly to voltage and uses an input attenuator. Fading is used to describe the volume of the signal after it has been amplified for output to a speaker.

Input devices connected to a mixer are assigned channels. Channel inputs can be either microphone level or line level inputs. A professional microphone is typically a low-impedance device. Line inputs are high-impedance devices. The mixer controls the incoming signal level and presents the appropriate impedance level.

Input signals are often processed using an array of effects that alter the characteristics of the sound. Equalization (EQ), reverberation (reverb), and compression are all aspects of sound that are processed using special effects processors. The mixing console may intrinsically support these effects, or they may be routed to external effects processing devices before returning to the mixer for output.

A mixer will often provide phantom power. Remember that microphones (particularly condenser microphones) require a nominal power source. Phantom power is provided to the microphone from the mixer using a balanced cable between the two.

Mixers have several signal lines that are used to route signals internally or to other devices. A signal line is called a bus. The master bus provides the pathway for signals for primary mixing. Auxiliary buses, sometime called sends, are used to route signals to external devices for special signal processing.

Mixers are described by the inputs and outputs to their bus structure. For example, a 24

× 8 mixer has 24 input channels and 8 output channels. This is sometimes referred to as a matrix. Mixes can be routed to each output, in variable degrees. For consoles with stereo outputs, you might see 24 × 8 × 2 as a console description, which indicates 24 inputs, 8 mixing, and 2 output channels. Many consoles allow inputs to be grouped into categories called subgroups or submasters. This allows control over multiple inputs such as percussion inputs, string inputs, and vocals.

2.6.0 Signal Processing

An equalizer allows specific frequencies to be filtered and enhanced. This plays a large part in the signal processing domain. There are other types of processing equipment, such as reverb and delay units, compressors and limiters, gates, and expanders.

2.6.1 Equalizers

The easiest way to think of an equalizer is a device that filters or enhances certain frequencies. The simplest equalizer is the tone control. There are controls for bass and treble on many audio devices. These controls allow the user to increase or decrease the level of low or high frequencies (the bass and treble, respectively). A unit with only these two controls would be called a two-band equalizer. Most equalizers provide more than just two bands, though, as shown in *Figure 24*.

Some equalizers allow sweeping of the center bands, which are the frequencies in the midrange. By sweeping, you can actually move the equalization into higher or lower frequencies. Rather than being fixed, the frequencies are selected by the sweep.

Graphic equalizers couple a linear slider control with a ⅓ to 1 octave bandpass filter. Typically, they are fixed at certain frequencies following ISO divisions, which are 62Hz, 125Hz, 250Hz, 512Hz, and so on. They are generally used to correct or alter the frequency response of the system. This allows the resonant peaks and dips of the speaker output to be adjusted and helps to better manage feedback. A parametric equalizer is an equalizer that permits selection of the center frequencies within these bands.

Figure 23 Mixing console.

33401-11_F24.EPS

Figure 24 Graphic equalizer.

Another means of providing equalization (EQ) is through the use of filters. Generally, there are three filters to consider: a high-pass, a low-pass, and a bandpass filter. A high-pass filter attenuates low frequencies, letting the high frequencies pass through. A low-pass filter does just the opposite; it attenuates high frequencies and passes through the lower ones. A bandpass filter attenuates frequencies on either side of a designated band of frequencies. All of these filters are often components of an equalizer. If not, however, they can be added as needed. For example, a high-pass filter with a 20Hz to 40Hz cutoff can protect speaker systems from the negative effects of low-frequency spikes. A low-pass filter can be used to strip off frequencies above those supported by a speaker system.

2.6.2 Reverb and Delay

Reverberations occur naturally in most indoor environments. They are the result of sound bouncing off reflective surfaces. Delay, on the other hand, is echo. An echo is a complete, distinct sound image, where reverb is a series of early reflections.

Reverb, in a reasonably modulated fashion, adds color and depth to the sound. It can enrich otherwise flat sounds. Delay is used to manipulate the sound. For example, in a large hall the delay may need to be adjusted so that sounds reaching the back of the hall arrive closer in time. These adjustments are in the range of milliseconds.

2.6.3 Compressors and Limiters

Compressors and limiters serve to reduce or level the dynamic range of a signal. A limiter sets a threshold for a given frequency using a ratio for input to output level. Simply stated, it allows a limited increase in output to result from a greater increase in signal input. This is measured in decibels and is often called the compression ratio. A limiter is usually applied to peaks with ratios of 8:1, 12:1, etc. If the ratio is reduced to 4:1 or below, and most of the waveform is affected, then the limiter is called a compressor.

2.6.4 Expanders

Expanders serve to enhance or extend the dynamic range. They are used to push down the noise floor, eliminating unwanted effects of hiss and other noise. They are particularly useful for the playback of compressed waveforms.

2.6.5 Gates and Downward Expanders

Noise gates, or just gates, are devices that turn off an input channel when the signal drops below a certain threshold. This keeps unwanted noise, hiss, and other distractions from entering the sound mix. They are particularly useful for managing microphones. A noise gate operates by scanning the input channel and opening or closing the gate based on a nominal signal threshold established by the device. For example, if the gate is set to 4dB, then any signal at 4dB or below is shut off.

A downward expander is similar to a gate. While a gate is basically an On-Off switch, a downward expander, on the other hand, adjusts the signal down when it drops below the designated threshold. However, it doesn't totally eliminate the signal as with a gate.

2.7.0 Amplifiers for Sound Reinforcement

Amplifiers are necessary in a sound reinforcement system to drive output signals to loudspeakers. Preamplifiers are needed to boost nominal signal levels from input devices, bringing them up to line levels as needed by mixing and processing equipment. This section provides a closer look at these devices and some of the factors that influence their use.

2.7.1 What Is an Amplifier?

An amplifier increases the voltage, current, or power wattage of a signal. The power rating of an audio amplifier determines the power that it will deliver. It is stated with reference to the distortion level and the frequency range.

Distortion is an effect resulting from unwanted signals. The most common form of distortion from an amplifier is harmonic distortion. Harmonics are integral multiples of the frequency of a fundamental tone. That means, given a frequency, its harmonic values result from multiples of whole numbers such as 1, 2, or 3. For example, the harmonics of a 1kHz sine wave occur at 2kHz, 3kHz, and so forth. Each of the increments is called an order. 2kHz is second order harmonic, and 5kHz is a fifth order harmonic. Harmonic distortions result when an amplifier creates harmonics for a waveform where they do not exist.

Harmonic distortions can cause clipping and other unwanted effects including popping, or the over-enunciation of consonants such as the letters S or P. Interestingly, harmonic distortions in even-order ranges such as 2, 4, or 6 are less disturbing to listeners than those in odd-order ranges such as 3, 5, or 7. What is important to keep in mind is that clipping can damage mid-range and high-range speaker systems by overpowering them. When selecting amplifiers, less than 0.5 percent harmonic distortion is desirable.

Amplifiers increase the amplitude of a signal. Amplifiers fall into two basic categories, voltage amplifiers and power amplifiers. A voltage amplifier increases the voltage of a signal, whereas a power amplifier increases the power of a signal. For sound reinforcement, the primary interest is in power amplifiers used to drive speakers, as well as wattage.

A power amplifier increases the power of a signal. The power output is measured in dBm. A dBm is a measurement in decibels, but with a special subscript that relates to the power of an electrical current. The power measurement is stated in terms of resistance and wattage. Resistance is the impedance property of the line, and wattage is the amount of electrical power. A single watt is the power dissipated by a current of 1 ampere flowing across a resistance of 1 ohm.

When connecting speakers to an amplifier, they can be connected in series or in parallel. Both schemes affect the impedance of the circuit. A series-connected set of speakers should be considered as additives. That is, the individual impedance ratings are added together. This is clearly seen in *Figure 25*. When speakers are being connected in parallel, it gets a bit more complicated. However, if all speakers are of the same impedance, the net impedance is that of a single speaker divided by the total number of speakers.

Every amplifier has a minimum impedance with which it can operate. Series-connected speakers can be used to stay above this level. Parallel-connected speakers (*Figure 26*) serve to lower the load and could bring it closer to the

minimum, but a single failure does not bring the whole circuit down.

A less common approach is to parallel-connect several series-connected banks of speakers. This allows a large number of speakers to be driven by a single amplifier. The total amplifier power is then shared among them. Parallel connections are the most desirable means used in sound reinforcement.

2.7.2 Preamplifiers

Electronic signaling equipment, particularly audio equipment, has input requirements across a range of varying signal levels. This is due in part to the signaling outputs of attached devices.

For example, a condenser microphone has a nominal voltage output, expressed in millivolts (thousandths of a volt), while the output from an auxiliary input device such as a CD player or tape player provides signaling at line levels of 0.5V to 2V. A preamplifier is used to boost low level signals from input devices, raising them to levels acceptable for line level input.

Some equipment, including mixing consoles and equalizers, provides built-in pre-amplification circuits. This allows them to accept inputs from microphones, phonograph pickup cartridges, and other low signal level inputs. The goal of a preamplifier is to increase the signal gain and keep noise and distortion levels low.

2.7.3 Headroom

Headroom is the ability of a sound system amplifier to get louder. It is the amount of available room between the nominal or base level and the highest level obtainable by the equipment. *Figure*

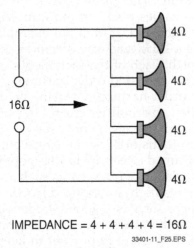

IMPEDANCE = 4 + 4 + 4 + 4 = 16Ω

33401-11_F25.EPS

Figure 25 Series-connected speakers.

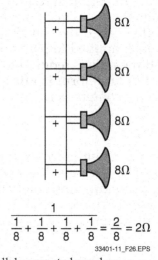

$$\frac{1}{\dfrac{1}{8} + \dfrac{1}{8} + \dfrac{1}{8} + \dfrac{1}{8}} = \frac{2}{8} = 2\Omega$$

33401-11_F26.EPS

Figure 26 Parallel-connected speakers.

The Invention of the Electronic Amplifier

In 1904, James Ambrose Fleming invented the vacuum tube rectifier, a device that converted alternating current to direct current. He called it an oscillator valve. His oscillator replaced the earlier methods of transmitting radio signals using a spark generator.

In 1906, Lee DeForest added a third wire to Fleming's oscillator, allowing it to act as an amplifier. Called a triode, it was later improved upon by Harold Arnold, who created a true amplifier circuit.

27 shows the concept of headroom. Headroom is measured in decibels and usually requires 10dB or more above the nominal or base level for public address systems. The goal is to provide enough room to avoid clipping problems. When the sound level is increased, it does not increase incrementally, but rather logarithmically.

2.8.0 Constant-Voltage Audio Distribution

Constant-voltage is the term used to describe the distribution of power between amplifiers and loudspeakers for special types of audio systems. These systems are used when driving a number of speakers from a single amplifier. Speakers are usually more than 50 feet and less than 1,000 feet away from the amplifier. As previously noted, speaker drivers have reasonably low voltage, but require high power inputs. The more speakers connected to an amplifier, the more difficult it is to maintain the power requirements. Remember, power is voltage times current ($P = I \times E$). As resistance increases, current decreases.

Most constant-voltage systems in the United States operate at 70.7 volts, and are commonly called 70V systems. (70.7 is the RMS value of a 100V peak waveform.) Because some municipalities and school systems require conduit for 70V systems, 25V systems were developed for use in such environments. This voltage level provides safety without the expense of conduit. These systems are particularly common in schools. However, they are not as efficient as 70V systems.

In a modern constant-voltage system, each loudspeaker has a built-in step-down transformer. This is shown in *Figure 28*. Speakers use taps to connect to the cabling. Taps provide differing output levels to drive the loudness of speakers. Amplifiers are designed to output the required 70.7V, eliminating the need for an output transformer. These modern systems are called direct-coupled audio systems, as the output from one component provides the input to the next.

Cabling for constant-voltage systems consists of a twisted pair of solid or stranded wires. These wires can be shielded or not. Typically, 18-gauge copper wire is used for runs as long as 700 feet; 16-gauge wiring has been used for runs of up to 1,100 feet. One of the original goals of a constant-voltage system was to keep the conductor size to a minimum, limiting installation costs. As cabling

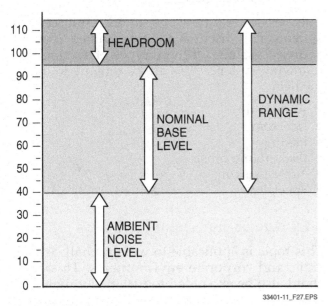

Figure 27 Headroom.

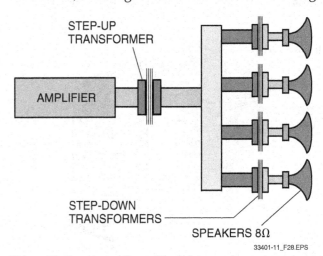

Figure 28 Constant-voltage direct-drive audio system.

lengths increase, the resistance increases, and the size of the wire would otherwise have to increase as well. By using constant voltage at 70.7V, smaller wires can be run farther, but the heaviest gauge of wire practical should always be used.

The requirements for constant-voltage systems that resulted in the 70.7V standard were partially driven by electrical codes. National electrical codes enacted in the late 1940s required all voltages above 100 peak volts to be run through conduit. The codes were enacted to help prevent the shock hazards presented by these high voltages. In some parts of the United States, codes were even stricter, and 25V systems were also standardized. In Europe, a 100V standard evolved.

> **NOTE**
> Most paging systems are 70V systems and most intercoms are 25V systems.

2.9.0 Audio Transformers

Transformers are devices that take a signal input in the form of AC and output the signal unmodified, also as AC. The input and output are not connected. Using principles of induction, a primary input is run through a conductor coiled around an iron core. This produces flux, or a force field. The secondary, or output coil, picks up this flux and generates an equivalent output signal. See *Figure 29*.

Transformers are described by their coil windings. The windings are inversely proportional to the output. A step-up transformer has more secondary than primary windings. The result is that electrical current is increased on the output side. A step-down transformer is the reverse; it has fewer windings on the output side, which generate a lower overall output. Transformer operation is expressed as a ratio of the input to output windings. For example, a step-down transformer may be described as 6.5:1 (ratio of 6.5 input windings to every 1 output winding).

Transformers can actually have more than one set of secondary windings. This allows a single transformer to provide different levels of output. Each set of output windings has its own leads. These transformers are called line matching transformers. Each set of leads, or taps, provides a different wattage. In an audio distribution system, this allows the desired output level to be selected for speakers in different locations. Output level selection means that some speakers can be louder or softer, regardless of the settings at the master volume control.

There are other uses for transformers besides stepping up or stepping down the signal voltage. They can be used to increase or decrease circuit impedance. They can block DC current, but let AC current flow. Electrically, they can isolate one device from another. Another audio application is to use a transformer to convert an unbalanced circuit to a balanced circuit and vice versa.

Transformers that block DC current and convert balanced to unbalanced circuits are called unity transformers. Unlike their step-up and step-down cousins, unity transformers have an equal number of windings for both input and output. Their ratio is expressed as 1:1 (one-to-one). They have a secondary effect of limiting RFI. Unity transformers are used for equipment isolation as previously described.

3.0.0 APPLICATIONS OF SOUND REINFORCEMENT SYSTEMS

Thee are several types of sound reinforcement systems. This discussion will be limited to public address systems. Of these types of systems, the following system types are commonly found in the field:

- Auditorium
- Intercom
- Paging
- Background music
- Noise masking
- Specialty

3.1.0 Auditorium Sound Systems

This topic is applicable to worship hall, school, civic, and corporate environments. This topic could also be expanded to include location-specific applications, such as a stadium or arena. This section will look at just a few typical uses and

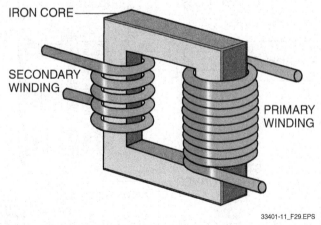

33401-11_F29.EPS

Figure 29 Basic transformer.

their requirements. The purpose here is to provide a sense of what a medium-to-large sound reinforcement system is, the equipment used, and the inherent challenges faced.

3.1.1 Requirements for an Auditorium Sound System

Auditoriums have requirements that differ from other sound reinforcement systems such as paging, intercom, or background music systems. Auditoriums are usually designed with acoustic needs in mind. That means walls may be covered in non-reflective materials, or the physical shape of the hall may enhance the overall quality of the sound. These are positive features. However, many auditoriums serve multiple purposes. They may be used for theatrical performances with an emphasis on vocal support. They may also be used for different types of musical performances ranging from orchestras to rock concerts, with opera or folk music filling the gaps.

Auditoriums may have different audience sizes. The number of people in a space has an effect on the loudness and quality of the sound. Related to this are various listening areas within the auditorium including the orchestra (main part of the auditorium); mezzanine (balcony sections); and spaces tucked away in pseudo-rooms such as those under a mezzanine, in lobby areas, or even concession areas. Combined with these production requirements are other needs. These include standard public address and paging in the event of an emergency.

3.1.2 Overview of Sound Reinforcement for an Auditorium

Sound reinforcement in an auditorium involves three critical factors: microphone selection and placement, loudspeaker selection and placement, and mixing. In theatrical productions, microphone selection is probably as important as placement. It is important to effectively isolate the sound coming from individual performers, but it is necessary to provide a rich sound level for all aspects of the performance. Lavalier mics provide excellent directionality and isolation for performers. It is important, however, not to pick up the rustling of costumes and other distractions. Boom mics provide directionality, but may limit movements of the performers. Stage microphones need cardioid pickup patterns to prevent audience sounds from entering the acoustical mix.

Speaker selection and placement play a crucial role in the reproduction of sound in an auditorium. Two key related elements should be

addressed: dealing with the Haas effect, and the negative effects of delay. The Haas effect, sometimes called the precedence effect, describes the physics of hearing as it relates to delay (*Figure 30*). The human ear will actually suppress, or tune out, secondary sounds such as echoes or reverberations. There is about a 40-millisecond delay built into our hearing. The sounds that are heard immediately after this brief delay actually enhance the overall quality of what is being heard. This effect has been capitalized on by the recording industry for years.

There are two speakers in the figure. One has a discernable direction at the front of the auditorium. In the back is a ceiling-mounted speaker. It acts to reinforce the sound coming from the front.

The problem is that the sound from the front takes longer to reach the listener at the back than the ceiling speaker. To compensate, a delay timer is set to delay the signal to the ceiling speaker, perhaps 20 to 30 milliseconds. This way, the sound from the two speaker sources arrive close enough in time that the listener cannot discern the difference. In fact, they may not even be aware of the ceiling speaker at all, though it is truly reinforcing the sound they hear.

Another way the Haas effect can be understood is through reflected sounds. Reflected sounds from speakers in front may actually reach areas such as under-balcony sections after locally placed speakers relay the sound, causing an echo effect. Though the human ear is accustomed to orienting towards the sound source, it may be important to administer appropriate delays at the console to compensate.

The spoken voice ranges in frequency from around 100Hz to 400Hz. Male voices are generally at the lower end of this spectrum and female voices toward the high end. Peaks in the 12kHz

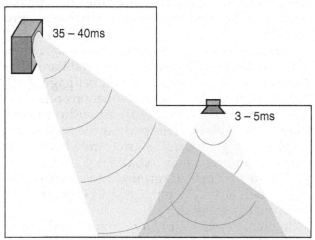

33401-11_F30.EPS

Figure 30 Haas effect.

range are possible, particularly in relationship to sibilants (s, p, and z sounds). Speaker systems for voice reinforcement are usually optimized in the 200Hz to 8kHz range. They should have a reasonable flat response to avoid sibilant peaks.

Another challenge in these environments is that as the distance from the sound source increases, there is a proportional decrease in sound pressure. For every doubling of the distance, a subsequent halving of the sound pressure level occurs. This is called the inverse square law. For this reason, speakers are often provided locally for hard-to-reach places such as under balconies.

3.2.0 School and Talkback Intercom Systems

Intercom systems are used to provide two-way communication. They can provide support between two offices or, on a much larger scale, two-way communication throughout an entire facility. Intercom systems consist of a master station and one or more remote units or substations. Some systems support special stations that are used to interface with third-party equipment such as telephone networks. Many intercom systems are constant voltage systems operating at 25V.

Intercom systems provide different methods for initiating a page. In some systems, a telephone interface provides access at the master station, while other systems use microphones. Many systems provide a PBX connection, allowing the intercom system to be connected to the company telephone network.

Remote stations may support station-to-station calling between remotes. Generally, communication is between the master station and the remote. Most systems are wired in a star configuration. Many remote stations are open-air transceivers. They have a microphone built into the speaker assembly. Pushbuttons or other switches are provided to initiate calls.

Intercom systems are installed both in residential dwellings and commercial establishments. Both installation types can support paging and music applications. Home installations often provide an AM/FM radio with an audio cassette player or CD player. This permits the homeowner to play a single musical source anywhere in the home. Commercial installations also have the ability to support background music. Often this is through auxiliary input of a CD player, radio receiver, or tape deck.

Intercom systems for large commercial and industrial installations may be modular in design. They consist of speakers mounted in ceilings or on walls. Electronic switching devices may be used to route intercommunications calls between a master station and the many remote stations. *Figure 31* shows intercom consoles. These systems provide interconnection for up to 10,000 potential nodes. Twisted-pair cabling is used for signaling and power. Each remote station is tapped into the system as part of a parallel-connected network.

Matrix intercom systems often appear to be more like small, private telephone exchanges. *Figure 32* shows a matrix intercom controller or switcher. Switchers provide for routing, with trunking applications that connect systems through telephone lines. These complex systems are used in areas such as television broadcast facilities and other real-time venues. They can be found in corporate environments as well as educational facilities.

3.3.0 Paging Systems

Paging systems are a specific application of sound reinforcement technology. Most paging applications are one way. They allow messages to be either broadcast or targeted, but do not necessarily provide for direct two-way communication such as that found in an intercom system. Paging technology is often used to provide emergency notification and communication within a defined area. Other uses might include the following:

- Locating people when they are not at their desks
- Providing call support after hours
- Announcing in-store special sales
- Providing background music

Paging systems are installed in stores, businesses, warehouses, industrial facilities, and hospitals. A typical system consists of a distributed network of loudspeakers, amplifiers, and one or more controllers. *Figure 33* shows a paging system controller. Historically, paging systems have been 70V or 100V direct-drive systems. A 70V system is centrally amplified and distributed. Newer 24VDC systems are self-amplified.

A 70V system uses a central amplifier. Individual speakers have transformers that match them to the 70V output. Speakers are tapped to allow selection of different output wattages. By selecting more or less power at the tap, speakers can cover areas with low to high levels of noise. Background music support is good. Wiring for 70V systems should not be run in close proximity to telephone wires.

Paging systems are often divided into zones. Each zone will have its own controller. The controller allows selection of a specific page mode,

such as one-way, talkback (for two way systems), or mixed support for both one-way and two-way communication. Many controllers allow connection into the company PBX for telephone initiated pages.

Paging systems use a variety of speaker types and enclosures. The most prevalent are ceiling-mounted speakers, wall baffles, and horns. Ceiling speakers typically have a wide dispersion rate, meaning they can cover a fairly large area. Turning the volume up on these speakers does not increase the coverage area. Volume controls

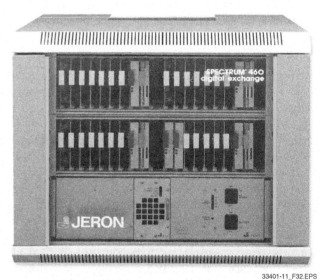

33401-11_F32.EPS

Figure 32 Matrix intercom controller.

33401-11_F31.EPS

Figure 31 Examples of workplace intercom consoles.

33401-11_F33.EPS

Figure 33 Paging system controller.

Valcom 24V Paging System

A 24V system uses speakers with a small amplifier that is either built-in or attached to each speaker. These types of systems require four wires to connect speakers: two wires for audio signal and two wires for DC power. Wiring can be routed with telephone cabling, or even across the same cable. Self-amplified systems do not work well for background music application because their amplifiers are usually designed for speech reproduction.

affect the loudness, allowing ambient noise levels to be overcome. To determine the number of speakers required for a specific area, the ceiling height times two rule can be used. It states that the best coverage is often achieved by spacing speakers twice as far apart as the ceiling is high.

Wall baffles are used where ceiling speakers are not practical, such as extreme ceiling height. Wall baffles are usually installed all facing in the same direction. Spacing is determined by the height of the speaker from the floor. For example, one manufacturer recommends that at 8 feet high, speakers should be spaced 20 feet apart. Each speaker can project up to 25 feet. Speakers mounted 16' high should be spaced 30 feet apart, with 45 feet of projection. Columns or posts can be used in this arrangement. If speakers have to be mounted facing each other on opposing walls, they should be staggered.

Horn speaker drivers are used in a number of special environments, such as outdoors or indoors where there are extremely high noise levels. They are also used in environments where extreme conditions exist, such as in freezers or heat-affected areas. Horns are often used in industrial applications where mounting aesthetics are not necessary. They are usually mounted at 15 to 20 feet from the ground and angled downward. Horns, like wall baffles, should be mounted facing in the same direction.

3.4.0 Background Music

Background music systems are designed to provide pleasant sound environments for listeners, workers, and shoppers. Background music systems are usually extensions of paging systems.

In fact, many paging systems include line-level inputs for adding music support. *Figure 34* shows a background music controller used in a paging system. Though most paging systems are optimized for speech, music is reproduced on them reasonably well.

3.5.0 Noise Masking

Noise (sound) masking systems are a relatively new application of sound reinforcement. The idea behind noise masking emerges within the modern office environment. Here, workers occupy cubicles with walls well below ceiling height. In this environment, privacy—particularly voice privacy—is at a minimum. The idea is to generate a non-distracting masking sound at a sound pressure level sufficient to reduce speech intelligibility.

A typical system consists of a noise generator, equalizer, amplifier, and loudspeakers. The noise generator is a special type of pink noise generator, which is filtered to provide a specific spectrum of noise. White noise is usually a single frequency, or several frequencies rolling from one to the other. Pink noise provides a complete, harmonic waveform and is more suitable for masking applications. The noise generator allows you to roll-off frequencies at the high end, approximately 4dB or so, for each octave of the frequency, and otherwise shape the sound through equalization. By applying this noise at sufficient levels, the ambient noise in the environment is raised and drowns out distracters such as office machinery and foot traffic. Conversations taking place at other locations within the space are also masked.

The goal of sound masking is to provide speech privacy. Other environments, such as doctors' offices and courtrooms, also benefit from sound masking. Speakers are mounted above a dropped ceiling and face up. Paging and background music system speakers are mounted in the ceiling facing down. With a masking system, the objective is to generalize the masking sound, including its directionality.

33401-11_F34.EPS

Figure 34 Background music and paging amplifier.

To avoid phase shifts between two speakers, many masking systems use a two-channel approach. They connect the output of each channel to a different masking generator. Another approach is to package the noise masking circuitry and the loudspeaker into the same enclosure. These types of systems are often distributed.

> **NOTE**
>
> The Health Insurance Portability and Accountability Act (HIPAA) of 1996 requires that sound masking be provided in some environments where privacy is important. Such environments include doctor's offices, clinics, pharmacies, hospitals, and a variety of other locations. Anyone involved in designing audio systems should be familiar with this law.

3.6.0 Room Combining Systems

Room combining systems are a special type of audio control system. They are used in applications such as hotels, conference centers, corporate office buildings, and educational facilities. The need for room combining results from the constant rearrangement of space in these types of facilities. For example, in a hotel or conference center, large spaces are often subdivided for specific purposes. A ballroom may be split into four smaller rooms for individual meetings. Later in the same day, two rooms may be combined for a larger meeting.

Room combining systems are often computer controlled. At a minimum, they may require computer software to configure the various parameters of individual sound systems. These systems primarily support speech applications, though they may be called on for music reinforcement as well. Additionally, they will often provide paging inputs or connection to site paging systems. They also may include non-audio controls for lighting.

Room combining technology can be broken into three key components or subsystems. Typically, there is a central control system, a distribution system, and individual room systems. With the exception of individual room controllers, these systems are often rack-mount systems.

The central control system for a room combining system consists of a master controller (sometime called a mainframe) a master volume control, and an interface to room systems.

The mainframe may house line inputs in the form of pluggable printed circuit boards. Each room supported would require one of these cards. The controller communicates with the cards using a special audio protocol, PA-422. Many residential and commercial audio components support this industry-approved interface standard.

A master volume control is needed to support combined rooms. Individual rooms have their own volume level controls, but when combined they need a single controller to ensure usable volume levels.

The interface controller is needed to connect individual room controllers to the system. When sectioned off, each room has its own console. That is, each room requires its own sub-master station to support microphone selection, audio mixes, and the like. The room console is often called the head table. Because each room is an individual, almost virtual, zone, the system interface controller is used to combine the requirements from each of the rooms. The interface unit in turn talks to both the master controller and the room head-table setups.

The distribution system consists of one or more amplifiers connected to 25V or 70V distributed speaker systems. These are similar in all aspects to the standard distributed, constant-voltage systems discussed earlier. Output from the master controller is routed to the distribution system based on determined volume levels as set by the master controller. Once audio signals are received at the individual rooms, they may be subject to local volume controls established by the room's controller, called the room panel.

The room panel provides for connections to the head-table console. It also generally provides volume control for that room or zone.

3.7.0 Distributed Residential Audio Systems

There is a trend in residential audio to provide professional-level support for home audio and hi-fi systems. Using state-of-the-art digital technologies, homes are wired to provide the benefits of intercom, paging, room combining, sound masking, and distributed audio throughout the home. This allows the homeowner to use a single set of audio and hi-fi components for the whole house, inside or outside.

The foundation for residential audio includes a structured wiring system. These wiring systems provide wiring for CATV, SATV, telephony, and audio. They provide the bus used to distribute audio. These wiring systems serve as the central hub for all audio connections throughout the home. Audio cables are then routed to each of the rooms. Hi-fi and audio equipment are plugged into special receptacles placed in various rooms.

Speakers are mounted in the ceiling, in the walls, and under eaves (outside). Home systems usually have a dedicated amplifier to support distributed speakers. Controls for volume and device selection are mounted in gang receptacles for easy access. Many systems provide infrared connections for remote control from anywhere in the room.

Like room combining applications, distributed home audio systems have a router. Some systems provide input to the router through a set of distributed input receptacles. In some cases, these input modules have their own amplifier.

4.0.0 AUDIO CABLING OPTIONS

Cabling for audio systems varies depending on the type of installation. This section introduces cabling requirements for various systems. It identifies the types of connectors used for various types of equipment. It concludes with an overview of structured wiring systems as found in conference systems and residential applications.

4.1.0 Impedance Versus Resistance

Resistance is the characteristic of a conductor to oppose the flow of electric current. Resistance as a measurement only applies to direct current (DC). Impedance is a characteristic of a device and is not dependent on signal. It is the resistance of a device that uses alternating current. Impedance has three dimensions: phase, magnitude, and frequency.

Source impedance determines how easily power will flow from the source. Load impedance determines how much power will be drawn from the source. The lower the impedance on a device connected to a source, the higher the load or draw on the source.

In modern equipment, the source impedance is kept low, and the load impedance is kept high. Some circuits have a load impedance that is ten times (10×) the source or input impedance. These circuits are called bridged circuits and do not ap-

preciably load the source; no power is transferred. In a matched circuit, the source and load impedances are equal. Matched circuits are typical of older systems, but are not common in new installations.

Circuit failure can occur if device impedances are mismatched. If the load is too low, the output equipment will try to deliver more power. This can result in blown fuses, tripped circuits, and an automatic shutdown.

4.2.0 Balanced Versus Unbalanced

The first thing to know about audio cabling concerns with the way cables are used as part of an overall grounding strategy. An audio system has either balanced or unbalanced lines (*Figure 35*). In an unbalanced line, the cable consists of a single conductor with shielding. The conductor is connected to the negative terminal, and the shield is connected to chassis ground. Because the shielding is now part of the circuit, the line is considered unbalanced.

A balanced line has two conductors and a shield. The two conductors are connected to the positive and negative leads. The shield is connected to chassis or earth ground. The shield serves to minimize unwanted RFI and EMI, both of which can cause significant distortions and hum. A balanced line carries the same signal on both conductors; they are just at different polarities. When an interfering signal is introduced, it is carried on both lines. However, they are out of phase, and the signal cancels itself out.

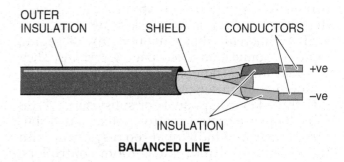

BALANCED LINE

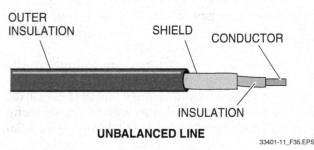

UNBALANCED LINE

33401-11_F35.EPS

Figure 35 Balanced and unbalanced lines.

Most signal processing, recording, and playback equipment uses balanced lines. Many microphones do not. Even those that are balanced are subject to being connected to unbalanced input lines.

There are several ways to convert an unbalanced to a balanced connection, or a balanced to an unbalanced connection. The best way might be to use a balancing amplifier. More likely, though, you will want to use a converter to accept unbalanced to balanced and balanced to unbalanced connections. *Figure 36* shows a device that converts cabling from balanced to unbalanced, or vice versa. You can also custom-terminate cabling, though this is the least desirable means.

Ground loops are a significant cause of line noise. Each component in a sound system has its own internal ground. This is often called signal ground or audio ground. When two devices are connected together, their grounds are often tied together via the conductors of the connecting cable. A ground loop occurs when the grounds are tied together at some additional place as well, such as power rails or the third wire on the AC cord. This allows current to flow in a loop from one device to the other and back, which can create a noticeable hum.

Normally with balanced circuits and interconnects the ground loop noise is rejected. Sometimes, though, the noise does affect the audio signal. This can often be traced back to improperly designed equipment that, even though balanced, is not really immune. In other cases, it can be traced to the connection of unbalanced equipment to balanced equipment.

4.3.0 Cabling and Wire

The rules for selecting appropriate cable and wire are similar to those for other electronic equipment. However, unlike many electronic installations, audio equipment—particularly microphone and speaker cable—is subject to a greater amount of flexing. This is important to consider. Excessive flexing can cause wire shielding to separate and even cause individual conductor wires to break. A host of problems can occur as a result. Cabling issues in general were covered in earlier modules.

This section identifies issues specific to audio systems.

4.3.1 Microphone Cable

As a general rule, microphone cables should be selected for their flexibility and noise handling capability. Microphone cables should not be used for long runs. Use shorter runs to connect them to a wall or stage box (*Figure 37*). Then use a snake or foil-shielded twisted pair to run them from there.

Other criteria for selecting microphone cables should include a high strand count for conductors. This will help ensure a greater flex life for the cable. Microphone cable is usually 22 gauge or lower, though there is a trend to use smaller 24-gauge cabling to reduce costs. Remember, the lower the gauge, the thicker the conductor. As the conductor size increases, resistance decreases.

Shielding is also an issue. Foil shielding provides the best protection from unwanted interference. However, foil is less flexible. Wire braided shielding provides adequate protection from interference, but is subject to unwinding and separation after repeated flexing.

4.3.2 Speaker Cable

Speaker cables should be low resistance. Large-gauge wire is better than small-gauge wire to enhance the low-impedance output from amplifiers. Smaller gauge wires attenuate high frequencies faster than larger gauge. The shielding of cables can increase the overall capacitance of the cable.

33401-11_F36.EPS

Figure 36 Two-way converter for balanced and unbalanced circuits.

33401-11_F37.EPS

Figure 37 Microphone snake and stage box.

When coupled with cable resistance, high capacitance in a cable can, in effect, create a type of low-pass filter. This contributes to signal loss over longer runs.

The loss of signal resulting from resistance can best be remedied with larger gauge speaker wire. It should not be twisted too much as twisting increases capacitance. Shielding is not as necessary for speaker cable as it is for a microphone. The biggest problem with speaker cable is that as impedance increases, the damping factor of the driver could decrease.

4.3.3 Distributed Speaker Systems Cable

With distributed, constant-voltage systems, the quality of the sound falls short of other types of sound reinforcement systems. This is primarily due to the fact that transformers are used throughout to connect speakers to cabling, and ultimately to amplifiers. Hundreds of speakers can be hung from these systems. Typically, 12-, 14-, 16-, or 18-gauge wiring is used.

4.4.0 Connectors

A wide variety of connectors are used in audio installations. Though no real standards exist, some types of connectors are used more often than others. This section identifies the most prominent and widespread connectors.

4.4.1 XLR Connectors

The XLR connector (*Figure 38*) is a three-pin connector used for balanced lines. XLR connectors are used frequently for microphone connections. There are two parts to the connector, male and female. Some connecting devices use a special chassis-mounted male connector port. Extension cables often use standard inline couplers.

4.4.2 Phone Plug Connectors

Phone plugs (*Figure 39*) get their name from when they were used exclusively for telephone switchboard cabling. There are two types of phone plugs. Both are ¼" in diameter. One is for an unbalanced circuit, called tip-sleeve (TS), the other for a balanced circuit, called a tip-ring-sleeve (TRS).

TRS plugs are used for balanced lines. Typically, the hot terminal is wired to the tip, and the return is wired to the ring. The sleeve is then wired to the shield for common. These types of connectors are sometimes called stereo jacks as they are usually found on stereo headphones.

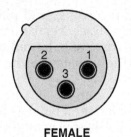
FEMALE

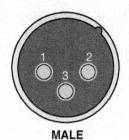
MALE

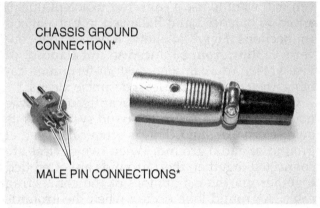

CHASSIS GROUND CONNECTION*

MALE PIN CONNECTIONS*

* The common arrangement is to solder the black lead to pin 3, the white lead to pin 2, and the shield to pin 1.

33401-11_F38.EPS

Figure 38 XLR connector.

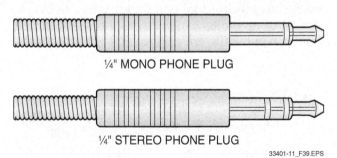
¼" MONO PHONE PLUG

¼" STEREO PHONE PLUG

33401-11_F39.EPS

Figure 39 Phone plug.

4.4.3 RCA Connectors

Sometimes referred to as a pin connector or phono connector, the RCA connector is the standard for many line-level components. RCA connectors are shown in *Figure 40*. These connectors are readily available prewired for consumer-grade audio device connections. They are sometimes used in professional setups because they are inexpensive, but be aware that these lines are unbalanced. One advantage is that they are smaller than other connectors, allowing more connection ports in a limited amount of space.

4.4.4 Banana Plugs and Connectors

Banana plugs and receptacles (*Figure 41*) are commonly used for making speaker connections. Their receptacles can be found serving double duty as binding posts.

4.4.5 Twist-On Connectors

Large speakers are often connected using twist-on connectors like the one shown in *Figure 42*. These connectors are designed to handle the larger currents used with modern speakers. The original

twist-and-lock connectors had to be inserted, then twisted to lock them into place. More recent versions do not require the twisting motion. They have a release tab that is used to lock them in place after they are inserted into the speaker connection.

4.4.6 Patch Bays and Patch Panels

Changing or connecting cables to the back of components can be challenging. Patch bays or patch panels (*Figure 43*) are designed to allow cabling

BANANA JACK

BANANA PLUG

33401-11_F41.EPS

Figure 41 Banana plugs and posts.

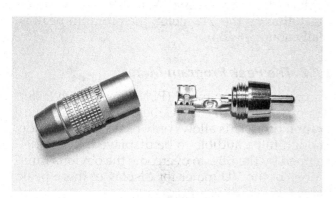

TWO PIECE

THREE PIECE

33401-11_F40.EPS

Figure 40 RCA plugs.

33401-11_F42.EPS

Figure 42 Twist-on connector.

requirements for both input and output at the same place. The various components are plugged into the back of the panel. Cross connections or signal routing is then accomplished on the front side.

Many patch panel ports are normalized. This means that if a plug is not inserted into a jack, then by default its output is connected to the input below it. Traditional patch panels use specialized phone plugs, either TS or TRS, which may be called Bantam or TT plugs. Some modern patch panels use D-shell and sub D-shell connectors.

5.0.0 INSTRUMENTATION AND TEST EQUIPMENT

Test instruments are invaluable tools for troubleshooting. It is important to understand them and, in some cases, be able to calibrate equipment and equipment meters.

Instrumentation is often an integral part of audio equipment, and test equipment also has its own instrumentation. This section examines various test instruments and discusses their uses and calibration requirements.

5.1.0 Understanding Console Instruments

Most consoles and mixers use one or more of three basic indicators. An indicator, though synonymous with the word *meter*, is used in this context for visual feedback. During the process of mixing and combining, the operator needs specific information in order to balance and equalize the sound mix appropriately. The three types of loudness metering examined here are as follows:

- Volume unit (VU) meter
- Peak program meter (PPM)
- Light-emitting diode (LED) meter

5.1.1 The Volume Unit Meter

The volume unit (VU) meter displays an average volume level. It responds similarly to the level-sensing capability of the human ear. A VU meter consists of a coil on which a needle is attached. As current passes through the coil, the needle moves accordingly. As an indicator, a VU meter is calibrated and used to measure the power level of the complex waveforms found in an audio signal.

When working with a VU meter, understand that the VU meter does not give an indication of peak signal levels. A VU meter, in a sense, samples the incoming signal over a period of about 300 milliseconds. Once the meter stabilizes, it provides a reasonable, though subjective, view of the average volume of the signal.

The VU meter is fairly slow, but its response closely parallels that of the human ear. A VU meter is driven from a full-wave averaging circuit defined to reach 99 percent full-scale deflection in 300ms. The VU meter should not overshoot more than 1.5 percent. This is why it is not used to measure peaks.

Figure 44 shows the scale of a VU meter. The scale starts on the left at –20dB and extends to 0dB. The meter then extends an additional 3dB to the right. Usually, values above 0dB are in red. Remember that the VU meter is an averaging meter. It does not give precise information about the peaks of the audio signal. When the peak signal hits 0dBm on the VU meter, the program peak is really about +6dBm.

5.1.2 The Peak Program Meter

The peak program meter (PPM) has a much faster response rate than the VU meter, with a 5ms integration time. This allows only peaks that are wide enough to be audible to be displayed. It was developed by the BBC to overcome the obvious limitations of the VU meter for display of these peak values.

The PPM shows how hard the channels are being driven. A major difference is in the scaling used. At least eight different scales may be encountered with a peak program meter. *Figure 45* shows two of these scales. This is because many European national broadcasters each have their own standards for the scaling factor. The differ-

Figure 43 Patch panels.

33401-11_F43.EPS

Figure 44 A VU meter.

33401-11_F44.EPS

ences are really determined by the reference point used. What is important to know is that the total meter range is typically 24 to 26dB.

5.1.3 The Light-Emitting Diode Meter

The light-emitting diode (LED) meter (*Figure 46*) evolved from the shortcomings of the two acceptable standards for loudness and peak measurement, the VU and PPM. The 10ms sampling rate of the PPM pretty much ignores average data, while the VU ignores the rapid, sharp peaks and transients of the waveform.

The advantages of an LED meter are numerous. It can be configured to use color for instant feedback. LEDs can be arranged to represent decibels change with a 1:1 correspondence. An LED of a different color can be used to reference peaks.

5.2.0 Audio Test Instruments

The professional audio technician will use several test instruments to successfully set up and configure a sound reinforcement system. This section introduces typical instrumentation found on the job, including the following:

- Real-time analyzer (RTA)
- Voltmeter
- Impedance bridge
- Spectrum analyzer
- Sound pressure level (SPL) meter

5.2.1 Pre-Installation Testing for a Sound Reinforcement System

There are several requirements for testing a sound reinforcement system. Some tests are conducted as part of the requirements specification and design of the system. Other tests are conducted during the installation of the system. Additional tests are conducted as part of overall systems test and, importantly, during the commissioning of the system.

You will likely participate in many, but not necessarily all, of the testing activities. For example, during requirements specification and design, acoustical tests are conducted to determine the

33401-11_F46.EPS

Figure 46 LED loudness meter.

sound characteristics of the environment. Under the supervision of an acoustical engineer, you may participate in determining these overall requirements. Testing most likely to be performed by a technician is known as real-time analysis.

The physical characteristics of a room can combine to attenuate certain frequencies while boosting others. The effect is a degrading of the overall frequency balance. A real-time analyzer (RTA) uses sophisticated fast Fourier transform (FFT) analysis to measure and display the energy content across the audio spectrum. This allows you to fine tune the system for sound equalization and determine the best locations for speaker placement.

An FFT, in essence, takes a discrete signal in the time domain and transforms that signal into its discrete frequency domain representations. A typical RTA can be used for testing of filters, sound systems, and room acoustics. An RTA is also used to monitor the audio spectrum from 20Hz to 20,000Hz.

5.2.2 Installation Testing for a Sound Reinforcement System

During the installation of a sound reinforcement system, you will perform several tests to verify correct function and operation of equipment. They also serve to support the alignment of loudspeakers for optimal sound reproduction. Many of these tests may be related, and include but are not limited to the following:

- *Sound pressure level (SPL)* – These tests determine the actual loudness of the sound as it is distributed throughout the location. The results of SPL tests are used to support sound equalization (EQ) among others. Sound pressure is measured in decibels.
- *Gain structure* – The gain structure represents the total gain produced by all equipment in a given audio signal path. Power amplifiers, signal processing, speakers, and speaker cabling all contribute to the overall gain structure. A sound system that is operating poorly may be subject to fluctuations in gain, with resulting pops, hum, hiss, and other noisy artifacts.

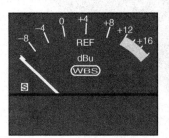 

33401-11_F45.EPS

Figure 45 Program peak meters.

- *Ground loops* – Ground loops occur when equipment is connected together in such a way that their grounds are tied together at two places, causing a form of feedback or noise. This is looped through the system, giving rise to distortions and other noise artifacts.
- *Absolute polarity and phase* – Polarity is inconsistently represented in electronic components. This can be seen with speakers where, for example, some manufacturers use red and black posts to determine polarity. Others use symbols such as plus and minus. Cables can be wired to connectors in a multitude of possible connections, such as balanced XLR to non-balanced phone connections or balanced phone to RCA. Problems resulting from a mismatch of polarity can cause phase cancellation. This means, for example, if one speaker in a bank of speakers is miswired based on polarity, it can cancel the output of another speaker. If there were four speakers in the group, only two would then produce useful output.
- *Impedance* – Impedance is the total AC resistance to current flow. Impedances vary between input devices. For example, there are both high-impedance and low-impedance microphones. Low-impedance microphones are usually selected for sound reinforcement. High-impedance microphones require a transformer when used with low-impedance connections, or even long cables. The technician will commonly test for these kinds of problems. Impedance testing is an important part of maintaining constant-voltage distributed systems as well.
- *Equalization* – Equalization means to change the strength of the audio signal for a given band of audio frequencies. Equalization affects the overall tone by changing the level of fundamental and harmonic frequencies. The result is to either boost or reduce the energy of the given frequency band. Every room has a unique set of characteristics that affect the quality of sound reproduction. You will test the system installation within the parameters of the room to ensure adequate sound reinforcement needs are met.

5.2.3 Post-Installation Testing for a Sound Reinforcement System

Once a sound reinforcement system has been installed and is operational, you may perform ongoing testing to ensure the successful operation of the system. These tests are similar in nature to the tests already discussed. Post-installation testing is driven by two requirements: keeping the system tuned and operational, and responding to problems.

When operating a sound system, it is important to ensure that the gain structure is maintained, that cables have not been damaged due to handling and use, and that there are no ground loop problems. In addition to test devices already discussed, there are two test devices used for routine maintenance: the cable tester and the microphone and line tester.

The cable tester is a valuable device that can be used to test most cable combinations, such as XLR, ¼-inch, or RCA phono plugs for shorts, opens, and polarity reverse. A cable tester is shown in *Figure 47*. Many cable testers can be used to locate intermittent flexation problems.

Use a cable tester to monitor a standard three-wire intercom feed or talkback from the main audio console, or to quick-check dynamic mics without firing up the entire system. It provides a handy mic or line-level tone for activating a signal path or identifying a specific cable in a pile of cable ends. Cable testers are powered by a single standard 9V alkaline battery and housed in rugged ABS.

Several all-in-one audio line testers are good for applications such as live sound, maintenance, and installation work. The Qbox from Whirlwind (*Figure 48*) is such a device. It includes a microphone, a speaker, a test tone generator, outputs for standard headphones, and a ¼-inch jack for line-in or a 2k ohm (telephone) earpiece out. It also has voltage presence LEDs for confirming phantom or intercom power.

The Qbox provides a built-in pink noise source used to send a tone or signal from the built-in condenser microphone at either mic or line levels. There is a volume control that lets you confirm that the incoming signals are at mic or line level.

33401-11_F47.EPS

Figure 47 Cable tester.

Figure 48 Microphone and line tester.

5.3.0 Real-Time Analyzer

As mentioned before, the physical characteristics of a room can combine to attenuate certain frequencies, while boosting others. The effect is a degrading of the overall frequency balance. A real-time analyzer (RTA) (*Figure 49*) uses sophisticated FFT analysis to measure and display the energy content across the audio spectrum.

The RTA measures the loudness of an audio signal, either through a flat-response microphone or a line input. Special microphones are used when sampling sound with an RTA. These microphones are generally omni-directional, but in special cases these microphones may be directional. They are usually air condenser or electret condenser microphones. Each individual RTA device will generally specify the microphone required for its operation. It is important to match the microphone as required by the RTA exactly.

On Site

The Qbox Cable Tester

The Qbox cable tester has some truly unique features. For an instant hands-free intercom, simply plug in headphones and hook up one Qbox on each end of a line.

Figure 49 Real-time analyzer.

A real-time analyzer usually displays frequencies in octave bands. An octave is the interval between two frequencies, where the second frequency is twice the frequency of the first. For example, 80Hz is a low-level frequency; 160Hz is an octave above 80Hz; 320Hz is the next octave, and so on. Octaves are expressed as a ratio of 2:1. An octave band is the range of frequencies in the octave. Octave bands are usually defined by their midpoint. Given a frequency range of 80Hz to 160Hz, the octave band would be referenced as the 120Hz band.

5.4.0 Impedance Bridges

An impedance bridge (sometimes called an impedance meter) measures resistance, conductance, inductance, and capacitance (*Figure 50*). In audio, it is used to troubleshoot the impedance of a circuit. For example, it can be used to calculate the impedance of speaker circuits. You can measure the amount of power that will be consumed on a circuit when power is applied. This enables you to verify the wattage of the speaker load. For example, 10 speakers tapped at two watts each in a constant-voltage system will consume approximately 20 watts at 1kHz.

An impedance bridge will measure the total impedance of each device, including insertion loss and cable loss at a preset frequency. This device is useful when troubleshooting a sound reinforcement system. It can help identify and locate opens, shorts, and overloaded circuits. The results of impedance testing provide appropriate documentation for close out.

5.5.0 Spectrum Analyzers

A spectrum analyzer (*Figure 51*) is an invaluable tool for measuring amplitude versus time. The spectrum analyzer shows what is happening to the sound energy as it propagates over time. It can be used to measure a particular spectrum or frequency band. It is commonly used to measure the frequency response of the system relative to the room. In typical applications, a pink-noise sound source is used to generate test tones.

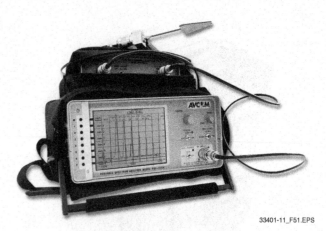

33401-11_F51.EPS

Figure 51 Audio spectrum analyzer.

The output of a spectrum analyzer is usually displayed on the device itself, though some newer models use a computer display to show information about the spectrum. In fact, one of the most popular approaches is to use a computer with an audio adapter (a sound card) specialized for spectrum analysis tasks. There are even spectrum analysis programs that use small handheld PCs or smartphones. The output is often used to adjust the location or alignment of speakers by first determining where the most desirable, or flattest, frequency response is in the room.

5.6.0 Sound Pressure Level Meters

An SPL meter is a device used to measure sound pressure levels. It consists of a special calibrated microphone, a small amplifier, and a meter. It presents sound level measurements in decibels. These measurements relate back to earlier discussions of loudness contours.

To manage different frequency requirements, an SPL meter (*Figure 52*) provides several weighting curves that contour the responsiveness of the meter to the frequencies being measured. This is necessary, as frequency plays an important role in perceived loudness. A 1kHz sine wave at 90dB is rather loud, but a 30kHz sine wave at 80dB is inaudible.

- The A curve measures sound waves with a fast roll-off of low frequencies. This weighting curve operates similar to human hearing and is used to measure sound pressure levels in the range of 20dB to 55dB.
- The B curve measures moderate SPLs between 55dB and 85dB.
- The C curve measures frequencies uniformly across the spectrum and is used for sound pressure levels of 85dB and greater.

33401-11_F50.EPS

Figure 50 Impedance meter.

- The linear response curve is similar to the C curve, but does not have a filter. It indicates the actual sound pressure levels present in the air. Some SPL meters do not have a linear weighting. In those cases, the C weighting should be used.

Sound pressure level meters usually allow you to choose between fast and slow response. This is accomplished with a switch. Fast response is used to get the peak SPL reading.

When using a sound pressure meter, it should be held at arm's length. Typically, it is pointed at 90 degrees to the sound source. Outdoor readings are usually taken with the microphone pointing up. These actions serve to minimize the effects of reflections from surfaces.

6.0.0 INSTALLING AUDIO SYSTEMS

The installation of a sound reinforcement system is driven by the specific requirements as determined for any individual system. However, there are general rules and guidelines applicable to all system installations. The topics in this section provide guidelines that can be applied to most installations.

33401-11_F52.EPS

Figure 52 Sound pressure level meter.

6.1.0 Overview of Installation Activities

A number of activities make up the overall installation of a given sound system. Each system is designed to serve specific goals for sound reinforcement. These goals are based on a variety of measurable criteria. The stages of the installation process can be summarized as follows:

- Design
- Planning
- Procurement
- Staging
- Performance
- Testing
- Commissioning
- Maintenance

6.1.1 Design

A sound reinforcement system is designed by various individuals and groups. The outcome of design is a set of specifications that describe the goals of the system. Included in the design are specific requirements needed to meet these goals. These requirements are further refined to identify subsystems and eventually the components and materials required.

The output of design is a set of documents that define the requirements, describe the solution, and most importantly, specify the measurements for success. Because the output of design is documentation, expect to find architectural drawings, equipment specifications, schematics, and justifications for key decisions.

You might participate in designing a sound reinforcement system. Typically, though, design is performed by engineers, consultants and—of course—the customer.

6.1.2 Planning

Once the design of a system is complete, planning activities can begin. It is acceptable to begin planning earlier, but until designs are finalized and approved, planning activities are usually limited.

The output of planning is documentation. Planning serves to identify and sequence individual tasks needed to complete the installation of the sound reinforcement system. Planning must take into account requirements for prerequisite and postrequisite tasks. For example, sufficient electrical power is required for the operation of the system. The prerequisite needs may include upgrading power panels and power distribution throughout the facility. Post-requisite tasks would then include fire and safety testing of the updated power distribution once the sound system loads are placed on it.

Planning also seeks to identify the labor and costs associated with installation. For example, you might be called upon to route cabling and connect components, but professional riggers may be required to fly loudspeaker arrays.

6.1.3 Procurement

Procurement centers on ordering and acquiring the materials and components for the system. In many cases, the components are determined as part of the specification and will include manufacturer, models, and costs.

However, in some cases, components may simply be specified according to design requirements. For example, rather than stating that a given amplifier from manufacturer X should be ordered, the specification may call for an amplifier that outputs 1,000 watts at 70.7V peak power, for a sustained period of Y minutes.

6.1.4 Staging

Staging is the act of preparing and organizing materials prior to, or during, the actual installation of the system. Many pre-installation tasks can be completed during the staging process. It is important that procurement, staging, and installation activities be coordinated. This is where electronic systems technicians often begin their work. During staging, activities such as prepping, individual component tests, and the like will take place.

6.1.5 Installation and Task Performance

Installation is where all of the materials and components are actually put together. It is important to be task and sequence sensitive. Quality issues must be dealt with throughout each performance step.

Often, changes to the design or installation plan will have to be made during these activities. These must be understood and approved by the customer, designer, and others.

During installation, certain testing procedures may be carried out. For example, grounding requirements should be monitored. Transformers in a distributed, constant-voltage system are brought online and verified, and wiring is tested for continuity. Many of the tests performed during installation are verification tests.

6.1.6 Testing

Once the system has been installed, system testing begins. Refer back to the design guidelines for determining and measuring specific system requirements, such as sound pressure levels, gain structure, or headroom. Additionally, system performance and operation tests are conducted. You will participate in most, if not all, of these testing activities.

6.1.7 Commissioning

Commissioning is the process of verifying and validating a system. Commissioning is an active set of testing procedures and often includes user training. The output of commissioning is documentation.

The ultimate goal of commissioning is to ensure that standards are met. Standards are determined by multiple groups or organizations. System performance standards are determined by the customer and design team. Additional standards are set by standards bodies, such as the *National Electrical Code®*. Others are established by state and local governments.

There are several levels to the commissioning process. At one level, the system should be validated. This means the system must be matched to the goals that were used to guide the design. At another level, the system is actually tuned and optimized. Troubleshooting activities may be triggered during this process.

6.1.8 Maintenance

Maintenance is the long-term activity of keeping the system in working order. There are two active processes: preventive and ongoing. Preventive activities involve inspection and testing. Inputs are provided that yield predictable results and are measured and tracked. Ongoing activities include upgrading of equipment and replenishing consumables, such as batteries. Troubleshooting activities are triggered throughout the process.

6.2.0 Speaker Rigging

Speakers come in all sizes. Some loudspeaker systems are built in floor-standing cabinets and enclosures. These are fairly easy to install; just stand them on the floor. Other loudspeakers and speaker systems are designed to be mounted into walls and ceilings. These are more challenging, but are reasonably easy to do. The loudspeakers and speaker systems that are the most challenging are those that need to be flown, or suspended.

The hardware and activities related to suspending speakers from ceilings are called rigging.

Loudspeakers or speaker systems must be designed for flying. Cabinets and enclosures for flown speakers are specially designed and constructed. Made of maple or birch multi-ply lumber, they are reinforced with internal bracing systems. Some speaker systems are mounted in enclosures made of engineered plastics. These newer speaker cabinets are often also rated for flying. Hardware is selected to support hanging load ratings. Hardware used to attach speakers to rigging must have established load ratings as well.

A load rating is a ratio. It represents the algebraic ratio of the minimum to maximum load stress. This is called the WLL, or working load limit. It measures the maximum allowable load versus the point where the system will fail. As a general rule of thumb, one manufacturer of speaker rigging systems suggests using a load ratio five times greater than the actual load (5:1). Another recommends ten times greater or 10:1. This means, for example, that if a wire rope has a WLL of 700 pounds, and breaks at 7,000 pounds, the load-to-failure ratio is 10 to 1. The inherent dangers of these types of systems make it important to never exceed working load requirements.

6.2.1 Speaker Arrays

A speaker array is a group or cluster of speaker systems and enclosures. The speakers within an array are oriented so that their combined response is more powerful and their coverage is generally greater than individual speakers. Speaker arrays are often flown.

Speaker arrays can be viewed as being either distributed or coupled. With a coupled array, the speakers are close together. With a distributed array, they are spread apart. When speakers are aligned in arrays, their acoustic patterns will converge either behind or in front of the array itself. Speakers in an array can be angled outward, inward, or along a plane. When speakers in an array are angled inward or outward, they are said to be splayed.

There are a number of acoustical challenges with arrays of any configuration. Of particular concern is an effect called comb filtering. Comb filtering results when the array of speakers lacks a uniform response. All of the speakers are aimed pretty much in the same plane, but the sound reaches the listener in such a way that they all have the same level, but with different delays. This causes gaps at various locations and severe sound distortions as the listener makes small position changes. A frequency plot for a comb filter has numerous equal amplitude peaks and gaps, and resembles a comb.

Another challenge with loudspeaker arrays has to do with their spacing. Ideally, speakers are separated from each other in such a way that their frequencies do not cross over and distort each other. The spacing of speakers is in part determined by their frequency: the higher the frequency, the smaller the waveform.

Three key speaker configurations are found in flown speaker arrays: cylindrical-section, spherical-section, and parabolic. Speakers are grouped in clusters to form any of these three configurations. The shape of these clusters has a significant effect on sound quality and speech intelligibility.

Cylindrical-section clusters (*Figure 53*) are formed by combining trapezoidal speaker enclosures edge to edge. With a cylindrical array, pan and tilt alignments are made to the cluster as a whole. You have to aim the entire cluster, as all of the speakers are locked together. Comb filtering effects are fairly restricted. Cylindrical clusters have high coherency. This means the sound waves that are produced all have similar direction, amplitude, and phase relationships and exhibit minimal interference.

Cylindrical speaker clusters can exhibit a noticeable degree of reduced performance. This is most pronounced when the shape of the venue is not well suited to the distribution pattern of the loudspeaker cluster. If a particular area of seats has poor sound, a single speaker or two cannot be adjusted to correct the problem, thereby limiting the usefulness of cylindrical-section clusters.

Spherical-section clusters (*Figure 54*) provide the best performance in many circumstances. These arrays or clusters are formed by aligning the speakers in several horizontal rows. Each row is tilted slightly below the one above. Speakers in individual rows are splayed outward, giving this array a spherical look. Each speaker in a spherical-section cluster can be individually aimed. This allows for the best possible coverage in all seating areas. In many circumstances, a spherical-section cluster provides the best performance when adequate coverage of all areas is the goal.

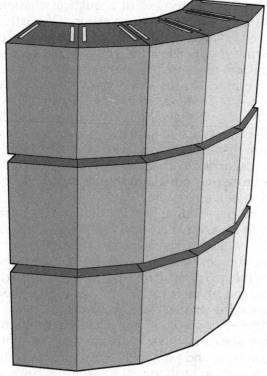

Figure 53 A cylindrical-section cluster of speakers.

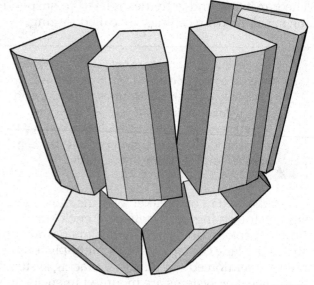

Figure 54 A spherical array.

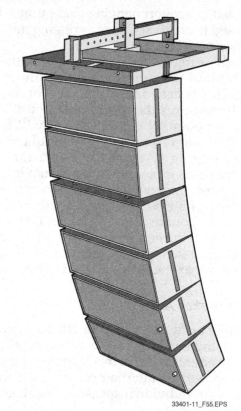

Figure 55 A parabolic array.

The third configuration is called a parabolic section (*Figure 55*). In technical terms, it is referred to as a vertically aligned, J-shaped array. It is also called a line array. Speakers at the bend of the J are focused downward towards the front rows. Each speaker (in its enclosure) generates a constant phase wavefront that combines to produce a single extended sound source. Each speaker is also designed so that the distance between speakers is defined as slightly less than one-half of the wavelength of the frequencies it reproduces.

Speaker alignments in a cluster or array are subject to rotation around three axes: horizontal, vertical, and rotational, or X, Y, and Z. One way to understand the positioning of speakers in three-dimensional space is to think about how an airplane orients itself in the air. An airplane uses the same three axes (X, Y, and Z) but they are called pitch, roll, and yaw. Refer to *Figure 56*.

When an airplane is level to the ground, its X-axis is parallel to the ground. The pitch changes as the airplane lifts higher in the air or as it approaches the ground. The nose of the aircraft lifts to increase or lowers to decrease the pitch. Speakers are referenced to the elevation angle, or the degree to which they are tilted relative to the ground. The X-axis can be measured using an inclinometer.

The Y-axis, or vertical plane, is used to measure roll. Roll is the degree to which a speaker is rotated relative to the horizontal plane. If an airplane is rotated 180 degrees, it is upside down relative to the ground. Its belly is facing up. The same is true with speakers. Speakers are rotated on the Y-axis in 90-degree increments. Zero degrees (0°) means the speaker is upright. If the speaker is aligned at 180 degrees, it is upside down. There is no common useful way to set or measure speakers on the Y-axis other than 0, +90, −90, or 180 degrees.

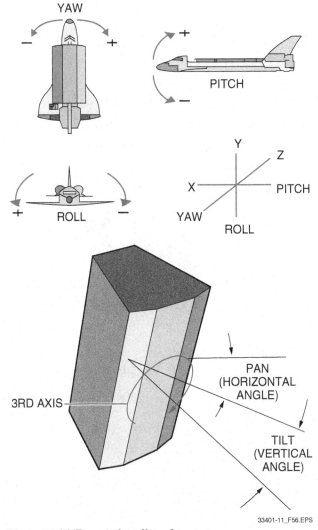

Figure 56 XYZ as pitch, roll, and yaw.

The Z-axis (yaw in aviation terms) is the panning angle used to aim speakers. The panning angle is the back and forth, or the left to right positioning of the speakers. These angles are measured with a compass or a transit (a surveyor's instrument). Mounting equipment may even have scales printed directly on the rigging equipment for easy alignment of the panning angles.

6.2.2 The Basics of Rigging Speaker Arrays

Loudspeakers that are intended for flying are designed with attachment points on the speaker enclosure. These points are arranged in a number of configurations, or patterns, as determined by the manufacturer. The ease-of-use of a given rigging system or methodology can be affected, as the rigging points may not align with many suspension products. Rigging costs, both in hardware and labor, are then affected.

As a general rule, the complexity of rigging a speaker can be determined from the data pro-

vided from the loudspeaker manufacturer. Speakers designed with rigging in mind often have specific types of rigging points in precise locations. The exact location of the center of gravity for the enclosure is identified. However, when the manufacturer provides little or no information about rigging points, their locations, or the center of gravity, the speakers will generally be more difficult to rig.

When flying horizontally aligned arrays, the top and bottom of individual speaker enclosures must be used. The sides are not available. The load placed on the rigging for horizontal arrays is significant and requires bracing through the individual enclosures.

When flying vertically aligned arrays, the top, bottom, and sides of the enclosures can be used to suspend the load. When using the sides, you actually brace through the enclosure. You may have the option (depending on the enclosure) of using the center of gravity to alleviate the need to brace through the enclosure.

Enclosures can have two-, three-, or four-point suspension (*Figure 57*). These are on the top and sides of the enclosure. With a three-point system, any one of the points must be able to support the load. With a four-point system, any two of the points must be able to support the load. All suspension points must still be used. Some speaker enclosures require a back chain to reach a suspension point on the bottom of the enclosure.

6.2.3 Rigging Configurations

The key to fast and effective rigging is to have control over each single enclosure before grouping the enclosures into clusters. Rotary spreader beams are popular rigging devices because they provide control over the pan or horizontal angle of flown speakers. The ZBeam®, manufactured by Polar Focus®, is an example of a rotary-spreader beam mount for speaker rigging.

To angle speakers downward, a pulley (*Figure 58*) is used to connect to the suspension point at the bottom of the speaker. For steeper downward angles, two cables can be connected to a suspension point.

Flown speaker configurations can be suspended from a steel beam (*Figure 59*). There are a number of ways to engineer a suspended or flown speaker array, including truss systems, grids, and tracks. The key principle to consider when combining speakers into an array is that if any single speaker is not under control, then the output of the entire array could perform poorly. This can result in comb filtering, lobing, and a number of listening areas with poor audio coverage.

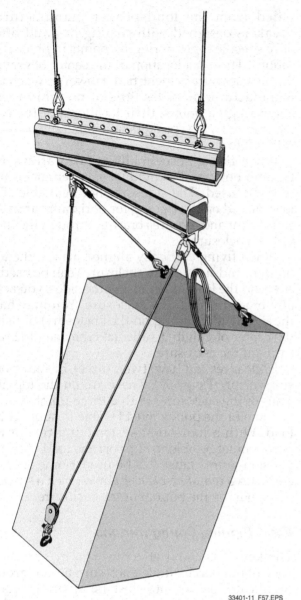

33401-11_F57.EPS

Figure 57 Three-point suspension for rigging flown speakers.

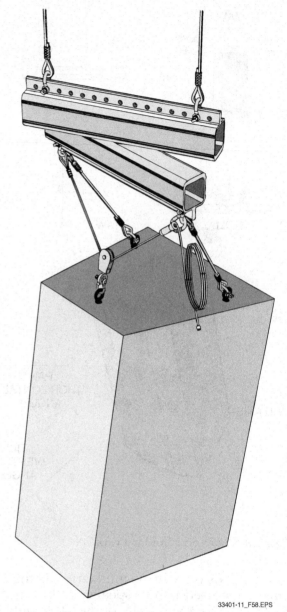

33401-11_F58.EPS

Figure 58 Angling speakers downward.

6.3.0 Troubleshooting an Audio Installation

Problem-solving has four primary, or high-level, steps that lead ultimately to a useful solution. The key to successful problem solving begins with calm and patience. It is extended by a methodical approach of divide and conquer. The end result is a well-formed solution. By following these steps, most problems can be resolved as they arise:

Step 1 Determine the problem.

Step 2 Determine the cause of the problem.

Step 3 Eliminate the problem cause.

Step 4 Verify that the system is operational.

Many problems can emerge in a sound system, regardless of whether it is a simple or complex system, but there are five primary problem areas:

• Power system failures or power quality problems
• Interconnection problems

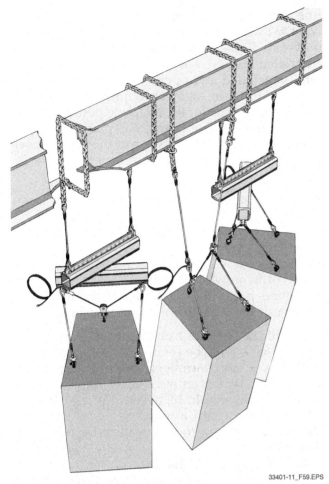

Figure 59 A speaker array suspended from a beam.

33401-11_F59.EPS

- RFI and EMI
- Headroom and SNR
- Operating limits of the equipment

By understanding the nature and causes of each of these problem areas, you will be prepared to identify and solve them. Power problems can be resolved by ensuring that the peak output of any set of devices doesn't exceed the capacity of the circuit from which they are drawing power. Interconnection problems are often avoided by using balanced lines wherever possible and using a direct input device when available. Grounding and power issues are explored in depth in the *Rack Systems* module.

RFI and EMI issues are often best dealt with through isolation; that is, isolating devices and providing line filtering. The increased use of wireless connections for musical instruments and microphones makes RFI an emerging problem area.

Headroom and signal-to-noise issues are often resolved using equalization, line filtering, and gating. Amplifiers, preamplifiers, and power distribution amps all contribute to improve the quality of the sound within the limits of the equipment.

One of the biggest challenges is to determine if a problem is electrical or acoustical. Electrical problems often generate noticeable noise and sound distortions. Feedback is an acoustical problem. Hum and hiss are electrical problems. Here are some common artifacts resulting from electrically related issues:

- Hum, particularly low-frequency hum, is often an indication of ground loop problems.
- A noticeable buzz can occur from poorly or improperly terminated cables and connectors, as can pops, clicks, and other static noise.
- Noises that sizzle or crackle can result from microphone placement or from RF interference derived from wireless microphones.
- High-pitched squeals may indicate a need to adjust for feedback, but more than likely point to cables being too close to power distribution lines.
- Waterfall sounds and hiss suggest system-generated noise, which can be traced back to the system gain structure.

A common acoustical problem is a loud screeching noise, which is a sure indication of feedback. Feedback is often more pronounced in reverberant environments. Drapes, wall coverings, carpeting, and acoustical tiles can reduce feedback, but not eliminate it. Some degree of equalization will be needed. This can be accomplished at the mixer or through the use of a feedback eliminator.

Another type of problem occurs when a signal doesn't appear at the speaker. This can result from simple errors, particularly cabling or routing errors. Make sure everything is turned on and has power. It is easy to mistakenly swap inputs for outputs when connecting components. Patchbays are often a source of these problems.

Once a problem has been identified, it must be resolved. Ground loops can be removed by lifting the cable shield ground on the receiving equipment. Cables can be re-terminated or replaced to manage termination problems. Feedback can be addressed by placement of microphones or through effective equalization. The important thing is to address the issues that cause the problem as a means of removing the problem.

Once the root cause of the problem has been addressed, make sure the system works. Don't wait for a program to start or an All Exit announcement to find out whether the problem was fixed. Even more important, make sure other problems have not emerged as a result of the fix. Systematic testing should always be used to verify the solution.

7.0.0 SYSTEM COMMISSIONING

The commissioning of an audio system is actually a two-phase approach to testing. The process of commissioning consists of a suite of test activities that take place at two key milestones: staging and installation.

The methods and activities of commissioning are common across all methods of sound reinforcement. However, each type of system will determine specific commissioning methods. For example, a constant-voltage audio system, such as an intercom system, will trigger different testing and verification activities than will an arena sound system.

7.1.0 General Commissioning Activities

During the commissioning of an audio system you will perform a number of testing and documentation tasks. General commissioning activities include the following:

- Factory acceptance testing
- Site acceptance testing
- Documentation review
- User training
- Customer sign off

7.1.1 Factory Acceptance Testing

Commissioning is part of a set of activities that range from design to installation. Before the actual installation there is a staging task, in which system components are collected and prepped for installation. Another term used for this staging process is *factory acceptance testing*.

The goal of this level of testing is to verify correct operation of equipment to ensure that everything works as designed before the system is actually brought online. Generally, this means testing everything except the loudspeakers. The focus of the tests is to ensure the electrical performance of the various components.

Until the system is actually cabled together, there are always unknowns. During the factory acceptance testing, external inputs will need to be simulated. For example, if a matrix switcher in a distributed constant-voltage system is used, it may accept alarm inputs that trigger fire or other evacuation procedures. These inputs will have to be provided in order to test the various functions. The factory acceptance test ensures that the desired outcomes are achieved.

Another suite of testing activities verifies that the equipment can recover from power problems. If a UPS or other backup power source is provided, you must make sure that it comes on-line within the allotted recovery time. You will want to connect the various devices to the backup power and ensure that they stop and start in a controlled fashion. You cannot wait until the system is installed and operational to find that the recovery time for a power amplifier is less than the response time for the alternate power source.

The overall goal of factory acceptance testing is to ensure that the installation of the system goes smoothly. To this end, it is necessary to ensure the following:

- Proper power requirements have been achieved.
- The phase integrity (polarity) of the system is verified.
- The system is free of noise, oscillations, hums, and other unwanted artifacts.
- Devices perform their desired functions.
- Devices work according to specifications.

In addition, any control systems that need to be customized should be set up and tested before installation. Some controllers have software, which will need to be configured and backed up.

All cabling should be tested. Lift all grounds and test them with a multimeter. Signal pads should be checked to verify they are free from distortion.

For rack-mounted hardware, the factory acceptance testing allows you to perform burn-in activities for components. Heat problems are also avoided by testing, measuring, and correcting any airflow and ventilation problems.

Once factory acceptance testing is complete, the job of installing the sound system begins in earnest.

7.1.2 Site Acceptance Testing

The site acceptance test is performed after installation. A variety of tests are performed. The ultimate goal of site acceptance test suites is to verify the acoustical performance of the system. Before turning the system over to operators or customers, you must ensure that the system works as designed.

During factory acceptance testing, the electrical performance of the equipment was verified, but speakers could not be tested. Site acceptance testing starts at the loudspeaker. To test acoustical performance, inputs are measured and verified at the speaker. Specific tests performed will include the following:

- Sound pressure levels
- Peak program levels
- Frequency response
- System gain structure

In addition to the testing procedures already discussed, additional activities are needed for a thorough site acceptance. For example, consumable materials, such as batteries for microphones, headsets, and belt packs, need to be inventoried. All of these devices should also be field tested to ensure they are operational.

External interfaces, such as fire alarm and evacuation systems, will need to be brought online. One of the goals of a modern sound reinforcement system is to provide public address capabilities that operate with high levels of speech intelligibility. External interfaces need to be verified as working within the established parameters.

All in all, site acceptance testing is the combination of systems tests, acoustical parameter tests, and ancillary systems tests. Once testing has been performed and the system performance is verified against the specifications, it is time for the documentation review.

7.1.3 Documentation Review

The documentation review is two-fold. It starts by first accumulating all relevant documentation. Once accumulated, it should be logged and recorded. Here is a list of documents that are provided for review:

- Scope of work
- Bill of materials
- Data sheets (manufacturers' spec sheets)
- As-built (hardcopy or CD)
- System commissioning reports
- System settings
- System operating procedures
- Operation manuals
- Service manuals
- Warranty documents
- Service and sales contracts
- Maintenance schedules
- Substantial completion forms (starts warranty)

Documentation requirements extend from the design to the commissioning of the system. The goal of the documentation review is to ensure that all relevant documents are ready to be handed over to the customer.

7.1.4 User Training

Not every system will require you to prepare and deliver training materials. If the audience is knowledgeable, you may just need to review the overall system setup and identify specific system settings and other key parameters. If the audience is not audio-savvy, you will have to prepare them to take ownership of the system. This may include providing a system overview, theory of operation, and hands-on exercises that develop specific operating skills.

User training can be a challenging undertaking, but it can be made easy by following a simple set of rules. The following will give you a sense of how to determine the requirements and match them to the needs of the audience:

- Determine user needs and what they want to be able to do with the system.
- Determine their knowledge and experience.
- Subtract what they know from what they need to know.
- Develop content that builds knowledge and skills that they lack.

As you develop content, the following actions will ensure that you have done a thorough job:

- Describe what the system is designed to do (scope, settings, operation).
- Cover bypass, reset, shutdown, and restart procedures.
- Demonstrate the system.
- Have the customer demonstrate.

Be sure to provide all relevant documents to support the actions taken. Once the training is complete, users will want to reference them.

When conducting the actual training, you can use the following guidelines to ensure a successful session. You may want to make yourself available for several days after the training to provide support.

- Target the audience.
- Use PowerPoint® or overheads.
- Introduce yourself and provide company contact info.
- Use a signup sheet.
- Provide manuals.
- Have a well-developed lesson plan.
- Videotape the session.
- Be patient.
- Watch the time.
- Don't ramble.
- Stay in control.

User training and session preparation activities are covered in detail in *System Commissioning and User Training*.

7.1.5 Customer Sign Off

Customer sign off is the final step in the commissioning process. It signals that the customer has accepted the system, knows how to operate and use it, and is ready to benefit from your many hours of work. The sign off should be accomplished with a formal document. At this point, any maintenance agreements will begin.

7.2.0 Commissioning a PA or Intercom System

As established earlier, many PA and intercom systems are constant-voltage systems. They use power amplifiers to distribute signals, and transformers to step-up and step-down the electrical current to provide audio signaling at individual speakers. Speakers are tapped to distribution wiring, allowing the volume of each speaker to be managed.

Collect all system schematics and blueprints. Be sure to include any changes and as-built documentation. You should have, at a minimum, the following tools:

- Walkie-talkie
- Impedance bridge or impedance tester
- Screwdrivers
- Pliers
- Crimpers
- VOM (volt-ohmmeter)
- 9V battery

Test each speaker. Make sure the speaker output is clear, with no rattles or distortion. Verify that the tap used by the speaker is appropriate for the sound level required. To make sure the speaker works properly, use a 9V battery and attach it to the speaker's input leads. This will cause the speaker cone to bow outwards, showing that the speaker is operational.

Test each speaker lead with the impedance tester or bridge. Make sure the impedance is as specified. Also, test each line in the system for ground shorts using a VOM.

As a final test, have a second party go to each speaker in the system as you initiate a page to it from the control center. The walkie-talkies are needed here if the system does not have a call-back function. Ensure that each room or zone is properly labeled.

7.3.0 Commissioning an Audio System

Collect all system schematics and blueprints. Be sure to include any changes and as-built documentation. You should have, at a minimum, the following equipment:

- VOM
- Signal generator
- Real-time analyzer (RTA)
- Spectrum analyzer
- SPL meter
- Impedance tester or bridge
- 9V battery

You will need to perform all of the testing procedures described for PA and intercom systems. In addition, you will want to ensure the phase relationship (polarity) with each point within the system. You will also need to test or verify the gain structure as well as the signal-to-noise ratio for the system.

Summary

This module has examined many facets of sound reinforcement, ranging from the science of human hearing to the physics of periodic waves and their various spectra. You learned about the various applications of technology and electronics as they support sound reinforcement. This final topic summarizes the many key points into three specific areas.

Sound is a waveform consisting of many frequencies. Human hearing operates in the frequency range of 20Hz to 20kHz. Human speech operates in the frequency range of 60Hz to 400Hz. The intelligibility of speech requires a clear audio signal at least 12dB above the noise level. Equalization of sounds within the range of hearing can result in clear and colorful sound.

Public address is a term used to describe a number of different sound reinforcement systems. These include PA systems, intercom, auditorium sound systems, room combining, and others. Systems can be direct-coupled with constant-voltage power, such as with an intercom or a room-combined system, or they can be powerhouses where the voice at a microphone is passed to a large array of speakers, such as those found in an arena. To work with these systems, an in-depth knowledge of microphones, speakers, and processing equipment is needed. In fact, it is the technologies themselves—amplifiers, transformers, effects processors, gates, and equalizers—that make this industry what it is for the electronic systems technician.

The process of installing and testing a sound reinforcement system is challenging. You must be knowledgeable in electronics and instrumentation and able to apply the theories of human hearing to the tuning of an installed system. Commissioning and training let you do more than just manage the complexities of the system, as you prepare manuals, documentation, and interact with users and operators.

Sound reinforcement is an art and a science. This module has tried to touch on the key points, but recognize that there is always more to learn. You are encouraged to do your own research, build your skills, and check the additional resources listed at the back of this module.

Review Questions

1. The loudness of sound is measured in _____.
   a. amperes
   b. joules
   c. decibels
   d. bels

2. The frequencies of a sound wave are measured in _____.
   a. amperes
   b. hertz
   c. decibels
   d. harmonics

3. A microphone is considered to be a type of _____.
   a. transducer
   b. amplifier
   c. booster
   d. transformer

4. A microphone that will pick up sounds equally from within a 360-degree circle is a(n) _____.
   a. radial microphone
   b. directional microphone
   c. omni-directional microphone
   d. shotgun microphone

5. In its most basic form, a speaker converts _____.
   a. waveforms into sound
   b. electrical current into sound
   c. sinusoidal energy into sound
   d. magnetic force into sound

6. The frequency response of a speaker is the ratio between the _____.
   a. waveform and its harmonics
   b. input power level and output sound pressure
   c. compression and rarefaction of its cone
   d. roll-off and slope

7. What types of sound waves have a greater tendency to cancel themselves out?
   a. High-frequency
   b. Mid-range frequency
   c. Low-frequency
   d. Supersonic

8. Phantom power is often provided to a microphone from the mixing console.
   a. True
   b. False

9. The device that allows you to filter or enhance specific frequencies is a(n) _____.
   a. amplifier
   b. equalizer
   c. mixer
   d. controller

10. A high-pass filter _____.
    a. attenuates low frequencies
    b. passes low frequencies
    c. amplifies high frequencies
    d. amplifies low frequencies

11. Harmonic distortions result from an amplifier creating harmonics for a waveform where they do not exist.
    a. True
    b. False

12. The minimum amount of headroom above the base level for a public address system is _____.
    a. 0.1dB
    b. 0.5dB
    c. 10dB
    d. 20dB

13. 18-gauge wire is suitable for runs up to _____.
    a. 700 feet
    b. 1,000 feet
    c. 1,100 feet
    d. 1,500 feet

14. The Haas effect describes a 40ms delay built into human hearing.
    a. True
    b. False

15. Under the inverse square law, doubling the distance from a sound source _____.

    a. has no effect on sound pressure level
    b. increases the sound pressure level by 25 percent
    c. reduces the sound pressure level by 25 percent
    d. reduces the sound pressure level by 50 percent

16. The best coverage for spacing ceiling speakers in a paging system is generally _____.

    a. one time the ceiling height
    b. two times the ceiling height
    c. three times the ceiling height
    d. four times the ceiling height

17. In a paging system, horns and wall baffles should be mounted facing _____.

    a. each other
    b. in the same direction
    c. down towards the floor
    d. upwards towards the ceiling

18. The goal of noise masking systems is to _____.

    a. provide speech privacy
    b. eliminate environmental noise
    c. lower the ambient noise level
    d. increase sound level for background music

19. Room combining systems have three key components: a distribution system, individual room systems, and a(n) _____.

    a. sound system
    b. central control system
    c. volume control system
    d. interface controller system

20. Tip-ring-sleeve (TRS) tipped phono cables are used for _____.

    a. unbalanced lines
    b. microphone connection
    c. balanced lines
    d. connecting large speakers

For Questions 21 through 24, match the appropriate word or phrase to each statement (not all apply).

21. Power amplifiers, signal processing, and speaker cabling contribute to _____.

22. Tests to determine loudness test for _____.

23. A form of feedback or noise is caused by _____.

24. Testing for _____ can help avoid phase cancellation.

    a. Sound pressure levels
    b. Microphone input
    c. Line levels
    d. Gain structure
    e. Ground loops
    f. Absolute phase and polarity
    g. Impedance

25. When measuring sound pressure levels, you should hold the meter 45 degrees to the sound source.

    a. True
    b. False

26. Pan alignments and tilt alignments are made to a cylindrical-section speaker array as a whole because all of the speakers are locked together.

    a. True
    b. False

27. Each speaker in a spherical-section cluster can be aimed individually.

    a. True
    b. False

28. The speakers in parabolic-section speaker clusters, or line array, combine to form a single, extended sound source.

    a. True
    b. False

29. The overall goal of factory acceptance testing is to _____.

    a. check external interfaces for integrity
    b. ensure that the system installation goes smoothly
    c. check that all the speakers work properly
    d. verify the sound pressure level

30. The goal of site acceptance testing is to _____.

    a. verify the acoustical performance of the system
    b. test each equipment component before installation
    c. commission the system
    d. run continuity tests on cables

Trade Terms Introduced in This Module

Acoustical: Audio properties of a device, space, or material.

Amplifier: A device that increases the power of an audio signal.

Amplitude: The strength of an audio signal, measured in decibels.

Bandpass filter: A filter which cuts off frequencies outside of an established range.

Bus: Signal lines in an audio mixer.

Clipping: An unwanted effect that occurs when the peaks of an audio signal are cut off by an amplifier or other device.

Compression: The forward movement of a speaker's cone.

Compressor: A device that reduces or levels the dynamic range of a signal, making loud parts softer and soft parts louder. A compressor is a limiter with a ratio of 4:1.

Crossover: A circuit found in loudspeaker systems that directs certain frequencies to specific speakers.

Decibel (dB): One-tenth of a bel. A decibel is a logarithmic measurement of electrical, acoustical, or power ratios.

Diaphragm: A thin membrane that vibrates, either in response to sound waves to produce electric signals, or the reverse, in response to electric signals to produce sound waves.

Driver: Another name for a loudspeaker.

Equalization: To modify the frequency range of an audio signal.

Expander: A device that enhances or extends the dynamic range. It is used to push down the noise floor, eliminating unwanted effects of hiss and other noise.

Fader: A control that allows a variable attenuation of the audio signal and, therefore, its waveform.

Feedback: A loud screeching noise that results when sound from an audio system is reintroduced into the system, forming a loop.

Frequency: Defines how often a periodic wave occurs. Frequency is measured in hertz, or the number of times per second. In audio, frequency relates to the pitch of the sound.

Frequency response: The ratio between the input power level and the output (sound pressure).

Gain: An increase in power or signal level provided by a device. Each device adds its gain to the overall strength of the signal.

Gate: A circuit that acts like a switch, either on or off, allowing frequencies above a certain threshold to pass.

Haas effect: A 40-millisecond delay built into human hearing. Delayed sounds are integrated by the sound equipment if they reach the ears within 40 milliseconds of the first sound.

Harmonic distortion: Harmonics of an input signal that are added to an output signal.

Harmonics: Similar to an overtone, harmonics are integral multiples of the frequency of a fundamental tone.

Headroom: The difference between the nominal operating level of a device and the maximum level that the device can pass without distortion.

Inverse square law: A decrease of 6dB for each doubling of the distance from a sound source.

Limiter: A device that limits the dynamic range of an audio waveform.

Lobe: A rounded energy projection. The lobe of a microphone is that portion of the microphone's three-dimensional sensitivity range where sound is picked up.

Masking: Using one sound to inhibit the hearing of another sound.

Mid-range: An audio driver designed to reproduce the middle range of audio frequencies.

Octave: An interval between two frequencies, where the second frequency is twice the frequency of the first.

Panning: Adjusting sound between two speakers, as in a stereo mix.

Period: For sine waves, the distance between two amplitude peaks.

Periodic: A repeating sine wave.

Phantom power: Using direct current electricity to power a condenser microphone over the same cable that connects the microphone to the equipment.

Phono-cartridge: A low-impedance transducer that converts the motion of a stylus on a phonograph record to voltage.

Pink noise: A complete, harmonic waveform resulting from random noise. Pink noise has a constant amount of energy in all octaves.

Rarefaction: The backwards movement of a speaker's cone.

Resonance: The intensity and continued expression of a sound vibration.

Reverb: Reflections of a sound wave that reach the ear with different delays. If the delay is too long, the sound is heard as an echo.

Roll-off: The portion of an audio waveform where high and low frequencies are attenuated.

Signal-to-noise ratio (SNR): A measurement of the amount of noise present after the value of a signal has been removed. It is expressed as ratio and measured in decibels. In audio, it relates to the strength of the signal over the amount of noise present.

Snake: A multi-cable microphone connector.

Sound damping: Removing energy from an audio signal.

Sound pressure level (SPL): A measurement of the amount of barometric change in the atmosphere due to a sound wave. SPL is expressed in decibels.

Transducer: A device that converts energy from one form to another.

Tweeter: An audio driver designed to reproduce the highest range of audio frequencies.

Waveform: The component frequencies of a complex sound wave.

White noise: Random noise with equal amplitude at each frequency.

Woofer: An audio driver designed to reproduce the lowest range of audio frequencies.

Additional Resources

This module presents thorough resources for task training. The following resource material is suggested for further study.

Audio Made Easy (OR How to be a Sound Engineer Without Really Trying). Ira White. Milwaukee, WI: Hal Leonard Corporation.

Audio Systems Design and Installation. Philip Giddings. Woburn, MA: Focal Press.

Guide to Sound Systems for Worship. Jon F. Eiche. Milwaukee, WI: Hal Leonard Corporation.

Live Sound Reinforcement. Scott Hunter Stark. Vallejo, CA: Mix Books.

Sound Reinforcement Handbook. Gary Davis and Ralph Jones. Milwaukee, WI: Hal Leonard Corporation.

The Audio Dictionary. Glenn D. White. Seattle, WA: University of Washington Press.

Figure Credits

NCCER CURRICULA — USER UPDATE

NCCER makes every effort to keep its textbooks up-to-date and free of technical errors. We appreciate your help in this process. If you find an error, a typographical mistake, or an inaccuracy in NCCER's curricula, please fill out this form (or a photocopy), or complete the online form at **www.nccer.org/olf**. Be sure to include the exact module ID number, page number, a detailed description, and your recommended correction. Your input will be brought to the attention of the Authoring Team. Thank you for your assistance.

Instructors – If you have an idea for improving this textbook, or have found that additional materials were necessary to teach this module effectively, please let us know so that we may present your suggestions to the Authoring Team.

NCCER Product Development and Revision
13614 Progress Blvd., Alachua, FL 32615

Email: curriculum@nccer.org
Online: www.nccer.org/olf

❑ Trainee Guide   ❑ AIG   ❑ Exam   ❑ PowerPoints   Other _____

Craft / Level: _____ Copyright Date: _____

Module ID Number / Title: _____

Section Number(s): _____

Description: _____

_____

_____

_____

Recommended Correction: _____

_____

_____

_____

Your Name: _____

Address: _____

Email: _____ Phone: _____

33402-12

Video Systems

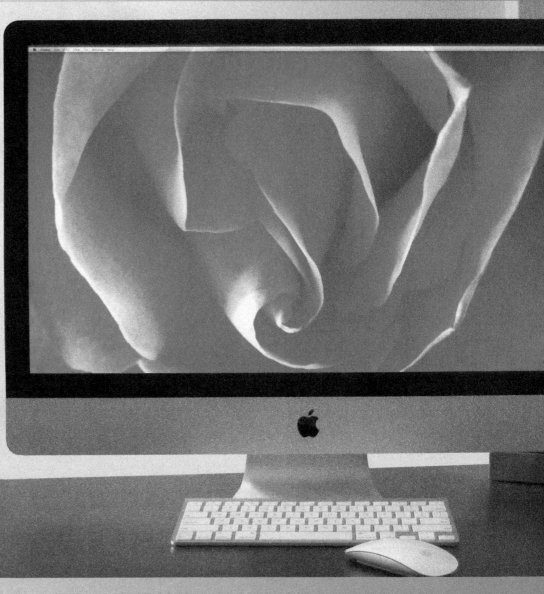

Endorsed by the Satellite Broadcasting & Communications Association (SBCA)

Module Two

Trainees with successful module completions may be eligible for credentialing through NCCER's National Registry. To learn more, go to **www.nccer.org** or contact us at **1.888.622.3720**. Our website has information on the latest product releases and training, as well as online versions of our *Cornerstone* newsletter and Pearson's product catalog.

    Your feedback is welcome. You may email your comments to **curriculum@nccer.org,** send general comments and inquiries to **info@nccer.org**, or fill in the User Update form at the back of this module.

 V.1 3/12

VIDEO SYSTEMS

Objectives

When you have completed this module, you will be able to do the following:

1. Describe what video is and the various ways in which it is produced and delivered.
2. Describe the characteristics of analog and digital video signals.
3. Describe video display technologies and video monitoring equipment.
4. Explain video processing methods and equipment.
5. Describe cabling and connectors used in video systems.
6. Identify the components of a video system.
7. Calculate the bandwidth of a video system.
8. Perform basic video system installation procedures, including:
   - Connect the components of a video system
   - Terminate an HD-15 connector
9. Isolate a fault in a video system.

Performance Tasks

Under the supervision of your instructor, you should be able to do the following:

1. Terminate a video connector.
2. Identify the components of a video system.
3. Connect a video system.
4. Calculate the bandwidth of a video system.
5. Set up a video display and verify proper operation.
6. Isolate a fault in a video system.

Trade Terms

3D TV
Aspect ratio
Advanced Television Systems Committee (ATSC)
Bandwidth
Blanking interval
Blu-ray
Chrominance
Commission Internationale de l'Clairage (CIE)
CMYK color
Color space
Composite video
Component video
Digital visual interface (DVI)
Frame rate
Gamma
High definition multimedia interface (HDMI)
High definition TV (HDTV)

Horizontal blanking interval
Hue
Interlaced video
Luminance
National Television System Committee (NTSC)
Progressive video
Resolution
RGB
RGBHV
RGB sync-on-green
Saturation
Standard definition television (SDTV)
S-video
Synchronization (sync)
Transition minimized differential signaling (TMDS)
Vertical blanking interval
Y/C video

Industry Recognized Credentials

If you're training through an NCCER-accredited sponsor you may be eligible for credentials from NCCER's Registry. The ID number for this module is 33402-12. Note that this module may have been used in other NCCER curricula and may apply to other level completions. Contact NCCER's Registry at 888.622.3720 or go to nccer.org for more information.

Contents

Topics to be presented in this module include:

Figures and Tables

Figures and Tables (*continued*)

1.0.0 INTRODUCTION

Video technology has become an important aspect of the communication process. The importance of visual information in general has increased over the last 120 years. Still photos from the nineteenth century and moving pictures from the early twentieth century can illustrate and bring to life the drama and character of a changing society. Television opens the world to scrutiny as events unfold before a global audience.

This module will explore video technology as it is applied to the public presentation of information. It will look specifically at the uses of video technology in the display of information for public, educational, and business applications. It will identify new technologies as well as existing, older video technologies and emphasize the integration issues that might be encountered by technicians.

Applications for video can be found in just about every business segment and enterprise. Some of the most common applications the electronics systems technician will encounter are as follows:

- Video conferencing
- Media management systems
- Banquet and meeting facilities
- Network operation centers
- Government and civic chambers
- Courtrooms and arraignment systems
- Distance learning classrooms
- Auditoriums
- Corporate boardrooms

Video applications can be found just about anywhere. It is important to keep in mind that video is usually accompanied with some form of audio. It is difficult to ignore the close relationship between audio and video; however, this module specifically emphasizes the video aspects. The only application in which audio is not likely to accompany video is in CCTV surveillance systems.

Video technology is found at the heart of a number of technologies. For example, CCTV surveillance is a typical application of video technology and so are home entertainment systems.

2.0.0 OVERVIEW OF VIDEO TECHNOLOGY

This section introduces video technology. It identifies the various devices likely to be found in the field. It starts with the concepts of a simple video presentation system and progresses to complex setups that include multiple video sources and outputs.

2.1.0 Simple Video Systems

A simple video system consists of a video source, such as a camera or video tape player, and a video output device. *Figure 1* shows this simple system using a DVD player and a monitor. The DVD player reads encoded video signals stored on a standard DVD, decodes them into electrical impulses, and sends them to the monitor. The monitor receives the electrical signals, translates them into images and sound, and displays them on the screen.

> **NOTE**
>
> Standard DVDs are not designed for playback on high-definition (HD) TV monitors. DVDs typically output data in a 480i or 480P format (740 × 480 pixels). HDTV pixel count, on the other hand, is 1280 × 720 (720P) or 1920 × 1080 (1080P). For that reason, a Blu-ray-compatible playback device is required for HDTV.

On Site

A Brief History of Television

The birth of television can be traced to the late nineteenth century when signals developed using a rotating disc were transmitted over wire by Paul Nipkow of Germany in 1884. The cathode ray tube, invented in 1897, paved the way for electronic television. For several decades until about 1937, electronic and electro-mechanical methods shared the television space.

Regularly scheduled network television broadcasts became available in the eastern United States in the 1946–1948 timeframe, and had reached the West Coast by 1951. By 1954, more than half the households in the United States had TV sets. TVs of that time were primitive by today's standards. Images were black and white and viewing screens were small. Color TV became generally available in the 1970s, when solid-state electronics began to replace vacuum tubes. This change also paved the way to electronic tuners and remote controls.

Plasma TVs came on the scene in the 1990s, in response to the need for a technology that could display high-definition TV. Liquid crystal display (LCD) TVs followed in the first few years of the twenty-first century. By 2010, the LCD TV had become the dominant player in the TV market and 3D TV was becoming commonplace.

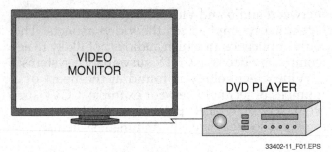

Figure 1 A simple video system.

2.2.0 Real-World Video Systems

Simple video systems can be found in a number of presentation environments, including classrooms, boardrooms, and even courtrooms. Such environments are likely to have multiple video inputs, such as DVD players, computers, and even satellite receivers. You will also find a number of output or presentation devices. These range from CRT monitors to rear projection screens, plasma or LCD video panels, and overhead LCD projection devices. There are even applications that transmit the video onto a computer network, such as a LAN or a WAN. *Figure 2* shows a fairly complex set of input and output devices as they might connect together for video presentation.

2.3.0 Video Connecting Equipment

A typical presentation environment, such as a corporate boardroom, will contain multiple input devices, but the output devices are usually limited. For example, there may be a single projection system that supports inputs from a computer and a DVD player. To accommodate the many-to-one relationship between input and output devices requires specialized connecting equipment. It would not be uncommon to find a scaler, a video switcher, and possibly even an amplifier between the input and output devices. These devices will be examined in greater depth in another section.

3.0.0 ANALOG VIDEO

Worldwide, there are three standard analog forms of broadcast video. There are additional digital forms of broadcast TV, including high definition. Before you look at these different formats, there are some general facts to learn about video signals.

Video signals are electronic signals. They carry three basic types of information:

- *Brightness* – Brightness is the amount of light and darkness. In essence, it represents the black and white aspects of an image. Brightness is referred to as the luminance portion of an image.

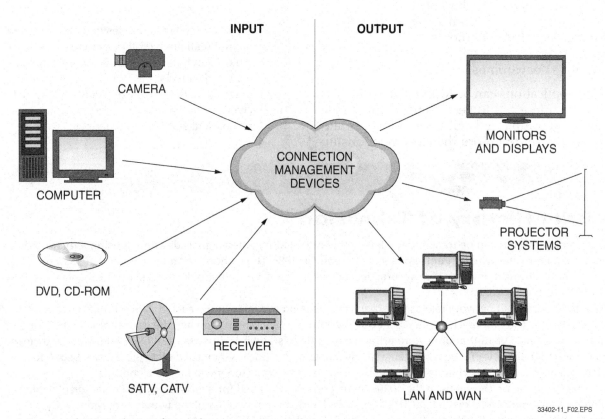

Figure 2 Input and output devices for a video presentation system.

- *Color* – Color represents two aspects of the color of an image: the hue and saturation. Hue is the actual color as it will appear on the display or monitor. Saturation represents the amount of gray in the color. Color information is referred to as the chrominance portion of an image.
- *Synchronization (sync)* – Synchronization (sync) signals are timing signals that allow the chrominance and luminance information to be matched into a coherent image on the screen.

3.1.0 National Television System Committee Analog Video

To understand video, it is necessary to review television broadcast standards. These standards not only determine the nature of broadcast video but also the requirements for various devices used for other forms of video presentation. These include monitors, displays, and televisions, as well as recording and playback devices such as VCR and DVD players, and of course, cameras.

The original broadcast standard in the United States, called the National Television System Committee (NTSC) standard, was formed in two parts. Originally, television was broadcast in black and white. Later, it was revised with color insertion. Keep in mind that a black and white image is different from a monochrome image. As the term suggests, a monochrome image has a single color. Elements of a monochrome image are black and white only, or white and the absence of white. A monochrome image is one-dimensional and contains only luminance information.

A black and white image, on the other hand, also contains chrominance information. A two-color palette, black and white is mixed to create varying shades of gray. Black and white images are two-dimensional; they have both hue and saturation. In this context, however, hue is not color but the amount of gray relative to other contrasting shades within the image.

The original NTSC standard established the rules for the broadcasting of black and white images. Each image is scanned into 525 individual lines. These lines are then grouped into two different fields and broadcast. The receiving device combines the two fields into a single frame on the television screen, recreating the images.

Each field of the NTSC standard consists of one-half of the scanned lines. The first field contains all of the odd-numbered lines, starting with one. The second field contains all of the even-numbered lines, starting with two. Each field is recreated every 60th of a second so that both fields are recreated in one 30th of a second.

An image is recreated by first tracing every odd-numbered line and then tracing every even-numbered line. This process of painting the screen is called interlaced scanning, or simply interlacing. *Figure 3* shows the scanning process on a CRT-type monitor. The odd-numbered lines are drawn or traced first. Then each even-numbered line is interlaced between the odd-numbered lines. As a technology, this process is referred to as interlaced video. The electron beam in the monitor has to make two complete, consecutive passes every second in order to recreate an image.

By tracing two fields of lines, or one frame, a single image is formed on the screen. There are a total of 525 lines carried in the signal, but only 480 of them actually contain image information. The original NTSC standard is often referred to as 525i, or 525-line interlaced video. A 1080-line version of interlaced scanning was developed in order to display high-definition TV (HDTV) on a CRT. It is commonly referred to as 1080i. It more than doubles the scan rate, resulting in an image of much higher quality.

This method of television broadcasting was prevalent in North and South America for many years. A single standard serves much of the western hemisphere. There are, however, several countries in the Caribbean and on the South American continent that use a variant of the NTSC standards.

3.2.0 Color Signaling for the Americas

In order to broadcast color images, the NTSC first reviewed the current broadcast standards. They decided that consumers with black and white receivers must be able to receive color signals but display black and white images. To accomplish this goal, they decided to maintain the bandwidth

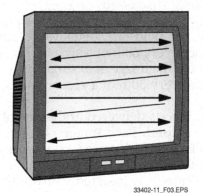

33402-11_F03.EPS

Figure 3 Scanning lines and horizontal blanking on a CRT.

---

requirements and overlay color onto the black and white images. This also meant that color receivers could receive and display black and white images.

To actually capture and broadcast color television would optimally require three cameras. Each camera would have a primary color filter. Primary colors are red, blue, and green. Each camera would be used to provide a full-bandwidth signal for one of the primary colors. However, to broadcast these color signals would require three full-bandwidth channels, one each for red, blue, and green. Each television channel is 6MHz wide. The FCC allocates channel frequencies, and they are limited. So to use 18MHz for color television transmission would severely limit the already constrained allocation of channels. It would also exclude existing black and white televisions from receiving the color broadcast.

The solution was to combine the full-bandwidth inputs for all three primary colors and subtract the elements that are common. The luminance information for brightness and detail are common to all three inputs and is already carried in the signal. The remaining color information is then gamma-corrected, or adjusted for the nonlinear relationship between the electrical voltage of the signal and the intensity of the image. The resulting color information is then transmitted on the chrominance signal with hue and saturation values for the red, blue, and green aspects of the image.

A single color camera contains three sensors: one each for the three primary colors. The luminance portion of each sensor signal is extracted. The remaining color values are normalized at 75 percent and gamma-corrected. The color information is then encoded onto the chrominance signal. The end result is a single composite signal that can be transmitted within the 6MHz-wide frequency band for a given channel and received by either a color or a black and white television.

3.3.0 Blanking Intervals

As previously discussed, analog video is displayed at two rates, both of which mean the same thing. One, the field rate, displays a single field of video once every 60th of a second. It takes two fields to create the image, so a single image is displayed once every 30th of a second and is called the frame rate.

Analog video is delivered at the rate of 30 images every second. Each image is composed of two fields. Each field is traced, one line at a time. To avoid unwanted screen artifacts, a horizontal blanking interval is inserted. During this brief moment, the electron beam repositions itself at the beginning of the next horizontal line. Likewise,

when the end of the first field is reached, the electron beam in the CRT has to be repositioned to the top of the display before tracing the next field. This is called the vertical blanking interval.

During both of these blanking intervals, a signal is still required. For both the vertical and horizontal blanking, the signal is encoded to send black.

3.4.0 Synchronizing Video Signals

There are several stages in the production, broadcast, and replay of video where synchronization is called into play. You will briefly learn the first two, as only a general understanding is required for this module. You will look at the third in closer detail because it is critical to the discussion regarding video signaling. Keep in mind during this discussion that it is covering composite video, where everything needed to create a video image is combined into a single, complex signal.

3.4.1 Audio and Video

Synchronization ensures that audio and video match up in such a way that when you see lips move, you also hear the sounds. You have most likely experienced video where the sound track and video images were not in sync, and it can be disconcerting.

3.4.2 Studio Equipment

Another area where synchronization is important is in the studio. Here there are video signals passing through multiple pieces of equipment. The video stream starts at the camera. It is passed through to the camera controller. From there it is channeled through a video mixer, special effects processors, time code corrector, and a host of other devices, before finally being recorded. It is critical that synchronization is maintained throughout this process of hand-off to each device in order to keep the myriad of signals correctly timed. The best way to do this is through a method called genlock. Here, a signal generator is used to create a timing signal in the form of regulated pulses. All equipment in the video chain is then synchronized (locked) to these timing pulses. These signals are often referred to as house black or house sync.

3.4.3 The Video Waveform

An important aspect of this discussion is the synchronization of the various elements in a video waveform. To ensure effective representation of

Vertical Blanking Interval

The vertical blanking interval consists of 21 lines out 525. It is also used to carry additional information. For example, closed caption text is encoded into the vertical blanking interval signal. All television receivers in the United States are required to have a special decoder for closed captioning. Manufacturers of receivers often opt to install additional decoders for proprietary information streams. Some broadcasters insert stock market data, TV listings, news reporting information, and even HTML data (used with a set-top box, digital TV, or a computer TV tuner card).

---

the images, the following aspects of the signal must be in sync:

- Horizontal scanning
- Vertical scanning
- Chrominance

In order to trace and interlace the individual lines contained in the two fields of the video frame, it is necessary to provide synchronization information. This allows the left and right edges of each line to be aligned, maintain the same level of contrast, and ensure that each line blends with the line above or below it.

The horizontal sync signal provides horizontal synchronization. Each line of the video field contains a horizontal sync pulse. This pulse signals the electron beam to go back to the left side in preparation for tracing a new line.

The vertical sync signal provides vertical synchronization. Each line in the video field contains a vertical sync signal. It is based on the width of the horizontal sync pulses (seven of them). This ensures that the signal is recognized.

Color burst is a special signal inserted into the composite signal. It is used to provide the color subcarrier, the frequency on which the color information is encoded. This allows all devices in the video chain to synchronize on that frequency and provide correct color information. A composite video signal is shown in *Figure 4*.

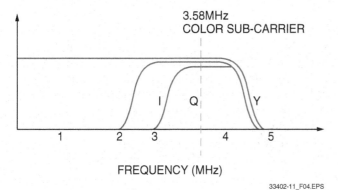

Figure 4 Diagram of a composite video signal.

3.5.0 Composite Video

The quality of a composite video image is influenced and limited by the way color information is generated, encoded, and transmitted. An important aspect of the video signal is the use of color information to recreate color images. This section introduces basic color principles and addresses the way color is encoded, decoded, and conveyed to a receiving device.

3.5.1 Color Spaces—Non-Video and Video

Different color spaces are found in day-to-day interaction with various media. For example, in the print world, there is the four-color space CMYK, consisting of cyan, magenta, yellow, and black. In video, there is the RGB color space, consisting of red, green, and blue.

The CMYK colors used for printing are considered subtractive colors. Subtractive color systems begin with light. When light strikes the surface of an object, wavelengths of light are subtracted, or absorbed from that light, giving the object color.

3.5.2 The RGB Color Space

The RGB color space consists of three primary colors—red, blue, and green. These colors are considered additive. Additive color systems start with the absence of light (black). New colors are created by adding the three primary colors together in some combination. Combining all three primary colors in equal intensities produces white. The RGB color space was selected for television and video components.

Interestingly, the human eye is most sensitive to the green range of the color spectrum. Also, color sensitivity is related to brightness. When color information is calculated for video, the green color composes almost 60 percent of the signal. This is clearly seen in the Commission Internationale de l'eclairage (CIE) (CIE is a standards body) color space shown in *Figure 5*. Red and blue are calculated by subtracting the luminance portion.

These two colors are then represented as R–Y and B–Y, meaning red minus Y or blue minus Y.

When the luminance portion of the signal is derived, it is based on the following formula:

$$Y = 0.299R + 0.587G + 0.114B$$

The green, or largest portion of the color spectrum, can be algebraically determined, given the R and B values. Therefore, only R–Y and B–Y are encoded on the color or chrominance portion of the signal. Green is encoded in the luminance portion.

For composite video, the chrominance information is added to the luminance information. Luminance is combined with horizontal and vertical synchronization signals and with blanking and color burst information to generate the waveform.

Composite video carries all video information in a single source signal. Prior to transmission, all aspects of the video stream—luminance, chrominance, and synchronization signals—are combined into the composite video signal. On the other hand, digital visual interface (DVI) and component video (including S-video and RGBHV) break the video signal into pieces. The individual signals are taken from the system before they would normally be combined into a composite signal. There are a number of component video formats.

3.6.0 S-Video

S-video provides better quality than composite video. Though S-video stands for separated video, it is often referred to as Y/C video and sometimes super-video. With S-video, the video signals are separated into black and white information (Y)

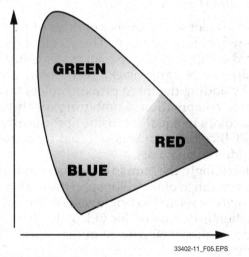

Figure 5 The RGB color space.

and color information (C). Synchronization signals are carried on the luma or brightness component.

S-video is the most common form of component video, particularly in consumer grade equipment. Most DVD players, CATV set-top boxes, and newer VCRs provide an S-video interface. When presented with a composite or S-video interface, the S-video connection should be chosen because it will generally provide the better image.

Most video encoding-decoding equipment uses comb filters to combine and separate color information on the luminance signal. S-video connections will separate the signal prior to encoding or immediately after decoding. The more processing involved with the signal, the more it is degraded. The Y signal contains the original black and white signal. The C signal carries the color information in a form prior to its encoding.

With composite video, the quality of the image is limited. This is partially the result of the original standard, which used a 19-inch monitor viewed at six times the distance. In this case, 19 inches multiplied by a factor of 6, which is the ideal viewing distance, equals 114 inches or just a little more than 9 feet. This distance takes advantage of the fact that the response of the human eye to color diminishes as distance increases.

3.7.0 Component Video YPrPb

Component video is the highest quality analog video connection available on consumer-grade devices. When available on both the source and destination equipment, component video should be used. Most DVD players support component video output because of the higher image quality.

The YPrPb labels refer to the signals carried on each one of the three cables. Y, for luminance, carries the brightness of the signal. Pr is the red portion minus the luminance signal. Pb is the blue portion minus the luminance signal.

3.8.0 YIQ

YUV is generally used in an NTSC system, but in some cases, YIQ is used instead. The U and V in YUV represent color information of blue-lumina (B-Y) and red-lumina (R-Y). The luma component (Y) is the same in YIQ, but the blue and red signals are in phase (I), and the quadrature phase (Q) is one that is modulated 90 degrees out of phase. YIQ signals can be either analog or digital.

Component vs. Composite Video

Component video is a video signal that has been split into two or more components. In common use, it describes analog video information that is transmitted or stored as three separate signals carried on three cables. Component video cables do not carry audio. Composite video, on the other hand, describes a video signal in which all the video information is combined into a single line level signal. It combines the luma (brightness), chroma (color), burst (color reference), and sync (horizontal and vertical synchronizing signals) into a single waveform carried on a single-wire pair.

3.9.0 RGB

Red, green, and blue can be carried on a three-wire cable, with each wire carrying an individual signal. RGB is the source for YUV, YIQ, and component video, so it is the best choice for color representation. Technically YUV, YIQ, and Y/C can be considered non-RGB color signals because they provide information needed to infer RGB colors but don't actually carry them.

3.10.0 RGB Sync-on-Green

RGB sync-on-green is an RGB component video stream carried across three wires. The three wires each carry a primary color signal. The third wire carries green as well as synchronization information. This is referred to as a three-wire system.

3.11.0 RGBS

RGBS translates to RGB with separate (composite) sync. The RGB components are each carried on a separate wire, and a fourth wire is used to carry synchronization information, both horizontal and vertical. This is referred to as a four-wire system.

3.12.0 RGBHV

This form of component video carries the RGB signals, each on a separate wire. Horizontal and vertical synch signals are also each carried on a separate wire, for a total of five wires. It is referred to as a five-wire system.

3.13.0 Other Video Standards

PAL and SECAM are two additional standards used for video. They share many aspects common to NTSC. Like NTSC, there are several variations of both used around the world. This topic discusses PAL and SECAM in general terms to introduce them to the electronic systems technician.

3.13.1 PAL

PAL, or phase-alternating line, is the analog video standard in Europe and other countries. It is similar to NTSC, with an aspect ratio of 4:3 and two fields of video that are interlaced to create a frame of video. Adopted in 1967, it supports a slightly higher resolution with 625 scanning lines. The frame rate is lower than NTSC, however, at 25 frames per second, and flicker is a bit more noticeable.

In PAL, the luminance signal (Y) is derived in the same fashion as NTSC, but a phase-alternating technique alters the subcarrier phase of the U and V components. Phase differences are cancelled out by reversing the phase (180 degrees) for V every other line. This eliminates color errors and improves contrast, but color saturation is diminished as a result.

3.13.2 SECAM

SECAM was adopted by France in 1967 and translates to sequential color with memory. It is a composite video standard that uses two color subcarriers, allowing it to maintain consistent color saturation as well as hue. This is accomplished by broadcasting each color sequentially, storing the previous line in memory, and then calculating the proper hue and saturation before displaying it. Like PAL, it has 625 scan lines and is subject to flicker.

4.0.0 DIGITAL VIDEO SIGNALING

All TV stations in the United States were required to begin digital broadcasts by the year 2002. They were also to begin full-time digital broadcasts by 2006 or when 85 percent of their broadcast areas had converted to digital. This requirement assumed that US households would upgrade televisions to digital or would install a set-top box that receives digital signals and converts them to acceptable analog inputs for display devices. On June 12, 2010, full-power television stations nationwide began broadcasting exclusively in a digital format. Anyone still using an analog TV receiver was required to connect it to a digital-to-analog converter box in order to watch digital programming.

To support the transition to digital TV, an overall standard for digital television was defined. The standard is called DTV and establishes the basis for the encoding, transmission, and decoding of digital TV signals.

The DTV standard does not define a standard display format. It specifies the requirements for the transmission, format, and compression of a digital television signal. Many different display types have emerged to receive these signals, and millions of older TVs are still in service. Providers have to deliver a full-time digital signal, so television stations were given an additional 6MHz channel band for digital broadcasts. This ensures older sets will continue to receive programming.

Just as analog video has come to be known by the committee that created the standard, digital video is often referred to by its committee. The Advanced Television Systems Committee (ATSC) has the responsibility for DTV.

The two types of DTV in the United States are HDTV and standard definition television (SDTV). Discussions of digital television usually concern HDTV.

4.1.0 Video Compression

Digital video files require a huge amount of storage space on a computer. Compression is a method of reducing the size of a video file by eliminating data bits that are not needed to recreate the image. In the 1980s, a group known as the Motion Picture Expert Group (MPEG), a sub-group of the International Organization for Standardization (ISO), began developing a standard for video compression. The initial effort, which became a recognized standard in 1992, was MPEG-1, with a 1.5 megabit/second (mbps) data rate. MPEG-2 was developed shortly afterward as a standard for full-broadcast video at data rates of 3 to 15 mbps. MPEG-2 is the compression technology used for DVDs. The initial release of MPEG-4 occurred in 1998. This version of the MPEG standard enables video streaming on the web to PCs and a variety of other devices, ranging from TV monitors to hand-held devices such as cell phones. The MPEG standards are divided into parts. The parts are different versions of the standard designed to support specific functions. To date, 27 parts have been released for MPEG-4. Part 27, released in 2009, establishes the requirements for 3-D graphics.

The purpose of compression is to make video images easier to store and use. Part of the compression process is the elimination of redundant images. For example, if the same item appears in many frames of video, the compression program will store it once, then retrieve it when needed. The assumption is that most content in a video scene does not change, or it changes slowly over time. Using MPEG, an initial frame is captured and serves as a reference frame. As the scene is monitored, only those elements that change are actually recorded. This continues until enough of the scene has changed to warrant capturing a new reference frame. This significantly reduces the data transmission requirements.

Compression algorithms can be classified as either lossy or lossless. Lossy compression is one in which data bits are permanently removed during the encoding process. They do not reappear when the file is recoded. This type of compression results in a smaller, more easily transmitted file. In lossless compression, fewer data bits are removed, and the ones removed are restored during decoding. Files compressed this way are of higher quality and can readily be edited. For example, if you wanted to shoot video and stream it directly to the web, you would choose a lossy compression program. If you were capturing raw footage to be edited for a TV production, you would choose a lossless program.

On Site

Video Compression

MPEG-2 and MPEG-4 are video compression algorithms that use the concept of change over time to compress images. The assumption is that most content in a video scene does not change, or it changes slowly over time. Using MPEG, an initial frame is captured and serves as a reference frame. As the scene is monitored, only those elements that change are actually recorded. This continues until enough of the scene has changed to warrant capturing a new reference frame. This significantly reduces the data transmission requirements. MPEG-2 is the compression technology commonly used for DVDs.

4.2.0 Benefits of High Definition TV

There are numerous benefits to high definition digital TV. They are best summarized as follows:

- Improved resolution
- Enhanced bandwidth
- Computer and network compatibility
- Clear signal transmission
- Improved audio
- Multicasting
- Aspect ratio

4.2.1 Improved Resolution

Digital television ushered in significant improvements in display resolution. NTSC has a resolution of 525 lines, of which only 480 are actually displayed. The NTSC non-digital pixel resolution can best be described as acceptable, though less than that of a VGA display. High definition televisions have several different resolutions, starting at 720p all the way to 1080i and 1080P lines of vertical resolution. Additional improvements are also achieved from progressive scanning at up to 60 frames per second. Also, the aspect ratio of HDTV is expanded to 16:9.

Interlaced scanning was suitable for analog video displayed on CRT monitors. When LCD monitors came along, they required a different approach. This resulted in the development of progressive scanning. In progressive scanning, each line, or row of pixels, is scanned in a sequential order, as opposed to the every-other-line method used in interlace scanning. A much sharper image is produced by scanning 60 lines per second, rather than 30. In the progressive scanning method, the screen is refreshed more often.

On Site

What is a Codec?

The term *codec* is shorthand for compression/ decompression. It is also interpreted as code/ decode. Regardless, a codec is a software implementation of an MPEG standard that is used to encode video and audio. There are hundreds of codecs; any file encoded by a particular codec must be decoded by a compatible codec. Most PCs have the most common codecs embedded in them.

4.2.2 Enhanced Bandwidth

DTV is still confined to the 6MHz-wide bandwidth provided with NTSC. Compression technologies, however, enable a much broader range of input signals to be compacted and delivered over the existing 6MHz bands. In fact, after decompression, a high definition TV signal is 27MHz wide.

4.2.3 Computer and Network Compatibility

DTV signals are digital and are therefore computer friendly. In earlier schemes, digital signals had to be converted to analog before being displayed. This occurred either before being delivered to the display device, or it was performed by the display.

4.2.4 Clear Signal Transmission

Unlike analog signals, where image quality is affected as signals get weaker over distance (the signal to noise ratio decreases), digital signals retain image and audio quality. With a digital signal, it is either there or it is not. There are two predominate digital broadcast protocols in place: 8VSB (8-level vestigial sideband) in the United States and COFDM (coded orthogonal frequency division multiplexing) in Europe.

4.2.5 Improved Audio

In the United States, Dolby® digital audio is used for audio information. There are two versions of Dolby® digital audio. Dolby® digital 5.1 carries five channels of high-quality digital audio. Front left and right channels, left and right side surround channels, and a center channel make up the five. An additional low-frequency special effects channel is also broadcast. This is sometimes referred to as 5.1 audio (pronounced *five-dot-one*). Dolby® digital 7.1 carries two additional side audio channels—left center and right center.

4.2.6 Multicasting

There are two primary types of digital broadcasts: HDTV and SDTV. An HDTV transmission carries a single channel of audio and video. SDTV (standard definition TV) transmissions allow broadcasters to package five different video

---

programs, or channels, into a single transmission signal delivered across a 6MHz wide band. This is called multicasting. SDTV is not limited to television programming. Providers can use the SDTV technology to deliver programming such as multiple views from different cameras or to provide interactive services.

4.3.0 HDTV Signaling

8VSB is an amplitude modulation (AM) scheme. It strips the sidebands of all but a set of desired frequencies. There are actually two key components to the digital signal. The first component is MPEG-2 video compression and encoding. The second, 8VSB, is actually the RF modulation scheme. Before a signal can be modulated, it must first be MPEG-2 encoded. This is accomplished with a piece of equipment called the MPEG-2 encoder. The signal modulation is performed using a device called the 8VSB exciter. The overall process for generating and transmitting the DTV signal can be seen in *Figure 6*. The bit stream from the MPEG encoder is passed through several levels of processing and is finally RF-modulated. The following sections describe the process in greater detail.

4.3.1 Synchronizing and Randomizing

The first phase of processing requires that the 8VSB exciter identify the beginning and end bytes of each MPEG-2 data packet. This packet is 188 bytes, and once identified, the MPEG-2 sync byte is removed, reducing it to 187 bytes. It will be replaced later with a segment sync byte.

The ultimate bitstream needs to appear as random noise. The randomizer changes each byte value using a known pattern of pseudo-random numbers that can be reversed by a receiver.

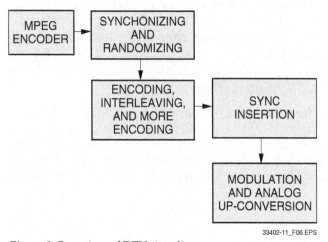

Figure 6 Overview of DTV signaling.

4.3.2 Encoding and Interleaving

There are two levels of encoding that take place at this phase. Both provide forward error correction. The data is first Reed-Solomon encoded, meaning it is manipulated in such a way as to derive a representation of the data that is used to predict errors. This is tacked onto the end of the packet, adding 20 bytes.

When the packet is received, the 20-byte signature is stripped off and compared to the received data. As you see later in the chain, each byte is transmitted as 8 tokens—hence, eight-level VSB—one level for each token. The receiver can select the reassembled packet with the minimal number of errors.

To help create these multiple tokens, the interleaver works to rearrange the bit sequences and creates new packets that consist of fragments of the original MPEG-2 data packets and their signatures. The method used here is called time diversity: it disperses the data across a bank of memory buffers during a 4.5-millisecond period.

A second encoding scheme called the trellis encoder is then brought into play. This is another forward error correction method. Each byte is converted into four 2-bit words. These in turn are compared to earlier 2-bit words to derive a special 3-bit value that represents the transitions between one 2-bit word and the next. These values are then inserted into the bitstream, replacing the original 2-bit values, and transmitted. In other words, every byte is processed and converted to 12 bits. A receiver can now use this information to predict errors as well as identify errors that have already occurred.

4.3.3 Sync Insertion

This phase of the signal processing chain adds signals that are used by the receiver to demodulate the 8VSB signal. A pilot signal is inserted to help the receiver recognize and maintain the correct synchronization of video frames.

A four-pulse ATSC sync symbol is inserted at the front of the MPEG-2 data packet. It replaces the MPEG-2 sync byte that was removed at the beginning of the chain. By this time, after the encoding phase, the data packet is now a stream of 828 eight-level symbols, and then the ATSC sync pulses are added. These pulses are used by the receiver to identify the data segment. *Figure 7* shows a representation of the ATSC data segment.

A third symbol inserted into the stream is the field sync symbol. It consists of an entire data segment of known pulses, and it acts to group a total of 313 data segments into a single field. *Figure 8* shows a complete field with a field sync segment.

SEGMENT SYNC	MPEG-2 DATA PACKET	REED-SOLOMON SIGNATURE
4 SYMBOLS (1 BYTE)	752 SYMBOLS (188 BYTES)	80 SYMBOLS (20 BYTES)

33402-11_F07.EPS

Figure 7 ATSC data segment.

4.3.4 *Modulation and Packaging*

The signal is now ready for RF amplitude modulation. During modulation, a large double sideband is generated that is too wide to be transmitted across the designated 6MHz band. A Nyquist™ filter is used to narrow-band filter the sidebands, eliminating the lower band and minimizing the upper. The signal now fits into the designated bandwidth.

The final step in the process is to take the intermediate frequency generated by the Nyquist™ filter and up-convert it to the assigned UHF or VHF band. It is then passed onto the DTV transmitter.

4.4.0 The Digital Color Space

Digital video uses two color spaces: YCbCr and RGB. The YCbCr color space is a digital representation of the YUV components discussed earlier in this module. Digital RGB is just the translation of the analog RGB signals into digital form.

Most video equipment, including digital television and DVD players, uses the YCbCr color space. Each of the color signals is sampled, and a digital value is assigned for each component.

There are several sampling formats used that contribute to the accuracy of the color reproduced on the display. Every sample contains a Y value, but the Cb and Cr components are sampled at varying rates to determine their values.

4.5.0 Digital Visual Interface

Digital visual interface (DVI) defines a standard digital interface for providing video signals to a computer display. There are two types of DVI: DVI-D is digital only, and DVI-I carries both integrated analog and digital signals.

Until recently, many displays were strictly analog devices. The computer's video sub-system had to convert the video to RGB or RGBHV before it was provided to a progressive video display. DVI, on the other hand, allows a pure digital signal to be presented to a display without intermediate steps for coding and decoding.

DVI uses a signaling method called transition minimized differential signaling (TMDS). Basically, TMDS works by converting 8-bit pixel data into 10-bit pixel data. This data is encoded for all three colors, which are then each carried on a separate line. Control data is carried on a fourth. The highest resolution supported by DVI is 1920 by 1080. Two TMDS channels can be used to increase the resolution, though at a slightly reduced bandwidth.

4.6.0 High Definition Multimedia Interface (HDMI)

High definition multimedia interface (HDMI) is an interface designed for the digital HDTV environment. It is used to transmit uncompressed digital video and audio between compatible devices such as DVD players, camcorders, PCs, set-top boxes, and game consoles. An important advantage of this technology is that it allows video and audio to be carried on a single cable. With HDMI, it is no longer necessary to have three video cables and two audio cables, as it is for component

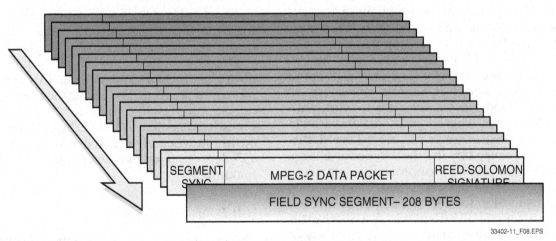

33402-11_F08.EPS

Figure 8 ATSC field.

video. Although HDMI delivers video as three distinct colors, it does so using three TMDS channels within a single cable. In HDMI, horizontal and vertical sync are added to the blue channel, and there are separate red and green channels. The HDMI cable can also carry eight digital audio channels. HDMI and DVI are very similar, and their devices are compatible, so an HDMI device and a DVI device can be connected without using a converter box.

HDMI was introduced in 2002 with Version 1.0. Since that time, there have been four additional versions, each providing additional features. Bandwidth has increased over time. Version 1 and 2 support 3.96 Gb of bandwidth, while Versions 3 and 4 provide 8.6 Gb. Versions 3 and 4 support 3D video, while Version 4 (2008) also provides support for an Internet channel.

4.7.0 3D TV

Three-dimensional (3D) video and movies allow the viewer to see depth on a flat screen, thus duplicating the three-dimensional image one would see in real life. For instance, a baseball hit in the recorded program would appear to be coming out of the screen and right at the viewer. The eyes are able to see in three dimensions because they receive two separate images that are processed by the brain to produce depth. 3D movies replicate this process by projecting two separate images onto the screen. That is why the 3D image looks blurry when viewed without the special 3D glasses. 3D TV achieves a three-dimensional view by alternately displaying a separate full-color image for each eye and sending a synchronizing signal to the 3D glasses.

Most 3D TVs on the market at this time (2011) are active-shutter displays. The glasses used in movie theaters cannot be used with these displays because the screens are not the same. The original 3D films that began reaching theaters in the 1950s used disposable cardboard glasses with one blue lens and one red lens. The glasses used to view newer technology 3D movies in theaters are passive polarized glasses designed to work with polarized screens. These glasses have relatively high quality plastic frames and are collected, disinfected, and reused by the theater. Because large TV screens cannot be economically polarized, different, more expensive glasses (sometimes ranging from $120 to $150) are required to view active-shutter 3D TV displays. Active-shutter glasses work by alternately blocking the view of one eye or the other in sync with the TV image. This can be 120 views per second, or 60 views per eye, or other rates, as long as the TV and glasses

are synchronized. Synchronization of the glasses is accomplished primarily with an IR signal, although an RF signal is sometimes used. Some projection TVs accomplish synchronization with digital light processing (DLP) technology, which uses a white signal on the screen to sync the glasses to the image. IR requires a built-in emitter, while RF uses a built-in transmitter. Glasses must match the IR emitter or RF transmitter. If an IR emitter is built into the TV, viewers will only be able to use the glasses supplied by the TV manufacturer, or other glasses that meet the manufacturer's specification.

Passive 3D TV has been introduced, but at this time does not provide the viewing quality of active-shutter systems. On the other hand, the cost of the glasses for passive systems is a fraction of that for active-shutter systems. It is possible to view passive 3D TV using the same glasses used in 3D movie theaters as long as they match the format of the program being viewed.

There are a number of 3D TV signal formats that can be delivered to the display device by source devices such as DVD players, satellite receivers, cable boxes, and game consoles. These devices must be designed for 3D viewing and any DVDs used with them must also be 3D compatible. An adapter may be needed to convert the signal from the source device to a format that is compatible with the display device. The most common signal formats used by source devices are as follows:

- *Frame packing* – Used primarily by Blu-ray players and game consoles
- *Side-by-side* – Commonly used by satellite receivers and cable boxes
- *Top-bottom* – Used primarily by cable boxes and some satellite receivers

The display device will convert the signal format to a frame-sequential format, which is the format required for active-shutter glasses.

The viewing angle for 3D TV is quite different than it is for LCD and plasma TVs and is much more of a concern. The recommended maximum horizontal viewing angle for 3D TV is ±30 degrees from center. In the vertical plane, 3D TV is optimum when viewed at eye level. A variance of ±15 degrees from eye level is the recommended maximum.

5.0.0 VIDEO DISPLAYS

The professional AV technician is responsible for the connection of display devices that match the presentation requirements of customers, the selected media, and the viewing needs of the audience. This section introduces the video display. It addresses issues regarding display characteristics such as aspect ratio, resolution, viewing angle, brightness, and other important attributes.

5.1.0 Overview of Display Technology

A display, in the context of these discussions, is any device upon which numbers, characters, graphics, or other data is presented. These devices can be CRT, LCD, LED, or other photo-luminescent panels. In regard to display technology, display monitors can be classified as follows:

- *Studio video* – Have a fixed scanning rate based on the TV standards in the country where they are used. Input is usually in the form of composite video or RGB.
- *Fixed frequency RGB high resolution* – Inputs are usually in the form of analog RGB. They often have multiple sync options and are generally used for computer workstation applications.
- *Multi-scan or auto-scan* – Support multiple resolutions and scan rates. Input is often provided as analog RGB using DB15 high density VGA connectors, although BNC jacks may be available as an auxiliary input.
- *VGA/SVGA/XGA* – Designed for use with PC video-graphic cards. SVGA and XVGA input is provided from an HD DB15 high-density VGA connector for analog RGB signals with separate horizontal and vertical sync. VGA uses a DB9 connector.

Note that many displays, whether CRT or otherwise, use video in an analog form. Digital video is converted to analog before being displayed. The conversion is done either by a device earlier in the video chain or directly at the monitor. It is important to know where this conversion takes place so the appropriate monitor can be selected.

Newer displays, primarily computer and high definition displays, are digital. These displays receive a digital signal and do not convert it into analog form before it is displayed.

5.1.1 Resolution

The term *resolution* is used to describe the level of detail presented by a display device. It describes how sharp the image appears to a viewer. As a measurement, it consists of variables from a number of different domains. For television or CRT displays, it usually means the number of scanning lines used to create a frame of video. The various facets of resolution are shown in *Figure 9*.

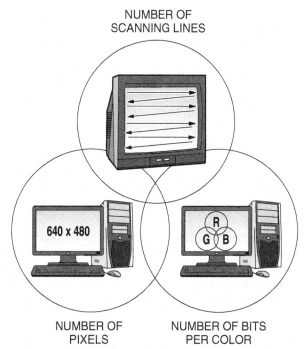

Figure 9 Three facets of resolution.

For computer displays, resolution represents the number of pixels, or picture elements, in the display. It is measured as the number of pixels in the horizontal axis times the number of pixels in the vertical axis. The higher the resulting number, the higher the resolution and the sharper the image is going to be. CRTs also have picture elements, but they are usually larger and fewer than found in a computer monitor. For example, a standard NTSC television monitor has 215,040 pixels per frame of video, while a VGA computer monitor has 307,200 pixels, and an SVGA monitor has 786,432 pixels.

Resolution is also used to describe the degree of color information provided by a display. Loosely related to the pixel count, the color resolution is determined by the number of bits needed to reproduce a single dot of color: the higher the number of bits, the greater the number of colors that can be reproduced by the display. For example, an 8-bit color monitor provides 256 different colors. A VGA monitor is 4 bit, and therefore provides only 16 different colors. A standard SVGA monitor is 16 bit, providing over 64,000 different colors. Newer monitors are 24-bit color devices and support over 16 million different colors.

5.1.2 Aspect Ratio

The aspect ratio is a way of describing the X and Y axes of a display. It relates the width of a display to its height. For example, in the United States, a sheet of paper used for photocopying or computer printing is called standard letter size. It has an aspect ratio of 8.5" × 11", referred to as eight and a half by eleven inches. There are other standard aspect ratios for paper, such as legal (8.5" × 14") or ledger (11" × 17").

Video monitors have different aspect ratios. A standard video monitor used for NTSC or PAL video has an aspect ratio of 4:3. This is referred to as four to three. The width of the screen is one-third wider than the height. This is often factored to the real value of 1.33 to 1, or 1.33:1. HDTV monitors have an aspect ratio of 16:9, or 1.78:1 (see *Figure 10*).

The aspect ratio becomes problematic when displaying output created for one aspect ratio on a device designed for another. This becomes apparent, for example, when converting film or movies to video or DVD formats. Originally, the 4:3 aspect ratio for NTSC video was selected because it closely matched the aspect ratio of movie films. Over the years, however, the movie industry has changed the aspect ratio of its theater screens. Wide screen movies have an aspect ratio of at least 1.85:1. Most wide screen formats stored using DVD formats have an aspect ratio of 2.35:1.

One solution for the difference in aspect ratios between movies created for wide screens but being displayed on 4:3 monitors is to use the letterbox format. Here the movie is resized to fit the horizontal aspects of a video display but with a large black border across the top and bottom. This retains the aspect ratio but at the expense of the vertical aspect and the overall size of the image.

As a result, the resolution of the image is somewhat diminished.

5.1.3 Viewing Angle

Viewing angle refers to the visibility and clarity of a displayed image based on the position of the viewer relative to the display device. It determines the locations in front of the display where an image can be seen without distortion. There are actually two different viewing angles to consider: vertical and horizontal. Vertically means the upwards or downwards angle. When looking at computer displays, the optimal viewing angle is approximately 30 degrees from center. This means that the most comfortable view of the displayed content is provided when the viewer's eyes are 30 degrees above the center of the screen, as shown in *Figure 11*.

The same human factors of vision used to determine the optimal viewing angle for a computer display can be applied to the viewing angle for presentation displays. However, because presentation displays are usually farther away, the effects of distance should also be considered. For example, the eye, head, and neck in particular, all participate in the viewing process. As the distance increases, the angle that is most comfortable for the eyes rises. This means the viewing angle actually gets smaller.

When looking at a computer display, it will typically be 20 to 40 inches away. The eyes are most comfortable looking down at that distance, and 30 degrees represents how low the center of the display is relative to the eyes. So a display being viewed from 30 feet may have a much smaller viewing angle although observers can see more of it. What you want to avoid is the head and neck tilting backwards as users look up. Over a small period of time, this can cause fatigue and therefore diminish the viewing experience.

Conversely, horizontal viewing angle refers to the side-by-side angles. This is where projection

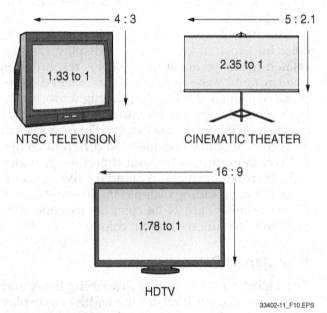

Figure 10 Aspect ratios.

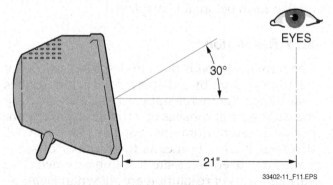

Figure 11 Optimum viewing angle for a computer display.

devices in particular have had problems in the past. Coupled closely with brightness and contrast (discussed later in this module), the horizontal viewing angle of many display devices is limited compared to that of a CRT. Throughout the 1990s, as flat panel displays entered and took over the market, continual improvement has progressed to the point where a 160-degree horizontal viewing angle is now the norm. However, keep in mind that contrast and brightness can significantly influence the horizontal viewing angle. Also, projection devices will typically have a narrower horizontal viewing angle compared with LCDs or CRTs.

Figure 12 shows a CRT with a horizontal viewing angle of 120 degrees. Viewers within the shaded region have an undistorted view of the image. The locations with a clear viewing space are defined by drawing a 60-degree line out from both edges of the television monitor.

5.1.4 Response Time

A CRT uses an electron beam to scan and retrace images. Two fields or one complete screen image is retraced every 30th of a second. With an LCD monitor, however, the image is created by thousands of pixel triads consisting of red, green, and blue pixels. The picture elements are actually diodes with a small capacitor associated with each, and there is an inherent limitation to the rate at which they can be turned on and off. This rate is called the response time.

The response time for LCDs varies from 30 milliseconds, which is slow enough to cause noticeable delays and other artifacts, to 20 milliseconds, which provides an acceptable refresh rate. LCDs are not subject to flicker caused by the 60Hz power cycle like CRTs are, but if the response time

is slow, an effect called smearing becomes apparent. Smearing occurs during the transition of one frame to another and is most apparent when video contains a high degree of motion.

5.1.5 Brightness and Contrast

Contrast relates to the difference between the black and white levels of a video waveform. As the difference between black and white levels increases, so does the contrast and its tolerance to extraneous room light. Contrast is often represented as a ratio. For example, the contrast of a CRT can approach 700:1. LCDs, on the other hand, may reach 500:1. This results from the fact that LCDs use reflected light while CRTs have a direct light source (the electron beam in the scanning mechanism). Lower contrast ratios cause darker shades to be displayed as black, reducing the amount of detail available in an image. Contrast ratios are more often used when describing LCD performance.

Brightness, on the other hand, is actually the overall DC voltage level of the video signal. When you adjust the brightness controls on a monitor, you are increasing or decreasing the DC voltage level of the signal being presented to the device. LCD brightness is measured in nits, which describes the intensity of the visible light. The greater the number of nits, the brighter the LCD display. LCDs commonly range between 75 and 300 nits. There is also a brightness signal, the luma portion that provides information about the amount of light for each point in the image.

5.2.0 Display Types

There are three primary types of display technology used. This section identifies, compares, and discusses the technology behind the LCD, plasma, and CRT displays, which are the types you are most likely to encounter.

5.2.1 LCD Display

The LCD flat panel TV has become the standard for TV sets up to about 65 inches. LCD technology uses the characteristic of certain materials that, when heated, create an almost plastic state. The molecules are solid, but they flow as if a liquid—hence the term liquid crystal.

An LCD forms an image by blocking light to an area that is energized with a small electrical charge. Many LCDs use reflective light, which makes them difficult to see. They actually have a mirror on the back to reflect incoming light. Back-

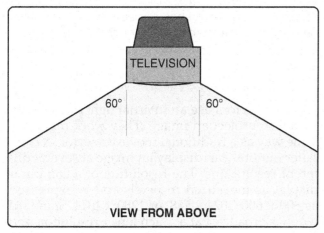

33402-11_F12.EPS

Figure 12 Undistorted image for a TV monitor.

lit models have a light source that is projected onto this mirrored surface.

Newer LCDs are created using miniature transistors laid out in a matrix on a thin film. These are called TFT, or thin film transistors. Collectively, they are called active-matrix TFT displays. There are also capacitors associated with each of these tiny transistors. Each transistor is addressable and is called a pixel. To turn on a particular pixel, a specific row is switched on. An electrical charge is then sent down a specific column. Because all of the other rows in that column are off, only the capacitor at the specific pixel receives the electrical charge. The capacitor holds the charge until the next refresh cycle. This is the origin of the word *active* in the phrase *active-matrix*.

A typical LCD TV has a depth of about 3 inches. LED-backlit versions of the LCD TV are much thinner at about 1 to 1.5 inches.

5.2.2 Plasma Display

A plasma display produces a very high-quality image. The image is significantly superior to that of a standard television monitor. The main advantage of a plasma display is its large image size. Plasma displays are available with display sizes up to about 105 inches.

A plasma display uses excited gases to produce light, much like a neon sign. This allows the display to be very thin. The big display, the shallow depth, and a wide viewing angle make this type of display ideal for mounting on a wall. Currently, a plasma display produces an image that is sharper than a projection high definition TV or a digital projector.

The plasma display shown in *Figure 13* has overall dimensions of 62 by 36 inches and a depth of 3.5 inches. The 160-degree viewing angle means that the image will be clear from almost any location in the room.

5.2.3 CRT Monitor

CRT monitors are commonly found in older installations and are rapidly being replaced with flat-panel LCD or plasma monitors. The main component of a CRT monitor is the cathode ray tube (CRT). The cathode generates an electron stream or beam. This beam is manipulated both horizontally and vertically by a magnetic field. This causes the electron beam to bend and scan across a phosphor-coated wire screen. As the beam strikes the screen, it excites the phosphor molecules, causing them to glow for a short period.

33402-11_F13.EPS

Figure 13 Plasma display.

Monochrome and black-and-white CRT monitors each have a single electron beam. Color CRTs have three beams—one for each primary color (red, blue, and green).

In a video monitor, the beam actually makes two passes across the screen, scanning from left to right, top to bottom. Based on NTSC standards, a video monitor displays 525 lines. As the beam scans, it traces every other line on the first pass and then fills in the missing lines on the next. As previously discussed, this is called interlacing. Most CRT video monitors produce an interlaced image. Computer monitors may use a CRT, but they are non-interlaced. Called progressive scanning, they scan from left to right and top to bottom, painting each line in succession.

Computer displays often have greater resolution but experience a side effect called flicker. This is a noticeable flashing of the picture based on the refresh rate—the amount of time that the phosphor coating takes to lose its luminance. A video monitor refreshes the screen 60 times per second. A computer display may have to refresh 75 times or more per second to retain an image.

5.3.0 Projection Systems as Displays

Video projectors use an internal light and display screen to project an image. They work much the same way as a traditional movie projector. A computer monitor can display an image at several different resolutions. The resolution of a computer display is measured in pixels and is expressed as 800 × 600, 1024 × 768, or 1280 × 1024. Standard video is broadcast at a much lower resolution and is measured in horizontal lines. As previously discussed, NTSC video, the broadcast standard in

LED Displays

In contrast to the LED-backlighting used on LDC displays, an LED display uses LEDs as the display medium. As discussed earlier, LCD and plasma displays are limited in size, whereas the LED display size is virtually unlimited. LED displays are used for the giant viewing screens found in sports stadiums and arenas. The largest LED display, which is located in the nation of Dubai, is 33 stories high. Another application of LED display technology is the organic LED (OLED) display. It uses organic material placed between glass plates and can be used to produce super-thin screens. With this technology comes the potential for flexible TV screens.

the United States and Canada, has 525 horizontal lines.

Video projectors have both an input resolution and an output resolution. The output resolution of a projector cannot be changed, so the video projector must convert the source image to its required output resolution. To work with a computer input source, some projectors require that the input resolution match the output resolution. That is, if a digital projector has an output resolution of 800 × 600, then the attached computer must also be set to display a resolution of 800 × 600. Higher quality projection equipment can convert any standard resolution input to the required output resolution. Common projector resolutions are 800 × 600 (SVGA) and 1024 × 768 (XGA). Higher-resolution projectors such as 1280 × 1024 (SXGA) and 1600 × 1200 (UXGA) are available, but are quite expensive. The 640 × 480 standard is no longer supported.

If the projector is to be permanently mounted, there should be a power outlet close by. Manufacturers have guidelines on the maximum power cord length. A 6-foot power cord is common.

Certain brands of video projectors allow for either front or rear projection. Rear projection (*Figure 14*) is obtained by reversing the image from left to right. With front projection, the image can be displayed on a screen, wall, or whiteboard. With rear projection, a translucent screen

is needed. The image is projected onto the back of the screen.

Projector placement is based primarily on the desired size of the projected image. A projector is typically mounted from 4 to 40 feet away from the display surface; the size of the screen determines the actual distance.

The brightness of the projector, which is measured in lumens, is also a factor. A 2,500-lumen projector will project farther to a larger screen than an 800-lumen projector. The brightness factor plays an additional important role in the presentation environment. The brighter the image projection, the easier it is to see with ambient room light. A 2,500-lumen projector will project a clear image at normal room lighting. An 800-lumen projector will require the lights to be dimmed or even turned off. If people are expected to take notes or consult printed material during a presentation, brightness is an important factor. *Figure 15* shows a typical range of image sizes and distances from the projection surface. Refer to the manufacturer's device specifications for exact guidelines on minimum and maximum image sizes based on mounting distance.

Ceiling-Mount Projectors

Many models of video projectors can be mounted from the ceiling. In such cases, the projector will be able to display the image upright while it is upside-down. Installing the projector upside-down allows users to have access to the controls located at the top of the projector. If ceiling mounting is desired, check with the manufacturer to see which models support this feature.

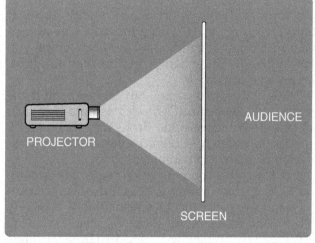

33402-11_F14.EPS

Figure 14 Rear projection setup.

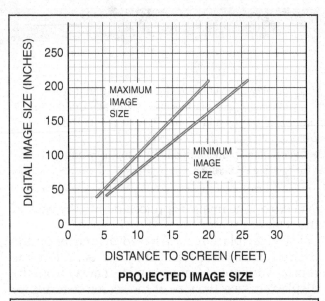

PROJECTED IMAGE SIZE

		DISTANCE TO SCREEN	
DIAGONAL SCREEN SIZE (INCHES)	IMAGE WIDTH (INCHES)	MAXIMUM DISTANCE (FEET)	MINIMUM DISTANCE (FEET)
40	32	5.1	3.9
50	40	6.4	4.9
60	48	7.7	5.9
100	80	12.8	9.8
150	120	19.2	14.8
200	160	25.6	19.7

RANGE OF DISTANCE TO THE SCREEN
FOR A GIVEN SCREEN SIZE

Source: InFocus® Corporation

33402-11_F15.EPS

Figure 15 Typical image size versus distance chart.

Many types of devices can be connected directly to a video projector. Projectors usually have, at a minimum, a composite video connector and a computer display port. Some models support both an input and an output for each type of connector, to allow for additional monitor equipment. For higher quality video, projectors may also have an S-video connector.

Most video projectors can be connected directly to a computer (*Figure 16*) using a standard 15-pin D-shell connector. Projectors will usually support a range of screen resolutions when connected directly to a computer.

Lower-end equipment often requires the computer screen resolution to match the output resolution of the projector. This resolution is typically 800 × 600, the standard for SVGA displays. More advanced equipment will support a broader range of input resolutions and will internally convert them to the required output resolution.

33402-11_F16.EPS

Figure 16 A laptop PC driving a digital projector.

Some projectors have speakers and connections for audio input and output. Depending on the size of the room and the placement of the unit, this may be enough to fulfill the sound requirements. A larger room will need a separate sound system.

5.3.1 LCD Display Projectors

LCD display projectors use an internal light and display screen to project an image. LCD display projectors use technology found in laptop computer displays but with some differences. For example, an LCD display uses a reflective surface at the back of the display. Light from one or more neon tubes is bounced off this surface and directed through the active matrix field. Projectors do not have this reflective surface; instead, there is a bright light source (a lamp) that is focused through the active matrix directly. *Figure 17* shows a digital projector. LCD projectors use three LCD filter layers—one each for red, green, and blue.

33402-11_F17.EPS

Figure 17 Digital multimedia projector.

Remote Control

Most projectors have a remote control. If the remote control also has keys with direction arrows, then the remote can be used like a mouse to control a computer connected to the projector. Therefore, a person giving a presentation at the front of an auditorium can also control a computer located in the back of the room. If a model supports this, then the mouse ports on the projector are output ports. That is, the projector acts as the mouse through the remote control. The mouse output from the projector should go to the mouse input on the computer, using either a PS/2 cable or a USB cable.

A computer monitor can display an image at several different resolutions, such as 640 × 480 or 800 × 600. Standard video is broadcast at a resolution of 320 × 240. Digital projectors have both an input resolution and an output resolution. The output resolution of a projector cannot be changed, so the digital projector must convert the source image to its output resolution.

5.3.2 Digital Micromirror Devices

Digital micromirror devices (DMD) use an array of small mirrors and a method sometimes referred to as digital light processing or DLP® (which is a trademark of Texas Instruments Corporation). The mirror array tilts forward and backward on a small hinge. Light is directed through color filters arranged on a spinning wheel and onto the mirror array. Each mirror, in essence, represents an individual pixel. The light is then passed through an optical system to merge the contents of each pixel and is projected outwards onto a screen. High-end DLP® projectors may have three DMD chips—one for each primary color. Lower end models will have a single chip.

Currently, these newer projectors are packaged as devices weighing less than three pounds and are very portable. They have a better contrast ratio than LCD projectors. DMD displays are a reflective technology whereas CRT and LCD projectors are transmissive. One drawback is that they are not as bright. Their color representation, however, is superior.

5.3.3 Liquid Crystal on Silicon Projectors

Liquid crystal on silicon (LCOS) projectors can be thought of as hybrid, possessing characteristics of both LCD and DLP® technologies. Instead of tiny mirrors, as in DLP® devices, they use liquid crystals. Unlike LCD projectors, however, they use reflective light. An LCOS projector will have three chips—one each for red, green, and blue—whereas DLP® technology uses a color wheel.

LCOS projectors have extremely high resolutions, however. LCD and DLP® projectors provide support for VGA (640 × 480) through SXGA (1280 × 1024). LCOS projectors typically provide resolutions starting at 1365 × 1024 and are generally more expensive as a result. Their contrast ratio is reasonably good, though, with performance at 500:1 up to 800:1.

5.4.0 Projection Screens

Video monitors provide their own viewing surface. Display projectors do not. All video projectors require a screen to project their image onto. There are two types of screens: front projection screens and rear projection screens. The screen selected for the video installation plays a major role in picture brightness and clarity. There are many different screen surfaces from which to select. Each screen is chosen to suit the particular projector.

5.4.1 Front Projection Screens

Front projection screens provide a reflective surface for the projection of video images. Screens are usually made of a heavy plastic or vinyl backing with a special reflective surface coating on the front. The coating, often having particulates such as silica, aluminum, or other material, is rough to the touch. Projection screens range in color from bright white to silver-gray. Projection systems such as LCD or DLP® achieve the best contrast and brightness with gray screens.

Projection screens are rated based on their gain, or ability to reflect the light that is projected on them. The higher the gain, the brighter the reflected image will appear.

Viewing angles are not affected by the gain of a screen, but each screen and screen type will have a different viewing angle. These two attributes need to be checked against recommendations from both projector and screen manufacturers when matching elements of a display system to the room or space where they are to be installed.

Mini-Projectors

Today, there are projectors small enough to fit in the palm of the hand. One model, developed by Dell, provides a 50-lumen output with 858 × 600 resolution and an image size ranging from 15 to 60 inches. These devices have built-in hard drives so that they can operate independently. For even greater convenience, mini-projector technology is also being incorporated into cell phones, although at this time the image quality is not as great as that of the mini-projector.

Screens come in different sizes and are packaged in several ways. They may be roll-up type screens, which are manually pulled down and positioned. They may be permanently installed, as in a cinema, or raised and lowered using a motor. Many permanently installed screens are laced to a frame, keeping the surface taut and ensuring consistent reflectivity across the entire surface. When installing screens, be careful not to damage the reflective surface: this diminishes its reflectivity and can cause visible shifts in image contrast and quality.

One of the key challenges encountered with pull-down screens is called keystoning. Here, the angle of projection is off, and the top of the projected image is wider than the bottom. To adjust for keystone effects, it is necessary to either adjust the angle of the projector or pull the bottom of the screen closer to the projection beam. Some suppliers provide an extension bracket that pushes the screen away from the wall on which it is mounted. Certain projectors also provide keystone adjustments. You should plan to manage these effects during the design and installation of the display system.

Large screens often have to be made from several smaller ones. For these applications, it is important to use materials that do not leave visible seams during projection.

5.4.2 Rear Projection Screens

Rear projection screens are translucent, which means they allow light to shine through them. There is always some loss of brightness, however, and it is important to match the optical properties of the rear projection screen to the ambient lighting conditions encountered during viewing.

Rear projection screens are available in different opacities, sizes, and portability. Some screen materials are darkened for improved contrast with high ambient light levels. Screens are made of glass or acrylic. Both offer good sound dampening qualities and thus minimize projector noise. Glass has the added advantage that it can be marked on with grease pencils or dry erase markers. Acrylic weighs less but is more subject to scratching.

One of the challenges of a rear projection system is managing the focal length of the projector. This often means that a larger room is required to have the proper distance between screen and projector. This can be mitigated with special mirror systems consisting of one or more mirrors, as shown in *Figure 18*. These systems reduce the distance between the projector and screen, allowing a smaller room to house them.

6.0.0 VIDEO PROCESSING AND DISTRIBUTION

There are numerous devices used in the video processing chain. This section will examine the various types of equipment used to process and distribute video for a presentation system. Processing equipment serves to alter or adjust the

33402-11_F18.EPS

Figure 18 Rear projection mirror system.

video signal. Distribution equipment is used to move the signal to displays and other devices.

6.1.0 Video Processing Equipment

In a presentation system, the goal is to provide video to a display device. In a simple video system, there may be a single video source being displayed on a single screen. In a more complex system, such as used in a courtroom or boardroom, one of several video sources is directed to one or more displays. In a very complex system, such as a command and control environment, there may be multiple inputs directed to one or more display devices simultaneously.

6.1.1 Video Scaler

A video scaler (*Figure 19*) is used to convert a video signal by adjusting it to match the requirements for a given display. This allows the video to be displayed on a host of output devices. For example, an RGB input can be converted to the resolution of a high definition plasma display. By adding lines, the scaler adjusts the video to fit the resolution of the target display and outputs a new video signal.

A scaler can address both the horizontal and vertical frequencies of the video image and can therefore change aspect ratios without losing image quality. It can also permit the display of video from diverse video sources, such as NTSC, PAL, or SECAM. Keep in mind that a scaler operates on either the video signal or the RGB signal. Some scalers operate on both, allowing the use of video and computer-video sources.

Many video scalers are programmable. They may contain presets for the assignment of parameters to specific output ports. Many have RS-232/RS-422 inputs, allowing them to be controlled or programmed from a network device.

6.1.2 Line Doubler

A line doubler takes an interlaced video signal and converts it to a progressive or non-interlaced signal. Because it adds the two video fields to-

gether, it doubles the scan rate, creating an image that can be displayed on a computer display. The quality of the image is enhanced a bit in the process, often to that of SVGA. Another type of line doubler is the line quadrupler, which adds four output lines of video for each input line for a much brighter and smoother image.

6.1.3 Video Switcher

The video switcher provides the means for directing an input signal to an output device. Often, a video scaler will be combined with a seamless switcher to select a given output device. This would be considered an integrated unit.

In other cases, the output of a video scaler or line doubler may be connected to a device called a video switcher (*Figure 20*). It accepts inputs from multiple sources and directs them to specific output devices.

Another type of video switcher is the matrix switcher (*Figure 21*). It accepts multiple inputs and supports multiple outputs. These types of switchers are typically nonlinear and can direct a signal to any output. When several matrix switchers are connected together, they can assign virtual addresses for physical input connectors as well as output connectors. They are controlled through either a series of presets via an RS-232/RS-422 interface or through some other controlling device such as a front panel controller. A matrix switcher will accept multiple video inputs, including composite, S-video, and component video.

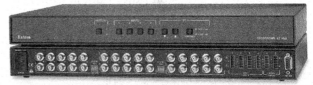

33402-11_F20.EPS

Figure 20 A video switcher.

33402-11_F21.EPS

Figure 21 A matrix video switcher.

33402-11_F19.EPS

Figure 19 A video scaler.

6.1.4 Video Transcoder

The video transcoder accepts an input signal and decodes it into its various components. It can then encode the resulting signal for transmission with different output signal characteristics. For example, a composite video signal can be decoded into its component signals, adjusted, and then encoded as S-video component video. Transcoders are used to convert between video sources, such as NTSC, PAL, or SECAM. These features are often integrated into switchers and scalers, or they can be packaged as separate units, as shown in *Figure 22*.

6.2.0 Video Distribution and Cabling

The distribution of video throughout an installation requires careful planning and attention to detail. This section discusses the equipment used to distribute video, the cabling requirements, and the various connectors you will encounter.

Initially, video was distributed across coaxial cabling. Currently, other cabling methods are also employed. Twisted-pair cabling, including Cat 5, Cat 5e, Cat 6, and other specialty cables, are now used to interconnect the various devices in a video system. As a result, several standard connectors have also emerged.

6.2.1 Twisted-Pair Cabling (UTP)

Cat 5 cable was originally introduced for computer networks. Cat 5 cabling consists of four pairs of twisted wires. Video is transmitted across them in parallel while computer data is transmitted serially. The ratio of twisting was determined by its ability to avoid cross talk and noise. Noise is generated by other electrical devices near the wir-

ing. Using a differential receiver, a complementary signal cancels out the noise.

When transmitting video on Cat 5, Cat 5e, and Cat 6 cabling, small differences in cable length can cause skew problems. Skew is a shift in color that results from two signals separated in time. For example, if a wire carrying one color is longer than another, the first color arrives, is displayed, and then the second color signal is displayed. This causes a color shift, as only one color is being drawn at a time. To solve this problem, equipment manufacturers use time delay circuitry to adjust for skew. Others include a special compensation cable (usually coaxial) that is attached to extend the length of the shorter cable.

Two other issues for UTP cabling are attenuation and impedance matching. UTP cabling can be used for runs up to 500 feet when using a transmitter and distribution amplifier like that in *Figure 23*.

UTP cabling usually has an impedance of 100 ohms. Without impedance matching, there is a loss in power, which will cause image ghosting and other convergence problems. The UTP video receiver (*Figure 24*) solves that problem by providing the proper signals to the display device.

6.2.2 Fiber-Optic Cabling

Fiber-optic cabling systems can be used to eliminate RFI, EMI, and other noise. It is used for longer distances such as throughout an arena or campus or between facilities in a command and control environment. Fiber optics can also be used to connect display equipment. *Figure 25* shows a DVI connector cable with a built-in optical transmitter and receiver.

Digital optical audio cable (*Figure 26*) is used to connect audio components within a system. This is a fiber optic cable that uses light pulses to transmit digital audio signals. Digital optical audio cables consist of a fiber core and an outer cladding. These cables can only be used on equipment that has optical jacks.

Figure 22 A video transcoder.

Figure 23 UTP transmitter and distribution amplifier.

Figure 24 UTP video receiver.

33402-11_F24.EPS

Figure 26 Digital optical audio cable.

33402-11_F26.EPS

Figure 25 DVI connector cable.

33402-11_F25.EPS

Figure 27 RGBHV cable terminated with BNC connectors.

33402-11_F27.EPS

6.2.3 Video Connectors

Several types of connectors are used in the installation of a video system. The choice of connector may be established by the equipment manufacturer or by company policy. The main types of connectors include the following:

- *BNC connectors* – BNC connectors can be found on many analog video cables, as seen in *Figure 27*. The cables themselves consist of several mini-coaxial cables, providing a shielded, dielectrically insulated single wire for composite and component video signals. They come in several configurations—three-wire, four-wire, and five-wire cables—and are sometimes called RGB, RGBS, or RGBHV cables. BNC connectors exist in both 50- and 75-ohm versions. Originally all were 50-ohm connectors

and were used with cables of other impedances. The small mismatch was negligible for lower frequencies. However, digital signals in video and telephony applications demand the use of 75-ohm connectors because mismatched 50-ohm connectors cause attenuation of the digital signal, resulting in slower rise time of the signal's square waves. This distortion of the signal can cause transmission errors. Also, impedance-mismatched connections cause reflections of the signal returning to the source. These reflections are known as return loss. One mismatched connection might not have a noticeable effect on system performance, but multiple mismatched connections in the link between the source and destination have a cumulative effect, causing possible distortion of the transmitted signal.

- *F connectors* – F connectors (*Figure 28*) are commonly used to connect video sources, such as cable TV sources and satellite TV antennas, to the premises receiving equipment. RG-6 cable with compression-type F connectors is the preferred method of hooking up cable and satellite TV. Although some manufacturers provide push-on F connectors, they should not be used for professional video work, as they are susceptible to interference from nearby RF sources, and do not provide the same level of performance as compression connectors.
- *RCA phono connectors* – RCA phono connectors are used for component video cables, as shown in *Figure 29*. They are often used for consumer and other unbalanced applications.
- *S-video connectors* – S-video connectors are of the mini-DIN variety. DIN stands for the German standards agency Deutsches Insitut für Normung. These connectors have four pins and are attached to dual 75-ohm coaxial cables packaged as a single cable (see *Figure 30*). S-video cables are available commercially in lengths from 3 to 50 feet. S-video does not carry audio signals, so it is necessary to provide audio connections for home theater and other applications. Some cables provide RCA-terminated audio cables as part of their assembly.
- *D-shell and D-sub connectors* – D-shell and D-sub connectors are used for multi-wire cable connections. These connectors are found on display units, splitters, and other video processing and distribution equipment. *Figure 31* shows a VGA to RGBHV cable.

> **NOTE**
>
> Care must be taken when terminating these connectors. Improper soldering can result in ghosting, color smear, and signal degradation.

- *Terminal blocks* – Terminal blocks are used to connect individual wires from UTP cabling. These blocks typically have small plated screws to lock the cable ends to the block.

6.2.4 DVI Connectors and Cables

DVI devices are connected using a 24-pin connector. There are two types of DVI connectors, DVI-I and DVI-D (*Figure 32*). As described earlier, the DVI-I can carry both digital and analog signals, while the DVI-D is strictly digital.

33402-11_F28.EPS

Figure 28 Compression-type F connector.

33402-11_F29.EPS

Figure 29 RGB component video cable.

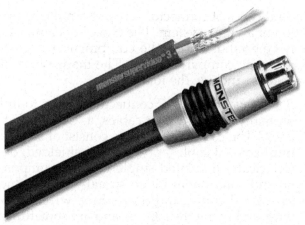

33402-11_F30.EPS

Figure 30 S-video cable and connectors.

The DVI specification does not address scaling: it is left up to the display to manage scaling operations. Initially designated for flat panel displays, DVI is now used for HDTV and for computer displays. DFP connectors are old style connectors. DVI cables must not exceed six feet. Length is critical.

6.2.5 HDMI Cables

HDMI cable (*Figure 33*) is designed to support true 1080P video resolution and high-end surround sound. As previously discussed, it carries three channels of video and audio. The cable has a 10.2 GB bandwidth. HDMI cables use a special 19-pin connector. The pinout is shown in *Figure 34*.

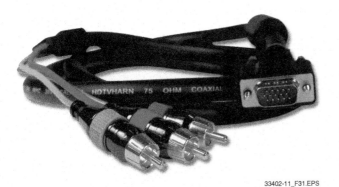

33402-11_F31.EPS

Figure 31 VGA to RGB component video cable.

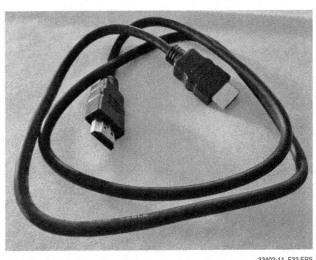

33402-11_F33.EPS

Figure 33 HDMI cable.

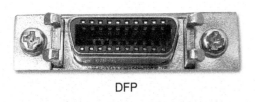

DFP

DVI-I

DVI-D

33402-11_F32.EPS

Figure 32 DVI connectors.

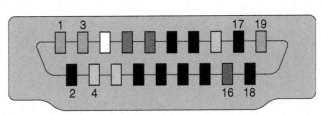

PIN	USE
1–9	TMDS data (video and audio)
10–12	Clock channel
13	CEC command channel
14	Reserved
15–16	Display data channel
17	Data shield
18	+5V power
19	Hot plug detection

33402-11_F34.EPS

Figure 34 HDMI connector pinout.

SUMMARY

Ever since regularly scheduled broadcast television began in the 1940s, television has been an increasingly important source of information and entertainment. Television serves as an important part of corporate information systems, and as a pivotal component in today's residential home entertainment systems. The transition from analog to digital video has added a level of complexity to the work of ESTs, who are required to upgrade or maintain existing analog systems and install newer digital systems. In order to do so, it is necessary to understand the nature of analog and digital video signals and the equipment and cabling systems used to process and display these signals.

1. A simple video system consists of a _____.
   a. DVD player and S-video connection
   b. video source and monitor
   c. CRT and monitor
   d. VCR and VHS tape

2. The luminance component of a video signal consists of _____.
   a. hue
   b. brightness
   c. amount of gray
   d. saturation

3. According to the original NTSC standard, before being broadcast, each image is scanned into 525 lines, then _____.
   a. changed to letterbox format
   b. changed into pixels
   c. grouped into two different fields
   d. passed through a color filter

4. The difference between black and white and monochrome video is that _____.
   a. monochrome video contains chrominance but no luminance
   b. monochrome video contains luminance and chrominance
   c. black and white video contains chrominance but no luminance
   d. black and white video contains luminance and chrominance

5. The bandwidth reserved for each NTSC channel is _____.
   a. 4.7MHz
   b. 5.0MHz
   c. 6.0MHz
   d. 6.3MHz

6. For NTSC video, the field rate of one field per 60th of a second and the frame rate of one frame per 30th of a second mean the same thing.
   a. True
   b. False

7. S-video is a _____.
   a. composite signal
   b. component signal
   c. four color signal for CRTs
   d. monochrome signal for CRTs

8. In an S-video system, the letter Y represents _____.
   a. yellow
   b. luma
   c. black and white information
   d. color information

9. The compression algorithm that supports video on cell phones is _____.
   a. MPEG-1
   b. MPEG-2
   c. MP-3 JPEG
   d. MPEG-4

10. The signal transmission protocol for DTV in the United States is _____.
    a. COFDM
    b. 8VSB
    c. HDTV
    d. SDTV

11. DVI defines a standard for providing digital video signals to a computer display.
    a. True
    b. False

12. The term *aspect ratio* is a way of describing _____.
    a. the X and Y axes of a display
    b. the viewing angle of a display
    c. the brightness of a display
    d. the distance from a display to a viewer

13. A normal horizontal viewing angle for flat panel displays is _____.
    a. 30 degrees
    b. 60 degrees
    c. 120 degrees
    d. 160 degrees

14. The term *contrast* in reference to a video wave-form relates to the difference between levels of _____.
    a. RGB
    b. black and white
    c. monochronicity
    d. YUV

15. What does a plasma display use to produce light?
    a. Miniature transistors
    b. Excited gases
    c. Phosphorus liquid
    d. Cathode rays

16. When using a front projection screen, one way to handle the problem of keystoning is to _____.
    a. switch the gray screen to a white screen
    b. change to a translucent projection screen
    c. project the image onto a different size mirror
    d. adjust the angle of the projector relative to the screen

17. A video scaler is used to _____.
    a. amplify a video signal
    b. match a video signal to the display device
    c. convert interlaced video to progressive
    d. direct an input signal to an output device

18. A matrix switcher accepts _____.
    a. a single input and provides a single output
    b. multiple inputs and provides a single output
    c. a single input and provides multiple outputs
    d. multiple inputs and provides multiple outputs

19. DVI devices are connected using a _____.
    a. coax cable with an F-type connector
    b. cable with a 24-pin connector
    c. HDMI cable with a 19-pin connector
    d. Cat 5e data cable

20. In an HDMI cable connector, video and audio data are carried on pin(s) _____.
    a. 1 through 5
    b. 6 through 12
    c. 1 through 9
    d. 12 through 19

3D TV: A video image in which three dimensions can be perceived.

Aspect ratio: The ratio between the horizontal and vertical axes on a display or monitor.

Advanced Television Systems Committee (ATSC): The committee that is responsible for DTV standards.

Bandwidth: The range of frequencies, expressed in MHz, over which the amplitude of a signal remains constant.

Blanking interval: The time between the end of one horizontal scanning line and the beginning of the next. Blanking occurs when a monitor's electron beam is positioned to start a new line or a new field.

Blu-ray: An optical disk storage format that allows playback of high-definition TV.

Chrominance: The color information in a television image. Chrominance has two properties relating to color: hue and saturation.

Commission Internationale de l'eclairage (CIE): A set of color matching functions and coordinate systems.

CMYK color: Cyan, magenta, yellow, and black are the four colors that make up the subtractive color space for printed materials.

Color space: A geometric representation of colors in space, usually of three dimensions.

Composite video: A video signal that contains all of the information needed to reproduce a color picture. Luminance and chrominance information is used to carry brightness and color. Composite video also contains horizontal, vertical, and color synchronizing information.

Component video: Video in the form of three separate signals, all of which are required to specify the color picture and sound. The signals are composed of luminance information, chrominance information, and sync.

Digital visual interface (DVI): A digital computer video transmission standard developed by the digital display working group (DDWG) consortium. The DVI standard involves two sub-types: a digital-only version (DVI-D) and an integrated analog and digital version (DVI-I).

Frame rate: The number of video frames per second. NTSC video consists of 30 frames per second. ATSC video consists of 60 frames per second but shows 30 frames twice each in succession.

Gamma: The nonlinear relationship between the electrical voltage of the signal and the intensity of a video image.

High definition TV (HDTV): High definition television.

High definition multimedia interface (HDMI): A digital HDTV interface that uses a single cable with 19-pin connectors to carry both video and audio.

Horizontal blanking interval: During this brief moment, the electron beam in a CRT repositions itself at the beginning of the next horizontal line.

Hue: The color that appears on the display screen or monitor.

Interlaced video: An image is recreated on a CRT by first tracing every odd-numbered line and then tracing every even-numbered line. The electron beam in the CRT makes two complete, consecutive passes every second in order to recreate a single field of video.

Luminance: The portion of a video waveform signal required for black and white images. Color systems use luminance as well, but it is obtained as the weighted sum of the red, blue, and green (RGB) signals.

National Television System Committee (NTSC): This committee established both the 525-scanning-lines-per-frame/30-frames-per-second standard and the color television system currently used in the United States. It is also the common name of the NTSC-established color television system.

Progressive video: Results from the raster of a CRT, which traces each line of a video image, one at a time, from the top to bottom. Sometimes called non-interlaced video.

Resolution: The sharpness or crispness of a picture. For a display device, it is measured based on three factors: the number of scan lines used to create each frame of video, the number of picture elements (pixels) contained in each line, and the number of bits per color.

RGB: A color space consisting of the three primary colors, red, green, and blue. Each color is carried on a separate wire in a three-wire RGB cable.

RGBHV: A form of component video where each color of the red, blue, green color space is carried on a separate wire, with synchronization signals for vertical and horizontal control also transmitted on individual wires.

RGB sync-on-green: A form of component video where the three colors of the red, blue, green color space are carried on separate wires; however, synchronization signals are combined with the green color signals and carried on the green signal wire.

Saturation: The amount of gray in a color (as opposed to hue).

Standard definition television (SDTV): Part of the DTV standards, SDTV is used to display video on a standard definition television (NTSC). It is also used to package up to five different NTSC television channels for transmission on a single 6MHz channel band.

S-video: A form of component video called separated video, commonly referred to as Y/C video. With S-video, the color information (C) for a video signal is separated from the brightness information (Y), and each is transmitted on a separate wire.

Synchronization (sync): Video signals are complex signals that are shared among different devices. Synchronization signals are timing signals that allow you to match the chrominance and luminance information into a coherent image on the screen.

Transition minimized differential signaling (TMDS): The signaling method or protocol used for the transmission of DVI video signals.

Vertical blanking interval: With a CRT, when the end of the first of two video fields is reached, the electron beam in the CRT is repositioned to the top of the display before tracing the next field. The amount of time needed to do this is called the vertical blanking interval.

Y/C video: A term used to describe the separation of an NTSC video signal into its luminance (Y) and chrominance or color (C) signals. Y/C video is used in S-video component video distribution.

Additional Resources

This module presents thorough resources for task training. The following resource material is suggested for further study.

How Video Works, Second Edition: From Analog to High Definition. Marcus Weisse. Burlington, MA: Focal Press.

Maxim Application Note 750, Bandwidth Versus Video Resolution, www.maxim-ic.com.

Video Bandwidth, www.extron.com.

Video Demystified: A Handbook for Video Engineers, Fifth Edition. Keith Jack. Burlington, MA: Newnes.

www.extron.com (video control equipment)

www.da-lite.com (projection devices)

www.draperinc.com (home theater systems)

Figure Credits

Topaz Publications, Inc., Module opener, Figures 16, 17, 26, 28, and 33

Samsung Electronics USA, Figure 13

InFocus Corporation, Figure 15

Da-Lite Screen Company, Figure 18

Photo courtesy of Extron Electronics, Figures 19–24

Ram Electronics Industries, Inc., Figures 25, 27, 29, and 31

Monster Cable Products, Inc., Figure 30

DVIGear.com, Figure 32

NCCER CURRICULA — USER UPDATE

NCCER makes every effort to keep its textbooks up-to-date and free of technical errors. We appreciate your help in this process. If you find an error, a typographical mistake, or an inaccuracy in NCCER's curricula, please fill out this form (or a photocopy), or complete the online form at **www.nccer.org/olf**. Be sure to include the exact module ID number, page number, a detailed description, and your recommended correction. Your input will be brought to the attention of the Authoring Team. Thank you for your assistance.

Instructors – If you have an idea for improving this textbook, or have found that additional materials were necessary to teach this module effectively, please let us know so that we may present your suggestions to the Authoring Team.

NCCER Product Development and Revision
13614 Progress Blvd., Alachua, FL 32615

Email: curriculum@nccer.org
Online: www.nccer.org/olf

❏ Trainee Guide ❏ AIG ❏ Exam ❏ PowerPoints Other _____

Craft / Level: _____ Copyright Date: _____

Module ID Number / Title: _____

Section Number(s): _____

Description: _____

_____

_____

_____

_____

Recommended Correction: _____

_____

_____

_____

_____

_____

Your Name: _____

Address: _____

_____

Email: _____ Phone: _____

33403-12

Broadband Systems

SBCA

**Endorsed by the Satellite Broadcasting &
Communications Association (SBCA)**

Module Three

Trainees with successful module completions may be eligible for credentialing through NCCER's National Registry. To learn more, go to **www.nccer.org** or contact us at **1.888.622.3720**. Our website has information on the latest product releases and training, as well as online versions of our *Cornerstone* newsletter and Pearson's product catalog.

Your feedback is welcome. You may email your comments to **curriculum@nccer.org,** send general comments and inquiries to **info@nccer.org**, or fill in the User Update form at the back of this module.

Objectives

When you have completed this module, you will be able to do the following:

1. Draw a block diagram of a selected CATV/SMATV/MATV system headend.
2. Describe the signal flow for selected processing paths in the headend of a CATV/SMATV/MATV system.
3. Identify the different assemblies and components used in CATV/SMATV/MATV systems and describe their function.
4. Select and terminate coaxial cables used for specific applications.
5. Calculate CATV/SMATV/MATV distribution system gains and losses.
6. Use selected test equipment to make measurements and checks in CATV/SMATV/MATV systems in order to evaluate system operation.

Performance Tasks

Under the supervision of your instructor, you should be able to do the following:

1. Install a video distribution system.
2. Use a signal level meter (SLM) to measure the strength and slope of a signal.

Trade Terms

Amplitude modulated (AM)
Automatic gain control (AGC)
Azimuth
Bandwidth
Baseband
Broadband
Broadcast television systems committee (BTSC) stereo
Carrier-to-noise (C/N) ratio
Decibel (dB)
Decibel-microvolt (dBμV)
Decibel-millivolt (dBmV)
Decibel-milliwatt (dBm)
Distortion
Downlink
Drops
Feeders
Frequency modulated (FM)

Geosynchronous orbit
Headend
Heterodyning
Hybrid fiber-coaxial (HFC)
Insertion loss
Isolation
Modulation
Modulator
Multiplexed
Noise figure
Orbital slot
Resonant dipole antenna
Signal-to-noise (S/N) ratio
Surface acoustic wave (SAW)
Transponder
Trunks
Uplink

Industry Recognized Credentials

If you're training through an NCCER-accredited sponsor you may be eligible for credentials from NCCER's Registry. The ID number for this module is 33403-12. Note that this module may have been used in other NCCER curricula and may apply to other level completions. Contact NCCER's Registry at 888.622.3720 or go to nccer.org for more information.

Contents ————————————————————————

Topics to be presented in this module include:

Figures and Tables

1.0.0 INTRODUCTION

Not that long ago all television signals were transmitted in analog format similar to how radio signals are transmitted. The video signal of analog television was transmitted in amplitude modulation (AM), while the audio signal was transmitted in frequency modulation (FM). Analog TV was subject to interference such as ghosting and snow, depending on factors like the distance and geographical location of the TV or other equipment receiving the signal. The amount of bandwidth assigned to an analog TV channel also restricted the resolution and overall quality of the image. The analog TV transmission standard in the United States was referred to as the National Television Systems Committee (NTSC). The NTSC standard was based on a 525-line, 60 fields/30 frames-per-second at 60Hz system for transmission and display of video images.

In recent years, digital television (DTV) was developed. DTV signals are transmitted as data bits—in the same way computer data, music, or other information is written to a hard drive or other storage media. The digital signal is either "on" or "off," so there is no gradual loss of signal as the distance between the transmitter and the receiver increases. If the receiver is too far from the transmitter, the viewer sees nothing at all.

Unlike analog TV, digital TV takes all the main factors of the television signal into consideration. Video and audio are transmitted as an interlaced signal that results in greater integrity and flexibility of the signal content. Also, a DTV signal that uses the same amount of bandwidth as an analog signal can provide a higher quality image and accommodate additional video, audio, and text signals. This allows broadcasters to provide more features like surround sound, multiple language audio, and text services. DTV also enables broadcasters to use a 16:9 widescreen format (versus a 4:3 aspect ratio used by traditional TV) so that viewers can see movies the way filmmakers intended.

In October of 2005, the US Congress passed the Digital Television Transition and Public Safety Act of 2005, which mandated that all full-power broadcast stations stop broadcasting in analog and start broadcasting only in digital format by June 12, 2009. This act changed how broadcasters transmitted TV signals and how customers received those signals. Customers with TVs capable of receiving digital signals did not need to make any changes. But customers with analog TVs were required to buy a DTV converter box to convert the digital signals to analog signals.

Despite the conversion from analog to digital, there are still situations where customers may be receiving signals from analog sources. For that reason, the characteristics of analog TV are still important.

This module is intended to introduce trainees to broadband cable television (CATV) systems, private broadband satellite master antenna television (SMATV) systems, and master antenna television (MATV) systems. CATV systems serve both rural and urban areas. They provide a large group of subscribers access to multi-channel TV programming distributed over a regional area by a network of coaxial and/or fiber optic cable. Private SMATV and MATV cable systems provide access to multi-channel TV programming to a much smaller group of users. These systems typically serve the occupants of an office building, industrial facility, school, hotel, or hospital. There, the TV programming signals are distributed to the building occupants via an intra-building network of coaxial or fiber optic cables.

2.0.0 EVOLUTION OF CATV SYSTEMS

In the past, television reception was accomplished through an individual antenna that was normally placed on the roof of the home. This method, although popular, had many pitfalls. Broadcasting stations and antenna receivers were often too far away to receive adequate signals. Another area of concern was adequate reception of the signal when there were hills or mountainous terrain in the surrounding area.

In the late 1950s, a better system for television reception was established—community antenna television (CATV). With CATV, transmitting stations using microwave technology could transmit far outside their local area, allowing the signals to be picked up at distant receiving stations and then routed by cable to individual homes. This provided community television systems with an expanded variety of program services.

The development of communication satellites made it possible for community systems to receive programs from around the world. The variety of programming available became great enough to be attractive to viewers in large cities, as well as the surrounding areas. At that point, community TV was no longer descriptive of the service, and cable TV or CATV became the accepted nomenclature. There are, however, still areas of the country in which cable television has not been installed, and in fact may not be practical. In these areas, direct satellite reception provides the only means of receiving the array of stations.

3.0.0 ARCHITECTURE OF CABLE SYSTEMS

CATV systems are designed to serve individual subscribers within a relatively large region or community, while private MATV and SMATV systems are designed to serve a smaller, restricted group of people typically in intra-building applications. Even though the CATV, SMATV, and MATV system applications are different, there are many similarities in their technology and architectures. A brief overview of the CATV, SMATV, and MATV architectures is provided here. Detailed information about the specific equipment used in these systems is given later in this module.

3.1.0 CATV Architecture

A basic CATV system can be divided into two main parts. One part is called the headend (often written as head end or head-end). The headend is the main station, which is sometimes called the master facility. The other main part of a CATV system is the distribution system. It consists of trunks, feeders, and drops to the subscribers. These basic elements are shown in *Figure 1*. Collectively, they provide for the acquisition, processing, and distribution of the desired CATV channel signals.

3.1.1 Headend

The headend of a CATV system is where incoming signals are received and where the final control of the broadcasted content takes place. Some examples of sources for incoming signals include satellite, microwave, computer, and off-air signals. Another source is locally originated programming. Most franchises require CATV companies to provide programs originating from production facilities at the CATV company studios or from other local sources like municipalities. For instance, many cities now air their city council meetings on a cable channel.

The headend equipment is typically housed in a dedicated building. National networks and movie channels, such as WTBS, USA Network, CNN, HBO, Showtime, and others, are received by the headend via satellite antennas (*Figure 2*). Local network and community broadcast station signals are received by standard off-air TV receive-only (TVRO) antennas, through microwave links from the local stations, or point-to-point fiber optic cable. Additionally, special services, such as pay-per-view events and movies, can be placed on the cable system. Special events such as boxing matches are received from a satellite, but pay-per-view movies may be received either from a satellite or played internally and placed on the system.

The carrier frequencies of the off-air, satellite, and microwave signals are converted and/or processed and combined in the headend. These carrier frequencies correspond to standard broadcast channels for viewing on cable-ready television sets. The pay-per-view and premium movie channels are scrambled or masked so that they cannot be viewed. Viewing of these channels by the subscriber is allowed by the cable company using one of several technologies. The first and oldest technology is a converter box, sometimes called a converter/descrambler, supplied by the cable company.

As the name implies, the converter operates to convert a selected basic programming signal, or to unscramble and convert a selected premium program signal, into a single broadcast channel frequency (typically Channel 3 or 4) that can be viewed on the TV. The conversion is necessary because many CATV channel assignments are

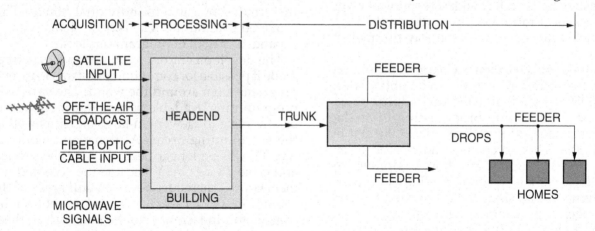

Figure 1 Basic cable television system architecture.

C-BAND ANTENNAS

DBS ANTENNA

33403-11_F02.EPS

Figure 2 Typical CATV headend satellite receiving antennas.

different from those supplied from off-air broadcast sources. For example, a channel designated as Channel 11 in the CATV system may be locally broadcast for off-air reception as Channel 24. When a converter is used, all channel selection is done via the converter, while the TV set remains tuned to Channel 3 or 4. It should be pointed out that cable subscribers with older non-cable ready television sets also must use a converter box in order to view even basic CATV programming.

A technology called interdiction also has become widely used. In this method, jamming or signal scrambling is injected at the subscriber's location instead of at the CATV headend. Control signals superimposed on the coaxial cable from the cable company turn the jamming off for each applicable channel, so that the channels that are transmitted in the clear from the headend can be viewed by the subscriber. With this technology, a converter box is not normally used.

Another common way to make specific programming available is through the use of inter-

active addressable converters. These devices are authorized for specific programming by the cable TV company using digital information sent along with the programming signals on the cable. Since most cable systems now have two-way capabilities, the set-top unit can communicate directly with the cable company either through an immediate or delayed response mode allowing the customer to change programming packages or purchase pay-per-view events and features simply by using their remote control. This process requires a set-top box, a two-way data stream, and usually access to interactive services.

In addition to television services, most cable TV customers today have access to Internet and/or telephone services directly from their program provider.

3.1.2 Distribution System and Subscriber Drops

Cable TV signals processed in the headend are applied to the distribution system (*Figure 3*) consisting of a combination of trunk, feeder, and drop cables. These can be run overhead, underground, or both in a community. Trunk cables supply the signals from the headend to large portions of the cable TV system network, such as an entire neighborhood. There, the trunk connects to multiple feeder cables that further distribute the signal along the paths of the individual streets, and then to drops that feed the individual homes that have subscribed to the cable service.

In the past, distribution systems used coaxial cable. Modern systems are more likely to use a combination of fiber optic cable and coaxial cable in a network that is often called a hybrid fiber-coaxial (HFC) system. In such a system, fiber optic cable is used for the trunk and feeder portions of the network, while coaxial cable is used for the subscriber drops. Fiber optic cables have less loss, can carry more signals, and have the capability to carry other communication signals on the same fiber as the CATV signals.

Ideally, a signal injected into the system network at the headend and transmitted through the entire distribution cable network would emerge at the subscriber's home without losing signal strength or quality. However, no practical system can be perfect. Fiber optic cable provides the least loss and attenuation, but all signals carried through coaxial or fiber optic cable can suffer from various signal losses and attenuations as they pass along the system. Therefore, it is often necessary to compensate for these inherent factors in such a manner that the signal tapped off the network at any point can supply a high-quality television signal.

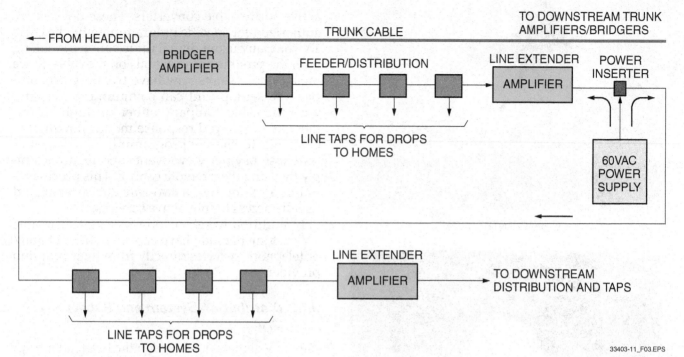

Figure 3 Simplified distribution system.

33403-11_F03.EPS

Signal loss in CATV distribution systems that use coaxial cable is compensated for by the use of line amplifiers, as shown in *Figure 4*. The amplifiers are usually placed at intervals of several thousand feet in a distribution system covering a typical neighborhood. The amplifiers are generally powered by 60VAC or 90VAC that is injected into the coaxial cable by a power supply.

Filters prevent the voltage from the power supplies from traveling down the drops to the homes. They also prevent the voltage from interfering with the signals carrying the programming. If a filter is not functioning properly, it creates an interfering carrier called hum. This interference generates a line that rolls through the picture on every TV in that line.

In practice, only a segment of the system is energized by a single power supply. Power block devices are used to prevent cross-connection between power supplies. *Figure 5* shows a typical pole-mounted power supply for use in cable distribution systems. The power inserter allows the power supply to feed the cable distribution system. Suspended line drops to homes or commercial buildings are commonly made using CATV-rated or radio-grade (RG)-6 coaxial cable. Cable used for aerial drops from the pole to the house has an integral steel messenger to provide adequate support. Without a messenger to support the cable, the tension on the cable would cause the cable jacket, the braid, and the dielectric to stretch and the center conductor to pull out. Cable

On Site

Converter Boxes

The converter box pictured here is typical of the ones used in CATV systems to convert and/or unscramble CATV signals for application to the TV set. Modern converters provide both Standard Definition TV (SDTV) and HDTV services and have a variety of video and audio outputs to maximize a TV's capabilities. This converter has radio frequency (RF), separate video (S-video), composite video, component video, left and right audio, high-definition multimedia interface (HDMI), and optical digital audio outputs.

33403-11_SA01.EPS

DISTRIBUTION AMPLIFIER WITH TWO BRIDGING LEGS

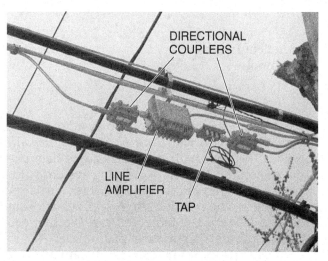

DISTRIBUTION AMPLIFIER WITH ONE OUTPUT

33403-11_F04.EPS

Figure 4 Typical line amplifiers.

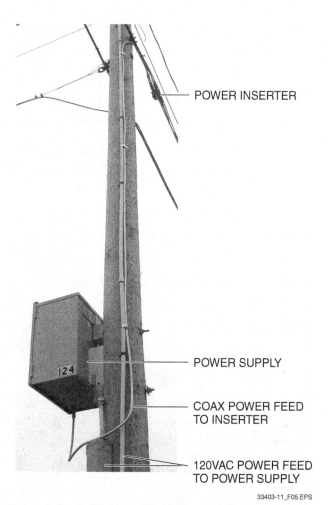

POWER INSERTER

POWER SUPPLY

COAX POWER FEED
TO INSERTER

120VAC POWER FEED
TO POWER SUPPLY

33403-11_F05.EPS

Figure 5 Typical line amplifier power supply.

used for routing on the exterior or throughout a house does not have an integral messenger. There is also a special flooded RG-6 used for underground drops. Although the flooded RG-6 cable and the regular RG-6 cable without a messenger look alike, the flooded cable should never be used inside or on the exterior of a home beyond the ground block. Likewise, standard (non-flooded) RG-6 cable should never be buried underground or placed in an underground conduit.

3.2.0 MATV and SMATV Architecture

Private SMATV and MATV systems can range from basic residential installations to large, professional installations similar to CATV installations. The professional systems are used primarily in commercial or institutional installations, including schools, hospitals, prisons, and hotels/motels. Many exist in a multi-building campus; such as

a university, major medical center, or apartment complex. *Figure 6* shows the architecture for a typical professional SMATV/MATV system.

As shown, the architecture for an MATV system is the same as for the SMATV system, with the exception that the headend of the MATV system does not receive the satellite antenna input. *Figure 7* shows typical receiving antennas for a SMATV system mounted on a large building. In addition to the antenna inputs, a feed from a CATV system may be included. Locally generated programming signals may also be applied in the headend for processing. These programming signals can be generated by devices such as video cameras, VCRs, DVDs, laserdisc players, and character generators.

SMATV/MATV systems consist of two main sections—the headend and the distribution system, consisting of trunks, feeders, and building room taps. In these systems, the headend and distribution components are typically housed within the same building. The TV signals processed in the headend are distributed to the building room taps via the trunk and feeders, in the same way as pre-

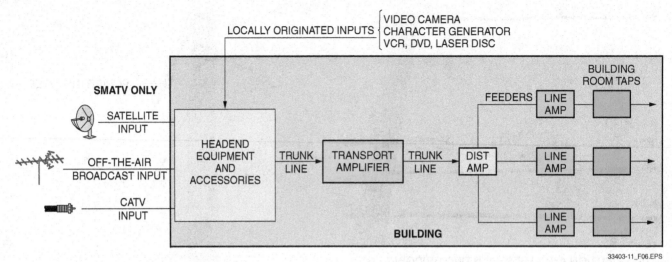

LOCALLY ORIGINATED INPUTS { VIDEO CAMERA / CHARACTER GENERATOR / VCR, DVD, LASER DISC

33403-11_F06.EPS

Figure 6 Basic SMATV/MATV system architecture.

33403-11_F07.EPS

Figure 7 Typical SMATV system receiving antennas.

viously described for a CATV system. Amplifiers are installed where needed in the system to boost signal strength in order to ensure a high-quality television signal. The amplifiers and other equipment used in the distribution system of SMATV/ MATV systems perform the same functions as like equipment used in a CATV distribution system.

4.0.0 BROADBAND SYSTEM BASICS

Before an in-depth examination of CATV/ SMATV/MATV system operation and equipment occurs, it is helpful to review some fundamental background information that applies to these systems. This information includes the following:

• Metric system prefixes
• Scientific notation
• The frequency spectrum

• TV channels
• Units of measure
• Symbols

4.1.0 Metric System Prefixes

The metric system and metric system prefixes were covered in the *Craft-Related Mathematics* module. Because metric prefixes are widely used in cable systems when expressing system values and component parameters, a brief review is provided here. The metric system prefixes are listed in *Table 1*. From the table you can see that each prefix represents an order of magnitude.

The most common metric system prefixes used in broadband cable work are giga (G), mega (M), kilo (k), milli (m), micro (µ), and nano (n). For example, 1GHz = 1,000,000,000Hz, 1MHz = 1,000,000Hz, 1kHz = 1,000Hz, 1mW = 0.001W, 1µV = 0.000001V, and 1nsecond = 0.000000001 second.

Table 1 Metric System Prefixes

Prefix	Unit	Power of Ten	
Pico- (p)	$\frac{1}{1,000,000,000,000}$	0.000000000001	10^{-12}
Nano- (n)	$\frac{1}{1,000,000,000}$	0.000000001	10^{-9}
Micro- (µ)	$\frac{1}{1,000,000}$	0.000001	10^{-6}
Milli- (m)	$\frac{1}{1,000}$	0.001	10^{-3}
Centi- (c)	$\frac{1}{100}$	0.01	10^{-2}
Deci- (d)	$\frac{1}{10}$	0.1	10^{-1}
Deka- (da)	10	10	10^{1}
Hecto-(h)	100	100	10^{2}
Kilo- (k)	1,000	1,000	10^{3}
Mega- (M)	1,000,000	1,000,000	10^{6}
Giga- (G)	1,000,000,000	1,000,000,000	10^{9}
Tera- (T)	1,000,000,000,000	1,000,000,000,000	10^{12}

4.2.0 Scientific Notation

Scientific notation and how to perform calculations using scientific notation were covered in the *Craft-Related Mathematics* module. As a reminder, scientific notation can be used to simplify calculations by expressing very large or very small numbers as a power of ten. In cable TV work, it is commonly used to express frequencies and numerous other units of measure. For example, a frequency of 1,000Hz can be expressed as 1×10^3Hz, a frequency of 1,000,000Hz as 1×10^6Hz, or a frequency of 1,000,000,000Hz as 1×10^9Hz. Similarly, a computer's hard drive with a storage capacity of 7,500,000,000 bytes can be expressed as 7.5×10^9 bytes. As shown in *Table 1*, the powers of 10^{-9}, 10^{-6}, 10^{-3}, 10^3, 10^6, and 10^9 relate to the metric prefixes commonly used in electronics of nano (n), micro (µ), milli (m), kilo (k), mega (M), and giga (G), respectively.

4.3.0 Frequency Spectrum

Designated portions of the RF frequency spectrum ranging from 5MHz to 1GHz are allocated by the Federal Communications Commission (FCC) for TV channel assignments used for over-the-air broadcast TV and cable TV. Each assigned channel has a bandwidth of 6MHz. The FCC has also allocated specific frequencies in the frequency range from 0.95GHz to 23.0GHz for use by other communications systems and platforms. *Figure 8* shows a broad overview of the channel assignments for broadcast TV, cable TV, and satellite TV. Specific channel assignments can be obtained from the FCC website.

It should be pointed out that the transmission of adjacent channels is not permitted in the same geographic location. This is because TV receivers may not be able to discriminate between a weak desired signal and a strong undesired signal on an adjacent channel. However, in CATV systems, adjacent-channel signals are allowed because the signal levels of all channels can be controlled to the same level.

CATV broadband cable channel frequencies are listed under three columns: Standard, Incremental, and Harmonic. The frequency assignments listed under the Standard column are the frequencies used by most cable TV systems. However, many cable systems use carrier frequencies for some of the channels that are shifted in frequency from the standard frequencies. These systems use frequencies designated as Incrementally Coherent Related Carrier (IRC) frequencies or Harmonically Related Carrier (HRC) frequencies. These designations apply to schemes that reduce interfering carriers in the third harmonic or composite triple beat (third order distortion).

For the systems that use the IRC channel frequencies, the frequencies for Channels 5 and 6 are shifted up +2.0MHz, while all the other channels remain at the assigned standard frequencies. For the systems that use the HRC channel frequencies, all the channels except for Channels 5 and 6 are shifted down in frequency by –1.25MHz, while Channels 5 and 6 are shifted up in frequency by +0.75MHz. In both the IRC and HRC schemes, an additional channel, designated as A8, is created between Channels 4 and 5.

Use of the IRC and HRC channel frequencies by CATV systems is done to reduce a form of interference called composite triple beat (CTB) that can cause interference lines in the TV picture and can interfere with nearby aeronautical radio services. CTB interference is covered later in this module. The equipment used in CATV systems that use the IRC or HRC channel frequencies must be capable of tuning to and processing the IRC or HRC frequencies. With some types of equipment, this is easily done by selecting either standard, IRC, or HRC channel frequency operation via front panel switches, or by using an on-screen menu or internal feature. It is also possible to purchase converters that are designated as standard, IRC, or HRC.

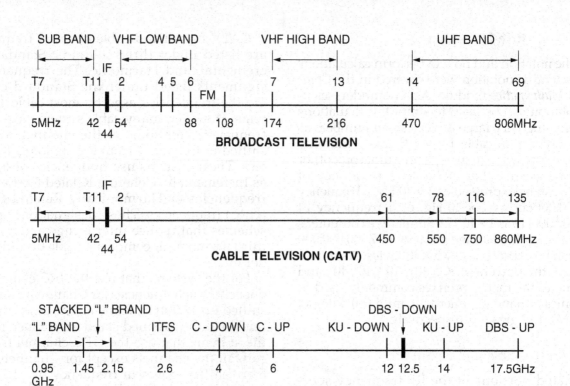

Figure 8 TV frequency spectrum.

4.4.0 TV Channels

In the United States, each 6MHz-wide TV channel is formatted in accordance with the NTSC standard. Multiple digital channels can be carried within the 6MHz range (*Figure 9*). As shown, the color burst and sound carriers are 3.58MHz and 4.5MHz, respectively, above the video carrier. The sound carrier is an FM 50kHz (±25kHz) signal. The video bandwidth is 4.2MHz.

4.4.1 High-Definition TV Channels

Federal law required all TV broadcasters to convert from analog to digital broadcasting on or before June 12, 2009. Congress mandated this conversion because digital broadcasting freed up frequencies for public safety communications used by the police, fire departments, emergency rescue personnel, and advanced commercial wireless services. Digital broadcasting is also a more efficient transmission technology. It allows broadcasters to provide better picture and sound quality and offer additional programming options.

More than two years prior to the analog-to-digital conversion, all television receivers shipped interstate or imported into the United States were required to have a digital tuner. Most DTVs and digital television equipment have labels or markings on them, or statements in the printed manuals for the equipment, to indicate that they contain digital tuners. Some common terms used to describe equipment that has a digital tuner are integrated digital tuner or digital tuner built in. In some cases, the word receiver might be used in place of tuner, and DTV, ATSC, or HDTV might be used in place of digital.

It should be noted that equipment labeled digital monitor, HDTV monitor, digital ready, or HDTV ready might not contain a digital tuner. For these and analog TVs, a separate digital-to-analog set-top converter box is required to view over-the-air digital television signals. The only additional equipment needed is some type of regular antenna.

Digital HDTV provides for a TV presentation with higher resolution and clarity, wider screen pictures, and better sound when compared to analog NTSC systems. HDTV broadcast signals are made up of coded digital instructions that are decoded by an HDTV receiver or an HDTV converter. To be a true HDTV signal, the signal format must meet the Advanced Television Systems Committee (ATSC) standards for HDTV. Since the transition to HDTV affected broadcasters, cable companies, and consumers, some important points about the transition to HDTV are as follows:

- Broadcasters had to purchase many new items of equipment such as cameras, remote broadcast units, control rooms, cables, and sound equipment, rendering most of their video production equipment obsolete.

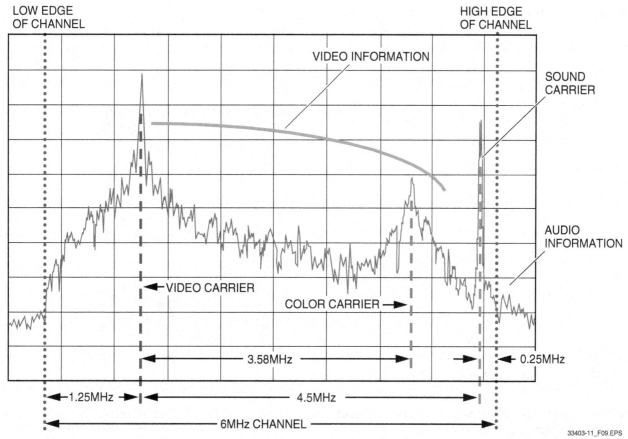

LOW EDGE OF CHANNEL

HIGH EDGE OF CHANNEL

VIDEO INFORMATION

SOUND CARRIER

AUDIO INFORMATION

VIDEO CARRIER

COLOR CARRIER

3.58MHz

0.25MHz

1.25MHz

4.5MHz

6MHz CHANNEL

33403-11_F09.EPS

Figure 9 Standard NTSC analog television broadcast channel.

- Cable operators had to convert their headend equipment and all of their TV set converter boxes.
- Consumers had to buy either a digital HDTV set or a special digital-to-analog converter in order to convert received HDTV digital signals into analog signals for viewing on an analog TV set. It should be pointed out that converted HDTV programming on an analog TV set loses the resolution, format, and other benefits that are obtained by using a TV set specifically designed to display HDTV signals.

4.5.0 Units of Measurement

The design, component selection, testing, and alignment of broadband cable systems requires the ability to understand and work with different units of measurement, especially the decibel (dB) and decibel referenced to 1 millivolt (dBmV).

4.5.1 Decibel

The bel is a basic unit in a logarithmic scale used for expressing the ratio of two amounts of power. The scale is named after its originator, Alexander Graham Bell. The decibel, which is equal to $\frac{1}{10}$ of a bel, is the more commonly used unit of measure. The decibel is used to compare two power levels or the power ratio of two voltage levels on a logarithmic scale. It can also be used to compare the power ratio of two current levels, but this is not commonly done in cable TV system work. The gains of antennas and amplifiers and the losses in components, such as splitters and directional couplers, are expressed in decibels. The advantage of using the decibel is that it allows the gains and losses within a system to be easily calculated using only addition and subtraction.

Decibel values are easily calculated using the common log (log) function of a scientific calculator. When comparing the power ratio of two power levels, the decibel gain or loss is found by multiplying the common logarithm of their ratio by 10. Expressed mathematically:

$$dB = 10 \times \log_{10} P1/P2$$

Where:

P1 = the power level being considered, typically the output signal level
P2 = an arbitrary reference level, typically the input signal level

To illustrate the calculation of decibel values when comparing the power ratio of two power levels, a few examples are provided here.

Example 1

If an amplifier has an input signal level of 200 milliwatts (mW) and an output signal level of 800mW, what is the gain of the amplifier in decibels?

Solution:

dBGain = 10 × log₁₀ (P1/P2)
dBGain = 10 × log₁₀ (800/200)
dBGain = 10 × log₁₀ (4) = 6
dBGain = 6dB

Example 2

If an attenuator has a signal input of 800mW and an output signal of 200mW, what is its attenuation loss in decibels?

Solution:

dBLoss = 10 × log₁₀ (P1/P2)
dBLoss = 10 × log₁₀ (200/800)
dBLoss = 10 × log₁₀ (0.25) = –6dB
dBLoss = 6dB

When comparing the power ratio of two voltage levels when the impedances across which the signals being measured are equal, the decibel gain or loss is found by multiplying the common logarithm of their ratio by 20. Expressed mathematically:

$$dB = 20 \times \log_{10}(V1/V2)$$

Where:

V1 = the voltage level being considered, typically the output signal level
V2 = an arbitrary voltage reference level, typically the input signal level

To illustrate the calculation of decibel values when comparing the power ratio of two voltage levels, a few examples are provided here.

Example 3

An amplifier has an input signal level of 1,000 microvolts (μV) and an output signal level of 2,000μV. What is the gain of the amplifier in decibels?

Solution:

dBGain = 20 × log₁₀ (V1/V2)
dBGain = 20 × log₁₀ (2,000/1,000)
dBGain = 20 × log₁₀ (2)
dBGain = 6dB

Example 4

A coaxial signal line has an input signal of 800μV and an output signal level of 100μV. What is the attenuation loss of the line in decibels?

Solution:

dBLoss = 20 × log₁₀ (V1/V2)
dBLoss = 20 × log₁₀ (100/800)
dBLoss = 20 × log₁₀ (0.125)
            = 18.062 rounded to –18dB
dBLoss = 18dB

4.5.2 Decibels and Decibel Conversions

Decibel power levels encountered in electronic systems can be expressed in several different ways, including the following:

- Decibel-milliwatt (dBm) – This is the abbreviation for decibels referenced to 1mW across a 50-ohm impedance. It is commonly used when working with microwave, satellite, and fiber optic systems. It is also commonly used in audio systems. However, in audio systems the dBm is assumed to be across a 600-ohm impedance, instead of a 50-ohm impedance. Milliwatts can be converted to dBm using the equation: dBm = 10 log₁₀ (x milliwatts). For example, 2mW converted to dBm = 10 log₁₀ (2) = 3dBm. Power levels expressed as +dBm indicate a signal that is greater than 1mW. Those expressed as –dBm indicate a signal that is less than 1mW.

Channel Bandwidth Spacing

In addition to the NTSC standard used to encode TV signals in the United States, two other standards are in use to encode and transmit analog TV signals worldwide. These standards are the Phase Alternation Line (PAL) and the Sequential Color With Memory (SECAM) standards. These standards both use a channel bandwidth of 8MHz rather than 6MHz. Use of an 8MHz bandwidth per channel yields better audio and video quality. However, fewer TV channels can be made available within the usable frequency spectrum. In addition to being used in the United States, the NTSC format is used in Japan, Canada, and several European countries. The PAL format is used in Germany, the United Kingdom, and Australia; the SECAM format is used in France, the French colonies, and Russia.

- **Decibel-millivolt (dBmV)** – This is the abbreviation for decibels referenced to 1mV across a 75-ohm impedance. It is used when working with cable TV systems. Millivolts can be converted to dBmV using the equation: $dBmV = 20 \log_{10} (\times \text{ millivolts})$. For example, 2mV converted to $dBmV = 20 \log_{10} (2) = 6dBmV$. Power levels expressed as +dBmV indicate that the signal is greater than 1mV. Those expressed as –dBmV indicate that the signal is less than 1mV.
- **Decibel-microvolt (dBμV)** – This is the abbreviation for decibels referenced to 1μV across a 75-ohm impedance. It is commonly used when working with off-air TV signals and conventional TV systems. Microvolts can be converted to dBμV using the equation: $dBμV = 20 \log_{10} (\times \text{ microvolts})$. For example, 2,000μV (2mV) converted to $dBμV = 20 \log_{10} (2,000) = 66dBμV$. Power levels expressed as +dBμV indicate that the signal is greater than 1μV. Those expressed as –dBμV indicate that the signal is less than 1μV.

It is often desirable to convert signals expressed as dBm to dBmV and vice versa. This is commonly done using conversion tables included by cable equipment manufacturers in the technical reference section or appendix of their product catalogs. An example of this type of conversion table, which gives the conversions between corresponding mV, dBmV, dBμV, and dBm voltage and power measurement values, is given in *Appendix A*.

The conversions from dBm to dBmV and vice versa can also be calculated using the log and y^x functions of a scientific calculator. A power level expressed in dBmV can be converted directly to power in dBm when the impedance (Z) is known, using the following formula. Note that for calculations pertaining to cable TV, the impedance is always 75 ohms.

$$dBm = (x - 30) - [20 \log_{10} (\sqrt{Z})]$$

Where:

x = value of dBmV to be converted
Z = known impedance

Example 5

Convert 48.75dBmV to dBm for cable TV.

Solution:

$$dBm = (x - 30) - [20 \log_{10} (\sqrt{Z})]$$
$$dBm = (48.75 - 30) - [20 \log_{10} (\sqrt{75})]$$
$$dBm = 18.75 - [20 (.9375)]$$
$$dBm = 18.75 - 18.75$$
$$dBm = 0$$
$$48.75dBmV = 0dBm$$

Conversely, a power level expressed in dBm can be converted directly to power in dBmV when the impedance (Z) is known, using the following formula:

$$dBmV = (x + 30) + [20 \log_{10}(\sqrt{Z})]$$

Where:

x = value of dBm to be converted
Z = known impedance

Example 6

Convert 0dBm to dBmV for cable TV.

Solution:

$$dBmV = (x + 30) + [20 \log_{10}(\sqrt{Z})]$$
$$dBmV = (0 + 30) + [20 \log_{10}(\sqrt{75})]$$
$$dBmV = 30 + [20 (.9375)]$$
$$dBmV = 30 + 18.75$$
$$dBmV = 48.75$$
$$0dBm = 48.75dBmV$$

4.6.0 Common CATV Symbols

As with most industries, the cable TV industry uses symbols on installation and other drawings to represent the various system components and their configuration. Ideally, all organizations in the CATV industry would use a standardized set of symbols. However, this is not always the case. For this reason, whenever using an installation or other drawing, always refer to the associated legend to determine exactly what each symbol being shown on the drawing means. *Appendix B* contains

---

On Site

Understanding the Decibel

It is important to understand that the decibel does not represent a specific power level. The value expressed in decibels only denotes the ratio between two discrete power levels. Several amplifiers with different input and output levels can all have the same gain expressed in decibels. For example, an amplifier with an input signal level of 1mW and an output level of 2mW has a gain of 3dB, while another amplifier with an input signal level of 2W and an output signal of 4W also has a gain of 3dB. Because the power ratio for both amplifiers is 2, the gain for both amplifiers expressed in decibels is 3dB ($3dB = 10 \times \log_{10} 2$).

---

a set of CATV symbols typical of those used in the industry.

5.0.0 SATELLITE TECHNOLOGY

Because both CATV and SMATV systems receive and process signals from satellites, an overview of satellite technology related to TV is covered here. Satellites were first used in the 1970s for the transmission of television broadcasting to and from networks, and then later to cable systems and directly to individual subscribers. Today, satellites are indispensable for the transmission and distribution of video programs, as well as for industrial applications including security and surveillance.

In a basic satellite system (*Figure 10*), Earth stations uplink (transmit) a video signal to a satellite by means of a narrow radiated beam. The satellite receives this signal, amplifies it, shifts the frequency to a downlink frequency using a transponder, then retransmits the downlink signal to Earth at the same time.

5.1.0 Classification of Satellites

Satellites are classified by their usage and technical characteristics. For television, the FCC defines two usages. One is the fixed satellite service (FSS) that includes most commercial communication satellites. The other is the broadcast satellite service (special FSS sub-classification), usually called direct broadcast service (DBS) for analog signals or digital direct broadcast service (digital DBS) for digital signals. The direct broadcast services are operated at higher power and wider spacing. This allows direct-to-home broadcasts using much smaller TVRO dish antennas. The technical characteristics of satellites used for television include their frequency spectrum bands, orbital locations, and service areas.

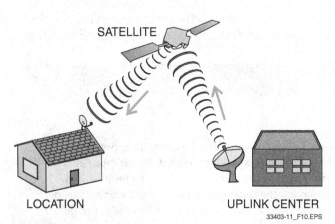

Figure 10 Basic satellite system.

SATELLITE

LOCATION

UPLINK CENTER

33403-11_F10.EPS

5.2.0 Television Satellite Frequency Spectrum Bands

Four frequency spectrum bands, designated as C-band, Ku-band, DBS, and Ka-band, are used for commercial television satellite communications. The downlink frequencies are used for TVRO purposes, while the uplink frequencies are used only by commercial network services to provide programming signals.

- C-band FSS:
  - Downlink = 3.7 to 4.2GHz
  - Uplink = 5.925 to 6.425GHz

- Ku-band FSS:
  - Downlink = 11.7 to 12.2GHz
  - Uplink = 14.0 to 14.5GHz
- DBS:
  - Downlink = 12.2 to 12.7GHz (Ku-band)
  - Uplink = 17.3 to 17.8GHz

- Ka-band FSS:
  - Downlink = 18.3 to 18.5GHz, 18.6 to 18.8GHz, and 19.7 to 20.2GHz
  - Uplink = 28.4 to 28.6GHz and 29.3 to 30.0GHz

Ka-band, which is the newest satellite broadcast band, is used by satellite programming and satellite Internet providers. For instance, WildBlue™ uses Ka-band to deliver satellite broadband services, and DirecTV® used Ka-band to supplement their existing Ku-band channel capacity.

DBS is a sub-band of the Ku-band. The DBS band is reserved for high-powered satellites intended for the delivery of digital television programming. DBS satellites are spaced nine degrees apart. This wide spacing allows for very high-powered transmissions. As a result, small parabolic dish-type antennas that are less than one meter in diameter can be used to receive the signals.

In addition to the higher-power DBS frequencies, lower-power C-band and other Ku-band frequencies are used for direct-to-home broadcast. In fact, the early home satellite systems used large (8' to 12' diameter) C-band antennas. The C-band service can be subject to problems associated with nearby microwave installations. In commercial applications that use the uplink as well as the downlink, the Earth station must be coordinated (cleared) with nearby microwave stations to prevent interference.

5.3.0 Orbital Positions of Satellites

Satellites used for communication travel around Earth at a speed that matches that of Earth's rotation. For that reason, satellites appear to be stationary in space. This is called a geosynchronous, or geostationary, orbit. The geosynchronous orbit is the plane at which the property of centrifugal force, pushing an object outward into space, equals the gravitational force (that force which pulls an object downward to the Earth). *Figure 11* shows a geosynchronous orbit.

The geosynchronous orbit surrounding Earth is divided into a series of arcs called orbitals. These arcs, or orbitals, are assigned worldwide via an international agreement for use by the different countries. Each individual satellite located within a specific orbital is assigned a fixed position, called an orbital slot. The orbital slot is identified by its longitude. Within these orbitals, slots are assigned by the regulatory bodies of the country to licensees. In the United States, the regulatory authority is the Federal Communications Commission (FCC). Only a specific number of satellites may be accommodated within the geosynchronous orbit. In order to maximize this number, the distance between orbital slots is critical and engineered to be minimal. The spacing is then determined by uplink and downlink antennas, as well as directivity. *Figure 12* shows an example of the North American C-, Ku-, and Ka-band satellite orbital slots.

5.4.0 Downlink Signal Parameters

A downlink signal covers an area known as the satellite's footprint. There are three general types of uplink/downlink configurations in use that determine the footprint of the downlink:

- *Point-to-point* – This configuration is used primarily for commercial communication carrier systems.
- *Point-to-multipoint* – This configuration is the one primarily used for satellite network broadcast TV or radio purposes. Various commercial network signals are transmitted from

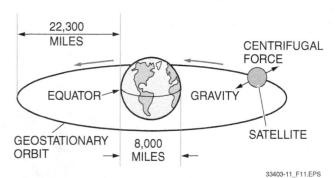

Figure 11 Geosynchronous orbit.

specific spots to a satellite. The signals are then rebroadcast to broad areas on Earth for reception by commercial television over-the-air broadcasters, commercial cable television (CATV) systems, private master antenna television (MATV) systems, home direct broadcast service (DBS), or other television receive-only (TVRO) satellite antennas and receivers.

- *Multipoint-to-point* – A configuration used for a very small-aperture terminal (VSAT) where mobile or fixed station point-of-sale terminals or monitoring equipment are linked to an office via small to medium satellite antennas. Such systems are typically used by gas stations, stock brokers, commodity traders, and other businesses.

5.5.0 Areas of Service

Satellite service can vary greatly depending on the downlink antenna size of the satellite. The coverage of a single satellite may incorporate a large portion of the hemisphere or a single metropolitan area. The greater the beam transmitted by the satellite, the larger the service area can be. The larger the downlink antenna, the narrower the transmitted beam is going to be, and the resulting service area to receive the signal is going to be smaller. The following terms are used to identify different beams:

On Site

Geostationary Orbit

A communication satellite must have an orbital time (period) equal to that of Earth's rotation in order for it to appear stationary above a given point on the ground. This period is equal to 23 hours, 56 minutes, the time it takes for one complete revolution of the earth, based on a sidereal day. (A sidereal day is defined as the interval between two successive transits of a star over the meridian. It is equal to 23 hours and 56 minutes.) Because the orbital period is a function of altitude, this means that the geostationary orbit for communications satellites must be located about 22,300 miles above the equator. The geostationary orbit is sometimes referred to as the Clarke belt, so named after Arthur C. Clarke, a writer who first envisioned the concept of having a network of satellites that formed a worldwide communications system.

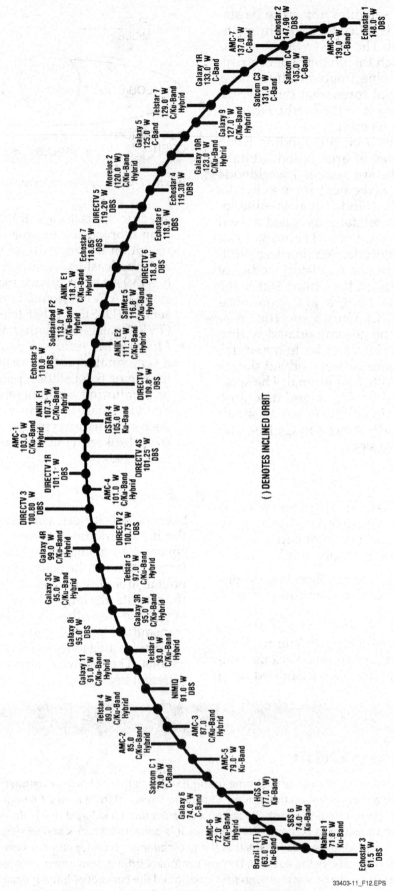

( ) DENOTES INCLINED ORBIT

Echostar 1
148.0° W
DBS

AMC-8
139.0° W
C-Band

Echostar 2
147.90° W
DBS

Satcom C4
135.0° W
C-Band

AMC-7
137.0° W
C-Band

Satcom C3
131.0° W
C-Band

Galaxy 1R
133.0° W
C-Band

Galaxy 9
127.0° W
C/Ku-Band
Hybrid

Telstar 7
129.0° W
C/Ku-Band
Hybrid

Galaxy 5
125.0° W
C-Band

Galaxy 10R
123.0° W
C/Ku-Band
Hybrid

Morelos 2
(120.0° W)
C/Ku-Band
Hybrid

Echostar 4
119.30° W
DBS

DIRECTV 5
119.20° W
DBS

Echostar 6
118.9° W
DBS

Echostar 7
118.85° W
DBS

DIRECTV 6
118.8° W
DBS

ANIK E1
118.7° W
C/Ku-Band
Hybrid

SatMex 5
116.8° W
C/Ku-Band
Hybrid

Solidaridad F2
113.0° W
C/Ku-Band
Hybrid

ANIK E2
111.1° W
C/Ku-Band
Hybrid

Echostar 5
110.0° W
DBS

DIRECTV 1
109.8° W
DBS

ANIK F1
107.3° W
C/Ku-Band
Hybrid

GSTAR 4
105.0° W
Ku-Band

AMC-1
103.0° W
C/Ku-Band
Hybrid

DIRECTV 4S
101.25° W
DBS

DIRECTV 1R
101.1° W
DBS

AMC-4
101.0° W
C/Ku-Band
Hybrid

DIRECTV 3
100.80° W
DBS

DIRECTV 2
100.75° W
DBS

Galaxy 4R
99.0° W
C/Ku-Band
Hybrid

Telstar 5
97.0° W
C/Ku-Band
Hybrid

Galaxy 3C
95.0° W
C/Ku-Band
Hybrid

Galaxy 3R
95.0° W
C/Ku-Band
Hybrid

Galaxy 8i
95.0° W
DBS

Telstar 6
93.0° W
C/Ku-Band
Hybrid

Galaxy 11
91.0° W
C/Ku-Band
Hybrid

NIMIQ
91.0° W
DBS

Telstar 4
89.0° W
C/Ku-Band
Hybrid

AMC-3
87.0° W
C/Ku-Band
Hybrid

AMC-2
85.0° W
C/Ku-Band
Hybrid

AMC-5
79.0° W
Ku-Band

Satcom C 1
79.0° W
C-Band

Galaxy 6
74.0° W
C-Band

HGS 6
(77.0° W)
Ku-Band

SBS 6
74.0° W
Ku-Band

AMC-6
72.0° W
C/Ku-Band
Hybrid

Nahuel 1
71.8° W
Ku-Band

Brazil 1 (T)
(63.0° W)
Ku-Band

Echostar 3
61.5° W
DBS

33403-11_F12.EPS

Figure 12 Example of North American C-, Ku-, and Ka-band satellite orbital slots.

- *Spot* – Spot beams cover a limited area and are used principally for point-to-point voice and data communication. Spot beams are used heavily by satellite TV providers to distribute local programming.
- *National (multipoint)* – National beams cover all or a significant portion of a single country. Most domestic US satellites are in this category. Dish Network®, DirecTV®, and standard satellite TV all fall within this category.
- *Regional* – Regional beams cover a group of countries (e.g., western Europe).
- *Global* – Global beams cover the entire area visible from the satellite. They are used for international communication.

6.0.0 HEADEND SIGNAL PROCESSING

The headend consists of receiving and processing equipment required to electronically prepare selected off-air, satellite, and locally generated signals for application to the distribution system. Headend systems sometimes use modular-type equipment to conserve space. In such systems, one or more rack chassis are populated with a variety of modular components, as shown in *Figure 13*. Information about rack assemblies was covered in detail in the *Rack Systems* module.

Processing of the signals in the headend involves all of the following:

- Amplification of the off-air, satellite, or local signals
- Rejection of co-channel and other spurious signals
- Frequency conversions
- Combining and/or splitting of signals
- Generation of local signals
- Signal gain control and balance
- Control of in-channel audio carrier levels relative to the video carrier levels
- Scrambling of premium program signals
- Descrambling of satellite program signals

The specific types of equipment used in the headend are determined by the source and frequency of the input signals and the output requirements. The equipment configuration can range from simple to complex. *Figure 14* shows a simplified block diagram of a CATV or professional SMATV headend.

With the exception of cases in which there is no satellite input or related equipment, the headend of a professional MATV system is similar. An overview of the processing activities within the headend is given here. More information about the major components used for headend process-

Figure 13 Rack-mounted headend equipment.

ing is provided later in this module. The signal processing activities within the headend typically include the following:

- VHF off-air signal processing
- UHF off-air signal processing
- Satellite signal processing
- Locally originated channel signal processing

6.1.0 VHF/UHF Off-Air Signal Processing

Aside from the frequencies involved, processing of off-air VHF and UHF signals is done in the same way. As shown in *Figure 14*, VHF frequency (54 to 88MHz and 176 to 216MHz) and UHF frequency (470 to 806MHz) off-air broadcast channels are received by high-gain VHF and UHF antennas, respectively. These antennas can be either single-channel or broadband antenna types. Low-level signals received by each of the antennas are applied via coaxial cable to the input of a low-noise preamplifier. Its function is to amplify and improve the signal-to-noise (S/N) ratio of the received signals to the level required for input to the processor. The preamplifiers are normally mounted on the related antenna mast, as close as possible to the antenna. Amplified VHF or UHF signals at the outputs of the preamplifiers are applied via coaxial cables to the associated processor units. Power for each of the preamplifiers is provided by a stand-alone

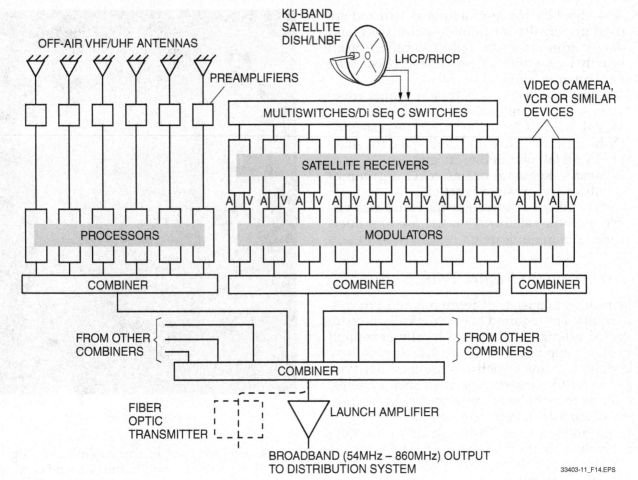

Figure 14 Simplified block diagram of a headend.

In the figure: OFF-AIR VHF/UHF ANTENNAS, PREAMPLIFIERS, KU-BAND SATELLITE DISH/LNBF, LHCP/RHCP, VIDEO CAMERA, VCR OR SIMILAR DEVICES, MULTISWITCHES/Di SEq C SWITCHES, SATELLITE RECEIVERS, PROCESSORS, MODULATORS, COMBINER, FROM OTHER COMBINERS, FROM OTHER COMBINERS, COMBINER, FIBER OPTIC TRANSMITTER, LAUNCH AMPLIFIER, BROADBAND (54MHz – 860MHz) OUTPUT TO DISTRIBUTION SYSTEM, 33403-11_F14.EPS

DC power supply (not shown) usually located in the headend building. The power supply output is simultaneously applied via a power inserter onto the same coaxial cable that carries the VHF or UHF signal from the preamplifier to the processor. This arrangement eliminates the need for running a separate power feed cable from the power supply to the preamplifier.

The processors that receive the off-air VHF or UHF signals from the preamplifiers perform several functions. These functions include signal amplification, rejection of unwanted adjacent channel signals received by the antenna, precise signal level control, audio carrier level control, and generation of output signals at the desired broadcast or CATV channel frequencies.

There are many different types of processors. The type used depends on the system requirements. Processors can range from the simplest type, called strip amplifiers, to the most sophisticated types, called frequency-agile processors. Strip amplifiers are single-channel amplifiers designed to receive a single VHF or UHF channel input and provide a single-channel output with no change in frequency. Frequency-agile processors have the capability to select any input VHF or UHF channel and provide an output at any other broadcast or CATV channel frequency. Other processors have channel selection capabilities ranging between the two. Each processor has output gain adjustments and input automatic gain control (AGC) circuitry that control the level of its output signal. During system alignment of the off-air processing equipment, all the processors are adjusted so that their video RF carrier output signals are equal to within ±1dB of each other. The signal levels need to be balanced in order to combine them together and with the signals produced in the satellite and local origination processor equipment. Each processor is also adjusted so that the selected channel's sound (aural) carrier level is set to about –15dB below its video carrier level.

The balanced signal outputs from all of the VHF or UHF processors are applied to combiners, where they are combined with each other and with the signals being generated by the satellite and local origination equipment. This produces the broadband output signal for application to the distribution system.

If the headend output is applied to a distribution system with a coaxial trunk cable, the broadband signal output from the final combiner is applied to a post amplifier, sometimes called a launch amplifier. There, it is further amplified to a level sufficient to drive the trunk cable. If the headend output signal is applied to a distribution system with a fiber optic cable, the output from the final combiner is applied to a fiber optic transmitter in order to drive the fiber optic cable.

6.2.0 Satellite Signal Processing

The type of satellite receiving antenna and headend equipment used to process satellite signals is determined by the programming to be received and processed by the headend. The transmission format used by early satellite systems was analog. The analog format used the standard C-band FSS. Today, most satellite systems use the digital format, as evidenced by digital DBS systems. DBS systems use multiple satellite locations in both the Ka- and Ku-bands. Moreover, many of the television signals transmitted by satellite, especially digital signals, are encoded using sophisticated scramblers to prevent unauthorized signal reception. For the purpose of this examination, processing for DBS-band satellite signals is described. Processing for C-band satellite signals is done in a similar manner.

DBS-band transponder downlink signals in the 12.2 to 12.7GHz range are received by a high-gain dish antenna (refer to *Figure 14*) and applied via its feedhorn to a dual low-noise block downconverter (LNBF). The LNBF, commonly referred to as LNB, is controlled by a voltage or a digital code sent from the receiver. The transponder signals provide for 32 channel frequencies. Each transponder has a bandwidth of 30MHz. Sixteen of the transponders are right-hand circularly polarized (RHCP) and 16 are left-hand circularly polarized (LHCP). In order to process all 32 transponders, a dual LNBF is used. If a single LNBF were used, the channel availability would be limited to only half of the available transponders, those being either the 16 RHCP or 16 LHCP channels. Each transponder signal can contain as many as 12 channels of programming, depending on the digital compression techniques used.

The function of the dual LNBF is to amplify and convert the DBS-band signals into L-band (950 to 1,450MHz) signals at the level required for input to the satellite receivers. Downconversion to the lower L-band frequencies is done because they encounter less signal loss in the coaxial cables through which they are applied to the satellite receivers. The LNBF shown in this example has two independent signal outputs. Each output is capable of supplying a signal of either polarity via multiswitches or digital satellite equipment control (DiSEqC) switches that are controlled by a voltage change or a digital signal. Polarity selection is based on the receiver program selection. Normally, the selection is such that one output provides LHCP signals and the other RHCP signals for input to the switches.

The multiswitches or DiSEqC switches also enable multiple satellite receivers to be connected to the dual LNBF. This enables each of the satellite receivers to independently select either the RHCP or LHCP signal for output from its related switch on a single cable for input to the satellite receiver. Each satellite receiver sends a control voltage to its LNBF or switch that selects the proper polarity based on the receiver program selection. The operation and polarization selection voltages applied to the LNBF/switches are provided via power inserters from two separate power supplies (not shown). In some systems, these voltages are supplied directly from the satellite receivers.

The selected satellite signal output from each multiswitch or DiSEqC switch is applied to its related satellite receiver. Satellite receivers are made in many different designs. All operate to tune the desired transponder signal and then demodulate it to provide baseband audio and video (A/V) outputs. Most satellite receivers also have the capability to descramble encoded received signals. In C-band systems, these receivers are called integrated receiver descrambler (IRD) receivers. The A/V outputs from each satellite receiver are applied to a related modulator. It accepts the baseband audio and video signals and amplitude modulates (AM) them onto a carrier with a frequency for one channel within the CATV frequency range. Each modulator has gain adjustments that affect the output level and carrier-to-noise (C/N) ratio of its output signal. During system alignment of the headend equipment, all the modulators are adjusted so that their output signals are balanced within ±1dB of each other, and with the signal outputs from the VHF/UHF processors and local origination modulators.

The balanced signal outputs from all the modulators are routed through combiners to be combined with those processed in the VHF/UHF and local origination signal-processing equipment. The broadband output signal from the final combiner is then applied to the distribution system.

6.3.0 Locally Originated Channel Signal Processing

Processing of signals from local origination equipment, such as TV cameras, VCRs, DVDs, character/graphic generators, and similar devices, is often done to provide additional channels of prerecorded programming (refer to *Figure 14*). Baseband audio and video signals generated by each of these devices are applied to related modulators. They accept the baseband audio and video signals and amplitude modulate (AM) them onto a carrier with a frequency for one channel within the CATV frequency range. The output signals from the modulators are equalized and multiplexed with those from the VHF/UHF and satellite signal-processing equipment, via combiners, for application to the distribution system.

7.0.0 HEADEND COMPONENTS

This section describes the main types of equipment commonly used for receiving and processing signals in the headend of CATV, SMATV, and MATV systems, including the following:

- Television broadcast receiving antennas and preamps
- Satellite receiving antennas and downconverters
- Off-air processors
- Demodulators and modulators
- Satellite receivers
- Stereo encoders
- Combiners and splitters
- Filters

7.1.0 Television Broadcast Receiving Antennas and Preamps

Broadband and single-channel television antennas are the two main types of off-air television receiving antennas. The typical residential receiving television antenna is the broadband type (*Figure 15*). This antenna is optimized to receive low-band VHF (Channels 2 through 6), FM, high-band VHF (Channels 7 through 13), and UHF signals. Antennas of this type usually consist of two antennas mounted on the same bracket—the VHF frequency band of 55 to 88 and 176 to 216MHz on one set of elements and the UHF frequency band of 470 to 890MHz on the other, making up the all-band antenna. This type of antenna is more or less directional, depending on the number of antenna elements used. If stations are located more than 90 degrees apart, an electric rotation system is needed to turn the antenna toward the station.

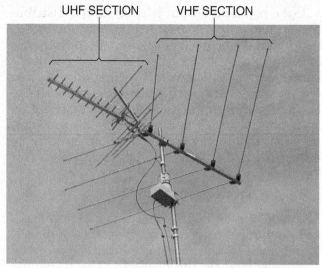

UHF SECTION   VHF SECTION

33403-11_F15.EPS

Figure 15 Residential VHF/UHF broadband receiving antenna with rotor.

Commercial receiving antennas used for off-air broadcast TV signal reception in CATV, SMATV, or MATV installations are heavy-duty versions of residential-type single-channel or broadband antennas, usually Yagi or log-periodic types (*Figure 16*). These antennas are made of thicker and heavier materials and are mounted on square beams that can withstand up to 125 mile per hour winds and substantial ice loads. The orientation of commercial antennas remains fixed once they are initially positioned for best signal reception. Various mounting frames are also available to accommodate different tower and mast mounting configurations.

Professional single-channel VHF antennas are designed to optimally receive one selected low-band VHF, low-band VHF and FM, or high-band VHF signal. UHF antennas are typically made to optimally receive selected groups of UHF channel frequencies. For example, a particular UHF antenna may be designed to receive Channels 14 through 19 or Channels 20 through 26. Broadband VHF or UHF antennas are designed to receive all the VHF or UHF channel frequencies, respectively.

7.1.1 Receiving Antenna Signal Reception

Factors that affect the quality of broadcast TV signal reception by a television off-air receiving antenna include the following:

- Type of receiving antenna
- Height of the receiving antenna
- Terrain between the transmitting and receiving antennas
- Types and number of obstacles between the transmitting and receiving antennas

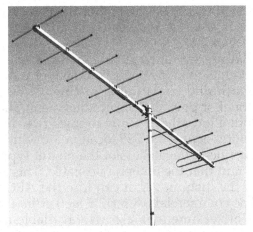

SINGLE-CHANNEL YAGI ANTENNA   BROADBAND LOG-PERIODIC ANTENNA

33403-11_F16.EPS

Figure 16 Typical professional receiving antennas.

- Distance between the transmitting and receiving antennas
- Height of the transmitting antenna
- Transmitter power
- Transmitter frequency
- Weather

The proper mounting and placement of receiving antennas for different frequencies mounted on a tower or mast is crucial to satisfactory signal reception. Proper spacing must be maintained so the tuned antenna elements are not affected by adjacent antennas.

Generally, the rules for antenna placement are as follows:

- The weakest station antennas should be mounted at the highest section of the tower. Place the highest frequency (smallest) antenna on top when equal signals are received.
- Always follow the procedures outlined for the proper minimum spacing. Refer to *Appendix C* for antenna spacing information and tables.
- Orient the antenna so that it faces the direction of the broadcasting station(s).

7.1.2 Antenna Grounding

In order to protect the antenna and headend equipment from the effects of lightning, proper installation of a good electrical grounding system must be done in accordance with *NEC Article 810*.

7.1.3 Antenna Specifications

In order to compare one antenna versus another for a particular application, you must compare their specifications. Common terms used by an-

tenna manufacturers to describe the characteristics of their products, regardless of antenna type, include the following:

- *Gain* – Indicates the amount of received signal level in a given direction when compared to a reference antenna, typically a resonant dipole antenna.
- *Bandwidth* – The range of frequencies (TV channels) for which the designed antenna is optimized.
- *Impedance* – The impedance of an antenna must be matched to the connecting equipment. Commercial and residential-type antennas typically have an impedance of 75 ohms for connection with 75-ohm coaxial cable.
- *Directivity* – The property of the antenna to receive more energy in some directions than in others.
- *Beamwidth* – Describes the width of the main (front) lobe of the antenna at the points (–3dB points) where the voltage is 0.707 times the maximum voltage (*Figure 17*).
- *Front-to-back ratio* – The front-to-back ratio of an antenna is the ratio of how much signal is picked up by the front lobe versus the back lobe. It is expressed in decibels.

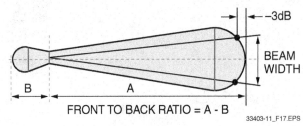

33403-11_F17.EPS

Figure 17 Basic antenna beam pattern.

7.1.4 Broadband VHF and UHF Preamplifiers

Because of long cable runs or weak received signals, an appropriate VHF or UHF preamplifier (*Figure 18*) is normally used in conjunction with a high-gain antenna to boost the received signal strength. As mentioned earlier, preamplifiers are usually mounted at or near the antenna. Most are fed DC power that is diplexed on the coaxial lead-in cable from a power supply located at the end of the cable inside a building. VHF/UHF preamplifiers are made in both single-channel and broadband models. Most have a test port that allows the output signal to be easily monitored without interrupting service.

All the information needed to select a preamplifier is given in the manufacturer's specifications. Preamplifiers typically range in gain from 20 to 30dB, and some are adjustable. In addition to the gain, the preamplifier noise figure is an important characteristic to take into consideration. It determines how much noise is added to the system when a very weak signal is being amplified. The lower the noise figure, the better the performance of the preamplifier is going to be. Typical noise figure values for preamplifiers like the one shown range from 3 to 5dB.

7.2.0 Satellite Receiving Antennas and Downconverters

The most common receiving antennas used with satellites have parabolic (circle) or quasi-parabolic (oval) reflector surfaces focused on a feed horn. Some are made from fiberglass, aluminum, or galvanized steel material; others are made of aluminum or steel mesh. The antenna size is mainly determined by the signal frequency or frequencies it is designed to receive. Normally, the larger

the antenna diameter is, the higher the gain. An example of a C-band and Ku-band antenna is shown in *Figure 19*.

These larger size antennas can have fixed mounts, mounts that have either manual or powered azimuth and elevation adjustments, or a single manual or powered polar adjustment. The polar mount allows the antenna to track two different satellites across the FSS location arc with a single adjustment. Regardless of the mount type, it is important that the antenna be stable. This is because if the dish moves, it can lose the ARC/ signals. The commercial 3.8m (12.5 feet), 5.0m (16 feet), and offset antennas are also available as folding transportable versions. To mount or place a reflector for standard satellite television use, a base or platform must be installed. In most cases, this base is made of concrete for stable support in order to protect the antenna. Base specifications are provided by the manufacturer.

The small DBS and digital DBS satellite dishes (*Figure 20*) are lightweight and do not require movable mounting structures, as they are aimed at one or more satellites that can be received from the same fixed antenna position. The dish is mounted on a single mounting pole or fastened to a wall, window, or roof of a structure. To prevent snow and ice buildup, a thin heating membrane can be applied to the surface of the dish, or the dish can be purchased with an internal heater.

7.2.1 Low-Noise Block Downconverters

Low-noise block downconverters (LNBF) amplify and convert received satellite signals into L-band (typically 250 to 2,150MHz) signals. *Figure 21* shows an example. It is called a block down-

33403-11_F18.EPS

Figure 18 Typical VHF/UHF preamplifier.

33403-11_F19.EPS

Figure 19 C-/Ku-band 1-meter TVRO antenna.

Figure 20 Typical DBS satellite dish.

33403-11_F20.EPS

33403-11_F21.EPS

Figure 21 LNBF.

converter because the entire received satellite bandwidth signal (not a single-channel) is converted. This block of frequencies contains signals from all the satellite transponders, thus allowing two or more satellite receivers to simultaneously tune any channel in the block of frequencies. As described earlier, downconversion is done because the lower frequencies encounter less signal loss in the coaxial cables through which they are applied to the satellite receivers.

Frequency conversion in the LNBF is done by a process called heterodyning, where two frequencies are mixed together in order to produce two other frequencies. One of the frequencies is equal to the sum of the two mixed frequencies; the other is the difference. For example, if DBS-band 12.2 to 12.7GHz signals are heterodyned with a fixed 11.25GHz signal generated by a stable local oscillator (LO), sum frequencies from 23.45 to 23.95GHz and difference frequencies from .95 to 1.45GHz (950 to 1,450MHz) are produced. Bandpass filtering is used to reject the sum frequencies while allowing the difference frequencies to pass for output from the LNBF. Frequency conversion in LN-

BFs designed for input signals in other frequency bands to the same 950 to 1,450MHz L-band output is done by using a local oscillator that has a different fixed frequency. For example, a fixed 2.75GHz LO signal would be used to downconvert C-band 3.7 to 4.2GHz signals to produce the 950 to 1,450MHz L-band signals.

Selection of the polarity signal in a DBS antenna is done electronically by a solid-state switch housed within the LNBF and controlled by the receiver. Selection of a single polarization signal in C-band systems is done at the antenna by an electrically controlled mechanical switch that causes a probe in the antenna feed to move to the desired position. As described earlier, this arrangement allows only half of the transponder channels to be used. The important characteristics to consider when selecting an LNBF include the input frequency, local oscillator frequency, noise figure, and supply voltage. Conversion gain expressed in decibels is the ratio of the available IF power output to the available RF power input. As with VHF/UHF preamplifiers, the lower the noise figure, the better the LNBF is going to perform.

7.2.2 Satellite Antenna Grounding

In order to protect the antenna and converter from the effects of lightning at the feed point, proper installation of a good electrical grounding system is required in accordance with *NEC Article 810*.

7.2.3 Satellite Antenna Alignment Procedures

To properly install a satellite system, the antenna must be located so that there are no obstructions in its line of sight for any satellite to be detected. Before the antenna is installed, a site survey should be performed to determine the best installation location. When erecting the antenna, it is important to make sure that the mount is stable to prevent unwanted movement. After the mechanical installation of the antenna is completed, it must be aimed to its specific azimuth and elevation angles. A compass, an inclinometer, and topographical maps can be used initially for this purpose. The inclinometer is used to see elevation, and the compass is used to determine right/left alignment. Satellite final antenna alignment is done by positioning the antenna to maximize the received signal using one of several types of available satellite signal level meters (*Figure 22*). It can also be done using a video monitor or spectrum analyzer. For DBS antenna alignment, the on-screen peaking meter of the related satellite receiver is used. It gives a relative intensity indication in the form of a bit error ratio, where the maximum signal relates to the cleanest picture.

When aligning a DBS antenna, observe the following guidelines:

- Make sure to use the proper heading (azimuth) and elevation data for the antenna location and satellite. These can be determined from charts or tables supplied with the equipment, from web sites, or from software programs where either a zip code or specific longitude/latitude values can be entered.
- Make sure the mounting post or bracket is stable, plumbed vertically with a level, and is not tilted. This ensures that initial elevation adjustments are reasonably close.
- After initial antenna orientation in azimuth and elevation, make sure that the line of sight from the dish to the sky location for the satellite is clear of all obstructions like trees, tree limbs or branches, bushes, poles, or buildings.
- When fine-tuning the dish, make sure that the movements are small and slow. Use 0.5° increments and hold it for several seconds before proceeding.
- The use of an alignment meter speeds the installation. If a meter is not available, most receivers have an audio indicator that helps to achieve maximum signal levels.
- A single cable run between the LNBF and the receiver is recommended. If a cable extension must be used, the extension connectors and cable must be completely under cover. Use under eaves is permissible.

7.3.0 Off-Air Processors

Several types of processors are available to process the off-air VHF and UHF signals. Processors can be grouped into three categories based on their electronic complexity:

- Strip amplifiers
- Single-channel converters
- Agile heterodyne processors

7.3.1 Strip Amplifiers

Strip amplifiers (*Figure 23*) amplify and stabilize the level of a single-channel VHF or UHF input channel signal for output at the same channel frequency. In addition to amplification circuits, strip amplifiers have AGC and bandpass filter circuits that collectively function to clean up and maintain the processor signal output at a constant level. They also have an adjustable aural carrier trap at the input and an adjacent channel intermodulation trap at the output. This allows their use for adjacent channel operation at high levels without the need for external sound traps or output filtering.

7.3.2 Single-Channel Converters

Single-channel converters (*Figure 24*) convert and amplify off-air single-channel VHF or UHF input signals to different single-channel VHF or UHF output signals. Models designed to convert VHF to VHF, VHF to UHF, and UHF to VHF channels are typical. It should be pointed out that some channel conversions are prohibited when using

33403-11_F23.EPS

Figure 23 Rack-mount strip amplifier.

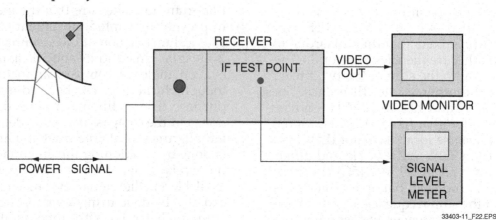

33403-11_F22.EPS

Figure 22 Satellite antenna alignment test setup.

Low-Noise Amplifiers and Downconverters

In older systems, the downconversion from the satellite downlink frequency to an IF is done using two separate assemblies: a downconverter and a low-noise amplifier (LNA). The downconverter can be a single-channel converter or a block converter. When using a single-channel converter, a control signal applied from the related satellite receiver is used to tune the downconverter to produce an IF signal (typically 70MHz) representing the desired single-channel frequency.

this type of converter. This is because certain beat frequencies generated during the conversion process cannot be filtered out of the final signal. The specific channel conversions that are prohibited for use are normally listed in the manufacturer's specifications for the converter. Like strip amplifiers, single-channel processors have band-pass filtering and AGC circuits that clean up and maintain the processor signal output at a constant level.

7.3.3 Agile Heterodyne Processors

Agile heterodyne processors (*Figure 25*) are the most widely used type of processor in many MATV and SMATV systems. Most cable TV systems have engineering standards that specify whether the headend should use frequency agile or fixed frequency units. While frequency agility is advantageous in making quick channel realignments, it is not as precise as a fixed frequency unit and, in most cases, is more expensive. Most CATV headends have frequency agile units as spare standby equipment in case a processor fails. They are the most electronically sophisticated of all the processors, enabling them to deliver the cleanest and most stable output signals. They are called agile because they can switch to any desired frequency. Agile heterodyne processors are commonly used to move any off-air VHF or UHF channel, or any other single-channel input source, to any unused broadcast or CATV channel.

Agile heterodyne processors are made by several manufacturers, and in many models. Depending on design, some are more agile than others. The most versatile models (agile input/agile output) are capable of converting any of the off-air VHF or UHF input channels to any of the broadcast or CATV output channels. Input channel and output channel selection is typically done by setting front panel dual inline package (DIP) switches to the desired channels.

Some models of agile heterodyne processors consist of a mainframe assembly into which a separate removable single-channel output filter module is installed. These are called agile-input, channelized agile-output processors. This arrangement allows frequency-agile input circuitry in the mainframe to select any off-air VHF or UHF channel for processing into the single-channel output. Input channel selection is typically done with front panel DIP switches. The specific filter module installed into the mainframe then determines the frequency of the single-channel output. Because the filter module can be removed and installed, output channel changes can be made easily by replacing the filter module installed in the mainframe with one designed for a different channel frequency.

Other models of heterodyne processors, called channelized input, channelized agile output processors, have agile input circuits that can select any off-air received channel within a limited group of channels. For example, it can select channels either in the VHF low-band, VHF high-band, or in the UHF band. The selected input channel signal is then processed and applied to frequency-agile output circuitry, where it can be converted into

33403-11_F24.EPS

Figure 24 Rack-mount single-channel converter.

33403-11_F25.EPS

Figure 25 Rack-mount agile heterodyne processor.

On-Channel Processor Operation

An off-air processor must sometimes be used to output the same channel as applied to its input. This is referred to as on-channel operation. When this is a requirement, a processor with an optional on-channel capability should be used. Failure to do so in the presence of strong off-air broadcast signals can cause flutter and a horizontal wavy pattern in the TV picture. These distortions are a result of inherent delay encountered by the signal as it passes through the processor relative to the same undelayed signal received directly off-air. When the same signal is received from multiple sources, it is often referred to as multipathing.

any single-channel for output. Selection of both the input and output channels is done with front panel DIP switches.

Most processors also have a standby carrier oscillator circuit that provides a blank-picture output signal should the input signal level applied to the processor drop below a usable level. Most also have an external IF loop that can be used to interface with signal scrambling systems and/or the emergency alert system (EAS). Many CATV franchises require an EAS that can be used by municipal government within the headend's specific coverage area. In the event of a disaster, all channels cease to broadcast their regular programming and are overridden by whatever is being transmitted over the EAS.

7.3.4 Agile Heterodyne Processor Frequency Conversion

Figure 26 shows a simplified block diagram of the frequency conversion circuits used in agile heterodyne processors. It also shows an example of the frequencies derived when a Channel 2 input signal is being converted to a Channel 12 output signal in the processor. As shown, the processor consists of a downconverter circuit where the off-air VHF or UHF input signal is heterodyned with a Channel 2 appropriate local oscillator (LO) signal to produce a downconverted (difference frequencies) IF signal. After downconversion, the IF signal is applied to an upconverter circuit. There, it is heterodyned with a Channel 12 appropriate LO signal to produce an upconverted (sum frequencies) Channel 12 RF carrier output. Examination of this circuit should make it apparent that selection of different analog input and output channel frequencies for processing requires only changing the upconverter and downconverter LO frequencies. This capability is why heterodyne processors are frequency agile processors. It should be pointed out that the heterodyne method for conversion of signals is also commonly used in other frequency-agile equipment.

The use of agile heterodyne processors in many MATV and SMATV applications is a cost-effective method for processing off-air received signals and provides options for quick, inexpensive channel lineup changes. This is because, in addition to their frequency agility, surface acoustic wave (SAW) and AGC filtering in the units provide for both excellent picture quality and adjacent channel rejection.

7.4.0 Demodulators and Modulators

Demodulator and modulator pairs are sometimes used to convert off-air received VHF or UHF channels to a different output channel in the broadcast or CATV band. This is done by connecting the outputs of an agile demodulator unit to the input of a modulator unit (*Figure 27*). Processing of the selected off-air received channel signal in the demodulator removes (demodulates) the carrier leaving the baseband (unmodulated) audio and video signals for output. The type of circuitry used to demodulate the signal depends on the method used to initially modulate the applied input signal. Typically, it is done by downconverting the channel input signal to an IF signal and then detecting the IF signal in separate video and audio detector circuits to remove the carrier frequency. The resultant baseband audio and video signal outputs from the demodulator are then applied to the modulator, where they are modulated onto a new carrier to produce an output signal at the desired broadcast or CATV channel frequency.

It must be pointed out that the modulator unit described here is commonly used separately in several other processing activities where baseband signals must be converted to a single broadcast or CATV channel. They are used to process the baseband audio and video signal outputs applied from satellite receivers, stereo encoders, VCRs, and similar equipment. Controls on the modulator provide adjustment for the video and audio modulation, audio/video ratio, and the output level. The audio and video modulation

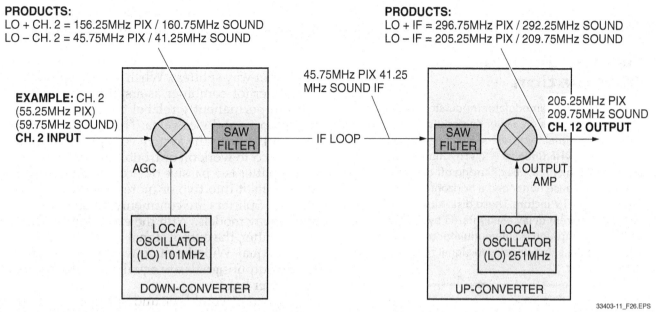

Figure 26 Frequency conversions in an agile heterodyne processor.

controls adjust the sound and picture brightness levels. The audio/video ratio control, sometimes called the aural level control, adjusts the level of the audio carrier relative to the video carrier.

7.5.0 Satellite Receivers

A satellite receiver (*Figure 28*) functions to select a single satellite transponder signal and convert it into baseband audio and video outputs for application to a related modulator. The signal applied to the receiver input is usually an L-band (950MHZ to 3GHz) signal derived from the downconversion of the received C-band, Ku-, or Ka-band signals by the LNBF at the antenna. The specific satellite receivers used depend on the desired satellite TV program services. This is because satellite receivers are designed so that they cannot process the signals transmitted from all the various satellites. For example, many digital receivers are designed specifically to receive one or more program formats, such as MPEG-1, MPEG-2, MPEG-4, DirecTV®, or DVB-compliant Dish Network® programs.

Satellite receivers are made by many manufacturers and in many digital models. Most com-

mercial satellite receivers are integrated receiver/descrambler (IRD) receivers. This means that they have a descrambler circuit used to decode signals that have been encrypted to prevent unauthorized reception. Because of incompatibility between the various program encryption methods, an IRD receiver must be used that is compatible with the encrypted program service or services it is meant to receive. Satellite receivers typically contain microprocessor and memory storage circuits that work in conjunction with factory-installed software to acquire the desired satellite signal. Many are preprogrammed with the locations and tuning parameters for all the available satellite TV services stored in memory. Digital receivers usually have a unique address stored on a communication card. They are capable of automatically handling virtual channel mapping using a programmer-transmitted satellite transponder lookup table.

Figure 28 Typical commercial satellite receiver.

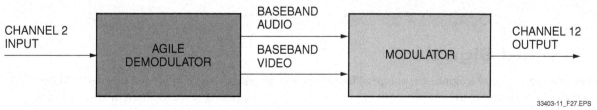

Figure 27 Demodulator-modulator frequency conversion.

Demodulation-Modulation

When using the demodulator-modulator method of channel conversion, never demodulate and modulate a signal to the same channel frequency, for example, VHF Channel 3 to VHF Channel 3. Doing so in the presence of strong off-air broadcast signals can cause flutter and a horizontal wavy pattern in the TV picture. These distortions are a result of inherent delay encountered by the signal as it passes through the demodulator-modulator relative to the same undelayed signal received directly off-air.

7.6.0 Stereo Encoders

Stereo encoders are used to modulate a TV channel in broadcast television systems committee (BTSC) stereo. Video, audio left, and audio right signals can be applied to the encoder input from a satellite receiver, VCR, or DVD player. A modulated second audio program (SAP) carrier input can also be applied. The stereo encoder processes these signals to provide baseband video and composite audio outputs for application to the input of a modulator.

7.7.0 Combiners and Splitters

Combiners (*Figure 29*) are used to take the balanced outputs from multiple processors and modulators and combine them into a single output signal for application to either an amplifier or the distribution system. Combiners commonly have 8, 12, or 16 inputs, called ports. Some can have frequency selective band filtering on the combiner inputs. The two types of combiners are passive and active. Passive combiners only combine the signals, while active combiners both combine and amplify the signals. Active combiners are designed to have unity gain (a gain of 1). This means that the amplification gain within the unit is only enough to overcome the loss of the combiner. It

should be pointed out that passive combiners are commonly used as splitters and splitters as combiners.

Figure 30 shows examples of two-way, four-way, and eight-way splitters. When using a splitter as a combiner (or combiner as a splitter), it is a good idea to permanently relabel the output ports as input ports, and vice versa. This can aid a person unfamiliar with the system configuration should they need to work on or troubleshoot the system.

A splitter is a passive device that takes a signal and splits it into two or more lower level output signals. Splitters are commonly made in 2-, 3-, 4-, and 8-way models. With the exception of a three-way splitter, the output signal levels from splitters are all equal. With the three-way splitter, only two of the output signals are equal; the third output is at a lower level.

Important combiner and splitter parameters include their insertion loss and port-to-port isolation. Insertion loss, or through loss, is a measure of the attenuation encountered by a signal as it passes through the combiner or splitter. The

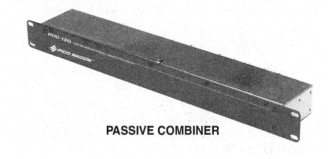

PASSIVE COMBINER

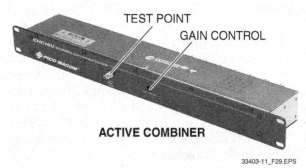

TEST POINT

GAIN CONTROL

ACTIVE COMBINER

33403-11_F29.EPS

Figure 29 Typical signal combiners.

Baseband Signals

For either a video or audio signal, the range of frequencies in the variations is called the baseband. These frequencies correspond to the actual visual or aural information, without any modulation or other encoding. Baseband audio frequencies range from about 15 to 20kHz; baseband video ranges from about 30kHz to 4MHz.

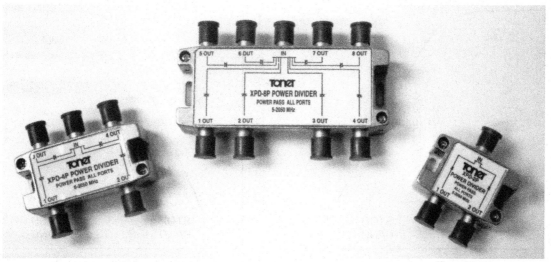

Figure 30 Splitters.

33403-11_F30.EPS

lower the insertion loss the better. Insertion loss is expressed in decibels. Isolation, also expressed in decibels, is a measure of the signal attenuation encountered by a signal from one port to another. The better the splitter, the higher the isolation. High isolation between ports reduces the amount of signal feedback from one port to the other. *Figure 31* shows some typical insertion losses and isolation losses for splitters.

7.8.0 Filters

Filters are frequency-selective devices used to pass or reject certain frequencies. Some filters are made as stand-alone assemblies; others are built into the circuitry of various equipment such as processors and receivers. There are four basic filter types: low-pass, high-pass, bandpass, and band rejection (*Figure 32*).

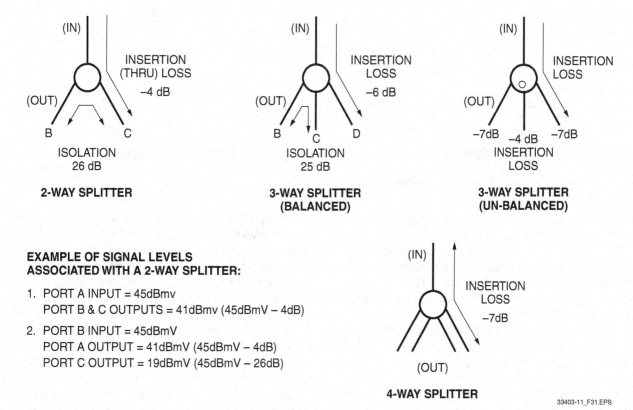

Figure 31 Examples of splitter insertion loss and isolation.

33403-11_F31.EPS

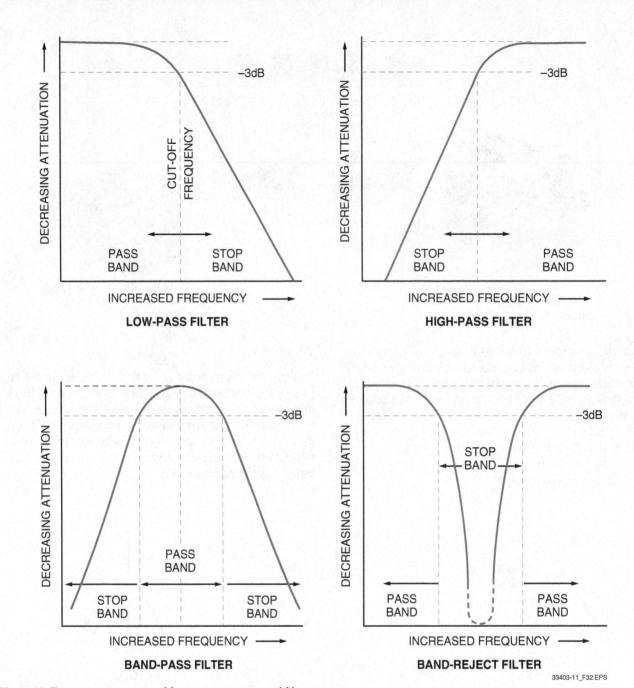

Figure 32 Frequency response of four common types of filters.

Low-pass filters permit all frequencies below a cut-off frequency point to pass, but reject all others. The tighter the filter response curve slope (skirt), the better the rejection of unwanted frequencies. For example, a low-pass filter can be used to select low-band VHF channel signals and reject high-band VHF channel signals.

High-pass filters reject all frequencies below a cross-over point, but pass all others. The tighter the skirt, the better the rejection of unwanted frequencies. For example, a high-pass filter can be used to select UHF channel frequencies while rejecting VHF channel frequencies.

Bandpass filters allow only a limited range of frequencies to pass. Again, the tighter the skirt, the better the rejection of unwanted frequencies. For example, a bandpass filter may be designed to pass one 6MHz-wide range of channel frequencies.

Band rejection filters permit all frequencies except those in a limited range to pass. For example, it may be designed as a channel-elimination filter used to remove one 6MHz-wide channel, allowing another channel to be re-inserted into the vacated channel's place. Channel elimination filters are designed to attenuate the audio and video carriers of the channel to be eliminated by a minimum of –50dB. A special type of rejection filter, called a notch trap, is designed to trap the video carrier signal in order to destroy or distort it for security purposes.

8.0.0 DISTRIBUTION SYSTEM COMPONENTS

This section describes the main types of components used in distribution systems to deliver signals from the headend of a CATV, SMATV, or MATV system to the customer. These include the following:

- Distribution amplifiers and line extenders (for coaxial distribution)
- Splitters
- Directional couplers and taps
- Attenuators and terminators
- Coaxial cable

8.1.0 Distribution Amplifiers and Line Extenders

Distribution amplifiers (*Figure 33*) and line extenders (line amplifiers) provide a high-level, low-distortion broadband signal used to drive distribution system trunk and/or feeder lines that use coaxial cables. Models are made with typical maximum gains of 15, 30, 43, and 50dB. They can have different output bandwidths, with frequencies typically ranging between 49MHz and either 450MHz, 550MHz, 750MHz, 860MHz, or 1,000MHz.

The rated output level from an amplifier is a function of its input signal and gain. For example, an amplifier may be rated as having a +44dBmV output and a gain of 36dB, based on an ideal flat signal level input of +8dBmV. Limiting the level of the signal applied to an amplifier is done because too high an input can cause the amplifier to overload and distort the output signal. It should be pointed out that if the level of a signal applied to an amplifier exceeds the specified maximum input level, an appropriate size attenuator is used to reduce the input signal to an acceptable level. This is done in most units by using a plug-in attenuator stage. In others, it can be done by attaching the attenuator externally to the input port of the amplifier.

Most amplifiers have built-in gain and slope controls used to adjust the maximum gain and slope of the output signal. Slope control circuits in the amplifier insert attenuation in the signal path to control the gain of the output signal so that the gain increases continuously starting with the lowest channel frequencies and moving toward the

Figure 33 Typical indoor distribution amplifiers.

highest channel frequencies. This is done to compensate for unequal signal losses of the amplifier's output signal as it passes through the distribution system coaxial cables. While passing through coaxial cables, the higher channel signals encounter greater losses than the lower channel signals. As the channel frequencies increase, so does the signal loss. This condition is called signal tilt. Signal tilt in coaxial cables is covered in more detail later in this module.

All amplifiers contain circuitry to minimize signal distortions that can be generated within the amplifiers themselves. Unless suppressed in the amplifier to acceptable decibel levels, typically in the –50 to –70dB range, these distortions can result in an undesirable TV picture. The types of distortions generated in amplifiers include:

- *Composite second order (CSO)* – Second order distortion can result from two signals mixing to produce sum and difference frequencies, or from the second harmonics of frequencies applied to the amplifier. CSO in VHF channel signals can result in diagonal lines in the TV picture.
- *Composite triple beat (CTB)* – Triple beat distortion results from the sum and difference products generated by the mixing of any three

---

video carriers that create a number of discrete beats within 20 or 30kHz of a video carrier. CTB makes a TV picture look grainy or wormy.

- *Cross modulation (XMOD)* – Results when a modulated signal from one channel crosses over into an adjacent channel. XMOD can cause a buzzing in the TV sound and sound lines in the picture.
- *Hum* – Results from the amplitude modulation of the carrier by a signal with a frequency that is usually a second or third harmonic of the 60Hz power line frequency. Hum can cause roll-up and wavy hum bars in the picture. One hum bar is 60Hz, two hum bars is 120Hz, and so on. Hum bars are usually caused by a bad filter capacitor in the power supply.

8.2.0 Splitters

Splitters are passive devices used to divide the signal into two or more outputs. A splitter divides the input signal into a number of equal but lower output signals, as described earlier in this module.

8.3.0 Directional Couplers and Taps

A directional coupler removes a specific portion of the main trunk line signal for output at one or more ports, typically to supply feeder or drop cables. Directional coupler parameters include insertion (through) loss, tap value, and isolation between the tap and through or output and through ports. *Figure 34* shows signal flow through a directional coupler. Note that the symbols for taps and couplers are often used interchangeably.

A tap is a passive device typically installed in a feeder cable. Like a directional coupler, it is used to split off a portion of the signal, usually for application to a subscriber's drop cable, while allowing most of the power to pass through to its output. Some taps are made with multiple tap ports with different isolation values. This is done to accommodate different drop distances along a trunk or feeder cable. As the distance from the headend or from a trunk line amplifier increases, the signal is attenuated more, resulting in the need to tap off a larger percentage of the signal. This is necessary in order to provide the subscriber with a signal level equal to that received by subscribers closer to the headend or line amplifier. Tap parameters include insertion loss, tap value, isolation between taps, and isolation between the tap port(s) and the output port.

8.4.0 Attenuators and Terminators

Attenuators, commonly called attenuation pads or pads, are used in applications where the strength of the signal must be reduced (attenuated). Attenuators can be inserted directly between two coaxial cables or between the output port of a device and a coaxial cable. Attenuators are made in both fixed and adjustable models. Fixed attenuators are made in several attenuation values, with 3, 6, 10, 12, 16, and 20 being typical.

Terminators used in cable systems are 75-ohm passive devices designed to provide an impedance match. They are installed on all unused splitter, directional coupler, and/or tap ports to eliminate signal reflections. It is especially important to terminate the output port of the last tap used in a feeder line.

8.5.0 Coaxial Cables

Many types of coaxial cables are made for different applications. Those used in CATV, SMATV, and MATV distribution systems all have a nominal impedance of 75 ohms. The specific type of coaxial cable used depends on factors such as the distances to be covered, whether the cable is run indoors or outdoors and whether it is aerial or buried. Semiflex (hard line) coaxial cables, typically used outdoors, are sized by the outside diameter of their metallic outer sheath: for example, cable size 0.500 inch (½ inch) or 0.750 inch (¾ inch).

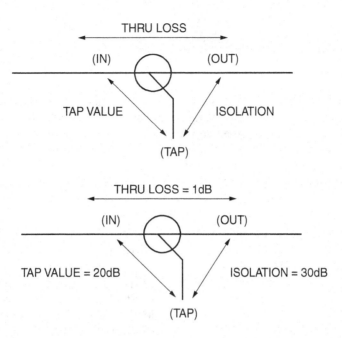

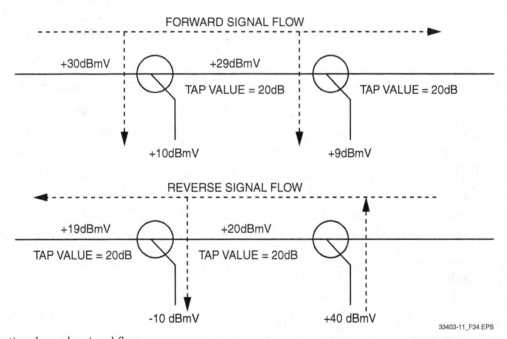

Figure 34 Directional coupler signal flow.

33403-11_F34.EPS

Non-semiflex (soft) coaxial cables, used both indoors and outdoors for CATV applications, are identified as RG-6 or RG-11, also known as series 6 and series 11. As shown in *Figure 35*, they usually consist of an inner conductor separated by a foam dielectric (core) from its outer conductor, which is typically a double-, triple-, or quadruple-aluminum foil and braid shield protected by either a PVC or plenum-rated outer sheath. This construction allows coaxial cable to conduct broadband signals without degradation from external inter-ference sources. The descriptions double-, triple-, and quadruple-shield refer to the number of layers of foil and braid used to form the shield. For example, the shield in quad-shield cable consists of four layers alternated as follows: (1) aluminum foil, (2) aluminum braid, (3) aluminum foil, and (4) aluminum braid. Cables are available with 60 to 95 percent braid outer shields. The difference between 60 and 95 percent shielding is that 60 percent shielding has a looser braid made of fewer strands of aluminum. For CATV/SMATV/MATV

Signal Leakage

Cable television systems and licensed over-the-air broadcasters use many of the same frequencies to transmit programming. Cable systems, for example, use TV, radio, and aeronautical radio channels. The FCC prohibits cable operators from interfering with over-the-air services that happen to be using the same frequencies as the cable operator and that are within the vicinity of the cable system. Such interference, especially on the emergency channels, can interfere with the communications of safety personnel or airplane pilots, potentially endangering lives or hindering rescue efforts. The FCC has set maximum individual signal leakage levels for cable systems, especially those that may interfere with aeronautical and navigation communications. The FCC also requires cable operators to perform periodic inspections to locate and repair leaks on their systems.

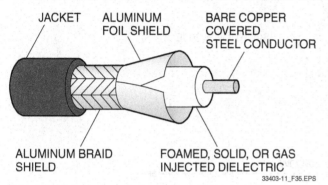

Figure 35 RG-type coaxial cable construction.

cabling, quad-shied cable with 100 percent foil and 60 percent braid is required.

Attenuation of broadband signals resulting from the length of a coaxial cable run is an important consideration. Further, the amount of attenuation experienced by a signal in any coaxial cable is always greater for the higher frequency channels than for the lower frequency channels. Cable manufacturers provide the specific attenuation values in relation to frequency for the different kinds of cables in their product literature. It should be pointed out that attenuation values given by one manufacturer may be different from those of another for the same type of cable (RG-6, for example). For this reason, always consult the cable manufacturer's specifications to determine the exact value of cable attenuation. *Table 2* shows some typical attenuation values for standard RG and semiflex coaxial cables. As shown, the attenuation values given are based on 100 feet of cable length at 68°F. It should be noted that the actual attenuation at each frequency is somewhat lower or higher with lower or higher temperatures, respectively.

Different types of 75-ohm cables, such as RG-6 and RG-11, are commonly mixed in CATV, SMATV, or MATV systems. It is important that cables with impedances other than 75 ohms, for example RG-62 at 93 ohms, never be mixed with

75-ohm cables in a system. Doing so causes an impedance mismatch, resulting in signal reflections that degrade the signal.

RG-6 cable can be run for longer distances, typically 150 to 200 feet, before the inherent cable loss dictates the need to use a line amplifier in order to reestablish the signal level. RG-11 cable has the lowest loss of the soft cables. That allows RG-11 to be run for much longer distances, typically 200 to 500 feet. Because of its low attenuation, semiflex ½-inch (0.500-inch) cable is used for runs longer than 500 feet.

8.5.1 Coaxial Cable Loss and Signal Tilt

Cable loss for coaxial cable is calculated based on the distance that a signal must travel, along with the lowest and highest frequency transmitted on the system. Because losses are greater at higher frequencies, cable loss should be determined by finding the loss for the lowest and highest frequency being carried by the cable. As mentioned

Attenuators

Shown here are examples of in-line DC voltage-passing and DC voltage-blocking attenuators.

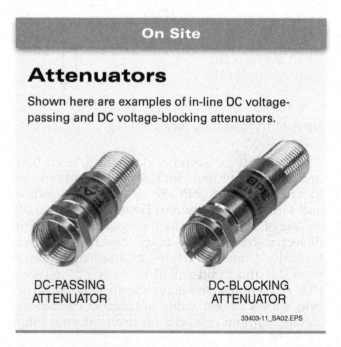

DC-PASSING ATTENUATOR

DC-BLOCKING ATTENUATOR

Table 2 Typical RG and Semiflex Cable Attenuation

Frequency (MHz)	Drop Cable			Semiflex Cable									
	RG6	RG7	RG11	412	500	625	750	875	1,000	565	700	840	1,160
5	0.57	0.56	0.36	0.20	0.16	0.13	0.11	0.09	0.08	0.14	0.11	0.09	0.07
55	1.50	1.22	0.95	0.68	0.55	0.45	0.37	0.32	0.29	0.47	0.37	0.32	0.24
211	2.87	2.29	1.81	1.35	1.08	0.89	0.73	0.64	0.58	0.93	0.74	0.64	0.48
250	3.12	2.49	1.98	1.49	1.19	0.98	0.81	0.70	0.64	1.03	0.82	0.70	0.53
300	3.43	2.74	2.17	1.64	1.31	1.08	0.89	0.78	0.72	1.13	0.90	0.77	0.59
350	3.72	2.98	2.36	1.78	1.43	1.18	0.97	0.84	0.78	1.23	0.98	0.84	0.65
400	4.00	3.20	2.53	1.91	1.53	1.27	1.05	0.91	0.84	1.32	1.05	0.91	0.70
500	4.51	3.61	2.85	2.15	1.73	1.43	1.18	1.03	0.96	1.49	1.19	1.03	0.80
600	4.98	3.99	3.16	2.37	1.91	1.58	1.31	1.14	1.06	1.64	1.31	1.14	0.89
750	5.62	4.50	3.58	2.68	2.16	1.79	1.48	1.29	1.21	1.85	1.49	1.30	1.01
870	6.09	4.87	3.90	2.90	2.35	1.95	1.61	1.41	1.33	2.01	1.62	1.41	1.11
950	6.39	5.11	4.10	3.03	2.49	2.04	1.72	1.50	1.35	2.15	1.75	1.51	1.15
1,000	6.54	5.25	4.23	3.13	2.53	2.11	1.74	1.53	1.44	2.17	1.75	1.53	1.20
1,200	7.18	5.77	4.71	3.44	2.83	2.32	1.96	1.72	1.55	2.45	2.00	1.72	1.33
1,450	7.89	6.34	5.29	3.81	3.12	2.61	2.16	1.90	1.81	2.66	2.13	1.90	1.52
1,750	8.74	6.93	5.95	4.23	3.47	2.92	2.41	2.13	2.03	2.96	2.36	2.13	1.71
1,850	8.99	7.13	6.12	4.36	3.60	2.97	2.52	2.22	2.07	3.13	2.57	2.23	1.74
2,000	9.34	7.41	6.36	4.55	3.76	3.12	2.64	2.32	2.11	3.27	2.69	2.33	1.82
Loop Resistance	39.6	26.8	19.5	2.5	1.7	1.1	0.8	0.4	1.3	0.9	0.9	0.6	0.3

Typical Cable Attenuation [dB/100' @ 68°F (20°C)]

Note: Loop resistance is shown in Ω/1,000'.

earlier, the characteristic of coaxial cables to attenuate signals at the higher frequency channels more than the lower frequency channels causes a tilt in the broadband signal response (*Figure 36*). The amount of tilt is expressed in decibels. The amount of cable tilt to expect for a particular signal can be approximated by subtracting the cable attenuation value at the lowest frequency from the attenuation value for the highest frequency.

Using the attenuation values provided in *Table 2*, 100 feet of RG-6 cable causes a tilt of approximately 12dB for frequencies ranging between 5MHz and 3GHz. Depending on the signal frequencies and the length of a cable run, the slope control circuits in distribution and/or line amplifiers may be able to completely or partially compensate for the tilt. When an amplifier's slope-controlled output is insufficient to correct for the tilt, a passive device, called an equalizer, can also

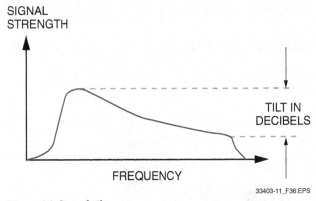

Figure 36 Signal tilt.

be used. When inserted into the coaxial signal line, the equalizer attenuates the low frequencies while passing the high frequencies through with little loss to provide a flat input to the amplifier.

8.5.2 NEC® CATV Coaxial Cable Classifications

NEC Article 820 governs the requirements for installing cable TV system equipment and coaxial cables in buildings. It classifies coaxial cables and their permitted uses as follows:

- *CATV plenum (CATVP) cable* – A coaxial cable for interior use in plenums and other environmental air spaces. It has good fire resistance and low smoke-producing characteristics.
- *CATV riser (CATVR) cable* – A coaxial cable for interior use in vertical shafts or in runs from floor to floor. It is resistant to fire to the extent that flames are not carried from floor to floor.
- *CATV cable* – A general-purpose coaxial cable for interior use, excluding use in risers or plenums. It is resistant to the spread of fire.
- *CATV limited use (CATVX) cable* – A coaxial cable for limited interior use in dwellings and raceways. It is resistant to the spread of fire.
- *CATV underground (CATVU)* – A coaxial cable suitable for underground applications.

8.5.3 Handling and Terminating Coaxial Cables

Many signal loss problems can be traced to coaxial cable damage and/or poor connections. Coaxial cable can be damaged by any activity that causes the cross-sectional area of the dielectric surrounding the center conductor to be reduced or otherwise distorted. Activities such as bending the cable too severely, installing the wrong type of connector, and fastening or stapling the cable

On Site

Equalizers

Shown here are examples of miniature in-line passive equalizers.

33403-11_SA03.EPS

too tightly all fall into this category. Distortion of the dielectric raises the impedance between the center conductor and foil. The resulting impedance mismatch can cause excessive signal reflections and attenuation losses to occur in the line. When bending coaxial cable, the radius of the bend should not exceed that specified by the cable manufacturer. If no manufacturer's specifications are available, a rule of thumb is to use a bend radius not to exceed 10 times the diameter of the cable. For example, when using a 0.500 semiflex cable, the bend radius should not exceed 5 inches (10 × 0.5).

It is critical that coaxial cable connectors be correctly installed on the coaxial cables. For CATV cables, you must use the proper type of compression connector designed for use with the particular type of coaxial cable to which it is to be attached. Connectors should always be installed in accordance with the connector manufacturer's instructions and using the tools specified by the manufacturer. This is because the same type of connector, for example, a male F connector, made by different manufacturers can be installed somewhat differently. The tools needed to terminate a cable with a connector normally include a cable cutter, cable stripper, and a compression tool designed to compress the type of connectors you are installing.

9.0.0 DISTRIBUTION SYSTEM TOPOLOGIES

The pattern or topology of the cabling arrangement used for cable distribution systems can take many forms. The form used is determined by the location, type, size, and complexity of the system, and its application. Three basic topologies are commonly used. However, most distribution systems actually use a combination of all three topologies.

- Home-run
- Loop-through
- Trunk-and-branch

9.1.0 Home-Run Cable Distribution Systems

In a home-run designed cable distribution system (*Figure 37*), all the cable runs start from one central location. The signal output from the headend is connected to a network of splitters or taps located in the headend equipment room or in a remotely located distribution box. From there, a separate cable is run to each subscriber outlet or wall plate. The main advantage of using a home-run system

Coaxial Cable Connectors

It is important that a connector used to terminate coaxial cable be designed for use with the particular type of coaxial cable to which it is to be attached, and it must be installed using the proper tools. About 90 percent of the cable problems encountered in a system can be traced to the use of wrong or improperly installed connectors.

is that it allows an operator to turn access to the system on or off without affecting anyone else.

9.2.0 Loop-Through Cable Distribution Systems

In a loop-through cable distribution system (*Figure 38*), one or more trunk cables exiting the headend are run as continuous vertical and/or horizontal risers through a building. The individual subscriber outlets or wall plates are then connected in series with one another in the riser cable at the required locations. The loop-through method uses less cable, making it relatively inexpensive when compared to the other designs. The main disadvantage of using it is that if the trunk cable is damaged, all those downstream of the damage are affected. In addition, it does not allow for control of subscriber access on an individual basis.

9.3.0 Trunk-and-Branch Cable Distribution Systems

The CATV system architecture shown earlier in this module represents a basic trunk-and-branch cable distribution system. In a trunk-and-branch cable distribution system that uses coaxial cable, the signal from a main trunk cable is typically applied to an amplifier. The amplifier has both trunk and distribution/feeder outputs. Inside the amplifier are two sections. One section controls the trunk amplification and tilt. The other section controls the feeder amplification and tilt. The signal level for the distribution/feeder is usually higher than that of the trunk.

The feeder output is used to further distribute the signal to a geographical location and/or building. Subscriber drop cables are tapped off the feeder cable as needed. A trunk-and-branch system typically includes an amplifier at the headend and other amplifiers where needed to boost the signal. This design is commonly used in CATV systems and in larger private SMATV/MATV systems.

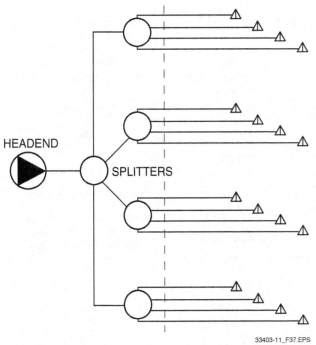

Figure 37 Simplified home-run cable distribution system.

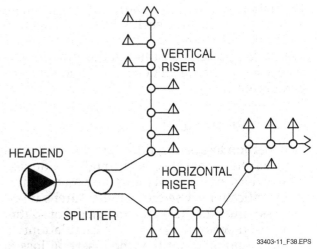

Figure 38 Simplified loop-through cable distribution system.

10.0.0 DISTRIBUTION SYSTEM GAINS AND LOSSES

The focus of this section is on how to calculate the gains and losses for a distribution system when you know the signal level output from the head-end. The ability to perform this task is necessary in order to select the proper directional couplers and/or taps needed to deliver the required signal levels to all subscribers. It is also necessary when troubleshooting so that you can evaluate the correctness of signal levels that are measured at the different points within the system. When calculating distribution system losses, the cable and component insertion losses must be considered.

10.1.0 Cable Losses

As discussed earlier, a certain amount of signal is lost as it travels through coaxial cable. The amount of signal loss depends on the type of cable, its length, and the frequency of the signal being carried. Signal losses increase as the frequency increases. Cable losses should be calculated at both the lowest and highest frequency being carried by the cable. The difference between these two loss values gives an indication of cable tilt. Technicians must compensate for cable tilt when it becomes excessive. This is usually done using the slope control adjustment provided in most amplifiers or by use of an equalization network.

To determine the cable attenuation values, use a cable attenuation chart provided by the manufacturer for the specific brand and type(s) of cable being used. If the manufacturer's data is not available, nominal values for coaxial cable attenuation can be used. Many cable equipment manufacturers include such tables in the technical or appendix sections of their product literature. *Table 2* presented earlier in this module also gives a set of typical values for RG and semiflex coaxial cable.

10.2.0 Splitter Losses

The insertion losses in splitters are the losses that occur as the signal passes through the device. These values are given in the splitter manufacturer's specifications. Use the manufacturer's insertion loss values for the brand and model of splitter being used when making your calculations. It should be pointed out that the insertion loss of splitters is different at different frequencies. For example, the splitter insertion loss for a two-way splitter may be 4dB at 10 to 470MHz, 4.5dB at 470 to 900MHz, and 5dB at 900 to 1,000MHz. *Table 3* lists some nominal values for splitter insertion loss that can be used if the manufacturer's data is not available. Note that the insertion loss of a splitter is the same no matter which direction the signal is traveling.

10.3.0 Directional Coupler/Tap Losses

Tap value and insertion loss must be considered when working with directional couplers and taps. Insertion loss occurs as the signal passes through the device. The tap value is determined by the rated tap value of the device. As the tap value of a directional coupler or tap decreases, its insertion loss increases. A rule of thumb is to use the highest tap value possible, while providing the correct signal level to the port at each location in order to keep insertion losses in the cable run to a minimum. Use the manufacturer's insertion loss values for the brand and model of device being used when making your calculations. *Table 4* lists some typical directional coupler/tap values and related nominal isolation losses that can be used if the manufacturer's data is not available.

> **NOTE**
> Always use equipment that is specified for the systems you are working on.

Table 3 Nominal Splitter Insertion Loss for Each Output Port

Type of Splitter	Nominal Insertion Loss (dB)
Two-way	4.0
Three-way	3.5 and 7.0
Four-way	7.0
Eight-way	10.5

Table 4 Typical Directional Coupler/Tap Values and Nominal Insertion Losses

Tap Value (dB)	Nominal Insertion (Through) Loss (dB)
30	0.5
27	0.5
24	0.5
20	0.6
16	0.7
12	0.9
8	1.2

10.4.0 Calculating Distribution System Gains and Losses

In order to calculate the gains and losses in a distribution system, and select the proper coupler/tap values, it is necessary to have the following information:

- Output signal levels from the headend at the system's lowest and highest channel frequencies
- The minimum signal level required at the subscriber taps
- A cable attenuation chart for the specific brand and types of cable used in the system
- Insertion and isolation loss values for the specific brand and models of splitters, directional couplers, taps, and similar passive devices used in the system
- Gain values for the specific brands and models of amplifiers and similar active devices used in the system
- A sketch showing the cable run distances and the type and location of each component in the run

Figure 39 shows an example distribution network for which the gains and losses have been calculated in order to select the proper tap values

needed to deliver the required minimum signal to the customer. The following parameters are assumed for this example:

- The lowest and highest frequencies processed are Channels 2 and 61.
- The headend output signal levels for Channels 2 and 61 are 40dBmV and 46dBmV, respectively.
- The minimum signal level delivered to the subscriber is +6dBmV.

The cable attenuation losses for RG-6 cable and the insertion losses for various tap values are also shown in *Figure 39*. Refer to *Table 3* for the splitter insertion loss.

A procedure for calculating the signal levels and selecting the tap values for the cable run shown on the left-hand side of *Figure 39* is described here. The signal levels and tap values for the run on the right-hand side of the figure

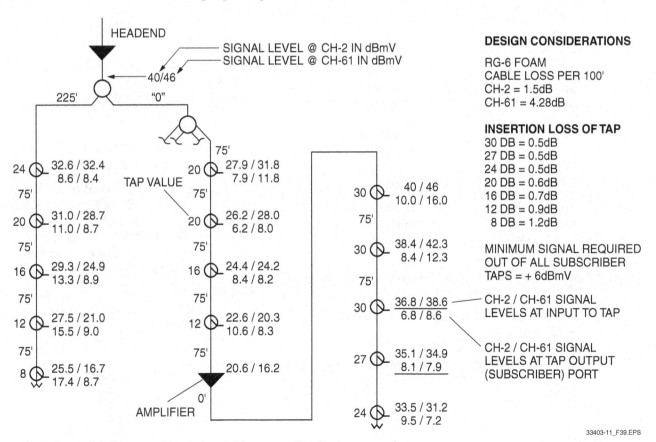

Figure 39 Example of gains and losses in a cable system distribution network.

were determined in the same way. After going through the procedure given here for the left-hand cable run, it is recommended that you do the calculations for the right-hand cable run to verify the accuracy of the values shown, and to make sure that you fully understand the tap selection process.

Step 1 Calculate the signal levels at the input to the first tap and determine the tap value needed to deliver a minimum of +6dBmV to the customer.

a. Calculate the Channel 2 signal level at the input to the first tap.

40.0dBmV – [(splitter insertion loss) + (225' cable loss)]
40 – 4 – (1.5dB per 100' × 2.25 hundred feet of cable)
40 – 4 – 3.375 = 32.625 rounded to 32.6dBmV

b. Calculate the Channel 61 signal level at the input to the first tap.

46dBmV – [(splitter insertion loss) + (225' cable loss)]
46 – 4 – (4.28dB per 100' of cable × 2.25 hundred feet of cable)
46 – 4 – 9.63 = 32.37 rounded to 32.4dBmV

c. Determine the highest tap value that can be used for the first tap.

Channel 61 level out of a 30dB tap = 32.4 – 30 = 2.4dBmV (level too low)
Channel 61 level out of a 27dB tap = 32.4 – 27 = 5.4dBmV (level too low)
Channel 61 level out of a 24dB tap = 32.4 – 24 = 8.4dBmV (level ok)
Channel 2 level out of a 24dB tap = 32.6 – 24 = 8.6dBmV (level ok)

Step 2 Calculate the signal levels at the input to the second tap and determine the tap value needed to deliver a minimum of +6dBmV to the subscriber.

a. Calculate the Channel 2 signal level at the input to the second tap.

32.6dBmV – (24dB tap insertion loss + 75' cable loss)
32.6 – 0.5 – (1.5 × 0.75)
32.6 – 0.5 – 1.125 = 30.98 rounded to 31dBmV

b. Calculate the Channel 61 signal level at the input to the second tap.

32.4dBmV – (24dB tap insertion loss + 75' cable loss)
32.4 – 0.5 – (4.28 × 0.75)
32.4 – 0.5 – 3.21 = 28.69 rounded to 28.7dBmV

c. Determine the highest tap value that can be used for the second tap.

Channel 61 level out of a 24dB tap = 28.7 – 24 = 4.7dBmV (level too low)
Channel 61 level out of a 20dB tap = 28.7 – 20 = 8.7dBmV (level ok)
Channel 2 level out of a 20dB tap = 31.0 – 20 = 11.0dBmV (level ok)

Step 3 Calculate the signal levels at the input to the third tap and determine the tap value needed to deliver a minimum of +6dBmV to the subscriber.

a. Calculate the Channel 2 signal level at the input to the third tap.

31.0dBmV – [(20dB tap insertion loss) + (75' cable loss)]
31.0 – 0.6 – (1.5 × 0.75)
31.0 – 0.6 – 1.125 = 29.275 rounded to 29.3dBmV

b. Calculate the Channel 61 signal level at the input to the third tap.

28.7dBmV – [(20dB tap insertion loss) + (75' cable loss)]
28.7 – 0.6 – (4.28 × 0.75)
28.7 – 0.6 – 3.21 = 24.89 rounded to 24.9dBmV

c. Determine the highest tap value that can be used for the third tap.

Channel 61 level out of a 20dB tap = 24.9 – 20 = 4.9dBmV (level too low)
Channel 61 level out of a 16dB tap = 24.9 – 16 = 8.9dBmV (level ok)
Channel 2 level out of a 16dB tap = 29.3 – 16 = 13.3dBmV (level ok)

Step 4 Calculate the signal levels at the input to the fourth tap and determine the tap value needed to deliver a minimum of +6dBmV to the subscriber.

a. Calculate the Channel 2 signal level at the input to the fourth tap.

29.3dBmV – [(16dB tap insertion loss) + (75' cable loss)]
29.3 – 0.7 – (1.5 × 0.75)
29.3 – 0.7 – 1.125 = 27.475 rounded to 27.5dBmV

b. Calculate the Channel 61 signal level at the input to the fourth tap.

24.9dBmV – [(16dB tap insertion loss) + (75' cable loss)]
24.9 – 0.7 – (4.28 × 0.75)
24.9 – 0.7 – 3.21 = 20.99 rounded to 21.0dBmV

c. Determine the highest tap value that can be used for the fourth tap.

Channel 61 level out of a 16dB tap = 21.0 – 16 = 5.0dBmV (level too low)
Channel 61 level out of a 12dB tap = 21.0 – 12 = 9.0dBmV (level ok)
Channel 2 level out of a 12dB tap = 27.5 – 12 = 15.5dBmV (level ok)

Step 5 Calculate the signal levels at the input to the fifth tap and determine the tap value needed to deliver a minimum of +6dBmV to the subscriber.

a. Calculate the Channel 2 signal level at the input to the fifth tap.

27.5dBmV – [(12dB tap insertion loss) + (75' cable loss)]
27.5 – 0.9 – (1.5 × 0.75)
27.5 – 0.9 – 1.125 = 25.475 rounded to 25.5dBmV

b. Calculate the Channel 61 signal level at the input to the fifth tap.

21.0dBmV – [(12dB tap insertion loss) + (75' cable loss)]
21.0 – 0.9 – (4.28 × 0.75)
21.0 – 0.9 – 3.21 = 16.89 rounded to 16.9dBmV

c. Determine the highest tap value that can be used for the fifth tap.

Channel 61 level out of a 12dB tap = 16.9 – 12 = 4.9dBmV (level too low)
Channel 61 level out of a 8dB tap = 16.9 – 8 = 8.9dBmV (level ok)
Channel 2 level out of a 8dB tap = 25.5 – 8 = 17.5dBmV (level ok)

Because the fifth tap is the last tap in the cable run, its insertion loss is not a factor. However, it must be terminated with a 75-ohm terminator to maintain an impedance match.

> **CAUTION**
>
> Failure to maintain an impedance match causes a portion of the signal called standing waves to be reflected back on the line toward the headend. These standing waves can cause ghosting and other TV picture distortions.

After analyzing the example in *Figure 39*, it should be apparent that tap values are selected by calculating the cable and insertion losses from tap to tap and by choosing the highest tap value that gives the desired output, which is +6dBmV in our example. The tap isolation values should decrease as the distance from the tap to the headend or a distribution amplifier increases. As a result, the last tap in the run should always have the lowest tap value because the signal level at this point is the lowest.

Calculating the gains and losses in the headend equipment is done in a similar manner to that described for the distribution system. Calculations for processing paths in the headend involve calculating the losses for the passive devices (cables, splitters, and/or combiners) and the gains for active devices, such as preamplifiers. The input to the processor must meet the processor requirements. After the processor, subtractions are made for devices such as combiners to arrive at a final output level. *Figure 40* shows a simplified example for the loss and gain calculations in a headend. In the example shown, the 0.200mV (–14dBmV) signal from the antenna is increased to a level of 28dB at the output of the four-way combiner calculated as follows:

Processor gain	+35
Combiner loss	– 7
	28dBmV

11.0.0 TEST EQUIPMENT

The alignment, balancing, testing, and troubleshooting of cable TV systems involves using the appropriate test equipment to make sure that the system and all its components meet both contractual and governmental requirements. In addition to the multimeter, several items of special test equipment are widely used for this purpose:

- Signal level meter (SLM)
- Spectrum analyzer
- Cable tone test set
- Satellite tester
- Portable color TV receiver

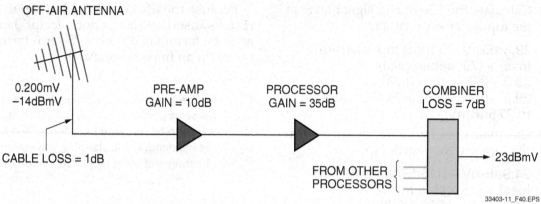

OFF-AIR ANTENNA

0.200mV
−14dBmV

CABLE LOSS = 1dB

PRE-AMP
GAIN = 10dB

PROCESSOR
GAIN = 35dB

COMBINER
LOSS = 7dB

FROM OTHER {
PROCESSORS {

→ 23dBmV

33403-11_F40.EPS

Figure 40 Simplified example of headend gains and losses.

11.1.0 Signal Level Meter

A signal level meter (SLM) is the most used item of test equipment in coaxial cable systems for maintaining and monitoring the signal levels. System measurements commonly made with an SLM include the following:

- Audio and video carrier levels on each channel
- Audio-to-video signal level ratio
- Carrier-to-noise ratio

SLMs detect RF energy and display the measurement in dBmV and/or dBµV. They are made by a number of manufacturers and in several models, ranging from relatively simple manually tuned models to microprocessor-controlled programmable models (*Figure 41*). Depending on the model, the SLM can be an analog or digital instrument. Today, digital SLMs are the most commonly used type of meter. The digital SLM usually shows the signal level and other measured parameters as text on a liquid crystal display (LCD). Typically, they have the capability to automatically tune by frequency and/or by channel, and to automatically switch in signal level attenuation as needed to give an accurate reading.

Ideally, the SLM is able to measure all the frequencies within the broadband CATV range of 55 to 890MHz. Most SLMs have a built-in speaker that provides demodulated audio for the channel under test. Sophisticated models have the capability to interface with a printer and/or computer, allowing the measured signal data to be downloaded to these devices.

11.2.0 Spectrum Analyzer

A spectrum analyzer (*Figure 42*) is used to determine the frequency and relative power levels of RF signals present in a specific area of the system. It should be pointed out that spectrum analyzers have many measurement capabilities in common

33403-11_F41.EPS

Figure 41 Signal level meter.

with signal level meters. Some system checks and measurements commonly made with a spectrum analyzer include the following:

- View groups of channels or individual channels within a distribution system
- Measure the frequency and amplitude of individual channel audio and video carriers
- View co-channel and other interference
- Determine tilt and flatness at various points in the system
- Determine the presence of composite triple beat and second order distortions

Spectrum analyzers are made by a wide variety of manufacturers and in several models. All

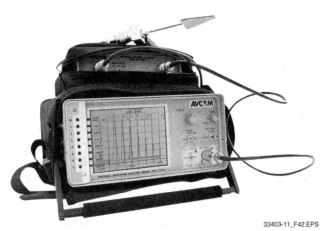

Figure 42 Spectrum analyzer.

33403-11_F42.EPS

consist of a scanning receiver that automatically tunes through a selected frequency spectrum and visually displays the signals present at its input on a cathode ray tube (CRT) as a plot of amplitude versus frequency (*Figure 43*). The frequency of the signal spectrum being viewed is represented by the horizontal axis increasing from left to right, and the signal amplitude is displayed vertically in decibels. The signal displayed horizontally across the bottom of the screen is the noise floor. The specific frequency range and power level gradient being used are selected by front panel controls. Spectrum analyzers used for making cable TV measurements typically have more than 70dB of dynamic range and are capable of making spectrum measurements in the 100kHz to 1,000MHz range.

11.3.0 Cable Tone Test Set

The cable tone test set (*Figure 44*) is used to generate and receive a tone signal through a cable, and any passive device along the cable path, for the purpose of locating and identifying a specific cable. It is also used as a continuity tester. A cable tone test set consists of a transmitter and receiver equipped with F-type connectors. Most provide both an audible and LED tone indication. As a continuity tester, it provides an indication of resistance and identifies the presence of AC or DC voltage on the cable under test. In addition to a direct-connection method of signal reception, most can also pick up signals by electrostatic pickup for non-terminated cables and electromagnetic pickup for terminated or shorted cables.

11.4.0 Satellite Signal Level Meter

There are several types of portable satellite signal level meters (*Figure 45*) that can be used to monitor the strength of the signal output from

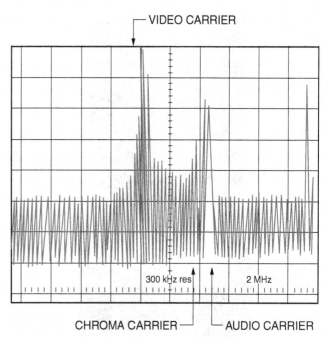

Figure 43 Spectrum analyzer display.

33403-11_F43.EPS

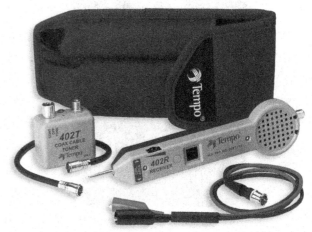

Figure 44 Cable tone test set.

33403-11_F44.EPS

the satellite antenna LNBF (low-noise block downconverter) as an aid when aligning a satellite dish. SLM-type testers provide a direct readout of signal strength in dBμV or dBmV. Antenna peaking-type meters have a bar graph indication that shows the relative strength of the signal, and a built-in variable frequency audio tone generator driven by the received input signal. This feature allows for no-look aiming of the antenna. Correct alignment of the antenna is indicated when there is a maximum signal indication on the meter and the highest audio pitch tone.

More sophisticated satellite signal level meters have other features to aid in evaluating and optimizing satellite receiver installations. Typically,

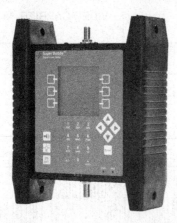

Figure 45 Satellite testers.

these features include the capability of measuring the DC voltage supplied to the LNBF, the intermediate frequency (IF), the local oscillator frequency, and the C/N ratio. Most have the capability of powering the LNBF in order to allow for independent evaluation of the antenna dish/LNBF part of the system. Some units have factory-programmed worldwide channel listings and provide for multiple channels of user-programmed memory. Models are also available that allow the focusing of two satellite dishes by being able to observe meter swings on two separate meters (one for each dish) at the same time.

11.5.0 Portable Color TV Receiver

A small portable color TV set is useful to provide a visual indication of picture quality at various points in the headend and distribution system. Sometimes ghosts and other picture distortions that are not indicated by an SLM measurement show up on the TV set picture. A common use of the TV receiver is to check the quality of the signal received by an off-air antenna. Also, when the cause of a problem is a malfunctioning subscriber's TV set, not the cable input signal, a good picture on the portable TV can be used to convince the subscriber that the cause of the problem is the subscriber's TV set.

11.6.0 Handling and Using Test Equipment

Proper calibration of the test equipment being used is important in order to obtain accurate measurement results. All instruments should be calibrated by qualified individuals at the interval recommended by the instrument manufacturer. Test instruments are precision devices and should be handled accordingly. Should an instrument inadvertently fall from a scaffold or ladder, or otherwise be subjected to severe shock, it should be calibrated before using it. When not being used, test instruments should be stored in their carry case or protective container to protect them from the environment and possible damage. Rough handling or improper storage of the instruments can cause them to become uncalibrated or damaged.

The items of test equipment used with cable systems are made by several manufacturers and in several models. Depending on the sophistication of the particular instrument being used, the procedures for operating the instrument can vary widely. For this reason, the instruments should always be operated in accordance with the applicable manufacturer's instructions.

12.0.0 HEADEND ALIGNMENT

Aligning and balancing the system involves using an SLM or spectrum analyzer. The object is to adjust the overall gains and audio and video carrier levels of the system's processors and modulators to provide a headend output signal with a level and balance that meet those specified in the system design. Before attempting a headend alignment, the following equipment/system conditions should exist:

- All off-air and satellite antennas are properly aligned.
- All frequency-agile processors and modulators are set to their assigned input/output channels.
- All satellite receivers are tuned to their chosen transponders.
- Input signal levels to the processors, modulators, and/or satellite receivers are within the range specified by the device manufacturer.

General guidelines for aligning the headend are as follows:

> **NOTE**
>
> If you are unfamiliar with the unit(s) to be adjusted, refer to the manufacturer's operation and maintenance instruction manual for the unit to familiarize yourself with the unit's capabilities and its controls and their functions.

Step 1 Set the level of the RF output (video carrier) from the highest frequency processor or modulator, allowing a minimum of 3 to 5dB margin below its rated output. Adjust the audio carrier (aural or A/V) for a level that is 15 to 17dB below the video carrier.

Step 2 Verify that the level of the highest frequency processor or modulator signal is within specifications at the final combiner output test port. Record this signal level for use as a reference level.

Step 3 Align each of the remaining channels to provide an output signal level from the final combiner equal to the reference level recorded in Step 2. This also includes adjusting their audio carriers for a level that is 15 to 17dB below their video carriers. Ideally, all channels should be balanced to within ±1dB of each other.

Step 4 The combined signal output level from the final combiner applied to the input of a launch amplifier should be flat within ±1dB across all channels.

Step 5 Record all pertinent signal levels. Leave a copy with the owner or responsible person and retain a copy for future reference.

13.0.0 TROUBLESHOOTING

While the reliability of electric/electronic systems, equipment, and cabling is quite high, failures do occur. These failures can be the result of defective parts, improper installation, manufacturers'

defects, and/or poor maintenance. To locate and correct a problem, you must do the following:

- Understand the purpose and function of each assembly or component.
- Observe and identify the symptoms that point to the improper operation of any part of the system.
- Use the proper procedures and test equipment to diagnose and correct the malfunction quickly and efficiently.

Troubleshooting can be defined as a procedure by which the technician locates the source of an equipment or cable problem, and then performs the necessary repairs and/or adjustments to correct the problem. A systematic approach to troubleshooting can be divided into five basic steps:

- Subscriber interface
- Physical examination of the system
- Basic system analysis
- Manufacturers' troubleshooting aids
- Fault isolation in system/unit problem area

> **NOTE**
>
> The vast majority of problems can usually be found quickly. Only the most stubborn and difficult problems require the use of all five troubleshooting elements.

13.1.0 Customer Interface

Troubleshooting should always begin with obtaining all the information possible regarding the equipment problem. Talking to and asking questions of a subscriber with first-hand knowledge of the problem is always recommended. This can provide valuable information on equipment operation that can aid in the troubleshooting process. The interview may sometimes determine that the source of a problem is the subscriber's TV set, thereby eliminating unnecessary troubleshooting of the cable system components. When interviewing the subscriber, ask the following questions:

- When did the problem begin?
- Can you describe the problem?
- Was anything changed right before the problem started?
- Are all channels affected?
- Is the television set new or has it recently been moved?
- Is the sound or picture most severely affected?
- Is the TV tuner set to Channel 3 or 4 if using a converter, VCR, or DVD player?
- Do other subscribers have the same problem?

13.2.0 Physical Examination of the System

Problems can sometimes be easily identified by checking for anything unusual with the equipment or cabling. When performing a physical examination of the system, do the following:

• Look for damaged, disconnected, or loose cables or power cords.
• Look for improperly installed connectors and/or missing or loose terminations.
• Look for power or other equipment switches being set to the wrong position.
• For operating equipment, check for odors of burning insulation, sounds of arcing, and/or abnormally warm units.

13.3.0 Basic System Analysis

Proper diagnosis of a problem requires that you know what the system and/or assembly should be doing when it is operating properly. If you are not familiar with how a particular assembly operates, you must first study the manufacturer's service literature to familiarize yourself with the assembly and its operation.

Next, you must find out what symptoms are exhibited by the problem. Carefully listen to the subscriber's complaints, and analyze the operation of the system or assembly yourself. This can mean making electrical or signal measurements at key points in the system. Compare measured values with a set of previously recorded values or with values given in the manufacturer's service literature to determine if there has been some degradation.

When troubleshooting, avoid making assumptions about the cause of problems. Always secure as much information as possible before arriving at your diagnosis. Due to the complexity of some systems and equipment, it may be necessary to seek help in locating the cause of a particular problem from a more experienced technician or from a manufacturer's service representative.

13.4.0 The Use of Manufacturers' Troubleshooting Aids

A good troubleshooter knows how and where to locate information, including vendor technical support lines, online resources, and the use of vendor-supplied manuals and other product documentation. Technical and service manuals for specific types of equipment are invaluable sources of information.

To aid in the isolation of faults, manufacturers' manuals typically include troubleshooting aids for their equipment such as troubleshooting tables and fault isolation diagrams. Troubleshooting tables are intended to guide you to a corrective action based on your observations of system operation. Fault isolation diagrams, also called troubleshooting trees, normally start with a failure symptom observation and take you through a logical decision-action process to isolate the failure. *Figure 46* shows a typical fault isolation diagram provided by one major manufacturer for troubleshooting distribution system problems. The approach shown on the figure is valid for all systems, regardless of the equipment manufacturer.

13.5.0 Guidelines for Troubleshooting the Distribution System

Troubleshooting should start by determining if the problem is isolated to one subscriber, a group of subscribers, or all subscribers. If one subscriber is affected, then the problem is isolated as applicable to the subscriber's splitter/tap, drop cable, converter (if used), or the VCR/DVD player and/or TV and related cabling. If a group of subscribers are affected, the cause of the problem can be a local distribution amplifier, splitter/tap, feeder cable, or power supply. The troubleshooting guidelines given in the fault location diagram shown in *Figure 46* describe how to diagnose and isolate such problems in a cable distribution system.

When all subscribers are affected, the problem usually exists in the headend equipment or the trunk cable exiting the headend. When troubleshooting distribution systems, start at the problem location and then work your way back upstream toward the headend.

It is important to point out that FCC Rules and Regulations, Part 76.605 defines the specific requirements for many of the signal parameters in CATV/SMATV/MATV systems. *Table 5* lists some typical values for the signal parameters at a subscriber tap. The actual values used when troubleshooting a system should be those specified for that particular system.

13.6.0 Guidelines for Troubleshooting the Headend

Troubleshooting in the headend can be accomplished in several ways. One approach is given here. Troubleshooting begins by determining if the problem is related to one channel, a group of related channels (VHF, UHF, satellite, or local), or all channels. It must also be determined if the problem is a result of interference.

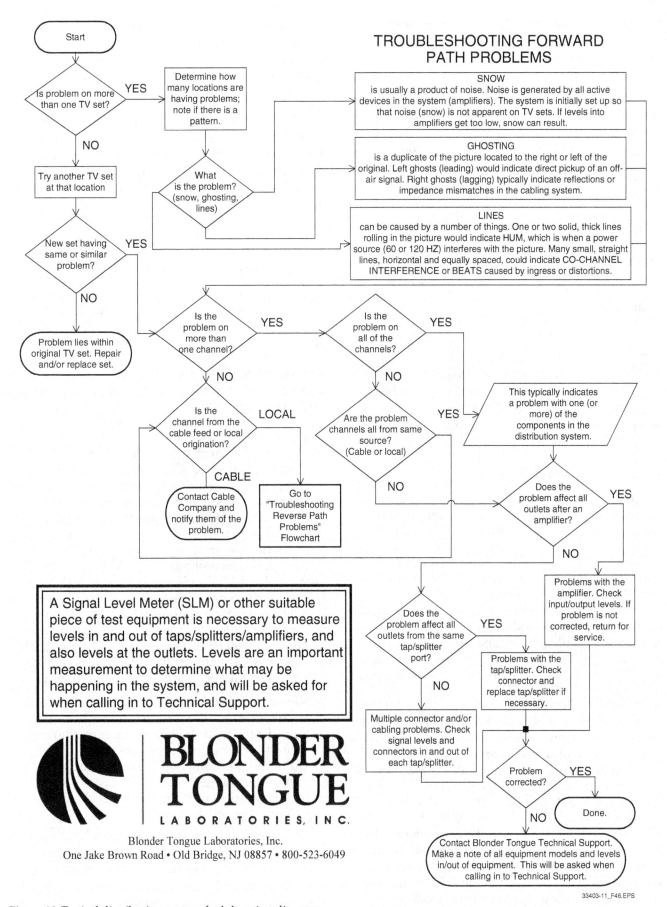

Figure 46 Typical distribution system fault location diagram.

Table 5 Typical Subscriber Tap Signal Parameter Values

Signal Parameter	Typical Value
Minimum video signal level	3dBmV
Maximum video signal level	10dBmV
Maximum difference between adjacent carriers	1dB
Maximum difference between any video carriers	7dB
Audio carrier levels relative to video carrier levels	–13dB to –17dB
Minimum carrier-to-noise ratio	42dB
Hum	1 percent
Reflections	–40dB

> **NOTE**
>
> The signal levels measured while troubleshooting should be compared with those previously recorded when the headend was operational. Any degradation in one or more of the measured signal levels from previously recorded values can point to the problem area.

13.6.1 Signal Level Problems

If a low or no channel signal exists in a single channel, the signal level at the output of the related processor (off-air VHF/UHF channel) or modulator (satellite or locally-generated channel) should be checked. If the output signal level is good, the problem is located downstream in the combiners or related cabling.

If the signal level is low or missing at the output of the processor or modulator, the input signal(s) to the unit should be checked. If the input signal level applied to a processor is within the level specified for the unit, the problem is with the processor itself. If the input signal level is bad, the cause of the problem is upstream in the cable run or equipment that supplies the received signal to the processor from the off-air antenna.

If the video and audio input signals to a satellite channel modulator are good, the problem is isolated to the modulator. If the video and/or audio input signal level is bad, the input signal level to the related satellite receiver should be checked next. If it is good, the cause of the problem is isolated to the satellite receiver and its related output cables. If bad, the cause of the problem is upstream in the cable run or satellite dish/LNBF that supplies the received signal to the satellite receiver.

A problem with a particular group of channels (VHF, UHF, or satellite, for example) that have low or no channel signals is caused by a component shared by the signals being processed by all of the affected channels. Typically, these are a common cable, splitter, combiner, or similar device located upstream or downstream from the affected units. A problem with all of the channels having degraded or no channel signals is caused by a downstream problem in the final combiner, launch amplifier, or their related cable. Another cause for a group of channels or all channels to be faulty is a malfunctioning primary power source that is common to all the affected units.

13.6.2 Interference Problems

If an interfering signal is present in one or more channels, its source can often be found by connecting a spectrum analyzer to view the output signal from the affected channel's processor/modulator. While observing the interference signal, the processor/modulator in all the other channels should be turned off, then on again, one at a time. If the interference signal in the affected channel disappears when a particular channel is turned off, the source of the interference is in that channel. Further troubleshooting in that channel should be performed to find and eliminate the source of interference.

If the above test fails to isolate the source of the interference to some other channel, the source can be in the affected channel itself or from some external source. Using the spectrum analyzer to identify the specific frequency and characteristics of the interference signal can provide the information needed to help identify the type and source of the interference. Often, the problem is the result of beat signals being generated because of improperly set levels in the affected channel and/or in the adjacent channels. For this reason, the processor/modulator in the affected and adjacent channels should be checked for the following:

- Video carrier level is balanced with the other channels
- Sound carrier of the lower adjacent channel is at least 13dB below the picture level
- Units are operating within their specified input and output ranges
- Spurious outputs from the units are at a sufficiently low level

When the above tests fail to isolate the source of the interference signal to the affected channel itself, or to some other channel, the interference is probably caused by an external source. The different kinds of interference and their sources are too numerous to describe here.

Unit Replacement

Before replacing a processor, receiver, or modulator suspected of being bad, always make sure that the frequency and mode selection switches have not been inadvertently placed in the wrong position(s). Make sure that the unit output level, aural level, and/or other signal parameter controls have not recently been incorrectly adjusted. If they have, readjusting the unit controls to obtain a signal that meets specifications may correct the problem.

14.0.0 TWO-WAY TRANSMISSION

Up to this point in the module, the description of CATV/SMATV/MATV systems has been about the one-way (downstream or forward) transmission of the channel signals from the headend to the last subscriber on the distribution system. However, most cable systems have the capability for two-way (bi-directional) transmission. This allows program selection or TV programming signals to be sent back to the headend via the distribution system cabling.

There are two types of signals commonly transmitted back to the headend via the distribution system cabling. First, on some systems, when subscribers select pay-per-view events, digital acceptance signals are sent. These are then processed to allow standard broadcast distribution of the content. In the case of pay-per-view, limited bandwidth is required. On the other hand, full bandwidth audio/video signals can be transmitted to the headend as well. For example, a dean of a college may need to provide a campus-wide announcement from her office. Audio and video signals can be transmitted then to the headend, where they are processed and re-transmitted on a specific channel via the distribution system.

The capability of two-way transmission is made possible by the allocation of the sub-band Chan-

nels T7 through T11 or T12 for the transmission of signals in the reverse direction. Sub-band Channels T7 through T12 encompass the frequency spectrum just below Channel 2 in the VHF low band. Two-way transmission is also made possible by the use of distribution system and headend components that are designed to handle bi-directional signals.

Figure 47 shows a simplified example of an amplifier designed to handle bi-directional signals. As shown, the key element in its design is the use of two-way (low bandpass-high bandpass) filters. Downstream going high-frequency signals (those above the sub-band) applied to the common port of the filter are passed through the high-bandpass portion for application to the downstream amplifier. From there, the amplified signal is applied to the distribution system through the high-bandpass portion of the second two-way filter. Similarly, upstream going sub-band signals applied to the common port of the two-way filter are passed through the low-bandpass portion for application to the upstream amplifier. From there, the amplified signal is applied to the headend through the low-bandpass portion of the second two-way filter. If desired, the sub-channel frequencies can be applied to a processor in the headend for up-conversion to a desired channel frequency suitable for redistribution.

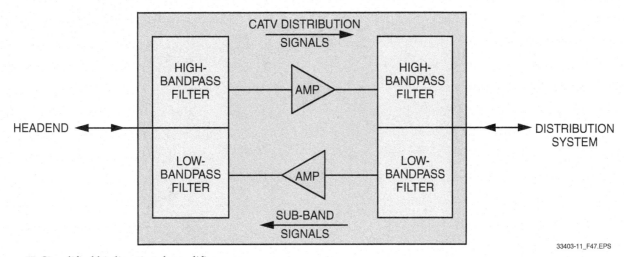

Figure 47 Simplified bi-directional amplifier.

SUMMARY

Broadband cable television (CATV) systems serve both rural and urban areas. They provide a large group of subscribers access to multi-channel TV programming distributed over a regional area by a network of coaxial and/or fiber optic cable. Private broadband satellite master antenna television (SMATV) systems, and master antenna television (MATV) systems provide access to multi-channel TV programming for a much smaller group of users. These systems typically serve the occupants of an office building, industrial facility, or school.

CATV/SMATV/MATV systems consist of two main parts: the headend or main station and the distribution system. Collectively, they provide for the acquisition, processing, and distribution of the desired CATV/SMATV/MATV channel signals to system subscribers. Designated portions of the RF spectrum ranging from 5MHz to 1GHz are allocated by the FCC for TV channel assignments used for over-the-air broadcast TV and cable TV. Each assigned channel has a bandwidth of 6MHz.

In the United States, each 6MHz-wide TV channel is formatted in accordance with the National Television System Committee (NTSC) standard. Most of the analog NTSC standards were replaced by the Advanced Television Systems Committee (ATSC) standards during the digital switchover that took place on June 12, 2009. ATSC standards apply to digital television transmission over terrestrial, cable, and satellite networks.

The ability to calculate gains and losses in a distribution system and in the headend is a necessary skill when working on cable systems, in order to select the proper directional couplers and/or taps needed to deliver the required signal levels to all subscribers. It is also necessary so that you can evaluate the accuracy of signal levels that are measured at different points within the system.

Troubleshooting of cable systems requires that you use an organized and systematic approach.

Review Questions

1. When a converter/descrambler box is used in conjunction with a TV set, the TV set tuner is normally set to _____.
   a. Channel 2 or 3
   b. Channel 3 or 4
   c. Channel 4 or 5
   d. any desired channel number

2. The architecture of an MATV system is the same as that of an SMATV system, except the headend of an MATV system cannot receive input signals from a(n) _____.
   a. off-air antenna
   b. CATV source
   c. satellite antenna
   d. character generator

3. The portion of the broadcast television frequency spectrum that ranges from 174 to 216MHz is assigned to the _____.
   a. VHF high channels
   b. VHF low channels
   c. UHF channels
   d. CATV channels

4. In the 6MHz-wide NTSC channel format for a TV channel signal, the sound carrier is 4.5MHz _____.
   a. below the video carrier
   b. above the video carrier
   c. below the color carrier
   d. above the color carrier

5. The formula used to calculate the power ratio in decibels for two voltages is _____.
   a. $dB = 10 \times \log_{10}(V1/V2)$
   b. $dB = 10 \log_{10} [Z \times 1000 \times 10(x/10)]$
   c. $dB = 10 \log_{10} [10 (X/10) \div (Z \times 1,000)]$
   d. $dB = 20 \times \log_{10}(V1/V2)$

6. In the headend of a CATV system, output signals from the VHF/UHF processors, the satellite, and local origination equipment are combined to produce the _____.
   a. broadband output signal for the distribution system
   b. L-band signal for an agile heterodyne splitter
   c. feedback signal for the demodulator
   d. broadband input signal for the stereo encoder

7. Processor units in the headend of a CATV system are used to process _____.
   a. satellite antenna signals
   b. VHF/UHF off-air signals
   c. combined modulator signals
   d. locally generated signals

8. In the headend of a CATV system, satellite transponder signals are converted to baseband audio and video signals by a _____.
   a. modulator
   b. processor
   c. satellite receiver
   d. baseband encoder

9. A device that is used to amplify and convert received satellite transponder signals into L-band signals is a _____.
   a. modulator
   b. VHF/UHF preamplifier
   c. demodulator
   d. low-noise block downconverter

10. A type of processor unit that is capable of converting any off-air VHF or UHF input channel to any of the broadcast or CATV output channels is called a(n) _____.

    a. strip amplifier
    b. single-channel converter
    c. agile heterodyne processor
    d. channelized input, channelized output processor

11. Which of the following statements regarding splitters is correct?

    a. Splitters and directional couplers are the same thing.
    b. Splitters can be used as combiners.
    c. Splitters are classified as active devices.
    d. All splitter outputs are equal.

12. Frequency-selective devices categorized as low-pass, high-pass, bandpass, and band rejection are all types of _____.

    a. demodulators
    b. filters
    c. amplifiers
    d. encoders

13. The type of coaxial cable suitable for installation in vertical shafts is _____.

    a. CATVU
    b. CATVR
    c. CATVX
    d. CATV

14. If the basic rule of thumb about bending coaxial cable is followed, then the maximum bend radius for a 0.500 semiflex cable should not exceed _____.

    a. 5" (10 × 0.500)
    b. 2.5" (5 × 0.500)
    c. 50" (100 × 0.500)
    d. 25" (50 × 0.500)

Refer to *Figure 1* to answer Questions 15 through 17.

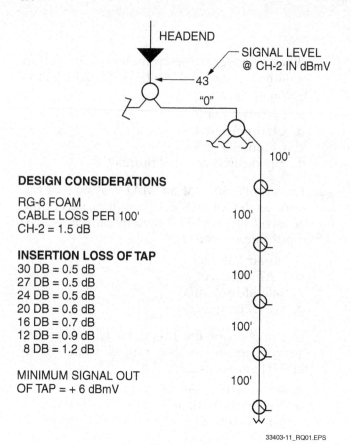

DESIGN CONSIDERATIONS

RG-6 FOAM
CABLE LOSS PER 100'
CH-2 = 1.5 dB

INSERTION LOSS OF TAP
30 DB = 0.5 dB
27 DB = 0.5 dB
24 DB = 0.5 dB
20 DB = 0.6 dB
16 DB = 0.7 dB
12 DB = 0.9 dB
 8 DB = 1.2 dB

MINIMUM SIGNAL OUT
OF TAP = + 6 dBmV

33403-11_RQ01.EPS

Figure 1

15. In *Figure 1*, the signal level for Channel 2 at the input to the first tap is _____.

    a. 30.9dBmV
    b. 43.7dBmV
    c. 46.4dBmV
    d. 50dBmV

16. In *Figure 1*, the tap value for the first tap needed to deliver a minimum of +10dBmV to a subscriber is _____.

    a. 16dB
    b. 24dB
    c. 27dB
    d. 30dB

17. The signal level for Channel 2 at the input to the last tap in *Figure 1* is _____.

    a. 24dBmV
    b. 29.2dBmV
    c. 31.4dBmV
    d. 33.1dBmV

18. A commonly used piece of test equipment in a coaxial cable system that detects RF energy and displays the measurement in dBmV and/or dBμV is a _____.

a. portable color TV receiver
b. cable tone test set
c. spectrum analyzer
d. signal level meter (SLM)

19. During headend alignment, all of the processing channels should be aligned so that their output signals at the output of the final combiner are balanced to within _____.

a. ±0.5dB of each other
b. ±1.0dB of each other
c. ±2.0dB of each other
d. ±3.0dB of each other

20. When using a systematic approach to troubleshooting, the first step should be to _____.

a. interview the customer about the problem
b. perform a basic system analysis
c. physically inspect the system
d. consult the manufacturer's troubleshooting aids

Trade Terms Introduced in This Module

Amplitude modulated (AM): A process by which the amplitude (size or strength) of a carrier signal is varied in accordance with the information being sent.

Automatic gain control (AGC): A type of circuit that maintains the output signal level of an amplifier or receiver constant within a range, regardless of variations in the signal strength of the applied input signal.

Azimuth: Angular position in a horizontal axis, usually related to true North.

Bandwidth: The range of usable frequencies processed by an electronic circuit or piece of equipment.

Broadband: A term used to describe radio frequency (RF) systems or equipment that process a relatively broad range of frequencies.

Baseband: The frequency band occupied by a signal used to modulate a carrier. For example, baseband audio frequencies range from about 15 to 20kHz and baseband video frequencies from about 30kHz to 4MHz.

Broadcast television systems committee (BTSC) stereo: A television sound system that allows simultaneous transmissions of a 15Hz to 15kHz stereo audio signal along with a completely second audio program (SAP) signal, all within the 6MHz television channel bandwidth.

Carrier-to-noise (C/N) ratio: Ratio of amplitude of a carrier to the noise power that is present in that portion of spectrum occupied by the carrier, expressed in decibels. The more negative the number, the better.

Decibel (dB): A logarithmic measure of signal power. Decibels indicate the ratio of power output to power input, expressed as follows: $dB = 10 \log_{10}(P1/P2)$.

Decibel-microvolt (dBμV): This is the abbreviation for decibels referenced to 1μV across a 75-ohm impedance.

Decibel-millivolt (dBmV): This is the abbreviation for decibels referenced to 1mV across a 75-ohm impedance.

Decibel-milliwatt (dBm): This is the abbreviation for decibels referenced to 1mW across a 50-ohm impedance.

Distortion: Undesired changes in the waveform of a signal so that a spurious element is added. All distortion is undesirable.

Downlink: The transmission of signals sent from a communications satellite to a ground receiving station.

Drops: In CATV/SMATV/MATV systems, the cables that connect individual ports to the feeder/trunk cable.

Feeders: In CATV systems, the cables that run in front of homes, from which individual homes are connected by drops. Also see drops.

Frequency modulated (FM): A process by which the instantaneous frequency of a carrier signal is varied by an amount proportionate to the amplitude of a modulating wave.

Geostationary orbit: A satellite orbit that matches Earth's rotation, causing the satellite to remain over the same point on Earth (22,300 miles above the surface).

Headend: In CATV/SMATV/MATV systems, the equipment that creates and amplifies the signals to drive the distribution network. This is the starting point in all networks.

Heterodyning: A process by which two frequencies are mixed together in order to produce two other frequencies that are equal to the sum and difference of the first two.

Hybrid fiber-coaxial (HFC): A common configuration in a CATV system in which the trunk cables and feeder cables are fiber optic and the subscriber drop cables are coaxial.

Insertion loss: The signal loss encountered between the input and output ports of a passive device. Insertion loss is expressed in decibels. The lower the insertion loss, the better.

Isolation: The signal loss encountered between adjacent and non-adjacent ports of a passive device. Isolation is expressed in decibels.

Modulation: The process by which a baseband signal is added to or encoded onto a carrier wave.

Modulator: A device in which the modulation of a signal occurs.

Multiplexed: The simultaneous transmission of two or more signals over a common transmission line or other medium.

Noise figure: A measure of the level of thermal noise at the output of an amplifier, receiver, or system, expressed in decibels. It is the ratio of the measured output noise to the noise generated by an ideal noise standard. It is assumed to equal the thermally generated noise appearing across an ideal resistor.

Orbital slot: The arc of the orbit allocated to a single geostationary satellite. They are identified by their longitude.

Resonant dipole antenna: The most elementary type of antenna, consisting of two $\frac{1}{4}$ wavelength elements that are insulated from each other. Each element is connected by a transmission line to a receiver. It is resonant when receiving a signal that has a wavelength that precisely matches that of the antenna.

Signal-to-noise (S/N) ratio: The ratio of signal power to noise power in a specified bandwidth. It usually is expressed in decibels.

Surface acoustic wave (SAW): The abbreviation for a type of solid-state device that processes signals via a technology that excites and detects minute acoustic waves that travel over the surface of the device substrate, much like earthquake waves travel over the crust of the earth. These devices are typically used for broad bandwidth signal delay, custom-designed filters, and complex signal generation and correlation applications.

Transponder: A microwave repeater. In a satellite, it amplifies and downconverts received uplink signals, and then retransmits the signals as downlink signals back to Earth.

Trunks: In CATV systems, the cables that carry the TV signals from the headend to major portions of the service area, where they connect to feeders. Also see drops.

Uplink: The transmission of signals sent from a ground transmitting station to a communications satellite.

Appendix A

TABLE OF CONVERSIONS

The following table lists the conversions between voltage and power measurements for the range of signal levels commonly encountered in broadband networks.

mV	dBmV	dbμV	dBm	mV	dBmV	dbμV	dBm
0.0010	-60	0	-108.75	0.0447	-27	33	-75.75
0.0011	-59	1	-107.75	0.0501	-26	34	-74.75
0.0013	-58	2	-106.75	0.0562	-25	35	-73.75
0.0014	-57	3	-105.75	0.0631	-24	36	-72.75
0.0016	-56	4	-104.75	0.0708	-23	37	-71.75
0.0018	-55	5	-103.75	0.0794	-22	38	-70.75
0.0020	-54	6	-102.75	0.0891	-21	39	-69.75
0.0022	-53	7	-101.75	0.1000	-20	40	-68.75
0.0025	-52	8	-100.75	0.1122	-19	41	-67.75
0.0028	-51	9	-99.75	0.1259	-18	42	-66.75
0.0032	-50	10	-98.75	0.1413	-17	43	-65.75
0.0035	-49	11	-97.75	0.1585	-16	44	-64.75
0.0040	-48	12	-96.75	0.1778	-15	45	-63.75
0.0045	-47	13	-95.75	0.1995	-14	46	-62.75
0.0050	-46	14	-94.75	0.2239	-13	47	-61.75
0.0056	-45	15	-93.75	0.2512	-12	48	-60.75
0.0063	-44	16	-92.75	0.2818	-11	49	-59.75
0.0071	-43	17	-91.75	0.3162	-10	50	-58.75
0.0079	-42	18	-90.75	0.3548	-9	51	-57.75
0.0089	-41	19	-89.75	0.3981	-8	52	-56.75
0.0100	-40	20	-88.75	0.4467	-7	53	-55.75
0.0112	-39	21	-87.75	0.5012	-6	54	-54.75
0.0126	-38	22	-86.75	0.5623	-5	55	-53.75
0.0141	-37	23	-85.75	0.6310	-4	56	-52.75
0.0158	-36	24	-84.75	0.7079	-3	57	-51.75
0.0178	-35	25	-83.75	0.7943	-2	58	-50.75
0.0200	-34	26	-82.75	0.8913	-1	59	-49.75
0.0224	-33	27	-81.75	1.0000	0	60	-48.75
0.0251	-32	28	-80.75	1.1220	1	61	-47.75
0.0282	-31	29	-79.75	1.2589	2	62	-46.75
0.0316	-30	30	-78.75	1.4125	3	63	-45.75
0.0355	-29	31	-77.75	1.5849	4	64	-44.75
0.0398	-28	32	-76.75	1.7783	5	65	-43.75

33403-11_A01.EPS

TABLE OF CONVERSIONS

mV	dBmV	dbμV	dBm	mV	dBmV	dbμV	dBm
1.9953	6	66	-42.75	158.4893	44	104	-4.75
2.2387	7	67	-41.75	177.8279	45	105	-3.75
2.5119	8	68	-40.75	199.5262	46	106	-2.75
2.8184	9	69	-39.75	223.8721	47	107	-1.75
3.1623	10	70	-38.75	251.1886	48	108	-0.75
3.5481	11	71	-37.75	273.8420	48.75	108.75	0
3.9811	12	72	-36.75	281.8383	49	109	0.25
4.4668	13	73	-35.75	316.2278	50	110	1.25
5.0119	14	74	-34.75	354.8134	51	111	2.25
5.6234	15	75	-33.75	398.1072	52	112	3.25
6.3096	16	76	-32.75	446.6836	53	113	4.25
7.0795	17	77	-31.75	501.1872	54	114	5.25
7.9433	18	78	-30.75	562.3413	55	115	6.25
8.9125	19	79	-29.75	630.9573	56	116	7.25
10.0000	20	80	-28.75	707.9458	57	117	8.25
11.2202	21	81	-27.75	794.3282	58	118	9.25
12.5893	22	82	-26.75	891.2509	59	119	10.25
14.1254	23	83	-25.75	1000.0000	60	120	11.25
15.8489	24	84	-24.75	1122.0185	61	121	12.25
17.7828	25	85	-23.75	1258.9254	62	122	13.25
19.9526	26	86	-22.75	1412.5375	63	123	14.25
22.3872	27	87	-21.75	1584.8932	64	124	15.25
25.1189	28	88	-20.75	1778.2794	65	125	16.25
28.1838	29	89	-19.75	1995.2623	66	126	17.25
31.6228	30	90	-18.75	2238.7211	67	127	18.25
35.4813	31	91	-17.75	2511.8864	68	128	19.25
39.8107	32	92	-16.75	2818.3829	69	129	20.25
44.6684	33	93	-15.75	3162.2777	70	130	21.25
50.1187	34	94	-14.75	3548.1339	71	131	22.25
56.2341	35	95	-13.75	3981.0717	72	132	23.25
63.0957	36	96	-12.75	4466.8359	73	133	24.25
70.7946	37	97	-11.75	5011.8723	74	134	25.25
79.4328	38	98	-10.75	5623.4133	75	135	26.25
89.1251	39	99	-9.75	6309.5734	76	136	27.25
100.0000	40	100	-8.75	7079.4578	77	137	28.25
112.2018	41	101	-7.75	7943.2823	78	138	29.25
125.8925	42	102	-6.75	8912.5094	79	139	30.25
141.2538	43	103	-5.75	10000.0000	80	140	31.25

33403-11_A02.EPS

Appendix B

COMMON CATV SYMBOLS

COMMON CATV SYMBOLS

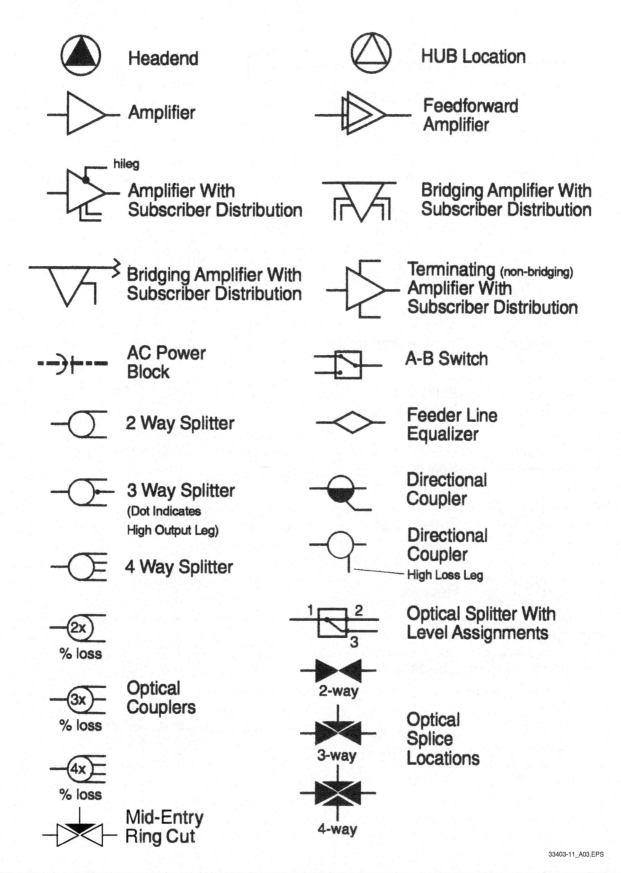

Headend		HUB Location	
Amplifier		Feedforward Amplifier	
Amplifier With Subscriber Distribution		Bridging Amplifier With Subscriber Distribution	
Bridging Amplifier With Subscriber Distribution		Terminating (non-bridging) Amplifier With Subscriber Distribution	
AC Power Block		A-B Switch	
2 Way Splitter		Feeder Line Equalizer	
3 Way Splitter (Dot Indicates High Output Leg)		Directional Coupler	
4 Way Splitter		Directional Coupler — High Loss Leg	
2x % loss	Optical Couplers	Optical Splitter With Level Assignments	
3x % loss		2-way	Optical Splice Locations
4x % loss		3-way	
Mid-Entry Ring Cut		4-way	

33403-11_A03.EPS

 33403-12 Broadband Systems

COMMON CATV SYMBOLS

———————	0.412 Inch Cable	⊡ (circle with square)	Standby Power Supply
- - - - - - -	0.500 Inch Cable		
- · - · - · -	0.750 Inch Cable	━■━	Power Inserter
- · · - · · -	1.000 Inch Cable	⊡ (circle with square)	Power Supply
———→	Termination		Interdiction Unit Symbol
Fixed Attenuator			

Interdiction Unit Symbol

Tap   # Ports

14	16
8	3

Eq.   Att.

–⬡14⟩	Terminating Tap	–◇#◇–	1 Output Tap
Stand Alone Status Monitor		–○#○–	2 Output Tap
Splice		–□#□–	4 Output Tap
Transmitter		▽#	6 Output Tap
Receiver		–⬡#⬡–	8 Output Tap
Male Connector **Female**		◯◯	Optical Figure Eight
Fiber Cable Representations		–⊘ F#–	Optical Cable With Number Of Fibers

33403-11_A04.EPS

Appendix C

TV Off-air Receiver Antenna Spacing Data

TV OFF-AIR RECEIVER ANTENNA SPACING DATA

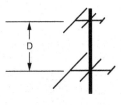

VERTICAL

D = MIN. ½ λ OF LOWER CHANNEL
OPTIMUM IS ⅔ λ OF LOWER CHANNEL

HORIZONTAL

D = 0.12 λ MIN. OF
LOWER CHANNEL

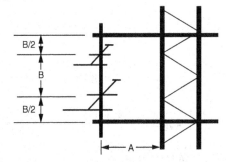

TOWER MOUNTING

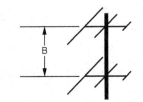

VERTICAL

B = ⅔ λ

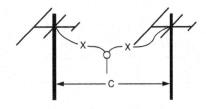

HORIZONTAL

C = 1 λ

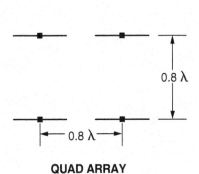

QUAD ARRAY

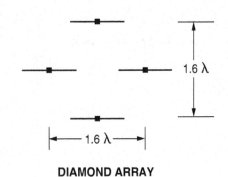

DIAMOND ARRAY

33403-11_A05.EPS

DIMENSION NOTES:

A) THE MINIMUM HORIZONTAL SPACING BETWEEN THE TOWER STRUCTURE AND THE ANTENNA CROSSBAR.

B) THE RECOMMENDED VERTICAL SPACING FOR A GAIN OF 3dB.

B/2) THE MINIMUM VERTICAL SPACING BETWEEN THE ANTENNA CROSSBAR AND ADJACENT MECHANICAL STRUCTURES.

C) THE RECOMMENDED HORIZONTAL SPACING FOR A GAIN OF 3dB.

D) THE MINIMUM SPACING BETWEEN ANTENNAS OF DIFFERENT CHANNELS AND IS THE FIGURE GIVEN FOR THE ANTENNA WITH THE LOWEST FREQUENCY.

SPACING CHART

CHANNEL NO.	A	B ⅔ λ	C 1 λ	D ½ λ
2	113	138	208	104
3	101	125	188	94
4	91	115	172	86
5	78	100	150	75
6	72	93	139	70
FM	72	80	120	60
7	40	44	67	33
8	39	43	65	32
9	37	42	62	31
10	36	40	61	30
11	35	39	59	29
12	34	38	57	29
13	34	37	55	28

DIMENSIONS IN INCHES

FORMULAS:

ONE WAVELENGTH IN SPACE .... $\lambda \text{ (INCHES)} = \dfrac{11811}{\text{FREQ. (MHz)}}$

ONE WAVELENGTH IN 75Ω COAX (SOLID) .... $\lambda \text{ (INCHES)} = \dfrac{7783}{\text{FREQ. (MHz)}}$

ONE WAVELENGTH IN 75Ω COAX (FOAM) .... $\lambda \text{ (INCHES)} = \dfrac{9565}{\text{FREQ. (MHz)}}$

ANTENNA NULLING (FINDING H) .... $d (\lambda) = \dfrac{1}{2 \sin \phi}$

33403-11_A06.EPS

Additional Resources

This module presents thorough resources for task training. The following reference material is suggested for further study.

National Electrical Code Handbook, Latest Edition. Quincy, MA: National Fire Protection Association.

Wireless Cable and SMATV. Steve Berkoff and Frank Baylin. Boulder, CO: Baylin Publications.

Manufacturer or distributor product literature, available from various cable equipment manufacturers and/or distributors.

Figure Credits

Ray Edwards, Module opener, Figures 3, 20, 21, and 45B

Topaz Publications, Inc., Figures 2, 4, 5, 7, 15, 19, and SA01

Blonder Tongue Laboratories, Inc., Figures 8, 9, 12, 16, 18, 25, 28, 31, 33, 34, 39, 46, and Appendices A–C

Pico Macom, Inc., Figures 13, 23, 24, 29, and SA02

Toner Cable Equipment Company, Inc., Figure 30 and SA03

Applied Instruments, Inc., Figures 41 and 45A

AVCOM of VA, Inc., Figure 42

Greenlee/A Textron Company, Figure 44

NCCER CURRICULA — USER UPDATE

NCCER makes every effort to keep its textbooks up-to-date and free of technical errors. We appreciate your help in this process. If you find an error, a typographical mistake, or an inaccuracy in NCCER's curricula, please fill out this form (or a photocopy), or complete the online form at **www.nccer.org/olf**. Be sure to include the exact module ID number, page number, a detailed description, and your recommended correction. Your input will be brought to the attention of the Authoring Team. Thank you for your assistance.

Instructors – If you have an idea for improving this textbook, or have found that additional materials were necessary to teach this module effectively, please let us know so that we may present your suggestions to the Authoring Team.

NCCER Product Development and Revision
13614 Progress Blvd., Alachua, FL 32615

Email: curriculum@nccer.org
Online: www.nccer.org/olf

❏ Trainee Guide   ❏ AIG   ❏ Exam   ❏ PowerPoints   Other _____

Craft / Level: _____   Copyright Date: _____

Module ID Number / Title: _____

Section Number(s): _____

Description: _____

_____

_____

_____

Recommended Correction: _____

_____

_____

_____

Your Name: _____

Address: _____

_____

Email: _____   Phone: _____

33404-12

Media Management Systems

SBCA

Endorsed by the Satellite Broadcasting &
Communications Association (SBCA)

Module Four

Trainees with successful module completions may be eligible for credentialing through NCCER's National Registry. To learn more, go to **www.nccer.org** or contact us at **1.888.622.3720**. Our website has information on the latest product releases and training, as well as online versions of our *Cornerstone* newsletter and Pearson's product catalog.

Your feedback is welcome. You may email your comments to **curriculum@nccer.org,** send general comments and inquiries to **info@nccer.org**, or fill in the User Update form at the back of this module.

Objectives

When you have completed this module, you will be able to do the following:

1. Explain the functions of a media management system.
2. Identify the major components of a media management system and explain their functions in the system.
3. Describe the database and operating software used to control a media management system.
4. Describe the various devices used to store media in a media management system.
5. Describe the types of playback and display devices used in media management systems.

Performance Tasks

This is a knowledge-based module; there are no performance tasks.

Trade Terms

Flicker
Hot-swappable
Interlacing
KVM

Local control unit (LCU)
Media management system (MMS)
Relational database
WORM

Industry Recognized Credentials

If you're training through an NCCER-accredited sponsor you may be eligible for credentials from NCCER's Registry. The ID number for this module is 33404-12. Note that this module may have been used in other NCCER curricula and may apply to other level completions. Contact NCCER's Registry at 888.622.3720 or go to nccer.org for more information.

Contents ———————————————————————

Topics to be presented in this module include:

Figures and Tables

Figures and Tables (*continued*)

1.0.0 INTRODUCTION

A benefit of the digital age is the ability to store, manage, and distribute all types of electronic media. One way to do this is with a centralized storage and retrieval system known as a *media management system (MMS)*. With a media management system, all media resources are stored in a central location, called a media library. Resources can include digital photographs, text, video, computer-based presentations, and audio recordings.

A person no longer needs to walk to a traditional library to check out a movie. With an MMS, a web browser can be used to request that the movie be played in a specific room at a given time. At the appropriate time, the media coordinator can start the playback, or it can be scheduled to start automatically. With some systems, the MMS can automatically turn on the television and play the movie. In addition, some systems allow real-time control of media playback.

This module introduces the concepts and principles of media management systems. It identifies the various technologies used in MMS, both hardware and software. You will learn what media management systems are, how they are used, and the many benefits they offer their users.

The term *media management system* has several interpretations, depending on the environment within which it is installed. For example, in the corporate information technology world, a computer tape library is called an MMS. A television production studio also uses an MMS, but to store and retrieve media objects in varying degrees of completion as they move through the production process.

In this module, we are looking at media management systems primarily as they apply to educational institutions. Media management systems in these environments involve the integration of digital media storage and retrieval with broadband video distribution systems. They can be installed in a given school, or even across an entire school district.

2.0.0 OVERVIEW OF MEDIA MANAGEMENT SYSTEMS

A media management system consists of the hardware and software needed to manage the distribution of stored content. The content can be in either analog or digital format. The MMS provides control capabilities, which can be made available to an individual location, such as a classroom, or within multiple locations across a campus. The system provides both scheduling and content-on-demand capabilities.

An MMS is primarily a video distribution system, and therefore contains headend equipment like that found in CATV systems. Video network infrastructure elements, such as coaxial and network cabling, distribution amplifiers, and other broadband system components, are included as part of the overall system.

An overview of a typical MMS is provided in *Figure 1*. Media resources are stored and broadcast from a central media center. Headend equipment for video broadcasting is located nearby. Individual viewing rooms have some form of video monitor equipment, such as a TV monitor. For some types of media, a networked computer must be located in the viewing room. Notice in

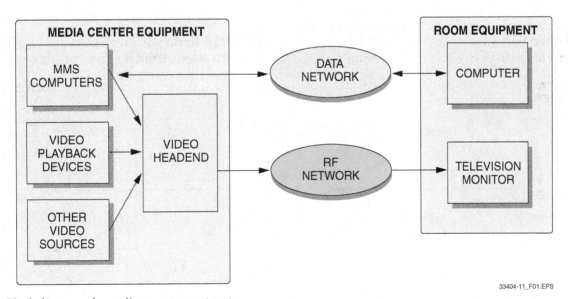

Figure 1 Block diagram of a media management system.

the diagram that there is both a data network and a video network connecting the viewing room with the distribution center.

Many media resources already exist in a format suitable for distribution by an MMS; for example, video tapes, audio CDs, and DVDs. Without an MMS, a video monitor and one or more playback devices must be placed in any room where viewing is needed. Often, this means moving mobile equipment from room to room.

In addition to finding and moving equipment, the process of identifying, locating, and obtaining a particular media program can be more challenging without an MMS. Schools and libraries often have limited copies or licenses to an individual media resource. Scheduling and tracking of the media can be difficult and, in some cases, can actually cause a media resource to be used less often. With an MMS, a single resource can be provided to multiple locations at the same time, reducing or eliminating the need for multiple copies. An MMS also simplifies the job of the librarian. All the media is in one location and is never physically removed from the media center, so it will not be lost.

With an MMS, the retrieval of information about available media is managed with specialized software, which can be accessed from any networked computer (*Figure 2*). It is not necessary to go to the media center to request a specific resource. A user can search for media and schedule playback from any computer on the network.

Specialized software is used to manage all functional and administrative aspects of an MMS. The software provides control capability for equipment management, as well as identification and scheduling activities. Using search capabilities, users can identify and locate media based on a variety of criteria, including subject, title, and author. Software is used to control the operation of video display equipment throughout the installation. Using the MMS software, a user can schedule a room for playback at the same time the media resource is requested.

Some type of video display is required for viewing. In the simplest of installations, the most desirable display device would be a cable-ready television. Many installations are equipped with television monitors or digital projectors. Some form of audio amplification may be required as well. *Figure 3* shows how the MMS interacts with video display equipment. As a media resource is broadcast over the video network, it is viewed in a room using either a television monitor or a projector and screen, which can be permanently installed.

An MMS uses many different playback devices. In *Figure 4*, the MMS controls a DVD player located in the media center. The DVD player may be controlled either by a direct serial link to an MMS computer, via a telephone handset, or by infrared signals. The output of the DVD player is broadcast to a specific room over the broadband network.

3.0.0 TYPES OF SYSTEMS

Media management systems are typically customized for a given installation, and therefore come in a variety of different forms. Often, the ideal system is built by mixing and matching components from several manufacturers to address a client's specific needs. An MMS can be designed for use in applications such as library management, video conferencing, and corporate content retrieval systems. Each application requires specific equipment and software.

3.1.0 Digital Library Systems

A digital library is an electronic storage and retrieval system. Content such as audio recordings,

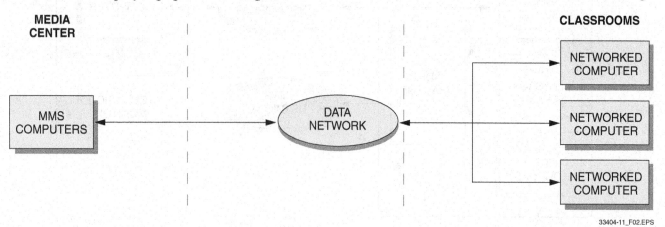

Figure 2 Media resources can be requested from any network computer.

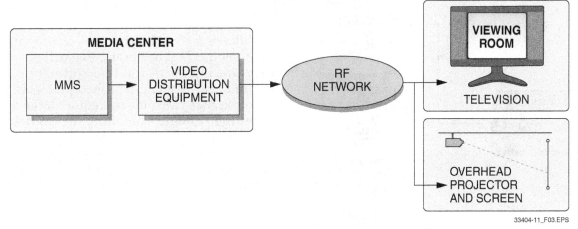

Figure 3 Video display equipment.

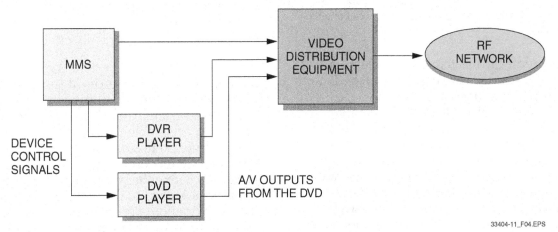

Figure 4 An MMS controls playback devices in the media center.

video, and other complex media are stored in a digital form. On request, the content is provided to an output device. Text, graphics, and multimedia data can be directed from online storage to a video display.

A digital library is much like a physical library. The digital library stores all the available media inside its system, much as all books in a library are located within the library facility. In a traditional library, the librarian is familiar with the layout of the library and can help a visitor find a specific resource. Similarly, a digital library is aware of all available resources and where they are stored. It can thus map a user's request for a media resource to a specific file, which is then made available for playback.

In a digital library, media resources, or at least information about them, are stored in a database. A record is created for each resource (*Figure 5*). This record contains information, such as the title of a resource, the author, and media type or format. It also contains information about the location of the resource. When a media request is made, the digital library searches all available

records. External sources for media may also be included in a search.

A database is a software application designed to manage large amounts of information. A relational database stores information in separate tables. Each table provides a specific view of the information. For example, a table may be used to organize information based on topic, media, or curriculum. Each row in a table is called a record. Each record in the table has a unique identifier used to identify it and relate it to other records. This permits information from a single record to be included in other views. Records in one table may refer back to records in another table using the unique identifier.

Consider a credit card transaction, for example. The transaction itself is one record in a database. This record also contains a reference to a specific account number (the credit card number). The account number then refers to another table that contains information about the customer, such as current and previous addresses. The account number is also used to locate payment information, credit balances, and more. To generate a

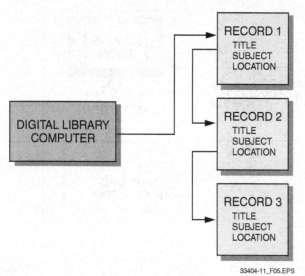

Figure 5 A digital library stores information about available media.

monthly credit card statement, information from several tables in the database is searched, combined, and then printed.

A relational database is used in a similar fashion as part of a digital library system. For instance, a table in the database may contain information about a media resource, such as the title, author, copyright notice, and creation date. The database may also use another table to track how many times a particular resource has been viewed. This information can be used to make decisions about buying new media resources. It may also be used to make payments to the publisher for licensed content.

The database in a digital library may also impose access rights. In an educational setting, it may be that only the English department has received permission from the publisher to show a specific movie. Thus, the database ensures that no other users are allowed access to the resource.

Some vendors of digital library systems integrate a database into their software. In that case, the user purchases what is known as a run-time license for the database as part of the library software acquisition. Other vendors simply sell the digital library system software and recommend one or more database choices.

Databases use industry standards. This allows the backend database to be changed without affecting the operation of the digital library. The two most common standards are structured query language (SQL) and open database connectivity (ODBC). SQL is often pronounced as sequel. Microsoft® SQL, IBM's DB2, and Oracle are very popular database solutions.

In some applications, the digital library software is installed on one computer and the data-

base software on another, with the two systems talking together over the data network. In other installations, the database runs on the same computer as the digital library. Regardless of where they are installed, the two software packages communicate using established protocols.

Audio and video media require large amounts of online storage. Therefore, the digital library provides various storage options. These options include high-capacity disk drives and optical drives, such as CD-ROM and DVD drives. These storage devices attach to the system using standard interfaces such as the small computer system interface (SCSI), the standard protocol used with high-performance storage devices. These devices can be located either inside the computer or in external expansion cases.

3.2.0 Content-on-Demand Systems

A content-on-demand system combines the media management and retrieval functionality of a digital library with an information delivery system. These systems are often used in business to make data accessible to employees, dealers, and clients, while exercising control over the data. Using a networked computer, a user requests a media resource for either immediate or future access. The software used to make the request may be a proprietary client-server application, requiring software to be installed on each networked computer that accesses the system. On many digital library systems, resource requests can be made using a standard web browser.

A digital library system stores static content, such as electronic documents, pictures, and short video clips. A content-on-demand system also incorporates dynamic content, including faculty presentations. Video files may be as long as feature-length motion pictures. The information delivery portion of a content-on-demand system adds video broadcasting capabilities. A content-on-demand system uses both a computer network and a broadband video network.

A content-on-demand MMS may also include features for video paging or video announcements. The MMS can activate all monitor equipment from a central location, allowing school announcements or emergency procedures to be broadcast over the system. The components of a content-on-demand MMS are shown in *Figure 6*. Content players are used to convert media into video signals, and video display equipment is used to view the resources. Computers used to control and manage the media, along with related software, tie all the pieces together to provide a complete solution.

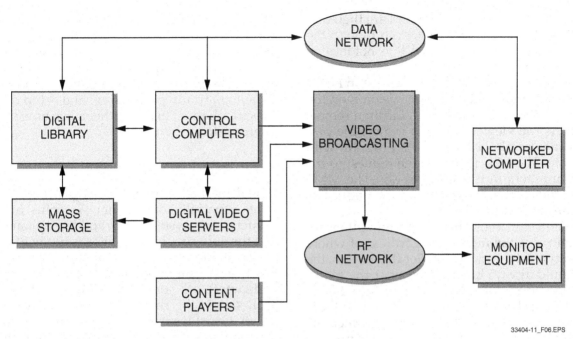

Figure 6 Content-on-demand system components.

3.2.1 Protection of Administrative and Personnel Records

It is likely that the administrative and personnel records are stored on the same network as the media library. For example, a school media library may share the network with the computers and storage devices used by school and district administrators. These devices would likely contain both student and personnel records. For this reason, access to these records must be protected against unauthorized access.

The federal government has established a number of laws that deal with protection of records. For example, the Disclosure of Non-Public Personal Information law, under its Financial Privacy Rule, requires financial institutions to disclose to customers how their private information is used and how it is protected. It also requires, under the Safeguards Rule, that these institutions have a written plan for protecting the private information of consumers.

4.0.0 VIDEO DISPLAY EQUIPMENT

Video display equipment is used to view and, in some cases, interact with the media delivered by the MMS. Although it is convenient to store media resources in electronic form, viewing of the electronic format requires special equipment. The selection of video display equipment is based on the needs of a particular system, but may be limited by cost considerations.

Video display equipment is located at the end of the video network (*Figure 7*). Most viewers will interact with the MMS only through this equipment. Television monitors and digital projectors are most often used for group viewing, although individual computer displays can be used where single user access is required. Plasma, LCD, and other large-format displays may be used in conference rooms and specialized classrooms.

4.1.0 Local Control Units

A local control unit (LCU) allows the central MMS computers to communicate with the display equipment. The LCU sends user signals back to the main MMS server to control media playback.

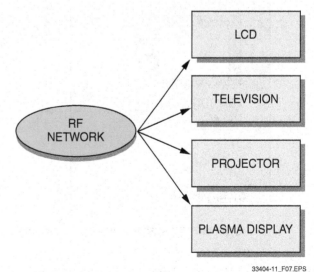

Figure 7 The video network drives the monitor equipment.

Functions such as pause and rewind are therefore available in the local classroom, even though the actual media player, such as a DVR, is not located in the classroom.

A typical LCU installation can be seen in *Figure 8*. The LCU is mounted near a television monitor. An LCU is often supported by an infrared remote control. As the viewers watch a movie, the presenter can control playback of the program with the remote. The LCU forwards the request to the MMS computers in the server room. The MMS matches the DVR to the room and then issues its own control signals to the DVR.

Presenters are accustomed to pointing their remote control at the monitor, regardless of where the local control unit is placed. For this reason, the LCU should always be mounted near the monitor. Handheld remote controls use infrared light to pass signals back to the receiving unit, so there must be an unobstructed line of sight between the presenter and the LCU for the remote control to function properly. Consider the room layout and where people will be located when positioning the LCU.

Communication between the LCU and the server in the media room is done through the data network. The LCU works best when it is placed near the monitor equipment. Therefore, a data network jack should be placed near the video network jack. Some systems may communicate over a two-way video network or a wireless network.

The MMS can control all monitor equipment that is connected to an LCU. This feature offers several advantages. One benefit is energy savings. All monitor equipment can be automatically powered off after normal operating hours. This prevents a remote monitor from being left on over a weekend. Another advantage is access control. If the power to the monitor equipment is hardwired into the LCU, then the equipment cannot be powered up without permission from the MMS. Another important benefit is video paging. During an emergency, all monitor equipment can be automatically powered on and set to display a specific channel to play either a live broadcast or a prerecorded message.

4.2.0 Television Monitors

Television monitors are ideal in educational settings, especially for smaller classrooms. In a video conferencing system, a TV and a telephone are often the only required monitor equipment.

Installation of a television monitor is also relatively easy. The unit should be connected to the video network controlled by the MMS. A video jack should be installed near the front of the room. Consider whether the monitor is going be permanently mounted when installing the jack. Equipment that is carted into a room is more easily connected to a jack placed lower on the wall. Permanently mounted equipment, such as a wall-mounted flat-panel TV, should have the jack placed near the mounting location, which will likely be near the ceiling.

TVs are less expensive than other types of monitors due to the lower quality of the image. TV monitors update their image at a slower rate than some other types of monitor equipment. The scan rate is a combination of both horizontal and vertical scanning. The vertical scan rate is also known as the refresh rate. A TV monitor refreshes the entire screen 60 times per second (60Hz). However, on each pass it refreshes every other line. This is called interlacing, and it takes two complete passes to complete the picture. Thus, the

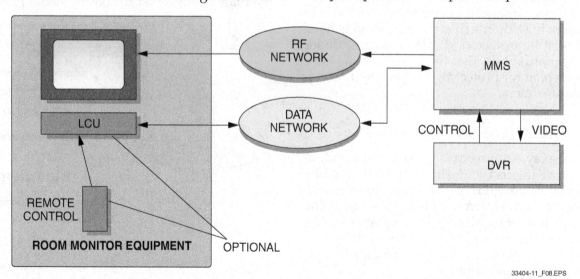

Figure 8 The LCU communicates with the DVR.

33404-11_F08.EPS

image is actually updated 30 times per second. A computer monitor, on the other hand, refreshes the entire image on each pass, progressively, or one line at a time. Computer displays typically operate at 75Hz or higher to reduce a phenomenon called flicker.

Interlacing does not create any noticeable effects when watching video. However, it does cause flicker when displaying certain types of text. In particular, a very thin horizontal line causes noticeable flicker when displayed on a television monitor. The type of content displayed should be considered when choosing a television monitor over other display equipment.

Flat-screen television monitors have become commonplace. Curved screens, such as those found on CRT-type monitors, cause more reflections than a flat screen. A flat screen monitor also has a larger viewing angle than a monitor with a curved screen. The viewing angle specifies the locations in front of the television monitor where the image can be seen without distortion. *Figure 9* compares the viewing angles of CRT and flat screen TVs. Viewers within the shaded region

have an undistorted view of the image. The locations with a clear viewing space are defined by drawing a degree line out from both edges of the television monitor, then dividing it in half.

4.3.0 Speakers

The television monitor or other monitor equipment may not have speakers large enough to fill the room with sound. In that case, a speaker system should be installed.

As shown in *Figure 10*, the volume of sound decreases with the square of the distance from the source. This is called the inverse square law. That is, for every doubling of distance from the source, there is a six-decibel decrease in sound levels, or four times the volume loss. This effect, along with distortions and reflections in the sound waves, make a multiple speaker system ideal for larger rooms.

A ceiling-mount, distributed speaker system is particularly well suited for classrooms and auditoriums because distortion is reduced when the speaker is placed above the listener. A single ceiling-mount speaker is shown in *Figure 11*. This approach also allows a clear line of sight between the speaker and the listener. In addition, overhead mounting reduces sound reflections and helps to compensate for background noise in the room.

4.4.0 Video Projectors

Video projectors (*Figure 12*) use an internal light and display screen to project an image. They work much the same way as a traditional movie projector.

If the projector is to be permanently mounted, there should be a power outlet close by. Manufacturers have guidelines on the maximum power cord length. A six-foot power cord is common.

Certain brands of video projectors allow for either front or rear projection. With front projec-

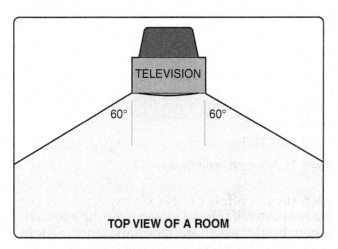

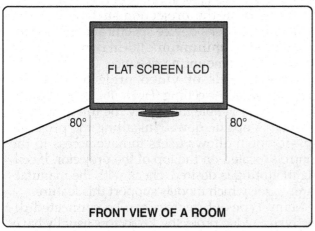

33404-11_F09.EPS

Figure 9 Undistorted image area for TV monitors.

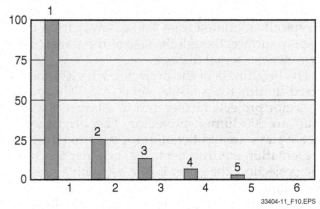

33404-11_F10.EPS

Figure 10 Volume-distance relationship.

Figure 11 Ceiling-mount speaker.

Figure 12 A multimedia projector.

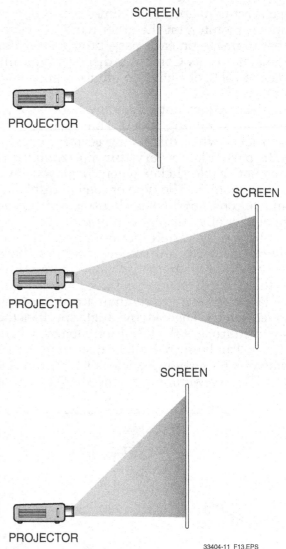

Figure 13 A rear-projection setup.

tion, the image can be displayed on a screen, wall, or whiteboard. With rear projection, the image is projected onto the back of a translucent screen. When using rear projection, the image must be reversed from left to right. A rear projection arrangement is shown in *Figure 13*.

Projector placement is based primarily on the desired size of the projected image. A projector is typically mounted from 4' to 40' away from the display surface, though the size of the screen determines the actual distance.

The brightness of the projector, which is measured in lumens, is also a factor. A 2,500-lumen projector projects farther and to a larger screen than an 800-lumen projector. The brightness factor plays an additional, important role in the presentation environment. The brighter the image projection, the easier it is to see with ambient room light. A 2,500-lumen projector shows a clear image at normal room lighting. An 800-lumen projector requires the lights to be dimmed or

even turned off. If people are expected to take notes or consult printed material during a presentation, brightness is an important factor. *Figures 14* and *15* show a typical range of image sizes and distances from the projection surface. Refer to the manufacturer's device specifications for exact guidelines on minimum and maximum image sizes based on mounting distance.

Many models of video projectors can be mounted from the ceiling (*Figure 16*). In such cases, the projector is able to display the image upright while it is upside down. Installing the projector upside down allows users to have access to the controls located on the top of the projector. If ceiling mounting is desired, check with the manufacturer to see which models support this feature.

Many types of devices can be connected directly to a video projector. Projectors usually have, at a minimum, a composite video connector and a computer display port. Some models support

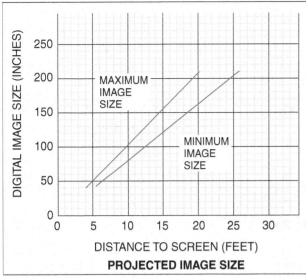

DIGITAL IMAGE SIZE (INCHES)

MAXIMUM IMAGE SIZE

MINIMUM IMAGE SIZE

DISTANCE TO SCREEN (FEET)

PROJECTED IMAGE SIZE

Source: InFocus® Corporation

33404-11_F14.EPS

Figure 14 Image size vs. distance chart.

DIAGONAL SCREEN SIZE (INCHES)	IMAGE WIDTH (INCHES)	DISTANCE TO SCREEN	
		MAXIMUM DISTANCE (FEET)	MINIMUM DISTANCE (FEET)
40	32	5.1	3.9
50	40	6.4	4.9
60	48	7.7	5.9
100	80	12.8	9.8
150	120	19.2	14.8
200	160	25.6	19.7

RANGE OF DISTANCE TO THE SCREEN FOR A GIVEN SCREEN SIZE

Source: InFocus® Corporation

33404-11 F15.EPS

Figure 15 Size vs. distance table.

both an input and an output for each type of connector to allow for additional monitor equipment. For higher quality video, projectors may also have an S-video connector.

Most video projectors can be connected directly to a computer (*Figure 17*) using a standard USB cable. Projectors usually support a range of screen resolutions when connected directly to a computer.

Lower-end equipment often requires the computer screen resolution to match the output resolution of the projector. Typical resolutions are 800 × 600, the standard for SVGA displays, and 1024 × 768, the XGA standard. Higher resolutions such as 1280 × 1024 (SXGA) and 1600 × 1280 (UXGA) are available, but at a significant price hike. The more

33404-11_F16.EPS

Figure 16 A ceiling-mounted projector.

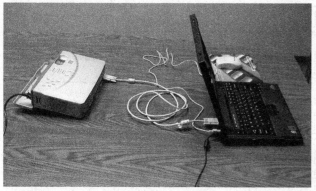

Figure 17 A laptop computer driving a digital projector.

advanced equipment supports a broader range of input resolutions, and internally converts them to the required output resolution. VGA (600 × 480) is no longer supported by projectors currently on the market.

Some projectors have speakers and connections for audio input and output. Depending on the size of the room and the placement of the unit, this may be enough to fulfill the sound requirements. A larger room needs a separate sound system.

4.5.0 Computers

Computers can also be used as monitors in an MMS. The computer serves multiple purposes. First, it allows the user to interact with the MMS to schedule media resources for playback. This is done using the software provided with the MMS. When computers are to be used, network connections should be installed near podiums and in conference rooms.

Second, the computer may be used as a playback device. Digital media is usually compressed. A computer is needed to decompress and view the content. In an MMS, the content is decompressed at the server for distribution over the video network. It may also send the compressed file directly to the computer for local playback. In this case, the computer serves as a monitor device. For this type of distribution, the data network is used instead of the video network.

The presenter can use local content if the room is configured so that the computer can be used as a monitor. This configuration is shown in *Figure 18*. Most digital projectors can switch between a video-input signal and a computer-input signal without any physical changes to the connectors. The remote control is used to select the active input. If the presenters write their own presentations, the room equipment can be used to show the presentations without adding the file to the MMS.

4.6.0 Displays

Plasma and LCD TVs have been around long enough that most installations that use video displays in favor of projectors have one or the other. One of the debates that has ensued since LCD TVs came on the scene is whether plasma or LCD is better. The answer seems to be that within a certain size range size, the differences appear to be negligible. It depends more on the manufacturer and

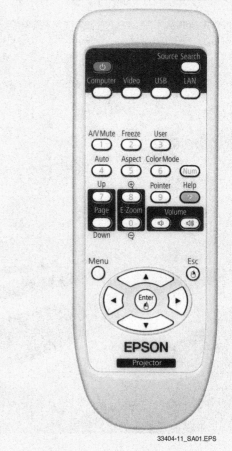

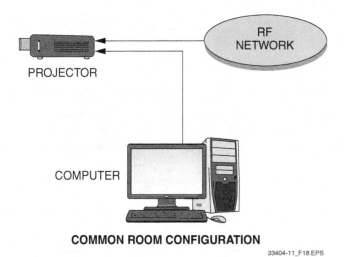

COMMON ROOM CONFIGURATION

33404-11_F18.EPS

Figure 18 A digital projector configured to receive inputs from a computer and a video network.

model than the display technology. Plasma TVs like the one shown in *Figure 19*, are made in larger sizes, going up to about 105 inches, although larger units do exist. High quality LCD screens run up to about 65 inches. At that point, the quality of LCD in comparison to plasma tends to fall off. Of course, all of this is constantly in a state of flux; what is true today will be old technology tomorrow.

5.0.0 STORAGE, RETRIEVAL, AND PLAYBACK EQUIPMENT

Storage, retrieval, and playback equipment makes up the next major section of an MMS. As these devices are used to manage various media resources, they can be referred to as source equipment. Video display and monitor equipment are used to view the media resources. Source equipment is used to store digital media and convert

33404-11_F19.EPS

Figure 19 Plasma display.

it to a usable form. It is then either transmitted across the data network or provided as input to a broadband video-distribution network.

An MMS has at least one computer to act as the main server for the system. The server for an MMS is typically located in the library or media center. The computer, digital storage, and broadcasting equipment are required elements of the MMS. Source connections to the data and video networks are also required. The equipment should be rack-mounted in an area with environment control.

Most media management systems use several high-performance workstations. These computers are located in the server room. Some systems require three to five computers located together in the server room. These computers can be rack-mounted. They can share a common monitor and keyboard by using a keyboard, video, and mouse (KVM) switch.

5.1.0 Local Digital Data Sources

The MMS stores most of its resources locally. Many of these resources are stored in a digital format directly on computer disk drives. Other types of computer storage equipment may also be used. Digital files are converted to video using several types of equipment, including digital video servers and presentation players.

5.1.1 Digital Video Servers

A digital video server is a computer that translates the stored digital media resource into video. The video is then sent for transmission over the video network. This computer has dedicated hardware to allow for a set number of simultaneous video outputs. Four outputs are quite common. The number of digital video servers required depends on the number of expected simultaneous digital video broadcasts.

A digital video server may also work with a loading station to provide the ability to convert analog video sources into digital video. The loading station may be part of the digital video server or may be a distinct workstation. An MPEG-encoding expansion card is required in the workstation to perform this task. By storing resources as digital files, heavily accessed media does not deteriorate with use. Digital storage also allows for simultaneous access by multiple users.

The digital video server is capable of playing multiple video files at one time. With the use of a multi-channel decoder, the output can be sent by an analog audio/video signal into an RF modulator for distribution onto the broadband video network.

LED TV

At the time this module was developed, so-called LED TVs had become a popular consumer TV technology. The LED TV is technically an LED-backlit LCD, rather than a pure LED device. True LED TVs exist, but are used only in large displays, such as those found in arenas and stadiums. LED backlighting replaced the cold-cathode fluorescent light (CCFL) backlighting that had been used in LCD TVs. LED backlighting provided the following improvements:

- Brighter display with better contrast
- Slimmer package
- Less power consumption
- Less heat production

5.1.2 Presentation Players

A presentation player is a dedicated computer located in the server room. The player can read and display presentation files, such as Microsoft® PowerPoint® or Apple® QuickTime®. This computer outputs the video to a scan converter that allows the presentation to be broadcast over the video network.

A digital presentation requires a computer to decode the file. A presentation player is required for rooms that are equipped with only a television monitor. The presentation player works like a local computer. It reads the presentation file and broadcasts it over the video network. The presentation player works with the local control unit. The local control unit (LCU) sends a signal back to the presentation player when the next page should be displayed. In this way, the presentation player works like a local computer would work.

The use of a presentation player in a classroom is shown in *Figure 20*. A local control unit is connected to the computer network. When the teacher presses the forward button, the control unit sends a message to the MMS to advance the presentation to the next page. The system for-

wards this signal to the presentation player. The player then converts a PowerPoint® presentation to video for delivery to a TV monitor or projector. When another signal is received from the LCU, the presentation is advanced to the next page, and a new video screen is created and broadcast over the video network.

5.2.0 Digital File Formats

There are several types of digital file formats. Some of the most common file formats are discussed below. Each format has particular advantages and disadvantages.

MPEG is a very common compression protocol for digital video transmission. MPEG is a set of standards used to encode and compress digital video for a variety of purposes. Digital cable TV and satellite broadcasts use versions of MPEG for content delivery. MPEG files are encoded with a bit rate. The bit rate specifies the amount of content per second. The higher the bit rate, the higher the video quality and the larger the source file will be.

There are several designations within the MPEG standard. MPEG1 produces video that is

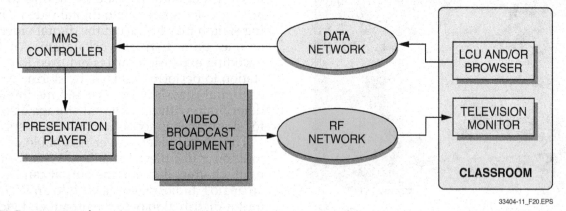

33404-11_F20.EPS

Figure 20 Presentation player setup.

slightly below the quality of a VHS videotape. MPEG4 is used for Internet video. MPEG2 supports video recorded at higher resolutions. This is the standard used for digital video broadcasting over cable TV and satellite systems. It is also the standard used for DVD. MPEG2 produces television monitor images of higher quality than normal analog transmission. MPEG4 trades quality for size. It is used for low-bandwidth situations. The video quality in MPEG4 shows degradation from the compression. However, it is still suitable for compressing animated pictures or incorporating text with low-quality video. Most MPEG players work with all of the MPEG formats.

The audio video interleave (AVI) file format was invented by Microsoft® to play video clips on a PC. There are many AVI resources available. For many years, AVI was the most common video format available on the PC. Many online reference books use the AVI format.

QuickTime® is a media presentation format developed by Apple®. It is more than a video format like MPEG or AVI. It combines video with scripting capabilities. This allows the viewer to interact with the video. While viewing a QuickTime® presentation, a presenter may use a mouse to select a link embedded within the file. This allows for a custom viewing experience. The use of QuickTime® also allows a presenter to play portions of the video at different times during the presentation.

Common intermediate format (CIF) is a set of specialized formats designed for use in video teleconferencing. It standardizes the horizontal and vertical YcbCr sequences in video signals. There are various versions of the standard, with resolutions ranging from 176 × 144 (QCIF) to 1408 × 1152 (16 CIF).

5.3.0 Local Digital Data Storage

Digital data can be stored using a variety of methods. The available storage methods vary in cost, retrieval speed, and storage capacity. Some of the storage devices are fixed, while others require an operator to install and remove them.

5.3.1 Hard Disk Drives

Hard disk drives are the most common device used for storing digital content on a computer. Hard drives can transfer vast amounts of information quickly. As with all magnetic storage, the data degrades with time. Hard disk drives are not appropriate for permanent archiving of media resources. They are also subject to complete failure

and data loss. Irreplaceable content should not be stored exclusively on hard disk drives.

Hard disk drives designed for server use now come in capacities up to about two terabytes, but breakthroughs in design are likely to multiply that capacity in the next few years. Of course, the ability to use stacked drives makes the capacity for a system nearly limitless.

Hard drives have a higher cost per megabyte than other types of storage. However, their rapid data access makes them a good choice for many applications.

Hard disk drives are also available in hot-swappable versions. These drives are typically placed in a rack-mounted external storage case. With this type of external hard disk drive case, additional drives can be added when required simply by sliding them into the case. They are automatically added to the system and used for storage.

5.3.2 RAID Systems

The term *RAID* is short for Redundant Array of Inexpensive Disks. A RAID array is shown in *Figure 21*. RAID is a data storage scheme in which data is divided among multiple disk drives. A RAID system is often used in digital library installations because the large size of many media resources creates the need for significant storage capacity. A RAID system combines high storage capacity with high data reliability. A RAID system works by putting a portion of each file on

33404-11_F21.EPS

Figure 21 Example of a dual RAID array.

---

several different hard disks. This is done in such a fashion that if any one of the hard disks fails, the missing data can be calculated and recreated. Thus, a RAID system allows the unit to continue functioning even during a disk failure.

RAID drives are hot-swappable. That is, a drive can be removed while the unit is powered and a read operation is in progress. Multiple simultaneous drive failures are necessary before data is lost. Data is automatically recreated on a replaced disk using the redundant information on the remaining drives.

5.3.3 Magneto-Optical Storage

Magneto-optical drives provide very high capacity and high performance storage of digital content. These drives are very similar in function to traditional hard disk drives. They are used when a large amount of storage capacity is needed or when data needs to be transferred off of the storage device at very high rates, such as when retrieving digital video.

The magneto-optical drive interfaces with a computer using the same protocols as high-performance hard disk drives. Many devices use the industry-standard SCSI protocol. To install a magneto-optical drive, either install it inside a computer that has a SCSI controller, or install the unit in an external SCSI expansion case.

Magnetic storage, such as hard disk drives and tape, begins to deteriorate after only a few years. A major advantage of a magneto-optical drive is the long shelf life of the storage media itself. This is important for sites that need to archive their existing media resources. Another advantage of magneto-optical storage is that some disks can be erased and rewritten. Removable media for a magneto-optical drive is currently available in sizes up to 9.1GB. A library of these removable cartridges, much like a floppy disk or CD-ROM, can store a large number of digital video titles.

5.3.4 WORM Storage

A write once, read many times (WORM) drive is a variant of a magneto-optical drive. WORM media works with many types of magneto-optical drives and provides permanent data storage. Because this media can only be written one time, it is not subject to the same data deterioration that affects magnetic media. Its very robust storage allows it to be used in situations where media preservation is critical; for example, where the data needs to be preserved for legal or auditing reasons.

Courtesy of: KINTRONICS

33404-11_F22.EPS

Figure 22 Magneto-optical jukebox storage units.

Figure 22 shows a range of optical jukeboxes. The highest capacity unit is capable of holding up to 3,360 optical discs in 96 magazines, for a maximum data storage capacity of 403TB. The HP 220mx is an optical jukebox (*Figure 22*). It stores up to 24 WORM or rewriteable optical cartridges. It has two active drives and a mechanical system to switch cartridges on demand. The unit provides up to 218 gigabytes of storage by selecting between the available cartridges.

5.3.5 DVD

DVDs are available in read-only and read-write versions. The most common DVD used for data storage is the 4.7 GB single-sided, single-layer disk. Other versions of the DVD include the single-sided, double-layer disk (8.7 GB), and the double-sided, single-layer disk (9.4 GB). There is also a double-sided, double-layer version (17.08 GB), but it is not generally available.

This technology is appropriate for storing digital video files or for general-purpose data storage and backup. Due to the general availability of DVD players, storing information on DVDs is very economical. Most PCs purchased today include a DVD drive. Because DVD drives can also play CDs, a separate CD drive is not necessary.

5.4.0 Internet Digital Data Sources

Most MMSs cannot handle content that is located on the Internet. To view this type of content, a local computer needs to be connected to the monitor equipment in the room. A presenter uses the local computer to access the Internet content and display it to the group.

5.4.1 Streaming Audio and Video

Digital audio and video files are quite large. Significant storage capacity, network bandwidth, and download time are required to download these files from the Internet for local play. Streaming has helped to overcome these problems and therefore is a popular way to display video and audio from the Internet. In streaming, the media is downloaded in segments that play while additional segments are being downloaded. Streaming is used extensively for Internet radio stations, and is also used for rebroadcasting live video onto the Internet. For a radio station, perhaps only one minute of content is downloaded at a given time. As the content is played, a bit more is downloaded. This eliminates the need for large amounts of local storage and reduces the long delays involved in downloading media.

Although storage requirements are reduced, a significant amount of bandwidth is required on the data network. Most installations do not have a need for extensive streaming audio and video capabilities. If a site does have a need for multiple streaming data sources to be played at the same time, a minimum of a 100-megabit data network should be installed.

5.4.2 Web Pages

An increasing amount of information is available on the World Wide Web. The need for web access continues to grow, particularly in classrooms. The typical MMS does not handle web pages directly. A local computer is required to view this type of content. If web pages will be viewed regularly, the local computer should be configured so that it can display on the local monitor equipment.

5.5.0 Content Players

Content players are located in the server room. These players may be preloaded with commonly accessed media resources or loaded by the system operator. The operator reviews scheduled media requests on a regular interval and loads tapes or DVDs into the appropriate player.

The players are managed by a headend controller unit. This unit manages the content players using infrared control signals. This enables a user to control the desired device using a network-based remote. *Figure 23* shows a bank of content players that are controlled by an MMS. Here, two VCRs and two DVD players are controlled. The number of content players is typically limited by the capacity of the video network. Many more players can be added to this configuration. A configuration such as this allows the using organization to retain existing VHS video tapes. Combination DVD/VCR units have become more common than individual VCRs.

On Site

Playback Software

RealOne™ player from RealNetworks is PC-based software. This software package can play many media types, including CD and DVD. It can convert audio CDs to MP3 files. It is also used to play streaming audio and streaming video from the Internet.

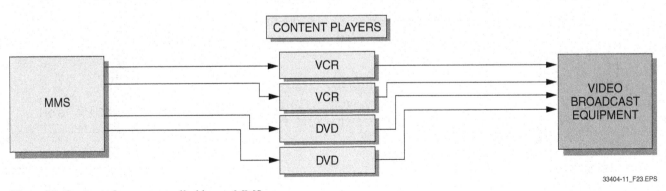

Figure 23 Content players controlled by an MMS.

5.5.1 Compact Disc (CD) Players

Audio CDs are the most common form of digital media. CD players may be incorporated into the server room to broadcast the content over the video network. One CD player is required for each simultaneous audio CD playback request.

Audio CDs may be compressed and stored as digital files. This eliminates the need for operator involvement when requesting audio resources. It also allows for simultaneous access of the same resource by different users.

The most common digital music file format is MP3. MP3 files are based on the MPEG standard, but they only support audio. These files preserve the quality of the original CD and use a fraction of the storage space. Only one CD player is required to convert an audio CD into the MP3 format. If all of the audio CDs are converted to digital formats, then CD players are not required as playback equipment.

5.5.2 Video Cassette Recorders (VCRs)

VCRs are used to play VHS videotapes. Although most movies and video programs are available on DVD, some libraries may have a large collection of programs on VHS tape that would be too costly to replace. A VCR is required for each simultaneous cassette played. Thus, if three teachers each want to watch a different movie in their classroom, at least three VCRs are required. The number of VCRs installed in the server room should be determined by the number of rooms that are likely to use a VHS resource at the same time.

Media management systems control the VCRs. The system includes a computer that outputs to an infrared transmitter. *Figure 24* shows this configuration. Each transmitter is mounted near a VCR so that it only controls one device. Since multiple infrared transmitters may be installed in the server room, care must be taken to ensure that each

transmitter controls only one playback device. In this way, the system is in control of all VCR functions except for changing tapes. Through the local control unit, the presenter can press the pause button. This information is sent over the data network to the MMS. The MMS coordinates a particular VCR with a particular room. It then sends the correct infrared codes to cause the specific VCR to pause. This same technique is used for other playback devices that are controlled by infrared or serial link.

Additionally, many MMSs support telephone touchpad (DTF) access in the same fashion. The teacher or presenter calls the MMS and controls selection, playback, and other actions by pressing the appropriate keys on the telephone.

Some playback devices can also be controlled by a serial link. The MMS can send commands over the serial link instead of an infrared. The serial link is connected to a serial port on an MMS computer. This wire is also connected to the playback device through either an RS-232 connector or a proprietary connector. When using a serial link, potential problems with infrared interference are eliminated.

A VCR can also be used as a tuner. A tuner, or demodulator, converts an input signal broadcast on a given channel frequency into baseband audio and video outputs. A VCR can convert a channel from the community cable TV (CATV) system into a format that can be broadcast over the local video network. An MMS may be able to use a VCR as either a player or a tuner, depending on demand.

5.5.3 High-Definition DVD Players

Standard DVDs are not designed for playback on high-definition (HD) TV monitors. DVDs typically output data in a 480i or 480P format (740 × 480 pixels). HDTV pixel count, on the other hand, is 1280 × 720 (720P) or 1920 × 1080 (1080P).

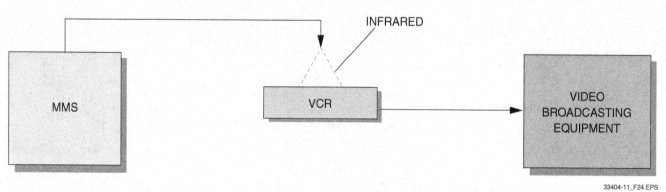

Figure 24 The MMS controls playback using infrared.

Initially, there were two competing and incompatible formats for HDTV playback—HD-DVD and Blu-ray. In 2008, Blu-ray became the universally-accepted format. Blu-ray uses what is known as blue-laser technology. The disk used for recording is the same physical size as a DVD, but has a capacity of 25 Gb or 50 Gb (dual-layer version). It can hold a feature-length movie or it can be used to record two hours of HD video.

An MMS controls the DVD player using an infrared transmitter. As with the VCR, the system controls all aspects of the player except for loading and unloading discs, which may require an operator in the library.

5.5.4 Digital Video Recorder (DVR)

The introduction of the DVR in the 1990s has allowed home users to record and play back scheduled TV programs at their convenience. It has become the defacto replacement for VCRs in the CCTV industry as well. The DVR uses a computer hard-disk drive to store recorded video. There are actually two types of DVRs. One, for general use, is similar to analog recorders. It takes a composite video source, digitizes and optionally compresses it, and then stores it on digital media. A network DVR, on the other hand, receives a digital video stream from network cameras connected to a LAN and stores it directly to digital media.

DVRs provide a number of controls in addition to those provided by a traditional VCR. For example, a DVR provides the ability to search stored video based on time, date, or even camera number. They also provide some form of video compression, usually based on either MPEG or motion JPEG (M-JPEG) standards.

5.5.5 Video Cameras

A video camera is an essential tool for local production of video. A school may wish to start a weekly news program, for example. A reasonable quality video camera is required for this project.

A consumer-grade camcorder or mini-digital video recorder can be used for this purpose. However, analog camcorders tend to record at or below television resolutions. Signal degradation occurs when the recording is edited and processed for broadcasting. Because a camcorder starts at the minimum resolution required, any degradation is noticeable.

Higher quality video cameras record the image at a resolution higher than the broadcast resolution. This allows the signal to be broadcast at the maximum resolution of the video network, even after editing.

By taking advantage of the two-way capability of coaxial cable, you can connect a video camera in a classroom to the broadband video network. This requires three components. First is a sub-channel modulator. It sits in the classroom and connects to the network. Its role is to transmit the video signal from the camera back to the headend. The video camera in turn connects to the sub-channel monitor. To actually move the video signal requires one diplexer at each end of the transmission. At the headend, a modulator receives the transmission, converts it to the proper frequency, and then passes it to the combiner, where it is then sent across the network to monitors and display devices.

5.6.0 Broadcast Sources

One of the advantages of an MMS is that all audio-visual (A/V) resources are centrally located. A broadband video network is necessary for an MMS. This means that as part of the MMS, a broadband video headend and distribution network are integral components and are installed as part of the system. This video network is similar to a CATV headend and it is a natural progression to use it to distribute broadcast content.

A CATV, MATV, or satellite signal must be routed to a tuner or demodulator in order to be used. In *Figure 25*, two channels from a local CATV source and two channels from an external antenna

Video Camera

The Canon XL-h1 digital video camera is designed to record high-definition video. Because it records in a digital format, no signal degradation occurs while editing.

33404-11_SA02.EPS

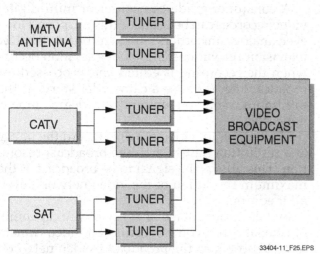

Figure 25 Adding broadcast content to an MMS.

are added to the local video network. The tuner serves to convert the signal into baseband audio and video outputs. These outputs are then sent to modulation equipment for local transmission.

5.6.1 CATV

Cable TV is available from a local provider in most communities. The transmission equipment required at the server room is well suited to add channels from a CATV source. The MMS may also be used to selectively filter certain CATV content. Only channels explicitly tuned and modulated at the headend are available on the local video network. Many schools want to broadcast a limited number of channels from CATV.

5.6.2 MATV

A master antenna television system uses an aerial antenna to receive broadcast content from local over-the-air broadcasters. The signal from the antenna is sent to the headend room, where the desired channels are tuned and modulated.

5.6.3 Satellite

A satellite system requires an externally mounted dish and a dedicated receiver for each channel that is modulated onto the video distribution network. Newer satellite systems broadcast digital video content and require much smaller dishes to receive their signal. Digital satellite systems do not always work well between different manufacturers. Therefore, it may be necessary to obtain all the receivers from the satellite service provider. Consult the product literature to determine if a single sat-

ellite dish is sufficient for the number of channels that are to be placed on the local video network.

5.6.4 Radio

A teacher may want to have a class listen to a radio broadcast. Perhaps a news commentary will be used to seed a discussion about current events. The video network can be used to broadcast a radio station within the site. This frees the teacher from using portable stereo equipment.

A radio signal is added to the video network the same way as a CATV signal is added. A tuner is used to convert the RF broadcast into a baseband signal. This signal is then sent to the video broadcast equipment for placement on a TV channel. If desired, a video source could be added to the signal as well. The video source could be a static image that is always broadcast on the channel. Many scan converters have the ability to grab and hold an image, thus freeing source equipment for other purposes.

6.0.0 NETWORK INFRASTRUCTURE

The network infrastructure is used to transmit both video and data between the source equipment and the monitor equipment. The network is a critical component of the MMS. The broadband video network is used to transmit the video from the headend to the monitor equipment. In most installations, the video is distributed over a local CATV system.

6.1.0 Broadband A/V

Broadband refers to a transmission medium that can carry a signal at many different frequencies. The frequency range is divided into multiple independent channels. In the case of a video network, each frequency range corresponds to one TV channel. These channels are combined and modulated onto an RF carrier and then transmitted over a wire, or by using wireless technology. A local site normally uses a wire for transmission. Wireless transmission introduces higher costs and a greater chance of interference on the broadcast.

6.1.1 Coaxial (RF)

The industry standard for distributing broadband video is 75-ohm coaxial cable. Most television monitors and VCRs have a 75-ohm coax input. These devices can demodulate the signal and tune each of the many channels available. When wiring a site, a 75-ohm coax connector should be wired in each room that is to be connected to the MMS. Both audio and video are transmitted over coaxial cable.

6.1.2 Fiber

When transmitting the video signal over long distances or between remote sites, fiber optic cabling may be appropriate. The RF signal carried over a coaxial cable can be converted to light pulses. These light pulses are then sent over fiber optic cabling. At the receiving end, the light is converted back into an RF signal and placed back on a separate coaxial network.

6.2.0 Baseband A/V

Baseband is a transmission method in which digital signals are sent without frequency shifting. With baseband transmission, only one channel is available on each wire.

In the headend, baseband transmission is used when dealing with individual signals. After all of the signals are created, each one is modulated onto a given channel for broadband distribution. This setup is shown in *Figure 26*. The technologies discussed in the following sections are used within the headend to transfer media from various playback devices to the broadband distribution network.

6.2.1 Stereo RCA

Baseband audio and video are usually transmitted together, but can be sent over several different types of wires. When transferring audio output from a receiving device, such as a VCR or TV monitor, it can be routed to speaker systems using cables that have two RCA connectors on each end (*Figure 27*). One wire is used for the left channel

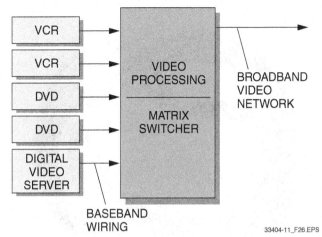

Figure 26 Both baseband and broadband are used in video distribution.

and one wire is used for the right channel. If the connectors are color-coded, the red connector is for the right channel and the white connector is for the left channel.

6.2.2 Composite Video

A single shielded wire with an RCA connector on each end is used to transmit composite video. If the cable is color-coded, each video connector is yellow. Composite video provides a signal quality superior to modulated coax. However, for baseband video transmission, composite video is the lowest quality.

6.2.3 S-Video

S-video stands for separated video, a form of composite video. Here the video is separated into both chrominance signals and luminance signals. An S-video connection is capable of transmitting a higher quality image than a composite video con-

Figure 27 Audio cable.

33404-11_F27.EPS

nection. If both the source and destination equipment have an S-video connection, it should be used instead of the composite video connection.

6.2.4 Component Video YPrPb

Component video is the highest quality video connection available on consumer-grade devices. When available on both the source and destination equipment, component video should be used. Most DVD players support component video output because of the higher image quality.

The YPrPb labels refer to the signal that is carried on each one of the three wires. Y, for luminance, carries the brightness of the signal. Pr is the red portion minus the luminance signal. Pb is the blue portion minus the luminance signal.

6.3.0 Data Network

The data network is used for almost all communication within the MMS. The LCU sends messages to the media management server to pause and resume playback. The server then converts the message into the appropriate infrared signals to control a playback device. The server also sends messages to the LCU using the data network. These messages may turn on a television monitor and tune it to a specific channel.

The data network is also used for communication with the scheduling software. In this case, access to a networked computer is required. The media management software is an application that the presenter may need to run, in which case a computer should be in the same room. For other installations, the management software may be accessed using a special web page. In either case, communications with the server occur over the data network.

6.3.1 LAN

The local portion of the data network in most installations is an Ethernet. For web-based software, the TCP/IP protocol is always used. Most media management software packages provide TCP/IP support. Some proprietary software may have unique protocol requirements. *Figure 28* shows a LAN-based MMS.

6.3.2 WAN

When data leaves the local site, it is usually transmitted across a wide area network (WAN). The most common WAN is the Internet. A local site may connect with a WAN using a variety of techniques. *Figure 29* shows a WAN-based MMS that can serve buildings in three locations.

For a connection between two sites, cabling or fiber optics may be run between locations. This ensures a very secure connection and eliminates any network dependency on outside providers.

Digital subscriber line (DSL), or cable modem is used to connect to the Internet through a service provider. ISDN may also be used. DSL and ISDN are point-to-point services in which data is delivered over telephone lines. A cable modem is a broadcast-based service used to deliver data over the CATV network. Two or more distinct LANs can transmit data to each other over a WAN.

6.4.0 Video Transmission Equipment

Video transmission equipment is used to convert baseband A/V signals into a form suitable for broadband distribution. Each particular source is placed on a single television channel of a broadband network. Many types of equipment are used to convert between baseband and broadband.

Figure 30 shows a common setup. Each baseband video output is sent to a modulator. The modulator moves the signal to the correct channel. The outputs of each of the modulators are combined. This result is a broadband signal suitable for transmission.

6.4.1 Modulators

Modulators convert standard audio/video inputs to a frequency suitable for local CATV distribution. A modulator is required for each simultaneous channel that is broadcast over the

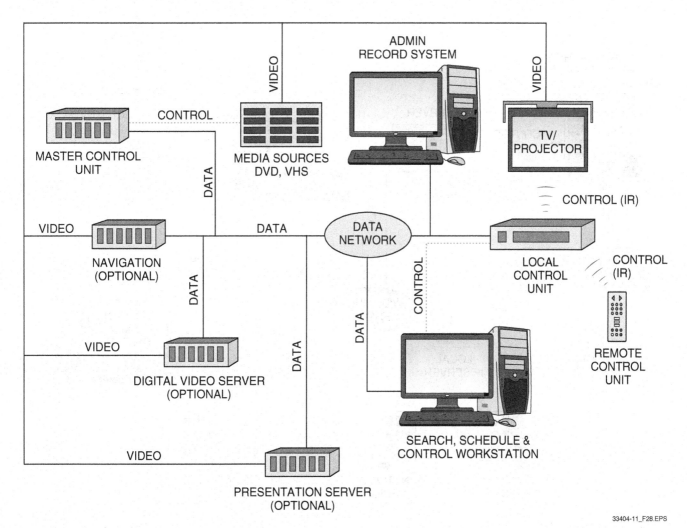

Figure 28 LAN-based MMS.

network. That is, if two DVDs, one VCR, and five electronic slide presentations are occurring simultaneously, a system with at least eight signal modulators is required. A modulator may also provide the ability to incorporate an additional language into the signal. This is known as second audio programming (SAP). It allows remote selection of the desired language through the monitor equipment.

6.4.2 Demodulators

An MMS may incorporate content from the local cable or television franchise, or from an antenna. For this content to be rebroadcast over the local video network, it must first be converted to a baseband signal and then remodulated onto the video network. A demodulator converts a broadband source input—for example, channel 12 from some other source—to baseband output. A/V equipment, such as a VCR with a tuner, may be used as a demodulator. Dedicated hardware may be installed as an alternative.

6.4.3 Combiners

Each modulator puts out a signal at a specific frequency range for the CATV standard desired. If a television monitor with a tuner was placed directly on this output, only the specific channel could be tuned. The signals from each of the modulators need to be combined so that all channels are available on the video network. A combiner performs this operation.

6.4.4 Processors

A processor performs several functions. These functions include signal amplification, rejection of unwanted signals, audio carrier level control, and generation of output signals at the desired frequencies.

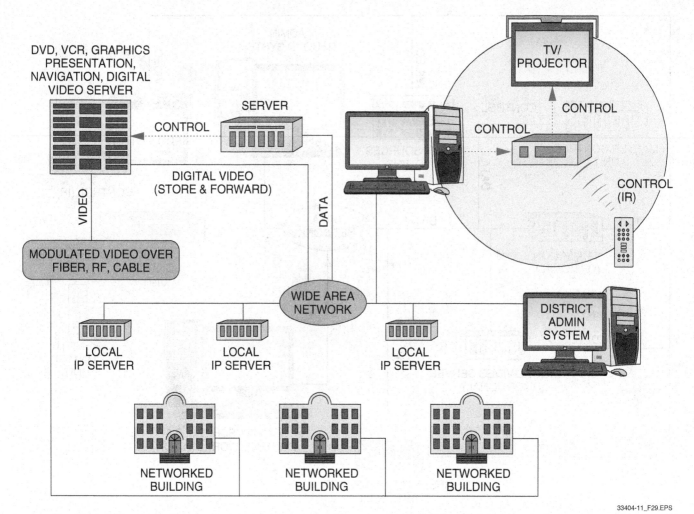

Figure 29 WAN-based MMS.

33404-11_F29.EPS

Figure 30 diagram labels:

VCR → AUDIO / VIDEO → MODULATOR

DVD → AUDIO / VIDEO → MODULATOR

ANTENNA → DEMODULATOR → AUDIO / VIDEO → MODULATOR

CATV → MODULATOR → AUDIO / VIDEO → PROCESSOR

DIGITAL VIDEO SERVER → SCAN CONVERTER → MODULATOR

COMBINER → RF VIDEO NETWORK

33404-11_F30.EPS

Figure 30 Video transmission equipment in the headend room.

6.4.5 Scan Converters

When playing a digital video file, the output signal is suitable for display on a computer monitor. However, this signal is not suitable for broadcast over the local video network. The signal must be converted before it can be passed to a modulator for assignment to a specific channel. A scan converter changes a signal that is conditioned for a computer monitor to a baseband audio/video output. The output signal is at a different resolution and a different refresh rate than the input signal. A scan converter is required for each simultaneous video playback from the digital video server.

7.0.0 MMS SOFTWARE

An MMS needs to be told what to do. This is accomplished using the software provided by the vendor. The software controls all aspects of system operation. It is responsible for scheduling content players; it provides an interface to the users so that they can choose from available resources; and it enables communication between local control units and the associated content player.

7.1.0 User Interface

The MMS software should allow the user to search for content and to specify the delivery method. The software also allows the user to specify the room where the content is viewed.

The media resources are accessed using a variety of criteria. At a minimum, the search system should allow for media selection based on title. A system may also offer search criteria such as author or subject.

Selecting the media resource is only half of what is needed when dealing with an MMS. After the resource is selected, the time and method of delivery should be specified. The software should handle the details of routing the specified content to the local television monitor, projector, or computer for viewing.

7.1.1 Web-Based Systems

Many schools and offices use a variety of computer hardware. Media management software may not be available for PC, Macintosh®, and UNIX® systems. However, each of these systems provides a web browser. Because of this, many media management systems offer a web-based user interface. This approach allows access to the system using any networked computer, and allows an organization to use its existing computer hardware.

7.2.0 Content Scheduling Issues

All media management systems involve the playback of media resources. Occasionally, there may be more requests for service than can be satisfied at a given time. When designing a system, tradeoffs are made between the number of VCRs, DVD players, and digital video servers, so there may be occasional conflicts. In order to minimize the conflicts, high-demand resources can be obtained in more than one medium. This approach ensures that there will be enough playback devices available to meet the demand.

Consider the following problem. A site has designed an MMS with three VCRs and three DVD players. About half of its resources are on videotape and the other half are on DVD. Up to six different resources could be played at the same time with this configuration. However, if four requests are scheduled at the same time, and the resource is only available on videotape, then the last request cannot be filled because there are not

---

On Site

Scan Converter

Extron VSC 700 and VSC 700D high resolution computer-to-video scan converters accept computer-video signals at resolutions up to 1920 × 1200 and simultaneously output a scan-converted video signal as NTSC or PAL composite video, S-video, and component video or RGB video. In addition, the VSC 700D includes a fourth, SDI serial digital video output. The VSC 700 and VSC 700D can be used in a variety of applications, including videoconferencing, video recording, and viewing of computer-video images on an NTSC or PAL monitor or other video display device.

33404-11_SA04.EPS

---

enough VCRs located at the headend to play the fourth videotape.

The videotapes and DVDs also need to be physically loaded into the players. For this reason, many resources are not available for immediate broadcast, although commonly requested titles might be digitized and stored on a digital video server.

An MMS requires an operator. Using the scheduling software, the operator would periodically check to see what content has been requested. In many cases, the MMS informs the operator that a specific title needs to be loaded into a specific player, in order to be available at a designated time. The operator then needs to physically locate the media resource and load it into the proper player. Some MMS software can be configured so that a report is printed out when media needs to be changed. For this reason, consider locating a printer in the server room or near the operator's work area.

SUMMARY

Media management systems contain a variety of components that must all work together. At the core of an MMS, computers provide the digital library services and headend control services. These computers and their software are the components provided when purchasing a media management system. However, a media management system is more than just the computers and software purchased from a provider. To be functional, additional equipment is required. Monitor equipment is needed to view the media resources. Source equipment is used to store the resources and convert them into video. A data network is required for control purposes, and a video network is needed for broadcasting.

There are several uses for an MMS. They are particularly well-suited for schools. They help to solve media storage and distribution problems. An MMS also eliminates the need to maintain multiple copies of the same resource. It makes tracking media resources much easier because the physical media stays in the media library. Therefore, there is much more flexibility in providing media content throughout a site.

The savings associated with an MMS are improved when an entire school district uses a centralized system. The media center may be located at any one of the district buildings. Multiple elementary schools would no longer need to maintain their own copy of the standard resources; all copies would be stored at a central location. Ultimately, the cost savings associated with installing a media management system can be significant. Equipment damage is reduced, since the need for cart-mounted media equipment is eliminated. The increasing problem of managing a resource library is also simplified.

1. In a school media management system, the headend equipment is located in the media library, but playback equipment, such as VCRs and DVD players, are located in each classroom.

   a. True
   b. False

2. Which of the following is a function of the software in a media management system?

   a. Loading DVDs
   b. Handling content located on the Internet
   c. Scheduling a room for playback
   d. Selecting the media to be played back

3. A digital library uses a database to store information about all available resources.

   a. True
   b. False

4. The database for a digital library might use _____.

   a. MPEG
   b. SQL
   c. KVM
   d. AVI

5. What type of system would be needed in order to broadcast a video announcement for emergency procedures?

   a. Content-on-demand
   b. Local control unit
   c. Hot-swappable
   d. Video conferencing

6. The device that allows the central MMS computers to communicate with the display equipment is the _____.

   a. video server
   b. local control unit
   c. video projector
   d. scan converter

7. A flat screen television has a larger viewing angle than a curved television.

   a. True
   b. False

8. According to the inverse square law _____.

   a. sound volume decreases with the square of the distance from the speaker
   b. speakers should be installed in the corner of a room
   c. sound volume is reduced by half for every 6 feet the listener is from the speaker
   d. sound volume is inversely proportional to the height of the ceiling

9. In an MMS, the equipment designated as source equipment is used to _____.

   a. display media
   b. convert media to a usable form
   c. transmit data between storage and display devices
   d. retrieve streaming video from the Internet

10. Which of the following is used to broadcast files such as Microsoft® PowerPoint® or Apple® QuickTime® onto the video network?

    a. VCR
    b. Multimedia projector
    c. Video server
    d. Presentation player

11. In streaming video, the media is downloaded in segments that _____.

    a. are compressed for data storage
    b. are stored for later retrieval
    c. are displayed in real time
    d. play while other segments are downloaded

12. Which type of equipment can be controlled by the headend controller unit using an infrared transmitter?

    a. Content players
    b. Digital file servers
    c. Video cameras
    d. LCU

13. The industry standard for distributing broadband video is _____.

    a. 60Hz coaxial cable
    b. 60-ohm coaxial cable
    c. 75Hz coaxial cable
    d. 75-ohm coaxial cable

14. Baseband is a transmission method in which digital signals are sent _____.
    a. across a coaxial cable
    b. across a fiber-optic cable
    c. without frequency shifting
    d. with frequency modulation

15. A scan converter is used to _____.
    a. convert a signal from baseband to broadband
    b. shift a video signal from one channel to another channel
    c. change the resolution of a video signal
    d. convert analog signals to digital signals

Trade Terms Introduced in This Module

Flicker: A distortion on a video display due to interlacing at slower refresh rates.

Hot-swappable: A hard disk drive system in which additional drives can be added by simply sliding a new drive into the rack. The new drives are automatically added to the system for storage.

Interlacing: A process in which every other line on a monitor is refreshed on every pass.

KVM: Keyboard, video, mouse. A KVM switch allows high-performance workstations to share a common monitor, keyboard, and mouse.

Local control unit (LCU): A piece of hardware that allows the central MMS computers to communicate with remote display equipment.

Media management system (MMS): A centralized system to store, retrieve, and distribute all types of electronic media.

Relational database: A software application designed to manage large amounts of information stored in separate tables.

WORM: Write once, read many times. This method of data storage allows for permanent storage without data deterioration.

Figure Credits

Peerless Industries, Inc., Module opener and Figure 16

TOA Electronics, Inc., Figure 11

Topaz Publications, Inc., Figures 12, 17, 27, and SA02

InFocus Corporation, Figures 14 and 15

Epson America, Inc., SA01

LaCie USA, Figure 21

Kintronics Blu-ray Library, Figure 22

ARRIS CHP Max5000® Converged Headend Platform, SA03

Photo Courtesy of Extron Electronics, SA04

NCCER CURRICULA — USER UPDATE

NCCER makes every effort to keep its textbooks up-to-date and free of technical errors. We appreciate your help in this process. If you find an error, a typographical mistake, or an inaccuracy in NCCER's curricula, please fill out this form (or a photocopy), or complete the online form at **www.nccer.org/olf**. Be sure to include the exact module ID number, page number, a detailed description, and your recommended correction. Your input will be brought to the attention of the Authoring Team. Thank you for your assistance.

Instructors – If you have an idea for improving this textbook, or have found that additional materials were necessary to teach this module effectively, please let us know so that we may present your suggestions to the Authoring Team.

NCCER Product Development and Revision
13614 Progress Blvd., Alachua, FL 32615

Email: curriculum@nccer.org
Online: www.nccer.org/olf

❏ Trainee Guide   ❏ AIG   ❏ Exam   ❏ PowerPoints   Other _____

Craft / Level: _____   Copyright Date: _____

Module ID Number / Title: _____

Section Number(s): _____

Description: _____

_____

_____

_____

_____

Recommended Correction: _____

_____

_____

_____

_____

Your Name: _____

Address: _____

Email: _____   Phone: _____

33405-12

Telecommunications Systems

Trainees with successful module completions may be eligible for credentialing through NCCER's National Registry. To learn more, go to **www.nccer.org** or contact us at **1.888.622.3720.** Our website has information on the latest product releases and training, as well as online versions of our *Cornerstone* newsletter and Pearson's product catalog.

Your feedback is welcome. You may email your comments to **curriculum@nccer.org,** send general comments and inquiries to **info@nccer.org**, or fill in and the User Update form at the back of this module.

V.1 3/12

Objectives

When you have completed this module, you will be able to do the following:

1. Explain common trade terms relating to telephone systems.
2. Briefly describe the history of telephones and the operation of the plain old telephone service (POTS).
3. Describe the operation of analog telephones.
4. Identify the main types of business telephone systems and describe their differences.
5. Identify the components used in key systems and traditional private branch exchange (PBX) systems.
6. Describe the differences between analog and digital telephone systems.
7. Describe the commonly used optional features for key systems and traditional PBX systems.
8. Describe emerging technologies.

Performance Task

Under the supervision of your instructor, you should be able to do the following:

1. Install a fully operational phone system.

Trade Terms

Alternate mark inversion (AMI)
Basic rate interface (BRI)
Bit rate
Channel bank
Digital subscriber line (DSL)
Foreign exchange service (FX)
Framing bit
Glare
Integrated services digital network (ISDN)
Key telephone system (KTS)
Local subscriber loop
Multiplexing
Off-premise extension (OPX)
Pair gain system
Plain old telephone service (POTS)
Primary rate interface (PRI)

Private branch exchange (PBX)
Private line automatic ringdown (PLAR)
Public switched telephone network (PSTN)
Pulse code modulation (PCM)
Pulse dialing
Ring
Ring generator
Telco
Tie line
Time division multiplexing (TDM)
Tip
Toll center
Touch-Tone®
Trunk
Voice over Internet Protocol (VoIP)

Industry Recognized Credentials

If you're training through an NCCER-accredited sponsor you may be eligible for credentials from NCCER's Registry. The ID number for this module is 33405-12 Note that this module may have been used in other NCCER curricula and may apply to other level completions. Contact NCCER's Registry at 888.622.3720 or go to nccer.org for more information.

Contents

Topics to be presented in this module include:

Figures and Tables

1.0.0 TELEPHONE HISTORY

Since its invention by Alexander Graham Bell in 1876, the telephone and related switching systems have evolved greatly. The very first telephones were hardwired together. Later, operators connected telephones, such as wall-mounted hand-cranked units (*Figure 1*) through manual telephone switchboards called cord boards. These boards used numerous patch cords and telephone jacks to route calls (*Figure 2*). Later, desktop telephones outpaced wall-mounted units and electromechanical relays replaced manual switchboards. Today's modern electronic telephone switching systems not only connect telephone lines, they can also provide data such as numbers dialed and the durations of conversations. They also provide services such as automatic answering and call forwarding. Modern systems pass data through the public switched telephone network (PSTN). This network is made up of telephone companies connected together to form an international telephone service and switching system.

Telephones have gone through numerous upgrades and changes over the years, shrinking in size over time. The cell phone of today is tiny compared with early telephones. The same is true of telephone switching systems (*Figure 3*).

2.0.0 PLAIN OLD TELEPHONE SERVICE (POTS)

The plain old telephone service (POTS) that exists today is a part of an all-analog copper-wire landline system that was formed over 100 years ago. It is the standard wire line service with dial tone and standard features package. The features

Figure 2 Manual switchboard.

Figure 3 Electronic switching system.

Figure 1 Early telephone.

Did You Know?

Western Union's Mistake

Shortly after the telephone was patented, Bell tried to sell the patent to Western Union and Western Electric. Western Union rejected the offer because they thought the telephone was impractical and that the telegraph was a better technology. Western Union was wrong. Bell Telephone later acquired Western Electric and a controlling interest in Western Union.

Hand-Cranked Magneto Telephone Operation

The hand-cranked magneto wall-mounted telephone shown is a Western Electric Type 317R set that was introduced in 1907. It was used primarily for rural party lines with eight or more parties. The hand-cranked magneto was used to generate an AC ringing voltage over an open two-wire telephone circuit to other parties on the circuit and a switchboard operator. The high-resistance ringer coils were normally connected (bridged) across the two-wire circuit on each set and rang whenever a magneto was cranked by any party on the circuit. The set could also be rewired internally to permit divided ringing when many parties were on the circuit. This divided the ringer load for the magnetos because each ringer used only half of the AC ringing voltage and current that was available across the lines.

When the receiver was lifted off the hook switch, three 1.5-volt carbon dry cell batteries connected in series and located in the battery box were connected to the two-wire circuit, along with anti-sidetone induction coils, to power the receiver and transmitter for talk. The anti-sidetone induction coils suppressed inherent noise on the two-wire circuit.

In use, the receiver was first lifted off the hook switch to make sure that no conversation was occurring on the two-wire circuit. Then it was replaced on the hook switch and the magneto was cranked to produce an AC ring voltage on the circuit in a prescribed sequence (one long, one long and one short, one long and two short, etc.) to signal the operator or another party on the circuit. Then the receiver was lifted from the hook switch to place DC battery voltage on the circuit for conversation using the transmitter and receiver. Hopefully, the signaled party would answer the call; otherwise the procedure would have to be repeated.

Note that this particular set was modified by a Western Electric employee with the addition of a standard dial telephone set in the battery box area so that the receiver and transmitter could be used to receive a call (and if necessary dial a call) over the current system.

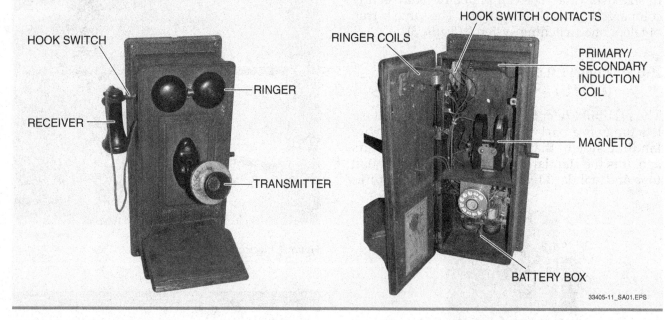

33405-11_SA01.EPS

depend on the utility company that is providing the service.

2.1.0 Local Subscriber Loop

The local subscriber loop is the individual hardwired telephone line that connects that subscriber to the telephone company (Telco) central office switch or switching system. In early systems, these wires were carried on cross arms attached to telephone poles. This system, called open wiring, cluttered the landscape in large cities. The development and use of multi-pair cables greatly reduced that clutter. A basic subscriber loop is a complete circuit consisting of a pair of wires that make a complete loop from the central office through the premise telephone, and back to the central office. When the phone is on-hook, the loop and circuit are open. When the phone is off-hook, the loop is closed and the circuit is complete. The central office monitors each subscriber line, and does all the switching to connect the individual subscriber loops.

In rural areas, where the one-way subscriber loop may be several miles long, the inter-electrode capacitance of the telephone cable pairs can degrade the voice signal. To counteract this capacitive reactance, load coils are used to balance out and null the capacitive reactance. Load coils are also installed on inter-switch tie lines and trunks that may be as much as 50 miles or more apart. Early tests with load coils were performed on some long-distance lines between New York City and Chicago. Load coils were placed at 2.5-mile intervals, and all amplifiers were removed from the lines. The goal was to see how much difference there was between the amplified lines and the coil loaded lines. There were some minor complaints about the volume of the load coil lines, but no noise complaints. The amplified lines still had noise complaints and some volume complaints. Over all, the load coil circuits performed with only a very slight reduction in customer satisfaction. Engineers were surprised that the load coil circuits, with no amplification, were still intelligible and viable.

2.2.0 Local Exchange Switch

There may be only one local exchange switch within a city, or there may be a number of local exchange switches that are connected with trunks (*Figure 4*). A trunk is a communication path between central offices and local centers. Any time a long distance number is dialed, the local exchange switch connects that call through a toll center to make the long distance connection. The lines between the local exchange and the toll center are called trunks. There is a fine distinction between tie lines and trunks. Tie lines indicate there are no time limits and charges involved.

A trunk indicates a long distance connection where the calling party is charged for that connection. Tie lines and trunks serve the same function of connecting two switches, but one is for toll charges and the other is not. Tie lines and trunks are covered in more detail later in the module.

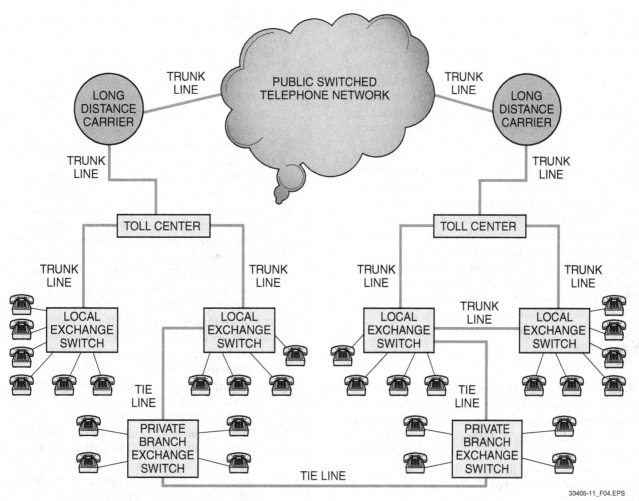

Figure 4 Public switched telephone network (PSTN).

2.3.0 Central Office Services

This section gives a broad overview of the central office services commonly used with telephone systems in the United States and Canada. Books referenced in the *Additional Resources* section of this module can provide more detailed information on the subject.

2.3.1 Analog Loop-Start Signaling

Figure 5 shows an analog loop-start circuit between a central office (CO) or private branch exchange (PBX) and an analog telephone in the on-hook (idle) condition. In this condition, the tip line is grounded at the CO and a nominal –48VDC steady-state voltage from the CO is present on the ring line. When the telephone handset is lifted to make a call, the hook switch closes and the –48VDC voltage is connected back through the tip wire to ground at the CO or PBX, causing current flow. This action seizes the line and the current flow is detected at the CO or PBX, which opens the ground path and applies a nominal –20VDC voltage followed by connection of a digit receiver and a dial tone generator to the tip wire. The nominal –20VDC on the tip wire reduces the nominal –48VDC on the ring wire to about –28VDC. This results in a 6 to 8VDC voltage difference between the two wires at the telephone and a lower current flow. This is the voltage used for normal off-hook telephone operation.

> **NOTE**
>
> In the on-hook condition, an analog telephone or device must have a DC resistance greater than five megohms for an applied voltage of 100VDC, and greater than 30 megohms for an applied voltage of 200VDC. The telephone companies use automated insulation tests as part of their routine line tests. If the line resistance is too low, the line is interpreted as being damaged and repair procedures are initiated. The prescribed DC resistance for telephones and devices prevents false service actions.

With a required internal DC resistance of less than 400 ohms (nominally 200 to 300 ohms) across the telephone in an off-hook condition, current flow during normal telephone operation can range from 20 to 50 milliamps (nominally 23 to 26 milliamps). In addition, in the off-hook condition, the AC impedance of the telephone or device must nominally match the AC line impedance of 600 ohms.

After a dial tone is present, the telephone number to be called can be dialed (a minimum of seven digits). At the start of the first digit dialed, the dial tone is removed. At end of the last digit dialed, the digit receiver is disconnected and the voice (audio) circuit is established at the caller end as the call is switched through the CO or PBX (and PSTN to another CO or PBX if necessary) to the lines of the called analog telephone. At the time of call, the destination telephone connected to a CO or PBX is in the same idle condition as shown in *Figure 5*. When the CO or PBX switch connects to the called telephone, a ring generator is connected to the ring line and a train of nominal +90VDC ringing voltage pulses is periodically applied. The periodic simulated AC-ringing signal generated is typically 20Hz in the United States. The resulting AC current passes through the ringer coil or transformer and back through the tip wire to ground, causing a periodic ringing sound to occur at the called telephone.

> **NOTE**
>
> The on-hook AC impedance of an analog telephone/device ringer coil or transformer ranges between 1,000 and 1,600 ohms depending on the frequency used for ringing. This is done to limit the DC ringing current to 3 milliamps or less. If more than one analog telephone or device is connected to the line, excessive current may still be drawn. To prevent this, the Federal Communications Commission (FCC) requires that a ringer equivalence number (REN) be labeled on any analog device that is connected to a POTS line. The amount of current drawn by the old Western Electric 500-type telephone with a ringer coil is used as a standard and is assigned an REN value of 1. A device that draws twice as much current is labeled with an REN of 2. A device that draws half as much current is labeled with an REN of 0.5 and so on. Ringer equivalences of parallel devices add, and the total REN allowed for most POTS lines is 5.0. The REN is followed by a letter that corresponds to the type of ringer signal frequency applied during testing of the analog device. Under FCC rules, either of only two ringer signals (A or B) can be used. A frequency of 20 or 30Hz is assigned the letter A and frequencies of 15 to 68Hz are assigned the letter B. The minimum AC impedance for ringer signal A is 1,000 ohms at 30Hz and 1,400 ohms at 20Hz. The minimum AC impedance for ringer signal B is 1,600 ohms at 15 to 68Hz.

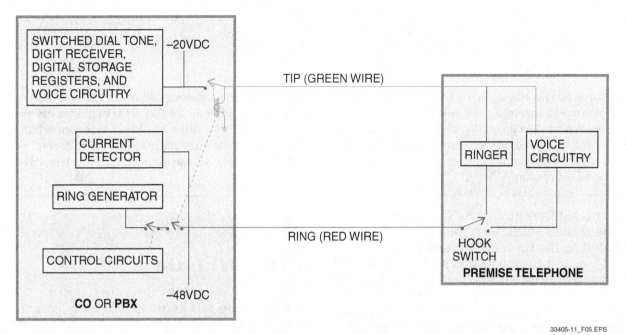

Figure 5 Analog loop-start circuit.

33405-11_F05.EPS

While the called telephone is ringing, a ring-back tone is sent by the CO for that telephone back to the caller to indicate that the telephone is ringing. If the telephone is in use, the local exchange sends a busy signal back to the caller. If the handset of the called phone is lifted, the hook switch closes, seizing the line and causing a different value of DC current to flow that is detected by the CO or PBX. The CO or PBX disconnects the ringing generator and completes the voice circuit back to the caller.

> **NOTE**
>
> For the billing protection of a Telco, the FCC requires automatic equipment that answers a call to conform to certain requirements that include a built-in two-second delay before sending usable data over the line. This does not apply to signals sent for establishing synchronization or mode of operation. Another requirement is that within the first five seconds after a call is answered, the device must draw at least as much current as a 200-ohm resistor or exhibit a current overshoot of less than 25 percent. This prevents the device from seizing the line and taking advantage of the differing current levels used for seizure and billing to receive a free call.

2.3.2 Telephone Numbers

The telephone numbers for North America conform to the North American Numbering Plan (NANP) and consist of 10 digits. The first three digits are the area code and the next three digits are the CO code, sometimes just called the office code. The remaining four digits are the station code, sometimes called the subscriber code, for the subscriber's telephone. The first digit of the area and CO codes can be any number from 2 to 9 and the remaining two digits can be any number from 0 to 9. When a 1 is dialed before the area code, a long distance call is signaled.

International numbering is based on a standard that all countries follow. The telephone number in any country cannot exceed 15 digits. The first three digits represent the country code, although not all three must be used. The remaining 12 digits are for the country's specific number code, although not all must be used. For example, when calling a number in North America from another country, the country code 1 must be dialed to access the NANP; then, the remaining 10 digits are dialed. Some countries use all 12 digits. In addition, some countries may use a set of digits to indicate an outgoing international call. The digits 011 are used within the United States to place an outgoing call to another country.

2.3.3 Analog Ground-Start Signaling

Analog ground-start signaling was originally used for party lines. *Figure 6* shows the analog ground-start circuit between a central office and a party line of two subscribers. It is shown in an on-hook (idle) condition. Remember, both subscribers are using the same pair of wires. In this condition, –48VDC is applied to the ring lead of the subscriber wire-line pair, and the tip lead is

Tip and Ring

The terms *tip* and *ring* for POTS lines is a carryover from the days of older corded switchboards. The cord that an operator used to connect a call was equipped with a plug that had three different contact areas that were insulated from each other. The tip section and the ring section were for the pair of POTS wires. The sleeve was connected to an earth-grounded braided shield around the tip and ring wires for interference suppression. Some of the interference was caused by the application of magneto-generated AC ringer voltage used for magneto telephones and other old phones.

disconnected from ground for the first subscriber. For the second subscriber, the ring generator is relocated to the tip wire connection at the central office, and the tip lead is still disconnected from ground. The tip lead is monitored by the central office switch for a ground. While the telephone set is on-hook, the ring lead is grounded through the telephone ringer and the tip lead is open. When the phone goes off-hook, the ground is removed from the ringer and connected to the tip wire. That grounded tip wire at the central office switch closes a relay and causes the line to be seized. It also connects the digit receiver and dial tone to the tip lead. The remaining processes are nearly identical to the operation of loop-start signaling. For modern operations, ground-start signaling is used mostly for connections between the local exchange switch and a private branch exchange (PBX).

2.3.4 Analog Ground-Start Signaling (PBX)

The primary advantage of a ground-start circuit between a PBX and central office is the ground connection to the ring lead. When the ring lead is open, false grounds are sometimes caused by long subscriber loops to a PBX that may make the cen-

The Whistle

In the 1960s a breakfast cereal maker placed a whistle in the cereal box as a prize for children. The whistle had a frequency of 2,600Hz. After the whistle was introduced, people began to complain that their long distance phone calls were dropping for no apparent reason. An investigation revealed that the frequency of the whistle was the same as that of the wink tone used to end and disconnect phone calls.

tral office feel there is an off-hook condition when the PBX is still in the on-hook state. This condition is known as glare. The use of a grounded ring detector circuit and an off-hook ground on the tip wire greatly reduces or completely eliminates glare (*Figure 7*).

2.3.5 E&M Tie Line and Trunk Signaling

E&M signaling is a form of supervisory signaling that is only used between switches. The tie lines, or trunks, between switches are always connected and ready for traffic. E&M signaling tells

Telephone Number Names

From the 1930s until the 1950s, as dial telephones and automatic switching became widespread, telephone numbers began to increase in the number of digits that had to be dialed. Because prior telephone numbers usually consisted of only three or four digits, names were used as part of longer telephone numbers as an aid in remembering the number. For instance, early branch exchanges used names like CEdar 4321. Later, as more digits were added, numbers with the familiar seven digits were used, like GRanite 5-4321. In both cases, the first two letters represented the first two digits of the phone number.

   This name association was the original reason for the alphabetical letters adjacent to the numbers 2 through 9 on the rotary dial face and on later keypads. This method of listing all telephone numbers was used into the early 1960s. However, the practice was abandoned because the Telcos ran out of viable names due to the increased number of lines and the need to use all available number combinations for CO codes. In addition, long-distance direct-dialing using area codes became widespread. Today, the letters are sometimes used by businesses and organizations in advertisements to construct unique 1-800 telephone numbers to aid in remembering the business or organization.

Name Number Exchange

The Name Number Exchange (NNX) dates back to when each local exchange switch had a name in which the first two letters of the name was part of the number and one other digit was used to separate the local exchanges within an exchange group. For example, ALpine-4 NNX was dialed as 254-XXXX; BUtterfield 8 was dialed as 288-XXXX, etc. A small town may have only two local exchanges, such as ALpine 4 and ALpine 5. Larger cities may have several local exchange groups such as BUtterfield 1-9 (281-xxxx through 289-xxxx), FRemont 1-9 (371-xxxx through 379-xxxx), and other name-number exchanges.

33405-11_SA02.EPS

the opposite side of the tie line when it is on-hook or off-hook (idle or busy). The E-lead is always a receive function and the M-lead is always a transmit function. Here is an easy way to remember E&M signals: E = ear = sound receiver; M = mouth = sound transmitter.

The M-lead is usually switched between –48VDC and zero volts DC, but not ground. The E-lead is switched between ground and open. The M-lead at one switch becomes the E-lead at the opposite end; the E-lead at one switch started as the M-lead at the opposite end. When one switch connects to a tie line, it will connect the –48VDC to the M-lead, which is received as an E-lead at the other end.

The E&M leads are supervisory and do not carry any voice signals. There is another pair (or two pairs) for voice signals. The –48VDC M-lead causes line seizure, which causes the digit receiver to be connected to the tip lead of the voice circuit. As dialed digits come across the voice circuit, they are interpreted by the switch to make a connection. If the switch is connected to another local exchange switch, where time and charges

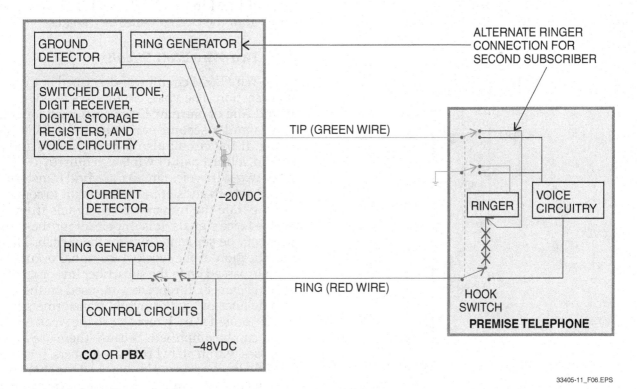

Figure 6 Analog ground-start circuit.

33405-11_F06.EPS

are not applicable, the area code is stripped off and only the last seven digits are used to make internal connections within that switch.

On Site

AC Ringer Voltage

Ringer voltage today is not a true AC voltage. It is a simulated AC voltage generated by a +90VDC pulse superimposed on the steady –48VDC of the ring wire at a 20Hz rate. As shown in the figure, an entire cycle of simulated AC voltage is 50 milliseconds long. It is made up of a 25-millisecond +90VDC pulse superimposed on the ring-line steady-state –48VDC voltage. This creates a +42VDC pulse followed by a 25-millisecond interval where the ring line returns to the –48VDC steady-state condition before the next +90VDC pulse occurs. The combination of both 25-millisecond periods results in current reversal every 25 milliseconds, which is the AC used to activate the ringer coil or transformer.

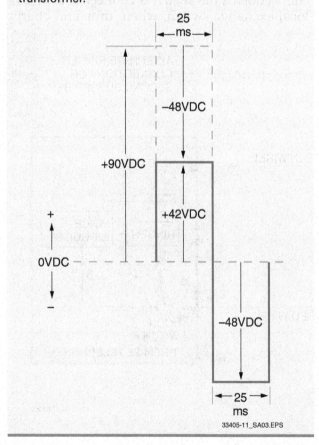

33405-11_SA03.EPS

The stripping of a dialed digit, or digits, is always a process performed at the transmit side of a switch. If the switched call connects to a toll center, it searches for a path to the distant local exchange switch, then routes the call and forwards all the dialed digits across the trunk or tie-down voice circuit.

Once the call is connected and the called party goes off-hook, the distant switch sees the line seizure and sends –48VDC over its M-lead back toward the caller. When both the called party and caller E-leads become a voltage, the charges timer starts and the automatic route selection tables apply the rate to the timer for charges.

For some applications, there is no –48VDC, or ground reference, available at the distant switch on which the E&M signals can operate. In those cases, M-battery and E-ground leads are added to the circuit. E&M requires one pair of wires; however the M-battery (signal battery) and E-ground (signal ground) use another pair of wires. M and M-battery are one pair, and E and E-ground use a second pair of wires. The M-battery and E-ground provide the supply/return path for the E&M signaling in locations where those references are needed. *Table 1* summarizes the off-hook and on-hook M and E lead conditions for both Type I and Type II interfaces.

A Type I interface only uses E&M signals. A Type II interface uses both E&M signals and M-battery and E-ground signals. There are five types of E&M interfaces. Types III, IV, and V are not used in the United States or Canada.

2.3.6 Two-Wire E&M Circuits

A two-wire E&M circuit was described in the previous section. The voice circuit uses two wires, a tip and ring of one pair. One pair is used for voice signals and a second pair is used for E&M signaling. If the circuit also uses M-battery and E-ground, a third pair of wires is required.

For two-wire voice circuits, a hybrid transformer is used at both ends of the voice circuit to combine and separate the transmit voice signals from the receive voice signals. It is important for these two signals to be separated within the switch. There, amplification of the weak voice signal occurs before it is passed on to the subscriber line or another switch (*Figure 8*). If repeaters are used on the two-wire tie lines or trunks, hybrid transformers must also be located at each repeater so the weak receive signal can be amplified. Because there are inherent losses within all hybrid transformers, two-wire E&M tie lines and trunks are used for short connections of only a mile or less. For longer tie lines, four-wire E&M lines are preferred.

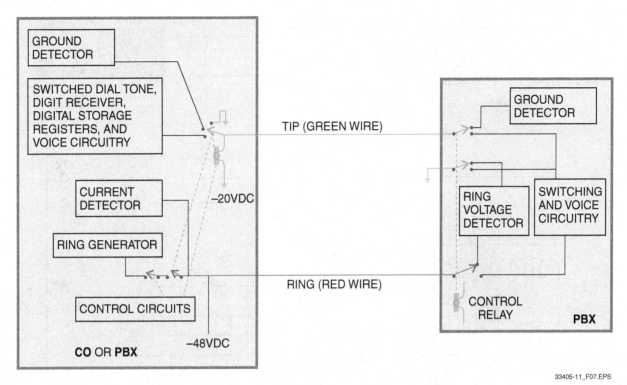

Figure 7 Analog ground-start signaling (PBX).

Table 1 Type I and Type II E&M Signaling Parameters

TYPE I		
Condition	Lead	Lead
On-hook	M = ground	E = open
On-hook	M = battery	E = ground
TYPE II		
Condition	Lead	Lead
On-hook	M = ground	E = open
On-hook	M = battery	E = ground

2.3.7 Four-Wire E&M Circuits

A four-wire E&M circuit uses two pairs of wires for the voice signals, resulting in four wires. The first voice pair (transmit) is listed as tip and ring. The second voice signal pair (receive) is listed as tip 1 and ring 1 (T&R and T1&R1). The E&M signaling is not changed, but the transmitted and received voice signals are separated. This separation of transmit and receive signals eliminates most of the noise and signal loss that occurs with two-wire hybrid circuit operations. Some tie lines and trunks can use as many as four pairs, or eight wires, for a single voice-grade analog tie line or trunk.

2.3.8 Foreign Exchange Service

A foreign exchange service (FX) is a feature used by businesses that have small numbers of employees based at remote locations or in foreign countries. It provides local exchange switch service from a central office at the company's primary office so that offices that are located away from the remote subscriber's actual location can use the primary office local exchange switch. The remote office can make local calls at the primary office location, reducing long distance charges. The system allows two separate remote offices to make calls between them without incurring long distance charges.

With foreign exchange service, a subscriber pays the Telco a fixed rate for the use of one or several private FX lines between the subscriber's local central office and the remotely located central office. Many times, the remote central office is located hundreds of miles away. Foreign exchange service lines are used by businesses to provide a local phone number from cities where the business has a remote office. When you call tech support and reach a technician at an overseas location, that technician may be using an FX service line. The advantage of foreign exchange service is that phone calls made between distant locations are billed as local calls rather than long distance calls.

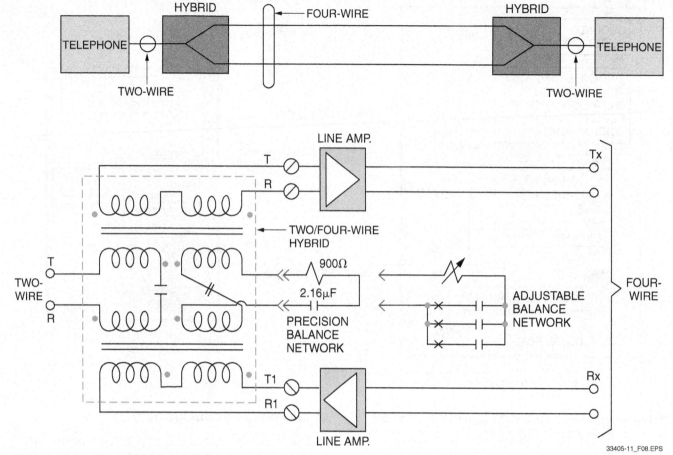

Figure 8 Two-wire to four-wire hybrid.

33405-11_F08.EPS

2.3.9 Off-Premise Station Lines (OPX)

An off-premise station, also called an off-premise extension (OPX), refers to a telephone that is installed at a location remote from the business premises where the main telephone system equipment is housed. A common application of this service is for employees who work at home. It is also used by answering services. The subscriber pays the local telephone company for a dedicated off-premise line off a PBX switch. It connects from the business location or the local central office, then continues through as many central offices as needed until it connects to the remote telephone. An off-premise telephone can be configured to have many or all of the features of the in-house phones. Telephones in off-premise locations are often analog phones. Analog phones are used because digital telephone signals used between a PBX and in-house telephones are often not compatible with the PSTN switches. However, newer PBX systems are better able to deal with compatibility issues.

2.3.10 Private Line Automatic Ringdown Circuits (PLAR)

A private line automatic ringdown (PLAR) circuit is a direct-connected line that will ring if the phone on the opposite end of the line is picked up. It is used for mostly for emergency circuits. If a guard at a main gate picks up the phone, the central security office telephone rings. Nothing has to be dialed and no special access code has to be entered; the guard just picks up the phone. If central security picks up the phone for the main gate, the main gate phone rings. The phones used in these systems often have no dial or keypad, just a cradle and a handset (*Figure 9*). They are frequently called hot lines.

2.3.11 Pulse Dialing

Old-fashioned rotary dial phones (*Figure 10*) use pulse dialing. When the rotary dial turns, the line is broken and reconnected to the −48VDC on the ring wire. Those pulses are interpreted at the

Figure 9 Private line automatic ringdown phone.

33405-11_F09.EPS

mechanical switch, causing stepper relays to step one level for each pulse received. If only one pulse is received, the stepper relay will step one level. If the number nine is dialed, the line is broken momentarily and reconnected nine times, causing the stepper relay to step nine levels. A zero is represented by 10 breaks. During the lull between dialing each number, a next level relay is connected and then steps with the dial pulses. Pulse dialing was the original dialing method. It was used because all switches were electromechanical relays. Modern electronic switching systems have to interpret the dial pulses. Pulse dialing is not the preferred dialing method when using an electronic switching system.

2.3.12 Touch-Tone® Dialing

The term Touch-Tone® was the name given by Bell Telephone for a new way to dial a telephone. To dial a number, a caller touches a numbered button or key and a tone is sent across the line to the

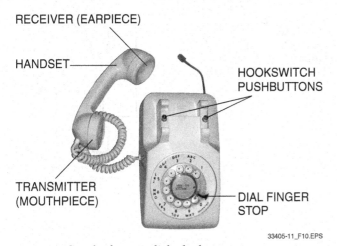

RECEIVER (EARPIECE)

HANDSET

HOOKSWITCH
PUSHBUTTONS

TRANSMITTER
(MOUTHPIECE)

DIAL FINGER
STOP

33405-11_F10.EPS

Figure 10 Standard rotary dial telephone.

digit receiver. In reality, each button has two tones. Each column of the buttons uses one tone and each row of buttons uses another. When a numbered button is pressed, two tones are sent at the same time (*Figure 11*). The dial pad looks like a calculator keypad, with three numbered rows going from one to nine. The bottom row contains the * (star) button, and the # (pound) button. The bottom row also contains the operator (0) button. If the caller presses the one (1) button, a combined tone made up of both 697Hz and 1,209Hz tones is sent to the central office. For the number nine (9), a combined 852Hz/1,497Hz tone is sent. Often the pound (#) button (941Hz/1,477Hz) is used as a hook-switch flash for things like call conferencing.

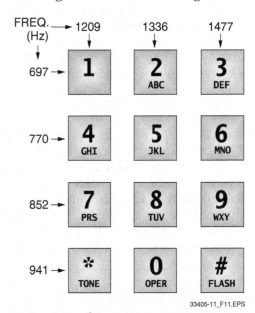

33405-11_F11.EPS

Figure 11 Touch-Tone® keypad layout.

Pulse dialing and Touch-Tone® dialing methods both send dialing information to the switch. However, the star and pound buttons used in Touch-Tone® systems allow more features within the central office switch or PBX. The method of interpreting the dialing information is what is different. Electromechanical switches are more compatible with dial pulses, while electronic switching systems are more compatible with Touch-Tone® dialing.

3.0.0 TELEPHONE SWITCHING SYSTEMS

There are several telephone system options available for businesses. For a large company with many employees, a privately owned switch may be appropriate. For a small company with only a few employees, the cost to own and maintain a telephone switch may not be worthwhile. In that case, a key system may make more sense.

3.1.0 PBX Systems

A PBX (PBX) or private area branch exchange (PABX) is a privately owned or leased telephone switch. They are common in large companies that have many employees who need to be able to make calls within the company without using lines that connect to an outside telephone company (*Figure 12*). At the same time, the company needs to have access to the PSTN to make calls

outside the company. Large companies may have a number of privately owned telephone switches in their facilities all over the country or even the world. These types of companies usually have high-capacity tie lines between their company switches. Each of the switches has access to one or more local exchange switches for local and long distance connections.

A PBX works like a central office switch, in that it provides a dial tone, busy signals, ringing voltages, and connections for calling parties. Many PBXs can also provide time and charges to each line so the cost of operating the system can be apportioned to each office or division. PBXs have subscriber lines, tie lines, and trunks. The subscriber lines are for individual telephones within each facility; the tie lines are for connections to other switches and facilities within the company. The trunks connect to the local exchange switches and the PSTN.

A PBX may connect to a local exchange switch using loop-start or ground-start analog lines, or it may connect using a high-capacity digital service. When connected to a local exchange switch using loop-start signaling, there can be a problem with an open ground connection for the –48VDC during an on-hook or idle condition. This situation tends to reflect the open back to the central office and cause the false off-hook condition called glare. With ground-start signaling, glare is greatly reduced or eliminated because there is al-

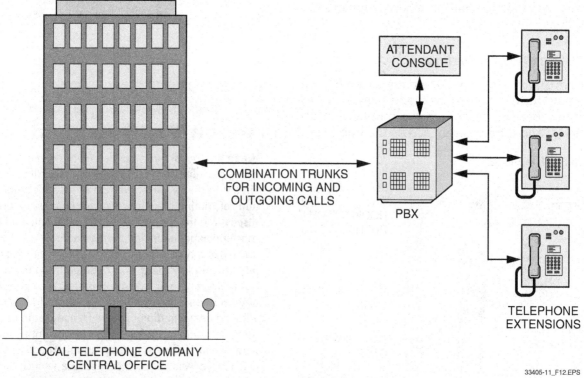

COMBINATION TRUNKS
FOR INCOMING AND
OUTGOING CALLS

ATTENDANT CONSOLE

PBX

TELEPHONE EXTENSIONS

LOCAL TELEPHONE COMPANY CENTRAL OFFICE

33405-11_F12.EPS

Figure 12 Traditional PBX telephone system.

ways a ground connection between the central office and the PBX. For most systems, ground-start analog central office lines are preferred.

A PBX is designed for internal communications. To allow calls outside the local PBX, a special dialing code is often required to reach the local exchange switch or other switches at distant locations. The number nine is widely used as the access code to the local exchange switch or PSTN, but that is not a requirement. Whatever access code is used, the PBX uses that code to route the call to a tie line, trunk, or trunk group. A trunk group is a group of trunk lines or tie lines that are all going to the same place, such as a local exchange switch or another company switch in another location. Once the PBX has directed the call to a trunk or trunk group, the access code is stripped off and only the required dialed digits are sent on to the connecting switch. Calls dialed within the PBX may require from three to seven digits.

Calls going through the local exchange switch and the PSTN may require as many as fifteen digits. The following are examples of calls placed through a PBX:

- A local internal company call may be four digits such as 8900. Three-digit dialing is very common, but cannot handle more than 1,000 internal subscribers.
- Access to the local exchange may require the access code of nine (9) and then the phone number (9-555-1212) for a local exchange switch.
- Access to long distance through the local exchange may require the access code of nine (9), a one (1) for long distance, an area code, and the called party's local exchange switch and subscriber line (9-1-800-555-1212).
- Access to other company phone systems, in a different city, state, or country may require an access code of 80 for one location, 81 for another location, 82 for another location, and so on (80-234-567-8900).

To reduce personnel costs, it is very common for a PBX to use an automated attendant instead of a human operator. If an automated attendant is used, a recorded message provides the incoming caller with a menu of dialing options, enabling that person to reach the desired party. The following are some of the options that a PBX may offer:

- Call waiting allows the caller to hear a false ring signal. This indicates an idle line, even though the line is busy. At the same time, an alert signal is sent to the called party, informing that person there is another call coming in.

- Caller ID lets the called party see the number of the incoming call so that he or she can decide how to respond to the call.
- Call forwarding allows incoming calls to be forwarded to another telephone. This feature prevents important calls from being missed.
- Voice messaging automatically answers incoming calls and gives a pre-recorded message to callers.
- Three-way or conference calling allows one person to call and have conversations with two or more people. An access code is often required to join the conversation.
- Redial allows the caller to automatically and continuously redial the last number called by pressing a redial button on the phone.
- Speed dialing allows a single number or short series of numbers to be dialed instead of a longer series of numbers. It is designed to save dialing time.

Because many PBX systems are equipped with voice mail and/or automated attendant features, they are set up differently. Many may have direct inward dial (DID) trunks (*Figure 13*) that allow each telephone extension to have a separate telephone number so that callers can reach the extension without going through the attendant console operator. If the extension is not answered or is busy, the call rolls over to voice mail. The caller can then leave a message that activates a message-waiting indicator on the extension telephone or may dial 0 to connect to the attendant console operator for immediate assistance. On smaller systems, the voice mail is usually a separate unit; however, on larger systems it may be internal to the PBX.

3.2.0 Key Telephone Systems

Small businesses with a few employees may need some of the services of a switch, but the purchase or leasing of a PBX may be too expensive. In those cases, a key telephone system (KTS) (*Figure 14*) may be an attractive option because of its lower initial cost. A key system has subscribers and trunks. The trunks are subscriber lines from a local exchange switch. Key system subscribers have full access to the local exchange switch by pressing the correct button on the telephone set.

If a person presses one of the local exchange switch line buttons and picks up the handset, he or she will hear a dial tone from that local exchange switch, not from the key system unit as a PBX would do. The only time a key system unit switches calls within the unit itself is when the in-

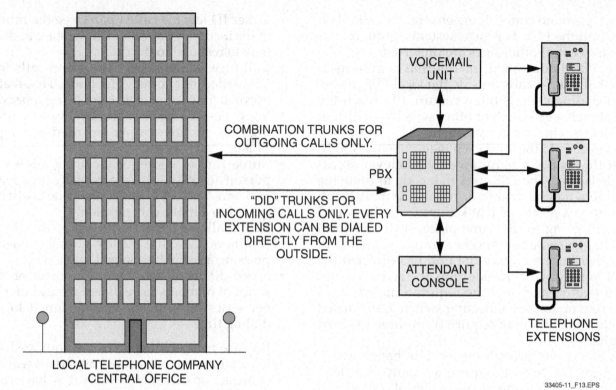

COMBINATION TRUNKS FOR OUTGOING CALLS ONLY.

VOICEMAIL UNIT

PBX

"DID" TRUNKS FOR INCOMING CALLS ONLY. EVERY EXTENSION CAN BE DIALED DIRECTLY FROM THE OUTSIDE.

ATTENDANT CONSOLE

TELEPHONE EXTENSIONS

LOCAL TELEPHONE COMPANY CENTRAL OFFICE

33405-11_F13.EPS

Figure 13 PBX system with DID trunks and voice mail.

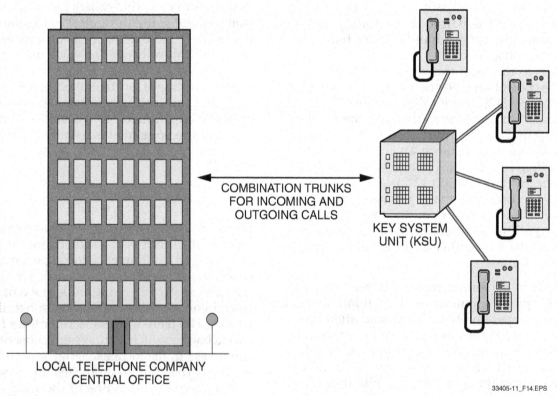

COMBINATION TRUNKS FOR INCOMING AND OUTGOING CALLS

KEY SYSTEM UNIT (KSU)

LOCAL TELEPHONE COMPANY CENTRAL OFFICE

33405-11_F14.EPS

Figure 14 Basic key system.

tercom button is pressed. When making intercom calls, they can only be connected within key system unit subscribers and cannot be connected to the local exchange switch or the PSTN. Intercom numbers are usually only one or two digits.

One advantage of the key system unit is that a person can make or receive a call, put that call on hold, call another employee on the intercom line, have all parties interconnected, or let another employee take charge of the call. This feature allows an assistant to make a call for a manager. The feature then lets the manager take charge of the call after the called party answers.

3.3.0 Electronic Key Service Units

An electronic key service unit (EKSU) is similar to a conventional electromechanical key service unit but with more features (*Figure 15*). Those features allow it to handle more lines both on the local exchange switch side and on the electronic key service unit subscriber line side. An electronic unit is more expensive to buy than a conventional unit.

Depending on the options and features available, modern electronic key service units can have as many features as a PBX, but on a smaller scale. It is often difficult to distinguish between the two. For some manufacturers, it may be possible to grow an electronic key service unit into a PBX by replacing components and software. Most modern electronic key service units have the ability to use digital voice or analog voice for their operations.

3.4.0 Hybrid Systems

Hybrid systems (*Figure 16*) have the capability to operate as proprietary or standard digital systems, or as analog systems. This capability may be on either its subscriber lines, or its tie lines and trunks. It is very common for hybrid systems to automatically sense whether the line is analog, proprietary digital, or standard digital transmission. A hybrid is a blend of PBX, electronic key service unit, and key system unit switching systems. Most hybrid systems use electronic switching. There are no electromechanical relays used for switching.

Like a PBX, a hybrid system does not normally need the larger 12- or 25-pair cable to the phone or subscriber lines. However, it is common for them to require at least two wire pairs for each phone, one for voice and one for control. Some may require as many as four wire pairs, one for voice signals, one for control signals, one for power, and one more for intercom.

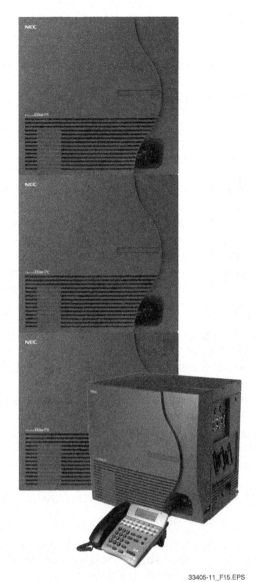

33405-11_F15.EPS

Figure 15 Electronic key system components.

33405-11_F16.EPS

Figure 16 Hybrid system control unit and telephone.

4.0.0 MULTIPLEXING

Multiplexing is the mixing of multiple signals on or in a single communications path during transmission. It is not uncommon for a number of communications wires or pairs to be placed into a single duct or conduit, or be contained in a single cable. Although each cable or cable pair may be carrying completely different information, they share a single path or route from point A to point B.

All radio stations are assigned a frequency. That frequency is used to carry voice and music signals. With amplitude modulated (AM) radio, the voice or music is used to change the height (amplitude) of the station's assigned frequency. That change in amplitude is a result of modulation. *Figure 17* shows the stages of amplitude modulation.

With frequency modulated (FM) radio, the voice or music is used to vary the frequency of the station's assigned frequency. That change in frequency is a result of modulation. *Figure 18* shows the stages of frequency modulation.

Modulation is not the same as multiplexing. For multiplexing, voice signals are mixed on a single carrier for transmission. The signals are divided into channels of different frequencies for multiplexing.

4.1.0 Analog Voice to Digital Voice Conversion

Most voice circuits are converted into digital signals by the time they reach, or are within, the local exchange switch. Exceptions include some electromechanical exchange switches, PBXs and

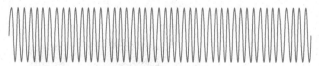

ORIGINAL FREQUENCY

AUDIO SIGNAL

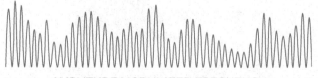

AMPLITUDE MODULATED FREQUENCY

33405-11_F17.EPS

Figure 17 Amplitude modulation.

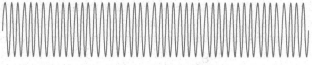

ORIGINAL FREQUENCY

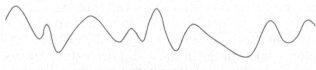

AUDIO SIGNAL

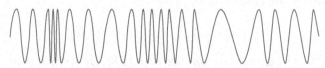

FREQUENCY MODULATED FREQUENCY

33405-11_F18.EPS

Figure 18 Frequency modulation.

key service units. From there, the digital voice signals, and digital data signals are intermixed and switched throughout the PSTN until they re-enter a subscriber loop or telephone.

There is a good reason why analog voice signals are converted to digital voice signals. The human voice emits sounds in the 100Hz to 4kHz frequency range. Sounds constantly change in both amplitude and frequency during conversation. A normal human ear can hear sounds in the 20Hz to 20kHz frequency range. Sounds can also be heard outside the frequency range of a voice. Analog voice circuits are designed to transmit and receive signals in the normal speech frequency range. All equipment in telephone systems must be capable of transmitting signals in the voice frequency range.

For voice-grade signals to be transmitted over long distances by way of twisted pair wiring, amplification of these signals must be done at cable length intervals to keep the signal audible. A big problem is that electrical noise is picked up and transmitted with the voice signals. As the voice is amplified, so is the noise. The farther the voice signals travel over the lines, the more noise gets picked up and amplified. At some point, the noise may be as loud, or louder, than the voice signal. This is known as signal-to-noise ratio. Digital signals are not as prone to noise as analog signals, and are generally much clearer for that reason. That is why most analog voice signals are converted into digital signals. The conversion is done at some point in the path where it is easiest to do. That point is usually the local exchange switch or PBX.

4.2.0 Pulse Code Modulation (PCM)

The most common form of analog to digital conversion is pulse code modulation (PCM). Amplitude and frequency modulation of a signal were described earlier. Most interference on audio signals affects the amplitude of that signal. For that reason, AM signals are more prone to noise interference than FM signals.

For PCM, the voice signals are sampled 8,000 times per second. When analog voice signals are sampled, an 8-bit digital signal is produced. It represents the voltage level of the voice signal at the point in time when it was sampled. At the receiving end, those 8-bit voltage levels are recreated to reproduce the original voice signal. Because the digital voice signal is 8 bits, and those 8 bits are produced 8,000 times per second, the digital voice signal is 8 times 8,000 or 64,000 bits per second (64kbps).

A problem with a digital voice signal is how to tell if the circuit is on-hook (idle) or off-hook (active). A process of bit robbing is done. Every sixth time an 8-bit byte is produced, the least significant bit, the one that is $\frac{1}{256}$ of maximum, is changed to a zero for on-hook indications or to a one for off-hook indications. This effectively takes the place of an M-lead. At the receiving side, it becomes an E-lead. Bit robbing introduces a new problem. If data is being transmitted, the loss or changing of this least significant bit corrupts the data in every sixth data byte, or every sixth character. That makes the data link unusable. For that reason, phone companies would not offer that least significant eighth bit for data transmissions. That leaves only 7 bits times 8,000 sampling times per second, which equals 56,000 bits per second (56kbps). For many years, 56kbps (or a multiple of that number) was the highest speed a Telco would offer for data transmissions.

In digital-transmission schemes, a 64kbps digital voice signal is referred to as a digital signal level zero, or DS-0. It is the zero level because it is the basis for all other digital voice transmissions. It is also the starting point for digital multiplexing. Each DS-0 has 256 different combinations of ones and zeros.

4.3.0 Digital Transmission Level 1 Signals (T-1)

A multiplexed digital transmission level one, called a T-1, starts with 24 PCM signals. It then multiplexes these 24 digital signal level zero (DS-0) signals into a single digital bit stream signal (*Figure 19*). A T-1 is also referred to as digital signal level one (DS-1), as it is the first multiplexed level of a digital voice signal. A DS-0 has not been multiplexed; it is only the analog to digital conversion. At the creation of the 64kbps PCM signal, 24 separate PCM signals are combined to form a single bit stream of 24 eight-bit channels. The bit stream is 1,536kbps (one 64kbps signal × 24 channels = 1,536kbps).

The problem is finding the starting and stopping point for each channel. Once Channel One is located, it is just a matter of counting 8-bits, 24 times, to define each individual channel. Here is how to find Channel One. At sampling time, one more bit, called a framing bit, is added to the T-1 bit stream (*Figure 20*). This identifies Channel One. Remember, this framing bit is added at the sampling time, which happens 8,000 times per second. In effect, 8kbps is added to the 1,536kbps signal, which equals 1,544kbps (1.544Mbps). This value is the bit rate of a T-1 or DS-1, not the frequency. There is a difference between frequency and bit rate. Frequency refers to how many cycles per second there are in an analog signal. Bit rate refers to the number of ones and zeros there are in one second. With any frequency, polarity changes twice during each cycle. A bit signal of all ones or all zeros does not change polarity during a one-second period. A digital signal that is a repeating ten-ten pattern changes direction 1.544 million times per second or 772Hz.

Each T-1 frame is located by a framing bit (*Figure 21*). The individual DS-0s are located by counting bytes. Use an 8-bit counter, 24 times, to identify each DS-0 after the framing bit. Each DS-0 has its own time space within the T-1. For that reason, this type of multiplexing scheme is called time division multiplexing (TDM). Time slot one is Channel One; time slot two is Channel Two and so on. Time slot 24 is Channel 24 and the end of the frame. The next bit is the framing bit that starts a new 24-channel T-1 TDM frame.

The 24-channel T-1 frame starts when the framing bit starts; it ends when the last and least significant bit of Channel 24 is received. There are 8,000 frames per second, just like the sampling pulses that developed the 8-bit PCM channel. Twenty-four 8-bit channels (8-bit bytes) plus one framing bit per frame equals 193 bits. Multiply 193 bits per frame by 8,000 frames per second to obtain 1.544Mbps. This value (1.544Mbps) may be called baud rate, speed, or bit rate.

> **NOTE**
>
> Never confuse baud rate with a frequency.

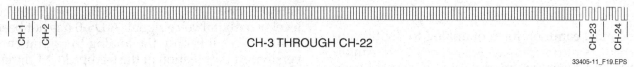

CH-1　CH-2　　　　CH-3 THROUGH CH-22　　　　CH-23　CH-24

33405-11_F19.EPS

Figure 19 DS-0 digital bit stream.

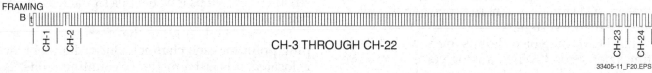

FRAMING
B t

CH-1　CH-2　　　　CH-3 THROUGH CH-22　　　　CH-23　CH-24

33405-11_F20.EPS

Figure 20 Framing bit added to T-1 frame.

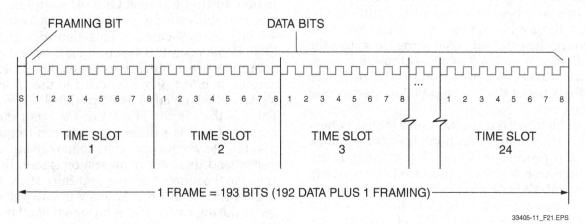

FRAMING BIT　　　　　　　　　DATA BITS

S 1 2 3 4 5 6 7 8 | 1 2 3 4 5 6 7 8 | 1 2 3 4 5 6 7 8 | ... | 1 2 3 4 5 6 7 8

TIME SLOT 1　TIME SLOT 2　TIME SLOT 3　TIME SLOT 24

◄────── 1 FRAME = 193 BITS (192 DATA PLUS 1 FRAMING) ──────►

33405-11_F21.EPS

Figure 21 T-1 frame organization.

Transmitted T-1 signals need to have embedded timing signals, but there is no space left in the bit stream to add the timing signals except for the framing bit. To aid in timing, a unique transmission scheme has been developed called alternate mark inversion (AMI).

A mark equates to a one in digital data. A mark is a one or voltage, and a zero is a space or no voltage. For a T-1 and alternate mark inversion, a mark (1) may be a positive or negative voltage. If the first one or mark is a positive voltage, the next mark or voltage is a negative voltage. Each time a mark is transmitted, it has the opposite polarity of the previous mark pulse. That is why it is called alternate mark inversion.

In *Figure 22*, the top signal represents two DS-0s in a T-1 and the lower signal represents the same

T-1 bits as an alternate mark inversion signal. Notice that only the marks, or ones, are pulses. All the zeros represent a no-voltage condition. The signals have marks as both negative and positive pulses because ones alternate polarity. Note that the marks are of shorter duration and appear as pulses rather than the normal even-width bits of most digital data signals. It is these pulses that do most of the timing reconstruction within the receiver of a T-1 signal. Many were installed in controlled environment vaults in residental and industrial locations to free up valuable cable pairs for other use. From the vault, individual subscriber loops were connected to individual service locations. This is still in use today for older electromechanical switches.

Another aspect of a T-1 shown in *Figure 22* is bipolar eight zero substitution (B8ZS). When multiple zeros are transmitted, the timing circuits tend to drift off frequency. To prevent this when the T-1 transmitter notes eight consecutive zeros, the least significant bit (bit 8) is replaced with two pulses of the same voltage polarity to keep the timing recovery circuits of the receiver in time with the transmitter that is sending the T-1 signal. When two sequential T-1 pulses have the same polarity, it is called a bipolar violation (BPV).

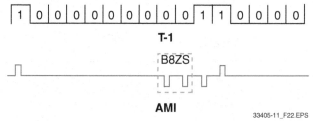

Figure 22 Alternate mark inversion.

33405-11_F22.EPS

4.4.0 Digital Transmission Level 3 Signals (T-3)

Digital transmission level three (T-3) is the next level of multiplexing above a T-1 or primary rate interface (PRI). This level is also called digital signal level three or DS-3. The bit rate of a T-3 is about 45Mbps. It contains 28 digital signal level ones (DS-1). Each DS-1 is 1.544Mbps, but is allotted 1.6Mbps of time in the frame (*Table 2*).

5.0.0 OTHER TELECOMMUNICATIONS TECHNOLOGIES

Various technologies and PBX solutions have emerged to handle combined analog voice, digital voice, digital data, and digital video transmissions. In reality, digital voice, digital data, and digital video transmissions that are routed through the PSTN are all handled as digital signals. Standard digital voice is always handled as a 64kbps signal. Digital data and digital video may have a number of channels grouped together to form a single path in multiples of 56kbps or 64kbps.

5.1.0 Digital Services

Telephone companies have dial-up digital services available. Subscribers use a telephone number to dial into their digital service access. A dial-up con-

---

On Site

T-1 History

When the T-1 was developed, it was aimed at high-density cities such as New York where there were numerous local exchange switches. At that time, the telephone company noted that local exchange switches in the city were never more than a mile apart. The Empire State Building needed at least one local exchange switch just to service that building. The T-1 was designed for transmission over twisted pair for up to 6,000 feet to meet that city's requirements. The T-1 would replace 24 two-wire or four-wire E&M circuits. That freed up as many as 96 cable pairs by using a T-1 that only used two pairs of wires. Central office technicians quickly called the T-1 a pair gain circuit because they could gain 92 cable pairs for each T-1 they installed. Engineers started referring to the T-1 as a high-capacity circuit because of the large number of voice lines two pairs of wires could handle.

33405-11_SA04.EPS

Table 2 Relationship of DS Levels to T-Trunk Levels

Signal Level	Trunk Level	Number of Voice Channels	Level Speed	Equivalent Number of T-1 Trunks
DS-0	None	1	64kbps	1/24
DS-1	T-1	24	1.544Mbps	1
DS-1C	T-1C	48	3.152Mbps	2
DS-2	T-2	96	6.312Mbps	4
DS-3	T-3	672	44.736Mbps	28
DS-4	T-4	4,032	274.760Mbps	168

nection is nothing more than a voice circuit telephone call. Once the dial-up connection is made, modems connected to each end of the line communicate with each other using audible-tone frequencies. The audible-tone frequencies are nothing more than modulated digital signals. In most cases, the modem makes the off-hook connection, dials the number for access, and answers the line when it rings. It can even mute the speaker when a connection is complete and the digital signals are communicating. When the digital session is complete, the modem closes the connection and hangs up the telephone line. Dial-up services are still in use today. Most dial-up systems are limited to 56kbps or less. This limits the speed and quantity of information that can be carried over the line. Dial-up is considered too slow by most computer users.

5.2.0 Digital Subscriber Line (DSL)

A digital subscriber line (DSL) is designed to handle both voice and data at the same time and over the same pair of wires. The analog voice operates at frequencies below 4kHz. Digital signals are carried at frequencies above 5kHz. DSL consists of 247 different channels that are constantly being monitored by the modem for usability. The term *DSL* is a generic term for a host of DSL services that may be offered.

5.2.1 Voice Over DSL

Voice over DSL is the normal service delivered to a customer on a wet DSL line. The term *wet* means there is voice signal on the telephone line or wire line pair. The first 4kHz of bandwidth is used for analog voice service and the remaining bandwidth, starting at 5kHz, is used for data.

5.2.2 Naked DSL

Naked DSL is a service that is offered over a pair of wires (called a dry pair) that has no voice service. Naked DSL can be provided by the telephone company. It can also be delivered by a service provider. That can lead to situations where a customer pays the phone company for a voice service line with no dial capability and pays a service provider for digital data DSL service.

5.2.3 Symmetrical DSL (SDSL)

Symmetrical DSL (SDSL) is service that provides transmit speed that is the same as the receive speed. This service is not often used. Digital voice circuits are symmetrical, and the transmit speed is the same as the receive speed.

5.2.4 Asymmetrical DSL (ADSL)

Asymmetrical DSL (ADSL) allows the customer to receive data at a higher speed than it is capable of transmitting. It is the most common DSL service provided. For most digital customers, it is more desirable to download (receive) large files than to upload (transmit) large files. Asymmetrical DSL provides that capability.

5.2.5 Rate Adaptive DSL (RADSL)

Rate adaptive DSL (RADSL) is a service where the customer can change either or both the transmit or receive speed. This service is normally used by businesses that have one service speed but require a higher speed for a short period. An example is a company that has to quickly transfer a lot of sales data at the end of a quarter.

5.2.6 High-Speed DSL (HDSL)

High-speed DSL (HDSL) is DSL service that starts at or above 1Mbps and goes up to 20 to 40Mbps. This service may be ADSL, SDSL, or RADSL, but is much higher than the normal transmit speed of 56kbps to 256kbps. This service often has forward error detection and correction built into the transmitted data.

5.2.7 Digital Subscriber Line Access Multiplexer (DSLAM)

Digital subscriber line access multiplexer (DSLAM) is the service provider's equipment. Individual subscribers are connected to channels of a DSLAM. These are then multiplexed and sent to the service provider's facilities where they are connected to the Internet or other facilities for internal operations.

5.2.8 DSL Modem

The term *modem* is an acronym for modulator and demodulator. A DSL modem (*Figure 23*) is subscriber equipment that converts the customer's digital signals into signals that are suitable for a wire line. It also converts the wire line signals back into digital data that can be used by the customer's equipment. Within the modem, a high-pass filter prevents voice signals that may be present from corrupting the digital data being received. A low-pass filter placed on each telephone jack prevents the hiss of the high-speed data from interfering with voice conversations. One common low-pass filter can be installed to protect all phones from data hiss. The speed at which DSL operates is limited by the quality of the telephone line carrying DSL signals. The modem constantly monitors all channels or frequency bands for quality. When a modem first makes a connection, it chooses the channel or combination of channels with the best qualities and use it for its transmissions. During transmission, the modem continues to monitor channels to see if the wire line is degrading. If a channel degrades to a point that it might affect digital data service, the modem switches to a better channel. There are many things that can affect DSL service, including:

- *Line losses* – Cable resistance restricts DSL cable length to less than 18,000 feet from the DSLAM to the customer's modem. Most any DSL speed of 1Mbps or less is available at distances of 6,000 feet or more. However, as the line gets longer, less signal strength is available and higher speeds do not function.

- *Load coils* – Load coils are designed to offset the capacitive reactance of frequencies below 4kHz. They adversely affect all DSL channels and degrade the signal of any channel. Because of line length and the presence of load coils, DSL is not available in all areas, even within the short-line subscriber loop cables of a city.

- *Multiplexers* – The multiplexing equipment used for analog and digital voice circuits is not compatible with DSL. If a particular subscriber loop is serviced by a multiplexer, between the local exchange and the premise, DSL is not available.

When DSL was first developed, the electronics of the day made the modem very large, which limited its application. Modern electronic circuits are tiny by comparison. DSL modems today are small enough for home use and are very compatible with commercial applications.

6.0.0 DIGITAL VOICE SERVICES

Before the introduction of electronic switching systems, T-1 services were in widespread use. Some of the other services now available include:

- Pair gain system
- Channel banks
- Integrated services digital network (ISDN)
- Integrated services digital network (BRI)
- Integrated services digital network (PRI)

6.1.0 Subscriber Line Carrier Pair Gain System

Prior to the introduction and installation of electronic switching systems, the subscriber line

33405-11_F23.EPS

Figure 23 DSL modem.

carrier was the mainstay for tie lines, trunks, high-density buildings, and other high volume distributions.

If one building needs 180 individual subscriber loops, it would take at least a 200-pair cable to provide service. However, a pair gain system could be installed to provide service to 192 subscribers using a smaller number of pairs out of the cable. In many cases, this would eliminate the need for the telephone company to invest in the installation of costly new cable to provide service to the building. If the same building were already serviced with a 100-pair cable, the pair gain system installation would eliminate the need to install a new or larger cable.

The pair gain system may be used as a high-capacity tie line or trunk group between switches. This eliminates the need to install multiple E&M circuits between two switches. The pair gain system can be configured with a number of different types of line cards including:

- Two-wire E&M
- Four-wire E&M
- Analog loop start
- Analog ground start
- Digital data service at speeds up to 56kbps
- FXO card (central office)
- FXS card (subscriber)
- OPX
- PLAR

6.2.0 Channel Banks

A channel bank is nothing more than a termination point for a T-1 that performs the analog to digital conversion for analog communications. It connects to a T-1 on the central office or PBX side and provides 24 individual subscriber or E&M circuits as its output. Most channel banks have one or two cards that process the T-1 signals and 24 other cards for the individual telephone line outputs. The variety of cards it can use is identical to the type of cards that may be installed in a pair gain system, but they may not be interchangeable.

6.3.0 Integrated Services Digital Network (ISDN)

Integrated services digital network (ISDN) is a total digital service that can be used for voice, video, and data at the customer's location (*Figure 24*). For voice service, the analog to digital conversion is the standard 64kbps PCM scheme, but the conversion is performed inside the customer's telephone. POTS phones do not work on an ISDN line. There are two standard interfaces used with

ISDN: basic rate interface (BRI) and PRI. The BRI interface is more appropriate for consumers and small businesses. The PRI interface is a step up in capacity and is more applicable to large businesses.

6.3.1 Integrated Services Digital Network (BRI)

BRI is the first level of service in ISDN. The secret to ISDN is how the on-hook and off-hook information, along with dialing information, is conveyed across the lines between the local exchange switch and the premise telephone or subscriber loop. Remember, the analog to digital conversion is performed inside the telephone, not at the local exchange switch. Dialing and hook switch information still has to be communicated between the premise telephone and the local exchange switch in digital format. To do this, a separate data channel, called the D-channel, is established between the local exchange switch and the premise telephone. This channel is always connected to the local exchange switch or PBX and is constantly reporting its on-hook and off-hook status. When the telephone is off-hook, it also communicates all dialing information and its connection status with the far end. The D-channel is a 16kbps channel containing all the connection, dialing, and hook status information. If the D-channel is disconnected, the premise telephone has to resynchronize with the local exchange or PBX before it can operate.

The digital voice is contained in a standard 64kbps PCM channel. There are two 64kbps channels that bear the voice information. They are called bearer channels or B-channels. A standard ISDN subscriber loop contains two 64kbps B-channels, one 16kbps D-channel (2B+D), and 48kbps of framing.

> **NOTE**
>
> The framing, timing, and synchronization information contained in the 48kbps signal is often referred to as overhead. This information cannot be seen or captured.

6.3.2 Integrated Services Digital Network (PRI)

For ISDN, the PRI is much like a T-1, except it handles bearer channels (B-channels) and D-channels. A PRI contains 23 B-channels and one D-channel (23B+D). For the BRI, the D-channel is 16kbps and it controls two B-channels. In a PRI, the D-channel is 64kbps and it controls 23 B-channels.

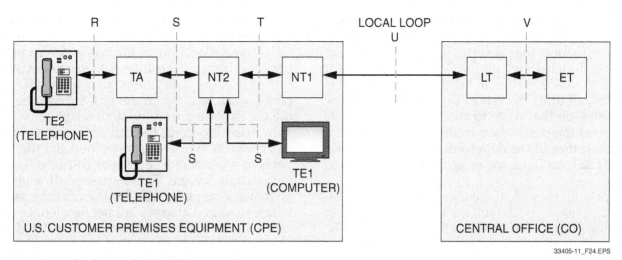

Figure 24 Example of a simple US ISDN network.

The T-1 contains 24 64kbps DS-0 channels. Both the PRI and T-1 are 1.544Mbps signals and have one framing bit that starts the frame. A PRI and T-1 are similar, except for the D-channel and bit robbing. A PRI does no bit robbing. For that reason, a DS-0 or B-channel is suitable for 64kbps digital voice or 64kbps data, but there are only 23 of them. A PRI is also transmitted as an AMI signal and B8ZS is still in use.

6.4.0 Voice Over Internet Protocol (VoIP)

Voice over Internet Protocol (VoIP) is not a switched-circuit service like POTS or ISDN. VoIP is a packet-switched service that uses the Internet for its connections. It is not routed from telephone switch to telephone switch. Instead, it is sent between service provider routers and is treated as data packets throughout its network routing. Just as data is often transmitted over switched telephone lines, voice can be transmitted over data lines and the Internet. Although some Telcos may offer this service, it is usually offered through their data transmission services rather than voice services.

With data over the Internet, data bytes are placed in packets of data, and a header is added within the packet. The header, called an address header, tells where the data is coming from and where it is going. With Internet protocol (IP), the header on the data packet is read by the router and then sent onto a route that connects to that destination address. If the route is down or overloaded, it automatically routes the packet onto an alternate route that has a path to the destination address. The individual packets may route through a number of routers and then be reassembled at the final destination.

Routing and alternate routing can create problems when data packets do not arrive at the destination address in the proper order or when they get lost in transmission. This is not a problem with normal data packets, because the destination equipment stores and reassembles the packets in the proper order to retrieve the data in its proper format. Voice service does not have the luxury of holding packets until all the previous packets arrive for reassembly. Each voice packet must be received and reassembled simultaneously in order to retrieve the voice signal.

To keep each voice packet on the same route and keep them arriving at the destination in the proper order, a quality of signal (QOS) modifier is added in the header. The signal modifier uses the H.323 standard that all routers must comply with for the voice signals to be routed and received in the proper order.

QOS (H.323) is used for all voice and video transmissions over the Internet, but the voice is more critical than the video. When a number is

---

On Site

International ISDN Services

Internationally, the ISDN interface terms *BRI* and *PRI* used in North America are called BRA for basic rate access and PRA for primary rate access, respectively. In Europe, an ISDN PRA interface consists of 30 64kbps B-channels and two 64kbps signaling channels. It is referred to as 30B+2D and is called an E-1.

---

dialed on a VoIP telephone, that telephone number is translated into an IP address that the routers use to route the packets. This can only be done with a Touch-Tone® phone. Dial phones cannot convert analog to digital. When the first packet is routed, all other packets of that voice data follow and stay on that route to ensure that all packets arrive at the destination in the proper order. If a single router in the data route does comply with the H.323 standard, the voice data may not be usable.

For VoIP to work, there must be Internet access for the phone. Special software is needed for both analog/digital conversion and IP routing. The technology is still fairly new and developers are continuously making improvements. Its major advantage is that there are no long distance charges.

Phone calls are not routed through telephone switches where time and charges can be applied. Calls to any location in the world are routed as packets through the Internet. Because voice is transmitted as data, the wiring for a VoIP telephone must be a minimum of Cat 5e rather than the Cat 3 cabling used for other telephone wiring.

In many locations, the call destination ends up as a connection to a subscriber loop and then connects to a telephone service or BRI lines as analog or digital voice. This bypasses all telephone switches except for the local exchange switch where time and charges are not applied. In those cases, the destination address is translated back into a telephone number that can be used by the local exchange switch.

On Site

Voice Over Internet Devices

Making a phone call using VoIP technology often requires an adapter between a USB port on the computer and the telephone line. The adapter uses a VoIP protocol to convert audio signals into a digital signal that can be sent over the Internet. These devices are heavily promoted on television. Two of the most popular devices/services are magicJack™ and Vonage®.

SUMMARY

Plain old telephone service (POTS) is the basic analog dial telephone service that has been around for over 100 years and expected to be around for many years to come. Newer technologies are changing how people email, text, tweet, or call others. What started out as a large wooden box on the wall has turned into a pocket-sized cell phone that can be carried anywhere. It can do much more than just allow people to talk.

Early telephone service was manually switched and required an operator to make telephone connections. The advent of electromechanical switches, area codes, and direct-dial long distance reduced or eliminated the need for operators.

Today's businesses have their own telephone systems. They handle internal calls, external calls, and voice mail without the need for an operator. At one time, a key system was the ultimate for a small business. That technology is becoming obsolete. Large corporations have systems that enable them to stay in contact with their offices around the country and around the world without incurring long distance charges. Cell phone service has reduced the revenues of the mainstay long distance carriers. Modern cell phone service providers may use as few as five electronic-switching systems nationwide, with multiple optical remote modules that connect to the cell towers.

Multiplexing is a technology that allows multiple signals to be sent over a single communication path. Various multiplexing schemes are available. New technologies such as digital subscriber line (DSL) have emerged to handle analog and digital voice, digital data, and digital video. Integrated services digital network (ISDN) is a total digital service that can be used for voice, video, and data. Voice over Internet Protocol (VoIP) is a service that allows phone conversations to be carried over the Internet, avoiding long distance charges.

1. Many phone companies are connected together to form an international telephone system that is known as the _____.

   a. International Telephone Union (ITU)
   b. Public Switched Telephone Network (PSTN)
   c. International Telephone Exchange (ITX)
   d. Public Telephone Association (PSA)

2. The hard-wired line that connects the individual telephone to the telephone central office is the _____.

   a. local line
   b. hard circuit
   c. local subscriber loop
   d. premise loop

3. When the premise telephone set is on-hook, the circuit between the central office and premise telephone is _____.

   a. closed
   b. grounded
   c. powered
   d. open

4. The telephone lines that connect toll centers and local exchanges are called trunks.

   a. True
   b. False

5. The normal off-hook voltage of a local analog telephone is about _____.

   a. –48VDC
   b. +48VDC
   c. +90VAC
   d. 6 to 8VDC

6. The first three numbers of a ten-digit dialing number are known as the _____.

   a. exchange ID
   b. number exchange
   c. area code
   d. name exchange

7. Analog ground-start operations are widely used for connections between a local exchange switch and a(n) _____.

   a. PBX
   b. operator
   c. DSL
   d. patch cord

8. In E&M signaling, the E-lead is used to _____.

   a. transmit
   b. receive
   c. modulate
   d. oscillate

9. The purpose of a hybrid transformer in a two-wire E&M circuit is to boost line voltage.

   a. True
   b. False

10. With a rotary dial phone, how many pulses will be produced when the number zero (operator) is dialed?

    a. Zero
    b. One
    c. Five
    d. Ten

11. An advantage of a key telephone system over a PBX is that it _____.

    a. has a lower initial cost
    b. allows easy Internet access
    c. can handle many more trunks
    d. can handle many more subscribers

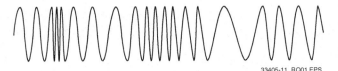

Figure 1

33405-11_RQ01.EPS

12. The waveform in *Figure 1* represents _____.

    a. amplitude modulation
    b. frequency modulation
    c. pulse code modulation
    d. 24kHz

13. What is the frequency range of analog voice signals?

    a. 100Hz to 4kHz
    b. 1,000Hz to 4,000Hz
    c. 16Hz to 18kHz
    d. 56kbps to 64kbps

14. How many times per second are analog signals sampled during their conversion to digital signals?

    a. 600
    b. 4,000
    c. 8,000
    d. 12,000

15. In digital transmissions, a digital 64kbps digital voice signal is referred to as _____.

    a. PRI
    b. DS-0
    c. PCM
    d. DS-64

16. The 24 channels in a T-1 are separated by _____.

    a. framing bits
    b. time slots
    c. counting bytes
    d. an alternate mark inversion

17. The term DSL stands for _____.

    a. dual safety line
    b. digital signal loss
    c. dual subset link
    d. digital subscriber line

18. The maximum distance that DSL service can be delivered without affecting the signal is _____.

    a. 18,000 feet
    b. 12,000 feet
    c. 10,000 feet
    d. 8,000 feet

19. The ability to increase the number of subscribers beyond the capacity of an existing system is provided by a(n) _____.

    a. channel bank
    b. PBX
    c. pair gain system
    d. ISDN subscriber loop

20. What technology allows a person to make a telephone call over the Internet?

    a. PBX
    b. VoIP
    c. PCM
    d. SLC-96

Trade Terms Introduced in This Module

Alternate mark inversion (AMI): A method for adding timing signals to the bit streams of transmitted T-1 signals.

Basic rate interface (BRI): The first level of service in an integrated services digital network (ISDN).

Bit rate: The number of ones and zeros contained in one second.

Channel bank: A termination point for a T-1 that performs the analog to digital conversion for analog communications.

Digital subscriber line (DSL): A subscriber line that is able to handle both voice and data over the same line.

Foreign exchange service (FX): A telephone service that provides local exchange switch service to a business office that may be located hundreds or thousands of miles from the main business office.

Framing bit: A bit added to a T-1 bit stream that identifies channel one.

Glare: A condition in which the central office senses an off-hook condition when the PBX is still in an on-hook state. Glare is caused by a false ground in long subscriber loops.

Integrated services digital network (ISDN): A total digital service that can handle voice, video, and data at a customer's location.

Key telephone system (KTS): A telephone system in which multiple telephones share multiple predetermined Telco central office phone lines to make or receive calls. Selection of an unused outside line to make a call, or to answer a ringing line, is made by the individual telephone users.

Local subscriber loop: The hard-wired lines that connect the subscriber to the telephone company central office.

Multiplexing: The mixing of multiple signals on or in a single communications path.

Off-premise extension (OPX): A telephone that is installed in a location that is remote from the business where the main telephone system equipment is located. A typical application is for a worker who has his/her office at home.

Pair gain system: A technology that allows a large number of individual subscriber lines to be installed in a building without investing in new cables.

Plain old telephone service (POTS): A basic telephone service involving analog telephones, telephone lines, and access to the public switched network.

Primary rate interface (PRI): The next higher level of service in an integrated services digital network (ISDN).

Private branch exchange (PBX): Privately owned switching equipment that allows communication within a business and between businesses. It is connected to a common group of lines from one or more Telco central offices.

Private line automatic ringdown (PLAR): A system of two directly connected phones. If either phone is picked up, the other phone rings. Phones used in this system do not require a rotary dial or numbered keypad.

Public switched telephone network (PSTN): The various telephone companies that are connected together to form an international telephone service and switching system.

Pulse code modulation (PCM): A modulation scheme used to convert analog voice signals to digital voice signals.

Pulse dialing: The type of dialing used with rotary dial telephones. The rotary dial breaks and reconnects a voltage on the ring line based on the number dialed. For example, dialing five breaks and reconnects the voltage five times.

Ring: In an analog subscriber line, it is the wire that connects to the ring of the telephone patch cord.

Ring generator: A pulsating +90VDC signal that causes a called party's telephone to ring.

Telco: A shortened name for a telephone company.

Tie line: A line that connects two switches that have no time and charges applied.

Time division multiplexing (TDM): A multiplexing scheme in which time slots are assigned to different T-1 channels.

Tip: In an analog subscriber line, it is the wire that connects to the tip of the telephone patch cord.

Toll center: A telephone switch in which time and charges are assigned to a call.

Touch-Tone®: A dialing method that creates a series of tones when numbered buttons are touched on a keypad on the phone. Each number or key function has its own distinct tone.

Trunk: A telephone line between the local exchange and the toll center.

Voice over Internet Protocol (VoIP): A technology that allows phone conversations to take place over the Internet.

Additional Resources

This module presents thorough resources for task training. The following resource material suggested for further study.

Business Telecom Systems: A Guide to Choosing the Best Technologies and Service. New York, NY: CMP Books.

Newton's Telecom Dictionary. San Francisco, CA: CMP Books of CMP Media LLC.

Next Generation Phone Systems: How to Choose a Voice and Data System for E-Business. New York, NY: CMP Books of CMP Media LLC.

The Telecom Handbook. New York, NY: CMP Books of CMP Media LLC.

Voice Over DSL. New York, NY: CMP Books of CMP Media LLC.

Figure Credits

Topaz Publications, Inc., Module opener, Figures 1, 10, 23, and SA01

www.sandman.com, Figure 2

Courtesy of BROADREACHNETWORKS.com, Figure 3

VintageRotaryPhones.com, SA02

NEC Unified Solutions, Inc., Figures 15 and 16

Dumond Chemicals, Inc., SA04

NCCER CURRICULA — USER UPDATE

NCCER makes every effort to keep its textbooks up-to-date and free of technical errors. We appreciate your help in this process. If you find an error, a typographical mistake, or an inaccuracy in NCCER's curricula, please fill out this form (or a photocopy), or complete the online form at **www.nccer.org/olf**. Be sure to include the exact module ID number, page number, a detailed description, and your recommended correction. Your input will be brought to the attention of the Authoring Team. Thank you for your assistance.

Instructors – If you have an idea for improving this textbook, or have found that additional materials were necessary to teach this module effectively, please let us know so that we may present your suggestions to the Authoring Team.

NCCER Product Development and Revision
13614 Progress Blvd., Alachua, FL 32615

Email: curriculum@nccer.org
Online: www.nccer.org/olf

❏ Trainee Guide   ❏ AIG   ❏ Exam   ❏ PowerPoints   Other _____

Craft / Level: _____   Copyright Date: _____

Module ID Number / Title: _____

Section Number(s): _____

Description: _____

_____

_____

_____

_____

Recommended Correction: _____

_____

_____

_____

_____

Your Name: _____

Address: _____

_____

Email: _____   Phone: _____

33406-12

Residential and Commercial Building Networks

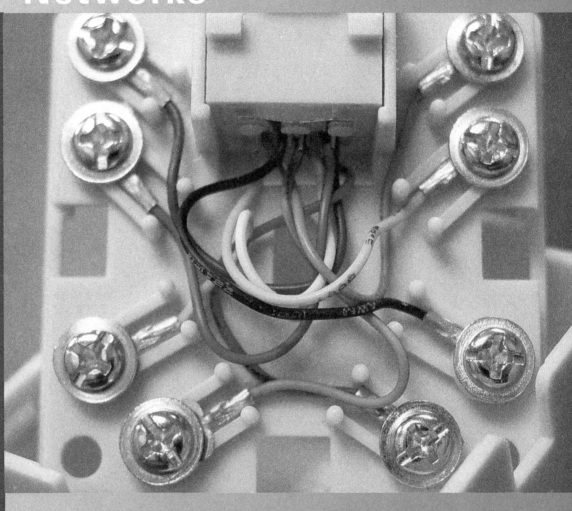

Trainees with successful module completions may be eligible for credentialing through NCCER's National Registry. To learn more, go to **www.nccer.org** or contact us at **1.888.622.3720**. Our website has information on the latest product releases and training, as well as online versions of our *Cornerstone* newsletter and Pearson's product catalog.

Your feedback is welcome. You may email your comments to **curriculum@nccer.org,** send general comments and inquiries to **info@nccer.org**, or fill in the User Update form at the back of this module.

Objectives

When you have completed this module, you will be able to do the following:

1. Describe the various configurations of residential and commercial networked systems.
2. Explain the various connection options and protocols commonly used for integration.
3. Describe network configurations.
4. Describe the various user interfaces used for integrated systems monitoring and control.
5. Explain the methods of communication between devices and controllers.
6. Explain how integrated systems can be remotely accessed and controlled.
7. Set up components on an Ethernet-based network that can be controlled remotely.

Performance Task

Under the supervision of your instructor, you should be able to do the following:

1. Set up components on an Ethernet-based network that can be controlled remotely.

Trade Terms

Acknowledgement byte (ACK)
Convergence
Dynamic host configuration protocol (DHCP)
High-level data link control (HDLC)
Logical address
Logical Link Control (LLC)
Media Access Control (MAC)
Non-acknowledgement signal (NAK)
Null modem cable

Parity bit
Simple mail transport protocol (SMTP)
Simple network management protocol (SNMP)
Static address
Systems controller
Telnet
Wi-fi protected access (WPA)
Wired equivalent privacy (WEP)

Industry Recognized Credentials

If you're training through an NCCER-accredited sponsor you may be eligible for credentials from NCCER's Registry. The ID number for this module is 33406-12. Note that this module may have been used in other NCCER curricula and may apply to other level completions. Contact NCCER's Registry at 888.622.3720 or go to nccer.org for more information.

Contents

Topics to be presented in this module include:

Figures and Tables

1.0.0 INTRODUCTION

When two or more forces are combined and the combined benefit is greater than the sum of the individual forces, synergy is created. One of the most exciting and rewarding aspects of being an electronic systems technician is the ability to create synergy by interconnecting multiple stand-alone electronic systems and then discovering that this integration of systems has resulted in a product that is better than the individual systems with which you started. Systems integration is the process of connecting two or more stand-alone systems together in a manner that improves the capabilities of each system.

You experience integrated systems in many areas of your life. For example, museums often use integrated systems in multimedia displays. When you push a button to activate a display, the surrounding lights dim, a spotlight focuses your attention on an artifact, and an audio program begins. The lighting and audio systems are integrated together and are automatically controlled in a manner that provides greater education and enjoyment for the visitor.

A number of systems at a typical movie theater are integrated and automatically controlled. As the movie begins, the lights slowly ramp down, the multi-channel audio system is engaged, the curtains covering the projection screen are withdrawn, and the projector begins showing the movie. Each of these systems (lights, audio, curtains, and video) is an individual stand-alone system, and each can be operated individually. However, once they are integrated, they can be automatically controlled with a systems controller. The resulting integrated system is much more efficient and enjoyable.

Some large department stores have monitors displaying videos of models wearing the latest fashions. Although it is not readily apparent, a large amount of systems integration is going on behind the scenes to make the advertising possible. Advertisement videos are created at corporate headquarters many miles away and stored on a media server at that facility. At a predetermined time, the advertisements for the new day are automatically sent over the corporate network from the media server to a video server in a local store. As the store opens, the monitors are automatically turned on, and the video server automatically begins playing the new advertisements. In this example, the corporate media server, the remote monitors, and the remote video server are all integrated on the store's corporate network. Once again, the resulting integrated system is an efficient and cost-effective way of integrating systems.

One large fast food restaurant chain automatically reads each cash register at predetermined times during the day. Doing so calculates how much money is on-site, how much product (by food type) has been sold, and how much product must be reordered. At the same time, the corporate office can help the local restaurant manager determine the required staffing for different times of the day and how many total employees and assistant managers are needed. From information drawn off cash registers at different restaurants, the corporate needs for an entire region can be determined and orders placed to suppliers. The corporate office may even be able to view the closed circuit TV cameras inside each restaurant for emergency response and tracking of employee behavior. This type of integration requires more than integrating an audio and video system. It requires the integration of communications, banking, accounting data, staffing, suppliers, and shipping.

Ultimate examples of systems integration are found at theme parks and entertainment centers. The electronic systems are integrated and precisely controlled to enhance the visitor's experience. Systems that are controlled include the following:

- Lighting
- HVAC (heating, ventilating, air conditioning)
- Audio
- Video
- Fireworks
- Rides
- Security
- Access control
- Animated robots
- Curtains, drapes, shades
- Clocks
- Fountains and water jets

This module helps put some concepts such as the open systems interconnection (OSI) reference model, networking protocols, and network communication into perspective, allowing you to see how these concepts apply to real-world applications.

2.0.0 REASONS FOR SYSTEM INTEGRATION

In order for these various electronic systems to be integrated, the products that make up the system must be able to communicate with each other and receive commands from an external source. It is also beneficial if the products can return acknowledgement of commands received and send status messages.

As a convenience to the user or installer of their products, manufacturers of these types of equipment provide methods for the user to communicate with the product. For example, the manufacturers of projectors, alarms, switches, and lighting provide software to configure and control them. However, even a small facility might have five to ten different systems, potentially requiring the client to learn how to operate five to ten different software programs. One of the greatest advantages of systems integration is the ability to provide clients with a single, unified, easy-to-operate user interface. Successful systems integrations merge the communications details of each system so that the client can easily use all of the systems. Integration of multiple systems using a single user-friendly system makes all the individual systems less complicated to operate. The key to the integration is the communication between users and between components of the integrated systems.

2.1.0 Convergence

Systems integration is the natural result of the convergence of many different technologies. Convergence can be viewed as the point where different technologies merge, such as telephony, computers, security, audio, and video, as well as many mechanical and electrical controls or devices. The level of convergence of these technologies and the growth of systems integration directly correlate.

Ethernet has been an important enabler of systems integration, allowing users to interconnect devices from different industries. Building automation systems (BAS), security systems, video distribution systems, surveillance systems, and audio systems are increasingly being integrated into local area networks (LANs). The level of integration and convergence is not limited to the LAN. Manufacturers from each of these industries are continually releasing Internet-capable products. Voice over Internet Protocol (VoIP) now gives users the ability to place telephone calls from any location where they can connect to the Internet, without paying long distance charges. These calls can be placed to other computers or to telephones.

At the same time, cell phones can send instant messages, check email, and send and receive text messages and photos. Many newer cell phones used for viewing TV and listening to radio. They can also give a driver turn-by-turn directions using global positioning satellite (GPS) technology. Several automobile manufacturers give the owner the capability to lock and unlock their vehicle, check the vehicle's location, and check tire pressure remotely. These are just a few examples of convergence of computers, networks, telephony, audio, and video. For this convergence to work seamlessly and with ease, a method of powering the convergence is required.

3.0.0 THE OSI REFERENCE MODEL

The International Organization for Standardization (ISO) has developed a seven-layer OSI model for communications between platforms or systems. They range from basic to complex. It subdivides a communications system into smaller parts called layers. In a complete communications system, each layer performs specific functions that provide services to layers above and below. Each layer depends on the information and format of information from the layers beneath it. The seven layers are:

- *Layer 1* – Physical Layer
- *Layer 2* – Data Link Layer
- *Layer 3* – Network Layer
- *Layer 4* – Transport Layer
- *Layer 5* – Session Layer
- *Layer 6* – Presentation Layer
- *Layer 7* – Application Layer

3.1.0 The Physical Layer

The first layer, the physical layer, defines the electrical and physical specifications for the actual connection to a device. The physical connector may be a screw terminal, an optical or light interface, an electrical plug and receptacle, or a radio antenna. It is a media interface. It also defines the physical electrical interface. There are multiple physical interfaces and most have certain recognized standards for that interface. The physical interface is a method of connecting one device to another across a distance so that they can communicate.

3.1.1 Recommended Standard 232 (EIA/TIA RS-232)

The recommended standard, RS-232, identifies a specific 25-contact subminiature D-shell connector (also called a D-sub connector) with 20-gauge contacts. It defines which contact is used for a signal and what that signal is called. *Table 1* provides the name and pin connection for RS-232 and compares it to other connectors and standards. In addition, it defines the voltage levels for the signals. Although RS-232 signals may be as high as 25VDC, the level most often used is about 5 to 6VDC. While a 25-pin D-sub is specified, many manufacturers have designated a 9-pin D-sub,

Table 1 Connector Configurations

Mnemonic	RS-232 Signal Name	RS-232 25-Pin Contact	RS-232 9-Pin	RS-232 8C8P	RS-422 Signal Name	RS-422 37-Pin Contact	RS-530 Signal Name	RS-530 Contact	V.35 Signal Name	V.35 34-Pin Contact
SD	Transmitted Data	2	3	6	Send Data-A	4	Transmit Data-A	2	Transmit Data-A	P
SD					Send Data-B	22	Transmit Data-B	14	Transmit Data-B	S
RD	Received Data	3	2	5	Receive Data-A	6	Receive Data-A	3	Receive Data-A	R
RD					Receive Data-A	24	Receive Data-B	16	Receive Data-B	T
ST	Transmitter Signal Element Timing	15			Send Timing-A	5	Transmit Timing-A	15	Serial Clock Transmit-A	Y
ST					Send Timing-B	23	Transmit Timing-B	12	Serial Clock Transmit-B	AA
RT	Receiver Signal Element Timing	17			Receive Timing-A	8	Receive Timing-A	17	Serial Clock Receive-A	V
RT					Receive Timing-B	26	Receive Timing-B	9	Serial Clock Receive-B	X
TT	Terminal/External Timing	24			Terminal Timing-A	17	Terminal Timing-A	24	External Clock-A	U
TT					Terminal Timing-B	35	Terminal Timing-B	11	External Clock-B	W
RTS	Request To Send	4	7	8	Request To Send-A	7	Request To Send-A	4	Request To Send	C
RTS					Request To Send-B	25	Request To Send-B	19		
CTS	Clear To Send	5	8	7	Clear To Send-A	9	Clear To Send-A	5	Clear To Send	D
CTS					Clear To Send-B	27	Clear To Send-B	13		
DTR	Data Terminal Ready	20	4	3	Terminal Ready-A	12	DTE Ready-A	20	Data Terminal Ready	H
DTR					Terminal Ready-B	30	DTE Ready-B	23		
DSR	Data Set Ready	6	6	1	Data Mode-A	11	DCE Ready-A	6	Data Set Ready	E
DSR					Data Mode-B	29	DCE Ready-B	22		
DCD	Data Carrier Detected	8	1	2	Receiver Ready-A	13	Received Line Signal Detector-A	8	Data Carrier Detected	F
DCD					Receiver Ready-B	31	Received Line Signal Detector-B	10		
SQ	Signal Quality	21			Signal Quality	33				
IC	Ring Indicator	22	9		Incoming Call	15				
SI	Data Signal Rate Selector	23			Signal Rate Indicator	2				
SG	Signal Ground	7	5	4	Signal Ground	19	Signal Ground	7	Signal Ground	B
FG	Frame Ground	1			Shield Ground	1	Shield Ground	1	Frame Ground	A
LL	Local Loopback	18			Local Loopback	10			Local Loopback	J
RL	Remote Loopback	21			Remote Loopback	14			Remote Loopback	BB
TM	Test Indicator	25			Test Mode	18			Test Indicator	K

and even an eight-contact, eight-position (8C8P) modular connector for use with RS-232.

An RS-232 uses unbalanced signals. As shown in *Table 1*, the transmitted data signal has only one contact (Pin 2) for that signal. Transmitted data is a signal that originates at a data terminal and terminates at the communications equipment. Notice in the same table that received data also has only one contact (Pin 3) for that signal. Received data is a signal that is terminated in the data terminal and has its origin in the communications equipment. The two signals are going in opposite directions and both signals use a common contact (Pin 7) for the return path to make a complete circuit. Depending on polarity, the voltage on Pin 7 (signal ground) may be going in either direction between the communications equipment and the data terminal. Pin 7 is used to balance the voltage levels of each of the unbalanced signals. Since RS-232 is an unbalanced interface, it is seldom used at a baud rate of more than a few hundred bits per second. At higher speeds, the signals interfere with each other and the data becomes corrupted. Because of the unbalanced signals, the length of an RS-232 cable is normally limited to no more than 50 feet. For speeds below 56kbps and short distances, this is a cost-effective option. At one time, almost all mainframe computer terminal connections used the RS-232 standard for data transfer. RS-232 cables typically use a 12½-pair cable with an over shield, or a 25-conductor, non-paired cable.

3.1.2 Recommended Standard 422 (EIA/TIA RS-422)

Data transfer rate requirements increased to support the growing amount of data that needed to be carried over long distances. To handle more traffic and faster transfer speeds, a new balanced interface was needed. Standard RS-422 was created by EIA/TIA to solve the problem. In *Table 1* under Send Data, notice that two contacts are used. This allows each signal a complete circuit without using the signal ground connection to carry any unbalanced load of that circuit. By using a balanced circuit, transmission distances are increased to over 300 feet. Transfer speeds for large files are also faster. The RS-422 standard specifies a 37-contact subminiature D-shaped connector (37-pin D-sub) with 20-gauge contacts. It may also be used with a screw terminal connector. In those cases, the signals are marked as TD minus and TD plus, instead of TD-A and TD-B, respectively.

An RS-422 output can also be connected to up to 10 termination devices at a time. The output amplifiers must be able to drive 10 termina-

tions; however this is not a common practice. It is common for RS-422 devices to be separated by as much as 300 feet or more. Because of the nature of balanced signals, cable losses are lower than when using an unbalanced interface. The voltage levels of RS-422 signals are bi-polar and vary from about -2VDC for a zero, to about +2VDC for a one. Because of lower voltage and current levels, RS-422 is less likely to interfere with timing and data signals that might be within the same cable. A typical RS-422 cable uses a 13-pair cable with an over shield, or a 13-pair, individually shielded pair cable. Because it is a balanced interface, it is important that the individual signals are connected onto paired cables. The speed of RS-422 may be anywhere from 19.2kbps to 10Mbps, and often higher on very short cable runs. The original design allowed for cabling of up to nearly 4,000 feet at 10 Mbps. The standard does not specify a maximum cable length. It does provide loss graphs based on cable length and transfer speed.

3.1.3 Recommended Standard 530 (EIA/TIA RS-530)

Several manufacturers have developed equipment with plug-in circuit cards that are interchangeable and have different standard outputs, such as RS-232, RS-422, V.35, etc. They use a standard 25-pin D-sub connector for all interface card connections on the back plane. Some of this equipment may have 96 or more 25-pin D-subs on the back plane for interconnections to equipment. This requires special cables for equipment compatibility. In an effort to standardize the outputs between manufacturers, the RS-530 standard was developed. It uses the 25-pin D-sub for any other standard interface. If the RS-530 interface is used, special cables are still required. However, they are not always manufacturer-specific. *Table 1* provides these interface contact connections. That way, a single back-plane connector can be used for any standard interface without rebuilding the back-plane of the equipment. An RS-530 cable typically uses a 12½-pair cable with an over shield, or a 12½-pair, individually shielded pair cable.

3.1.4 Universal Serial Bus (USB)

The universal serial bus (USB) is one of two high-bandwidth data communication standards that are now commonly used in computer and control systems. A single USB interface can support up to 127 devices, with a data transfer rate of 12Mbps, which is much faster than previous forms of serial communication. Multiple devices may be daisy-chained to the USB interface, or a hub (or multiple hubs) may be used to create a mini-network of devices. USB 2.0 increased the speed of the USB interface to 480Mbps. USB 3.0, with a 5.0 gigabit speed, became available in 2010. USB 3.0 devices are backward-compatible with USB 2.0. The length of USB 2.0 cables is limited to 5 meters (16.4 feet). The cable length for USB 3.0 is 3 meters (9.8 feet), although this length can be extended using a repeater extension cable or a USB bridge.

3.1.5 ITU-T V.35 Interface

The V.35 standard is an international standard of the International Telecommunications Union – Telecom (ITU-T) Standardization Section. It is not an ANSI-sanctioned standard. It was originally submitted to ANSI/EIA/TIA as a submission for the RS-422 standard. The standard was rejected because the connector was too bulky and not all V.35 signals are balanced. In *Table 1*, notice that the send data, receive data, send timing, receive timing, and terminal timing are balanced. All other signals are unbalanced. Data and timing signals are more susceptible to interference and corruption because of their high speed. The control or handshaking signals get set to a high or low and remain there as long as the communications link is intact and operating. They may remain in one state for up to several minutes at a time.

> **NOTE**
>
> While ANSI, the EIA, and the TIA were debating the balanced submissions for RS-422, this partially balanced interface was submitted to the ITU-T for consideration. The ITU adopted the standard intact.

The V.35 standard combines much of the RS-422 voltage level standard for its balanced data and timing signals of –2VDC and +2VDC. At the same, time it uses much of the RS-232 standard for the control and handshaking signals of up to 6VDC. However, the current levels of V.35 may be as much as 200 milliamps versus the 20 to 50 milliamps of RS-422 and RS-232, respectively. The connector is not a D-sub. It is a 34-contact M-series rectangular connector with 12-gauge contacts. The larger contacts allow for the higher current needs of V.35. A V.35 cable typically uses an 8- or 9-pair cable with an overall shield, or an 8- or 9-pair individually shielded pair cable. Although V.35 is not an ANSI/EIA/TIA standard, it is still in wide use throughout the United States and Canada for high-speed interfaces up to 2Mbps.

3.1.6 Recommended Standard RS-485

The RS-485 standard was developed for equipment that may have to be daisy-chained together. Each output amplifier can drive as many as 32 input devices. The voltage levels are close to the RS-422 levels. In some cases they may be interfaced together. Although there is no specified connector for an RS-485 interface, a typical RS-485 connector has both a male and female interface stacked upon each other on each end of the cable. This allows the cables to be stacked one on top of the other. The cable exit is on one side of the connector instead of out of the back of the connector as with RS-232, RS-422, RS-530, and V.35. The contact of this type RS-485 connector is a side-mounted strip called a Centronics® connector. It is similar to the connector used with PBX or key telephone system cables. The Centronics® connector is the norm. However RS-485 may be applied to a 25-pin D-sub, a 9-pin D-sub (*Figure 1*), a screw terminal connector, and even a modular 8C8P connector. A typical RS-485 cable uses between four and 15 pairs that may be individually shielded, or have an overall shield on the cable. Although it is possible to connect RS-485 devices with long cables, most RS-485 cables are closer to 30 feet long. On balanced cable, the interface is designed to operate at up to 100kbps on a 4,000-foot cable, and up to 35Mbps on a 30-foot cable.

RS-485 is designed as a multi-point bus-type architecture where a single transmitter is designated the master, and the remaining devices are designated as slaves. Although the slaves primarily receive and process information, they may also transmit data to the master and to other slaves in the daisy-chain bus. The voltage levels of RS-485 are based on differential voltages of –5VDC and +5VDC.

3.1.7 Universal Service Ordering Code (USOC)

The Universal Service Ordering Code (USOC) was developed by Bell Labs to define the user interface to the public network. There are a number

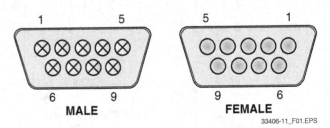

Figure 1 Pin layout of a 9-pin RS-485 connector.

of USOC specifications that all use a miniature modular connector. Some are used for telephone service and some are used for high-capacity services. The modular connectors come in six- and eight-position versions (*Figure 2*). Modified and keyed versions are also available (*Figure 3*). The internal wiring determines the registered jack (RJ) configuration. Registered jacks, regardless of their configuration, have a similar outward appearance.

The following is a list of common registered jacks and their purpose:

- *RJ-11* – A six-contact, six-position (6C6P) connector that can be used for up to three separate telephone lines (*Figure 4*). It is commonly installed for telephone service and also used for most DSL service. It is also found in PBX and electronic key systems.
- *RJ-31X* – An 8C8P connector used with alarm systems (*Figure 5*). It has provisions to interrupt an existing telephone connection and seize the line to contact public safety personnel.
- *RJ-45S* – An 8C8P connector for up to two telephone lines with termination resistors.
- *RJ-48C* – An 8C8P connector for use with high-capacity full duplex circuits such as a T1.
- *RJ-48S* – An 8C8P connector for low-speed full duplex circuits such as 56K digital data service. It is not widely used today.

Figure 4 RJ-11 connector.

- *RJ-48X* – An 8C8P connector that has an automatic loop-back connection when the connector is unplugged from the jack. It is normally used as the network jack for T1 service to keep the line active. It does this even when the terminal equipment is disconnected from the network at the network jack. The automatic loop-back feature is not in effect if the terminal equipment is not operating but still connected to the network.

> **NOTE**
> Many 8C8P connectors are mistakenly referred to as RJ-45 connectors.

3.1.8 T568 Interface

The T568 interface is defined by the ANSI/EIA/TIA 568 standard for commercial building cabling. This standard is an evolution of a Bell USOC RJ-48C that was designed for high-capacity circuits. Originally used with T1 and other high-speed circuits, it was modified for use with Ethernet and other data/telephone networking systems. It uses an 8C8P modular connector and a four-pair, unshielded twisted pair cable. There are two versions of the T568 standard, T568A and T568B.

Both interfaces locate pair one on contacts three and six, with pair four on contacts seven and eight. Both standards have the tip lead on the lower numbered contact except for pair one, which has the tip lead on contact five.

All USOC and T568 wiring has the tip lead of pair one as the higher numbered contact of the two contacts in the center of the connector. All other pairs are reversed from that and have the tip on the lower numbered contacts. T568A lo-

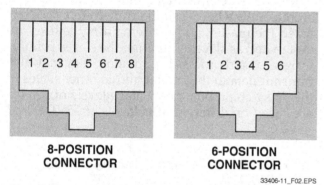

8-POSITION CONNECTOR   **6-POSITION CONNECTOR**

33406-11_F02.EPS

Figure 2 Registered jack positions.

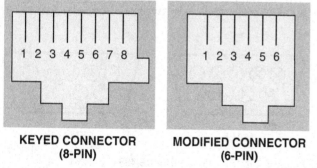

KEYED CONNECTOR (8-PIN)   **MODIFIED CONNECTOR (6-PIN)**

33406-11_F03.EPS

Figure 3 Modified and keyed registered jack positions.

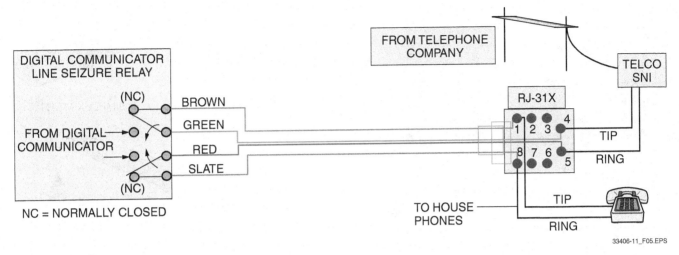

Figure 5 RJ-31X connector.

cates pair two on contacts one and two, with pair three on contacts three and six. T568B locates pair two on contacts three and six, with pair three on contacts one and two.

3.1.9 ST Fiber Connector

The standard termination (ST) fiber connector (*Figure 6*) was originally designed by Bell Labs for use with their fiber distribution systems within a dial central office. The connector uses a bayonet connection where the connector is placed inside a keyed connector. The exterior barrel is then twisted about 15 degrees to lock it in place. The connector uses a 2.5mm ferrule with a 125-micron clad fiber permanently attached inside the ferrule.

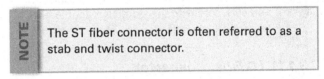

NOTE

The ST fiber connector is often referred to as a stab and twist connector.

3.1.10 SC Fiber Connector

The standard connector (SC) is placed into a chassis or bulkhead connector and pressed in place until it clicks and locks in (*Figure 6*). It was the first fiber optic connector approved by ANSI/EIA/TIA. The SC connector uses the same 2.5 mm ferrule as the ST connector, but the body has a rectangular cross-section and it uses a snap-in-place latch rather than a twisting latch. As with most fiber optic connectors, the ferrule has a 125-micron clad fiber permanently attached inside the ferrule.

On Site

Integration through Wiring

Integrating the signals between controllers built by different manufacturers may, in many cases, be accomplished directly without any modifications. This, of course, requires that the circuitry in the respective input and output channels adhere to the same communications standard.

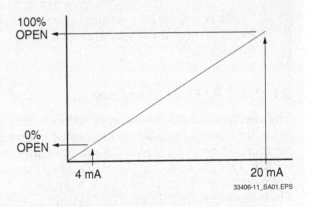

The output channel of one controller may produce a 4 to 20mA signal, and the input channel of the other controller may be designed to receive a 4 to 20mA signal. The values assigned to the signal range in the receiving controller may be matched to the sending controller by using standard controller software configuration. For example, 4mA may correspond to 0-percent open, and 20mA may correspond to 100-percent open in the sending controller. During the installation and programming phase of the receiving controller, the value of 0 percent would be assigned to 4mA, and 100 percent would be assigned to 20mA. For signals between 4 and 20mA, the software in the receiving controller would interpolate in a linear way to arrive at the value being transmitted from the sending controller.

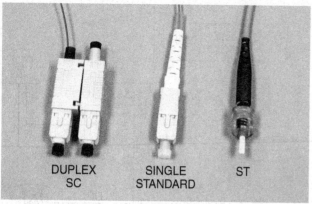

Figure 6 ST and SC fiber connectors.

33406-11_F06.EPS

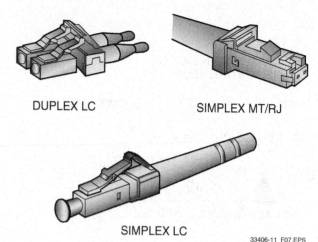

Figure 7 LC and MT-RJ fiber connectors.

33406-11_F07.EPS

3.1.11 LC Fiber Connector

Several manufacturers sell smaller fiber optic connectors to allow for higher densities within a distribution or patch panel. The local connector (LC) is popular because it is small and has a lower loss per mated pair of connectors. LC connectors (*Figure 7*) have a 125-micron clad fiber permanently attached inside the ferrule. However, the ferrule is 1.25 mm, half the size of ST and SC ferrules. This allows for much higher density within distribution and patching facilities. Because of the smaller size ferrule, the variations in fiber locations within a connector are much less, meaning that the losses between mated pairs is much lower. Manufacturing tolerances are much tighter.

3.1.12 MT-RJ Fiber Connector

The mechanical transfer-registered jack (MT-RJ) is another small form factor fiber optic connector designed to terminate two 125-micron clad fibers within a single housing (*Figure 7*). The fibers are not permanently attached inside a single ferrule. They are permanently attached within a rectangular connector that has the same overall dimensions as an eight-contact, eight-position registered jack. The MT-RJ connector fits into an RJ-45 or RJ-48 jack, but does not attach in place. It provides for higher density on bulkhead or chassis mounting and is often used on terminal equipment for duplex operation. One drawback to the MT-RJ connector is that one end within a mated pair requires alignment pins to ensure that the end faces of the fiber optic are aligned.

3.1.13 MPO Fiber Connector

The multiple-fiber push-on, pull-off (MPO) fiber optic connector is designed to terminate multiple 125-micron clad or 250-micron buffered fibers within a single connector. The connector may permanently house as few as six, or as many as 48 individual fibers. It is designed to be used with fiber optic ribbon cable that has as many as 24 single fibers connected side-by-side.

3.1.14 MTP Fiber Connector

The mechanical transfer push-on, pull-off (MTP) fiber optic connector (*Figure 8*) is a version of the MPO connector. These connectors are interchangeable. Connectors such as the LC and MTP/MPO do not have an ANSI/EIA/TIA standard applied to them, but they are becoming the standard in the fiber optic industry.

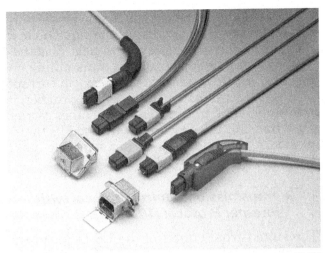

Figure 8 MTP/MPO connectors.

33406-11_F08.EPS

3.1.15 Hardened Fiber Optic Connectors (HFOC) and Hardened Fiber Optic Adapters (HFOA)

Hardened fiber optic connectors (HFOC) and hardened fiber optic adapters (HFOA) are passive components used outside the plant. They are designed to withstand temperature extremes and extreme weather conditions. They may be located inside various enclosures or terminals at the customer premises. They can be mated in the field and are highly desirable for fiber to the premises (FTTP), fiber distribution hubs (FDH), and industrial applications. At the present time, they are not in widespread use.

> **NOTE**
>
> Most fiber optic connectors do not have ANSI/EIA/TIA standards applied to them. However, they are sanctioned by ANSI/EIA/TIA as viable fiber optic connectors.

3.2.0 Layer 2 – The Data Link Layer

The data link layer defines the physical addressing of the first layer. Sometimes multiple devices are connected using a single electrical path for communications. Each device needs only the information that is designated for that device. Just as an address is placed on an envelope to designate where it is to go, an address is placed within each device to identify it. An address placed within the data defines which device is to receive what information. Data is seldom sent as just bits. It is normally sent as data bytes or data packets. A DS-0 is the lowest level of digital telephone communications. Within each DS-0, each of the eight bits represents a voltage level. The most significant bit has the highest voltage value, and the least significant bit has the

lowest voltage value. This process is the framing of the DS-0. When 24 DS-0s are joined together, they form a T1 or ISDN primary rate interface where each DS-0 has its own time slot. The T1 or primary rate interface (PRI) is started by a framing bit to tell the receiving end which DS-0 is number one and which is number 24. That is how the T1 or PRI is framed. Regardless of which framing method is used, it is necessary for the data to be framed to determine what bit is what, within it. For older style framing, a start bit or start baud is inserted to define the starting point for the data byte. At the end, a stop bit is added to tell the receiver where the byte ends.

Layer 1, the physical layer, defines the physical connection and electrical specification. Layer 2, the data link layer, defines how the data is to be framed during transmission. If the physical layer does not provide the connection, the data link layer has nothing with which to work. If the framing is not correct, the physical layer can de-

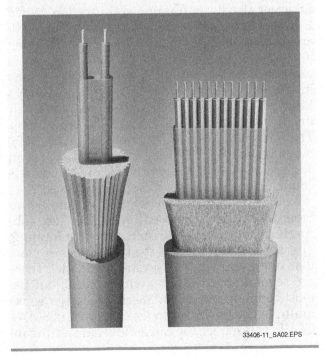

liver the bits and bytes but the data link layer cannot interpret the information for the upper layers to process. There are a number of communication techniques for framing information for transmission and submission to upper layers. They include the following:

- RS-485
- Ethernet IEEE 802.3
- Transmission control protocol with Internet protocol (TCP/IP)
- High-speed transmission level one (T1) and integrated switch data network (ISDN)
- Token ring
- Modem to terminal handshaking
- Error detection
- Flow control

3.2.1 RS-485

For RS-485 or other multipoint connections, an address is inserted within the data that identifies the device that is to receive the information. If two or more devices have the same address, they both will try to respond to the original transmitter at the same time. That can confuse the original transmitter. With two or more devices trying to reply at the same time, the data of all devices becomes corrupt and unusable. Some communication standards address more than one layer in the OSI model. RS-485 defines the electrical specifications of the physical layer and the framing specifications of the framing layer.

3.2.2 Ethernet (IEEE 802.3)

Ethernet also has a framing specification. When Ethernet was first designed, it used a large-diameter coaxial cable that had to be tapped along its length for devices to be attached. The coaxial cable and the tap were the physical layer and the interface to the tap provided the framing. Today's Ethernet is usually transmitted over four-pair unshielded twisted pair using T568 connections. The frame itself may be anywhere from about 50 bytes to about 1,500 bytes. Within the frame is an address section that identifies where the data is coming from and where it is supposed to go.

IEEE Specification 802.3 defines the framing for Ethernet. Within that specification, there are subspecifications that define more than just framing. This specification crosses the boundaries of more than one layer in the OSI seven-layer model. There are several specifications within 802.3, but the most common in North America is carrier sense multiple access with carrier detection (CSMA/CD). In this model, the carrier for transmitting information is started before any device (node) puts information on the network. At the same time, it monitors the network for carrier signals. That prevents two or more devices from transmitting on the network at the same time, which can corrupt data. If a second device starts sending carrier, it ceases operation until it senses the other carrier is not transmitting. Once it has finished its transmission, it shuts down its carrier, allowing other devices to use the network.

3.2.3 Transmission Control Protocol with Internet Protocol (TCP/IP)

IP resulted from efforts of the US Department of Defense to design a communications network that would remain operational if two or more major communications hubs were destroyed during a war. Most of the Internet protocol has nothing to do with the physical layer and only a slight amount to do with the data link layer. However, it does have some framing information that has to be operated in the data link layer, such as the addressing scheme.

> **NOTE**
>
> Prior to 1989, a person had to obtain permission from the US Army to access the Internet.

3.2.4 High-Speed Transmission Level 1 (T1) and Integrated Services Digital Network (ISDN)

In the public switched telephone network (PSTN), both T1 and PRI are referred to as DS-1. Within each DS-1 there is a framing bit to show where each frame starts. That is the data link layer information. The telephone wiring is the physical layer and the voice information within each DS-0 is part of another layer in the seven-layer OSI model. DS-0, T-1, and T-3 are strictly point-to-point, with one transmitter and one receiver in each link.

3.2.5 Token Ring

Token ring is another transmission medium that requires addressing. This specification defines a specific cable (IBM Type 1 or Type 2), a universal connector, and a specific addressing scheme that belongs to the data link layer. Token ring does not use the carrier signals of Ethernet. Instead, the host sends a token to the ring. If a node (device) on the network receives a token and has data to transmit, it does so and then releases the token back onto the ring for other nodes. A node cannot transmit any information until it receives a token from the host. As the name implies, this is a ring network. All nodes are interconnected to form a ring that starts and ends at the host, with all other nodes being attached onto the ring.

3.2.6 Modem to Terminal Handshaking

Within communications links, there are devices that operate over distances, as well as devices that process input and output data. The devices that operate over distances are called data communications equipment (DCE). Devices that are designed to process data are called data termination equipment (DTE). Both types of equipment may operate over the physical layer in one of three modes: simplex, half-duplex, or full-duplex.

Simplex devices are one-way devices. Transmitters will only transmit and the receivers will only receive. Broadcast TV and radio are examples of simplex devices. Both send signals out for all to use. The station sends signals but cannot receive them. Your TV set receives signals but cannot transmit them.

Half-duplex communication is where the devices both transmit and receive, but not at the same time. A walkie-talkie is an example of half-duplex communications. You can send or receive information with a walkie-talkie. However, you cannot send and receive at the same time.

Full-duplex communication is a communications link where information can be sent and received at the same time. A telephone is a full-duplex communication device. Both parties can talk at the same time.

A modem is a DCE device. A terminal is a DTE device. Modems and terminals must communicate with each other to ensure that each is operating properly. Handshaking signals are used for that purpose. Handshaking signals include the following:

- *Request-to-send (RTS)* – Used when the terminal is ready to send data over the physical layer. It sends a signal to the modem telling the modem that it has data to send. In some older half-duplex equipment, this signal was used to get the modem to change from a receiver to a transmitter so the terminal could send data over the link.
- *Clear-to-send (CTS)* – Used when the modem is ready to transmit the data over the communications link. It sends the clear-to-send signal back to the terminal so it can put its information on the physical layer.
- *Data terminal ready (DTR)* – Used when the modem is telling the terminal that it is operational and ready to transmit or receive data over the

physical link. This signal is the terminal telling the modem that it has power and is ready for operation. This signal may be called DTE-ready or terminal-ready.
- *Data set ready (DSR)* – Used when the modem is telling the terminal that it is operational and ready to transmit or receive data over the physical link. This signal is the modem telling the terminal that it is available for operation. This signal may be called data mode, or DCE-ready.
- *Carrier detected (CD)* – The modem monitors the communications link. When there is a receivable carrier signal present, it sends a carrier-detected signal to the connected terminal. For half-duplex modems, this signal is used to tell the modem to send a clear-to-send signal to the terminal. For full-duplex modems, this signal allows the terminal to hold its request-to-send signal until the carrier-detected signal is ready for communicating over the link. This signal is also known as received line signal detector and receiver ready.
- *Incoming call (IC) and ring indicate (RI)* – Both serve the same function. The modem tells the terminal that a call is coming in and needs to communicate with it. It is based on the telephone ring signal. The modem and/or terminal often go into sleep mode after it has been idle for a length of time. When the modem sees a ringing voltage on the phone line, it wakes up and sends this signal to the terminal telling it to wake up and get ready for operation. This sequence commonly occurs when a fax is sent over telephone lines.

Here is an example of how handshaking signals are used. Assume a half-duplex modem is connected to a dumb terminal. Data is entered using a keypad or other device. Then an Enter or Send key is pressed. The terminal first looks to see if data set ready is active so that the terminal knows that the modem is operational. If data set ready is active, it sends a request-to-send signal to the modem. The modem then sends a carrier signal on the communications link and waits for the distant receiver to get ready to receive. The modem then sends a clear-to-send signal to the terminal. The terminal forwards the information it wishes to transmit. The modem remains in the transmit mode until the other end starts sending back its own carrier. At that point, the modem holds all clear-to-send signals until the modem can again return to the transmitter operation. All handshaking signals depend on the physical layer, but they operate in the data link layer.

On occasions, a terminal must connect to a terminal, or a modem must connect to a modem. In those cases, a null modem cable is required because the handshaking signals will not be complete. There are only three terminal signals to replace all five modem signals that the terminal may need for proper operation. *Table 2* gives the pin connections for a null modem cable using three different configurations.

3.2.7 Error Detection

Error detection determines if the information received is correct. Data that is received should be the same data that was sent. Since termination equipment is the source of that information, it must be in proper working order to ensure that the information it is sending is correct. A parity bit is used for error detection. The parity bit tells if there is an even or odd number of ones in the data byte. If the parity bit does not match when the byte is received, it indicates that the information has been corrupted and is not usable. If the parity matches, the receiver sends an acknowledgement byte (ACK) back to the transmitter. If a byte is not a parity match, the receiver sends a non-acknowledgement signal (NAK) back to the transmitter and throws away the erroneous byte. When an NAK signal is received by the transmitter, it retransmits the original byte until an ACK byte is received. This method is one of many types of forward error control that work over the physical layer but operate in the data link layer.

3.2.8 Flow Control

Many transmission algorithms provide special flow control measures within the transmitted data. One example is X.25. If the received data is suspected of being corrupt (parity error), the receiver sends the transmitter an NAK signal. It will then have the transmitter send the data again until the receiver gets correct data. With no parity error, an ACK byte is sent. If the transmission media is of poor quality, it can cause throughput

Table 2 Null Modem Cable Pin Connections

25-Pin to 25-Pin RS-232 Null Modem		9-Pin to 9-Pin RS-232 Null Modem		25-Pin to 9-Pin RS-232 Null Modem	
2	3	2	3	2	2
3	2	3	2	3	3
4 (& 22)	5	7	8	4 (& 22)	8
5	4 (& 22)	8	7	5	7
6 & 8	20	6 & 1	4	6 & 8	4
20	6 & 8	4	6 & 1	20	6 & 1
7	7	5	5	7	5

delays and slow the transmission rate to a crawl. For transmission by way of satellite, the time delay between uplink and downlink causes the link to disconnect. Other algorithms, such as high-level data link control (HDLC), provide a numbering sequence while transmitting. The receivers acknowledge all good data and reject suspect or bad data using the same numbering sequence. Once the receiver has sent an acknowledgement for a bit stream, the transmitter restarts its numbering sequence. Continuing with a numbering sequence that is larger than the data itself reduces throughput to a trickle.

Within Ethernet, there are two sub-layers to handle this. The Media Access Control (MAC) sub-layer allows information to go only to the device or devices that are specifically addressed. Within each device, each manufacturer assigns a MAC address; similar to the way the postal service assigns street addresses for each house and building within a city. Each MAC address is unique and not duplicated within a manufacturer. Since part of the address also identifies the manufacturer with a unique code, no two addresses are alike. Within a MAC, errors are detected but not corrected. Corrections take place in higher layers of the seven-layer OSI model.

The next sub-layer in Ethernet is the Logical Link Control (LLC). This sub-layer deals with addressing and multiplexing when multi-access media is required. It tends to address a route plus an address for the data's end location or locations. It is used more for bus networks rather than point-to-point connections.

3.3.0 Layer 3 – The Network Layer

Not all data is transmitted using a single 8-bit byte, byte-by-byte data transfer method. Most transmissions are sent as a number of bytes of data, packaged within headers and trailers to form a packet. The packets may be of a specific length. Some packaging is done with expanding and contracting packets that do not have a specific length. These are variable length data sequences like Ethernet. They are normally transmitted from a source host on one network to a destination host on a different network.

Within the network layer, routing information such as a destination address, source address and packet length are added. Within this layer, routing information may be added to the packet by a device that is not part of the sender or receiver. An in-route router adds specific route information along the entire path the data travels. A single router may send one packet on one route, and one or more subsequent packets on a different route. This is especially true if the first path becomes degraded or overcrowded. The packet sequencing numbering scheme tells the receiver how the sequencing of the packets is performed. That is done so the receiver can reassemble the packets in the proper order, even if the packets are not received in the same order in which they were originally sent. This is important if the original data becomes fragmented between packets.

Within the network layer, addressing is not always a device address; it may be an interface address. For example, a 48-port data switch has each port with its own address. The device connected to that port may or may not have an address. The original transmitter addresses the port interface and allows the device connected to that port to receive the data. This is a common connection for layer one RS-232 and RS-422, and layer two Logical Link Control protocol.

There are a number of ways addresses are assigned to ports or devices on a network. One way is to have an address permanently assigned to a specific port on a device. This is called a static address. If the Ethernet port of a PC has an IP address permanently assigned to it, that address never changes without going into the setup of the computer and assigning a different address. Another method is to let the dynamic host configuration protocol (DHCP) assign a computer an address when a connection is made within its network. In this protocol, the host has a number of addresses that it can assign to any computer at will. Each time a connection is made to that host, the host assigns an address that identifies the computer to the host. A different address may or may not be assigned each time a connection is made to that host. The host controls the address and identifies each device connected to that host.

The network layer not only contains the addressing information and packet definitions, it also contains some error control schemes. One example is an expansion of the parity bit used in layer one. The basic parity bit is considered a horizontal error control. Think of adding a column of bytes with the sum being totaled at the bottom. That is how vertical parity works. A number of bytes are summed and then parity is assigned to that sum.

The sum and parity sum are transmitted to double-check the data bytes within the packet. If there is a vertical parity error, there often is a horizontal parity error. Some algorithms require the entire packet to be retransmitted. Others discover the bad byte and have it retransmitted. The algorithm is a function of layer four, but this error control is part of layer three.

With horizontal and vertical parity, there is sometimes a counter within the packet that tells the receiver how many data bytes are inside the packet. This is especially true with varying length packets in media like Ethernet. If a packet is received and the proper number of data bytes is not present, the receiver rejects the packet. When a packet is received with both the horizontal and vertical parity correct, and the correct number of data bytes is present in the packet, it can be concluded that the packet is correct. If that occurs, the receiver sends an acknowledgement back to the original transmitter. A numbering scheme for each packet of information determines which packets are acknowledged and which ones are rejected. When the receiver has all good packets, the numbering scheme restarts to prevent the packet numbering system from overloading the packet with the numbering scheme. If this is not done, throughput of data is lost.

Internet protocol version four (IPv4) is one of the most common network three layers in use today. This protocol uses a series of eight numbers in a base set of numbers to form an Internet address. The IPv4 protocol is currently in the process of being replaced with Internet protocol version six (IPv6) which uses a series of 16 numbers in a base set of numbers to form an Internet address. The advantage is more Internet addresses for worldwide use.

The network layer is network-specific but can be translated by the data link layer for use between differing networks. It depends on the data link layer presenting correct data over the communications link. The data link layer depends on the physical layer to make the physical and electrical-optical connection.

3.4.0 Layer 4 – The Transport Layer

The function of the transport layer is to provide a common interface to the communications network. It translates whatever unique requirements the higher layers might have into data the network can understand. The transport layer detects and corrects errors in transmission and provides for the expedited delivery of priority messages. It also checks the data bytes, puts them into the proper order when necessary, and usually sends an acknowledgement back to the originating transport layer. Layer 4 also attempts to re-establish contact in the event of a network failure. The transport layer checks the data to make sure it is correct and in the correct order to be used at a higher level. This is where parity bits are added and sequence numbers are placed onto each packet. The receiving data link layer checks the individual packets for the proper parity and either puts them back into the proper sequence, or requests a retransmission of information using acknowledge (ACK) or non-acknowledge (NAC) characters for particular packets or characters.

3.5.0 Layer 5 – The Session Layer

The session layer establishes the communication link between computers. It buffers the information to the device or program that performs the presentation function. Layer 5 recognizes users and acknowledges both their arrival and departure. In some systems, the session layer can be a driving factor in system design; in others, it is a very small consideration. It is this layer where the DHCP or static IP address assignments are performed and monitored.

3.6.0 Layer 6 – The Presentation Layer

The presentation layer prepares the information for the application. For example, a file received from a computer using ASCII would be converted into the proper format for display on a system using rich text format (RTF) by Layer 6. Each system uses different codes to represent a character, letter, or number. The presentation layer must know the differences and provide for them.

3.7.0 Layer 7 – The Application Layer

The application layer interacts with the user's software programs. Layer 7 is where the information in the message and how well it serves the user is determined. This is where application programs call upon the communication services. If this function is not handled properly, the entire system will be useless, because the goal of the system is to serve the user. Typical protocols in Layer 7 include File Transfer Access Management (FTAM) and Virtual Terminal (VT). FTAM is primarily within the computer such as the CPU. VT usually handles input and output devices, such as the keyboard, mouse, and monitor.

4.0.0 COMMUNICATION BETWEEN SUBSYSTEMS

It is normal for each subsystem of an integrated system to have its own proprietary operations protocol. Although the application layer may connect

to a centralized controller of all the subsystems, it is the job of the central controller to provide control signaling to each of the subsystems. The topology of the control is the variable and the interpretive segment of the central control system. The central controller must communicate with each of the subsystems and provide the necessary inputs and outputs to control, monitor, engage, or disengage each subsystem.

4.1.0 Basic Topology

Most basic integrated systems have a hub topology. A topology is a map or plan of a control system and its associated subsystems. As a general rule, at the center or hub of every integrated system is a systems controller. The systems controller is typically a computerized piece of equipment with the dedicated purpose of connecting multiple stand-alone systems into a cohesive integrated system. *Figure 9* shows one example of a systems controller and how it interfaces all the subsystems. Each subsystem is a complete system in its own right, but is a subsystem of the controller system.

A typical controller has one or more user interfaces that operate the systems controller. It is the human or user interface. The systems controller may have one of several other controlling interfaces, such as a dial-in/dial-out access or Internet

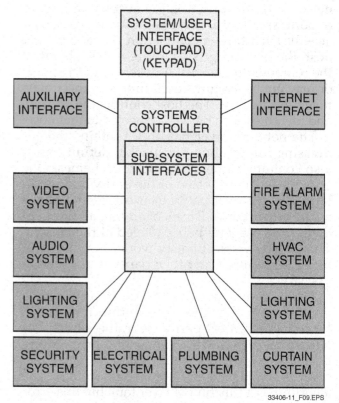

Figure 9 Systems controller and interfaces.

access to the systems controller programming. For fire and intrusion, it is important that the system controller be able to contact emergency personnel. It is desirable for the owner or operator to be able to access the systems controller to monitor or control the subsystems from another location.

The user interface, whether local or remote, often needs to have a specific access code to perform any control functions at the systems controller. The access code may be as simple as a dial string or as complicated as a user name and password.

4.2.0 Protocols

The systems controller has all the hardware and software to control or interpret the interface protocol of each of the subsystems. Subsystems protocols can be varied. Examples of subsystem protocols include:

- *RS-232* – Serial communications protocol
- *RS-422* – Serial communications protocol
- *RS-485* – Serial communications protocol
- *IR* – Infrared optical interface
- *RF* – Radio frequency interface
- *USB* – Universal serial bus interface
- *Firewire* – IEEE-1394 serial bus interface
- *Voltage-sensing* – Contact closures/changes in voltage
- *Voltage ramping*
- *TCP/IP* – Network protocols
- *Ethernet* – Network bus protocol
- *Contact closing/opening* – Providing on/off functions as control
- *Proprietary protocols* – Specific software languages that the systems controller must interpret and manipulate
- *Multimedia over Coax Alliance (MoCA)* – Standard for data communication and multimedia streaming over coax.

4.3.0 Network Configurations in Complex Systems

In a simple system, one controller may control a single facility, such as a movie theater or a meeting room. As a client's needs become more complex, integration of the systems required to fill those needs increases (*Figure 10*). For example, a hotel conference center may have one or more large meeting rooms plus several smaller meeting rooms. It is common for large meeting rooms to be divided into smaller rooms. Each of these smaller rooms may require separate controls. In such an arrangement, a master control serves the larger room, with separate controls for the smaller rooms. Some of the functions to be controlled include electrical power, audio, video, lighting, and HVAC.

Figure 11 shows how a single large room can be divided into smaller meeting rooms. As many as four systems have to be converged into a single control system. That system, in turn, controls all the systems within the four separately controlled spaces. During the course of a larger meeting, each smaller meeting in different rooms may have completely different needs, ranging from simple to complex. For example, one room may require a complex audiovisual control system, while the room next to it only requires a microphone.

4.4.0 Intersystem Connections

There are a number of connection methods available that ensure that intersystem connections perform properly.

4.4.1 Ethernet

Ethernet (*IEEE 802.3*) describes the physical characteristics for the transmission of data on four-pair twisted-pair cables, coaxial cables, and fiber optic cables operating at various speeds. Although Ethernet was originally designed as bus architecture, it is in common use for point-to-point connections via bus architecture and an addressing scheme. It has entries in the lowest four layers in the seven-layer OSI model.

4.4.2 Transmission Control Protocol/Internet Protocol (TCP/IP)

Transmission control protocol/Internet protocol (TCP/IP) is a suite of protocols used by devices on a network to communicate with each other. It has appearances in layers two, three, four, and five of the seven-layer OSI model. It is the predominant set of protocols used in LANs as well as the Internet. It is often confused with Ethernet. Ethernet defines the physical interface and the structure of the packet that contains the message or data to be sent. TCP/IP defines the structure of the messages transmitted on the network. TCP/IP messages are put into Ethernet packets and transmitted on Ethernet networks.

4.4.3 File Transfer Protocol (FTP)

The file transfer protocol (FTP) is used to transfer files from one machine to another. It may well be an attachment to an email. The programming of some systems controllers can be updated using FTP. Many systems controllers have a specific location in memory on a hard disk to store files that

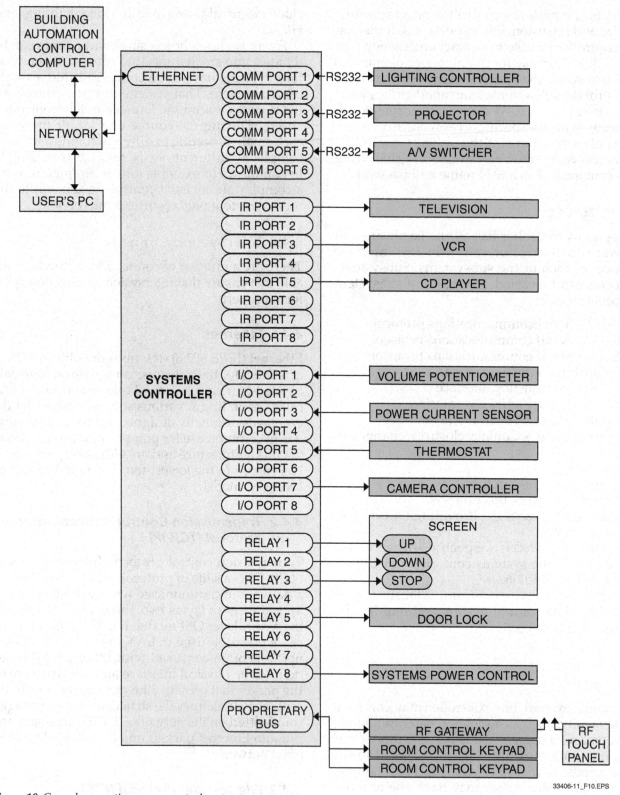

Figure 10 Complex meeting room control system.

are important for a project, such as source code, graphics, user interfaces, and plan files like Auto-CAD® drawings. Some systems controllers allow transfer of files to and from this memory location with FTP.

4.4.4 Simple Network Management Protocol (SNMP)

Simple network management protocol (SNMP) is a standard for gathering statistical data about network traffic and the behavior of network com-

33406-11_F10.EPS

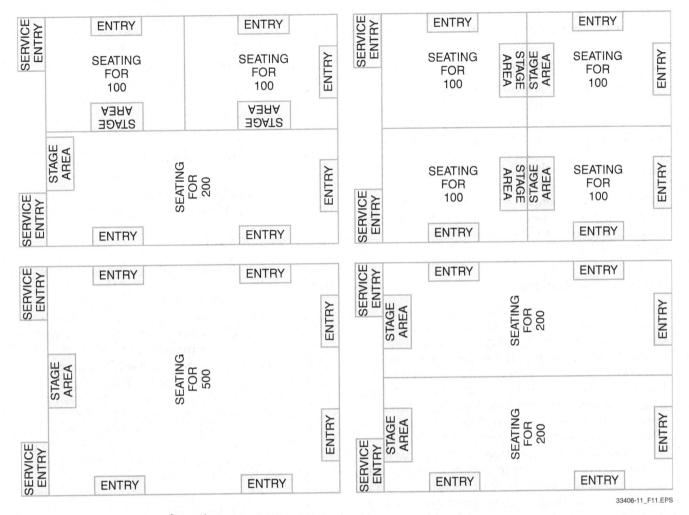

Figure 11 Meeting room configurations.

33406-11_F11.EPS

ponents. It is a protocol that allows a centralized application to monitor the status and health of a remote device. SNMP-enabled devices can send out special messages (traps) when certain events occur. For example, an SNMP-enabled amplifier can send out a power-off trap when it has been put into standby. It could also send a temperature-exceeded trap if it gets too hot. The SNMP-enabled computer (systems controller) listens for these traps and takes appropriate action when they are received.

4.4.5 Simple Mail Transport Protocol (SMTP)

Simple mail transport protocol (SMTP) is used to send email on the Internet. If a device is SMTP-enabled, it can send out email messages when certain events occur. Using the example from the preceding section, if the systems controller receives a message from an amplifier that the user-defined temperature threshold has been exceeded, it can send an email message to the A/V maintenance crew.

4.4.6 Telnet

Telnet is a protocol for remote computing on a LAN as well as on the Internet. It allows a computer to act as a remote terminal on another machine. When you Telnet to a remote computer, which must be configured to accept a remote connection, the remote computer (systems controller) accepts input directly from your computer. Output from the remote computer is directed to your screen. Many network-capable devices used in an integrated system can be configured and controlled through a Telnet session. Often, extensive system troubleshooting can be done remotely through use of the Telnet protocol.

4.5.0 Network Device Addresses

To send control signals, software updates, or commands between interconnected or integrated components, each device must be addressed properly so the messages can be received and verified by both the systems controller and the subsystems controller. These addresses fall into two categories—physical and logical.

4.5.1 Physical Address

Every device on a network has both a MAC address and an IP address. The MAC address is a physical address placed in the device by the manufacturer. Using a MAC address, a specific device can be identified on a network. In theory, no other device in the world can imitate it. For example, a security application can identify a specific IP-based camera on the network based on its MAC address. No other camera's video can take its place because it is the only camera with that MAC address.

4.5.2 Logical Address

A logical address is an address that must be assigned within a network, and if needed that address can be changed. A logical address on one network can be identified from another network. If the logical address is not on the network that is looking for the device, the seeking network uses a gateway to leave one network and enter another network. Every device on the Internet requires an IP address to uniquely identify it.

Within most large networks, the network is segmented into sections for easier access within each segment. A sub-net mask identifies each of the network segments for locating each item on each segment of the network. If network devices are on different logical networks, the IP configuration for each device must identify a default gateway as a logical route it can use to get out of its network and gain access to another network.

The default gateway normally needs a sub-net mask to locate the network to which it is attached. When a message is sent over the network, the network first determines if the destination address is within its own network. If the destination address is not in the local network, it forwards the packet to the default gateway for routing to another network. The router then locates the destination and sends the packet onto a route that reaches the destination address. Using this technique, a packet is sent from router to router to router over the Internet until it arrives at the destination network and then on to the destination device.

4.6.0 Networks, Hubs, Switches, and Routers

Networks, hubs, switches, and routers are designed to send and receive messages and packets. They do not originate or terminate messages or packets; they just send them on to another point.

4.6.1 Network Hubs

A network hub, often called a repeating hub, is an Ethernet device. Any packet that is received on any port of a hub will be repeated back out of each port of that hub. A hub can be looked at as a ported bus. Like a bus, all ports are connected and have the same information of every other port. It receives packets and resends those packets out of every port at the same time. A hub does not handle streaming audio or video very well. The hub repeats all the streaming audio or video out of every port, leaving the hub saturated with streaming data. No other device connected to the hub is able to send or receive any information until the streaming data has ended.

4.6.2 Network Switches

Compared to a hub, a network switch is a more intelligent Ethernet device. As devices are attached to the switch ports, the switch port gathers information about the device, including its MAC and IP addresses. That information is then stored in a table in the switch memory. When a switch receives a packet, it looks at the address information to determine if the device is connected to one of its ports. If the device is connected to one of its ports, it repeats the received packet only on the port to which the device is connected. Streaming video or audio is repeated only to the port that has the

On Site

NEC® Requirements for Communications Wiring

The types and ratings of cables for system integration must meet *NEC®* safety requirements, local codes, and equipment manufacturer requirements. Most of the requirements governing installation of building management system communications wiring are covered in *NEC Articles 725* and *800*. Other articles may apply as well, depending on the system design. *NEC Article 725* covers Class 1, Class 2, and Class 3 remote control, signaling, and power-limited circuits. *NEC Article 800* covers communication circuits.

address found in the packet destination address. No other ports are affected by the streaming data, so they transmit and receive packets as normal.

Network switches are either manageable or non-manageable. A non-manageable switch cannot be configured or monitored. For that reason, it is difficult to control what information should be sent through such a switch. A manageable switch can be configured and monitored.

Manageable switches are preferred since individual ports can be turned on or shut down to limit access and prevent unauthorized access to a network. In addition, manageable switches can be used with the following systems integration protocols:

- Spanning tree
- Internet group management protocol (IGMP)
- SNMP
- Load balancing
- Virtual local area network (VLAN) configurations

> **NOTE**
> Unauthorized access to a network enables criminals to gather information and to insert malicious programs (viruses).

4.6.3 Virtual Local Area Network (VLAN)

Many integrated systems have components that reside on the client's LAN. In some implementations, a significant portion of the integrated system is on the LAN. This can be a cause of concern to the client's network administrator for the following reasons:

- The network administrator might not be familiar with equipment being installed by a third party resulting in a low comfort level concerning the stability of the hardware.
- The type of traffic that the new hardware will be handling may be different from what is typical for the network. For example, the part of the network used by the accounting department may require protection from unauthorized access. The network administrator may not like adding hardware that may compromise network access.
- For security reasons, it may not be desirable to have data generated by the integrated systems available on the same network as other users.
- Applications such as streaming audio and video can affect bandwidth needed for other network traffic and cause a slowdown.

These problems can be overcome. In some installations, it might be desirable to install a separate data network that just handles the requirements of the integrated system. This solution isolates all data generated by the integrated system hardware. However, this solution can be expensive.

An alternative is the implementation of a virtual local area network (VLAN). While a LAN is a physical grouping of network devices, a VLAN is a logical grouping of network devices. In other words, all the network devices on a VLAN act as if they are all physically attached to the same network segment. In reality, they may be physically connected to separate networks or network segments. Using this technique, a single physical LAN can be logically configured to appear as if it was two or more separate LANs. Each logical network is invisible to the other. This technique offers great benefits, including the following:

- It is much less expensive than running a separate network for the integrated system.
- Security is enhanced. A user may be on the same physical network segment as a device from the integrated system but has no access to the integrated system's network data, and vice versa. That is because they are on separate virtual networks.
- Network traffic load balancing is simplified. High-bandwidth applications on one VLAN do not affect the traffic on a different VLAN, even though they may be physically located on the same network segment.

A virtual network is typically created by configuring network switches. Non-manageable switches do not have the ability to be configured into VLANs. If a virtual network is being planned, the designer must verify that any switches to be used in the integrated system are manageable and have VLAN capability. An example where a VLAN could be used involves transferring high-fidelity audio over an Ethernet network. It is critical that there are not obstacles to the transmission of data over this high-bandwidth network because if Ethernet packets containing audio data are delayed or lost, the users hear a popping noise instead of the audio. It is also critical that this high-bandwidth transmission not interfere with other applications on the network. In this situation, the network switches in the path between the audio source and the audio destination can be configured into separate network traffic from the audio device into a VLAN. This isolates the audio data from the rest of the network and keeps it from interfering with the data from other applications. It also keeps other data from interfering with it.

4.6.4 Quality of Service (QoS) and Transmission Algorithm H.323

When transmitting data over a network or between networks, packets are often transmitted across different routes from the sender to the receiver. Because of the varying paths and delays along the way, packets often do not arrive at the destination in the same order in which they were transmitted. With normal data, this is not a problem because the end equipment can realign the packets and display or record the data in the same way it was transmitted. When streaming audio and video packets are received in an incorrect order, the end equipment cannot reassemble them in the proper order without having a delay in the received signal. It is important that all packets are received in the same order as when they were sent. The only way to ensure that is to make sure each packet follows the same route from sender to receiver. Transmission algorithm H.323 adds a tag within the header for each router along the path to keep each tagged packet on the same path as the previous tagged packet. This may cause some minor delay in the streaming data, but it keeps all packets in the proper order so that the received audio or video is not distorted. If both audio and video are sent at the same time, audio remains in sync with the speaker's lips. VoIP networks are often set up with both H.323 protocol and assigned to a specific VLAN. This keeps the VoIP traffic separate from other data and provides quality of service with each packet of voice-over-Internet data.

4.6.5 Wireless Networking (IEEE 802.11)

Wireless networking is a very popular option for networks. Many schools have a wired LAN for internal operations and a wireless LAN for persons to access the Internet. The use of wireless networks in integrated systems is also common. Wireless devices offer flexibility to the installer, the integrator, and the client.

Here is an example of an advantage of using a wireless network in an integrated system. Once an audio system for an auditorium has been installed and configured, it must be balanced for optimal performance. A computer with digital signal processing software is used to optimize the audio system. Because the digital signaling processing equipment and each speaker are part of a LAN segment, the system can be tuned with a PC. A wireless-capable laptop PC can be used as a person moves from location to location within the hall to balance and equalize the system. The wireless network allows this to be done. The tech-

nician does not have to deal with dragging a network cable around the hall.

The operation of a wireless device in relation to a wireless access point transceiver is a function of distance. Within a building there also are materials that can block or absorb wireless signals, limiting the range of reception. There are several specifications within *IEEE 802.11* that can be used to help solve some of these problems. Devices can operate at different frequencies, use different modulation schemes, and have different bandwidths that can accommodate different data rates. The different standards provide a wireless operating range that varies from a little over 100 feet indoors to nearly 800 feet outdoors. As a wireless device moves around a wireless network, it can move out of range for one transceiver and come within range of another. The network automatically hands over the connection from one to another so the device continues to operate seamlessly. The number of transceivers required for a facility depends on the anticipated number of users, distance limitations of hardware, and physical obstacles to the wireless signal.

A big problem with wireless operation is unauthorized use. In congested areas, it is common for one user to access the Internet using someone else's wireless access without that person knowing that his or her wireless system is being accessed. Anyone with a wireless device may be able to access any wireless transceiver, and thus the wireless network. This can lead to unauthorized access to a person's private information. One method of limiting unauthorized use is to assign encryption codes within the network to limit access into the wireless network. Most encryption codes such as wired equivalent privacy (WEP) are susceptible to hacking because they use 40-bit code that is easy to defeat. Some 128-bit encryption codes can also be defeated by serious hackers.

> **NOTE**
>
> The National Security Agency (NSA) has developed a number of very secure encryption codes. These codes are strictly controlled by the US government.

A newer encryption called Wi-fi protected access (WPA) is more difficult to defeat. The Internet Security Association and Key Management Protocol (ISAKMP) is a protocol defined for establishing security associations and cryptographic keys in an Internet environment. ISAKMP only provides the framework for authentication and key exchange and is designed to be key exchange independent.

Another method of access control is to monitor each wireless access point for MAC addresses, and only allow authorized MACs to operate on the wireless network. This is effective in detecting and stopping unauthorized use. The method requires a table of all known authorized MACs and constant updating of equipment. It is a difficult system to deal with. Multiple wireless local area networks (WLANs) can operate in the same space, yet be totally separate and independent. Each WLAN must have its own service set identifier (SSID), and each WLAN must be configured with the service set identifier. Using this method, users can connect to the correct WLAN by selecting the appropriate identifier.

4.6.6 Unicast, Multicast, and Broadcast

It is very common to stream audio and video over data networks. Systems integrators are often asked to provide clients with the ability to view programming at the desktop as it happens. This provides unique challenges to the systems integrator as well as to the network administrator. Streaming data can be a bandwidth-intensive application. If the network is not configured properly, streaming can seriously affect the network's performance.

One of the factors to consider when configuring a network for audio and video streaming applications is to determine if the transmission will be broadcast, multicast, or unicast. In the unicast transmission of a stream, the packets of the stream are destined for only one address. If there are to be multiple recipients of a unicast transmission, there will be multiple streams, each with a distinct communication from the sender to the recipient.

In the multicast transmission of a stream, the multiple recipients of the stream become members of a group. In this case, only one stream is transmitted by the sender and it is addressed to the multiple recipients in the group by using a special multicast destination address. Those willing to be part of the group may then receive the stream. Internet group management protocol (IGMP) must be enabled on a switch to allow nodes attached to it to receive multicast transmissions. IGMP is the TCP/IP protocol that permits Internet hosts to take part in IP multicasts.

Broadcast transmission of a stream is what is more commonly called video on demand (VoD) or Internet protocol TV (IPTV). In reality, it consists of a video server where the video is stored and then forwarded either as a unicast or multicast stream to a single user of a user group.

Streaming data over coaxial cable is supported by the MoCA protocol. MoCA is a standard for using existing coax cabling to connect devices such as TV set-top boxes, DVRs, wireless access points, and game consoles to support data communication and streaming multimedia. MoCA version 2.0 offers basic and enhanced modes, with 400Mb per second and 800Mb per second throughputs. In point-to-point WAN applications, performance can be optimized, with basic mode delivering a 500Mb per second throughput and enhanced mode providing 1Gb per second throughput.

5.0.0 SYSTEM PROGRAMMING

Proper programming is what makes systems integration happen. The power, functionality, and flexibility of the system depend largely on the programming done by the systems integrator. The programming used in a specific system depends on the following factors:

- The systems controllers being used
- The subsystems programming languages and connections
- The complexity of the system
- The level of integration with the client's databases
- The necessity of interfacing with the client's network authentication scheme

By definition, systems controllers are computer devices with an operating system that allows programming by the systems integrator. Each manufacturer's operating system is proprietary and is designed to make the most of the capabilities of their controller. Therefore, the programming language developed by each manufacturer is proprietary and can only be used with their systems controllers. The amount of programmability varies from model to model. Because the functionality built into the simplest systems controllers is very limited, the programmability of those devices is very limited. On the other hand, systems controllers with high levels of functionality and flexibility are programmed using very powerful programming languages and tools.

Major manufacturers of systems controllers offer training classes for their customers. Some manufacturers also offer programmer certification programs to ensure that those persons doing programming for their products are properly trained. A systems integrator may choose to hire a third-party programmer rather than do programming in-house.

5.1.0 Program Development Tools for Systems Controllers

Each systems controller manufacturer has its own proprietary software. For that reason, each manufacturer has developed its own programming language and program development tools. These tools are user-friendly PC applications designed to streamline the programming task. Some development tools use point-and-click programming, some use script methods, and some have point-and-click combined with script methods (*Figure 12*) to create the programs. Each of these programming environments is powerful and flexible, and each has advantages and disadvantages. The process of choosing the correct system controller for a project should include an evaluation of the strengths of the program development tools for the system controller to be used.

5.2.0 Graphical User Interface (GUI) Development Tools

In addition to program development tools, manufacturers of systems controllers also provide tools that can be used to create attractive graphical user interfaces (GUI). These tools are written for use with that manufacturer's system only. Sometimes, the files created by these tools work only on that manufacturer's touch panel or keypads and can only be used with that manufacturer's system controller. Many of the subsystems that the systems controller control also use a GUI to provide the initial setup of that subsystem.

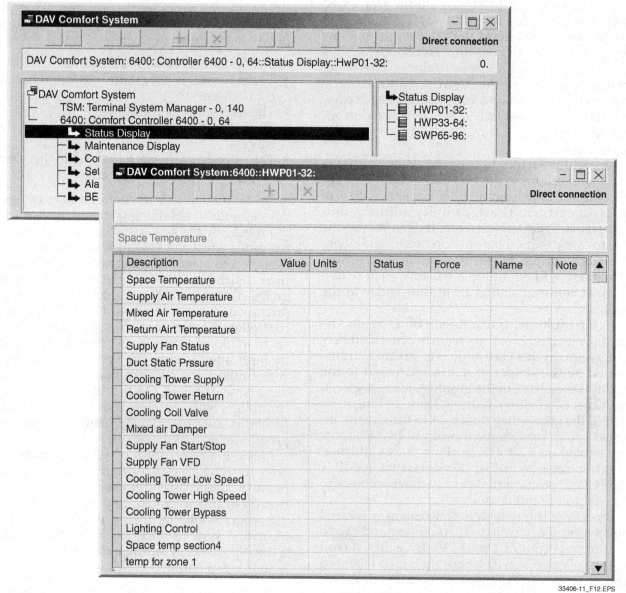

33406-11_F12.EPS

Figure 12 Sample of programming environment that uses point-and-click combined with script methods.

Some manufacturers have also developed GUI tools the programmer can use to create web pages that can be uploaded to a web server. Many of these tools allow the programmer to create user interfaces that can be uploaded to a client's PC and run as a standard executable application. Again, these web pages and applications provide a user interface for only that manufacturer's systems controller.

5.3.0 Other Specialized Development Tools

Some integrated systems require the control of IR or RF controlled devices. The systems controller controls these devices by use of a wired IR or RF emitter. For each controlled device, control codes must be uploaded into the memory of the systems controller. Some manufacturers have a library of control code files for the many devices that can be downloaded into their equipment. If a control code file for an IR or RF device is not available, the systems integrator must create one. Systems controller manufacturers have created tools that allow systems integrators to capture the codes from

controls and save those codes in the proprietary file format specific to that manufacturer. That file can then be uploaded into the systems controller.

5.4.0 Advanced Systems Controllers

As is typical in the computer industry, the capabilities of systems controllers continue to expand. As capabilities advance, the programming capabilities also advance to support the additional functionality. Some manufacturers now integrate web servers into their advanced systems controllers. Using these controllers, the programmer is able to upload created web pages directly into the systems controller using the user interface development tools. This approach allows their clients to access and control their integrated system from anywhere on the LAN. As shown in *Figure 13*, if the appropriate network security is applied, the client can also access and control their system from anywhere on the Internet. The programming and functionality designed into some systems controllers allow a user to take control over an external PC and operate it from a touch panel. This approach enables clients to expand the usefulness of their help desk, allowing their staff to

On Site

Graphical User Interface

Here is an example of a GUI for a building subsystem. The operator can review system status, change the system configuration, alter operating parameters, and run diagnostics by clicking active elements on the screen.

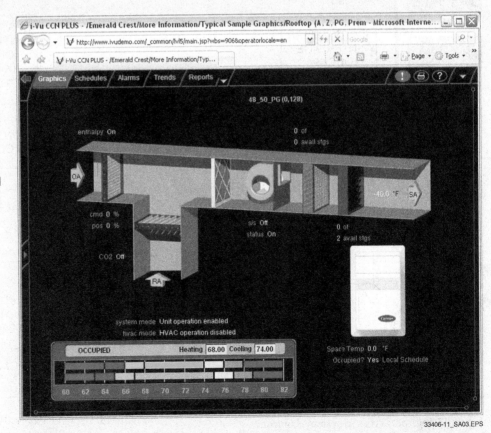

33406-11_SA03.EPS

control not only the subsystems in the remote environment but to also control the remote PC.

5.5.0 External Computer as Systems Controller

Up to this point, the traditional systems controller has been seen as the hub or center of the integrated system. However, this is not always the case. *Figure 14* shows a system where the external computer serves the functions of the central systems controller. This topology is very common in large-scale building automation systems. These systems depend on the ability to perform tasks that traditional systems controllers are unable to handle, such as accessing databases and printing reports. Many of the algorithms necessary to accomplish these tasks are unable to be performed by traditional systems controllers.

In a building automation system, it is possible for a traditional systems controller to be one of the subsystems dependent on the controlling computer system. In these integrated systems, the traditional systems controller can draw from the central computer such data as temperature and lighting scenes for display on the integrated system touch panel. Us-

ing the touch panel, the client can give commands to the systems controller to change the temperature in a room. The systems controller forwards this command to the building automation system computer, which in turn commands the HVAC system to increase the temperature in the room.

The automation system computer can log this change into a database and create energy management reports that reflect the change. With building automation systems, the integrated system becomes very complex. Building automation systems are generally package solutions and are configured rather than programmed. The manufacturer of the system generally does all of the programming, and the systems integrator configures the system according to the client's needs. Using this type of system may require the client to either alter their method of operation to match the capabilities of the building automation system or to accept a level of functionality that they do not want. A creative systems integrator has all the tools available to develop a building automation system tailored to the specific needs of each client.

6.0.0 USER INTERFACES

While it is programming that makes systems integration happen, the user of the system never sees that programming. The client's only interac-

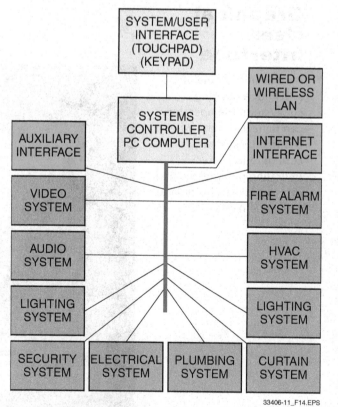

33406-11_F13.EPS

Figure 13 Internet access to systems controller.

33406-11_F14.EPS

Figure 14 Using an external computer as the systems controller.

tion with the system is with the user interfaces provided by the systems integrator. From the client's viewpoint, the user interface is the system. Because of this perception, it is critical that user interfaces be attractive, well labeled, logical, and easy to use.

6.1.0 Feedback

When clients interact with a user interface, they expect to get some sort of reinforcement that the command just issued is being acted upon. Feedback is a response, generally visual or audible, returned to a user indicating that the action just taken was recognized by the system. For example, if a user presses a button on a touch panel, the system recognizes that action by changing the appearance of the button or making a sound.

Feedback is also used to provide the user with device status or environment information. For example, as the temperature in a room changes, a digital indicator on a wall plate or touch panel can change to reflect the new value. Also, in an audio teleconferencing session, a graphical element on the touch panel can change to reflect the telephone line status. One icon can represent a telephone on-hook, which can be replaced by another icon representing a telephone off-hook.

Providing accurate feedback to users is critical. It increases their confidence in the operation of the system and makes the interface easier to understand and use. If accurate and immediate feedback is not provided to users, they may become impatient and press buttons until something happens. Unfortunately, when this occurs, the event that happens is often an undesirable one.

The two types of feedback are true feedback and simulated feedback. True feedback is returned to the user to indicate that an action has positively occurred. For example, when the user presses the projector-on button on a touch panel, the systems controller sends a power-on command to the projector through an RS-232 communications port. The projector turns on and then sends an acknowledgement back to the systems controller that the command was successfully executed. Upon receipt of the positive acknowledgement, the systems controller in turn changes the appearance of the projector-on button on the touch panel to provide positive feedback to the user that the projector is, in fact, running.

In this example, the communications between the systems controller and the projector were two-way RS-232 communications, allowing both a command and an acknowledgement.

Simulated feedback is returned to the user to indicate that the systems controller, in effect, thinks that an action has occurred. Assume that the projector in the example accepted only one-way IR commands. The systems controller would be able to issue the power-on command, but the projector would be unable to send an acknowledgement that it actually executed the command. In this case, it is still important that feedback be given to the user that the button press was recognized, and that the projector was turned on. Therefore, the systems controller acts as if the projector really turned on (even though no acknowledgement was received) and in turn changes the appearance of the projector-on button on the touch panel to let the user know that the projector is now on.

Because true feedback is more desirable than simulated feedback, most systems integrators prefer using devices that have two-way communications. If it is not possible to have two-way communications, the systems integrator typically uses a companion device to sense the status of the device being controlled. In the example, a power current sensor might be used in conjunction with the projector. The power current sensor can evaluate the amount of current being drawn by the projector and, in turn, determine if the device is on or off. The power current sensor provides the power-on acknowledgement to the systems controller.

6.2.0 User Interface Types

Many different types of user interfaces can be used in an integrated system. This section will focus on two of the most common: wall plates and touch panels.

6.2.1 Wall Plates

Probably the simplest example of a wall plate user interface is a standard light switch. When the switch is up, the light is on, and when the switch is down, the light is off. This is not done by accident. If the light being controlled is out of sight from the user, the user is able to determine the status of the light by looking at the switch. If the switch is up, the light should be on. This is an example of simple feedback.

In many integrated systems, a simple switch is replaced by an electronic control that offers the user more options than just on and off. *Figure 15* shows an electronic lighting control plate. In this case, there are five lighting scene selection buttons, along with level up/down buttons. Each button is labeled. The scene selection buttons have LED indicators next to them that provide feedback to the user. The user can select lighting scenes and levels and, at a glance, can determine the currently selected scene.

Many systems controller manufacturers also offer multi-button wall-mounted control panels that connect to the systems controller's proprietary communications bus. The communications bus is able to support many user interface devices, giving the systems integrator the flexibility of installing multiple control locations within a room or within a facility.

Typically, the wall plate is not programmable. It simply communicates with the systems controller, notifying the controller that a button was pressed. All of the programming is done on the systems controller. The systems controller identifies each event associated with each button, such as a button press and a button release (two events). It can then take appropriate action based on the event.

6.2.2 Touch Panels

In many integrated systems, the touch panel is the single most important component of the entire system to the user. Often, the client's impression of the touch panel comprises their entire impression of the system. There are many different types of touch panels:

- Wall-mounted, lectern-mounted, or tabletop (*Figure 16*)
- Wired, IR wireless, or RF wireless
- Communications via TCP/IP, RS-232, or proprietary bus
- Monochrome or color
- Multiple screen sizes
- One-way communications or two-way communications
- No video, composite, S-video input window, or RGB input window
- Multiple pixel resolutions
- Various aspect ratio displays
- Simple static graphics displays or complex motion graphics engines

In selecting the appropriate touch panel, you must carefully consider each of the following often-conflicting factors:

- What functionality is needed?
- What will the client's budget allow?
- What impression do you want the client to have of the system?

Manufacturer-provided design tools are used to create the basic touch panel GUI design. The design is saved as a computer file and is then uploaded into the touch panel. As with wall plates, the touch panel is not programmed. All of the programming is done on the systems controller.

When considering the design of touch panels, evaluate the graphic design skills required for a project in relationship to the graphics design skills of the programmer. Some firms have programming staff with adequate skills to do both the GUI and the programming. If in-house design skills are lacking, a systems integrator may choose to hire a third-party programmer and/or graphic designer to complete the job.

Using the touch panel design application, a GUI designer is able to create touch panel pages. On each page, the designer can then draw buttons and assign to each button various attributes, such as color, text, border, an icon, and a numeric ID value.

Like a wall plate, the touch panel is assigned a unique device address. This combination of device address and button identification allows the systems controller to identify each button on each touch panel in the system. Using this method, the touch panel is able to operate in the same fashion as a wall plate. However, the advantages of a touch panel over a wall plate are obvious:

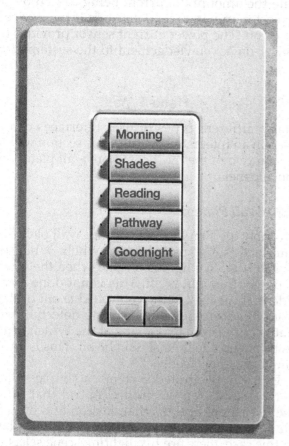

©2011 Lutron Electronics Co., Inc.

33406-11_F15.EPS

Figure 15 Electronic lighting control plate.

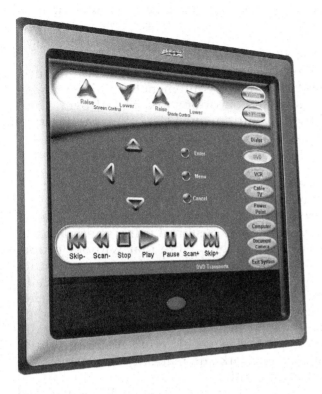

33406-11_F16.EPS

Figure 16 In-wall and tabletop style touch panels.

- The configuration of a wall plate is fixed. The layout of a touch panel page can be any configuration the designer chooses. It can also be easily changed.
- The wall plate can have only one layout. Touch panels can have many pages and many layouts.

The touch panel design tools provided by some manufacturers are very complex with many features. In many cases, attractive and functional touch panel designs can be created completely in these applications. It is also possible to use other graphic design tools outside of the manufacturer's design application to assist in the design.

7.0.0 FAULT TOLERANCE PROCEDURES

Appropriate fault tolerance procedures provide a number of benefits:

- System reliability is increased.
- Using preemptive service, system life can be extended.
- The client's confidence in the system is increased.

While extensive fault tolerance can greatly impact the price of a project, these benefits should be considered as systems are designed. However, not all fault tolerance procedures are expensive. Appropriate planning and programming can go a

long way toward providing a reliable, affordable, fault-tolerant system.

There are multiple levels of fault tolerance in integrated systems. Fault tolerance must be considered for the following:

- Individual components
- Subsystems
- The entire integrated system

The computer and networking industries have taught much to the systems integration industry about fault tolerance and the development of solutions that provide high levels of stability and reliability to integrated systems. Many of the techniques used in integrated systems to provide for fault tolerance are implementations of techniques used in networks. This is particularly the case when the integrated system resides on the LAN.

7.1.0 Individual Components

Various manufacturers of integrated systems components take different approaches to the task of making their products fault tolerant. One solution used in a number of products is the SNMP protocol. With it, a piece of equipment can be configured to automatically send special messages, called traps, over the network to notify a system administrator that an event has occurred. For example, if a critical unit begins to overheat, an over-temperature trap would be sent prior to shutdown, allowing time for a system administrator or technician to remedy the problem before any damage is done.

Many components have protective procedures that are automatically implemented when certain out-of-tolerance conditions arise. Excessive temperature is a common problem. Manufacturers often embed procedures to protect their devices from it. These procedures are often much more than a protective thermal circuit. For example, some projectors go into standby mode if they get too hot, and do not function until they have cooled off to safe level.

Some computer-based products have redundant, hot-swappable components, such as power supplies and hard drives. A hot-swappable component is one that can be replaced while the unit is still operating. This is very important when a key unit in a system must be operational at all times, and it is not an option to bring the system down to replace a component.

> **CAUTION**
>
> Before swapping out a component, make sure it is designated as hot-swappable. This includes powered connections such as USB and HDMI.

To ensure a high level of fault tolerance, steps must be taken to monitor the status of individual components as well as provide preemptive service. One of the major advantages of systems integration is the ability of one central application to monitor each intelligent device in the system. The programmer can query a device for its status. If an inappropriate response is received, or if no answer is received, the program can send an alarm to the system administrator or technician.

This method can also be used to provide assistance in preemptive service. For example, the systems controller can be programmed to query a projector for the number of hours the projection lamp has been used and how many hours of lamp life remain. When the critical number of hours is reached, the system can send an email to the system administrator prior to lamp failure stating that it is time to replace the lamp.

7.2.0 Subsystems

One fault tolerance solution on the subsystem level is the ability to monitor the status of various key components of the system and, upon finding a failure, to automatically reconfigure the system to use redundant equipment. This method is embedded into products by the manufacturer and can also be programmed by the systems integrator.

Consider a situation where a manufacturer has built into the equipment the ability to monitor system conditions and to fail over to redundant equipment. Very often, the physical network is a key component of the integrated system, and the system does not operate properly if the network fails. Many manufacturers of network hardware have implemented fault tolerance measures, such as the spanning tree protocol, into their switches. This protocol allows a network to be designed using multiple paths between switches. Note that without proper configuration of the spanning tree or similar protocol, this would not be possible. Typically, multiple paths between two network switches causes a failure on the network. The redundant cables between switches are then run along different paths. Using this protocol, if one wire is cut, the network automatically fails over to the redundant path.

If three or more switches are in the system, and if the spanning tree protocol is properly implemented, much of the loss of functionality due to the failure of any one switch is automatically compensated for by redundant switches using alternate physical paths. The implementation of this technique is critical to systems where the audio is transmitted on the network using protocols, such as CobraNet®.

Now consider a situation where fault tolerance procedures are designed and implemented by the systems integrator. In a subsystem where a computer is a key component, two identical computers can be provided. One is provided for redundancy in the event the other fails. In this case, the systems controller can be programmed to continuously query the status of each computer. If it finds that the primary computer is not operating properly, it can automatically reconfigure the system to use the redundant computer. That changes the role of the redundant computer so it acts as the primary computer. This allows time for the original primary computer to be repaired and reinstalled, at which time it becomes the secondary redundant computer. This technique can also be used to provide fault tolerance for any component whose health can be assessed by the systems controller.

7.3.0 Integrated Systems

The fault tolerance techniques for individual components and for subsystems can be applied on a larger scale, taking in the entire integrated system. For example, a large integrated system may consist of several buildings on a campus. The system can be designed so that a master systems controller systematically queries each subsystem to assess its health, regardless of where the subsystem is located. When something wrong is found, the master systems controller can automatically take corrective action if such actions have been preprogrammed and can notify the system administrator of the situation.

8.0.0 RESIDENTIAL APPLICATIONS

In a home, the most common systems are the telephone and a fire alarm system. For home applications, the fire alarm system must announce to all occupants within the home that a fire may exist. The next step is to report the fire to an outside responder. Few residential systems are actually set up to do reporting. Most jurisdictions require annunciation, but few require reporting for a private residence. The residential system that is usually programmed for reporting is an intrusion (burglar) alarm system (*Figure 17*). When a possible break-in is detected, an alarm goes off in the home. At the same time, the alarm system seizes a telephone line and reports the break-in to an outside party (*Figure 18*). That party may be a 9-1-1 operator, or it may be a private security firm that notifies public safety officials after they verify that a true emergency exists. With many private security systems, the fire alarm is also connected into the intrusion alarm system for an integrated, monitored system. If any system is installed to monitor, annunciate, and report a possible fire, it is considered a fire alarm system and must follow the local codes for fire alarm systems. In such cases, the intrusion detection is a secondary

Figure 17 Residential intrusion keypad.

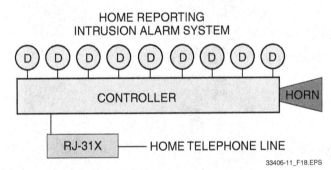

Figure 18 Reporting intrusion alarm system.

alarm, regardless of what the owner of the system considers to be its primary function.

If such a reporting system is installed, it must have at least one method for reporting the emergency. An RJ-31X telephone outlet (*Figure 19*) is a requirement for this reporting. This telephone jack allows the emergency reporting system to break any telephone connection that may be in progress. This opens the line for an on-hook condition. Once the telephone line has had enough time to clear the existing call, the reporting system takes control of the telephone line, go off-hook, and dial a preprogrammed number that connects to a private monitoring company or a public safety agency.

> **NOTE**
>
> In many areas, a false alarm can result in a penalty or fine. For that reason, intrusion alarm systems must be properly installed and maintained to prevent false alarms.

Figure 19 RJ-31X telephone jack.

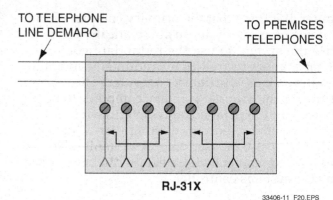

Figure 20 Internal wiring of an RJ-31X telephone jack.

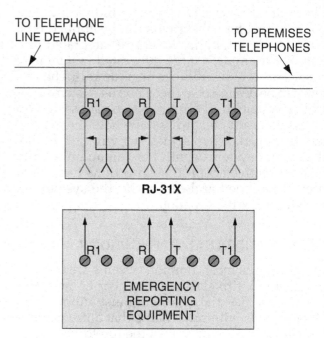

Figure 21 RJ-31X line seizure.

An RJ-31X telephone jack looks like a normal RJ-45S telephone jack. The difference is that it has an internal in and out connection (T to T1 and R to R1). If nothing is connected into the jack, the telephone line goes in and out with no break in the circuit (*Figure 20*). If a monitoring system is plugged into the jack, the reporting system controls whether the line goes into the premises wiring for normal telephone operations, or is used by the intrusion alarm system during an emergency (*Figure 21*). In this way, the intrusion alarm system has control of that telephone line and controls it in an emergency situation.

The intrusion alarm may be connected to a PC and the Internet. In the event of a break-in or other emergency, preprogrammed alerts can go out to emergency responders, the homeowner, or another party such as a neighbor or relative. In remote areas, it is common for the system to use a cell phone to contact the appropriate parties. There are some residential systems that send an alert to appropriate parties if the temperature

inside the home falls below its preset level. This feature prevents plumbing from freezing and bursting. Interior and exterior surveillance cameras may be monitored from an Internet connection if an intrusion alarm is activated. These same cameras may be used to view and/or record any fire or break-in that may occur. The system may even provide alerts or record attempts to tamper with the system. Most of these systems are simple and only use dial-out when there is an emergency. More elaborate systems provide remote control of heating, cooling, and lighting, as well as remote locking and unlocking of doors.

8.1.0 Integration of Other Home Systems

Home theaters (*Figure 22*) and smart home systems are good candidates for integration. A home theater usually involves the integration of audio, video, and lighting control systems using a controller that handles all three of these subsystems. A smart home system may be elaborate enough to control all lighting, entry door locks, interior room temperatures, and surveillance equipment within that home. A smart home integration may include access to an intercom system to allow someone at a remote location to talk with someone in the home using a PC or smart phone that has Internet access.

An audio system consists of an amplifier with speakers to project the audio into one or more rooms. Most audio systems have inputs to allow selection of a radio tuner, tape deck, compact disk player, TV audio, DVD audio, or a microphone input. The selection of the input may be from a keypad, remote control, or wireless touch screen. The video system may be as simple as a TV set and its tuner, or as complex as multiple video projections onto a screen. The most common is a single flat screen display located for the best visibility. There may be options for video inputs from cable or satellite television, a broadcast TV antenna, a DVD player, or home surveillance cameras. There also may be outputs to record what is being displayed using different recording devices. The lighting control system has a wired or wireless control to turn lights on and off or to provide dimming. Some lighting systems control blinds, drapes, or curtains. Systems controller features vary by manufacturer. Cost varies based on the features offered. Lower-end systems use a simple wireless

remote control with buttons to control the individual systems. A higher-end system may use a wireless touch screen with icons to select and control each aspect of each subsystem. Touch screen systems usually have a main screen to select the individual system, with additional screens set up in menu fashion to control features of each subsystem. The control icon may be a single button, a slide bar, or choice buttons. Choice buttons are connected so that only one can be picked from the selection. Choice buttons are very popular for channel selection, as well as audio and video input selection.

A smart home (*Figure 23*) starts with a dedicated PC with software that interfaces with each subsystem. The interface to the PC may be from an interconnection, a telephone connection, a wireless or radio connection, or a cell phone inter-

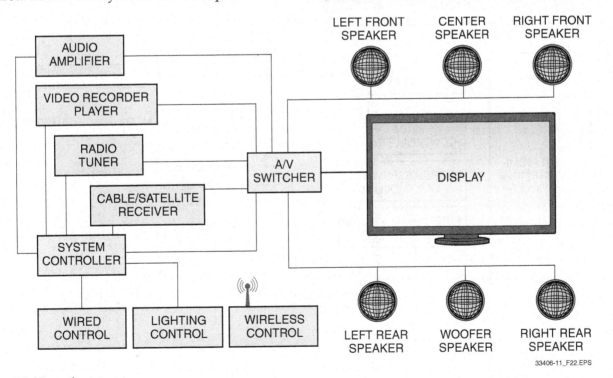

Figure 22 Home theater system.

face. Each subsystem of the smart home requires its own software and operating hardware that may be controlled by a separate system. Simple electromechanical room thermostats are not compatible with smart home systems. Digital room thermostats (*Figure 24*) are typically used for this purpose. Digital thermostats are designed to control many aspects of the indoor environment. They contain features that allow them to communicate with a smart home systems controller.

Home appliances are major consumers of energy. To conserve energy, smart home systems have the ability to control when certain appliances operate. For example, it may be cheaper to heat water in the early hours of the morning when electric rates are lower. The system may prevent the operation of some appliances at the same time to reduce overall power consumption. The control of lighting can also reduce energy use. Motion sensors can detect if a room is occupied. If no one is in a room, the smart home system does not allow lights to come on. If a room is flooded with sunlight, sensors may prevent lights from coming on. That same sensor may close blinds to prevent overheating of the room or open blinds in the winter to capture solar heat.

A common thread for most smart home systems is remote access from an Internet-compatible device, such as a computer or smart phone. Some of the functions that can be performed from a remote location with an Internet-compatible device include the following:

- Change the set-point on a room thermostat
- Unlock a door to allow a maid to enter
- Monitor security cameras
- Turn lighting on/off
- Turn on an appliance

There are innovative applications being introduced every day for use with smart home systems. For example, windows are now available that are made of liquid crystal display (LCD) material. By applying a small electrical charge by way of a smart home controller, the windows change from clear to opaque. A person could darken the home's windows and turn on interior and exterior lights using an Internet-compatible device before arriving home from work. That person would arrive at a home that is safer and more secure.

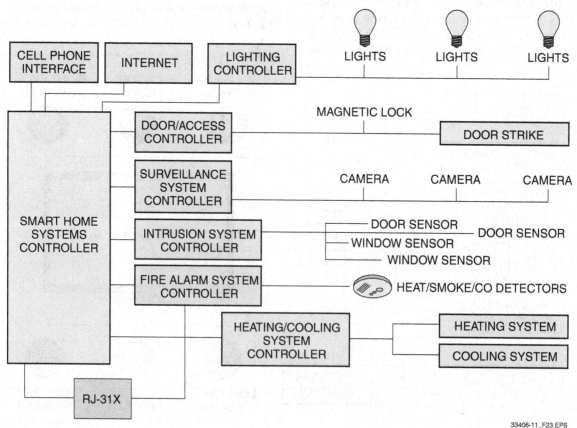

33406-11_F23.EPS

Figure 23 Smart home system.

HVAC Equipment and the Smart Home

HVAC equipment (furnaces, air conditioners, etc.) are the largest consumers of energy in most homes. The makers of this equipment have improved the efficiency of their products over the years. However, they realized that how the equipment is used in the home has as much to do with saving energy as the equipment itself. They now offer products that are fully compatible with smart home technology. For example, indoor temperature and thermostat set points can be monitored and changed from a remote location using a smart phone or a PC.

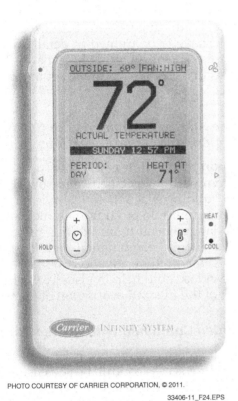

PHOTO COURTESY OF CARRIER CORPORATION, © 2011.

33406-11_F24.EPS

Figure 24 Digital thermostat.

9.0.0 COMMERCIAL BUILDING APPLICATIONS

A commercial building contains a web of subsystems whose number and complexity vary with the size and use of the building. From a building management point of view, system integration is the process by which various building control functions are integrated into one or more centralized, seamless human interfaces. Several common commercial building systems and a brief explanation of them are provided here.

HVAC – All commercial buildings require climate control to maintain space temperature, humidity, carbon dioxide levels, air cleanliness, and sound levels within industry standards. The complexity of the control system varies with the size and use of the building. Controls range from a simple thermostat in small buildings to a networked digital control system in larger commercial buildings. In a large digital control system, a central computer is interfaced through a network of digital controllers located at the building cooling and heating equipment. Digital control technology is used to schedule the movement of elevators, lock or prevent their use, and optimize the movement of elevators to meet the demand of people or supplies.

Fire alarm systems – All buildings must deal with the possibility of fire. In a fire emergency, a quick response by a fire alarm system can greatly reduce the chance of deaths and/or injuries.

Different types of buildings and applications have different fire safety goals. However, most automatic systems are set up to do the following:

- Protect life by detecting abnormal conditions and warning people
- Protect property by operating a variety of devices to reduce fire damage
- Provide early detection and response to avoid or minimize business interruption

To achieve these goals, sensors and fire alarm control units (FACUs) are arranged into systems. The complexity of these systems varies with building size and use. FACUs are further tied into local public safety agencies. Human interface varies from a single FACU to a centralized computer interfaced through a network with multiple monitoring panels throughout the building.

Most commercial and industrial applications require that all fire alarms be reported to emergency responders. This reporting is in addition to the annunciation requirement of any fire alarm system. During a fire emergency, exit doors must be unlocked so occupants can leave. Some interior doors may have to be closed in order to confine a fire to a certain area. These functions can be handled by a building automation system:

Building security – A building security system contains components designed to control building access and detect and report possible threats to property or human life. Security systems use building access control, badge readers, alarms, and video surveillance to help safeguard people and their possessions. To achieve these goals, components are networked together to create systems that comply with all applicable codes.

Lighting – Lighting can represent 20 to 40 percent of a building's energy budget. Commercial buildings require the control of lights during business hours and at other times to reduce energy use. Some systems even control lights to match the hours of daylight. Lights also need to be controlled during emergencies. Programmable controllers are available to interface with lighting fixtures.

Audio/video – Many commercial buildings require the ability to feed audio and video signals to different parts of the building. A large hotel requires flexibility in video and audio programming for its many rooms and common areas. Systems vary in their complexity based on the building size and use patterns. Audio systems are often integrated with fire alarm systems for voice evacuation notices.

Sound and communications – Whatever the application, manufacturers offer products that provide intercom, public address, and paging functions. Many of these systems are networked through the building phone system. A wide variety of speakers, horns, front-end interfaces, and networks are available to meet the unique needs of each building.

Facility and property management – Many large commercial buildings require coordination of property management, facilities management, and service providers to respond to problems that arise in the building. Software programs that use the Internet can respond to problems, maintain inventories, create purchase orders, track progress, register billing, and track payments.

Asset management – Building management teams often want to know where employees and/ or equipment are at a given time and the number of hours they have worked. A variety of time clock and badge-reading systems are used to track employee activities. Loss prevention applications help maintain equipment availability.

> **NOTE**
>
> Buildings such as hospitals and schools require specialized building automation systems to meet their unique needs.

9.1.0 Natural Combinations

Some different building systems combine naturally with others. The individual needs of each building determines which interfaces are required to operate the building successfully.

Rarely does one single operator have to interface with all building systems because the operator would have to be trained in detail on each of the systems. This is an unlikely scenario. The following sections contain examples of natural combinations.

9.1.1 Fire Alarms Combined with Security, Sound, and Communications

A common application is the integration of the fire system with a voice-activated sound system to aid in building evacuation. The voice activation can contain multiple programmed messages and multiple languages if required. In a prison, door access is a natural combination with the fire and sound system. In more advanced systems, the addition of video could help in visually tracking the progress of the fire and evacuation of people.

9.1.2 Video Combined with Access Control

The use of video imaging is a natural integration with building access control. Besides using badge access, a building can use video cameras to track possible entry by unauthorized people. The camera images can be recorded for later review. Improvements in cameras and recording equipment have enabled large video files to be stored and accessed.

9.1.3 Fire Alarms Combined with HVAC, Elevators, and Lighting

When a fire breaks out, the HVAC duct system can isolate and evacuate areas filled with smoke, while pressurizing other areas to exclude smoke. In addition, elevators need to be controlled so that people are not isolated on a floor above a fire. Finally, emergency lighting needs to be activated on floors affected by smoke or fire to assist in evacuation and fire control. *Figure 25* shows the architecture of a building management system that integrates these functions.

The smoke detector shown in *Figure 26* is designed to detect smoke in HVAC air duct systems. These units can be networked so that when one detector alarms, all the other detectors in the network react to control blowers and smoke dampers.

9.1.4 Traffic Control Combined with Video

One growing need in the transportation industry is the desire to integrate fire, police, snow removal, and emergency medical services in response to traffic accidents and weather conditions. Monitoring of highways and city streets is done with video cameras and interfaced to traffic lights, police, other city services, and dispatch personnel who activate and direct the service organizations to respond to emergencies.

9.1.5 HVAC Combined with Lighting

In an attempt to minimize energy cost, building management systems often integrate HVAC system energy users with lighting control. Under such a combination, all the major building energy users are controlled centrally, and power usage is optimized to limit the building's utility bill. Less critical energy users are cycled on and off, and building comfort limits are expanded to minimize energy use and avoid higher utility charges during peak demand periods.

Some companies specialize in developing custom systems integrations. They design and develop network configurations, software, and user interfaces that allow diverse systems to work with each other. Examples of systems integrations per-

---

On Site

Digital Controllers

Direct digital controllers are the core of all building management systems. All controllers receive input signals, act on them, and produce output signals. The controller is really a small computer that consists of input channels, output channels, and a processor that runs control routines, or algorithms. Input channels are wired to sensors, such as motion detectors and thermostats. These sensors measure analog or discrete variables required for the controller to perform its task. The sensor-input signals are translated into software values by special monitoring circuits. Input software values are then fed into an application-specific algorithm. The algorithm, based on the input values, requests that an output action take place. Special circuits translate the algorithm output command into analog or discrete signals. The output signals are then sent to the controlled devices, such as relays, valves, motors, alarm bells, and speakers, to initiate an action.

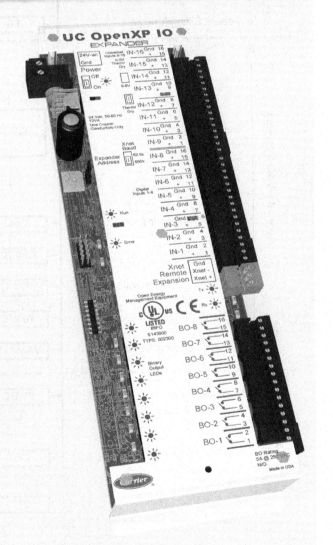

PHOTO COURTESY OF CARRIER CORPORATION, © 2011.

33406-11_SA04.EPS

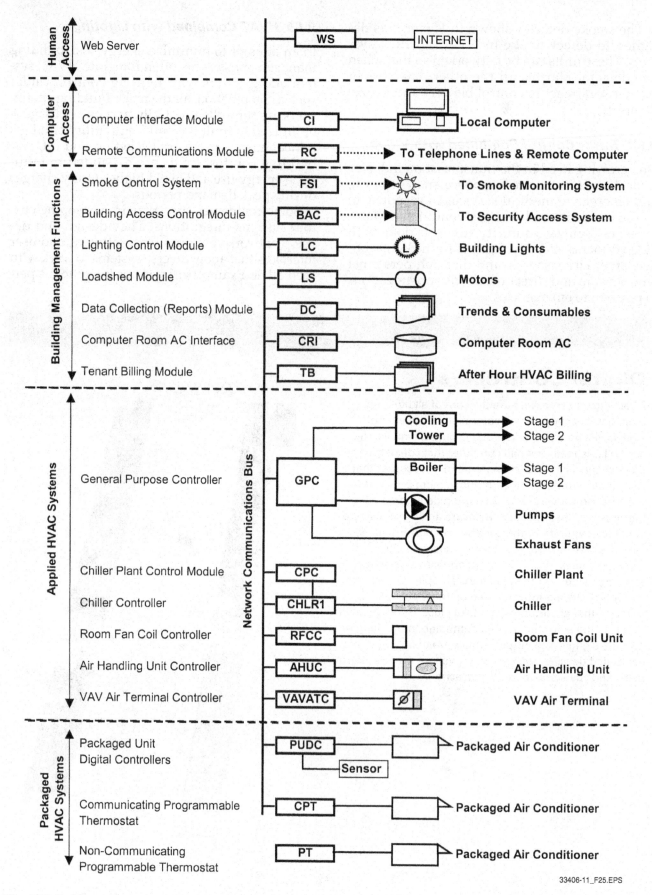

Figure 25 Building management system architecture.

33406-11_F25.EPS

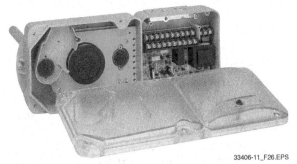

Figure 26 Duct smoke detector.

33406-11_F26.EPS

formed by such a company are shown and described in *Appendixes A* through *E*.

10.0.0 THE FUTURE OF RESIDENTIAL AND COMMERCIAL BUILDING NETWORKS

It is anticipated that TCP/IP will emerge as the predominant transmission, control, and routing protocol used in the systems integration industry. Preparing for this, current trends indicate that manufacturers will be putting greater emphasis on producing products that can reside on the data network. This shift in predominant protocols will require a change in the way integration is viewed by both IT and systems integration professionals. It will require a greater level of willingness to share the channel with other applications. In other words, the limited amount of bandwidth available on a client's LAN will need to be shared between traditional data applications, building automation applications, audio and video on the network, IP telephony, IP security cameras, and more.

To accommodate this increased traffic on the network, advances in technologies will include solutions providing greater bandwidth on the LAN. With the rapid acceptance and deployment of Ethernet, the availability of bandwidth has regularly and rapidly increased. Consider the advancement to 10Base-T, followed by 100Base-T Fast Ethernet, followed by gigabit Ethernet. At the time of this writing, the deployment of 100-gigabit Ethernet is imminent. As networking and systems integration technologies continue to converge, the need for greater available bandwidth also increases. Networking technologies must and will advance rapidly to fill this need.

Currently, one of the greatest obstacles to the acceptance of wide-scale systems integration is the lack of common standards for interconnection, communications, and performance. Techni-

cal advances are rapidly outpacing the ability to establish such standards. While plug-and-play is the goal of the computer and networking industries, currently it is hardly even considered in the industries that comprise systems integration. In the future, industry pressures will force manufacturers to participate in alliances whose purpose will be the establishment of codes and standards for systems integration. These alliances will consist of representatives from manufacturers, systems integrators, end users, and possibly government.

As systems integration codes and standards are developed, training and certification programs will evolve rapidly. As these programs develop, systems integrators, on both an individual and corporate level, will need to accept greater demands to become more highly trained and certified.

The future promises greater acceptance of systems integration by corporate clients. As clients become more knowledgeable of the technologies and methods used in systems integration, their level of acceptance will grow. Having clients who are more knowledgeable will force systems integration firms to provide higher levels of customer service.

As systems integration technologies increase, integration firms will recognize that the traditional roles they have played in the past are changing. Integration firms—particularly smaller firms—will recognize the difficulty of developing in house all of the skills necessary to be successful in all areas of systems integration. They will find that they can develop certain core competencies upon which they can depend. However, they will more often need to partner with other firms having complementing core competencies that they lack, all for the benefit of their client and the successful completion of their project.

Technicians will also find that their traditional roles are changing. They will recognize that the responsibility of education, training, and certification is on their own shoulders. While on-the-job training will remain a critical portion of their education, they must gain knowledge from other sources outside of normal employment, just as is done in the computer and IT industries. While successful past experience will continue to be the most important factor on a potential employee's resume, the importance that employers put on formal training and certifications will grow tremendously.

SUMMARY

Systems integration helps different technologies merge into a simple unified interface. It is useful for both residential and commercial systems, including fire and intrusion systems, lighting, telephony, HVAC, and audio and video systems. The key to systems integration is the communication between both the users and the components of the system.

The seven-layer OSI model divides communication between parts that serve a specific function for data transfer and control. Network topology determines how the systems controller manages the subsystems. Network protocols and devices, such as hubs, switches, and routers, ensure that the systems are properly interconnected. Programming is also important to the functionality of the systems being integrated.

1. Convergence can be looked at as the point where different technologies merge.

    a. True
    b. False

2. How many layers are there in the OSI Model?

    a. 3
    b. 5
    c. 7
    d. 9

3. Which statement best describes an RJ-11 connector?

    a. It has an automatic loop-back connection.
    b. It is an eight-contact, eight-position connector.
    c. It can be used for up to three separate phone lines.
    d. It is used with high-capacity duplex circuits.

4. Which of the following fiber optic connectors uses a snap-in-place latch to make a connection?

    a. LR
    b. SC
    c. ST
    d. MT-RJ

5. Which of the following fiber optic connectors is designed to terminate two 125-micron clad fibers within a single housing?

    a. LC
    b. SC
    c. ST
    d. MT-RJ

6. What layer in the OSI Model defines the physical addressing of the first layer?

    a. Data link layer
    b. Transport layer
    c. Physical layer
    d. Session layer

7. What layer in the OSI Model controls connections between computers?

    a. Data link layer
    b. Transport layer
    c. Physical layer
    d. Session layer

8. How are TCP/IP messages sent over the Internet?

    a. They are placed in Ethernet packets.
    b. They are placed in FTP packets.
    c. They must be routed through SNMP devices.
    d. They are routed by way of Telnet.

9. Which of the following addresses is considered a physical network address?

    a. Logical address
    b. Default address
    c. MAC address
    d. VoIP address

10. A network hub is designed to handle streaming audio and/or video.

    a. True
    b. False

11. When a virtual local area network is being built, any switches used in the integrated system must be _____.

    a. virtual
    b. optical
    c. manageable
    d. reversible

12. What is wired equivalent privacy (WEP)?

    a. An encryption code
    b. Anti-virus software
    c. Password protection software
    d. A VoIP address list

13. In a multicast transmission of streaming video, how many streams does the sender transmit?

    a. 1
    b. 2
    c. 4
    d. up to 12

14. A systems controller is a device that _____.

    a. allows a system to be programmed
    b. contains the system on/off switch
    c. displays system status
    d. serves as the primary customer interface

15. What type of feedback occurs when an operator hears music after touching a speaker icon on a touch screen systems controller?

    a. Positive
    b. Delayed
    c. Simulated
    d. True

16. A multi-button wall plate is an example of a(n) _____.

    a. positive feedback device
    b. programmable interface
    c. user interface
    d. simulated feedback device

17. What protocol is used to send trap messages?

    a. SNMP
    b. VoIP
    c. TCP/IP
    d. SMTP

18. Which of the following conditions would result in a trap message being sent?

    a. An unauthorized user is on the network.
    b. A wrong address was used to send a command.
    c. Several people are using the systems controller.
    d. A video projector is overheating.

19. What allows a telephone call in a home to be interrupted so that an emergency signal can be sent to a public safety agency?

    a. A fire alarm override relay
    b. An RS-232 telephone jack
    c. A 9-1-1 operator
    d. An RJ-31X telephone jack

33406-11_RQ01.EPS

Figure 1

20. The device shown in *Figure 1* is a(n) _____.

    a. HVAC thermostat
    b. duct smoke detector
    c. network switch
    d. automatic lighting control

Trade Terms Introduced in This Module

Acknowledgement byte (ACK): A signal sent back to a transmitter indicating that parity bits match. See *non-acknowledgement signal*.

Convergence: The point where different technologies such as telephony, computers, security, audio, and video merge together.

Dynamic host configuration protocol (DHCP): A protocol that allows centralized management and automation of the assignment of IP addresses on a network.

High-level data link control (HDLC): An algorithm that provides a numbering sequence while data is being transmitted.

Logical address: An address assigned within a network that can be changed.

Logical link control (LLC): A sub-layer in Ethernet that deals with addressing and multiplexing when multi-access media is required.

Media access control (MAC): The hardware address of a device connected to the network. The equipment manufacturer assigns the address.

Non-acknowledgement signal (NAK): A signal sent back to the transmitter indicating that parity bits do not match. See *acknowledgement byte*.

Null modem cable: A cable required when terminal-to-terminal or modem-to-modem connections are required. It is used because handshaking signals will be incomplete.

Parity bit: Used to tell the number of odd or even ones in a data byte. It is used for error detection. If parity bits do not match, it indicates that information has been corrupted and cannot be used.

Simple mail transport protocol (SMTP): The protocol used to send email on the Internet.

Simple network management protocol (SNMP): A standard for gathering statistical data about network traffic and the behavior of network components.

Static address: An address assigned to a specific port or device within a network.

Systems controller: A configurable and programmable computer that acts as the hub of an integrated system.

Telnet: A protocol for remote computing on a LAN as well as on the Internet.

Wi-fi protected access (WPA): A newer encryption code for wireless Internet access that is more secure than older encryption codes.

Wired equivalent privacy (WEP): An early wireless Internet encryption code. It is not as secure as newer encryption codes.

AUDITORIUM A/V

The following is an example of a 600-seat auditorium audio, video, and video control installation that took place in 2003. The auditorium was under construction when a sound contractor was contacted to design the sound system. The only provisions for the system were four 2-inch conduits for the placement of cables between the sound room in the back of the auditorium and the storage room behind the stage.

The architect asked for a sound system to accommodate choir, band, orchestra, and theater productions. The auditorium will also be open to the public for town meetings, available for rental by community organizations, and used for teaching by the school faculty. The contractor suggested a projector system to enhance the use of the building for these additional functions. The architect agreed and increased the budget to include the projector and screen.

As the design developed for the sound system and projector system operation, control of the projection system was discussed, and a control system was added. It allows the entire system to be turned on by a control panel on stage right, with inputs of RGB and S-video stage right, RGB and composite video stage left, and RGB and S-video in the control booth in the rear of the auditorium.

After reviewing the design, the school staff wanted to add an intercom system for theatrical plays and a sound system for the foyer areas, so people would hear announcements to return to the auditorium and music during intermission. As a result, speakers in the foyer area and a two-channel intercom were included in the design.

The sound systems included the following:

- A speaker cluster was designed, using three main speakers and two subwoofers, to accommodate a full range of audio performance, including music, band, theater, speaker presentations from a podium, and other voice inputs. The building was far along, so the woofers were installed with the main speakers because there was no place for them near the floor.
- Four monitor speakers were chosen for the stage area.
- Amplifiers were chosen to handle the following:
  - Three main speakers in the cluster
  - Two subwoofers in the cluster
  - Four monitor speakers
  - The speakers in the foyer area

- Mixers/preamplifiers were chosen for the following:
  - The main cluster
  - The subwoofers
  - The monitors
  - The speakers in the foyer area

- For audio control (*Figure A-1*), an automatic system was installed to handle the functions of crossover.
- For mixing, a 32-channel, 8-bus mixing board was installed in the control booth at the rear of the auditorium to accommodate the outputs to the mixers/preamplifiers and the microphone inputs.

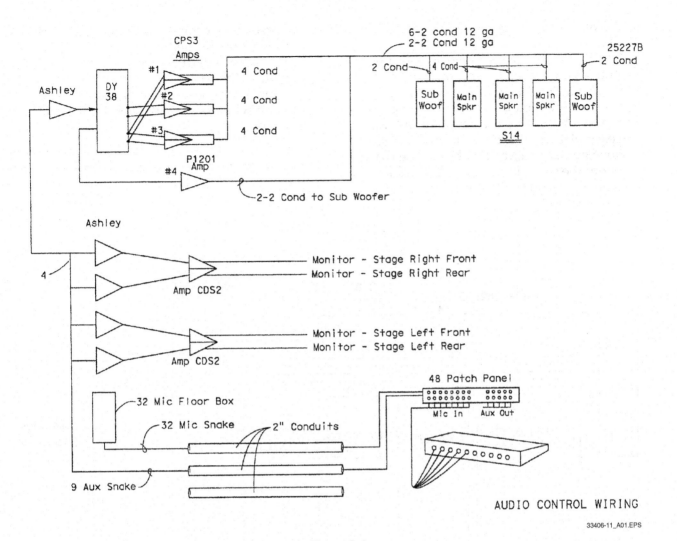

AUDIO CONTROL WIRING

33406-11_A01.EPS

Figure A-1

- A total of 45 microphone jacks were included as follows:
  - 28 on the stage area, to accommodate any setup, accomplished by providing eight microphone outlets (four each in two gang plates), 16 microphone outlets (four each in two gang plates with a monitor out jack in each plate for the monitors), two microphone outlets in a Mystery box in the wood floor of the stage, two microphone outlets in a single gang box on the front of the stage (see *Figure A-2*)
  - Three in the ceiling, for choir use and other events that dictated the use of these speakers
  - Two in the control booth
  - Four in the back wall, for setup of microphones for questions and answers
  - Eight wireless microphones, for various purposes
- A patch panel was installed in the control room rack to facilitate setting up 32 of the 45 microphones. (See *Figure A-3*.)

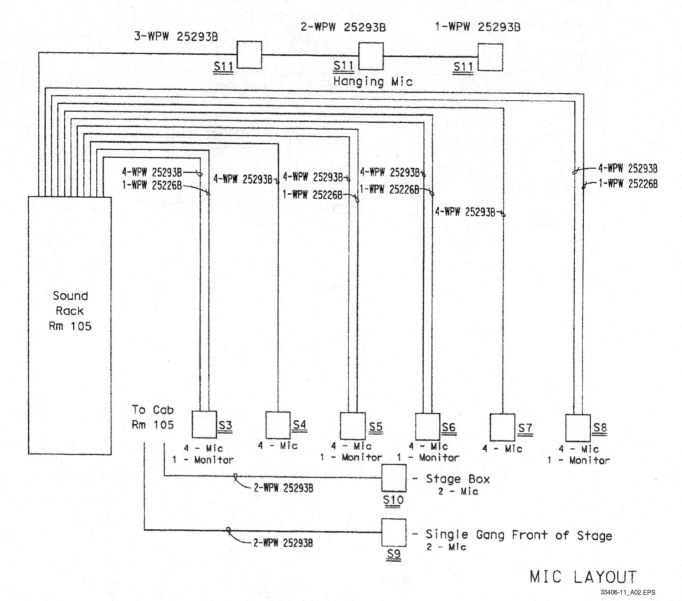

MIC LAYOUT

33406-11_A02.EPS

Figure A-2

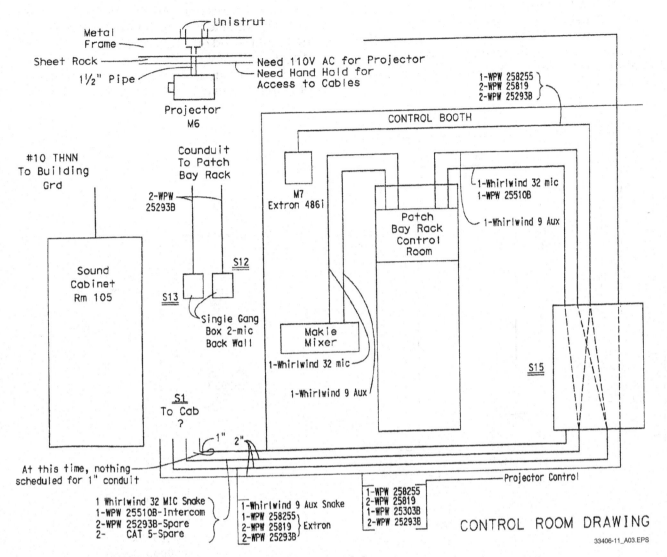

Figure A-3

- An assistive listening system was installed per ADA requirements. (See *Figure A-4*.) An antenna is installed in the storage room near the equipment rack and housed in plastic surface raceway to minimize damage by persons in the storage room.

- Two direct boxes were provided with the system to accommodate line level outputs from amplified guitars and other musical instruments. The direct box converts line level to microphone level, so one can use microphone inputs instead of using line level inputs to the mixing board.

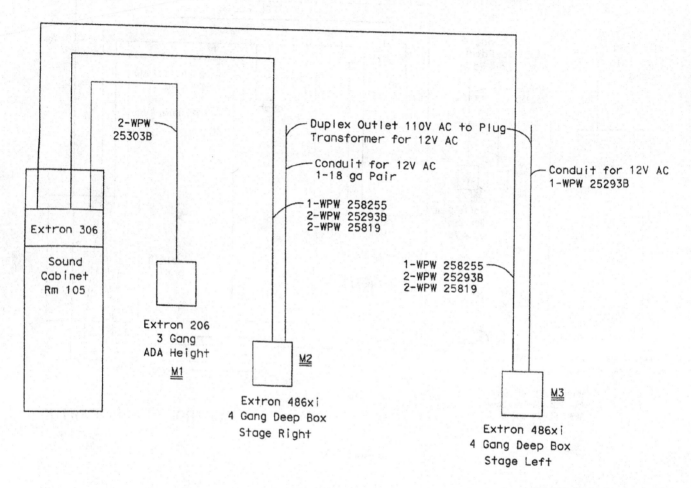

Figure A-4

The intercom system was wired as follows (*Figure A-5*):

- A two-channel wired intercom system was chosen because of the requirements for spotlights for stage plays. One channel is used for the stage production, and the second channel is used by the mixer board operator to advise the lighting crew when to change lighting effects without interrupting the stage production people with this information.
- The control booth was wired for a master intercom unit that is not. The master allows the operator to communicate via headset or hands-free device.

- The male and female dressing rooms were each wired for a substation intercom unit that is not yet installed. This substation allows users of the room to monitor the activity of the production hands-free device and communicate with users of the intercom system.
- To communicate with the production people running the play, wall jacks with channels A and B were installed in the hallways behind the stage.
- To communicate with the people running the lighting system, wall jacks with channels A and B were installed next to the lighting units in the corners of the auditorium.

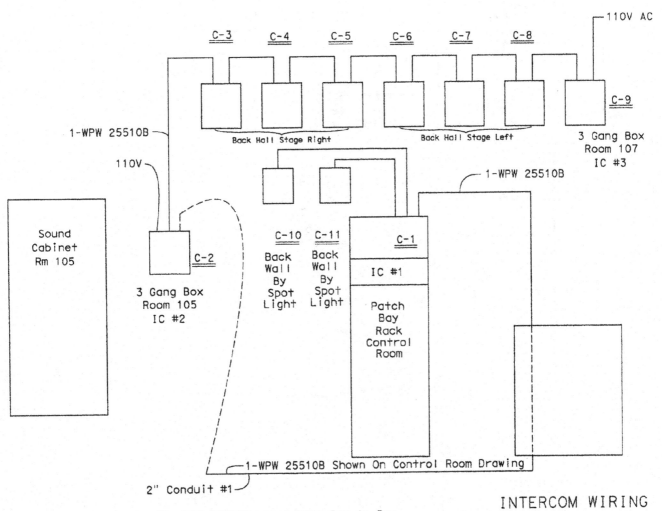

INTERCOM WIRING

33406-11_A05.EPS

Figure A-5

The projection system includes the following:

- A 12' wide by 9' high screen was installed to meet the criteria of this particular auditorium. The screen is square video format. It is an electrically operated screen and is controlled both manually and automatically from the control system (*Figure A-6*).

- Based on the type of lighting the school wants to use during video presentations, a 3,300 ANSI lumens projector was installed.

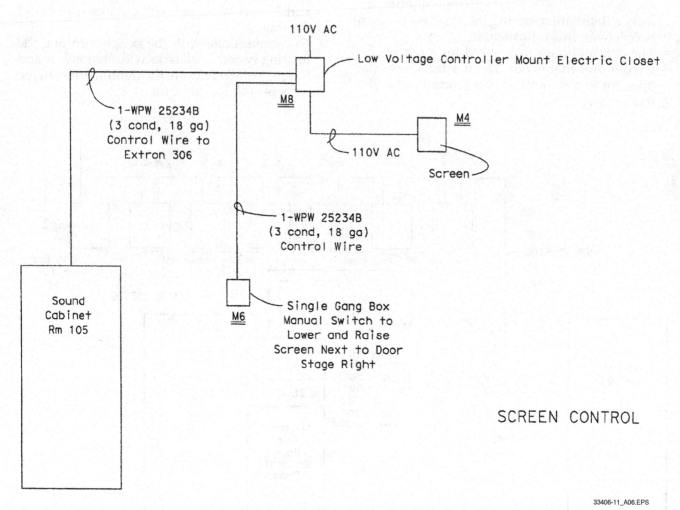

Figure A-6

- The following inputs to the projector would be used (*Figure A-7*):
  - RGB inputs, available from stage left, stage right, and the control room, for computer inputs from laptops brought in for Pow-erPoint® presentations and/or classroom teaching events
  - S-video inputs, available from stage right and the control room, for presentation of this media using DVD devices
  - Composite video input, available from stage left, for presentations of this media using VCR devices

Due to the fact there will be many users of the system, a control system was installed that allows each user to select inputs by merely pushing buttons. A reasonably inexpensive user-friendly control system was installed, allowing the user to perform the following:

- Turn on the entire sound and projection system by pressing one button on the control panel, which accomplishes the following:
  - Lowers the screen
  - Turns on the projector
  - Dims the lights, if desired
  - Sets up the control panel to activate the six projection system inputs described earlier
- Select one of the six projection system inputs described previously. It is not necessary to use the projector remote device to change from RGB to S-video or composite video. The control system commands the projector to switch to the proper input when the button on the control panel located stage left is pressed to change the input device.

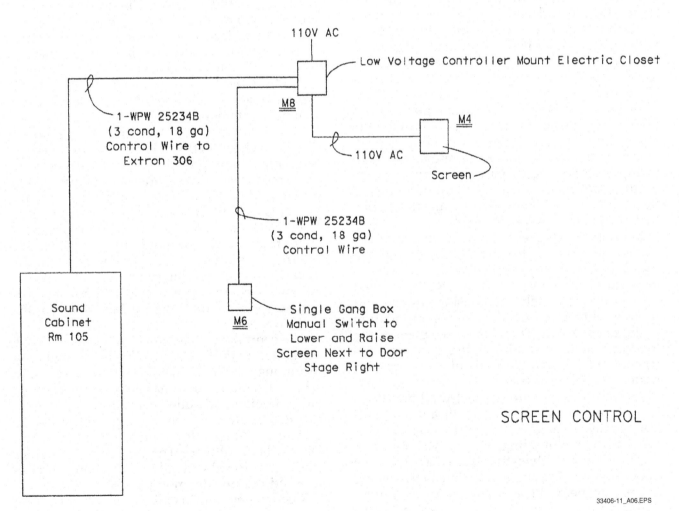

SCREEN CONTROL

33406-11_A06.EPS

Figure A-7

Appendix B

D-WAN AND A/V TECHNOLOGY FOR SCHOOLS, AVON COMMUNITY SCHOOLS

Avon Community Schools, located in suburban Indianapolis, has for many years been recognized as a leader in the practical application of educational technology in the classroom. Starting in 1993, through a multi-year general renovation project, every school has been equipped with their own data network, telephone system, and video distribution system, bringing the full range of voice/video/data technologies into the classroom. When Avon Schools made the decision to build a new $65 million high school, Avon's director of technology, Karl Illg, wanted to ensure that the technology suite for this new building reflected Avon's leadership position with the newest technologies yet would still fully integrate with the existing technologies already in place in the other schools. At Karl's direction, a team was assembled that included representatives from the consulting firm of Fanning/Howey, Cisco Systems^{SM}, and South Western Communications (SWC). The goal of the team was to provide the new high school with high-tech, yet practical, voice/video/data, sound, and presentation technology systems; and, as part of the same project, to integrate these systems with all buildings in the district into a complete, fully functional, district-wide area network (D-WAN).

In order to achieve the D-WAN functionality, it was first necessary to establish connectivity between all seven schools as well as the administrative and maintenance buildings. Karl reviewed several options, including leasing circuits and wireless options, before deciding on installation of the school's own fiber optic cable plant as the most cost effective and future-proof solution. After coordinating with the local utility on easement issues, a cabling infrastructure was installed. The new infrastructure connected all buildings in a star topology, whereby a centralized operations center (located in the district's center) connects all school buildings via twelve strands of fiber. Of these twelve strands, two are used for data, two for voice, and two for interactive video, leaving 100 percent spare capacity for the future. Although sometimes located miles apart, the total cost per school for installation of the fiber infrastructure was less than one year's connection fee for a leased circuit for that same school. In both the short and long term, installation of the school's own fiber infrastructure was, by far, the most cost effective and flexible solution for the district.

The centralized operation center is the nerve center of the D-WAN, housing the centralized head-end components of each of the main systems. This includes a fully managed 100MB Cisco Systems^{SM} data network, a district-wide digital telephone, voice mail, a homework hot line, and a two-way interactive Rauland Telecenter® IP video system. The video system, installed by SWC, provides 110 channels of broadband video—essentially providing the school district with their own campus cable TV system. Utilizing the Telecenter® IP system, users may view and control video from over 50 sources, including local and campus-wide bulletin boards, outside cable TV channels, satellite sources, VCRs, laser discs, as well as internal broadcasts from each school's dedicated channels from classrooms equipped for local origination. There is even a provision for a tie to the local cable provider's local origination outlets to be able to broadcast to the community at large. The Telecenter® IP user interface is a standard web browser that allows the teacher, either from their classroom or from home via the Internet, to search and schedule items from the media library. By having centralized the D-WAN equipment and media library, one operator is able to manage and control the media systems for the entire district.

The centralized Telecenter® IP system is fully integrated with the state-of-the-art presentation technology systems installed at the new high school. Each classroom has an RCA multimedia video monitor and outlets that allow access to local video sources as well as the centralized video system via either I/R remote control or Telecenter® IP client software. The 16 science/computer labs are equipped with NEC^{SM} LCD video projectors and multimedia monitors. Each projector is I/R controlled and installed on a disappearing motorized mount. In addition to the centralized system, the high school has its own mini-system, which allows local selection of cable TV channels and video sources, to supplement the main campus centralized system.

A showcase facility like Avon High School would not be complete without an equally impressive distance learning facility. Avon's distance learning classroom, located just off the high school media center, is equipped like a university level distance learning facility or a Fortune 500 boardroom application, with the exception that there is no dedicated connectivity cost. By utilizing the integrated services digital network (ISDN) technology inherent to the digital phone system, the system accesses otherwise unused bandwidth in the district's digital telephone service, resulting in an estimated $30,000 in annual savings. Seating approximately 24 people, the distance learning classroom is equipped with a wide range of presentation technology equipment that bring first-rate audio quality to the video conferencing/distance learning experience: a 72" SmartBoard™, an NEC℠ LCD video projector, four RCA multimedia monitors, ParkerVision® Cameraman™ instructor and student cameras, Renkus Heinz speakers, Crown® student voting microphones, and a Tandberg Codec mated with a Gentner™ audio processing system. All systems are integrated via a Panja control system and controlled via Panja touch panels built into the SWC teaching station. A second touch control panel is located in the back of the room to provide full control of the system to a dedicated moderator during complex multipoint conferences.

The SWC teaching station is a custom cabinetry unit housing a VCR, computer, document camera, and various other presentation sources. It allows the instructor one-touch selection and control of A/V and computer sources, both within the room and at remote locations. Through the use of the teaching station, the instructor can choose the level of presentation technology with which they feel most comfortable—from a simple PowerPoint® presentation to a multi-point video conference for communicating among several schools or universities. The campus D-WAN is integrated into the room as well, allowing full access to video and data sources from either the centralized operations center or from anywhere in the world via the Internet. This combination of connectivity, presentation technology, and integration makes Avon's distance learning classroom perfect for virtually any type of meeting or presentation.

Similar presentation technology systems were installed in the school's 999-seat auditorium, including a professional sound system, an NEC℠ video projector with motorized disappearing mount, and Panja touch screen controls. Engineered sound systems were installed in the school's gymnasium, natatorium, and several athletic fields, as well as an overall school intercom system.

Although all systems installed are leading technology, care was taken to ensure the design will accommodate emerging technologies. Next on the agenda is digital video streaming. SWC is currently working with the district on plans to install a Rauland digital video server and connect it to the D-WAN. The current library of VHS tapes will be digitalized and stored on a video server, allowing instructors to sort, preview, play, and control all the media without needing human intervention. Streaming the video on the school's broadband TV systems minimizes bandwidth demands, as only tiny data control packets are sent over the data network. This system is a simple plug-in to the existing technology, replacing a bank of VCRs in the centralized operations center. The digital video server is a very powerful presentation tool, bringing the Internet into the classroom in an easy-to-use and practical format. The system is cost-effective as well, with the cost of the server offset by the labor savings in media staff time.

TRAINING LAB A/V SYSTEM INTEGRATION

Integrating audio/video equipment involves the use of hardware components and cable connections that exist in most people's homes. Such systems should be familiar to most trainees. The TV in your home made by one manufacturer must be connected in such a way that it can work with the VCR and DVD players made by other manufacturers. Every home is thus a system integration project.

The following example system integration project is based on the integration of commercial audio and video equipment. In this example, it should be easy to relate to standard cable connections between individual controllers and the requirement for customized software to solve system integration problems. The example reflects a system integration project that was implemented by General Communication of Salt Lake City, Utah.

Interpreter Training Program Lab at Salt Lake City

This project involves the design of an interpreter training program (ITP) lab for Salt Lake City Community College. The lab (*Figure C-1*) consists of 12 student workstations and one instructor workstation and provides an automated teaching environment for student instruction in the American Sign Language (ASL) program.

Customer Requirements

Conferring with Salt Lake City Community College representatives, the ITP lab requirements were defined as follows:

- All A/V system users (students and instructors) must provide log-in information to gain access.
- Student access to the A/V system must be limited while instructor access must be unlimited.
- Instructors must be able to limit student access to various areas of A/V system functionality.
- Students must be able to control their own A/V equipment located within their workstation.
- Instructors must be able to control all A/V equipment in all workstations and switch any A/V source to any A/V display.
- Instructors must be able to access and maintain student information pertaining to the ALS program.
- Instructors must be able to encode any A/V source as an MPEG file. All MPEG files must be maintained on the A/V system server. Sources to be available for encoding include existing training videos, new training material to be encoded during live teaching sessions, and student testing.
- Students must have access to the ITP lab during off hours for testing.
- Students must be able to save their tests as video files on the server for subsequent grading by the instructor.
- Instructors must be able to access the A/V system and saved video files from both local and remote locations.

Figure C-1

33406-11_A08.EPS

System Integration Needs

On the surface, this project appears to be a typical A/V application and looks like a good candidate for a traditional A/V control system. Traditional A/V control utilizes a controller manufactured by companies such as AMX® or Crestron® and requires a system integration company to write control software to link all signal sources and destinations. However, four unique customer requirements went beyond the capability of a traditional A/V solution and required the integration of course administration functions with normal A/V system control functions.

- *Protected access* – It was very important to the college that each system user be authenticated by the A/V system prior to use. This authentication process had to be kept separate from the authentication used for the general computer network. Thus, students could access the general network, but their level of access to the A/V system could be controlled (or denied) by the ASL department staff and instructor.

- *Remote access* – The college wanted the ability to control the ASL system from remote locations. This would allow faculty to configure the system from their offices without having to physically be in the lab. It also would allow a staff member to assist instructors and students in a help desk capacity.

- *Test administration and records* – In order to take tests, the college wanted to allow students to access the system when the lab is not staffed during the evening. The system had to be able to record and save all tests as individual files on the server. Instructors could then review tests at their convenience. The college also wanted the instructors to be able to view test files from remote locations, such as their offices. It was also necessary to allow instructors to view the files from the instructor's home; however, due to bandwidth limitations that cause excessive transmission times, this feature is seldom used.

- *Course content library* – The college has an extensive library of ASL training materials that had been previously recorded on VHS videotapes. In order to make this content accessible to more students simultaneously, the new system had to be capable of encoding and saving videotape material on the ASL server as MPEG-encoded files. Due to the high costs associated with a third party performing the encoding work, the system design had to allow the ASL staff to encode the files themselves as their time permitted, to avoid paying additional fees. The system design also had to allow the encoding of a live training session as it happened, permitting the college to quickly and easily build its library of training materials.

The Solution

Due to these requirements, the A/V control system was designed using a client/server computer model running in Windows 2000 where clients are required to authenticate to a dedicated server. The client/server network is equipped with 13 client workstations: 12 for use by students and one for use by an instructor. It also includes one central server computer that controls the entire system. The 13 workstations (*Figure C-2*) are identical, and each includes the following:

- Computer with keyboard, monitor, mouse, and network connection
- Video monitor with an S-video input
- Camera
- VCR
- Headset with microphone and volume control

All workstation cameras, microphones, VCRs, and client computer screens are wired to a central audio/video switcher (*Figure C-3*) using standard A/V cable connections. The client computer screen image (RGB signal) is routed through a scan converter into the A/V switcher. Each workstation is equipped with picture-in-picture (PIP) hardware. The PIP hardware integrates two video images into a single picture-in-a-picture image, which in turn is routed back to the A/V switcher as a visual signal source. The A/V switcher, under control of the server computer, can route audio signals to the workstation headphones and input video signals to the workstation TV monitor.

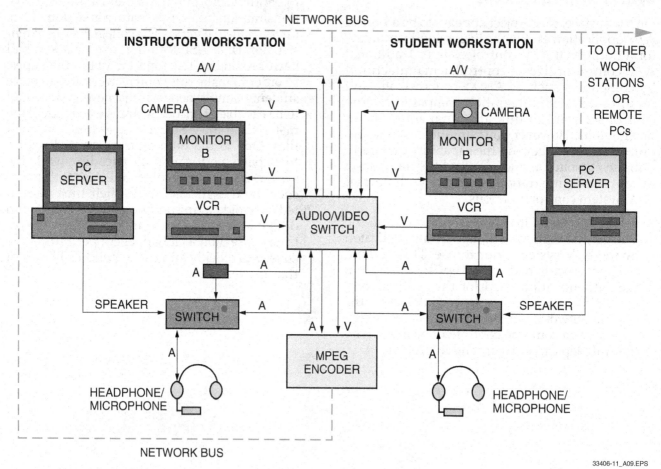

33406-11_A09.EPS

Figure C-2

A central network switch (*Figure C-4*) connects all 13 client workstation computers with the server computer. The server computer, in turn, is wired through an RS-232 cable to the A/V switcher. With this arrangement, a central server computer is used to authenticate all users and control all signal routing through the A/V switcher. Any A/V source can be directed to any A/V destination through the audio/video switcher. In addition, an MPEG hardware encoder is connected to the network, and MPEG decoder software is installed on each of the workstation computers. A central rack houses the central server computer, network switch, A/V switcher, MPEG encoder, and 13 PIP devices.

Finally, a custom graphical user interface (GUI) was written and installed on each of the workstation computers. The GUI (*Figure C-5*), programmed in Visual Basic®, communicates with the server computer using TCP/IP network protocol. The server computer, in turn, communicates with the A/V switcher equipment using the RS-232 protocol. This control arrangement allows the A/V switcher and MPEG hardware to easily interpret commands sent on the network by the customized GUI software.

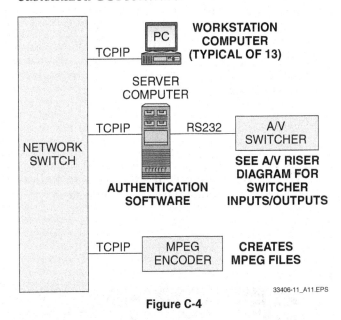

Figure C-4

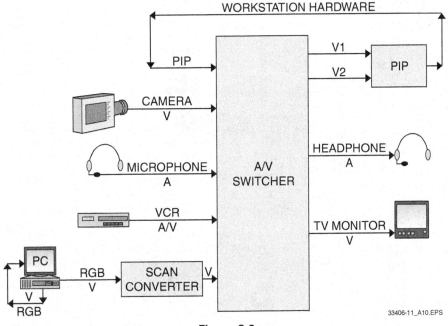

Figure C-3

Using the customized GUI software, students and instructors are able to select an A/V source, and drag and drop that source into display devices. Students are able to route their own audio and video sources to the display devices in their own workstation. Instructors are able to drag and drop any A/V source into any display device in the system. All signal routing flows through the audio/video switcher.

The GUI software provides the following functions:

- *Authenticates users of the A/V system* – Using the GUI software, each user logs into the server. The server authenticates the user and allows the user to interact with the A/V system.

- *Controls the A/V switching equipment* – Using the GUI software, the client (student) may request data and submit commands to the A/V system. However, each of these requests is submitted to the server for evaluation. If the request is valid, the customized server software issues commands to the A/V switcher to route the requested source signals to the appropriate destination devices.

- *Maintains student records related to the ASL system* – The GUI records and maintains a list of all ASL program students, their course enrollment, and test grades (*Figure C-6*).

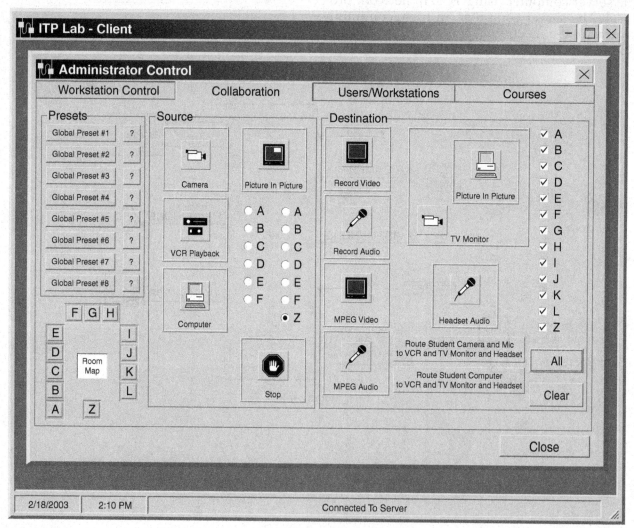

Figure C-5

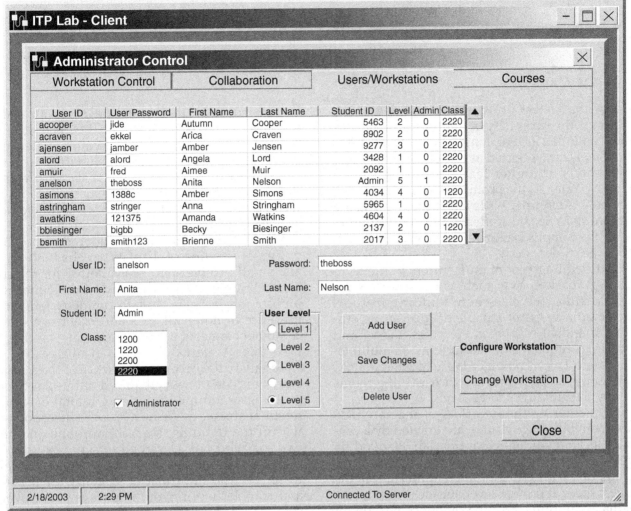

Figure C-6

33406-11_A13.EPS

- *Provides global presets* – The GUI allows instructors to initiate global presets that establish often-used configurations for each workstation in the A/V system. This feature helps in getting the lab ready for the next group of students.
- *Manages MPEG files* – Using the MPEG hardware encoder, MPEG decoder software loaded on each computer, and the GUI software, the system can perform the following:
  - Encode existing video-based training materials, and save them as MPEG files on the server

  - Encode new training sessions as they happen, and save them as MPEG files on the server
  - Allow users to select and view previously saved MPEG video files
  - Allow students to view video files sent to their workstation by an instructor
  - Encode and save student tests on the server for subsequent grading by instructors

The system has been in operation since the beginning of 2002 and is providing all the functionality the college requested.

Appendix D

COLLEGE CAMPUS

One of the fastest growing universities in the country is the University of Central Florida in Orlando. Founded in 1963, it has also earned the distinction of being selected as one of the most wired institutions of higher learning in the United States by Yahoo!®, the prominent provider of Web services. Currently, with enrollment hovering at more than 56,000 students and continuing to climb, the campus is gearing up to meet the needs of a large student body.

Outfitted by major manufacturers in the technologies marketplace, such as Crestron®, Sony, and Proxima, the classroom building features multimedia systems capable of connecting students to global resources via voice, video, and data links. In total, two auditoriums, 18 multimedia classrooms, a computer lab, two notebook PC classrooms, and a pair of distance learning rooms occupy the building's three levels.

Upon entering the first floor of the new UCF classroom building, visitors are greeted by a centrally located student computer lab. Flanked on both sides and to the rear by 11 multimedia classrooms of various sizes, the computer lab is the largest space at ground level outside of two multimedia auditoriums residing on opposing sides of the floor plan in what amount to their own separate wings.

A faculty multimedia center serves as the central hub of activity on the second floor, where the facility's notebook PC classrooms are also joined by studio classrooms. The site of more multimedia classrooms, the third floor is additionally home for distance learning rooms and an anatomy lab.

Each of these spaces houses a ceiling-mounted video/data projector from Proxima, a large projection screen, computer with a CD/DVD player and network connection, a videocassette deck, connections for laptop computers, a high-resolution document camera, and a quality JBL® sound system powered by Crown® amplifiers. Control for all of these systems is unified and standardized via a Crestron® network accessible from Crestron® VideoTouch VT-3500 color touch screens dedicated to each room.

Serving as the university's liaison with the architectural design team, UCF faculty support expert and academic facilities designer Gerry Ewing was instrumental in defining the true nature and scope of the project's technical needs and then creating appropriate spaces to accommodate these needs within the architectural blueprint. As a result, all design efforts were aimed at ensuring that the classroom building's infrastructure would be well-equipped to deal with today's technologies, as well as the technologies of tomorrow.

Lighting was carefully considered to provide optimum viewing of display screens both large and small. Ample space was allotted for the sizable cabling raceways required of the building's multimedia systems, and, to compensate for the presence of notebook PCs and other necessary computer connections, a detailed map was drawn up of logical interface locations in all areas.

To satisfy the necessary presence of the myriad electronic components required of the classroom building's multimedia systems, a total of six interstitial spaces, or Techways, were incorporated within the building, stacked atop one another, two to each floor. Designed to serve as centralized service corridors that allow 24-hour access to any of the multimedia systems for service and troubleshooting, the Techways house every component that doesn't require hands-on adjustment, such as the Crestron® controllers, switching equipment from Extron®, and Crown® amplifiers. Linked above and below with conduit stubs, the Techways additionally include the classroom building's electrical panels plus data and fiber optic hubs. Also inhabiting each Techway are two equipment racks to secure each space's inventory of gear, an uninterruptible power supply, and all appropriate power strips, rack hardware, and accessories.

In each of the classroom environments, teaching consoles supplied by a custom cabinetmaker in Orlando keep all multimedia components and the controlling Crestron® VideoTouch touch screens within easy reach of faculty members. From any of the consoles, commands are sent via the VideoTouch touch screens to each room's Techway-inhabiting Crestron® CNMSX-PRO modular control system. Featuring Internet/Ethernet/LAN compatibility, each CNMSX-PRO network includes hardware and software that links all of the controlling VideoTouch touch screens,

---

58   NCCER – *Electronic Systems Technician Level Four*   33406-12

room computers, and the diverse range of components found in each classroom.

Serving as the "brains" behind the technical functions of the entire classroom building, the Crestron® network does more than simply offer faculty members one-touch access to any multimedia systems function at their disposal. The industry standard for Internet-based control, Crestron's e-Control® places a comprehensive level of remotely-accessible system control and monitoring functions within the reach of a standard Web browser. Compatible with virtually any electronic device, e-Control® is operable with any Crestron® CNX system using its LAN connection and/or proprietary Internet/Ethernet card. With e-Control®, control graphics can be used directly from a PC monitor to maintain a centralized source of help desks, resource sharing, and networking. Effectively offering a universal systems link capable of being spread across any number of rooms or locations, the technology has allowed Ewing and his staff to construct a multimedia help desk that permits a single operator to troubleshoot and diagnose operational problems anywhere in the building.

"From the help desk, we can mirror the look and feel of any VideoTouch touch screen in any classroom," Ewing explains. "This capability lets us assist in a number of ways, because we can literally monitor what's happening on a systems level in any given area."

"While we most certainly could, it's not our intention to become operators for faculty members from the help desk. The systems are simple and intuitive on their own, and most of our faculty members have experience with technology in the classroom."

"The help desk is primarily a resource that allows us to quickly respond to problems of any nature. We can help change media selections, run volumes up and down, prepare a projector for presentation, or perform any other multimedia function that can be done locally. The levels of support we now provide using one person used to require a sizable staff."

Centrally controlling the classroom building in this fashion has other benefits as well. Because the network can be accessed via the Internet, some of the software programming was performed remotely from Crestron's Atlanta office. Internally, Ewing also has plans to use the network to introduce software program upgrades in an area-wide fashion to components of all description. Also, with a clock feature found in each classroom environment's main Crestron® controller, UCF is able to automatically program individual systems to power on and off as needed, thereby saving equipment life. Projector lamp life can also be monitored automatically, thereby ensuring that potential failures can be discovered and remedied before they occur.

Like the architectural portion of the project, technical design credits are given to a talented team of individuals led by UCF's Gerry Ewing, and Schenkle-Schultz is the firm responsible for the actual systems integration.

"Technology like this, combined with the innovative architectural design needed to accommodate it, is changing the face of teaching forever," Ewing adds on a final note.

"With the integration of the PC, large-screen video projection, document cameras, high-end audio, Internet accessibility, and all-encompassing control networks, the classroom has been outfitted with teaching tools that exceed the capabilities of anything imaginable just five or six years ago."

"Implemented with mindful care and thought, systems like these are testimony to the notion that technology really is here to help us. Today, we can actually do more with less and free our minds to develop a new generation of ideas."

FIRE ALARM

A fire alarm system is often required to interface with other building systems for the safety of the occupants in the building. Integration for fire systems is typically a one-way signaling system, which means that the fire system sends signals to other building systems, but the other systems do not send signals back to the fire system.

As an example, imagine a seven-story high-rise building. The building has an elevator hoistway with smoke detectors located in each elevator lobby and at the top of the shaft. Access control, HVAC, and elevator control systems are also located in the building. The fire alarm needs to be able to signal all of these systems during a building alarm condition. Once signaled, these independent systems respond by taking action on the systems they control.

If the fifth floor elevator smoke detector senses smoke, a signal is sent to the fire alarm control unit (FACU) that an alarm is present (*Figures E-1* and *E-2*). The FACU then processes this input and turns on the evacuation signals on floors four, five, and six as programmed in the FACU for floor above, floor below signaling. This alerts the occupants on those floors that a fire condition exists, and there is a need to evacuate. At the same time, the FACU signals the elevator controller, access control system, and the HVAC system of the fire condition.

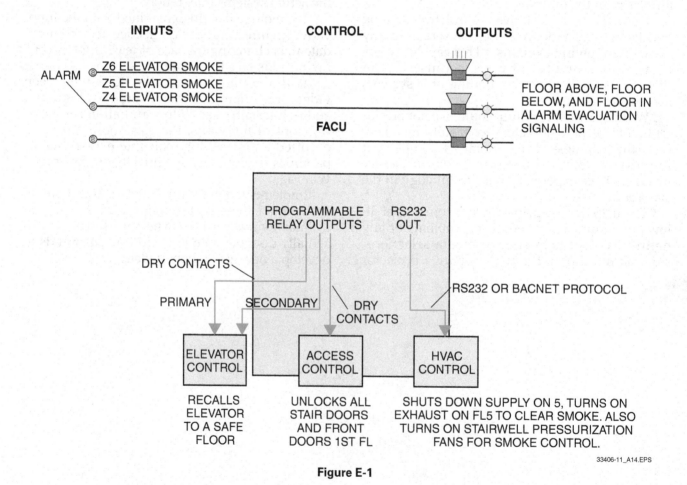

Figure E-1

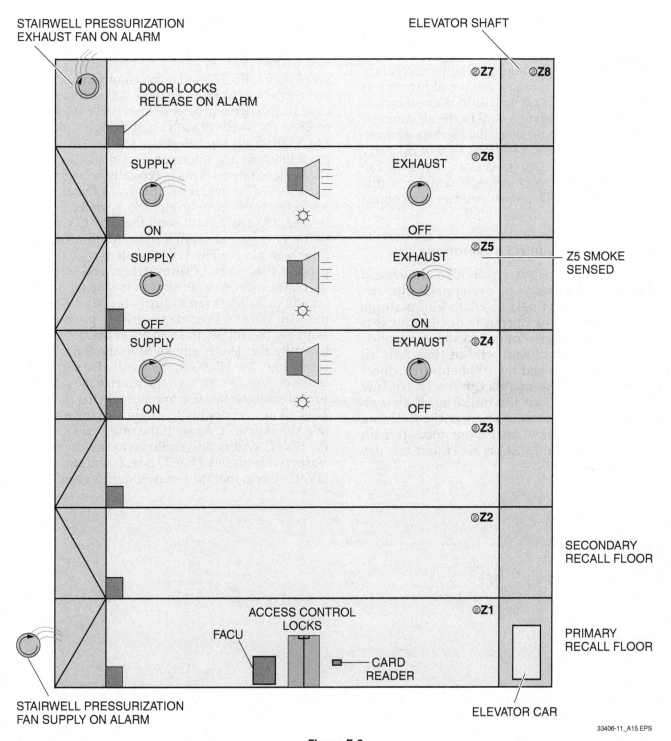

STAIRWELL PRESSURIZATION
EXHAUST FAN ON ALARM

ELEVATOR SHAFT

DOOR LOCKS
RELEASE ON ALARM

SUPPLY · EXHAUST · Z6 · ON · OFF

SUPPLY · EXHAUST · Z5 · OFF · ON

Z5 SMOKE
SENSED

SUPPLY · EXHAUST · Z4 · ON · OFF

Z3

Z2

SECONDARY
RECALL FLOOR

ACCESS CONTROL
LOCKS

FACU

CARD
READER

Z1

PRIMARY
RECALL FLOOR

STAIRWELL PRESSURIZATION
FAN SUPPLY ON ALARM

ELEVATOR CAR

33406-11_A15.EPS

Figure E-2

Building Elevator Control Integration

The elevator controller is typically interfaced to the FACU by means of two-pair #14 THHN in conduit terminated to two programmable relay outputs. The elevator controller responds to one of two possible contact closures from the FACU and processes the signal as either a primary or secondary recall signal. The determination of which contact is closed is processed by the FACU according to the alarm activation zone or device address. Typically, primary recall is activated by any smoke detector in the lobby of the main floor of egress from the building. In the example, floor one is the primary recall floor. Secondary recall is activated by any smoke detector in the lobby of all remaining floors and includes the top-of-shaft detector. In the example, floor two is the secondary recall floor. The purpose of any recall function is to send the elevator car to a safe floor of egress for any occupants who are still in the elevator car during a fire. Once recalled, the elevator controller will lock out the cab buttons so that no party other than the fire department can move the cab to a fire floor. The elevator will remain in this state until the FACU is reset and the dry contact restores.

Building Access Control Integration

The access control system is also interfaced to the FACU by means of dry contact closure. Typically, only one pair of wires and a single programmable relay output is needed for this function. The access control system responds to the dry contact closure and in turn tells all the stairway doors and the front entrance doors to unlock so that occupants can freely exit into the stairwell and exit the building. It also allows the fire department to enter the building during an emergency. Again, the doors remain unlocked until the FACU is reset and the dry contact restores.

Building HVAC System Integration

The last external system integrated to the fire system is the HVAC system. The interface in the example is via RS-232 serial communications over a twisted shielded pair of wires. As each event occurs in the fire alarm system, the HVAC system receives the status of each point change within the FACU through the interface. The HVAC system must interpret the information and take action on the status changes. This is typically accomplished in the HVAC provider's software. In the example, when the fifth floor elevator smoke detector is activated, the FACU will send the HVAC system a string of information that indicates that the fifth floor is in alarm. The HVAC system will, in turn, process that piece of information and will then signal the fifth floor supply fan to shut off and the fifth floor exhaust fan to turn on. This will cause the fifth floor to become negatively pressurized, stopping the supply of fresh air to the fire and exhausting the deadly smoke from the floor. At the same time, the HVAC system will also signal the stairwell fans to turn on to pressurize the stairwell so that smoke is not able to travel into the stairwell. This allows occupants to safely exit from a floor into the stairwell. As with the other two systems, the HVAC system will continue to run the fans in alarm mode until the FACU is reset and signals the HVAC system that the fire condition is clear.

Additional Resources

This module presents thorough resources for task training. The following resource material is suggested for further study.

Network Architecture & Design: A Field Guide for IT Professionals. Jerome F. DiMarzio. Indianapolis, IN: Sams Publishing.
Smart Home Automation with Linux. Steven Goodwin. New York, NY: Apress.
Smart Home Hacks: Tips & Tools for Automating Your House. Gordon Meyer. Sebastopol, CA: O'Reilly Media.
Top-Down Network Design, 3rd Edition. Priscilla Oppenheimer. Indianapolis, IN: Cisco Press.

Figure Credits

NCCER CURRICULA — USER UPDATE

NCCER makes every effort to keep its textbooks up-to-date and free of technical errors. We appreciate your help in this process. If you find an error, a typographical mistake, or an inaccuracy in NCCER's curricula, please fill out this form (or a photocopy), or complete the online form at **www.nccer.org/olf**. Be sure to include the exact module ID number, page number, a detailed description, and your recommended correction. Your input will be brought to the attention of the Authoring Team. Thank you for your assistance.

Instructors – If you have an idea for improving this textbook, or have found that additional materials were necessary to teach this module effectively, please let us know so that we may present your suggestions to the Authoring Team.

NCCER Product Development and Revision
13614 Progress Blvd., Alachua, FL 32615

Email: curriculum@nccer.org
Online: www.nccer.org/olf

❏ Trainee Guide ❏ AIG ❏ Exam ❏ PowerPoints Other _____

Craft / Level: _____ Copyright Date: _____

Module ID Number / Title: _____

Section Number(s): _____

Description: _____

_____

_____

_____

_____

Recommended Correction: _____

_____

_____

_____

_____

Your Name: _____

Address: _____

_____

Email: _____ Phone: _____

33407-12

Intrusion Detection Systems

Objectives

When you have completed this module, you will be able to do the following:

1. Identify and describe intrusion detection system sensing and notification devices.
2. Describe the control equipment and methods used with intrusion detection systems.
3. Configure an intrusion detection system to meet a specified need.
4. Describe system and equipment installation practices.
5. Describe the inspection, testing, maintenance, and troubleshooting practices associated with intrusion detection systems.
6. Install and wire an intrusion detection system consisting of sensors, notification devices, and a control panel.
7. Program a control panel and describe the different components, inputs, and programming options used in controlling intrusion detection systems.
8. Test and troubleshoot an intrusion detection system.
9. Wire an RJ-31X connector for line seizure.

Performance Tasks

Under the supervision of your instructor, you should be able to do the following:

1. Identify types of security sensors, notification devices, and control panels.
2. Select the correct sensors, notification devices, and control panels for various applications.
3. Install, wire, and program an intrusion detection system.
4. Troubleshoot an intrusion detection system.

Trade Terms

24-hour zone
Access control
Active infrared (photoelectric beam) sensor
Active sensor
Alarm verification
Americans with Disabilities Act Accessibility
   Guidelines (ADAAG)
Armed
Attack
Audio detection system
Audio discriminator
Bollard
Built-in siren driver
Burglar alarm screen
Buried line intrusion sensor
Central Station system
Channel detection
Circuit response time
Circumvent
Cold-flow
Combination system
Combined technology sensor
Control unit

Controlled zone
Delay zone
Detection device
Digital alarm communicator receiver (DACR)
Digital alarm communicator system (DACS)
Digital alarm communicator transmitter (DACT)
Digital communicator
Electric field sensor
End-of-line (EOL)
Entry delay
Exit delay
False alarm control team (FACT)
Glass-break detector
Hardwired intrusion system
Holdup alarm (HUA)
Home and away
Instant zone
Interior
Interior follower zone
Intrusion detection system (IDS)
Key switch
Kiss-off tone
Labeled

Trade Terms (*continued*)

Line carrier system
Local system
Microwave detector
Monitored intrusion system
Monitoring site
Motion (space) detector
Multiplexing
Notification device
Open collector output
Passive sensor
Perimeter
Perimeter follower zone
Photoelectric beam detector
Photoelectric detector
Piezoelectric sensor
Pressure mat
Protected distribution system (PDS)
Protected premises
Proximity sensor

Repeater
RJ-31X
Secured area
Security system
Seismic sensor
Shock (vibration) detector
Silent alarm
Spot detection
Stress sensor
Structural-attack piezoelectric sensor
Switched telephone network
Transmitter
Trap detection
Trouble signal
Visible notification device
Volumetric detection
Walk test light
Zone

Note: *NFPA 70*®, *National Electrical Code*® and *NEC*® are registered trademarks of the National Fire Protection Association, Inc., Quincy, MA 02269. All *National Electrical Code*® and *NEC*® references in this module refer to the 2011 edition of the *National Electrical Code*®.

Industry Recognized Credentials

If you're training through an NCCER-accredited sponsor you may be eligible for credentials from NCCER's Registry. The ID number for this module is 33407-12. Note that this module may have been used in other NCCER curricula and may apply to other level completions. Contact NCCER's Registry at 888.622.3720 or go to nccer.org for more information.

Contents

Topics to be presented in this module include:

Contents (*continued*)

Contents (*continued*)

Contents (*continued*)

Figures and Tables

1.0.0 INTRODUCTION

An intrusion detection system (IDS) is a combination of components designed to detect and report the unauthorized entry or attempted entry of a person or object into the area or volume protected by the system. Intrusion systems are commonly incorporated into total security systems or combination systems along with fire alarm and/or holdup alarm (HUA), panic, or emergency systems. This module focuses on those components and devices used in intrusion systems. Components and devices used with fire alarm systems and for building access control are covered in the modules entitled *Fire Alarm Systems* and *Access Control Systems*, respectively.

An intrusion system does not function properly if it is not designed, installed, and maintained according to specifications. It might fail to detect an intrusion, or it could activate when no intrusion has occurred to create a false alarm. No two system installations are exactly alike. To help regulate the different applications that exist, codes and standards have been established that dictate the specifics of design and installation for electronic systems. Always seek out and abide by any National Fire Protection Association (NFPA), Electronic Security Association (ESA), *National Electrical Code®* (*NEC®*), federal, state, and local codes and requirements that apply to the installation of each system. Underwriters Laboratory (UL) guidelines are not legally binding unless the alarm system is installed to meet UL Certificated requirements. Failure to comply with all applicable codes and requirements places both the electronic systems technician and the employer at risk of civil and even criminal liability.

2.0.0 INTRUSION SYSTEM OVERVIEW

A functioning intrusion system (*Figure 1*) consists of the following six components connected via wires and cables or through wireless technology:

- *Detection (initiating) devices* – Devices that include the system's various sensors and contacts
- *Annunciation (notification) devices* – The means through which the system announces to personnel within the secured area any perceived threats
- *Control panel* – The device that controls the intrusion detection system and which the other components are connected to as inputs or outputs
- Control units – The means through which security personnel operate the system

- *Communication devices* – The means through which the system reports to personnel outside the secured area about any perceived threats
- *Power supplies (primary and backup)* – Devices that provide the system with its operating power

Access control in an intrusion system includes any means of monitoring and restricting human traffic through doors, gates, and elevators. This includes using video cameras and monitors, time and attendance reporting, badges and access passes, pedestrian and vehicular traffic control, intrusion detection devices, and mechanical or electronic locks. For details about the various access control devices, refer to the *Access Control* module.

Intrusion systems commonly use closed-circuit television (CCTV) video cameras to monitor a given location. The video display may be located nearby to allow a person to watch an adjacent room or an inaccessible corner of the same room. The display might also be located in a central observation area or security office. A switcher connected to the equipment permits security personnel to monitor multiple areas from one location. Some CCTV systems have recorders for reviewing events or for legal purposes. For details about CCTV system components and operation, refer to the *CCTV Systems* module.

2.1.0 Local

Intrusion systems designed to alert only the occupants of a secured area or only those within range of an alarm bell, light, siren, or voice warnings in the secured area are local systems. The first alarm systems, consisting of little more than a bell designed to wake a house's occupants (or frighten

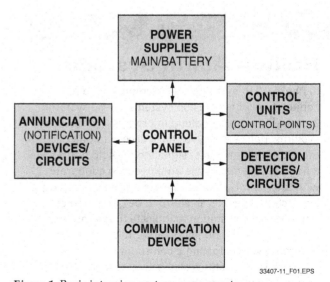

33407-11_F01.EPS

Figure 1 Basic intrusion system components.

Video Motion Detection (VMD)

Video Motion Detection (VMD) is a method that allows a video camera to serve as an initiating device. VMD can be implemented in the camera itself or in the video processing equipment through the use of specialized software. When implemented through the camera, video recording can be limited to situations in which motion is detected. This capability eliminates the need for continuous recording. The software can be programmed to perform a variety of functions, including triggering an alarm, sending alerts via phone or email, activating lights, and opening or locking doors.

and drive off an intruder), were local systems. There are two disadvantages with local systems. First, someone must be nearby to hear or observe the warning device. Second, skilled intruders have been known to quickly disable the system's notification devices so that no one is aware of the intrusion.

2.2.0 Monitored

An intrusion system connected to a monitoring function or designed to alert property owners, police, or security personnel outside the secured area is a monitored intrusion system. Monitored systems provide more protection than local systems and are less likely to be disabled before alerting the appropriate personnel.

2.3.0 Types

While specific codes and standards must always be followed when designing and installing an intrusion system, many different types of systems with different types of technology can be used. Most of these systems are categorized by their means of communication between the sensors,

Holdup Alarm Systems

In banks, HUA systems are commonly used in conjunction with intrusion systems. These systems are used to report a holdup in progress. In HUA systems, bank employees have access to one or more discreet means of signaling for help during a holdup. This allows the employees to cooperate with the robber while pressing a button or tripping a switch that causes an alarm to be generated at a remote central station to indicate that a robbery is in progress. Responding authorities then react accordingly, anticipating a confrontation with a criminal. Most systems also activate video cameras to photograph the robber.

or detectors, and the control panel. The three major types of systems are hardwired intrusion systems, multiplex systems, and addressable systems.

2.3.1 Hardwired System

A hardwired system using conventional sensors (mechanical and electrical switches and sensors) and conventional notification devices (such as bells, horns, and lights) is the simplest of all intrusion systems. Large buildings or secured areas are usually divided into zones to identify the area where a potential threat is perceived. A simple hardwired system is limited to zone detection only. It has no means of identifying the specific sensor that activated the alarm unless each sensor is on a separate zone.

2.3.2 Multiplex System

Multiplex systems are similar to hardwired systems because they rely on zones for detection. The difference, however, is that multiplexing allows multiple signals from several sensors to be both sent and received over a single communications line. Each signal can be uniquely identified. This results in reduced control equipment, less wiring infrastructure, and a distributed power supply.

2.3.3 Addressable System

Addressable systems have device-specific identifiers, not just a zone of devices. They use advanced technology and sensors that allow discrete identification of alarm signals at the individual sensor level. The system can pinpoint an alarm to the precise physical location of the activated sensor. The basic idea of addressable systems is to provide identification or control of individual sensor, control, or notification devices on a common circuit. Each component on the signaling line circuit has an identification number or address. The addresses are usually assigned using switches or similar means.

3.0.0 Types of Intrusion System Sensors

Sensors are devices designed to detect human intrusion. The intrusion may be detected either actively or passively through motion (space) detection.

- Motion (space) detectors respond to changes in the environment caused by the movement or presence of human beings.
- Photoelectric beam detectors use an infrared light transmitted to a photocell. A break in the beam activates the sensor.
- Active sensors include ultrasonic, microwave, and photoelectric detectors. Active sensors transmit energy into an area and monitor the area for changes caused by motion.
- Passive sensors include passive infrared detectors, passive sonic detectors, and audio discriminators. Passive sensors monitor the environment and respond to changes in it.
- Combined technology sensors use two technologies, usually one passive and one active, to prevent false alarms and verify that an intrusion is caused by human motion or presence.

Pressure sensors and stress sensors detect an intruder after entry has been achieved. Pressure sensors detect the weight of the foot. Stress sensors react to flexing of the floor caused by the weight of the intruder.

The two major categories of sensors are perimeter and interior.

3.1.0 Perimeter

The perimeter is defined as the outer limits of the secured area. It could be the walls, floor, and ceiling of a building, or it could be a fence or the outside boundary of the intrusion system's sensors. Perimeter sensors detect unauthorized access to the secured area from outside the area.

3.1.1 Magnetic Switch Sensors

Magnetic switch sensors use magnets to detect whether a door, window, or other means of accessing a secured area is open or closed. The magnetically actuated switch is positioned on the stationary side of the opening, and a magnet is positioned on the moving side. When aligned, the magnet exerts a force on the switch. When the door or window is opened, the magnet moves away from the switch, removing the magnetic force. The removal of this force activates the sensor switch.

Although they are widely used, generally reliable, and easily understood, magnetic switch sensors should not be used alone to protect high-value items. They do not detect an intrusion if, to gain access, the intruder cuts through the door or window to the secured area without opening the protected entry point. Normally open contacts are used on normally closed (closed loop) circuits and vice versa.

There are three basic types of magnetic switch sensors (*Figure 2*), including the following:

- The magnetic mechanical switch consists of a magnetically actuated switch and a magnet. Switches of this type are inexpensive and can handle relatively high levels of electric current. They are vibration sensitive and some are not sealed. The unsealed sensors are subject to contamination from dust, dirt, moisture, or other debris. They require relatively high levels of electric current to operate properly.
- The magnetic reed switch consists of contacts formed by two thin, movable, magnetically actuated metal vanes or reeds held in a normally open or normally closed position within a sealed glass envelope. (Normally closed switches are more common, but some have both normally open and normally closed contacts.) The contact points are activated with a magnet. A magnetic reed switch has a longer and more consistent operating life because the sealed contacts are protected from corrosion and the accumulation of dirt, dust, or other debris. Switches of this type are inexpensive and have relatively low contact resistance. But, they cannot handle high levels of electric current and they have small contacts that can freeze. Like magnetic mechanical switches, these switches are vibration sensitive.
- The balanced magnetic reed switch incorporates a balanced magnetic field. It is activated when it detects an increase or decrease in magnetic field strength. A balanced magnetic reed switch is more difficult to circumvent than standard magnetic reed or magnetic mechanical switches because it is sensitive to both the absence of its activating magnet and any attempt by an intruder to use another magnet on the switch. These switches are inexpensive and have relatively low contact resistance. Like standard magnetic reed switches, they are vibration sensitive, cannot handle high levels of electric current, and have small contacts that can freeze.

SWITCH SHOWN IN
NORMAL POSITION
WITH MAGNET AWAY
FROM SWITCH.

SWITCH SHOWN IN
ENERGIZED POSITION
WITH MAGNET AT SWITCH.

MAGNETIC MECHANICAL SWITCH (NORMALLY OPEN)

SWITCH SHOWN IN
NORMAL POSITION
WITH MAGNET AWAY
FROM SWITCH.

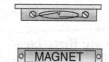

SWITCH SHOWN IN
ENERGIZED POSITION
WITH MAGNET AT SWITCH.

MAGNETIC REED SWITCH (NORMALLY OPEN)
CONTACTS ARE ENCLOSED IN HERMETICALLY SEALED TUBE.

SWITCH SHOWN IN
NORMAL POSITION
WITH MAGNET AWAY
FROM SWITCH.

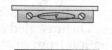

SWITCH SHOWN IN
ENERGIZED POSITION
WITH MAGNET AT SWITCH.

MAGNETIC REED SWITCH (NORMALLY CLOSED)
CONTACTS ARE ENCLOSED IN HERMETICALLY SEALED TUBE.

SWITCH SHOWN IN
NORMAL POSITION
WITH MAGNET AWAY
FROM SWITCH.

SWITCH SHOWN IN
ENERGIZED POSITION
WITH MAGNET AT SWITCH.

THIRD MAGNET
CHANGES SWITCH
TO OPEN POSITION.

BALANCED MAGNETIC REED SWITCH
CONTACTS ARE ENCLOSED IN HERMETICALLY SEALED TUBE. BALANCING MAGNET IS ADDED.

33407-11_F02.EPS

Figure 2 Magnetic reed switches.

3.1.2 Glass-Break Detectors

Glass-break detectors detect the sound, the frequency of breaking glass, or a combination of sounds and vibrations including the thump that immediately precedes the shattering of glass. How glass is constructed can determine how it breaks and what types of sounds or vibrations are produced when it does so. Always check the glass type (indicated by the installer's mark on the corner of the glass), and select a compatible sensor. Glass types include the following:

• Regular or plate glass is common in homes and older commercial buildings. When plate glass breaks, it makes a loud noise and breaks in hard, jagged pieces. It creates relatively little vibration because it is easily broken.

• Tempered glass is plate glass cut to size and fired in an oven at 1,400°F. It is found in sliding doors, in windows located next to doors, and in most commercial buildings. Tempered glass breaks into numerous small pieces. When it breaks, it is quieter than other types of glass and produces less vibration.

• Wired glass is plate glass with a wire mesh insert and is commonly found in industrial and fire doors. The wire holds the glass together and absorbs some of the shock when the glass is broken; the wire then breaks as the glass pieces separate.

• Laminated glass is a sandwich of glass and a plastic polymer filling and is used in commercial applications, such as offices and retail storefronts. If the glass is broken, the plastic ab-

Hall Effect Magnetic Switches

A magnetic switch like the one shown can be used as a perimeter sensor. This magnetic switch, called a Hall effect switch, is an electronic switch; therefore, it contains no moving parts. Like other perimeter magnetic switches, it has two parts: a switch unit that houses a Hall effect sensor and related electronic switch, and a magnetic unit. The voltage developed by the Hall effect sensor, which is proportional to the magnetic field strength, is applied to the electronic switch. A weak or absent magnetic field causes the electronic switch to open the circuit, initiating an alarm.

When a current-carrying semiconductor material is in the presence of an external magnetic field, a small voltage, called the Hall voltage, develops on two opposite surfaces of the conductor. The magnitude of the Hall voltage is directly proportional to the strength of the magnetic field.

The small voltage generated by the Hall effect must be amplified.

Hall effect sensors contain circuitry to convert the small Hall effect signal into useful voltage levels. Some sensors have an analog output that is proportional to the strength of the magnetic field. Others are designed with a digital output that switches logic levels when the strength of the magnetic field exceeds a certain level.

Hall effect magnetic switches typically provide for better protection against tampering and defeat than balanced magnetic reed switches. They have little susceptibility to pressure and temperature changes, provide solid-state reliability, and are more compact and more sensitive than reed switches.

HALL-EFFECT SWITCH UNIT

MAGNETIC UNIT

33407-11_SA01.EPS

---

sorbs most of the shock and the broken glass sticks to the polymer. Breaking laminated glass produces a loud thump with relatively little vibration.

- Sealed insulated glass consists of two panes of plate, tempered, or laminated glass separated by spacers and sealed airtight. The space between the panes normally contains a drying agent that helps to prevent fogging from condensation.

If necessary, contact the manufacturer to ensure the correct sensor for the window type. If the window type and sensor do not match, intrusions can occur undetected.

Piezoelectric sensors are in common use. These sensors use a piezoelectric crystal that vibrates and produces electricity in reaction to the frequencies produced by breaking glass. Piezoelectric sensors can be self-contained or wired to a processor. They operate without an external power source because they generate electricity. There are two types of piezoelectric sensors—on-glass and remote.

On-glass piezoelectric sensors mount directly to the glass. The adhesive used must not dampen or buffer the transmission of vibrations through the glass to the sensor. (The manufacturer should specify and supply the adhesive.) On-glass piezoelectric sensors can be used on any plate or sheet glass, but they cannot be used on tempered, wired, or laminated glass. Mount the detector at either a single corner or two diagonally opposite corners of a window. These sensors are small and unobtrusive. At least one on-glass piezoelectric sensor is required to cover each individual windowpane, which can be prohibitively expensive on multi-pane windows. Solar film on windows reduces the effectiveness of these sensors. On-glass piezoelectric sensors are also subject to false alarms if segments of cracked window glass rub together.

Remote piezoelectric sensors are also known as acoustic sensors or audio discriminators. Unlike on-glass detectors, remote detectors do not have to be mounted directly to the glass surface and therefore they are not visible to intruders. Tuned to listen specifically for the sound frequency produced by breaking glass, remote piezoelectric sensors differ in range, sensitivity, and physical size. They can be used on any plate or sheet glass. These sensors cannot be used on tempered, wired, or laminated glass, or on any surface that is likely to vibrate. Remote piezoelectric sensors are more reliable than most other detectors used to protect glass. These sensors can also be used to cover multiple panes of glass. Remote piezoelectric sensors are larger than on-glass sensors. They only respond to breaking glass.

All remote piezoelectric detectors use the piezoelectric effect as one means of sensing activity on the glass. There are several different means of processing the signal from the piezoelectric crystals, including the following:

- *Standard detectors* – These monitor the speed at which the higher frequencies generated by the breaking glass rise and the point at which the frequencies peak.
- *Dual-frequency glass-break detectors* – When glass is broken, this detector generates two frequencies. A low frequency is generated by the flexing of the glass when it is impacted. A high frequency is then generated when the glass breaks. Dual-frequency detectors monitor the points at which the low frequency and the high frequency generated by the breaking glass exceed predetermined levels. Both frequencies must exceed the predetermined level within a specific time period in order to trigger an alarm.
- *Three-frequency detectors* – These monitor the speed at which the higher, midpoint, and lower frequencies generated by the breaking glass rise, the duration of the sounds, and the ambient noise at the detector location.
- *Signal processing detectors* – These sensors use a microprocessor to analyze several factors to decide when glass has been broken. These factors include the points at which the low frequency and the high frequency generated by the breaking glass exceed predetermined levels. Both frequencies must exceed the predetermined level within a specific time period in order to trigger an alarm. The pattern (including the duration) of the frequencies must match a profile for glass breaking. The pattern of sound cannot match common ambient noises and the glass-breaking pattern of sounds must occur within a programmed time period. Advanced

signal processing detectors that evaluate high and low frequency, time, and ambient noise do not require mounting on the glass to detect intruders. This hides the detector from potential intruders who may attempt find a way to defeat the system and does not spoil appearance in high-end interiors.

- *Dual microphone detectors* – These are used to listen to sounds at the glass and within the room where the glass is located. Sound that occurs only in the room is ignored, but sound heard by both microphones is analyzed. If the sound comes from the room first, it is ignored; if the sound comes from the glass first, it may trigger an alarm. This feature is usually added as an additional check to a signal processing detector.
- *Advanced technology detectors* – Detectors are now available that detect breakage of all types of glass. They use sophisticated sound analysis of glass-break profiles. These detectors react to sound amplitude thresholds as well as preprogrammed statistical patterns of glass breaking.
- *Thump detection* – Most advanced audio discriminator detectors also listen for the thump sound that causes the glass to break. The detector must hear the thump first followed by the frequencies of glass breaking.

3.1.3 Burglar Alarm Screens

A burglar alarm screen consists of a fine-gauge wire woven (laced) through the fabric of a window screen, forming a complete electrical circuit (*Figure 3*). The wire is connected to a control device. If the screen (with its combined wire) is cut, the circuit is broken, and an intrusion is reported. Burglar alarm screens are a discreet, stable, easily understood, and reliable method of sensing intru-

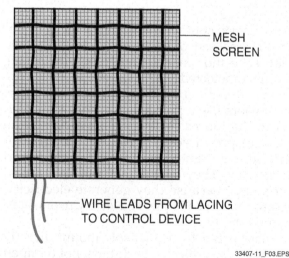

MESH SCREEN

WIRE LEADS FROM LACING TO CONTROL DEVICE

33407-11_F03.EPS

Figure 3 Burglar alarm screen.

sion through a window. They allow windows to be secured while remaining open for ventilation. Precise measurement of the window dimensions is required in order to install such screens properly.

3.1.4 Shock (Vibration) Detectors

Shock (vibration) detectors are inexpensive devices designed to detect the shock of attack before intrusion (*Figure 4*). The tension on the detector's contacts is adjustable. It is difficult to set the device's sensitivity, and false alarms may result. Shock or vibration detectors are used primarily to detect glass breakage or wall vibration. They cannot be used on windows subject to excessive vibration. Sensors of this type are extremely susceptible to false alarms caused by vibration unrelated to intrusion.

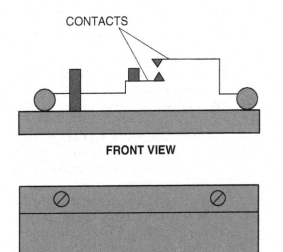

CONTACTS

FRONT VIEW

SIDE VIEW

33407-11_F04.EPS

Figure 4 Shock (vibration) detector.

3.1.5 Structural-Attack Piezoelectric Sensors

Structural-attack piezoelectric sensors may also be called vibration detectors. A wall or other barrier vibrating while under structural attack, such as hammering or drilling, generates frequencies much higher than normal background noises. These higher frequencies activate the structural-attack piezoelectric sensor.

Like all piezoelectric sensors, these sensors cannot be used on any surface that might vibrate randomly or excessively. They are very sensitive and difficult to defeat. Sensors of this type are relatively expensive and suitable for very limited applications, such as protecting safes, vault doors, and automated teller machines.

3.1.6 Electric Field (Capacitive) Sensors

An electric field sensor is a type of capacitive sensor used with fences. The sensors consist of parallel field and sensor wires, either free-standing or mounted to the fence. The field wire is connected to a generator that produces an electrostatic field. A processor panel monitors the field through the parallel sensor wire. The electric field sensor is activated by the change in capacitance created when an intruder approaches the fence.

There are three basic electric field sensors:

- *Single field wire* – Consisting of a field wire and a grounded post, it is the cheapest and least reliable type of electric field sensor.
- *Double wire* – This uses a field wire and a ground wire. It is more reliable than a single field wire electric field sensor.

- *Triple wire* – This uses a high wire, a low wire, and a central ground wire and is the most reliable electric field sensor system.

Electric field sensors can also be used with safes, wall tops, or roofs but should not be used where grounding is poor or vandalism and false alarms are likely.

While they take up little space and are difficult to defeat, electric field sensors are expensive and extremely sensitive. This sensitivity makes them very susceptible to false alarms. A fence with an electric field sensor should be enclosed within an exterior fence that can screen out small animals or blowing refuse that may trigger false alarms. Sensors of this type are rarely used today but may be used in special circumstances.

3.1.7 Coaxial Cable Systems

Coaxial cable (coax) systems, also known as ported or leaky coax, consist of a coaxial transmit cable and one or two coaxial receive cables buried under the ground. The transmit cable produces an electrical field that is monitored by a processor connected to the receive cables. An intruder moving in the energy field produces frequencies that will activate an alarm.

In some coax cable systems, the transmit cable has a special leaky or ported shield that radiates electrical energy. Coax cable systems can be adapted to terrain that is not level and will not interfere with ground-level activities. They should not be used to protect unmanned sites or in areas with a high water table, heavy snow melt, or frozen ground.

Coax cable systems are difficult to detect and can be used under concrete. They are very sensitive. Coax cable systems are unaffected by most environmental conditions, but they are complicated and expensive. A fence with a coax cable system should be enclosed within an exterior fence that can screen out things like small animals or blowing refuse to reduce false alarms. Like electric field sensors, coax cable systems are rarely used today but may be used in special circumstances.

3.1.8 Buried Line Intrusion (Seismic) Sensors

A buried line intrusion sensor, or seismic sensor, consists of a coaxial cable with piezoelectric ceramic disks located between the center conductor and the shield at equal intervals (*Figure 5*). Pressure from seismic motion on the disks causes them to generate a signal.

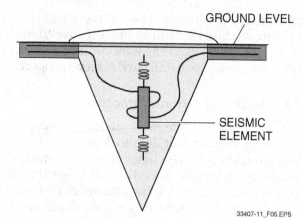

Figure 5 Buried line intrusion (seismic) sensor.

Buried line intrusion sensors are used under the ground around the perimeter of a secured area, sometimes around fences. They are difficult to detect. Their installation is labor-intensive. Like other similarly sensitive systems, a fence with this type of sensor should be enclosed within an exterior fence in order to prevent false alarms caused by small animals or blowing refuse. Sensors of this type are also rarely used today but may be used in special circumstances.

3.1.9 Outdoor Microwave Sensors

An outdoor microwave sensor uses microwave energy to form an invisible link between a microwave transmitter and a microwave receiver (*Figure 6*). The receiver is tuned to the strength of the signal reaching it. The presence of an intruder in the sensor area reduces the strength of the microwave signal reaching the receiver, which then activates the sensor.

The microwave receiver accounts for gradual changes in the strength of the signal. This means that only a rapid change in the signal strength activates the sensor, which helps prevent false

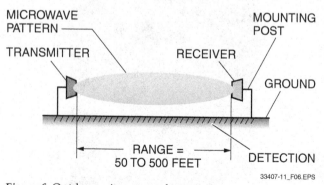

Figure 6 Outdoor microwave alarm sensors.

alarms. The effective distance between the microwave transmitter and receiver varies by the frequency of the system. X-band systems have a range of about 10 feet (3 meters) to 600 feet (183 meters). Y-band systems operate from 100 feet (30 meters) to 1500 feet (457 meters). A link that is too short creates too strong a signal, preventing a human from reflecting sufficient signal away from the receiver to create an alarm. A link that is too long creates a signal that is just above the alarm threshold, triggering false alarms.

Outdoor microwave sensors are extremely sensitive, which, depending on the application, can be an advantage or a disadvantage. They have a long range and can be used over hard surfaces like asphalt. They are not affected by air movement, temperature changes, fog, or rain, and have a relatively low current draw. They are difficult to defeat. Outdoor microwave systems are expensive and require labor-intensive clearing of the area to be protected. Preventing undesired spread of the microwave beam requires great care. Outdoor microwave systems are susceptible to false alarms caused by passing traffic, drifting snow, or even the movement of trees and tall grass caused by wind. Sensors of this type require line-of-sight transmission to the receiver.

3.1.10 Active Infrared (Photoelectric Beam) Sensors

An active infrared (photoelectric beam) sensor (*Figure 7*) transmits a beam of infrared light to a photocell. If an intruder interrupts the beam, the sensor is activated. Special circuitry can be added to the sensor to deal with blockage of the light beam by snow or rain. Heaters and fans may be

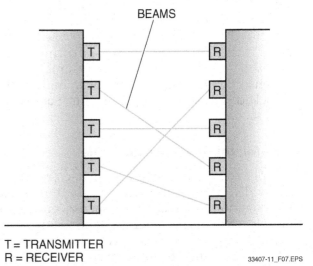

T = TRANSMITTER
R = RECEIVER

33407-11_F07.EPS

Figure 7 Outdoor photoelectric beam array.

Outdoor Microwave Sensors

Outdoor microwave sensors that use a separate transmitter and receiver and related antennas are called bistatic microwave sensors. Monostatic microwave sensors like the one shown here are also available.

A monostatic sensor combines the microwave transmitter and receiver, called a transceiver, in one unit along with a shared antenna. Pulsed microwave energy is transmitted from the transmitter via the antenna into the protected area. The receiver compares returned reflected signals against a known reflected signal reference established at installation and looks for a change in energy. The motion of an intruder causes the transmitted pattern of the reflected energy to change, resulting in an alarm.

Microwave detection systems are available in portable versions and are very useful for temporary perimeter security requirements.

33407-11_SA03.EPS

used to keep the unit at a temperature within the manufacturer's specifications.

While they are used primarily for flat terrain perimeter detection, active infrared sensors can be adapted to uneven terrain by vertically adjusting the light beam. Stacking active infrared sensors in series, or using mirrors, can provide barrier detection. These sensors are compact, easy to install, and have a potentially long range. Like outdoor microwave sensors, active infrared sensors require a labor-intensive clearing of the secured area and are also susceptible to false alarms caused by small animals, drifting snow, or wind-driven movement of trees and tall grass. The sensor lenses must be kept clean and cannot be blocked. Vibrations negatively affect active infrared sensors.

3.1.11 Microphonic Sensor Cable

Microphonic sensor cables are perimeter sensors designed to detect and analyze the noise patterns common to forced entry, such as cutting or climbing a fence. The systems can be strung along fences, roofs, and walls, and can also be used to protect vaults and strong rooms. Features built into the system allow it to distinguish the sounds caused by weather and small animals, thus reducing the incidence of nuisance alarms. Sensor cables for these systems typically are supplied in 1,000-foot lengths. The systems can be set up to monitor single or multiple zones. Systems can be set up to trigger local alarms through a relay contact or networked to feed information to a central monitoring location.

3.1.12 Fiber Fence Security

A fiber fence security system uses fiber optic cable strung along the length of the fence and secured with cable ties. Light is transmitted through the fiber. An activity, such as climbing or cutting a fence, disturbs the light pattern or power level. Special software algorithms are used to analyze the disturbance. These algorithms can distinguish between genuine alarms and nuisance events, such as those that might be caused by weather or small animals.

3.2.0 Interior

Interior sensors are any detection devices used within the secured area. They can be used to pinpoint the location of an intruder or to isolate a specific threat to property or human life.

The many types of interior sensors can be more broadly categorized as trap detection, spot detection, channel detection, and volumetric detection sensors (see *Figure 8*):

- *Trap detection* – Provides narrow or wide detection for high-traffic areas or for the path an intruder is expected to take to get from one point to another within a secured area
- *Channel detection* – Provides narrow detection in an area where an intruder is expected to cross
- *Spot detection* – Focuses on a particular object or on high-value areas, such as safes, vaults, and storage areas
- *Volumetric detection* – Provides wide detection in a defined area

---

On Site

Fiber Optic Mesh Sensors

Panels made of fiber optic mesh are available for use both as underground intrusion and perimeter fence intrusion sensors. When used as an underground sensor, a grid is formed by mechanically and optically connecting two or more mesh panels together then covering the grid with gravel or dirt. If someone steps on the ground above the mesh, the mesh fibers bend and cause a change in the supervised optical signal being transmitted through the mesh. This change causes an alarm to be generated.

When used as a perimeter fence intrusion detector, the mechanical and optically connected mesh panels are typically suspended from a taut-wire sensor at the top of the perimeter fence to form a combined fence barrier and intrusion sensor. A taut-wire sensor consists of a horizontal wire of high-tensile strength that is connected under tension to one or more transducers near the midpoint of the wire span. A fiber optic transducer generates a pulsed infrared (IR) signal for application through the mesh fibers, processes the signal from the mesh, and monitors the taut-wire sensor. If the mesh is cut or broken, an alarm is triggered. Any attempt to climb the mesh is detected by the top taut-wire sensor. Perimeter intrusion detection systems such as this are typically used in military applications.

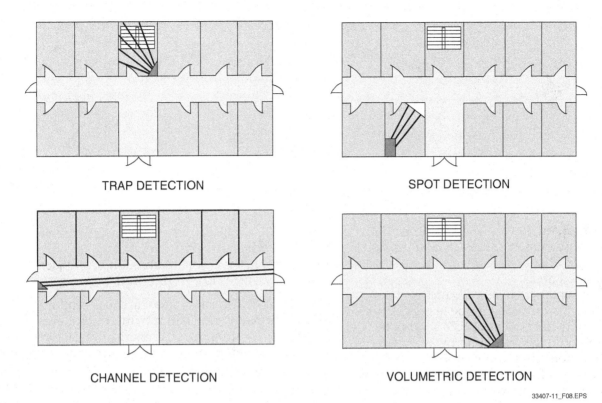

Figure 8 Trap, channel, and volumetric detection.

Interior sensors should be installed in addition to perimeter devices near all movable entryways, such as doors and windows. They provide a backup to the perimeter system and verify entry through a particular door or window. Interior sensors can also be used as augmentation to protect large glass windows, eliminating the need for foil or glass-break detectors. Interior sensors can detect intruders who hide in the secured area while it is being locked.

3.2.1 Microwave Detectors

An interior microwave detector (*Figure 9*) is an active volumetric detector that transmits microwave energy. The floor, ceiling, and walls of the secured area, as well as objects within the area, reflect a portion of this energy at a constant rate. The presence of an intruder alters the frequency of the reflected microwave energy. This activates the detector. Microwave-only detectors are now seldom used; interior microwave detectors are used as part of combined technology sensors instead.

Interior microwave detectors are used primarily in large rooms or long internal corridors. They are very sensitive, have long range, and are not affected by air turbulence, temperature change, ultrasonic noise, humidity, or sunlight. Reflections of the microwave energy can cause changes in the

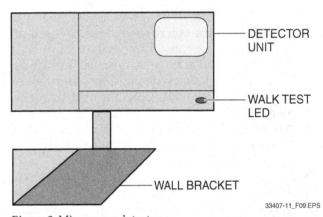

Figure 9 Microwave detector.

intended pattern. Penetration of building materials by microwave energy can extend the range of detection beyond the intended area, which is generally undesirable. It is also important to note that if more than one microwave sensor is used in an area, each unit must use a different frequency.

The beam height of a microwave transceiver pattern is approximately half its width (*Figure 10*). The pattern may range as long as 200 feet (approximately 61 meters), depending on the device.

Microwaves travel in a straight line and, unlike sound waves, do not require air as a medium. Theoretically, microwaves travel forever. They will penetrate most building materials to varying degrees. Glass transmits, rather than absorbing or

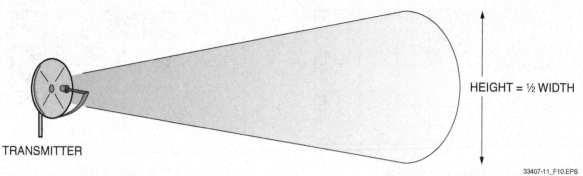

HEIGHT = ½ WIDTH

TRANSMITTER

33407-11_F10.EPS

Figure 10 Microwave transceiver pattern.

reflecting, microwave energy. Metal, however, reflects a portion of microwave energy, and metal ducts can channel microwave energy to and from unwanted areas.

The human body reflects microwave energy that is roughly equivalent to the energy reflected by a piece of metal 6 inches (15.24 centimeters) square. As previously stated, microwave reflections and penetrations may cause problems for microwave detectors. Reflections can alter the intended microwave pattern, and penetrations can move the pattern beyond the desired detection space.

Some other facts regarding microwaves and microwave detectors include the following:

- Vibration of the microwave detector's mounting surface can cause the detector to perceive an intruder when no intrusion has occurred.
- Moving metal signs, decorations, and fixtures activate a microwave detector and trigger an alarm.
- An intruder walking across a microwave detector's pattern at the far end of the detector's range may not be detected.
- Microwave detectors have difficulty perceiving a slowly moving target. Slowly moving targets may allow the detector enough time to adjust to the intrusion as part of the reflected microwave pattern.

3.2.2 Passive Infrared (PIR) Sensors

All objects at a temperature above absolute zero radiate infrared energy. The amount of infrared energy radiated depends on several of the following factors:

- The surface temperature of the object compared to surrounding surfaces
- The availability of energy within the object
- The surface emissions of the object (whether it tends to absorb or reflect)
- The physical size of the object

- The object's internal generation of energy (metabolism or activity)
- The emissions of surrounding objects (their tendencies to absorb or reflect)

If all surfaces in an area are stationary, and no human beings are moving in that area, a pattern of radiating infrared energy is established. If a human being moves or enters the room, the pattern is disturbed in two ways—the flow of infrared energy radiating from some surfaces is blocked, and the human body generates its own pattern of radiating infrared energy.

Passive infrared (PIR) sensors perceive and respond to these changes in infrared energy levels. They are used primarily in low- to medium-risk areas, such as offices, and to protect corridors and hallways. Passive infrared sensors can be triggered by sunlight, vibration, heat changes in stationary objects, and movement of small animals. Sensors of this type should not be used in any area where heating or cooling appliances, lighting, direct sunlight, or open fires can cause a rapid change in emission of radiant infrared energy. PIRs should not be directed at windows.

Passive infrared sensors are the most commonly used sensors today. They are easily understood and reliable if positioned correctly. Because they are passive, they have no detectable beam, no moving parts, and are not affected by audible noise. They are relatively inexpensive, and their range and pattern of detection are easily changed. Multiple passive infrared sensors can be used to protect a site and they do not interfere with one another.

There are four basic parts to a passive infrared sensor, listed as follows:

- Optics assembly
- Infrared transducer (pyro-electric element)
- Electronic amplifier
- Signal processing circuitry

The infrared energy pattern perceived by a passive infrared sensor is determined by the sensor's optics assembly. The optics assembly gathers infrared radiation and focuses it on the transducer, or infrared sensor. The optics assembly generally consists of a number of optical elements, each of which is aimed in a specific direction. This is done to cover the widest possible area. Some patterns are wide for full room coverage; others are narrow for hallway or corridor coverage. Various types of passive infrared patterns are shown in *Figures 11, 12,* and *13* and are listed as follows:

- Short range
- Medium range
- Long range
- Wide angle
- Narrow angle
- Curtain
- Barrier
- Pet alley
- Ceiling
- Beam
- Wall

There are different methods of processing signals in passive infrared sensors:

- *Multiple step logic (pulse count)* – One sensor must alarm more than once during a given time window. An alarm is triggered only after there has been a series of detections in a short period of time. A pulse count is used in difficult environments where false alarms are a problem.

- *Event verification* – The alarm signal must be verified by duration or repetition.
- *Signature detection* – The alarm must match the signature of an intruder. The signature is a record of the complex relationships between the amplitude, polarity, and timing of the signal received from the pyro-electric element.

There is no such thing as a volumetric passive infrared sensor. All wide-angle PIRs use segmented lenses. These segments create a smaller field of view to allow a human-sized target to be

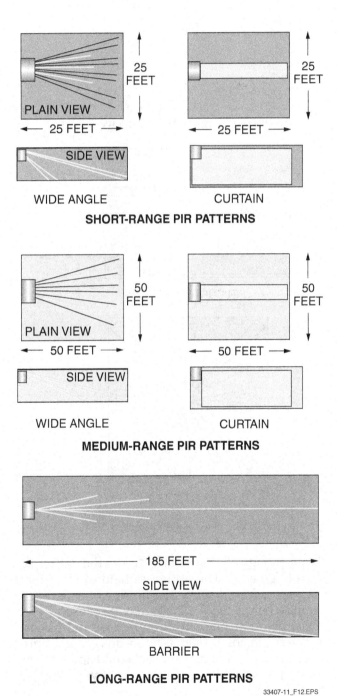

SHORT-RANGE PIR PATTERNS

MEDIUM-RANGE PIR PATTERNS

LONG-RANGE PIR PATTERNS

33407-11_F12.EPS

Figure 12 Short, medium, and long range focal patterns.

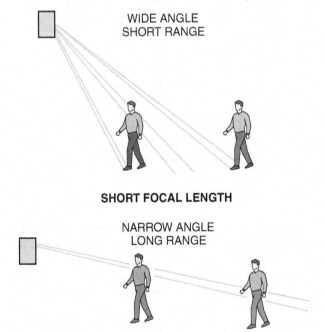

33407-11_F11.EPS

Figure 11 Short and long focal lengths.

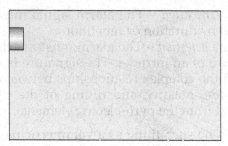

PIR BARRIER PATTERN

PIR PATTERNS WITH PET ALLEY

PIR CURTAIN PATTERN

CEILING PATTERN

33407-11_F13.EPS

Figure 13 Barrier, pet alley, curtain, and ceiling PIR patterns.

A TYPICAL ROOM

WHAT THE PIR "SEES"

A SPIDER WALKING ON THE PIR DETECTOR FACE
COULD CAUSE A FALSE ALARM.

33407-11_F14.EPS

Figure 14 PIR view of a room.

large enough to fill the field across the entire pattern (*Figure 14*). The optimum field of view for a passive infrared sensor is the size of an average human being. False alarms may be caused by insects or other pests walking across the passive infrared sensor. If there is no means of signal processing to prevent it, the insect could appear large enough to block the sensor's field of view, triggering an alarm.

False alarms from pets are prevented by installing the PIR sensor with the bottom of the infrared detector pattern higher than the height of the pet. A pet that is two feet tall does not cause an alarm if the bottom of the pattern is three feet above the floor. Pet-activated false alarms may still occur if the pet climbs onto furniture. If intruders see a pet in a secured area, the intruders can remain undetected by crawling under the pattern.

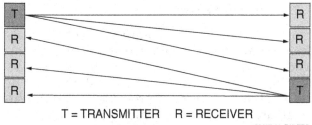

T = TRANSMITTER   R = RECEIVER

33407-11_F15.EPS

Figure 15 Multi-beam photoelectric wall.

3.2.3 Photoelectric Detectors

A photoelectric detector (also known as an electric eye) is an active infrared sensor that transmits infrared light. This infrared light beam is received by a photocell. If an intruder interrupts the light beam, the detector is activated.

Depending on the application, photoelectric beams can be classified as either perimeter or space detection devices. The space the beam covers, an invisible cylinder of light between the transmitter and receiver, is small but can cover a wide area. Photoelectric detectors are often used to cover long perimeter areas, but they can also be used to form a multi-beam wall for space detection (*Figure 15*).

Photoelectric detectors are designed to detect an interruption of a concentrated beam of infra-

red light that is pulsed at varying frequencies. An intruder would have to duplicate the exact pulse rate of the light in order to defeat the system by projecting light on the photocell. This makes photoelectric detectors difficult to defeat, although concealing the transmitter and receiver also contributes greatly to a successful detection.

Photoelectric detectors can be used to cover an area up to 1,000 feet (approximately 305 meters). They are used primarily across door and window openings, gates, skylights, storage racks, intersecting rooms, or to protect long hallways and corridors. They are compact, easy to install, reliable, and they activate if left blocked. They are not affected by air turbulence, heat changes, or noise.

Photoelectric detectors have certain disadvantages. They are susceptible to false alarms caused by vibrations and falling objects, and they require a clear line of sight between the transmitter and the receiver. It is sometimes difficult to align the transmitters with their receiving photocells. It is also important to keep the detector lenses clean.

Using mirrors with photoelectric detectors reduces the effective range of the infrared beams. Mirrors can be used to bend the beam around obstacles, but this practice is not recommended.

A practical alternative for using photoelectric detectors when a power source or mounting surface is not available for both sides of the detection area is a bounce-back active infrared detector (*Figure 16*). This is a self-contained combined transmitter and receiver with a reflector. The bounce-back unit is best suited for short-range notification purposes.

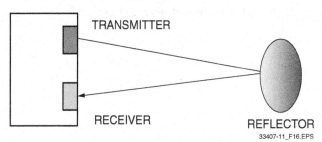

TRANSMITTER

RECEIVER

REFLECTOR

33407-11_F16.EPS

Figure 16 Bounce-back active infrared detector.

3.2.4 Pressure Mats

Pressure mats (*Figure 17*) are thin rubber mats that contain metal contact strips. When an intruder walks on the mat, the contacts close, activating an alarm. Pressure mats should not be used as the sole means of protection, in high-traffic areas, or in areas in which furniture will be moved. They can be concealed under carpeting, making them very discreet, particularly if a cavity is cut in the padding under the carpet so that the pressure mat area is indistinguishable from the rest of the carpet. Any pressure great enough to close the contacts of a pressure mat activates the alarm, making such devices vulnerable to false alarms from large pets. They are also easily circumvented if an intruder steps over them. They can be damaged or activated by furniture. A pressure mat is usually a normally open switch used on a closed loop circuit.

3.2.5 Stress Sensors

Stress sensors are devices that react to the flexing of the material to which they are mounted (*Figure 18*). Stress sensors can be attached to wood joists, stairs, decks, and fire escapes (often with epoxy) and usually are connected to a processor that enables the sensitivity of the sensor's response to be adjusted. Stress sensors are easily understood and generally reliable. They ignore normal vibrations and are unaffected by the presence or activity of small pets. They require precise mounting and should not be used on solid structures or in areas where the sensors are inaccessible.

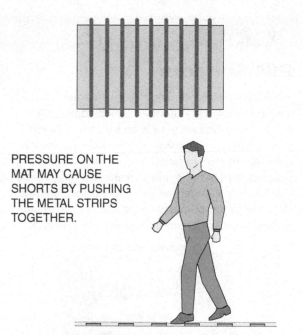

PRESSURE ON THE MAT MAY CAUSE SHORTS BY PUSHING THE METAL STRIPS TOGETHER.

33407-11_F17.EPS

Figure 17 Pressure mat.

3.2.6 Audio Sensors

Audio sensors, sometimes called audio processors or audio discriminators, are designed to detect the sounds or vibrations caused by intruders attempting to enter a secured area. A typical audio detection system consists of microphones and a control unit containing an amplifier, an accumulator, and a power supply. The unit's sensitivity is adjustable to prevent ambient noises or normal sounds from initiating an alarm signal until the noise rises above a preset level or until accumulations of impulses reach a specified total.

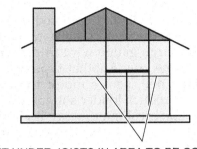

STRESS SENSORS ARE EFFECTIVE ON WOOD JOISTS, STAIRS, DECKS, AND FIRE ESCAPES.

MOUNT UNDER JOISTS IN AREA TO BE COVERED.

33407-11_F18.EPS

Figure 18 Stress sensors.

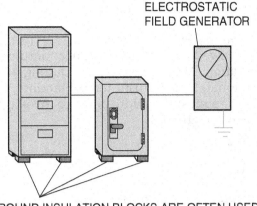

ELECTROSTATIC FIELD GENERATOR

GROUND INSULATION BLOCKS ARE OFTEN USED.

33407-11_F19.EPS

Figure 19 Capacitance proximity sensor.

Audio sensors often include a means for security personnel to listen in on what is happening at the alarm site. They are used primarily to protect vaults, safes, schools, galleries, museums, and similar facilities. Systems of this type should use concealed wiring or wiring enclosed in conduits. Audio sensors cannot be used in areas of high ambient noise or areas containing machinery that generates noise.

3.2.7 Capacitance Proximity Sensors

Any metal object, such as a safe or filing cabinet, can be electronically connected to a processor to form a capacitance proximity sensor (*Figure 19*). This is accomplished with a generator that produces an electrostatic field. The processor monitors the capacitance of the metal object. The presence of an intruder causes a change in capacitance and activates the sensor.

Capacitance proximity sensors can be used for 24-hour protection of an occupied space if no authorized person is likely to approach the protected object while the intrusion system is active. Systems of this type ignore normal vibration and are generally reliable. A sufficient earth ground is required for proper operation of a proximity sensor. The object being protected must be isolated from the ground and the walls or other objects that could become part of the sensor's pattern.

3.2.8 Protected Distribution System

A protected distribution system (PDS) prevents or detects physical access to alarm system communication lines. There are two types of PDS equipment, simple and hardened. Most of these systems are found in government applications where there is a need to protect sensitive data or communication circuits. In a simple distribution PDS, communication cable and sensor wiring are

in a protective carrier. The joints and access points are protected with sensors on a 24-hour zone.

In a hardened distribution PDS, communication cables and sensor wiring are located inside a physical barrier. Protection can be implemented in three forms: hardened carrier PDS, alarmed carrier PDS, and continuously viewed carrier PDS.

- *Hardened carrier PDS* – Locate data cables in a carrier constructed of electrical metallic tubing (EMT), ferrous conduit or pipe, or ridged sheet steel ducting. All joints are permanently sealed completely around all surfaces with welds, epoxy, or other such sealants. For underground cable installations, encase the carrier containing the protected cables in concrete. Visual inspections of all the joints are necessary to determine if there have been any intrusion attempts. Inspection frequency depends upon the threat level and the level of access control to the protected area.
- *Alarmed carrier PDS* – All cable carriers are wired with sensors, supervisory loops, or specialized optical fibers deployed within the conduit to sense acoustic vibrations caused by tampering to gain access to the cables. The cable carriers are on a 24-hour zone and monitored at the control panel, central station, or guard station. An alarmed carrier PDS has several advantages:
  - Continuous monitoring
  - Visual inspections unnecessary
  - Easier installation
  - The carrier can be above ceilings or below floors
  - No requirement for welding and epoxying of the connections or encasing them in concrete outdoors

– No need for locks on junction boxes and manhole covers

- *Continuously viewed carrier PDS* – A continuously viewed carrier PDS is under continuous observation, 24 hours per day (including when operational) by a security force. Standing procedures require the guard force to investigate all intrusion attempts.

4.0.0 ANNUNCIATION (NOTIFICATION) DEVICES

Annunciation, or notification, devices communicate information about an intrusion system to its user. The information can be communicated visually, audibly, or both. Concerns resulting from the Americans with Disabilities Act Accessibility Guidelines (ADAAG) have prompted the introduction of olfactory (sense of smell) and tactile (sense of touch) types of notification as well.

The information communicated may include whether or not the intrusion system is armed, the status of the system's alarms, and any system error messages. Visible notification devices include any means of transmitting information visually. They include LED displays, alphanumeric displays, graphic displays, and strobes. Audible notification devices produce noise, such as a bell, siren, or voice message.

4.1.0 Strobes

Strobes are high-intensity lights that flash when activated. They can be separate devices, or they can be mounted on or near audible devices. Strobe lights are more effective in residential, industrial, and office areas where they don't compete with other bright objects. Because ADAAG requirements dictate clear or white xenon or equivalent strobe lights, most interior visual notification devices are furnished with clear strobe lights. Strobe devices can be wall- or ceiling-mounted and are usually combined with an audible notification device. When multiple strobe lights are visible from a single location, the strobes must be synchronized to avoid confusing, random flashing.

4.2.0 Bells, Buzzers, Horns, Chimes, and Sirens

Noise-producing devices use different means to produce sound, including the following:

- *Bells* – Bells use a clapper to strike a gong. Some bells use an electrically vibrated clapper to re-

peatedly strike a gong. These types of solenoid-operated bells can also produce electrical noise, which interferes with controls unless proper filters are used. Solenoid-operated bells draw relatively high current (750mA to 1,500mA) and require large gauge wire to prevent low sound from line loss. Another type of bell, the motor-driven bell, uses a motor to drive the clapper and produce louder sounds than a solenoid bell. Electrical interference is eliminated in motor-driven bells, and less current is required to operate the bell.

- *Horns* – Horns usually require more power than bells. They can be polarized or non-polarized. They usually consist of a continuously vibrating membrane or a piezoelectric element.
- *Buzzers* – Buzzers use less power than horns. Like most horns, they consist of a continuously vibrating membrane.
- *Chimes* – Chimes are electronic devices that draw relatively low levels of current. Consisting of speakers and electronic sound equipment, they are similar to self-contained sirens.
- *Sirens* – A siren is a combination of speakers and sound equipment that may produce noise, relay voice messages, or both. Self-contained siren packaging saves installation time.

4.3.0 Voice Messages

Some systems have the capability of annunciating one or more prerecorded voice messages to alert personnel to detected intrusion. *Figure 20* shows a residential wireless voice alert system consisting of PIR motion sensor transmitters and a remote receiver-speaker unit. The users of the system record their own alert messages and are notified with their own voice message when activity in a protected area or zone is detected. This system allows a maximum of six user-recorded messages with each being specific to one protected zone.

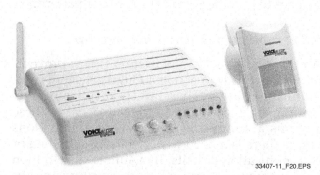

33407-11_F20.EPS

Figure 20 Residential wireless voice alert system.

5.0.0 Control Panels

The control panel coordinates and organizes all the components of an intrusion system. The control panel, connected to the system's sensors, detects possible threats through those sensors and reports the results. It is connected to a battery or outside power source and converts that power for use by the system. Through the control panel, the intrusion system operator may arm or disarm the system or individual zones, set operating preferences, override or acknowledge an alarm, and silence bells, horns, or other notification devices. If desired, the control panel can usually communicate possible threats to personnel outside the secured area.

In addition to zones, some control panels allow multiple system partitions that can be controlled from a master keypad or by separate user control points. A partition is a separate section of an alarm system that can operate independently. Partitions are used to allow for different hours of operation or to restrict particular individuals or groups of users from certain areas.

Control panels must be UL approved/listed for use in intrusion systems. UL Certificated systems require UL-approved/listed panels. A UL-approved panel is determined by the type of enclosure, method of grounding, method of connecting wiring, internal components used, standby capabilities, and marking or labeling. Additionally, local panels used at the premises should be UL Grade A or B.

UL Local Grade A panels have the following:

- 60 seconds of attack resistance
- Supervision of power supplies
- An automatic daily sounder test
- A separate bell mounting that is visible from the street

UL Local Grade B panels have the following:

- 45 seconds of attack resistance
- Continuous metering of power supplies
- A separate bell mounting that is visible from the street

5.1.0 Control Units and Combination Systems

A control unit provides a means of establishing user control over an intrusion system. Sometimes the control unit for a system is combined with other building system controls, such as those for fire alarms or climate controls. Input of user commands or receipt of system feedback is through operating panels, also called control points.

5.2.0 Operating Panels (Control Points)

Operating panels or control points provide the interface between the control panel and the system operator. Control points can provide different degrees of user input and feedback. They usually provide the user with controls to turn the system or system zones on or off (to allow exit or entry) as well as indications of system alarms and status. Status indications can show active sensors and open doors. User controls can also allow the reset of indicators for system events, such as alarms or trouble signals for system equipment, circuits, or external communications. Control points include such devices as keypads, key switches, touch screens, telephone controls, wireless controls, and computer controls.

5.2.1 Keypads

Two general types of keypads are in use today: LCD-alphanumeric and LED (*Figure 21*). Either type can be mounted at the control unit or separate from it. Both allow the entry of a numerical code to program various system functions. Alphanumeric keypads combine keypads, similar to a pushbutton telephone dial, with an alphanumeric display, which is capable of showing letters and numbers. Light-emitting diode (LED) keypads use a similar keypad but display information by lighting small LEDs.

Keypads can be of the self-contained, binary, serial data, or integrated-control variety:

- *Self-contained* – These keypads can be used at the control unit or at a remote location. They usually provide more than one function, such as panic, duress, and alarm arm/disarm. Self-contained keypads are vulnerable to tampering because decoding is done in the keypad. This requires protection of the keypad and of the wiring between the keypad and the control panel.

33407-11_F21.EPS

Figure 21 Typical keypad.

- *Binary* – In a binary keypad, a decoder unit decodes the numbers depressed. The decoder unit, located separately or in the control unit, then transfers the decoded signal to the control unit circuits for action. Binary keypads, connected by multi-conductor wiring to a decoder, have very limited vulnerability because the code is processed at a decoder unit located in the secured area. Multiple keypads can be connected to one decoder unit.
- *Serial data* – These keypads are similar to binary keypads in that the decoding is accomplished at a remote location. The data is transmitted in serial form from the keypad to the decoder by twisted pair or shielded twisted pair cable. Like binary keypads, serial data keypads have limited vulnerability to compromise.
- *Integrated-control* – These keypads are similar to self-contained versions and have all control functions, including the keypad, integrated into one unit. These units, like self-contained keypads, are somewhat vulnerable to tampering.

5.2.2 Key Switches

Key switches use mechanical keys to activate or deactivate some aspect of the intrusion system. They are simple to operate and easy to understand. Key switches are rarely used in newer systems. The two types of key switches are momentary and maintained: Momentary key switches are spring-loaded and return the key to its original position unless the operator applies pressure. The key must be inserted and removed in the same position. Maintained key switches are not spring-loaded. The key is turned to different positions for different functions and does not have to be inserted and removed in the same position.

5.2.3 Touch Screens

A touch screen is a video display that responds to the touch of a finger. Multiple customized displays of graphic and textual information are available. This allows the user to interact easily with the system.

5.2.4 Telephone Controls

Most alarm systems are connected to telephones, and telephones are often located with control points. As a result, telephones may be incorporated as control points for the intrusion system. Feedback and reports on alarm status are given audibly. Any authorized person with access to a telephone may control the system remotely.

5.2.5 Wireless Controls

Control points may use wireless technology, in which signals are carried to the system through the air rather than through cables. Wireless control points include cellular or digital phones, handheld radio transmitters (similar to garage door openers or even television remote controls), and key fob units (similar to the remote door lock controls sold with many vehicles).

5.2.6 Computer Controls

The elements of an intrusion system may be connected to a single computer or to a computer network. This connection can either be local or run through a telephone modem or high-speed data line. Through the computer, system users can input commands, control events, and receive information. Also, the Internet can provide access to intrusion alarm systems. The user can monitor their personal security system from any computer browser via the Internet. This is accomplished by providing the alarm system with an IP (Internet Protocol) address.

5.3.0 Control Unit/Panel Circuit Labeling

Effective system design and installation require that zones (or sensors) be labeled in a way that makes sense to all who use or respond to the system. In some cases, this may require two sets of labels: one for the alarm user and another for the police and emergency response authorities. Central station operators should be aware of both sets of labels because they may be talking to both the police and the alarm user. Labeling for the alarm user is done using names familiar to the user (such as Johnny's room and kitchen). Labeling for the police and emergency response authorities is done from the perspective of looking at the building from the outside (such as first floor east and basement rear).

5.4.0 Types of Control Unit Outputs

Most panels provide one or more types of outputs. These outputs are described in the following sections.

5.4.1 Relay (Dry) Contacts

Relay (dry) contacts are electrically isolated from the circuit controlling them, which provides some protection from external spikes and surges, but additional protection is sometimes needed. Always check the contact ratings to determine how much voltage and current the contacts can handle.

5.4.2 Built-In Siren Drivers

Built-in siren drivers use a transistor amplifier circuit to drive a siren speaker. They are not electrically isolated. Impedance must be maintained within the manufacturer's specifications. Voltage drop caused by long wire runs of small conductors can greatly reduce the output level of the siren.

5.4.3 Voltage Outputs

Voltage outputs are not isolated from the control; therefore, some protection may be required. Spikes and surges, such as those generated by solenoid bells or the back EMF generated when a relay is de-energized, can be a problem.

5.4.4 Open Collector Outputs

Open collector outputs are taken directly from a transistor. They often are used to drive a very low current device or relay. They are not electrically isolated. Filtering may be required. Use great caution to avoid overloading these outputs because the circuit board normally has to be returned to the manufacturer for testing if even a very brief overload occurs.

6.0.0 COMMUNICATION AND MONITORING

A communications interface is used to contact personnel who are too far away to directly see and hear an intrusion system's notification devices. Communication involves the transmission and reception of information from one location, point, person, or piece of equipment to another. Understanding the information that a system communicates makes it easier for the user to determine what is happening at the alarm site. Knowing how that information gets from the alarm site to the monitoring site is helpful if a problem occurs somewhere in between. There are several communication formats supported by alarm panels. Some of these formats are 3+1, 3+2, 4+1, and 4+2. These formats are pulse formats selected at the panel. Contact ID is a specialized format that uses tones rather than pulses. This allows information to be sent quickly and also allows more information to be sent at one time. SIA (Level 2) is a specialized format that communicates information quickly using frequency shift keying rather than pulses.

6.1.0 Communications Options

Electricity, light, and radio waves can be modulated to transmit sound, pictures, and other information. A receiver then converts the modulated information back into its original form. The three basic communications options are as follows:

- *Wire* – Electric current can carry a signal across a cable or wire. A device known as a digital communicator converts information into electric current. The current is sent over the wire to another device known as a digital receiver, which converts the current back into information.
- *Wireless* – A signal can be carried across the air in the form of radio waves. Radio transmitters use oscillators to convert information into radio waves. When the wave hits an antenna, it produces electric current. A radio receiver recovers the electrical signal and converts it back into information. This type of communication is found in cellular or digital communications systems.
- *Fiber optics* – A signal can be carried by light across special glass or plastic tubes called fiber optics. A modulator changes an otherwise constant light beam in response to information such as a closed-circuit television (CCTV) picture. The modulated light waves are sent across optical fibers, and, at their destination, the light pulses are converted back into information.
- *Network* – Communications can be transmitted through the computer network system. Software installed on a computer communicates in real time with the alarm control panel. This can provide the user or security forces immediate access and supervision of the alarm system.

6.2.0 Monitoring Options

There are a number of options for monitoring the signals of an alarm system:

- An answering service is a business that contracts with subscribers to answer incoming telephone calls after a specified delay or when scheduled to do so. The answering service may also provide other services like relaying fire or intrusion alarm signals to the proper authorities and personnel.
- A central station is a location, normally run by private individuals or companies, at which operators monitor receiving equipment for incoming intrusion and security system signals. The central station may be a part of the same company that sold and installed the system, but it is also common for the installing company to contract with another company to do the monitoring on its behalf. Certified Central Station systems are monitoring facilities that are constructed and operated according to a standard. They are inspected by a listing agency to verify

compliance. Several organizations, including UL, publish criteria and list central stations that conform to those criteria.

- A proprietary monitoring facility is similar to a Central Station system except that the notification devices are located in a constantly manned guardroom maintained by the property owner for internal security operations. The guards may respond to alarms, alert local law enforcement agencies when alarms are activated, or both.

6.3.0 Communication Methods and Systems

Unless an intrusion system is strictly local in nature, it must possess some means of communicating possible threats to security personnel or other individuals or organizations outside the secured area. There are a number of different methods for communicating this information (*Figure 22*):

- Internet provider
- Multiplex systems
- Fiber optic systems
- Long-range radio or satellite systems
- Digital communicators
- Cellular or digital wireless backup

6.3.1 Internet

An intrusion system may make use of a facility's Internet connection to communicate information through a computer control panel. One of the benefits of this method is that security personnel can check the status of the system from any computer that has Internet access.

6.3.2 Multiplex

Multiplexing is the transmission of multiple pieces of information over the same communications channel. Thousands of signals can be sent over a single transmission medium. Transmission media include radio waves, metal wires, and fiber optics. The three types of multiplex systems are frequency division multiplexing (FDM), time division multiplexing (TDM), and wavelength division multiplexing (WDM). These three systems are described in the *Wireless Communication* module.

Polling is one of the things that can be done with multiplexing. In TDM, polling is used to control the communication between the alarm sites and the monitoring site. Each alarm site is polled (told when it can send its information) at regular intervals. Computers can poll each site many times each second, and the alarm sites are set up to retain their information until it is their turn. The individual channels of FDM also can be polled at regular intervals. If a particular alarm site fails to respond to the poll, the monitoring site can be set up to indicate this as a communication failure. Monitoring equipment can be set up not only to indicate whether the signal is present but also to indicate whether that signal is weak.

6.3.3 Fiber Optic

Fiber optic cables transmit large quantities of information over long distances. They are immune to most types of interference and do not conduct electricity. Fiber-optic cables do not corrode, are unaffected by most chemicals, do not radiate electrical energy, and can handle a large signal band-

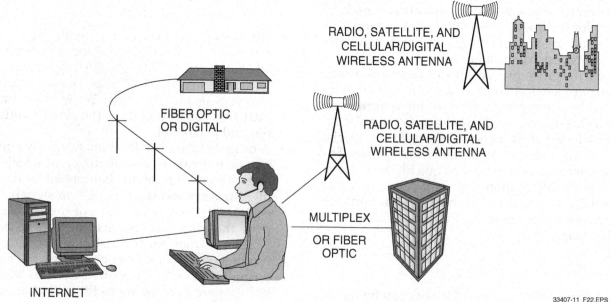

Figure 22 Examples of communication methods.

33407-11_F22.EPS

width. Refer to the *Fiber Optics* module for more information.

6.3.4 Long-Range Radio or Satellite

Long-range radio or satellite systems allow equipment in the secured area to communicate signals to the monitoring site through the air. The electrical signal from the system's control panel is combined with a carrier signal. An altered or modulated signal produces radio waves at the antenna of the system's transmitter. The waves spread in all directions, like ripples in a pond, and hit an antenna at the monitoring site. A radio receiver at the monitoring site, tuned to the frequency of the carrier wave, recreates the original transmitted information. Signals may be sent directly to the monitoring site or relayed through a repeater or satellite system. Refer to the *Wireless Communication* module for more information on radio systems.

6.3.5 Digital

Digital communicators use standard or wireless telephone service to send and receive data. Costs are low using this method because existing voice lines may be used, eliminating the need to purchase dedicated communication links. Standard voice-grade telephone lines are also easier to repair than other security communication links. Refer to the *Buses and Networks* and *Wireless Communication* modules for additional information on other digital communication protocols and systems.

Digital communicators are connected to a standard, voice-grade telephone line through a special connecting device, the RJ-31X (*Figure 23*). The RJ-31X is a modular telephone jack into which a cord from the digital communicator is plugged. The RJ-31X separates the telephone company's equipment from the security system equip, and is approved by the Federal Communica Commission (FCC).

While using standard telephone lines has se eral advantages, problems may arise if the cus tomer and the intrusion system both need to use the phone at the same time. The digital communicator can control or seize the line whenever it needs to send a signal. If the customer is using the telephone when the intrusion system needs to send a signal, the digital communicator disconnects the customer until the signal has been sent. Once the signal is sent, the customer's phones are reconnected.

The typical sequence that occurs when a digital communicator is activated is shown in *Figure 24*. When an alarm is to be sent, the digital communicator energizes the seizure relay. The activated relay disconnects the house phones and connects the communicator to the telephone line. After the communicator sends the signal and then detects a signal called the kiss-off tone, it de-energizes the seizure relay. This disconnects the communicator and reconnects the telephones.

When an intrusion and fire alarm system are both incorporated into a total security system, it is important to remember that in the event of both an intrusion alert and a fire alarm, the fire alarm signal supersedes the intrusion alert signal, even if the intrusion signal occurred first.

The RJ-31X is a modular connection, and, like a standard telephone cord, it can unplug easily. If the cord remains unplugged from the jack, the digital communicator is disconnected from the telephone line, and a local trouble alarm or indication normally occurs. Even if the intrusion system is disconnected from the telephone line, the local annunciator devices is activated in the event of a fire or intrusion.

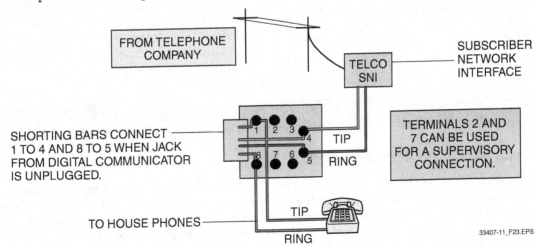

Figure 23 RJ-31X connection device.

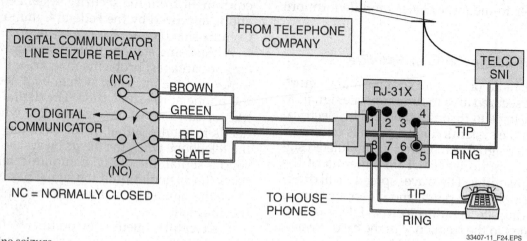

Figure 24 Line seizure.

The National Fire Protection Association (NFPA®) uses the following terms to refer to digital communications:

- Digital alarm communicator receiver (DACR) – A system component that accepts and displays signals from digital alarm communicator transmitters (DACT) sent over the public switched telephone network
- Digital alarm communicator system (DACS) – A system in which signals are transmitted from a DACT (located in the secured area) through the public switched telephone network to a DACR
- *Digital alarm communicator transmitter (DACT)* – A device that sends signals over the public switched telephone network to a DACR

Several conditions have been established by *NFPA 72* to apply to digital communicators:

- They can be used as remote supervising station fire alarm systems when acceptable to the authority having jurisdiction (AHJ).
- Only loop start and not ground start lines can be used.
- The communicator must have line seizure capability.

- A failure-to-communicate signal must be shown if 10 attempts are made unsuccessfully.
- They must connect to two separate phone lines at the protected premises. There is an exception to this rule that states the secondary line may be a radio system. This does not apply to household systems.
- Failure of either phone line must be annunciated at the premises within four minutes of the failure.
- If long distance telephone service, including wide area telephone service (WATS), is used, the second telephone number must be provided by a different long distance provider, where available.
- Each communicator must initiate a test call to the Central Station at least once every 24 hours. This does not apply to household systems.

There are several enhancements available for digital systems:

- *Listen-in option* – This option allows the operator to hear what is going on in rooms at the alarm site that have been equipped with microphones. The option enables the operator to pass on valuable information to the responding personnel or authorities.

On Site

Telephone Line Problems

One method of determining whether the intrusion system has caused a problem with the telephone line is to unplug the cord to the RJ-31X. This disconnects the intrusion system and restores the connections of the telephone line. If the customer's telephones function properly when the cord is unplugged, then the intrusion system may be causing the problem. If the problem with the telephone line persists with the cord to the RJ-31X unplugged, then the source of the problem is not the intrusion system.

- *Two-way voice systems* – These systems allow the operator to hear what is going on at the alarm site and to speak to the occupant of the room through microphones and speakers placed at the alarm site.
- *Video verification* – This option uses video technology to allow the operator to see several snapshots of activity before and after the alarm or to monitor real-time activity from the alarm site. Video systems can be separate from the rest of the security system or integrated into the control and communicator. Video may be viewed on standard computer monitors or through a separate monitor. Video received at the monitoring site can be recorded for future reference.
- *Caller ID* – Caller ID can be used to identify the phone number from which a signal came. This can be useful in dealing with signals sent in error by an installer before the installer has notified the monitoring site or in tracing a malfunction in a communications system.

Digital communicators may experience a runaway condition. A runaway occurs when a digital communicator cannot obtain either a handshake (the process in which predetermined arrangements of characters are exchanged by the receiving and transmitting equipment to establish synchronization) or a kiss-off tone from its receiver and continues to make attempts to dial. Most digital communicators can be programmed to limit the number of attempts made to dial.

If a runaway communicator calls in every minute around the clock, costs for wide area telephone service (WATS) lines and the interference with legitimate signals can become very serious. It can require lengthy, expensive telephone traces and investigation to identify the malfunctioning unit. If the message is garbled in the first few attempts, it remains garbled until the unit is repaired.

6.3.6 *Cellular or Digital Wireless Backup*

Some intrusion systems that rely on phone lines for communications have a cellular or digital backup system. These backup systems use wireless phone technology to restore communications in the event of a disruption in normal telephone line service (*Figure 25*).

7.0.0 SYSTEM DESIGN

The objectives of intrusion system design vary depending on the needs of the client and the perspective of the system designers. System design is a team activity that starts when a client con-

tacts the security consultant or security systems dealer. The process then works its way through dealers, distributors, installers, subcontractors, and service representatives. Each member of the team has something to say about the project, although some have more input than others. When properly coordinated, the observations and input of each party contributes to a successful and profitable project and a satisfied customer.

7.1.0 Applications

The manufacturer's operating specifications must be examined for all intrusion devices, including sensors, control panels, notification devices, and communicators, to determine if they are suitable for a given application. If devices are installed in environmental conditions outside the specified range or used with hardware or software that does not meet the manufacturer's requirements, the system may not operate properly.

Intrusion system designers and installers must understand and evaluate many different environmental factors to determine whether the devices to be used can operate within the manufacturer's specified range. These factors include the following:

- Local conditions, such as whether or not the facility to be protected is located on iron-rich soil, which might cause radio frequency interference (RFI) problems
- Community-wide problems, such as telephone line noise or frequent power interruptions
- Disaster potential, including hurricanes, earthquakes, and floods
- The effect on the facility of windstorms, sand, dust, grit, and other environmental conditions
- The short- and long-term effects of moisture on equipment in the secured area
- The difference in environmental conditions between day and night

In order to install the most effective devices and to troubleshoot problems that arise with the devices, it is important to understand the environment in which intrusion system devices operate. There are several factors to consider, including the following:

- Will the device be inside, outside, or in a building that is refrigerated, not insulated, or not heated?
- Will the device be subject to moisture, such as in a location near a fountain?
- Will the device be located near equipment that produces or circulates dust?

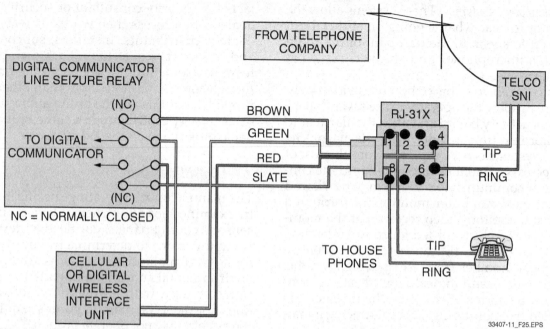

Figure 25 Wireless telephone backup system.

33407-11_F25.EPS

- Will the device be subject to heat or ultraviolet radiation from direct sunlight or to seasonal temperature fluctuations that are outside the manufacturer's operating specifications?
- Will the device be subject to interference from electrical equipment?

Intrusion system devices may be subject to problems specific to their environment that might make some devices more suitable than others. Factors that must be considered include the following:

- Extreme temperature shifts may cause devices to fail.
- Dust and grit may interfere with system devices.
- Direct sunlight or white light may interfere with a passive infrared device.
- Ultraviolet radiation from sunlight may harm plastic device components.
- Settling or shifting of the building may interfere with devices that depend on a precise fit with door and window frames.
- Sunlight and white light RFI can distort the low frequencies used by some system devices.
- Iron deposits in the soil, metal studs, reinforcement in the building, foil wall coverings, or other metal can block devices using radio frequencies (RF).
- Electromagnetic interference (EMI) from power lines, electric motors, or unshielded computer cables can interfere with system devices.
- Animals and insects may damage system equipment, particularly in rural locations or in areas where food is stored nearby.

- System devices installed in basements or near plumbing may be damaged by flooding or by plumbing failures.
- Air pressure from wind or from poorly balanced heating and air conditioning systems may push doors open that are connected to intrusion sensors, causing false alarms.
- Digging equipment used by contractors or public works personnel may damage system cables or underground devices.

7.1.1 Motion (Space) Detectors

Motion (space) detectors play various roles in the design of an intrusion system. Some common detector functions are listed here:

- *Primary detection* – Motion detectors should be installed in addition to perimeter devices near all movable entryways, such as doors and windows.
- *Backup to the perimeter system* – Motion detectors should be installed as protection in conjunction with a perimeter system to verify an entry through a particular door or window.
- *Augmentation of a partial perimeter system* – Motion detectors can be used to augment foil or glass-break detectors for large glass windows.
- *Detection of stay-behinds* – Motion or space detection is usually the first event to be sensed by a system when an intruder hides in a building while it is being locked.

System Design Objectives

Intrusion system design and performance are judged on three factors: probability of detection, false alarm rate, and susceptibility to defeat. Most intrusion alarm system specifications define the values of probability of detection and false alarm rates that a system must achieve. For an ideal system, the probability of detection (PD) is 1. However, no system achieves this value. Factors that affect the probability of detection include the following:

- Type and nature of targets to be detected
- Type and quality of sensor hardware
- Installation conditions and quality
- Weather

A nuisance alarm is any alarm not caused by an actual intrusion. The false alarm rate is the number of false alarms that occur over a given period. In an ideal system, the false alarm rate would be 0. Again, no system achieves this value. Common causes of false alarms can include noises from wind-driven vegetation, wildlife, and weather conditions. Other causes include vibrations, movement of debris, and electromagnetic interference. False alarms can be caused by the equipment itself, poor sensor design, poor installation, poor maintenance, or a faulty component.

Ideally, an intrusion system should be immune to defeat. To achieve this, the design must make it difficult for an intruder to bypass or deceive the system sensors.

7.1.2 Selecting Appropriate Sensors

The intrusion system installer must select the appropriate sensor for the application. Problems arise when sensors are used in environments, or to detect events, for which they are not designed. Disregarding the manufacturer's specifications for a sensor or ignoring environmental or hardware conditions that may negatively impact a sensor results in a poorly designed system.

7.1.3 Combined Technology Sensors

Combined technology sensors use two technologies—usually one passive and one active—to prevent false alarms and verify that an intrusion is caused by human motion or presence. Combined technology sensors contribute to the reliability of an intrusion system and should therefore be considered in its design.

7.1.4 Concealment

If an intruder is unaware of a sensor's presence, the sensor is much less likely to be circumvented or defeated. Concealing or hiding sensors is the best way of preventing an intruder from discovering them. Sensors can be concealed by or disguised as any number of common objects, fixtures, and appliances, including the following:

- Electrical outlets
- Thermostats
- Smoke detectors
- Speakers
- Light fixtures

7.1.5 Walk Test Lights

There is some debate regarding whether or not walk test lights (lights that indicate a sensor is functioning in response to human motion) should be left active or disabled. Disabling walk test lights helps to prevent an intruder from determining a sensor's pattern. Functioning walk test lights may discourage intruders by demonstrating that a secured area is well covered. Functioning walk test lights also verify proper operation of sensors for the security system customer.

> **NOTE**
>
> Hard-wired and wireless walk test lights are different. Wireless devices use a battery saving method and may not show a response for a period of time after the initial test.

7.2.0 Methods for Connection

An intrusion system's components, including control points, detection circuits, notification circuits, transmission circuits, and power supplies, may be connected to the control panel in different ways. The major methods for connecting these components include hardwire, line carrier, and wireless systems.

7.2.1 Hardwired Systems

Hardwired systems use concealed or exposed wiring to connect the components. Even when wireless techniques are used for parts of the system, sounding devices are usually connected to

---

33407-12 Intrusion Detection Systems

the control panel with wiring. While the time and cost of installing the wire can be greater, hard-wired systems are not generally subject to RFI. Individual batteries at each system device are not required because the devices can be powered from a central source. Hardwired systems are preferable to wireless systems in areas with high levels of RFI and in systems where maintenance of batteries in each transmitting device would be a problem.

7.2.2 Line Carrier Systems

Line carrier systems use the existing electrical wiring in the secured area to transmit messages between the system components. Signals between the system sensors and the control panel may be multiplexed on the secured area's power lines. A common use of line carrier technology is to send signals to lights or sounding devices in a building when an alarm is activated.

Line carrier technology is limited by power outages or other problems with the power lines that prevent signals from reaching the security system's components. Many secured areas have more than one phase or set of power lines, requiring special equipment to establish communication between the two phases.

7.2.3 Wireless Systems

Wireless systems use radio frequencies to connect sensors to the control panel (*Figure 26*). In wireless systems, hardwired or wireless methods may be used to connect the user control points to the control panel. Battery-powered radio transmitters are used to signal alarms to a radio receiver in the control panel. When the battery is depleted, it must be replaced, or the transmitter's signal cannot reach the control panel. Some wireless systems use common household batteries while others use special long-life batteries. Wireless systems are preferable to hardwired systems in areas where concealing wires is difficult or when portability of the system is desired.

Each wireless system has its own general radio frequency or facility code. Some systems assign each receiver and transmitter a specific sub-frequency, or path. In others, multiple transmitters can be connected to a single receiver. If all transmitters use the same signal, supervision is sacrificed. Many systems supervise the signal strength and verify operation by monitoring each sensor at preset intervals. If the transmitter fails to check in a specific number of times during the preset interval, an indication is made locally or remotely that the sensor is not functioning. This signal is

separate and distinct from an alarm signal. Batteries can also be monitored by a local or remote indication when they fall to a predetermined voltage, which is also a separate and distinct signal. Refer to the *Wireless Communication* module for additional information about line carrier and wireless communication systems.

> **CAUTION**
>
> Constant use of a walk test light on the unit may reduce its battery life and cause false alarms or a deactivated device.

7.2.4 Zones

A zone is another term for a detection circuit. A detection circuit is a portion of the detection or monitoring system that responds in a specific manner to sensed conditions. Zoning is dividing an intrusion system into a series of subsystems. Each zone, or subsystem, can consist of a single device or a group of devices in a given area. Zones usually have separate notification devices in the secure area and at the monitoring location. Zoning has several benefits, including the following:

- Zoning pinpoints the signal from a sensor or detector. Dividing an intrusion system into parts helps security personnel or authorities respond effectively and efficiently to an alarm.
- Zoning reduces service time by enabling the intrusion system users and service technicians to pinpoint areas needing corrective action or maintenance. Zoning also makes testing and troubleshooting easier by pinpointing the problem area. Some controls allow defective zones to be bypassed while the other detection circuits remain active.

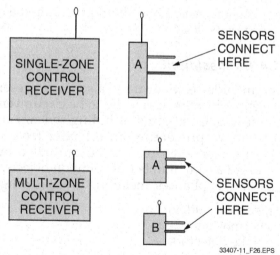

Figure 26 Single-zone and multi-zone wireless systems.

- Zoning allows flexible response. It permits individual sections of a security system to be set up or programmed to react to the same type of input in different ways. For example, a door contact on one door might activate an immediate alarm, but another door contact might activate a delayed alarm. Some controls allow the intrusion system user to bypass certain zones while arming other parts of the system.

There are several recommendations for properly zoning an intrusion system:

- *Limit the number of devices in a zone* – Adding too many devices to a zone makes it more difficult to pinpoint the location of an intrusion or alarm event and can add to troubleshooting time. Adhere to the manufacturer's specifications for the system's control panel regarding the number of devices that can be added to each zone. This is particularly important when devices are powered from the zone rather than from the control panel.
- *Label zones appropriately* – Assign certain parts of the secured area to each zone, and label the zones in a way that makes sense to the system user, security personnel, and the appropriate authorities.
- *Use point notification (if appropriate)* – Point notification is identifying each individual sensor or detector. It allows the system user or security personnel to identify the exact device and location where an alarm has been initiated. Controls and sensors that can be identified individually are also known as addressable devices.
- *Use cross zoning (where appropriate) to prevent false alarms* – Cross zoning is the practice of suppressing an alarm signal until two or more sensors or detectors in separate zones perceive alarm conditions. The system's control panel performs cross zoning if it is programmed for this feature.

7.2.5 Conditions

Zones can be used to monitor various conditions such as the following:

- Medical emergencies
- Low or high temperatures
- Failure of equipment
- Operation of equipment (such as generators and sump pumps)

It is important to specify how the central station should respond to these zones when the system is set up.

7.2.6 Zone-Programming Options

Installers can program zones with several options to customize the operation of circuits or zones. Among these options are the following:

- *Automatic reset* – Some systems have a feature that automatically silences the notification device and returns the system to its non-alarm condition after a specified interval.
- *Automatic restore* – Automatic restore is a feature that allows the system to automatically reset alarm system circuitry and sensors to prepare for an additional alarm signal (if necessary) after a specified interval.
- *Automatic zone shunting* – Zones faulted when arming are automatically bypassed in automatic zone shunting. Automatic zone shunting must not be used with 24-hour zones because 24-hour zones are not armed and disarmed and will not reset.
- *Chime zone* – This is a zone in which a chime sounds when the zone is violated while the system is disarmed. If the system is armed, an alarm sounds when the zone is violated.
- *Cross zoning* – As previously stated, cross zoning is the practice of suppressing an alarm signal until two or more detectors wired to separate zones register an alarm condition.
- *Day zone* – This is a feature of an intrusion system that continuously monitors an area even when the system is disarmed.
- *Manual shunt (bypass)* – A manual shunt or bypass allows the user to bypass a single zone or multiple zones through some control procedure.
- *Priority zones* – Problems in a priority zone prevent the system from arming.
- *Priority with bypass* – A problem in a priority with bypass zone prevents the system from arming, but bypassing can be implemented using a special code.
- *Swinger shutdown* – Swinger shutdown is a method used to prevent more than one set (or a programmable number) of alarms from being transmitted by a particular zone.
- *24-hour zone* – A 24-hour zone initiates an alarm regardless of whether the system is armed or disarmed.

7.2.7 Alarm Verification

Alarm verification is any means of determining whether an alarm signal from a sensor or detection circuit is false and should be ignored or is valid and should initiate an alarm. There are several methods for verifying an alarm:

- *Delay alarm verification control logic* – This assumes that the first signal from a sensor or a detection unit is false and should be ignored. The control panel resets the sensor or detection circuit and waits for a specified interval to retransmit the alarm. If the sensor continues to activate, the control panel transmits the alarm signal.
- *Audio sensors* – These often include a means for security personnel to listen in on what is happening at the alarm site. Security personnel use this information to determine whether an alarm is valid and to respond appropriately.
- *Video or CCTV cameras* – These may be used to monitor a secured area. A detection circuit may alert monitoring station personnel to pay closer attention to a specific zone or portion of a secured area in order to verify an alarm.
- *Telephone* – Some systems allow monitoring personnel to telephone the secured area to determine if an alarm is the result of legitimate intrusion or if a system user has initiated a false alarm. In some setups, authorized users have a code phrase or number that can be used to assure security personnel that no alarm should be reported.

7.2.8 Circuit Response Time

The *circuit response time* is the minimum length of time that an open condition or short must last before the circuit responds. The manufacturer's specifications should state a circuit's response time. If a sensor's minimum output signal is shorter than a circuit's response time, the detection circuit misses the alarm signal. It is important to match sensors with appropriate circuits to ensure that the security system functions properly.

Fast-response sensors include the following:

- *Mechanical vibration* – 15 to 50 milliseconds
- *Mercury glass-break* – 3 to 15 milliseconds
- *Non-powered glass-break* – 3 to 14 milliseconds
- *Older photoelectric beam (without built-in delays)* – 25 to 50 milliseconds

Pressure mats are considered fast response sensors. But if an intruder runs across an older pressure mat and steps on it lightly, it may not react.

Slow-response sensors include magnetic contacts. Magnetic contacts can produce very brief faults during windy conditions if they are installed on loose-fitting doors or windows.

7.3.0 UL Certificated Requirements

Underwriters Laboratories (UL) has developed commercial and residential extents of protection as a designation to describe the amount of detection installed in each detection area.

7.3.1 Commercial Extents

UL Commercial Extent Number 1 calls for complete detection on all windows, doors, transoms, skylights, and other openings leading from the premises. Complete detection is also required for ceilings and floors, as well as all hall, pantry, partition, and building walls enclosing the premises, except building walls that are exposed to the street or public highways, or any part of any building that is at least two stories above the roof of an adjoining building.

UL Commercial Extent Number 2 has three options: perimeter, volumetric, and beam.

- The perimeter option requires the following:
  - Complete detection on all accessible windows, doors, transoms, skylights, and other openings leading from the premises
  - Contacts only on all inaccessible windows
  - Detection on all ceilings and floors not constructed from concrete and all hall, pantry, partition, and building walls enclosing the premises
- The volumetric option requires the following:
  - Contacts on all movable openings leading from the premises
  - Provision of a system of invisible radiation to all sections of the enclosed area in order to detect four-step movement
- The beam option requires the following:
  - Complete detection on all accessible windows, doors, transoms, skylights, and other openings leading from the premises
  - Contacts only on all inaccessible windows
  - Provision of a network of invisible beams to subdivide floor space of each floor or separate section of the detected area into three approximately equal areas, adding more where necessary to provide at least one subdivision per 100 square feet (approximately 30.5 square meters) of floor space

UL Commercial Extent Number 3 calls for the following:

- Complete detection on all accessible windows, doors, transoms, skylights, and other openings leading from the premises
- Contacts on all movable accessible openings leading from the premises and providing one or more invisible rays or channels of radiation with the minimum overall length of the rays or radiation equivalent to the longest dimension of the area
- Contacts on all doors leading from the premises and providing a system of invisible radiation to all sections of the enclosed area in order to detect four-step movement

7.3.2 Residential Extents

The basic system designation requires either contacts on all accessible openings or contacts on all exterior doors and motion sensors, floor mats, interior contacts, or similar devices in selected areas. The extent system designation requires complete detection of all accessible openings and contacts on doors of selected areas, such as the master bedroom, dining room, or library.

7.4.0 False Alarm Prevention and False Alarm Control Teams (FACT)

When an intrusion system alarm is activated and no human intrusion or criminal activity is present, a false alarm has occurred. False alarms are a significant problem for system users, security personnel, and responding authorities. They cost money in the form of penalties or fines to the owner or customer, increased police budgets, and increased labor and equipment costs for the system's company. False alarms also reduce the credibility of the alarm owner with responding authorities.

False alarms can be caused by activation of the intrusion system detection circuits due to environmental factors or other unexplained actions. False alarms may also be caused by failure of telephone equipment or lines or by intrusion equipment failure. Overwhelmingly, however, it is the intrusion system subscriber who causes false alarms.

Rather than expressing alarm rates in terms of false alarms (the false alarm rate will always be relatively high), the appropriate measure of a system's operation is the alarm factor. The alarm factor is expressed as the number of false alarms divided by the number of alarm dispatches. Comparing the number of dispatches to the number of alarms yields a more useful number for purposes of analysis.

Security firm dealers and installers of intrusion systems must consider the problem of false alarms because they lead to reduction in the following:

- The public image of the security firm
- Referrals to the security firm
- The security firm's sales leads
- Upgrades and new system sales
- Productivity for all parties involved

False alarms also affect a security system firm's direct operating costs by increasing the following:

- Station signal traffic monitoring
- Follow-up costs of sales, for customer service, or of servicing equipment
- Management time and attention
- Labor diverted from income-producing projects

Many police departments no longer respond directly to intrusion system alarms because of the prevalence of false alarms. Negative perceptions or misconceptions about the security industry can hinder relations between security personnel and police. Some police departments have responded to the problem of false alarms in one or more of the following ways:

- Significant increase in response time in reaction to alarms
- Refusal to respond to alarms from sites that generate excessive false alarms
- Attempts to recover false alarm response costs by the use of fines, licensing, and permits

The most effective way to reduce false alarms is to educate the intrusion system's users, including building occupants and private security personnel. All commercial and multi-dwelling intrusion system users should form a *False Alarm Control Team (FACT)*. The following are general guidelines for starting a FACT:

- Assemble key personnel from all departments. If necessary, include police and end users in the FACT.
- Identify problem accounts. The monitoring service or accounting department usually supplies this information. The police departments involved can also point out problem accounts.
- The FACT must determine which department is responsible for eliminating the problems with each account. If multiple departments are responsible, one of them should be appointed as the account leader.
- Research the installation and service, as well as the monitoring, legal, and financial records. Make sure that the steps necessary to eliminate the false alarm problem do not cause more problems than they solve.
- Implement the solutions to fix each problem account.
- Review the steps taken to see how best to prevent future occurrences of the problems revealed during the investigation.

7.4.1 Programming Options

In order to be fully effective, an intrusion system should incorporate a variety of programming options. Desirable programming options include the following:

- AC reporting
- Dynamic battery testing
- Low battery reporting
- A minimum *entry delay* of 40 seconds
- A minimum *exit delay* of 60 seconds
- A bell time of five minutes
- Entry delay pre-warn
- Exit delay pre-warn
- Burglar zones programmed for audible notification
- Entry and exit delays on all burglar zones
- Cancel reports
- Exit error signals (if the control panel supplies them)
- Arming that is resistant to operator error
- *End-of-line (EOL)* supervised loops for holdups

An intrusion system should *never* have the following programming options:

- Silent burglar zones
- An entry delay shorter than 40 seconds
- An exit delay shorter than 60 seconds
- A duress signal code consisting of an operating code plus one digit
- Instant burglar zones
- Delays on holdup alarms (HUAs)

An intrusion system should incorporate the following hardware options:

- Supervised end-of-line loops
- An earth-grounded control panel (based on manufacturer specifications)
- Rechargeable batteries (except for premise RF systems)
- Indoor and outdoor audible notification devices

An intrusion system should *never* have the following hardware options:

- Normally open (NO) unsupervised detectors
- An electric current drain of 75 percent or more of the power supply's rated capacity

The Security Industry Association (SIA) publishes *ANSI/SIA CP-01, Control Panel Standard – Features for False Alarm Reduction*, which lists recommended control panel factory default settings for new equipment in order to reduce false alarms.

8.0.0 GENERAL INSTALLATION GUIDELINES

Proper equipment installation is the key to effective operation of an intrusion system. Although each manufacturer provides specific installation requirements for their equipment, this section contains general installation instructions that apply to all types of intrusion systems.

8.1.0 General Wiring Requirements

The *National Electrical Code®* (*NEC®*) specifies the wiring methods and cables required for intrusion systems. The cable types, as well as cable installation methods, termination methods, and voltage/power-drop calculations, have been covered in detail in the *Low-Voltage Cabling*, *Terminating Conductors*, and *Cable Selection* modules.

Various *NEC®* articles cover other items of concern for intrusion system installations, including the following:

- *NEC Article 310* – *Corrosive, Damp, or Wet Locations*

- *NEC Articles 700* and *800 – Ducts, Plenums, and Other Air-Handling Spaces*
- *NEC Article 550 – Locations Classified as Hazardous*
- *NEC Article 725 – Remote-Control and Signaling Circuits (Building Control Circuits)*
- *NEC Article 770 – Fiber Optics*
- *NEC Article 800 – Communications Circuits*
- *NEC Article 810 – Radio and Television Equipment*

In addition to the *NEC®*, some AHJs may specify requirements that modify or add to the *NEC®*. It is essential that a person or firm engaged in alarm work be thoroughly familiar with the *NEC®* requirements as well as any local requirements for alarm systems.

8.2.0 Workmanship

Intrusion system circuits must be installed in a neat and professional manner. Cables must be supported by the building structure in such a manner that the cable will not be damaged by normal building use. One way to determine accepted industry practice is to refer to nationally recognized standards, such as *ANSI/EIA/TIA 568, Commercial Building Telecommunications Wiring Standard*; *ANSI/EIA/TIA 569, Commercial Building Standard for Telecommunications Pathways and Spaces*; and *ANSI/EIA/TIA 570, Residential and Light Commercial Telecommunications Cabling Standard*. Refer to the *Pathways and Spaces* module for more information.

8.3.0 Access

Maintenance technicians must have clear access to equipment for testing and troubleshooting. Access to equipment must not be blocked by wires, cables, and other obstructions that prevent removal of any protective or concealing panels, including suspended ceiling panels.

8.4.0 Circuit Identification

Intrusion system circuits should be identified as such at the control panel and at the interior of all junctions. The interior of junction boxes should be clearly marked as intrusion system junction boxes to prevent confusion with commercial light and power. In the case of combination fire/intrusion alarm systems, the exterior of all junction boxes must be labeled as fire alarm system junction boxes. AHJs differ on what constitutes clear marking. Verify what the authority in your area expects before starting work. Examples of some AHJ acceptable markings for combination system equipment covers are as follows:

- Red-painted cover
- The words Fire Alarm on the cover
- Red-painted box
- The word Fire on the cover
- Red stripe on the cover

8.5.0 Power-Limited Circuits in Raceways

Power-limited circuits must not be run in the same cable, raceway, or conduits as high-voltage circuits. Examples of high-voltage circuits are electric light, power, and non-power-limited fire (NPLF) circuits. When NPLF cables are run in the same raceway, they must be run in accordance with the *NEC®*. A ¼-inch (6.35-millimeter) spacing from Class 1, power, and lighting circuits must be maintained.

8.6.0 Mounting of Detector Assemblies

There are special precautions you must follow when mounting detector assemblies such as PIR and microwave devices. These precautions include the following:

- Mount intrusion system devices according to manufacturer's instructions.
- Do not use circuit conductors to support detectors.
- Do not use plastic masonry anchors to mount detectors to drop ceilings, gypsum drywall, or plaster.
- At minimum, use toggle or winged expansion anchors for mounting to gypsum drywall.
- The best choice for mounting is an electrical box. Most detector assemblies are designed to be fastened to a standard electrical box.

National Electrical Code®

The *NEC®* specifies the minimum requirements necessary for protecting people and property from hazards arising from the use of electricity and electrical equipment. The safe installation and maintenance of intrusion detection electrical systems and equipment require that you be aware of how to use and apply the code on the job. The *NEC®* was first published in 1897 and is revised every three years. Always make sure you are using the most recent edition.

33407-11_SA06.EPS

8.7.0 Outdoor Wiring

The *NEC®* governs any intrusion system circuits that extend beyond one building and sets up the standards on size of cable and methods of fastening required for cabling. Some system applications or manufacturers prohibit aerial wiring; however, if aerial wiring is used, the *NEC®* also specifies clearance requirements from the ground for cable. Overhead spans of open conductors and open multi-conductor cables of no more than 600 volts, nominal, must be at least 10 feet (3.05 meters) above finished grade, above sidewalks, or from any platform or projection from which they might be reached (where the supply conductors are limited to 150 volts-to-ground and accessible to pedestrians only). Additional requirements apply for areas with vehicular traffic.

GOING GREEN

Saving Energy

When cables must pass through exterior walls, install sealant or insulation around holes to prevent heat or air conditioning from escape.

8.8.0 Fire-Stopping

Install intrusion system equipment and cables in a way that does not help the spread of fire. Maintain the integrity of all fire-rated walls, floors, partitions, and ceilings. Use an approved sealant or sealing device to fill all penetrations. Consider any wall that extends from the floor to the roof or from floor to floor a firewall. In addition, seal all raceways and cables that go from one room to another through a fire barrier. Check state and local jurisdictions and job-site specifications regarding certification requirements for fire-stopping technicians.

8.9.0 Air-Handling Spaces

In many commercial buildings, the space between the suspended ceiling and the roof deck is used as a return air plenum for the air conditioning system. Wiring in air-handling spaces requires using approved wiring methods, some of which are as follows:

- Plenum-rated cable
- Flexible metal conduit (FMC)
- Electrical metallic conduit (EMT)
- Intermediate metallic conduit (IMC)
- Rigid metallic conduit (RMC)

Standard cable tie straps are not permitted in plenums and other air-handling spaces. Ties must be plenum rated. All splices, equipment, and other devices must be in approved fire-resistant and low-smoke producing boxes.

Plastic Anchors

The prohibition on the use of plastic masonry anchors does not mean that a plastic anchor specifically designed for drywall cannot be used. Masonry plastic anchors should not be used because they do not open as far on the end as drywall plastic anchors.

8.10.0 Hazardous Locations

The *NEC®* includes requirements for wiring in hazardous locations such as those where explosive or flammable liquids or materials are used or stored, or where there is an increased danger due to building occupants. The following *NEC®* articles cover these locations:

- *NEC Article 511 – Commercial Garages, Repair and Storage*
- *NEC Article 513 – Aircraft Hangars*
- *NEC Article 514 – Motor Fuel Dispensing Facilities*
- *NEC Article 515 – Bulk Storage Plants*
- *NEC Article 516 – Spray Application, Dipping, and Coating Processes*
- *NEC Article 517 – Health Care Facilities*
- *NEC Article 518 – Assembly Occupancies*
- *NEC Article 520 – Theaters, Audience Areas of Motion Picture and Television Studios, Performance Areas, and Similar Locations*
- *NEC Article 545 – Manufactured Buildings*
- *NEC Article 547 – Agricultural Buildings*

8.11.0 Wet or Corrosive Environments

A minimum of ¼-inch (6.35-millimeter) spacing must be maintained between an exterior wall and the equipment or conduit. Conduits and ca-bles must be listed for the environment in which they are used. Outdoor applications require boxes that are labeled as rain-tight, rainproof, or outdoor type. Wet locations or areas where walls are washed down, dairies, laundries, swimming pools, and chemical storage areas are all classified as wet or corrosive locations.

8.12.0 Underground

Only conduit materials unaffected by corrosion may be used for direct burial. Plain steel EMT may last only six months in some environments and must not be used for direct burial. Use RNC (PVC), IMC, rigid (galvanized), or direct burial cable (*Figure 27*).

8.13.0 Remote Control Signaling Circuits

Building control circuits (*Figure 28*) are normally governed by *NEC Article 725*. However, *NEC Article 760* governs circuit wiring that is both powered and controlled by a combination burglar and fire alarm system. A common residential problem occurs when using a combination system. Many believe that the keypad is only a burglar alarm device. If the keypad is also used to control the fire alarm, it is a fire alarm device, and the cable used to connect it to the control panel must comply with *NEC Article 760*. In addition, motion detectors that are controlled by and powered by the combination fire alarm and burglar alarm power must be wired in accordance with *NEC Article 760*, using fire-rated cable.

8.14.0 Wiring Protection

The *NEC®* requires physically exposed wiring to be protected if it could be subject to damage. Protection may be accomplished as part of the building construction or by using conduit. Where

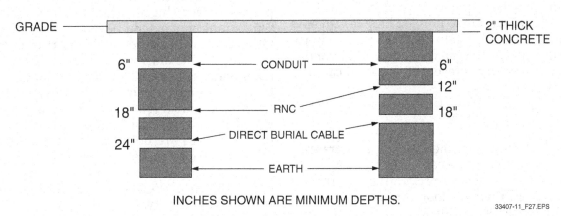

INCHES SHOWN ARE MINIMUM DEPTHS.

33407-11_F27.EPS

Figure 27 Minimum underground wiring burial depths under concrete.

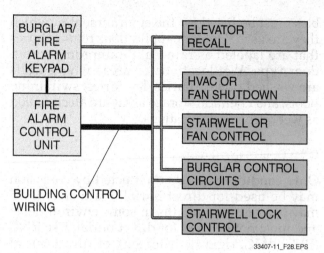

Figure 28 Combination burglar and fire alarm building control circuits.

exposed wiring is installed, cables must be adequately supported and installed in such a way that maximum protection against physical damage is provided by building construction (such as baseboards, door frames, and ledges). Protection must also be provided where cable is located within 7 feet (2.13 meters) of the floor. Cables must be securely fastened in an approved manner at intervals of no more than 18 inches (457 millimeters), except where fished.

The wiring protection described by the *NEC®* is the minimum required to prevent accidental damage. Intrusion system wiring that must be routed in the open and is exposed and accessible from outside the protected areas should be contained in conduit, preferably rigid metal or IMC. This conduit should not contain any separable joints, junctions, or junction boxes in the exposed run. This helps protect against external efforts to bypass or disable the system. In internal areas, it may be desirable or required, especially in high-security installations, to protect all exposed wiring in at least EMT conduit to minimize any covert tampering. In certain very high-security installations, all system wiring may be required to be placed in fully grounded metal conduit to provide EMI as well as signal radiation shielding. In any case, the conduit runs should appear the same as normal light and power conduit runs to prevent easy identification of the wiring's purpose.

8.15.0 Floor-to-Floor Cables

Riser cable is required when wiring penetrates more than one floor. The cable must be labeled to indicate it is capable of preventing fire from spreading from floor to floor. An example of a Listed riser cable is CMR-labeled cable. This re-

quirement does not apply to one- and two-family residential dwellings.

8.16.0 Cables in Raceways

All cables in a raceway must have insulation rated for the highest voltage used in the raceway. Power-limited wiring may be installed in raceways or conduit, exposed on the surface of ceiling or wall, or fished in concealed spaces. Cable splices or terminations must be made in Listed fittings, boxes, enclosures, or utilization equipment. All wiring must enter boxes through approved fittings. It must be protected against physical damage.

8.17.0 Raceways as Cable Support

The external surfaces of raceways may not be used to support power-limited alarm circuit cables. The *NEC®* specifically prohibits this practice because overloading the raceway might damage the fittings and/or cable.

8.18.0 Cable Spacing

Power-limited alarm circuit conductors must be separated at least 2 inches (50.8 millimeters) from any electric light, power, Class 1, or NPLF alarm circuit conductors. This is to prevent damage to the power-limited alarm circuits from induced currents caused by the electric light, power, Class 1, or NPLF alarm circuits.

8.19.0 Elevator Shafts

Wiring in elevator shafts must directly relate to the elevator and be installed in accordance with *NEC Article 620*.

8.20.0 Wiring Methods

All wires must be neatly trimmed and stripped at the terminals. In addition, wires should be placed under the terminals. Excess trimmed wire should not extend beyond the terminal lug. Otherwise, conductors could short circuit components on the alarm panel. The wiring for circuits using end-of-line (EOL) terminations must be done so that removing the device causes a trouble signal (*Figure 29*). Devices should be wired in and out, not T-tapped such that some of the devices are not in the wiring path of the EOL device (*Figure 30*). As specified by manufacturers, T-taps may be permitted in some wiring styles that use alternate means of supervising the wiring integrity.

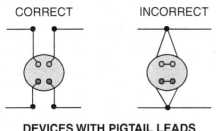

CORRECT   INCORRECT

DEVICES WITH PIGTAIL LEADS

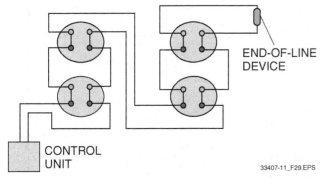

END-OF-LINE
DEVICE

CONTROL
UNIT

33407-11_F29.EPS

Figure 29 Correct wiring for devices with EOL termination.

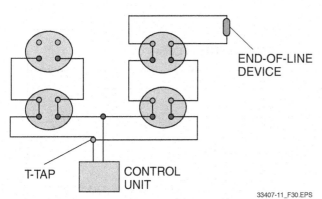

END-OF-LINE
DEVICE

T-TAP   CONTROL
UNIT

33407-11_F30.EPS

Figure 30 Incorrect application of T-tap in an EOL-terminated circuit.

8.21.0 Primary Power

If one 20-amp circuit cannot provide the necessary power, additional circuits may be used. However, no single circuit is permitted to exceed 20 amps. To help prevent system damage or false alarms caused by electrical surges or spikes, surge protection devices should be installed in the primary power circuits unless the alarm equipment has self-contained surge protection. Receptacles of dedicated kitchen, dining room, laundry, or other such circuits should not be used for alarm systems. Non-switched receptacles of lighting circuits or a separate dedicated alarm circuit should be used.

8.22.0 Secondary Power

An approved generator supply or backup batteries must be used to supply the secondary power to an intrusion system. The secondary power system must, upon loss of primary power, maintain the system in non-alarm mode for a minimum of four hours. In the case of combination intrusion and fire alarm systems, the fire alarm standby power requirements must be followed. Fire alarm systems must have a minimum of 24 hours of standby power in non-alarm mode.

Figure 31 shows two forms designed to help calculate the proper size standby battery for an alarm system. Additional blank copies of this form are included in *Appendix A*. In the Amps per Device column of the first form, the normal (standby mode) and alarm (or action) current draw shown for each device must be obtained from the device label or the device manufacturer's data sheet. The totals calculated in the first form are entered and calculated in the appropriate spaces on the second form.

- Multiply the current draw of each device by the number of devices installed for both the normal and alarm states.
- Insert the Normal and Alarm Amps into the second chart.
- Multiply the Normal Amps by 24 hours to obtain Total Normal Amps.
- Multiply the Alarm Amps by 5 minutes to obtain Total Alarm Amps. Use 0.08333 as the numeral for 5 minutes.
- Add the Total Normal Amps and the Total Alarm Amps together. This value equals the System Demand.
- Multiply the System Demand by 1.2 to add a battery de-rating factor to compensate for battery aging.

The result is the proper battery capacity (in amp/hrs) for whatever battery voltage the system requires (such as 12 volts or 24 volts).

Some equipment manufacturers specify a certain size battery. Using a smaller rated battery can cause premature battery failure since battery rating does not match charging current. Always follow the manufacturer's guidelines.

8.23.0 Grounding

Intrusion system control panels and enclosures must be properly grounded. Proper grounding of the panel and enclosure allows protection in case of a ground fault or accidental contact with higher

INTRUSION SYSTEM STANDBY BATTERY CALCULATION FORM

Job Name:

Job Address:

Account#	Phone#		Contact			
Device	Number of This Type of Device	Amps per Device		Total Amps		
		Normal	Alarm	Normal	Alarm	

INTRUSION ALARM STANDBY BATTERY CALCULATION FORM

Job Name:

Job Address:

Account#	Phone#	Contact
Total Normal Amps		
$\times$ Time (24 Hours)		$\times$ 24
= Total Normal Amps		
Total Alarm Amps for 5 minutes		
$\times$ Time (0.08333 Hours)		$\times$ 0.08333
= Total Alarm Amps		
Total Normal Amps		
+ Total Alarm Amps		
= System Demand		
$\times$ De-rating Factor (1.2)		$\times$ 1.2
Minimum Standby Battery		

Figure 31 Battery calculation form.

33407-11_F31.EPS

Alarm Wiring

Some recommended guidelines for intrusion system wiring include the following:

- When using alarm wiring 18 AWG or smaller, use stranded copper or tinned copper wire to reduce the risk of breakage from movement or impact by a foreign object.
- With the exception of control panel interiors, 24 AWG wire should not be used.
- Any cable containing 22 AWG wire or smaller and having more than three pairs of conductors (6 wires) should have, at a minimum, at least one additional pair of conductors for use in future expansion or breakage repair. In addition, for each additional six pairs in the same cable, an extra pair should be allowed.

voltage wires. It also drains leakage current and static charges to ground. The grounded panel and enclosure also act to shield internal electronic circuits from disruption by EMI.

9.0.0 SYSTEM AND EQUIPMENT INSTALLATION GUIDELINES

In all cases, the equipment manufacturer's installation requirements must be observed when installing intrusion system equipment. However, the following suggestions may also be considered, if they do not conflict with the manufacturer's requirements.

9.1.0 Minimum Secondary Power

Every alarm system must have standby power sufficient to operate the system in a non-alarm status for a minimum of four hours.

- After installation, test each secondary power system for the desired operational duration.
- Instruct all customers in the procedures to periodically test standby power.
- Every residential alarm system control unit must have either push on/off connectors to the battery and a description of the primary power transformer location permanently affixed to the inside of the control panel, or an Off/On switch that disconnects the battery and transformer from the circuits inside the control unit.
- Provide every alarm system with a supervised standby power supply that causes a local notification when standby power falls below the manufacturer's recommended specifications.

9.2.0 Control Units

Selection of a control unit should be based on the needs of the alarm user and the requirements of the alarm site. The challenge is to match the needs of the customer with the features of the panel. A system with features that are easily used and understood is preferable to a complicated system that exceeds the customer's needs. Success in this effort is measured in customer satisfaction and in reduced false alarms.

Carefully consider the location of the control unit by balancing security, convenience, and cost. When positioning the control unit, consider the following factors:

- The unit must be in an area that is secure under all conditions, even when the area is unsupervised. Unfortunately, controls are often placed near reception desks in open areas that are not secure after hours. Whenever possible, locate controls in a room with a lockable door that is wired to the alarm system.
- The unit must be in an area where it is not visible from the outside.
- The unit must be in an area where building occupants and visitors cannot easily tamper with the unit.
- Any alarm or trouble-signaling device for the control unit should be located in a constantly occupied area where it is likely to be heard.
- The unit should be within 50 feet (15.24 meters), preferably 25 feet (7.62 meters) or less, of both a good ground and a 24-hour unswitched power source.
- The unit must be as close as possible to the phone connection. Attempt to conceal the phone connection if possible.
- The unit must be in a temperature-controlled area. Extreme temperatures must be avoided.
- Avoid nearby magnetic and static fields.
- The unit must be at the economic center of the wiring, if possible. Although most intrusion devices do not suffer greatly from voltage drop, excessive cable distances increase the cabling costs.

The control unit must be protected internally or externally against transient voltages and surges. Transient voltages and induced surges from lightning or other causes can enter the control unit through the phone lines, power lines, alarm system circuits, or fault grounds. Surge protectors, if not part of the control unit, should be installed in the phone lines and power lines feeding the control unit. The control must also be connected to a unified ground system to stabilize voltage during operation, protect against ground faults, drain current leakage and static charges to ground, and block EMI.

Any control unit for an alarm system that has a touch pad or other device designed to allow the user to activate the alarm (even in a disarmed mode) must be configured for operation as follows:

- Panic, fire, and medical emergency alarms must be audible.
- Duress or holdup alarms can be silent alarms. Duress codes must be dramatically different from user codes.
- All holdup alarms requiring pushbutton activation must use simultaneous two-button activation or a keyed manual reset after activation.

9.3.0 Perimeter Sensors

The following guidelines may be used when installing remote piezoelectric sensors, foil, magnetic, and magnetic-reed switches.

9.3.1 Remote Piezoelectric Sensors

Remote piezoelectric sensors should be located at equal distances from all of the glass being covered. This reduces the possibility of the sensor being oversensitive for one window, yet not sensitive enough for another. Do not exceed the sensor's range. One unit should not be used to cover multiple rooms because high frequencies do not travel around corners. Most audio discriminators can detect various types of glass breaking up to 50 feet (15.24 meters) away if aimed directly at the glass, but only up to 35 feet (10.67 meters) away when installed at right angles (90 degrees) to the windows. All mounting must follow manufacturer's instructions. Examples of mounting considerations for remote piezoelectric sensors may include the following:

- Detectors should be aimed at the anticipated point of entry. Detectors should be on the opposite wall from, and in the direct line of sight of, the glass being covered.
- Mounting the detector 6 to 8 feet (1.83 to 2.44 meters) above the floor reduces the chance of it being blocked by furniture.
- Most sound discriminators offer the option of either a recessed or surface mount.

- Sound discriminators are passive and can be mounted on the wall or ceiling. If mounting the units in the ceiling, place them 6 to 8 feet (1.83 to 2.44 meters) away from the window to avoid an acoustical dead spot common to these types of sensors.
- Never mount the sensor outside the manufacturer's range, even if the tester shows that the unit will operate at a longer range.

9.3.2 Structural-Attack Piezoelectric Sensors

Sensors designed to detect a structural attack require special installation considerations. Structural-attack piezoelectric sensors should be mounted as follows:

- About 4.5 feet (1.37 meters) above the floor and up to 19 feet (5.79 meters) on center
- On safe doors 4.5 feet (1.37 meters) apart

9.3.3 Magnetic Switch Sensors

While magnetic switch sensors are used primarily for windows and doors, they can also be used for other openings, such as cabinets and drawers. The best magnetic switch installations use concealed wiring or wiring enclosed in a conduit. There are two ways to mount magnetic switches (*Figure 32*):

- Surface-mounted switches are installed on the surface of the door or window frame with a corresponding magnet on the surface of the door or window.
- Recessed or flush-mounted switch installations are mounted within the surface of the door or window frame and within the surface of the door or window. Concealing the magnet switch and magnet makes the magnet switch sensor harder for intruders to circumvent; however, recessed or flush-mounted switches must be precisely aligned when installed.

On Site

Sensitivity Adjustment

A sensitivity adjustment is available on most sound discriminators. This adjustment is used to calibrate the detector to the size of the room and to a variety of room conditions. After installation, use the manufacturer's recommended method of testing and calibrating the unit.

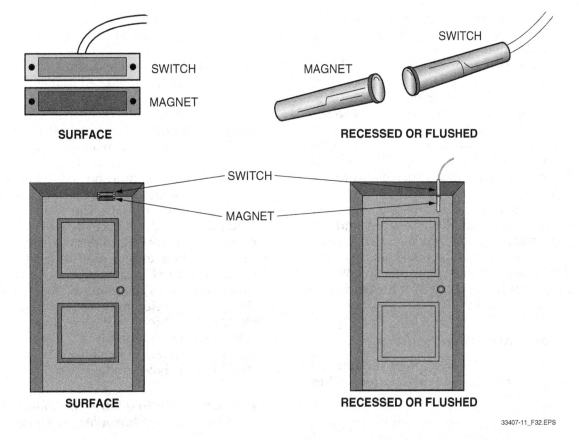

Figure 32 Surface mounting and recessed (flush) mounting.

Magnetic switch sensors should be used on reasonably heavy, well-fitting doors or windows that do not vibrate excessively. Vibration may not cause a false alarm directly or immediately, but it varies the magnetic pull on the magnetic switch contact, causing it to flex. This may cause fatigue and lead to premature failure of the sensor. *Figure 33* describes the proper placement of a magnetic switch sensor on a doorway.

9.3.4 Magnetic Reed Switch

Magnetic reed switches should be installed in accordance with the following requirements:

- Drill oversized mounting holes for press-fit switches to be installed in wood. Switches should be held in place by flexible rubber cement or RTV compound to allow for expansion and contraction of the wood.
- The magnet gap should be within the limit specified by the manufacturer. Typically, the standard gap rating for a reed switch is 0.75 inch (19.05 millimeters) or less, a wide gap rating is 0.75 inch (19.05 millimeters) or more, and a super-wide gap rating is 1.5 inches (3.81 centimeters) or more.

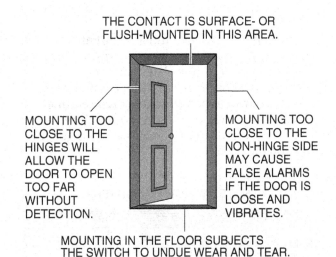

Figure 33 Magnetic switch mounting locations.

9.4.0 Perimeter Fence or Exterior Detection Systems

The following guidelines may be used when installing fence-mounted systems, outdoor microwave sensors, and outdoor photoelectric units. Observe the following precautions for all fence-mounted systems:

- Firmly secure fence fabric to all posts.

- Repair rusted or cut sections in the fence.
- Remove any large signs from the fence, and securely fasten small signs to the fence.
- Test gates to make sure they close and latch properly and securely.
- Chains should not bang against the fence. Secure rigging arms for barbed wire as well as the barbed wire itself.
- No metallic items, such as old cars or wire, should touch the fence. Remove all debris that touches or is near the fence, such as old tires or wood debris.
- Remove trees, tree limbs, brush, or other vegetation near the fence. Debris, trees, and vegetation touching the fence cause uneven sensitivity and false alarms. They may also provide a method for an intruder to enter the protected area by climbing over the fence without having to touch it.

9.4.1 Outdoor Microwave Sensors

Special care must be taken to prevent damage to outdoor microwave sensor installations. These sensors should be installed as follows:

- Locate units so that they are out of the way of vehicular traffic. Install curbing or bollards to protect units if necessary.
- Make sure gates do not strike units when opened.
- Terrain between units must be flat and without valleys, drainage ditches, or similar features that can result in non-detection areas.
- Grades must be on an even plane with no rises or valleys of more than 6 inches (15.24 centimeters).
- Bodies of water cannot be covered by microwave systems. Small puddles are not generally a problem.

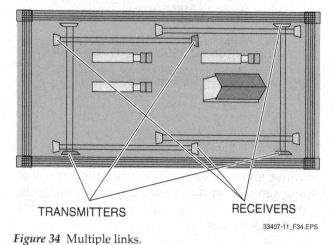

NOTE OVERLAPPED PATTERN

TRANSMITTERS   RECEIVERS

33407-11_F34.EPS

Figure 34 Multiple links.

- A clear path 6 feet (1.83 meters) wide centered on the link path must be maintained. Piles of dirt or debris must be removed. Snow buildup may lead to marginal operation.
- Multiple links may be used to form a continuous perimeter (*Figure 34*). The links should overlap a minimum of 30 feet (9.14 meters) when continuing links are installed, and 15 feet (4.57 meters) at the corners. The links can also be set up as zones, if desired.
- A wall or fence should be erected outside the coverage area to prevent accidental false alarms by people and animals. The fence should extend below grade as necessary to prevent animals from burrowing under the fence.
- Transducers are best mounted on building walls. The next best mounting is on 4-inch (10.16-centimeter) galvanized pipes or 4" × 4" (10.16 cm × 10.16 cm) pressure-treated posts mounted in 2-foot (60.96-centimeter) square concrete pads to a depth of 3 feet (91.44 centimeters). The average mounting height is 3.5 feet (1.07 meters) above grade (*Figure 35*).

9.4.2 Outdoor Active Infrared Pulsed Multi-Beam Photoelectric Units

The following instructions apply when installing outdoor active infrared pulsed multi-beam photoelectric units:

- Terrain between units must be relatively flat without valleys, drainage ditches, or similar features that can result in non-detection areas.
- Clearance for objects on both sides of the beam must be 2 to 3 feet (60.96 to 91.44 centimeters).
- A wall or fence should be erected outside the beam-path clearance to prevent accidental false alarms by people and animals.
- The beam path must be kept clear of drifting snow and vegetation.

9.5.0 Interior Intrusion Systems

The following guidelines may be used when installing passive infrared (PIR) sensors, single- or multi-beam photoelectric units, microwave transceivers, capacitance proximity sensors, and combined sensors.

9.5.1 PIR Sensors

PIR sensors should be installed as follows:

- Choose the correct optics pattern for the area that is to be protected. For example, if a long-range pattern is used across a wide room, some areas of the room may not be covered.

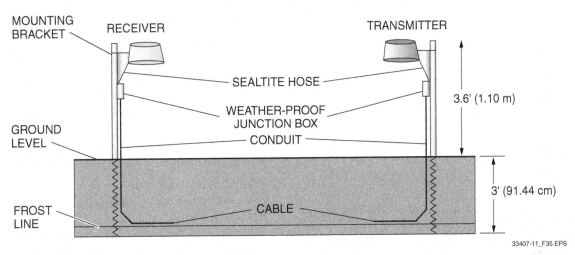

Figure 35 Typical microwave sensor installation.

- Locate the unit so that an intruder passes across the beam pattern rather than into or away from the unit. *UL 681* calls for a person taking three steps, at one step per second, and with a 30-inch (76.2-centimeter) stride to be detected in any three of four walk tests across the coverage area pattern of the unit.
- Do not install a PIR unit in an area that is either larger or 50 percent smaller than the unit's specified coverage.
- Always terminate the unit's coverage pattern on a solid wall, floor, or other object. Failure to do so can cause false alarms because temperature changes can affect the unit's range and result in unwanted detections. Typically, units are mounted 8 feet (2.44 meters) or higher and aimed downward so that their pattern terminates on a floor at maximum range.
- Avoid mounting the unit where sources of heated, cooled, and/or moving air, including windows, can blow directly onto the unit.
- Do not mount the unit so that direct sunlight can strike the face of the unit.
- Be aware of pets on the premises. Pet movements could cause false alarms unless you use the proper detector.
- Do not aim a unit at an outside window. Mount the unit above the window and aim it into the area being covered.
- Seal all knockout holes to prevent the entrance of moving air and insects, which can cause false alarms.
- Avoid mounting units in or on uninsulated buildings.
- Do not mount the units where flapping signs or moving reflective Mylar® materials are located.
- If it is necessary to mask part of the unit's coverage area, use a mask kit provided by the manufacturer. The mirrors in a mirrored unit require

a walk test to determine the appropriate mirror segments to mask.
- Aim units to avoid large areas (such as floors or walls) that are subject to sudden and intense heating by sunlight.
- Avoid reflections from unpainted metal.

9.5.2 Indoor Active Infrared Pulsed Single- or Multi-Beam Photoelectric Units

Active infrared pulsed single- or multi-beam photoelectric units may be installed up to 1,000 feet (304.8 meters) apart; however, aiming without an appropriate aiming tool is difficult. Single-beam units can be installed at reduced ranges using mirrors, as shown in *Figure 36*.

The effective range of the unit is dependent on the angle of reflection and the number of mirrors used:

- Without a mirror, the effective range of the photoelectric beam is 100 percent of its rated distance.
- One mirror reflecting at an angle of less than 60 degrees reduces the beam's effective range to 70 percent of its rated distance.
- One mirror reflecting at an angle of more than 60 degrees reduces the beam's effective range to 40 percent of its rated distance.
- Two mirrors, both reflecting at angles less than 60 degrees, reduce the beam's effective range to 30 percent of its rated distance.

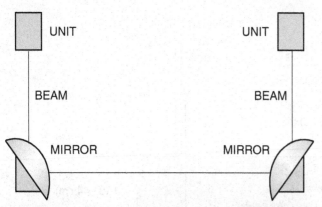

NUMBER OF MIRRORS	EFFECTIVE RANGE
NO MIRROR	100%
ONE MIRROR	
REFLECTED ANGLE <60°	70%
REFLECTED ANGLE >60°	40%
TWO MIRRORS	
REFLECTED ANGLES <60°	30%
ONE ANGLE <60° OTHER ANGLE >60°	25%
REFLECTED ANGLES >60°	20%

33407-11_F36.EPS

Figure 36 Single-beam mirror installation.

- Two mirrors, one of which is reflecting at an angle less than 60 degrees and the other of which is reflecting at an angle greater than 60 degrees, reduce the beam's effective range to 25 percent of its rated distance.
- Two mirrors, both reflecting at angles greater than 60 degrees, reduce the beam's effective range to 20 percent of its rated distance.

9.5.3 Microwave Transceivers

Microwave transceivers should be installed with the following in mind:

- Mount units so that the intrusion path is toward or away from the transceiver.
- Do not mount units on surfaces subject to vibration. False alarms can result.
- Moving metal signs, decorations, and fixtures cause false alarms and must be removed.

Various materials reflect or appear transparent to microwave radiation. Microwaves can transmit through solid materials, particularly metallic surfaces. Coverage is affected by the amount of energy reflected or transmitted by various materials. A typical coverage pattern for a wall-mounted unit is shown in *Figure 37*.

9.5.4 Capacitance Proximity Sensors

When installing capacitance proximity sensors, a good earth ground for the unit is required for proper operation. Device(s) being protected must be isolated from ground and from walls or other objects that could become a part of the sensor's pattern.

9.5.5 Combined Technology Sensors

When installing combined technology sensors, the unit must be carefully placed to ensure reliable detection for both types of sensors. Make sure that patterns of both sensors overlap in the entire area being covered so that detection occurs.

10.0.0 PROGRAMMING OPTIONS

There are several different programming options for intrusion systems. These options allow the user to set the system responses as appropriate for the system application and the user's needs.

10.1.0 Controlled and 24-Hour Zones

As previously stated, control points allow the user to arm or disarm the intrusion system. Circuits or zones that turn on and off through the control point are known as controlled zones. Circuits or zones that remain on 24 hours a day are known as 24-hour zones or day zones.

10.2.0 Entry and Exit Delays

The intrusion system user must be able to enter and exit the secured area without triggering an alarm. Placing the system controls outside the secured area leaves them vulnerable to tampering, so these controls are normally installed inside the secured area. Entry delays and exit delays give the user time to get to the control point to deactivate the system after entering or to arm the system and then leave the secured area without causing an alarm.

10.2.1 Entry

Entry delay allows the user to enter the secured area. An activated timer counts down for a specified interval during which the user must disarm the system. Normally a buzzer, sounder, or other pre-alarm is active during this countdown to remind the user to disarm the system. The pre-alarm and the intrusion system alarm, or alarms, are deactivated when the user successfully enters

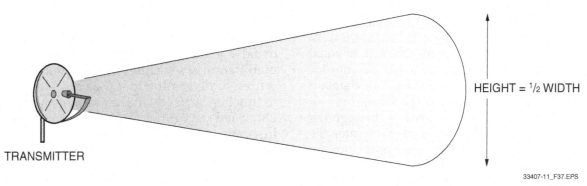

HEIGHT = $\frac{1}{2}$ WIDTH

TRANSMITTER

33407-11_F37.EPS

Figure 37 Typical microwave coverage pattern.

the disarm command or code. If an intruder or an authorized user fails to disarm the system during the pre-alarm countdown, the intrusion system alarm is activated.

10.2.2 Exit

Exit delay allows the user to leave the secured area after entering the arm command or code at the control point. Once armed, the intrusion system counts down for a specified interval during which the user must leave the protected site. Provided the user does not remain within the secured area after the countdown is completed, no alarm is activated. If the user remains within the secured area longer than the time allowed, the entry delay process is initiated. The user must then disarm the system during the entry delay countdown or the intrusion system alarm will be activated. If an intruder hiding within the secured area (a stay-behind) is detected when the system is armed, and the exit delay countdown has expired, the alarm will be activated.

10.3.0 Delayed and Instant Zones

Using an entry-exit delay can delay an alarm response to the activation of a sensor or detection circuit. Instead of using such a delay on all the system's detection circuits, however, control programming can be used to delay only certain parts of the system. Detection circuits using this delay are delay zones. The detection circuits that do not use the delay and trigger an alarm immediately when activated are instant zones.

10.4.0 Perimeter and Interior Zones

As previously stated, the perimeter consists of the outer boundaries of the intrusion system. The interior is the area within the perimeter. Sensors or detection circuits may be positioned on either perimeter zones or interior zones. These desig-

nations, perimeter or interior, help to identify a specific location within the secured area and are used to determine which sensors or detection circuits are active when the system's users are inside or outside the secured area.

10.5.0 Home and Away Feature

The home and away programming feature allows the user to arm an intrusion system that includes interior sensors while remaining inside the secured area (home). When a user arms the system in the home, or perimeter-only, mode, the perimeter detection circuits are active, but the interior detection circuits are not. In the away mode, all of the detection circuits (perimeter and interior) are active. The home and away feature allows the user the flexibility of having a portion of the system active while remaining within the secured area and prevents false alarms that would occur without this feature.

10.6.0 Interior and Perimeter Follower Zones

An interior follower zone is a zone that is located between an entry/exit zone and the alarm system control panel. This zone is temporarily ignored by the alarm system during entry/exit delay periods. This enables a person to walk (without causing an alarm) in front of a motion detector that is associated with the interior follower zone after entering through an entry zone on the way to the control panel, or when leaving the protected premises after system arming. Sensors connected to interior follower zones are delayed only after an entry-exit delay zone is activated. This enables the sensor to activate an alarm immediately in most situations, while allowing an authorized user to be present within the secured area.

For example, if the system control point in a building is located on a wall near the front door, a motion detector could cover the area between the control point and the door. In a system with normal entry-exit delay, this motion sensor must always be delayed. In a system with the interior follower feature, the motion sensor is delayed when the user leaves through the front door, or when the user (or an intruder) enters through the front door. If the user or an intruder activates the motion sensor before opening the front door, an alarm is immediately activated.

A perimeter follower zone is a non-entry/exit zone. It is typically a perimeter zone located on an entry/exit path, which is treated as an entry/exit zone during an entry/exit time. Essentially, these zones are instant unless an entry/exit delay zone is triggered first, causing them to become delay zones.

10.7.0 Panic, Duress, Medical, and Fire Zones

In some control panels, a specific code can be entered at the keypad to cause a programmable event. A switch connected to an alarm control panel can also generate these alerts. These events can be defined as a panic alarm, duress alarm, medical emergency, or fire emergency. These events can transmit information 24 hours a day and are not dependent upon arming of the control panel.

11.0.0 INSPECTION, TESTING, AND MAINTENANCE

After an intrusion system is installed or serviced, it must be inspected and tested thoroughly. For the duration of its service life, the system must be properly maintained. Proper inspection, testing, and maintenance are critical components of intrusion system applications.

11.1.0 Purpose of Testing

Testing is the only way to determine that the system is providing the level of security for which it is designed. Improper wiring, damage to the sys-

Climate Control

To control energy costs, monitor the building environment. Put the HVAC system on an alarm zone. The control panel can transmit an appropriate message if desired temperature and humidity parameters are exceeded.

tem, defective equipment, and damaged equipment often, but not always, cause trouble signals or false alarms. Testing in accordance with the manufacturer's written testing procedure is required to determine that the equipment is likely to function correctly in the event of an emergency. Qualified personnel must perform the required inspection, testing, and maintenance of an intrusion system.

11.2.0 Before Testing

Prior to any testing, it is extremely important to notify everyone who may be affected. This includes the monitoring station, building occupants, building engineer or owner, and anyone else who may become aware of the alarm and think it is an actual emergency instead of a test. Do not forget that people may continue to enter the building after an announcement that a test is being conducted. It may be advisable to post a notice at all entrances.

Review the results of activating a system alarm. Take appropriate action to prevent unnecessary operation of auxiliary systems. For example, automatic locking of all the doors during the test may create a safety hazard during the test unless extra measures are taken before activating the alarm system.

11.3.0 Precautions for Occupied Buildings

It is possible for an actual security problem to occur while a scheduled test is being conducted. Always be sure of the signal that is expected to result from the device being tested. If a signal from another device or zone is received, take immediate action in accordance with any established policy. If an established procedure does not exist, your company should have a policy that provides for the highest level of safety reasonably possible during a test. If you are uncertain what action you should take, notify the authorities immediately. Someone familiar with the policy should remain at the control panel any time the supervising station has been instructed to ignore alarm signals from the system.

11.4.0 Definitions

It is important to understand the following terms used in testing intrusion detection systems:

- *Test* – A test is an act intended to verify and confirm the functional operability of all or part of a device, circuit, or system.
- *Inspection* – An inspection is a visual evaluation to verify or confirm that something appears as desired. Is it in the proper location? Does it ap-

pear to be attached properly? Has a listing label been attached?

- *Maintenance* – Maintenance is an act or acts intended to prevent performance failure and to keep equipment, devices, and systems in proper working order.
- *Acceptance* – Acceptance is the point at which the system has been demonstrated and witnessed to function as specified.
- *Test methodology* – Test methodology is the procedure used to perform a test.
- *Test frequency* – Test frequency is the time interval at which a test is performed.

11.5.0 General Requirements

A 100-percent acceptance test should be performed in the presence of the alarm owner, user, or other representatives following an installation or any alterations. Inspection, testing, and maintenance programs should follow the equipment manufacturer's instructions.

Before connecting any devices to wiring, tests should be conducted for stray voltage, ground faults, short circuits, and loop resistance. Any voltage found on a pair of wires may be from an induced source. This may result in false alarms or may shorten the life of the system. Alarm wires should not be run parallel to power, lighting, or Class 1 circuits. A minimum of 2 inches (5.08 centimeters) of spacing is required. Short circuits result in an alarm signal or an overload when connected to the control panel.

Ground faults must be cleared before connecting wiring to the control panel. More than one ground fault occurring on a system may destroy part of the control panel. Ground faults are very serious problems and must be corrected immediately.

Record the loop resistance to use as an aid in future troubleshooting. Poor connections and corrosion may cause intermittent problems, which can often be found if the circuit resistance prior to the problem is known and can be compared.

11.6.0 Testing Methodology

Test in accordance with the manufacturer's instructions and the requirements of any applicable agencies or authorities. For example, a UL-Listed installation must comply with UL requirements. Test the following conditions, as applicable:

- *Open and grounds* – A single open and a ground fault must be created when testing each device. A trouble signal must be indicated at the re-

mote notification devices and monitoring station as applicable.

- *All sensors* – Activate each magnetic or mechanical sensor by opening and closing all protected openings. Use test devices to simulate breaking glass for glass sensors and other sounds for audio sensors. Use minimal intrusion to test beam-type sensors. Walk-test all infrared, microwave, proximity, and combination sensors. Use a minimal level of vibration to check all vibration-type sensors. Use a minimal amount of stimulation for other types of interior, perimeter, and fence sensors.
- *Control panels* – Illuminate all lamps and LEDs, remove fuses, and verify that trouble signals occur. Test the primary power by interrupting the secondary power and verify occurrence of trouble signals. Induce the maximum load for five minutes while the secondary power is disconnected. Test the secondary power by interrupting the primary power and verify that the trouble signal is activated. Measure the standby current per manufacturer's instructions and calculate adequacy for required standby time.
- *Control panel trouble silence* – Induce a trouble condition, and verify audible and visual signals. Silence the audible signals, and verify that visual signals remain on. Induce a supervisory signal, and verify the audible and visual signals. If a common sounder is used, it should sound. Silence and correct all trouble signals. If the silence switch is a maintained switch, verify that the trouble signal sounds until the switch is restored to normal.
- *Control panel zone and city disconnect* – Actuate each separately, and verify that a trouble signal is activated.
- *Control panel alarm silence* – Induce an alarm; then actuate a trouble signal, and verify that it is indicated. Induce additional alarms on another circuit, and verify that the alarm signal is reactivated. Reset all alarm and trouble signals (reset panel). Actuate and verify a trouble signal when no alarm condition is present.
- *Control panel supervisory silence* – Induce a supervisory signal, and verify that both visual and audible indications occur. Check that the signal is either visually or audibly distinct from the trouble signal. Silence, but do not restore, the supervisory signal. Activate another supervisory circuit, and verify indications.
- *Control panel-signaling circuits* – Induce an alarm, trouble signal, and supervisory condition, and verify that each is properly received at the monitoring center. Test every condition monitored.

- *Batteries and backup power supplies* – Test per manufacturer's instructions. Batteries are tested at different intervals depending on the type of battery used.

11.7.0 After Testing

After testing is complete, be sure to notify everyone involved. This should include the monitoring station, the building owner or representative, and the building occupants, as required. Do not forget to remove any signage posted to indicate the test. Failing to notify everyone may result in a serious lapse of coverage. The monitoring station may disregard any actual signal, and occupants may ignore the alarm.

Make sure to restore the system to its normal operation. Verify that all indicating lights are in their normal condition.

Complete a written report of your test, and provide the customer with a copy. The report should include who conducted the test, along with a notice of any problems, deficiencies, and professional recommendations. After any problems or deficiencies have been corrected and tested, fill out a verification form. Verification forms confirm that an intrusion system has been installed. Insurance companies sometimes require proof of the system's installation for rate reductions. (*Appendix B* contains a typical verification form that may be modified and used if desired.) Place a copy of the final inspection and testing report along with the verification form in the premise records, and file copies in your office in case the premise copies are lost.

12.0.0 INTRUSION SYSTEM TROUBLESHOOTING GUIDELINES

The troubleshooting approach to any alarm system is basically the same. Regardless of the situation or equipment, some basic steps can be followed to isolate problems:

- *Know the equipment* – For easy reference, keep specification sheets and instructions for commonly used and serviced equipment. Become familiar with the features of the equipment.
- *Determine the symptoms* – Try to make the system perform or not perform as it did when the problem was discovered.
- *List possible causes* – Write down everything that could have possibly caused the problem.
- *Check the system systematically* – Plan activities so that problem areas are not overlooked or that time is not wasted.
- *Correct the problem* – Once the problem has been located, repair it. If it is a component that cannot be easily repaired, replace the component.
- *Test the system* – After the initial problem has been corrected, thoroughly check all the functions and features of the system to make sure other problems that were masked by the initial problem are not present.

Troubleshooting charts provided by system manufacturers can be very helpful in isolating faults. *Figures 38, 39,* and *40* are examples of different troubleshooting charts. The following guidelines provide information for resolving problems for specific conditions.

Always check sensor power, connections, environment, and settings. Lack of detection can be caused by a loose connection, obstacles in the area of the sensor, or a faulty unit. Do not forget to check the battery. In some circuit designs, batteries are part of the power supply. A dead battery can cause a power supply to appear to have failed. Always recalculate standby battery capacity when a system is upgraded. If unwanted alarms occur, recheck the installation for changing environmental factors. If none are present, cover, disable, or bypass the sensor to confirm that the alarm originated with the sensor. If alarms still occur, then wiring problems, power problems, or EMI could be the problem. If the alarm stops when the sensor is covered, disabled, or bypassed, then the environment monitored by the sensor is the source of the problem. Replacing the unit should be the last resort after performing the preceding checks.

On Site

Troubleshooting

Your skill at troubleshooting intrusion systems can be enhanced by keeping pace with industry trends. This includes keeping in touch with equipment manufacturers to learn about any problems or field modifications pertaining to the particular sensors and other equipment used in the systems you service. It also means checking with other users of the same technology to learn from their experiences and reading trade magazines to stay current about new intrusion control equipment and techniques.

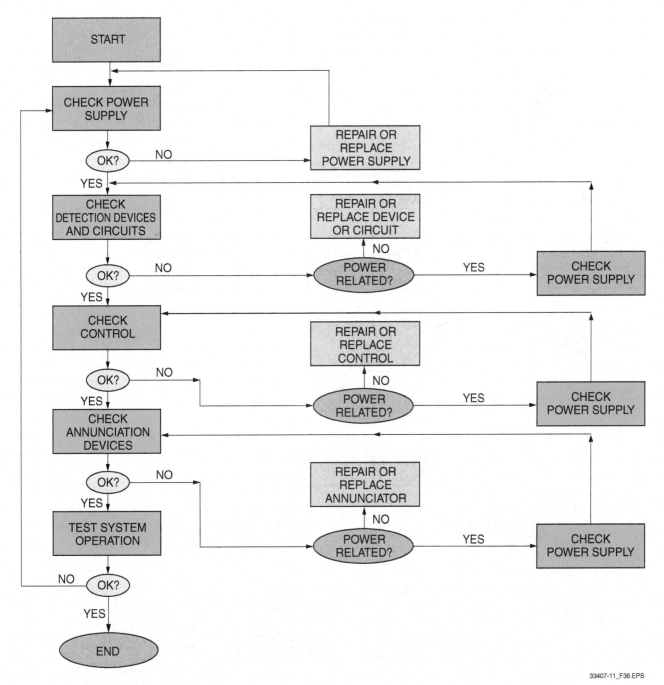

Figure 38 System troubleshooting chart.

Open circuits account for the largest percentage of faults. Major causes of open circuits are loose connections, wire breaks (ripped or cut), staple cuts, bad splices, cold-solder joints, wire fatigue due to flexing, and defective wire or foil.

Shorts or ground faults in a circuit cause alarms or trouble signals. Events occurring past a short are not be seen by the system. Some of the common causes of shorts or ground faults are staple cuts, sharp edge cuts, improper splices, moisture, and cold-flow of wiring insulation. Cold-flow is caused when wiring is under tension and routed around a corner of a conductive object or when the wire is under external pressure. The pressure of the wire on the corner of the object eventually causes the wire insulation to flow away from under the contact point of the wire until the wire touches the conductive object. If a pair of wires, primarily twisted pair, is under external pressure from objects like furniture, the pressure can cause the insulation to flow from between the wires until they touch. This causes a short. Cold-flow of

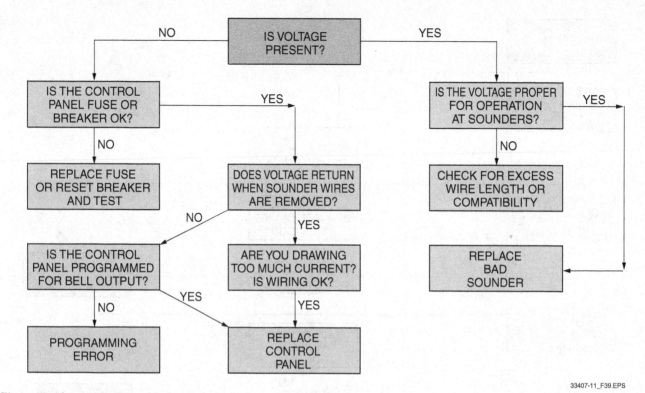

Figure 39 Alarm output troubleshooting chart.

33407-11_F39.EPS

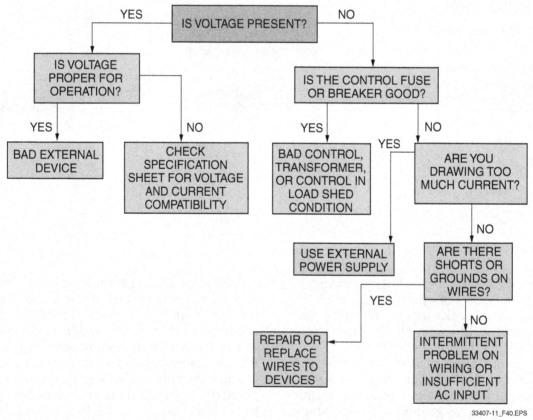

Figure 40 Auxiliary power troubleshooting chart.

33407-11_F40.EPS

wiring insulation can occur over a short or long period of time that depends on the amount of tension or pressure applied to the wire insulation. To eliminate staple cuts, use other types of fasteners, or leave wires under the staples loose. To eliminate insulation cold-flow or edge cuts, do not pull wires tight around corners. Make sure that heavy objects are not placed on under-carpet wiring.

Splices should properly insulate conductors from each other and from ground. Make sure soldered splices are properly taped. Make sure mechanical splices are properly crimped.

For RJ-31X problems, refer to *Figure 41* to troubleshoot various conditions of an RJ-31X interface. The voltages shown in the boxes represent correct voltages for the conditions stated.

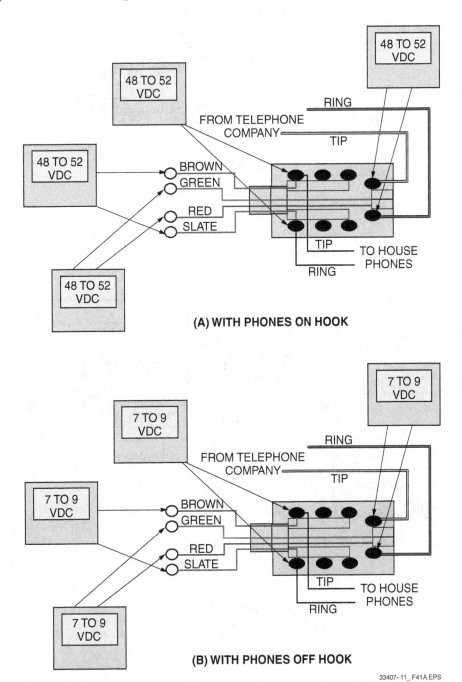

33407-11_F41A.EPS

Figure 41 RJ-31X troubleshooting chart (1 of 2).

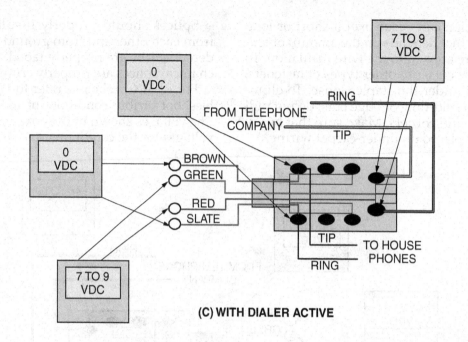

(C) WITH DIALER ACTIVE

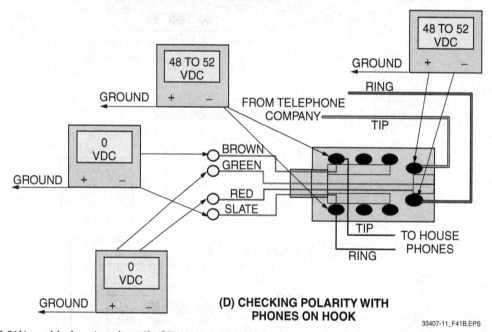

(D) CHECKING POLARITY WITH PHONES ON HOOK

33407-11_F41B.EPS

Figure 41 RJ-31X troubleshooting chart (2 of 2).

Summary

An intrusion system is a combination of components designed to detect and report possible threats to property or human life. A functioning system consists of the control panel, user control units, detection devices, annunciation or notification devices, communication devices, and power supplies. There are many different types of detection and notification devices, and each type has applications for which it is best suited. There are also multiple methods for connecting and supervising intrusion system components and for communicating system information.

An intrusion system does not function properly if it is not designed, installed, and maintained according to specifications. No two system installations are exactly alike. To help regulate the different applications that exist, codes and standards have been established that dictate the specifics of design and installation for electronic systems. The electronic systems technician must comply with all applicable codes and requirements or risk civil and even criminal liability.

1. In an intrusion system, the device that reports any perceived threats to personnel outside of the secured area is the _____.
   a. control unit
   b. annunciation device
   c. detection device
   d. communication device

2. The function of an annunciation device in an intrusion system is to _____.
   a. detect a threat
   b. notify occupants of a threat
   c. display system status information
   d. report threats to a monitoring facility

3. A monitored intrusion system _____.
   a. uses wireless communication instead of telephone lines
   b. is less likely to be disabled before alerting the appropriate personnel
   c. is connected to a monitoring facility
   d. alerts only the occupants of a secured area

4. Reduction of control equipment and wiring is a benefit of _____.
   a. hardwired systems
   b. zoned systems
   c. multiplex systems
   d. addressable systems

5. The type of intrusion system in which sensor devices return discrete identification information is _____.
   a. hardwired
   b. zoned
   c. multiplex
   d. addressable

6. The two major categories of sensors are _____.
   a. perimeter and interior
   b. perimeter and discrete
   c. perimeter and passive
   d. active and interior

7. On-glass piezoelectric sensors can be used on _____.
   a. tempered glass
   b. wired glass
   c. laminated glass
   d. sheet glass

8. A perimeter sensor that transmits infrared light to a photocell is a _____.
   a. PIR sensor
   b. photoelectric beam sensor
   c. piezoelectric sensor
   d. proximity sensor

9. Photoelectric beam sensors can provide barrier detection using _____.
   a. monostatic sensors
   b. microwaves
   c. mirrors
   d. X-band signals

10. An interior sensor that provides wide detection in a defined area uses _____.
    a. trap detection
    b. spot detection
    c. channel detection
    d. volumetric detection

11. In a PIR, the processing method in which the alarm signal must be verified by duration or repetition is _____.
    a. multiple step logic
    b. event verification
    c. signature detection
    d. single step logic

12. Active infrared pulsed single- or multi-beam photoelectric units can be placed up to _____.
    a. 1,000 feet (approximately 305 meters) apart
    b. 1,500 feet (approximately 457 meters) apart
    c. 2,000 feet (approximately 610 meters) apart
    d. 2,500 feet (762 meters) apart

13. The Americans with Disabilities Act Accessibility Guidelines dictate that strobe lights be _____.

   a. red or yellow
   b. yellow or orange
   c. clear or white xenon
   d. blue or green

14. A notification device that consists of a continuously vibrating membrane or a piezoelectric element is a _____.

   a. bell
   b. horn
   c. chime
   d. siren

15. UL Local Grade A control panels have how many seconds of attack resistance?

   a. 20
   b. 30
   c. 60
   d. 80

16. A control panel output that is isolated from the control and which may experience trouble from power surges is a(n) _____.

   a. voltage output
   b. relay or dry contact
   c. built-in siren driver
   d. open collector output

17. If a digital communicator receives an intrusion alert followed by a fire alarm, it will act on the intrusion alert first.

   a. True
   b. False

18. The practice of suppressing an alarm signal until two or more detectors wired to separate zones register an alarm condition is called _____.

   a. automatic reset
   b. automatic zone shunting
   c. cross zoning
   d. manual shunting

19. A circuit option that prevents more than one set (or a programmable number) of alarms from being transmitted by a particular zone is called _____.

   a. priority zone
   b. priority with bypass
   c. swinger shutdown
   d. 24-hour circuit

20. A zone that always initiates an alarm whether or not the system is armed is a(n) _____.

   a. 24-hour zone
   b. controlled zone
   c. instant zone
   d. interior follower zone

21. The most effective way to reduce false alarms is to _____.

   a. educate the intrusion system's users
   b. change the sensitivity of the detectors
   c. enlarge the perimeter dimensions
   d. merge many zones into fewer zones

22. An intrusion system should *never* have a hardware option that includes _____.

   a. indoor and outdoor audible annunciation devices
   b. a current drain of 75 percent or more of the power supply's rated capacity
   c. supervised end-of-line loops
   d. rechargeable batteries

23. An intrusion system should *never* have a hardware option or options that include(s) _____.

   a. supervised end-of-line loops
   b. normally open unsupervised detectors
   c. an earth-grounded control panel
   d. rechargeable batteries

24. An intrusion system should *never* include a programming option or options that call(s) for _____.

   a. low battery reporting
   b. silent burglar zones
   c. a minimum entry delay of 40 seconds
   d. a minimum exit delay of 60 seconds

25. All of the following materials are permitted for direct burial *except* _____.
    a. plain steel EMT
    b. RNC
    c. IMC
    d. rigid (galvanized) cable

26. Circuit wiring that is powered and controlled by a combination burglar and fire alarm system is governed by _____.
    a. NEC Article 725
    b. NEC Article 760
    c. NEC Article 770
    d. NEC Article 775

27. Which of these is the minimum required when wiring penetrates more than one floor?
    a. Rigid metallic conduit
    b. Plenum-rated cable
    c. Intermediate metallic tubing
    d. Riser cable

28. An act intended to verify and confirm the functional operability of all or part of a device, circuit, or system is a(n) _____.
    a. test
    b. inspection
    c. maintenance function
    d. acceptance function

29. Illuminate all lamps and LEDs, remove fuses, and verify that trouble signals occur when testing _____.
    a. audio sensors
    b. backup power
    c. ground faults
    d. control panels

30. Cold-flow of wiring insulation is a common cause of _____.
    a. open circuits
    b. short circuits
    c. sensor problems
    d. power supply problems

24-hour zone: A zone that remains active 24 hours a day without interruption.

Access control: Any means of monitoring and restricting human traffic through doors, gates, and elevators.

Active infrared (photoelectric beam) sensor: A sensor that transmits a beam of infrared light to a photocell. Interruption of the beam activates the sensor.

Active sensor: A sensor that continuously emits energy, such as low-power microwave or photoelectric beams, into a protected area.

Alarm verification: A feature of a security system control panel that allows for a delay in the activation of alarms after receiving an alarm signal from one of its detection circuits.

Americans with Disabilities Act Accessibility Guidelines (ADAAG): Building construction recommendations developed as the result of an act of Congress intended to ensure civil rights for physically challenged people.

Armed: An activated alarm system.

Attack: An attempt to burglarize or vandalize or an attempt to defeat a security system.

Audio detection system: A system that detects the sounds or vibrations caused by attempted forceful entry into a protected structure.

Audio discriminator: A sound detection and evaluation device capable of discriminating between different types of sounds, such as the difference between a passing truck and breaking glass.

Bollard: An upright wood, metal, or concrete post used as a vehicle barrier.

Built-in siren driver: A driver that uses a transistor amplifier circuit to drive a siren speaker.

Burglar alarm screen: A window screen with a fine gauge wire that is woven through the fabric and connected to a control device.

Buried line intrusion sensor: See seismic sensor.

Central Station system: A system or group of systems in which the alarm or supervisory signaling devices are received, recorded, maintained, and supervised from an approved central station.

Channel detection: Narrow detection in an area where an intruder is expected to cross.

Circuit response time: The minimum length of time that an open condition or short must last before the circuit responds.

Circumvent: To defeat an alarm system by avoiding its detection devices, such as by jumping over a pressure mat, entering through a hole in an unprotected wall (rather than entering through a protected door), or keeping outside the range of an ultrasonic motion detector.

Cold-flow: A condition in which pressure on a wire causes the insulation to flow away from the conductor, leaving it exposed.

Combination system: A local protective signaling system for security alarm supervisory, or guard's tour service, whose components may be used in whole or in part with a non-security signaling system, such as a fire alarm system, paging system, musical program system, or a process-monitoring service system, without degradation of or hazard to the protective signaling system.

Combined technology sensor: A sensor that uses two technologies, usually one passive and one active, to prevent false alarms and verify that an intrusion is caused by human motion or presence.

Control unit: The means through which security personnel operate a security system.

Controlled zone: Zones that turn on and off through a control point.

Delay zone: A detection circuit on which a delay has been applied that does not apply to all the parts of a security system.

Detection device: A sensor used to initiate a signal.

Digital alarm communicator receiver (DACR): A system component that accepts and displays signals from a digital alarm communicator transmitter (DACT) that are sent over public switched telephone networks.

Digital alarm communicator system (DACS): A system in which signals are transmitted through the public switched telephone network from a digital alarm communicator transmitter (DACT) located at the protected premises to a digital alarm communicator receiver (DACR).

Digital alarm communicator transmitter (DACT): A system component in the secured area to which initiating devices or groups of devices are connected. The DACT seizes the connected telephone line, dials a preset number to connect to a DACR, and transmits signals that indicate a status change of the initiating device.

Digital communicator: A device that uses standard telephone lines or wireless telephone service to send and receive data.

Electric field sensor: See proximity sensor.

End-of-line (EOL): A device used to terminate a supervised circuit; normally a resistor or a diode placed at the end of a two-wire circuit to maintain supervision.

Entry delay: A delay that allows a security system user to enter the secured area and disarm the system before an alarm is triggered.

Exit delay: A delay that allows a security system user to arm the system and exit the secured area before an alarm is triggered.

False alarm control team (FACT): An association of security system users, security personnel, and responding authorities, whose purpose and responsibility is to reduce false alarms.

Glass-break detector: A device used to detect the frequencies generated by breaking glass.

Hardwired intrusion system: An intrusion system that uses conventional sensors and conventional notification devices connected together by wires.

Holdup alarm (HUA): An alarm system that uses a device whereby signal transmission is initiated by the action of a bank employee or other person without alerting the intruder. Money clips and cash drawer alarms are examples of HUAs.

Home and away: Programming features that allow parts of a security system, specifically perimeter sensors, to remain active when the secured area is occupied, whereby false alarms within the perimeter are prevented.

Instant zone: A detection circuit that immediately triggers an alarm when activated.

Interior: The area inside the perimeter.

Interior follower zone: A zone that establishes how an entry delay is applied. This zone is temporarily ignored by the alarm system during entry/exit delay periods.

Intrusion detection system (IDS): A detection and alarm system for signaling the entry or attempted entry of an object or person into the area or volume protected by the system.

Key switch: A switch operated by the use of a mechanical key.

Kiss-off tone: A signal from a receiving device confirming that its transmission has been received.

Labeled: Equipment or materials to which a label, symbol, or other identifying mark of an organization acceptable to the authority having jurisdiction (AHJ) has been attached.

Line carrier system: A security system that uses existing electrical wiring in the secured area to transmit messages among the system components.

Local system: An intrusion system designed to alert only the occupants of a secured area or only those people within range of an alarm bell, light, or voice warning in the secured area.

Microwave detector: A sensor that transmits microwave energy and analyzes the reflection of that energy.

Monitored intrusion system: An intrusion system connected to a central station or designed to alert property owners, police, or security personnel outside the secured area.

Monitoring site: The location where a security system is monitored, also known as the central station.

Motion (space) detector: A device that responds to changes in the environment caused by the movement of human beings.

Multiplexing: A signaling method using wire path, cable carrier, radio, fiber optics, or a combination of these techniques and characterized by the simultaneous or sequential transmission (or both) and reception of multiple signals in a communication channel, including a means of positively identifying each signal.

Notification device: An electrically or mechanically operated visible or audible signaling device. Examples of audible signals are bells, horns, sirens, electronic horns, buzzers, and chimes. A visible signal consists of an incandescent lamp, strobe lamp, mechanical target or flag, meter deflection, graphical display, or equivalent. Also known as an annunciation or indicating device.

Open collector output: An output directly from a transistor that has limited current output.

Passive sensor: A sensor that detects natural radiation or radiation disturbances but does not emit the radiation on which its operation depends.

Perimeter: The outer limits of the secured area.

Perimeter follower zone: A non-entry/exit zone, typically a perimeter zone located on an entry/exit path, which is treated as an entry/exit zone during an entry/exit time.

Photoelectric beam detector: A type of sensor in which an infrared light is transmitted to a photocell. A break in the beam activates the sensor.

Photoelectric detector: An active infrared sensor that transmits infrared light.

Piezoelectric sensor: A device that uses piezoelectric crystals to generate electrical signals when the sensor is activated.

Pressure mat: A thin rubber mat containing metal contact strips. When an intruder walks on the mat, the contacts close, which activates an alarm.

Protected distribution system (PDS): Wiring and/or installation method that prevents or detects physical access to alarm system communication lines.

Protected premises: See secured area.

Proximity sensor: A device that detects a change in the capacitance of a metal object caused by the presence of an intruder.

Repeater: Equipment that relays signals between supervising stations, subsidiary stations, and secured areas.

RJ-31X: An interface that connects telephone line signaling devices to the telephone line.

Secured area: The physical location protected by a security system.

Security system: A combination of components designed to detect and report possible threats to property or to human life.

Seismic sensor: A sensor that responds to vibrations transmitted through the ground.

Shock (vibration) detector: A device designed to detect the shock of attack before intrusion.

Silent alarm: An alarm without an obvious local indication that the alarm has been sounded.

Spot detection: Detection focused on a particular object or on high-value areas such as safes, vaults, and storage areas.

Stress sensor: A device that reacts to the flexing of the material to which it is mounted.

Structural-attack piezoelectric sensor: A device that uses a piezoelectric crystal and is designed to detect the vibrations associated with forced entry into a secure area.

Switched telephone network: An assembly of communications facilities and central office equipment, operated jointly by authorized service providers, that provides the general public with the ability to establish transmission channels via discrete dialing.

Transmitter: A system component that provides an interface between signaling line circuits, initiating device circuits, or control units, and the transmission channel.

Trap detection: Narrow or wide detection for high-traffic areas or for the path an intruder is expected to take to get from one point to another within a secured area.

Trouble signal: A signal initiated by a security system or device that is indicative of a fault in a monitored circuit or component.

Visible notification device: A notification device that alerts by the sense of sight.

Volumetric detection: Wide detection in a defined area.

Walk test light: A light that indicates a sensor is functioning in response to human motion.

Zone: A defined area within the protected premises. A zone can define an area from which a signal can be received, to which a signal can be sent, or in which a form of control can be executed.

Appendix A

Battery Calculation Forms

INTRUSION SYSTEM STANDBY BATTERY CALCULATION FORM

Job Name:

Job Address:

Account#	Phone#	Contact			
Device	Number of This Type of Device	Amps per Device		Total Amps	
		Normal	Alarm	Normal	Alarm

33407-11_A01.EPS

INTRUSION ALARM STANDBY BATTERY CALCULATION FORM

Job Name:

Job Address:

Account#	Phone#	Contact
Total Normal Amps		
× Time (24 Hours)		× 24
= Total Normal Amps		
Total Alarm Amps for 5 minutes		
× Time (0.08333 Hours)		× 0.08333
= Total Alarm Amps		
Total Normal Amps		
+ Total Alarm Amps		
= System Demand		
× De-rating Factor (1.2)		× 1.2
Minimum Standby Battery		

33407-11_A02.EPS

Appendix B

VERIFICATION FORM

VERIFICATION FORM

Phone _____   Date: _____

To: _____

No. _____

RE: Name _____

        Address _____

        City, State, Zip_____

The property listed has the following protection:

Burglar Alarm System: Fire Alarm System:

[ ] Local Alarm   [ ] Local Alarm

[ ] Monitored   [ ] Monitored

1. [ ] Maintenance Contract   [ ] On-Call Repairs

2. Burglar Alarm Features:

                    [ ] Audible Exterior Alarm

                    [ ] Audible Interior Alarm

                    [ ] Exterior Doors Protected

                    [ ] All Movable Windows Contacted

                    [ ] Partial Windows Contacted

                    [ ] Interior Traps Such As Motion Detectors

                    [ ] Backup Power Supply

                    [ ] Visual Status Indicator(s)

                    [ ] System Will Not Alarm Unless All Zones Secured Or Bypassed

                    [ ] Can Cancel Accidental Alarm

                    [ ] Telephone Line Monitor (Sounds Alarm If Line Cut)

3. Method of Signal Transmission:

                    [ ] Digital Dialer Primary

                    [ ] Radio Backup

                    [ ] Radio Primary

                    [ ] Digital Dialer Backup

33407-11_A03.EPS

TO BE COMPLETED AT JOB SITE

Customer Name: _____   Date: _____

Location Address: _____

Location Phone: _____

Circle As Appropriate:

1.	Are all items installed?	Yes	No	NA
2.	Have all items been checked for proper operation?	Yes	No	NA
3.	Have all motion detectors been aimed/masked properly?	Yes	No	NA
4.	Have all drill holes been caulked properly?	Yes	No	NA
5.	Is the building completely restored and job debris removed?	Yes	No	NA
6.	Is the Owner's Manual filled in, and has it been given to the customer?	Yes	No	NA
7.	Have any abort or alarm cards been given to the customer?	Yes	No	NA
8.	Have decals been installed on all sides of the building?	Yes	No	NA
9.	Has a yard sign been installed?	Yes	No	NA
10.	Has the Master Security Code Label been removed from the control box, if applicable?	Yes	No	NA
11.	Is the system properly grounded?	Yes	No	NA
12.	Are all zones communicating to the central station?	Yes	No	NA
13.	Are zoning labels and proper activation decals placed on control points?	Yes	No	NA
14.	Is a "property of" label attached to the control box, if applicable?	Yes	No	NA
15.	Has all paperwork been completed?	Yes	No	NA
16.	Has programming data been uploaded to the office computer?	Yes	No	NA
17.	Has the customer received proper instruction on using the system?	Yes	No	NA

Control box location: _____

Telephone number to which alarm is connected: _____

Location of AC transformer: _____

Location of telephone tie-in point: _____

Which zones are group shunted? _____

Attic and under-house access areas (if not obvious): _____
_____

Siren/Bell location(s): _____

Splice point location(s): _____

Fire Cert. left in panel _____   Fire tag on panel _____   Floor plan copy left _____

Zone	Wire Numbers and Area	Zone	Wire Numbers and Area	Zone	Wire Numbers and Area
1.	_____	3.	_____	5.	_____
2.	_____	4.	_____	6.	_____

I (we), the customer (end user or representative), have: (check all that apply)

[ ] Received a copy of the owner's manual(s) and been provided with any keys/codes needed to access user controls

[ ] Been shown how to use all customer functions and controls

[ ] Been advised of how to contact the monitoring station and/or applicable responding authorities

[ ] Been advised of whom to contact for repair and maintenance service

[ ] Been advised of whom to contact for new employee training relating to the use of this system

[ ] Been advised of whom to contact for assistance in developing a false alarm reduction plan

[ ] Been advised of required testing frequency

[ ] Received an exact copy of this verification form

Signed _____   Print name _____

Witness _____   Print name _____

33407-11_A04.EPS

Additional Resources

This module presents thorough resources for task training. The following resource material is suggested for further study.

Security for Building Occupants and Assets. The Whole Building Design Guide/National Institute of Building Sciences. www.wbdg.org.

The Design and Evaluation of Physical Protection Systems, 2007. Mary Lynn Garcia. Boston, MA: Butterworth-Heinemann.

Figure Credits

NCCER CURRICULA — USER UPDATE

NCCER makes every effort to keep its textbooks up-to-date and free of technical errors. We appreciate your help in this process. If you find an error, a typographical mistake, or an inaccuracy in NCCER's curricula, please fill out this form (or a photocopy), or complete the online form at **www.nccer.org/olf**. Be sure to include the exact module ID number, page number, a detailed description, and your recommended correction. Your input will be brought to the attention of the Authoring Team. Thank you for your assistance.

Instructors – If you have an idea for improving this textbook, or have found that additional materials were necessary to teach this module effectively, please let us know so that we may present your suggestions to the Authoring Team.

NCCER Product Development and Revision
13614 Progress Blvd., Alachua, FL 32615

Email: curriculum@nccer.org
Online: www.nccer.org/olf

❏ Trainee Guide   ❏ AIG   ❏ Exam   ❏ PowerPoints   Other _____

Craft / Level: _____   Copyright Date: _____

Module ID Number / Title: _____

Section Number(s): _____

Description: _____

_____

_____

_____

Recommended Correction: _____

_____

_____

_____

_____

Your Name: _____

Address: _____

_____

Email: _____   Phone: _____

33408-12

Fire Alarm Systems

Supports the Skill and Knowledge Statements used as the basis for NICET's Fire Alarm Installer Certification Tests

Module Eight

Trainees with successful module completions may be eligible for credentialing through NCCER's National Registry. To learn more, go to **www.nccer.org** or contact us at **1.888.622.3720**. Our website has information on the latest product releases and training, as well as online versions of our *Cornerstone* newsletter and Pearson's product catalog.

Your feedback is welcome. You may email your comments to **curriculum@nccer.org**, send general comments and inquiries to **info@nccer.org**, or fill in the User Update form at the back of this module.

V.1 3/12

Objectives

When you have completed this module, you will be able to do the following:

1. Explain the terminology associated with fire alarm systems.
2. Describe the relationship between fire alarm systems and life safety.
3. Identify and explain the role that various codes and standards play in both commercial and residential fire alarm applications.
4. Describe the characteristics and functions of various fire alarm system components.
5. Explain and describe the different types of circuitry that connect fire alarm system components.
6. Explain the operation of conventional, addressable, and analog fire alarm systems.
7. Draw a two-wire and four-wire initiating circuit showing proper supervision.
8. Install and troubleshoot a four-wire initiating device circuit.
9. Wire either a conventional zone or a fire alarm system pull station.
10. Troubleshoot an instructor-induced ground fault of a fire alarm system.
11. Isolate a short circuit on a fire alarm system.
12. Isolate an open circuit on a fire alarm circuit.
13. Program a system.
14. Commission a system.
15. Correctly wire an RJ-31X telephone jack.
16. Complete an NFPA record of completion.

Performance Tasks

Under the supervision of your instructor, you should be able to do the following:

1. Draw a two-wire and four-wire initiating circuit.
2. Install a fire alarm system.
3. Commission a system.
4. Correctly wire an RJ-31X telephone jack.
5. Complete an NFPA record of completion.
6. Troubleshoot a fire alarm system.

Trade Terms

Addressable device
Air sampling (aspirating) detector
Alarm signal
Alarm verification
Authority having jurisdiction (AHJ)
Automatic fire alarm system
Backflow preventer
Candela (cd)
Ceiling surfaces
Certification
Class A circuit
Class B circuit
Digital alarm communicator receiver (DACR)
Digital alarm communicator system (DACS)
Digital alarm communicator transmitter (DACT)
End-of-line (EOL) device

Fault
Fire alarm control unit (FACU)
Flame detector
General alarm
Ground fault
Heat detector
Heat tracing
Indicating device
Initiating device
Initiating device circuit (IDC)
Light scattering
Listed
Multiplexing
Non-coded signal
Obscuration
Path (pathway)

continued

Trade Terms (*continued*)

Photoelectric smoke detector
Positive alarm sequence
Pre-action sprinkler system
Projected beam smoke detector
Rate compensation detector
Rate-of-rise detector
Remote supervising station fire alarm system
Signaling line circuits (SLCs)
Smoke detector

Spot-type detector
Stratification
Supervising station
Supervisory signal
System unit
Thermal lag
Trouble signal
Visible notification appliance
Zone

Industry Recognized Credentials

If you're training through an NCCER-accredited sponsor you may be eligible for credentials from NCCER's Registry. The ID number for this module is 33408-12. Note that this module may have been used in other NCCER curricula and may apply to other level completions. Contact NCCER's Registry at 888.622.3720 or go to nccer.org for more information.

Note: *National Fire Alarm and Signaling Code* and *NFPA 72*® are registered trademarks of the National Fire Protection Association, Quincy, MA. Some jurisdictions may not have adopted the 2010 edition of *NFPA 72*®. For that reason, tables from both the 2002 and 2010 codes have been provided.

Note: *NFPA 70*®, *National Electrical Code*® and *NEC*® are registered trademarks of the National Fire Protection Association, Inc., Quincy, MA 02269. All *National Electrical Code*® and *NEC*® references in this module refer to the 2011 edition of the *National Electrical Code*®.

Contents

Topics to be presented in this module include:

Contents (*continued*)

Contents (*continued*)

Figures and Tables

Figures and Tables (*continued*)

1.0.0 INTRODUCTION

Fire alarm systems can make the difference between life and death. In a fire emergency, a quick and accurate response by a fire alarm system can reduce deaths and injuries dramatically. However, practical field experience is necessary to master fire alarm installation and maintenance. Remember to focus on doing the job properly at all times. Failure to do so could have tragic results.

Always seek out and abide by any National Fire Protection Association (NFPA), federal, state, and local codes or standards that apply to each and every fire alarm system installation.

Different types of buildings and applications have different fire safety goals. Therefore, each building has a different fire alarm system to meet those specific goals. Most automatic fire alarm systems are configured to provide fire protection for life safety, property protection, or mission protection.

Life safety fire protection is concerned with protecting and preserving human life. In a fire alarm system, life safety can be defined as providing warning of a fire situation. This warning occurs early enough to allow notification of building occupants, ensuring that there is sufficient time for their safe evacuation.

Many times, fire alarm systems are taken for granted. People go about their everyday business with little or no thought of the behind-the-scenes safety devices that monitor for abnormal conditions. After all, no one expects a fire to happen. When it does, every component of the life safety fire alarm system must work properly in order to reduce the threat to life.

2.0.0 CODES AND STANDARDS

Fire alarm system equipment and installation are regulated and controlled by various national, state, and local codes. Industry standards have also been developed as a means of establishing a common level of competency. These codes and standards are set by a variety of different associations, agencies, and laboratories that consist of fire alarm professionals across the country. Depending on the application and installation, you need to follow the standards established by one or more of these organizations.

In the United States, local and state jurisdictions select a model code and supporting documents that detail applicable standards. If necessary, they amend the codes and standards as they see fit. Some amendments may be substantial changes to the code, and in certain situations, the jurisdictions author their own sets of codes. Once a jurisdiction adopts a set of codes and standards, it becomes a binding legal document within that jurisdiction.

The following list details some of the many national organizations responsible for setting the industry standards:

- *Underwriters Laboratories (UL)* – Establishes standards for fire equipment and systems.
- *The National Fire Protection Association (NFPA)* – Establishes standards for fire systems. The association publishes the *National Fire Alarm and Signaling Code* (NFPA 72®), the *Life Safety Code*® (NFPA 101®), and the *Fire Code (NFPA 1)*. These codes are the primary reference documents used by fire alarm system professionals.
- *FM Global* – Establishes standards for fire systems.
- *National Electrical Manufacturers' Association (NEMA)* – Establishes standards for electrical equipment.

On Site

Installing and Servicing Fire Alarm Systems

In certain jurisdictions, the person who installs and/ or services fire alarm systems must be a licensed fire alarm specialist.

On Site

Property and Mission Protection

The design goal of most fire alarm systems is life safety. However, some systems are designed with a secondary purpose of protecting either property or the activities (mission) within a building. The goal of both property and mission protection is the early detection of a fire so that firefighting efforts can begin while the fire is still small. Fire alarm systems with a secondary goal of property protection are commonly used in museums, libraries, storage facilities, and historic buildings in order to minimize damage to the buildings or their contents. Systems with a secondary goal of mission protection are commonly used where it is essential to avoid business interruptions, such as in hospitals, banks, security control rooms, and telecommunication centers.

- *The Federal Bank Protection Act* – Establishes fire alarm equipment and system standards for banks.
- *The National Institute for Certification in Engineering Technologies (NICET)* – Establishes standardized testing of fire alarm designers, installers, and service technicians.

2.1.0 The National Fire Protection Association (NFPA)

NFPA is responsible for setting the national standards and codes for the fire alarm industry. Consisting of fire alarm representatives from all areas of business and fire protection services, the NFPA responds to the ever-changing needs of society by using a process to form consensus standards that are acceptable to all members. The NFPA reviews equipment and system performance criteria and input from industry professionals to set an acceptable level of protection for both life and property.

2.1.1 NFPA Codes

The following five widely adopted national codes are specified by the NFPA:

- *National Electrical Code® (NFPA 70®)* – *The National Electrical Code® (NEC®)* covers all of the necessary requirements for all electrical work performed in a building. The Fire Alarm Signaling Systems portion of the code (*NEC Article 760*) details the specific requirements for wiring and equipment installation for fire protection signaling systems. Specifications include installation methods, connection types, circuit identification, and wire types (including gauges and insulation). The *NEC®* places restrictions on the number and types of circuit combinations that can be installed in the same enclosure.
- *National Fire Alarm and Signaling Code (NFPA 72®)* – The recommended requirements for installation of fire alarm systems and equipment in residential and commercial facilities are covered in *NFPA 72®*. Included are requirements for the installation of initiating devices (sensors) and notification appliances (visual or audible). Inspection, testing, and maintenance requirements for fire alarm systems and equipment are also covered.

> **NOTE**
> Not all jurisdictions have adopted *NFPA 72®* for one- and two-family residential occupancies.

- *Life Safety Code® (NFPA 101®)* – *NFPA 101®* is focused on the preservation and protection of human life, as opposed to property. Life safety requirements are detailed for both new construction and existing structures. Protection for unique building features and construction are detailed. In addition, chapters are organized to explain when, where, and for what applications fire alarm systems are required, the necessary means of initiation and occupant notification, and the means by which to notify the fire department. This code also details any equipment exceptions to these requirements.
- *Fire Code (NFPA 1)* – This code was established to help fire authorities continually develop safeguards against fire hazards. A chapter of this code is dedicated to fire protection systems. Information and requirements for testing, operation, installation, and periodic preventive maintenance of fire alarm systems are included in this portion of the code.
- *NFPA 3, Recommended Practice for Commissioning and Integrated Testing of Fire Protection and Life Safety Systems* – It should be noted that *NFPA 3* is a recommended practice rather than a code. Its purpose is to define a commissioning process to ensure that fire and life safety systems function as intended.
- *NFPA 5000®, NFPA Building Construction and Safety Code®* – *NFPA 5000®* provides minimum design regulations to safeguard life, limb, health, property, and public welfare. It provides information for regulating and controlling the permitting, design, construction, quality of materials, use and occupancy, location, and maintenance of all buildings and structures within the authority having jurisdiction (AHJ). It also regulates certain equipment within all buildings and structures.

2.1.2 NFPA Standards

The NFPA also publishes a large number of standards that are used by fire alarm system professionals. These include the following:

- *NFPA 13, Standard for the Installation of Sprinkler Systems*
- *NFPA 17, Standard for Dry Chemical Extinguishing Systems*
- *NFPA 17A, Standard for Wet Chemical Extinguishing Systems*
- *NFPA 20, Standard for the Installation of Stationary Pumps for Fire Protection*
- *NFPA 25, Standard for the Inspection, Testing, and Maintenance of Water-Based Fire Protection Systems*

NFPA Codes

Several NFPA code books are required for installing alarm systems. Nearly every requirement of *NFPA 101, Life Safety Code*®, has resulted from the analysis of past fires in which human lives have been lost. The code books shown are the current editions upon publication of this module. Code books are revised and changed periodically, typically every three or four years. For this reason, you should always make sure that you are using the most current edition of any code book.

33408-11_SA01.EPS

- *NFPA 37, Standard for the Installation and Use of Stationary Combustion Engines and Gas Turbines*
- *NFPA 75, Standard for Protection of Information Technology Equipment*
- *NFPA 80, Standard for Fire Doors and Other Opening Protectives*
- *NFPA 90A, Standard for the Installation of Air Conditioning and Ventilating Systems*
- *NFPA 92A, Standard for Smoke Control Systems Utilizing Barriers and Pressure Differences*
- *NFPA 96, Standard for Ventilation Control and Fire Protection of Commercial Cooking Operations*
- *NFPA 110, Standard for Emergency and Standby Power Systems*
- *NFPA 111, Standard on Stored Electrical Energy Emergency and Standby Power Systems*
- *NFPA 1221, Standard for the Installation, Maintenance, and Use of Emergency Services Communications Systems*

2.2.0 Other Codes

- *ANSI A17.1, Safety Code for Elevators and Escalators* – ANSI A17.1 covers the design, construction, operation, inspection, testing, maintenance, alteration, and repair of elevators and escalators, as well as power-driven walkways.

- *International Building Code (IBC)* – The *IBC* establishes minimum standards to protect life and safety by regulating and controlling the design construction, and quality of materials and practices used in construction.
- *International Fire Code (IFC)* – The *IFC* includes regulations governing the safeguarding of life and property from all types of fire and explosion hazards.

2.3.0 Labeled and Listed Equipment

Labeled equipment means that manufacturers label equipment with symbols or other identifying marks of various organizations, acceptable to the AHJ, to indicate compliance with organization standards and requirements for evaluation and periodic inspection. Listed equipment means that the manufacturers' equipment appears on the lists published by the various organizations for the same requirements. The difference is that the equipment may not be labeled. However, some organizations require Listed equipment to also be labeled. The AHJ should use the system adopted by the applicable organization to identify acceptable equipment.

3.0.0 FIRE ALARM SYSTEMS OVERVIEW

Fire alarm systems are designed for the protection of life, property, and mission. They are designed to detect and warn of abnormal conditions, alert authorities, and operate safety devices to minimize fire danger. Through a variety of manual and automatic system devices, a fire alarm system links the sensing of a fire condition with people inside and outside of the building. The system communicates to fire professionals that action needs to be taken.

3.1.0 Fire Alarm Circuit Designations

NFPA 72® requires that an initiating device, notification appliance, or signaling line circuit (SLC) be designated as a Class A circuit or Class B circuit and/or as one of the styles defined by the code under Class A or B for these types of circuits. This is necessary to define the circuit's ability to operate under fault conditions. Class A circuits and styles are regarded as more reliable than Class B circuits and styles. Because the purpose of a Class A circuit is survivability, the source and return legs of a Class A circuit are run in non-redundant paths. Fire system designers usually list only the style designation because the styles defined under a Class A or B designation define the performance of the circuit for specific failure conditions. If wire-to-wire short failures are not important to an application, then only the class designation may be listed. *Tables 1, 2,* and *3* show the minimum performance of Class A and B circuit styles for initiating device circuits (IDC), notification appliance circuits, and signaling line circuits. If desired, the specifier can require circuits to exceed these minimum requirements.

3.2.0 Types of Fire Alarm Systems

Although specific codes and standards must always be followed when designing and installing a fire alarm system, many different types of systems that employ different types of technology can be used. Most of these types of fire alarm systems are categorized by the means of communication between the detectors and the fire alarm control unit (FACU). The three major types of fire alarm systems are conventional hardwired, multiplex, and addressable intelligent.

3.2.1 Conventional Hardwired Systems

A hardwired system using conventional initiation devices (heat and smoke detectors and fire alarm boxes) and notification appliances (bells, horns, or lights) is the simplest of all fire alarm systems. Large buildings or areas that are being protected are usually divided into zones to identify the area where a fire is detected. A conventional hardwired system is limited to zone detection only, with no means of identifying the specific detector that initiated the alarm. A typical hardwired system might look like *Figure 1* with either two- or four-wire initiating circuits. In two-wire initiating circuits, power for the devices is superimposed on the alarm circuits. In four-wire initiating circuits, the operating power is supplied to the devices separately from the signal or alarm circuits. In either two- or four-wire Class B circuits, end-of-line (EOL) devices are used by the FACU to monitor circuit integrity.

3.2.2 Active Multiplex Systems

Active multiplex systems are similar to hardwired systems in that they rely on zones for fire detection. The difference is that multiplexing allows multiple signals from several sources to be sent and received over a single communication line. Each signal can be uniquely identified. This results in reduced control equipment, less wiring infrastructure, and a distributed power supply. *Figure 2* shows a simplified example of a multiplex system.

3.2.3 Addressable Systems

Addressable systems use advanced technology and detection equipment for discrete identification of alarm signals at the detector level. An addressable system can pinpoint an alarm location to the precise physical location of the initiating detector. The basic idea of an addressable system is to provide identification or control of individual initiation, control, or notification devices on a common circuit. Each component on the SLC has an identification number or address. The addresses are usually assigned using switches or other similar means. *Figure 3* is a simplified representation of an addressable or analog addressable fire alarm system.

The FACU constantly polls each device using an SLC. The response from the device being polled verifies that the wiring path (pathway) is intact (wiring supervision) and that the device is in place and operational. Most addressable systems use at least three states to describe the status of the device: normal, trouble, and alarm. Addressable smoke detection devices usually make the decision regarding their alarm state internally just like conventional smoke detectors. Output devices, like relays, are usually checked for their presence and, in some cases, for their output status. Notification output modules also supervise the wiring to the horns, strobes, and other devices, as well as the

Table 1 Performance of Initiating Device Circuits (IDCs) (1 of 2)

Data from NFPA 72®-2002, Table 6.5

Class	B			B			B			A			A		
Style	A			B			C			D			Eα		
	Alm	Trbl	ARC	Alm	Trbl	ARC	Alm	Trbl	ARC	Alm	Trbl	ARC	Alm	Trbl	ARC
Abnormal Condition	1	2	3	4	5	6	7	8	9	10	11	12	13	14	15
Single open	—	X	—	—	X	—	—	X	—	—	X	R	—	X	R
Single ground	—	X	—	—	X	R	—	X	R	—	X	R	—	X	R
Wire-to-wire short	X	—	—	X	—	—	—	X	—	X	—	—	—	X	—
Loss of carrier (if used)/channel interface	—	—	—	—	—	—	—	X	—	—	—	—	—	X	—

Alm = Alarm.
Trbl = Trouble.
ARC = Alarm receipt capability during abnormal condition.
R = Required capacity.
X = Indication required at protected premises and as required by Chapter 8.
α = Style exceeds minimum requirements of Class A.

Reprinted with permission from NFPA 72®-2002, *National Fire Alarm Code*. Copyright © 2002, National Fire Protection Association, Quincy, MA 02269. The reprinted material is not the complete and official position of the NFPA on the referenced subject, which is represented only by the standard in its entirety.

33408-11_T01A.EPS

Table 1 Performance of Initiating Device Circuits (IDCs) (2 of 2)

Data from NFPA 72®-2010, Table A.12.3(a)

NFPA 72-2007 Class	B			A		
NFPA 72-2010 Class	B			A		
	Alm	Trbl	ARC	Alm	Trbl	ARC
Abnormal Condition	1	2	3	4	5	6
Single open	–	X	–	–	X	R
Single Ground	–	X	R	–	X	R

Alm: Alarm. Trbl: Trouble. ARC: Alarm receipt capability during abnormal condition. R: Required capacity. X: Indication required at protected premises and as required by Chapter 26.

Reprinted with permission from NFPA 72®-2010, *National Fire Alarm Code®*, Copyright © 2009, National Fire Protection Association, Quincy, MA. This reprinted material is not the complete and official position of the NFPA on the referenced subject, which is represented only by the standard in its entirety.

33408-11_T01B.EPS

availability of the power needed to run the devices in case of an alarm. When the FACU polls each device, it also compares the information from the device to the system program. For example, if the program indicates device 12 should be a contact transmitter but the device reports that it is a relay, a problem exists that must be corrected. Addressable fire alarm systems have been made with two, three, and four conductors. Capacitance limits the SLC conductor lengths and the number of devices connected to the conductors. To accommodate additional detectors, some systems may contain multiple SLC circuits. These are comparable to multiple zones in a conventional hardwired system.

3.2.4 Analog Addressable Systems

Analog systems take the addressable system capabilities much further and change the way the information is processed. When a device is polled, it returns much more information than a device in a standard addressable system. For example, instead of a smoke detector transmitting that it is in alarm status, the smoke sensor device actually transmits the level of smoke or contamination present to the fire alarm control unit. The control unit then compares the information to the levels detected in previous polls. A slow change in levels (over days, weeks, or months) indicates that a device is dirty or malfunctioning. A rapid change, however, normally indicates a fire condition. Most systems or sensors have the capability to compensate for the drift caused by the dirt buildup. The system or sensors adjust the sensitivity to the desired range. Once the dirt buildup exceeds the compensation range, the system or sensor reports a trouble condition. The system or sensor can also perform self-checks to test the sensor's ability to respond to smoke. If the airflow around a device is too great to allow proper detection, some systems or sensors generate a trouble report.

Table 2 Performance of Notification Appliance Circuits (NACs) (1 of 2)

Data from NFPA 72®-2002, Table 6.7

Class	B		B		B		A	
Style	W		X		Y		Z	
	Trouble indication at protected premises	Alarm capability during abnormal conditions	Trouble indication at protected premises	Alarm capability during abnormal conditions	Trouble indication at protected premises	Alarm capability during abnormal conditions	Trouble indication at protected premises	Alarm capability during abnormal conditions
Abnormal Condition	1	2	3	4	5	6	7	8
Single open	X	—	X	R	X	—	X	R
Single ground	X	—	X	—	X	R	X	R
Wire-to-wire short	X	—	X	—	X	—	X	—

X = Indication required at protected premises.
R = Required capability.

33408-11_T02A.EPS

Table 2 Performance of Notification Appliance Circuits (NACs) (2 of 2)

Data from NFPA 72®-2010, Table A.12.3(c)

NFPA 72-2007 Class	B		A	
NFPA 72-2010 Class	B		A	
	Trouble Indications at Protective Premise	Alarm Capability During Abnormal Condition	Trouble Indications at Protective Premise	Alarm Capability During Abnormal Condition
Abnormal Condition	1	2	3	4
Single open	X	–	X	R
Single ground	X	R	X	R
Wire-to-wire short	X	–	X	–

X: Indication required at protected premises and as required by Chapter 26.
R: Required capability.

33408-11_T02B.EPS

The information in some systems is sent and received in a totally digital format. Others transmit the polling information digitally but receive the responses in an analog current-level format.

In most cases, the fire alarm control unit, not the device, determines the alarm state. In many systems, the LED on the detector is turned on by the control unit and not by the detector. This ability to make decisions at the control unit also allows the sensor's sensitivity to be adjusted at the panel. For instance, an increase in the ambient temperature can cause a smoke sensor to become more sensitive, and the alarm level sensitivity at the control unit or sensor can be adjusted to compensate. Sensitivities can even be adjusted based on the time of day or day of the week. Other sensor devices can be programmed to adjust their own sensitivity.

The ability of an analog addressable system to process more information than the three elementary alarm states found in simpler systems allows the analog addressable system to provide prealarm signals and other information. In many systems, five or more different signals can be received.

Most analog addressable systems operate on a two-conductor circuit. Most systems limit the number of devices to about 100. Because of the high data rates on these signaling line circuits, capacitance also limits the conductor lengths. Always follow the manufacturer's installation instructions to ensure proper operation of the system.

4.0.0 FIRE ALARM SYSTEM EQUIPMENT

The equipment used in fire alarm systems is generally held to higher standards than typical electrical equipment. A fire alarm system has seven main components:

- Alarm initiating devices
- Control panels
- Primary (main) power supply

Table 3 Performance of Signaling Line Circuits (SLCs) (1 of 2)

Data from NFPA 72®-2002, Table 6.6. 1

Class	B			B			A			B			B		
Style	0.5			1			2α			3			3.5		
	Alm	Trbl	ARC	Alm	Trbl	ARC	Alm	Trbl	ARC	Alm	Trbl	ARC	Alm	Trbl	ARC
Abnormal Condition	1	2	3	4	5	6	7	8	9	10	11	12	13	14	15
Single open	—	X	—	—	X	—	—	X	R	—	X	—	—	X	—
Single ground	—	X	—	—	X	R	—	X	R	—	X	R	—	X	—
Wire-to-wire short	—	—	—	—	—	—	—	—	M	—	X	—	—	X	—
Wire-to-wire short & open	—	—	—	—	—	—	—	—	M	—	X	—	—	X	—
Wire-to-wire short & ground	—	—	—	—	—	—	—	X	M	—	X	—	—	X	—
Open and ground	—	—	—	—	—	—	—	X	R	—	X	—	—	X	—
Loss of carrier (if used)/channel interface	—	—	—	—	—	—	—	—	—	—	—	—	—	X	—

Class	B			B			A			A			A		
Style	4			4.5			5α			6α			7α		
	Alm	Trbl	ARC	Alm	Trbl	ARC	Alm	Trbl	ARC	Alm	Trbl	ARC	Alm	Trbl	ARC
Abnormal Condition	16	17	18	19	20	21	22	23	24	25	26	27	28	29	30
Single open	—	X	—	—	X	R	—	X	R	—	X	R	—	X	R
Single ground	—	X	R	—	X	—	—	X	R	—	X	R	—	X	R
Wire-to-wire short	—	X	—	—	X	—	—	X	—	—	X	—	—	X	R
Wire-to-wire short & open	—	X	—	—	X	—	—	X	—	—	X	—	—	X	—
Wire-to-wire short & ground	—	X	—	—	X	—	—	X	—	—	X	—	—	X	—
Open and ground	—	X	—	—	X	—	—	X	—	—	X	R	—	X	R
Loss of carrier (if used)/channel interface	—	X	—	—	X	—	—	X	—	—	X	—	—	X	—

Alm = Alarm.
Trbl = Trouble.
ARC = Alarm receipt capability during abnormal condition.
M = May be capable of alarm with wire-to-wire short.
R = Required capability.
X = Indication required at protected premises and as required by Chapter 8.
α = Style exceeds minimum requirements of Class A.

33408-11_T03A.EPS

Table 3 Performance of Signaling Line Circuits (SLCs) (2 of 2)

Data from NFPA 72®-2010, Table A.12.3(b)

	NFPA 72-2007 Class B / Style 4 / NFPA 72-2010 Class B			NFPA 72-2007 Class A / Style 6 / NFPA 72-2010 Class A			NFPA 72-2007 Class A / Style 7 / NFPA 72-2010 Class X		
	Alarm	Trouble	ARC	Alarm	Trouble	ARC	Alarm	Trouble	ARC
Abnormal Condition	1	2	3	4	5	6	7	8	9
Single open	-	X	-	-	X	R	-	X	R
Single ground	-	X	R	-	X	R	-	X	R
Wire-to-wire short	-	X	-	-	X	-	-	X	R
Wire-to-wire short & open	-	X	-	-	X		-	X	-
Wire-to-wire short & ground	-	X	-	-	X	-	-	X	-
Open and ground	-	X	-	-	X	R	-	X	R
Loss of carrier (if used) / channel interface	-	X	-	-	X	-	-	X	-

ARC: Alarm receipt capability during abnormal condition. R: Required capability. X: Indication required at protected premises and as required by Chapter 26.

33408-11_T03B.EPS

- Secondary (standby) power supply
- Notification appliances
- Communications
- Monitoring

5.0.0 FIRE ALARM INITIATING DEVICES

Fire alarm systems use initiating devices to report a fire and provide supervisory or trouble reports. Some of the initiating devices used to trigger a fire alarm are designed to detect the signs of fire automatically (automatic detectors). Some report level-of-fire conditions only (sensors). Others rely on people to see the signs of fire and then activate a manual device. Automatic detectors or sensors are available that sense smoke, heat, and flame. Manual initiating devices are usually some form of a manual fire alarm box (pull box).

Figure 4 shows the application of various detectors or sensors for each stage of a fire. However, detection or sensing may not occur at any specific point in time within each stage, nor will the indicated detectors/sensors always provide detection or sensing in the stages represented. For example, alcohol fires can produce flame followed by heat while producing few or no visible particles.

Besides triggering an alarm, some fire alarm systems use the input from one or more detectors, sensors, or pull boxes to trigger fire suppression equipment, such as a carbon dioxide (CO_2), dry chemical, or water deluge system.

It is mandatory that devices used for fire detection be Listed for the purpose for which they are being used. Common UL Listings for fire detection devices are as follows:

- UL 38 – *Manual Signaling Boxes for Fire Alarm Systems*
- UL 217 – *Single and Multiple Station Smoke Alarms*
- UL 268 – *Smoke Detectors for Fire Alarm Systems*
- UL 268A – *Smoke Detectors for Duct Application*
- UL 521 – *Heat Detectors for Fire Protective Signaling Systems*

5.1.0 Conventional versus Addressable Commercial Detectors

Commercial detectors (*Figure 5*) differ from stand-alone residential alarms in that a number of commercial detectors are usually wired in parallel and connected to an FACU. In a typical conventional commercial fire alarm installation, a number of separate zones with multiple automatic

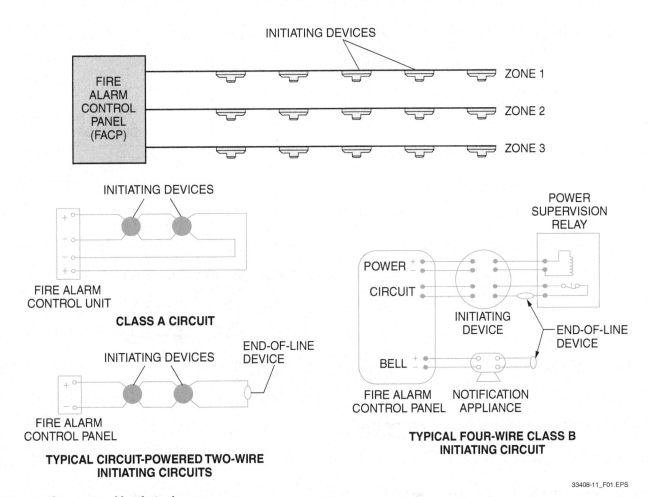

INITIATING DEVICES

ZONE 1

ZONE 2

ZONE 3

FIRE ALARM CONTROL PANEL (FACP)

INITIATING DEVICES

FIRE ALARM CONTROL UNIT

CLASS A CIRCUIT

INITIATING DEVICES

END-OF-LINE DEVICE

FIRE ALARM CONTROL PANEL

TYPICAL CIRCUIT-POWERED TWO-WIRE INITIATING CIRCUITS

POWER SUPERVISION RELAY

POWER

CIRCUIT

INITIATING DEVICE

END-OF-LINE DEVICE

BELL

FIRE ALARM CONTROL PANEL

NOTIFICATION APPLIANCE

TYPICAL FOUR-WIRE CLASS B INITIATING CIRCUIT

33408-11_F01.EPS

Figure 1 Conventional hardwired system.

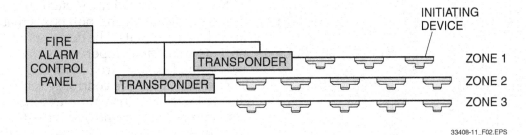

INITIATING DEVICE

FIRE ALARM CONTROL PANEL

TRANSPONDER

TRANSPONDER

ZONE 1

ZONE 2

ZONE 3

33408-11_F02.EPS

Figure 2 Multiplex system.

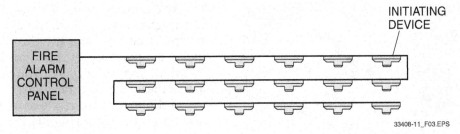

INITIATING DEVICE

FIRE ALARM CONTROL PANEL

33408-11_F03.EPS

Figure 3 Addressable or analog addressable system.

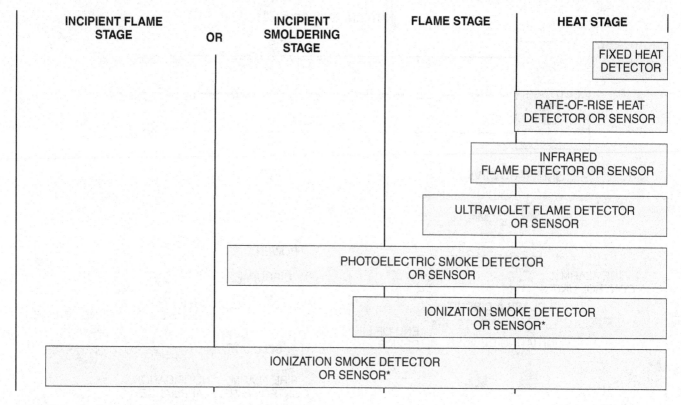

INCIPIENT FLAME STAGE	OR	INCIPIENT SMOLDERING STAGE	FLAME STAGE	HEAT STAGE

FIXED HEAT DETECTOR

RATE-OF-RISE HEAT DETECTOR OR SENSOR

INFRARED FLAME DETECTOR OR SENSOR

ULTRAVIOLET FLAME DETECTOR OR SENSOR

PHOTOELECTRIC SMOKE DETECTOR OR SENSOR

IONIZATION SMOKE DETECTOR OR SENSOR*

IONIZATION SMOKE DETECTOR OR SENSOR*

*IONIZATION DETECTORS REACT MORE QUICKLY TO INCIPIENT FLAME, BUT PHOTO DETECTORS ARE QUICKER TO REACT TO INCIPIENT SMOLDERING.

33408-11_F04.EPS

Figure 4 Detection versus stages of fires.

PHOTOELECTRIC AND THERMAL DETECTOR

ADDRESSABLE PHOTOELECTRIC AND THERMAL SENSOR

DUSTY/DIRTY ENVIRONMENTS

MULTI-CRITERIA SMOKE DETECTOR

33408-11_F05.EPS

Figure 5 Commercial automatic sensors.

detectors are used to partition the installation into fire zones.

Newer commercial fire systems use addressable automatic detectors and pull boxes that use coded identification signals. The detectors and pull boxes provide individual supervisory and trouble signals to the FACU, as well as any fire alarm signal. This provides the fire control system with specific device location information to pinpoint the fire, along with detector and pull box status information. In some cases, notification appliances may supply an individual status to the FACU as well. In many new commercial fire systems, analog addressable sensors are used. Instead of sending a fire alarm signal when polled, these devices communicate information about the fire condition (level of smoke, temperature, etc.) in addition to their identification, supervisory, or trouble signals. In the case of these sensors, the control unit analyzes the fire condition data from one or more sensors, using recent and older information from the sensors, to determine if a fire alarm should be issued.

> NOTE
>
> The descriptions of operation discussed in the sections covering automatic detectors, pull boxes, and notification appliances are primarily for conventional devices.

5.2.0 Automatic Detectors

Automatic detectors can be divided into the following types:

- *Line detectors* – Detection is continuous along the entire length of the detector. Typical line detectors may include certain older pneumatic rate-of-rise tubing detectors, projected-beam smoke detectors, and heat-sensitive cable.
- *Spot detectors* – These devices have a detecting element that is concentrated at a particular location. Typical spot detectors include bimetallic detectors, fusible alloy detectors, certain rate-of-rise detectors, smoke detectors, and thermo-electric detectors.
- *Air sampling (aspirating) detectors* – Air sampling (aspirating) detectors consist of piping or tubing distributed from the detector unit to the area(s) to be protected. An air pump draws air from the protected area back to the detector through air-sampling ports and piping or tubing. At the detector, the air is analyzed for products of combustion.
- *Addressable detectors or analog addressable sensors* – Addressable detectors provide alarms and individual point identification, along with supervisory/trouble information, to the control unit. In certain analog addressable sensors, adjustable sensitivity of the device can be provided. In these sensors, an alarm signal is not generated. Instead, only a level of detection signal is fed to the control panel. Alarm sensitivity can be adjusted due to drift, and the level of detection can be analyzed from the panel (based on older data) to reduce the likelihood of nuisance alarms in construction areas or areas of high humidity on a temporary or permanent basis. In newer sensors and detectors, software is used to analyze the signal, automatically adjust drift compensation and sensitivity, and perform other evaluations and adjustments to reduce nuisance alarms. Because all detectors are continually polled, a T-tap splice is permitted with Class B signaling line circuit styles. T-taps are not permitted with Class A signaling line circuits.

5.3.0 Heat Detectors

There are two major types of heat detectors in general use. One is a rate-of-rise detector that senses a 15°F per minute increase in room/area temperature. The other is one of several versions of fixed-temperature detectors that activate if the room/area exceeds the rating of the sensor. Usually, rate-of-rise sensors are combined with fixed-temperature sensors in a combination spot heat detector. Heat detectors are generally used in areas where property protection is the only concern or where smoke detectors are not needed.

5.3.1 Fixed-Temperature Spot Heat Detectors

Fixed-temperature heat detectors activate when the temperature exceeds a preset level. Detectors are made to activate at different levels. The most common temperature settings are 135°F, 190°F, and 200°F. The four types of fixed-temperature heat detectors are as follows:

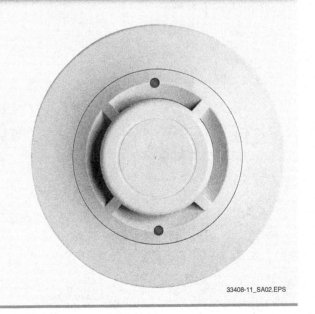

- *Fusible link* – The fusible link detector (*Figure 6*) consists of a plastic base containing a switch mechanism, wiring terminals, and a three-disc heat collector. Two sections of the heat collector are soldered together with an alloy that causes the lower disc to drop away when the rated temperature is reached. This moves a plunger that shorts across the wiring contacts and causes a constant alarm signal. After the detector is activated, a new heat collector must be installed to reset the detector to an operating condition.
- *Quick metal* – The operation of the quick metal detector (*Figure 7*) is very simple. When the surrounding air reaches the prescribed temperature (usually 135°F or 190°F), the quick metal begins to melt. This allows spring pressure within the device to push the top portion of the thermal element out of the way, causing the alarm contacts to close. After the detector has

been activated, either the detector or the heat collector must be replaced to reset the detector.
- *Bimetallic* – In a bimetallic detector (*Figure 8*), two metals with different rates of thermal expansion are bonded together. Heat causes the two metals to expand at different rates, which causes the bonded strip to bend. This action closes a normally open circuit, which signals an alarm. Detectors with bimetallic elements automatically reset when the temperature returns to normal.

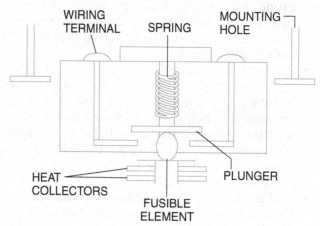

Figure 6 Fusible link detector.

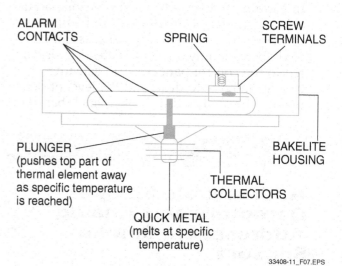

Figure 7 Quick metal detector.

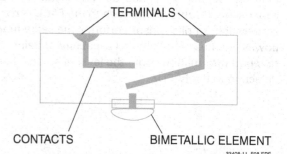
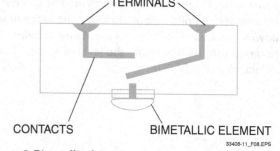

Figure 8 Bimetallic detector.

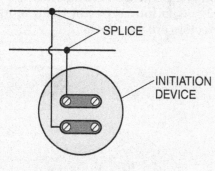
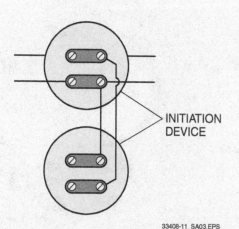

- *Thermistor* – Analog or analog-addressable heat detectors use a thermistor to sense heat. The resistance of a thermistor decreases as temperature increases. A circuit detects an increased current flow through a thermistor as the heat increases. The circuit sends a signal to the FACU that represents the temperature being monitored. The detectors reset when the temperature in the monitored space returns to normal. They can be powered from a two-wire circuit or by a separate power source in a four-wire circuit.

5.3.2 Combination Spot Heat Detectors

Combination spot heat detectors contain two types of heat detectors in a single housing. A fixed-temperature detector reacts to a preset temperature, and a rate-of-rise detector reacts to a rapid change in temperature even if the temperature reached does not exceed the preset level.

- *Rate-of-rise with fusible link detector* – (*Figure 9*) Rate-of-rise operation occurs when air in the chamber (1) expands more rapidly than it can escape from the vent (2). The increasing pressure moves the diaphragm (3) and causes the alarm contacts (4 and 5) to close, which results in an alarm signal. This portion of the detector automatically resets when the temperature stabilizes. A fixed-temperature trip occurs when the heat causes the fusible alloy (6) to melt, which releases the spring (7). The spring depresses the diaphragm, which closes the alarm contacts (4 and 5). If the fusible alloy melts, the center section of the detector must be replaced.

- *Rate-of-rise with bimetallic detector* – (*Figure 10*) Rate-of-rise operation occurs when the air in the air chamber (1) expands more rapidly than it can escape from the valve (2). The increasing pressure moves the diaphragm (3) and causes the alarm contact (4) to close. A fixed-temperature trip occurs when the bimetallic element (5) is heated, which causes it to bend and force the spring-loaded contact (6) to mate with the fixed contact (7).

- *Rate-of-rise and fixed-temperature thermistor detector* – Analog or analog-addressable heat detectors use dual thermistors to sense a rate-of-rise in temperature as well as a specific temperature. An electronic circuit sends a signal to the FACU when the rate of temperature rise that is sensed exceeds 15°F per minute and when a specific temperature has been sensed. These detectors self-restore when the space temperature returns to normal. They can be powered from a two-wire circuit or by a separate power source in a four-wire circuit.

5.3.3 Spot Heat Detector Ratings

As with all automatic detectors, heat detectors are also rated for a Listed spacing as well as a temperature rating (*Table 4*). The UL Listed spacing is typically 50' × 50'; however, for an FM listing, the spacing may be different. The spacing may also vary if the detector is a rate-of-rise or fixed-temperature type. The applicable rules and formulas for proper spacing (from the standards) are then applied to the Listed spacing. Caution is advised because it is difficult or impossible to distinguish the difference in the Listed spacing for two different types of heat sensors based on their appearance.

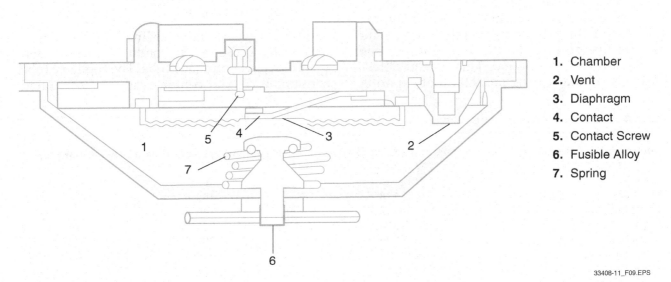

1. Chamber
2. Vent
3. Diaphragm
4. Contact
5. Contact Screw
6. Fusible Alloy
7. Spring

33408-11_F09.EPS

Figure 9 Rate-of-rise with fusible link detector.

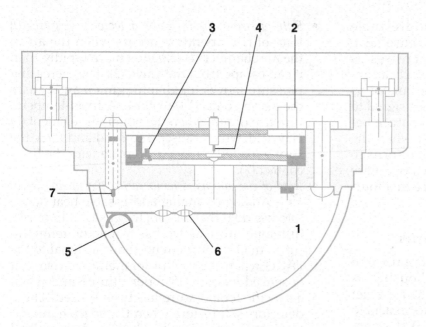

1. Air Chamber
2. Breather Valve
3. Diaphragm Assembly
4. Rate-of-Rise Contact
5. Bimetal Fixed-Temperature Element
6. Contact Spring
7. Fixed-Temperature Contact

33408-11_F10.EPS

Figure 10 Rate-of-rise with bimetallic detector.

The maximum ceiling temperature must be 20°F or more below the detector's rated temperature. The difference between the rated temperature and the maximum ambient temperature for the space should be as small as possible to minimize response time.

Heat sensors are not considered life safety devices. They should be used in areas that are unoccupied or in areas that are not suited for smoke detectors. To every extent possible, heat sensors should be limited to use as property protection devices.

5.3.4 Rate Compensation Detectors

A rate compensation detector (*Figure 11*) is a device that responds when the temperature of the surrounding air reaches a preset level, regardless of the rate of temperature rise. The detector also responds if the temperature rises quickly over a short period. This type of detector has a tubular casing of metal that expands lengthwise as it is heated. A contact closes as it reaches a certain length. These detectors are self-restoring. Rate compensation detectors are more complex than either fixed or rate-of-rise detectors. They combine the principles of both to compensate for thermal lag. When the air temperature is rising rapidly, the unit is designed to respond almost exactly at the point when the air temperature reaches the unit's rated temperature. It does not lag while it absorbs the heat and rises to that temperature. Because of their precision, rate compensation detectors are well suited for use in areas where thermal lag must be minimized.

5.3.5 Semiconductor Line-Type Heat Detectors

A restorable semiconductor line-type heat detector (*Figure 12*) uses a semiconductor material and a stainless steel capillary tube. The capillary tube contains a coaxial center conductor separated

Table 4 Heat Detector Temperature Ratings

Temperature Classification	Temperature Rating Range	Maximum Ceiling Temperature	Color Code
Low	100°F–134°F	80°F	No color
Ordinary	135°F–174°F	100°F	No color*
Intermediate	175°F–249°F	150°F	White
High	250°F–324°F	225°F	Blue
Extra high	325°F–399°F	300°F	Red
Very extra high	400°F–499°F	375°F	Green
Ultra high	500°F–575°F	475°F	Orange

*Gray ring around element for Fixed Temperature/Rate-of-Rise

33408-11_T04.EPS

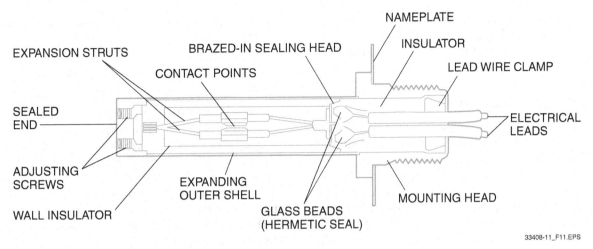

Figure 11 Rate compensation detector.

33408-11_F11.EPS

from the tube wall by a temperature-sensitive thermistor material. Under normal conditions, a small current (below the alarm threshold) flows. As the temperature rises, the resistance of the thermistor decreases. This allows more current to flow and initiates the alarm. When the temperature falls, the current flow decreases below the alarm threshold level, which stops the alarm. The wire is connected to controls that establish the alarm threshold level and sense the current flow. Some of these control devices can pinpoint the location in the length of the wire where the temperature change occurs. Line-type heat detectors are commonly used in cable trays, conveyors, electrical switchgear, warehouse rack storage, mines, pipelines, hangars, and other similar applications.

5.3.6 Fusible Line-Type Heat Detector

A non-restorable fusible line-type heat detector (*Figure 13*) uses a pair of steel wires in a normally open circuit. The conductors are held apart by heat-sensitive insulation. The wires, under tension, are enclosed in a braided sheath to make a single cable. When the temperature limit is reached, the insulation melts, the two wires contact, and an alarm is initiated. The melted and fused section of the cable must be replaced following an alarm to restore the system. The wire is available with different melting temperatures for the insulation. The temperature rating should be approximately 20°F above the ambient temperature. Special controls or modules are available that can be connected to the wires to pinpoint the fire location where the wires are shorted together.

5.4.0 Smoke Detectors

Photoelectric smoke detectors (*Figure 14*), which sense the products of combustion, and ionization detectors (*Figure 15*), which sense the presence of combustible gases, are two basic versions of spot-type smoke detectors. For residential use, both types are available combined in one detector. For commercial use, either type is also available combined with a heat detector. Newer commercial fire systems use addressable smoke detectors that signal an alarm, or analog addressable sensors, which do not signal an alarm. The analog addressable sensors return a signal that represents

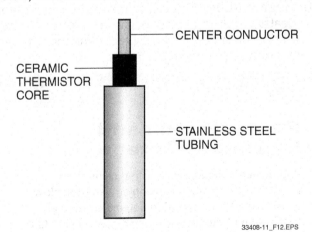

33408-11_F12.EPS

Figure 12 Restorable semiconductor line-type heat detector.

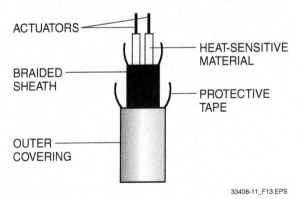

33408-11_F13.EPS

Figure 13 Non-restorable fusible line-type heat detector.

the level of detection to the FACU for further analysis. A smoke detector that uses photoelectric sensing and covers multiple areas is available as an aspirating smoke detection system.

5.4.1 Ionization Detectors

An ionization detector (*Figure 16*) uses the change in the electrical conductivity of air to detect smoke. An alarm is indicated when the amount of smoke in the detector rises above a certain level. The detector has a very small amount of radioactive material in the sensing chambers. As shown in *Figure 17A*, the radioactive material ionizes the air in the measuring and reference chambers, allowing the air to conduct current through the space between two charged electrodes. When smoke particles enter the measuring chamber, they impede current flow (*Figure 17B*). The detector activates an alarm when the conductivity decreases to a set level. The detector compares the current drop in the main chamber to the drop in the reference chamber. This allows it to avoid alarming when

Figure 14 Photoelectric smoke detector.

33408-11_F14.EPS

Figure 15 Ionization smoke detector.

33408-11_F15.EPS

the current drops due to surges, radio frequency interference (RFI), or other factors.

5.4.2 Photoelectric Smoke Detectors

Light-scattering detectors and beam detectors are two basic types of photoelectric smoke detectors. Light-scattering detectors use the reflective properties of smoke particles to detect the smoke. The light-scattering principle is used for the most common single-housing spot-type detectors. Beam detectors rely upon smoke to block enough light to cause an alarm (obscuration principle). Early spot-type smoke detectors also operated on the obscuration principle.

Light-scattering detectors are usually spot detectors that contain a light source and a photosensitive device arranged so that light rays do not normally fall on the device (*Figure 18*). When smoke particles enter the light path, the light hits the particles and scatters, hitting the photo sensor, which signals the alarm.

Projected beam smoke detectors that operate on the light obscuration principle (*Figure 19*) consist of a light source, a light beam focusing device, and a photosensitive device. Smoke obscures or blocks part of the light beam, which reduces the amount of light that reaches the photosensor, signaling an alarm. In some cases, mirrors are used to direct the beam in a desired path; however, the mirrors reduce the overall range of the detector. The prime use of the light obscuration principle is with projected beam-type smoke detectors that are employed in the protection of large, open high-bay warehouses, high-ceiling churches, and other large areas. A version of a beam detector was used in older style duct detectors.

5.4.3 Duct Detectors

After a number of fires where smoke spread through building duct systems and contributed to deaths, building codes began to require the installation of duct detectors. Duct detectors can be ionization or photoelectric types.

On Site

Photoelectric and Ionization Detectors

Photoelectric detectors are often faster than ionization detectors in sensing smoke from slow, smoldering fires. Ionization detectors are often better than photoelectric detectors at sensing fast, flaming fires.

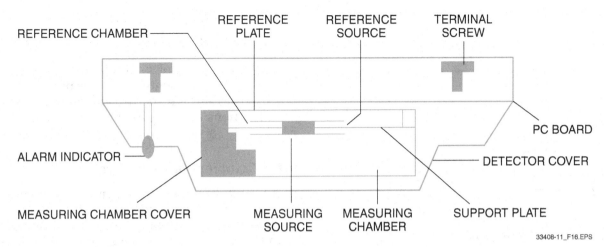

Figure 16 Ionization detector.

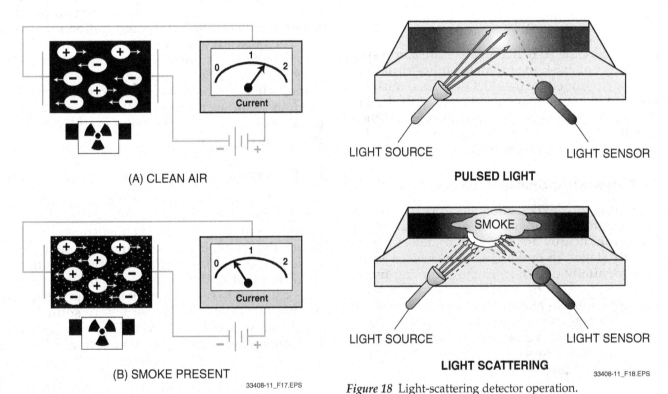

(A) CLEAN AIR

(B) SMOKE PRESENT

Figure 17 Ionization action.

PULSED LIGHT

LIGHT SCATTERING

Figure 18 Light-scattering detector operation.

Duct detectors enable a system to control the spread of smoke within a building by turning off the HVAC system, operating exhaust fans, closing doors, or pressurizing smoke compartments in the event of a fire. This prevents smoke and fumes from circulating through the ductwork. Because the air in the duct can be moving at a low or high speed, the detector must be able to sense smoke in either situation. A typical duct detector has a Listed airflow range within which it functions properly. It is not required to function when the duct fans are stopped. At least one duct detector is available with a small blower for use in very low airflow applications. Duct

detectors must not be used as a substitute for open area protection because smoke may not be drawn from open areas of the building when the HVAC system is shut down. Duct detectors can be mounted inside the duct with an access panel or outside the duct with sampling tubes protruding into the duct, as shown in *Figure 20*.

The primary function of a duct detector is to turn off the HVAC system to prevent smoke from being circulated. However, *NFPA 90A* requires that duct detectors be tied into a general fire alarm system if the building contains one. When tied into a fire alarm system, *NFPA 72®* and mechanical codes allow the duct detector to be a supervisory device

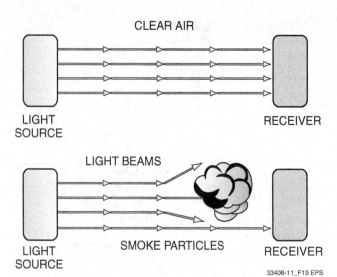

Figure 19 Light obscuration principle.

if allowed by the AHJ. If a separate fire alarm system does not exist, then remote audio/visual indicators, triggered by the duct detectors, must be provided in normally occupied areas of the building. Duct detectors that perform functions other than the shutdown of HVAC equipment must be supplied with backup power.

5.4.4 Aspirating Smoke Detection Systems

Aspirating smoke detection systems (*Figure 21*) use a single detector and one or more sampling tubes from nearby areas (zones) to detect fire at an early stage. An optional scanner causes air to be sequentially drawn from the sampling tubes by the aspirator. The sample air is then passed through a filter to remove dust and dirt before it enters the detection chamber. The second ultra-fine filter introduces clean air into the chamber to act as a barrier across the optical surfaces to pro-

tect them from contamination. The light source is usually a diode laser. These detectors are very sensitive and can measure very low to extremely high concentrations of smoke. By using a scanner, these systems can identify the zone of first occurrence to an FACU. If the fire spreads, additional zone locations are identified and reported to the FACU.

5.4.5 Combination Detectors

In addition to smoke, a fire produces heat and deadly carbon monoxide (CO). Heat or carbon monoxide alone can be deadly. In combination with smoke, either poses an even higher danger. To reduce this danger, combination smoke detectors are available. Typical combinations include smoke and heat, and smoke and carbon monoxide. The outward appearance of these combination detectors is similar to conventional smoke detectors. Installation is also similar.

Individual carbon monoxide detectors are also used. In accordance with *NFPA 72*®, signals from carbon monoxide detectors are permitted to be supervisory signals.

5.4.6 Multi-Criteria Detectors

NFPA 72® recognizes multi-criteria detectors as devices containing multiple sensors that separately respond to stimuli such as smoke, heat, CO, or fire gases. It has programmable sensitivity levels. If a sensor detects the presence of a condition that indicates a fire, the built-in software causes it to sample one of the other sensors in order to verify the threat before sounding an alarm. It is able to respond quickly, but at the same time, significantly reduce nuisance alarms.

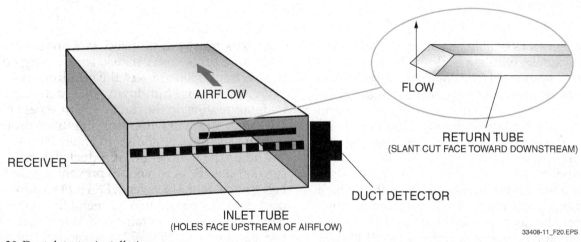

Figure 20 Duct detector installation.

Analog Addressable Laser Smoke Sensor

The spot-type analog addressable smoke sensor shown uses a microprocessor, diode laser, and precision optics to obtain an early warning capability comparable to aspiration technology. A highly focused beam minimizes reflection from accumulated dust in the optical chamber. In this particular sensor, built-in software performs independent analyses of the sensor's signal and includes dust spike rejection, drift compensation, smoothing, and multi-sensor alert and action levels. If spikes occur in the sensing circuit, this model can confirm the signal by checking the response from adjacent detectors to suppress nuisance alarms. In addition, the sensor's sensitivity level is programmable from the FACU. This model also has a fixed-temperature thermistor heat detector.

33408-11_SA05.EPS

5.5.0 Other Types of Detectors

This section describes special-application detectors. These include ultraviolet flame detectors and infrared flame detectors. In addition to those described here, CCTV cameras are also Listed as initiating devices.

5.5.1 Ultraviolet Flame Detectors

An ultraviolet (UV) flame detector (*Figure 22*) uses a solid-state sensing element, or a gas-filled tube. The UV radiation of a flame causes gas in the element or tube to ionize and become conductive. When sufficient current flow is detected, an alarm is initiated.

Figure 22 also shows typical sensitivity graphs for a detector. They show the response of a specific detector to various sizes of a gasoline fire-emitting ultraviolet radiant energy. Response time varies be-

cause of the fire size, distance to the detector, and type of fuel involved. The field of view for this radiant energy detector is also shown. The field of view for a radiant energy detector is defined by *NFPA 72*® as the cone that extends out from the detector along its center axis. The outer and furthest edges of the cone are defined as the point where the sensitivity of the detector is at least 50 percent of its on-axis, Listed, or approved sensitivity. Because these types of detectors are sensitive to any source of UV radiation, they are used indoors where direct sunlight or electric arc cutting and welding operations can be shielded or placed out of the detector's field of view to prevent nuisance alarms.

5.5.2 Infrared Flame Detectors

An infrared (IR) flame detector (*Figure 23*) consists of a filter and lens system that screens out

Disposal of Ionization Smoke Detectors or Alarms

Do not dispose of ionization type detectors or alarms in the trash. They contain trace amounts of hazardous radioactive material and must be disposed of in accordance with applicable federal, state, and local codes. Manufacturers must take these smoke detectors back for recycling.

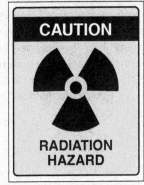

33408-11_SA04.EPS

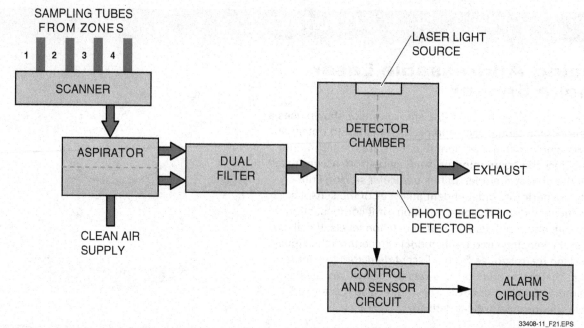

Figure 21 Simplified diagram of an aspirating smoke detection system.

unwanted radiant-energy wavelengths and focuses the incoming energy on light-sensitive components. These flame detectors can respond to the total IR content of the flame alone or to a combination of IR with flame flicker of a specific frequency. They are used indoors and have filtering systems or solar-sensing circuits to minimize nuisance alarms from sunlight.

The unit shown in *Figure 23* is a single-wavelength type designed to have peak sensitivity to radiant IR energy at the 4.45-micron wavelength. Hydrocarbon fires emit strong IR energy from excited carbon dioxide (CO_2) molecules in a specific frequency band. This band is called the CO_2 spike and is the major IR emission from hydro-carbon fires. The flicker effect of these fires is also used to help discriminate against unwanted responses by employing an electronic filter. This particular unit can be used indoors and outdoors because it has a relatively high immunity to nuisance alarms caused by hot objects, lightning, cutting and welding, X-rays, artificial light sources, and sunlight. Other units are available with dual-wavelength response in the CO_2 spike. The difference in intensity between the two wavelengths is used to help discriminate against false signals. Dual-wavelength units can be identified by the two sensors that are visible in the optical window.

Projected Beam Smoke Detectors

The transmitter and receiver for a projected beam and reflected beam smoke detector are shown here. Projected beam units are designed for use in atriums, ballrooms, churches, warehouses, museums, factories, and other large, high-ceiling areas where conventional smoke detectors cannot be easily installed.

33408-11_SA06.EPS

Duct Sensors

This is a typical analog addressable duct smoke sensor with the cover removed. This four-wire model for Class A circuits includes a tamper indicator that signals a trouble alarm after 20 minutes if the cover is not replaced after unit testing or maintenance. It also has two sets of 10-amp contacts operated by a relay to control connected fans, blowers, and dampers. A set of 2-amp relay-operated supervisory contacts is also provided.

33408-11_SA07.EPS

5.5.3 Fire-Gas Detectors

Fire-gas detectors sense the presence of various fire gases. The infrared detector shown in *Figure 23* is an indirect type of fire-gas detector. Fire-gas detectors designed for CO detection must not be confused with CO warning equipment used in homes to provide alerts for CO gas.

5.5.4 Video Image Smoke Detection

In remote locations, a video camera installed with a flame or smoke detector allows the alarm event to be viewed from a central monitoring station. The images can be used to help evaluate the extent of the fire so that an appropriate response can be generated. The images are also recorded to allow the event to be studied after the fact to help determine the cause of the fire. The cameras are designed to resist fire and explosions, and are often installed with multi-purpose sensors that can detect UV and/or IR. Video image smoke detection is covered in *NFPA 72®*, *Section 17.7.7*.

5.6.0 Manual Fire Alarm Boxes (Pull Boxes)

When required by code, manual pull boxes must be distributed throughout a commercial monitored area so that they are unobstructed, readily accessible, and in the normal path of exit from the area. Outdoor pull boxes must be rated for outdoor use. Examples of pull boxes are shown in *Figure 24*.

- *Single-action pull boxes* – A single action or motion operates these pull boxes. Pulling the handle that closes one or more sets of contacts generates the alarm.
- *Glass-break pull boxes* – In these devices, a glass rod, plate, or special element must be broken using a handle or hammer that is an integral part of the station. When the alarm is activated, one or more sets of contacts are closed and an alarm is actuated. Usually, the plate, rod, or element must be replaced to return the unit to service, although some boxes operate without the rod or plate.

Programmable Aspirating Smoke Detector Unit

The unit shown here operates on the obscuration principle using a laser light source. Over time, this reduces the unwanted or nuisance alarms caused by accumulating dust. It includes an optional programmer and display module. The unit can be locally programmed for sensitivity and has four configurable alarm levels (alert, action, fire 1 and 2, and day/night/weekend/holiday) and seven software-configurable relays (alert, action, fire 1, fire 2, maintenance, fault, and isolate).

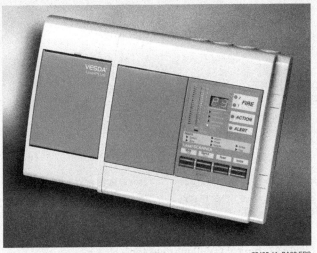

33408-11_SA08.EPS

- *Double-action pull boxes* – Double-action pull boxes require the user to lift a cover or open a door before operating the pull box. Two discretely independent actions are required to operate the box and activate the alarm.

- *Tamper-cover pull boxes* – This type of double-action pull box has a cover designed to discourage unauthorized alarms. An alarm sounds if the tamper cover is lifted. There have been cases in which people have confused the sound of the tamper alarm with the sound of fire alarm activation. The sounding of this tamper alarm has caused them to fail to activate the pull box.

- *Key-operated pull boxes* – Applications for key-operated pull boxes (*Figure 25*) are restricted. Key-operated boxes are permitted in certain occupancies where facility staff members may be in the immediate area and where use by other occupants of the area is not desirable. Typical situations would include certain detention and correctional facilities and some mental health care facilities.

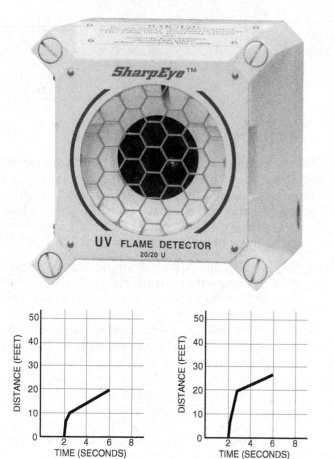

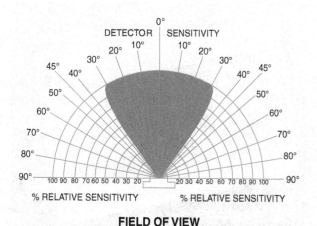

FIELD OF VIEW

33408-11_F22.EPS

Figure 22 Typical UV flame detector.

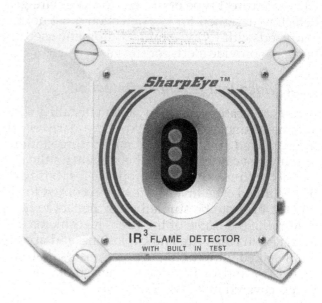

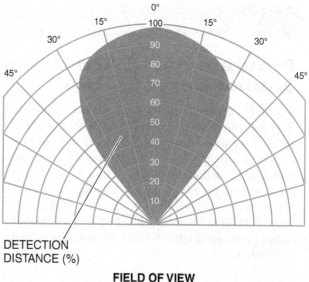

FIELD OF VIEW

33408-11_F23.EPS

Figure 23 IR flame detector.

| SINGLE-ACTION PULL BOX | GLASS-BREAK COVER FOR PULL BOX | TAMPER COVER PULL BOX | DOUBLE-ACTION PULL BOX |

33408-11_F24.EPS

Figure 24 Pull boxes.

5.7.0 Sprinkler System Fire Alarm and Supervision Equipment

This section describes various types of sprinkler system supervision and fire alert equipment. It covers wet and dry sprinkler systems and water flow alarms.

> **NOTE**
>
> Check the local jurisdiction for licensing required in order to operationally test sprinkler systems.

KEY LOCK
33408-11_F25.EPS

Figure 25 Key-operated pull box.

5.7.1 Wet Sprinkler Systems

A wet sprinkler system (*Figure 26*) consists of a permanently piped water system under pressure, using heat-actuated sprinklers. When a fire occurs, the sprinklers exposed to high heat open and discharge water to control or extinguish the fire. They are designed to automatically detect and control a fire and protect a structure. Once a sprinkler is activated, and depending on whether an alarm valve or shotgun riser valve is used, some type of water flow or pressure sensor signals a fire alarm. A problem with wet systems is that they can freeze if a power failure or loss of heat in the protected space occurs. Wet systems can be protected against freezing using antifreeze, heat tape, or temperature sensors.

Sprinkler pipes can be filled with a non-toxic antifreeze solution. If this is done, a backflow preventer must be installed in the piping to prevent the antifreeze from contaminating the potable water system. Thermostatically controlled electric heat tapes may be wrapped around or placed along sprinkler piping. This process is called heat tracing. The tape energizes if temperatures in the space approach freezing. Temperature sensors in

> **On Site**
>
> ## UV/IR Detectors
>
> Ultraviolet and infrared sensors are available as a combined detector. Microprocessor based signal processing within the detector allows discrimination against nuisance alarms by using both infrared and ultraviolet signals.

Carbon Monoxide Warning Detectors

These detectors are used to sense unacceptable levels, or level changes, of CO in a monitored, occupied space. Odorless CO is produced as a byproduct of combustion, either from a fuel-powered heating or cooking device or from an open fire. Most units have a multi-stage indicator warning system. It is important that these devices are installed in residential occupancies in addition to smoke alarms and that the devices meet *UL 2034*.

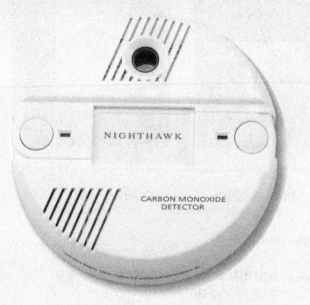

33408-11_SA09.EPS

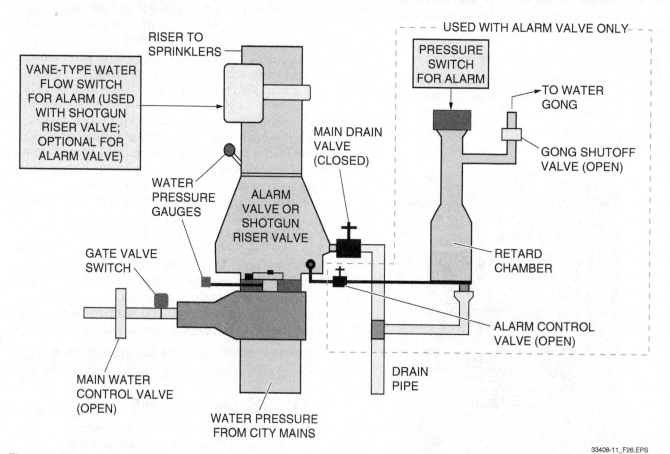

Figure 26 Wet sprinkler system.

33408-11_F26.EPS

Addressable Manual Pull Boxes

The addressable key-lock reset pull box shown here can be wired in parallel using a single SLC. Like other pull boxes, it is available as a single- or double-action box or as a glass-break or key-operated box. When a key-lock reset box is pulled, a latching mechanism is engaged that prevents the pull cover from being returned to the closed position. The only way the station cover can be reset to the closed position is by using the appropriate reset key.

33408-11_SA11.EPS

the protected space can signal if the temperature drops to unsafe levels. The device can be as simple as a normally open bimetal switch that closes on a drop in temperature. It can also be in the form of a thermistor that can measure actual temperature. The sensors can be tied into an alarm or security system to provide automatic alerts of a low temperature condition.

> **NOTE**
>
> Some codes prohibit heat tracing.

Glass-Break Pull Box

When installing a glass-break pull box, consider where the glass will go when broken and the danger it may pose to children and the disabled.

33408-11_SA10.EPS

5.7.2 Dry Sprinkler Systems

A dry sprinkler system (*Figure 27*) consists of heat-activated sprinklers that are attached to a piping system containing dry air or nitrogen under pressure. Normally, low air pressure in the pipes holds a differential dry-pipe water valve closed, keeping high-pressure water out of the piping system. When heat activates a sprinkler, the open sprinkler causes the air pressure to be released. This allows the water valve to open, and water flows through the pipes and out to the activated sprinkler. Once a sprinkler is activated and high-pressure water starts to flow, a water pressure switch signals a fire alarm. Water flow switches are not used in a dry sprinkler system. Because the pipes are dry until a fire occurs, these systems may be used in spaces subject to freezing.

Fire Extinguisher Monitoring

Fire extinguishers must be placed in an accessible location and must be charged to the correct pressure for proper operation. In the past, fire extinguishers were checked manually to ensure they were in compliance. The 2010 edition of *NFPA 72®* now allows fire extinguishers to be monitored electronically. Pressure is monitored by an electronic pressure gauge that sends pressure signals to an FACU. A tether switch signals the control unit if the extinguisher is removed from the wall. A photocell placed near the extinguisher signals the control unit if the extinguisher becomes obstructed.

5.7.3 Wet or Dry Sprinkler System Water Flow Alarms and Supervisory Systems

When a building sprinkler is activated by the heat generated by a fire, the sprinkler allows water to flow. As shown in *Figures 26* and *27*, water pressure switches or vane-type water flow switches are installed in the sprinkler system along with local alarm devices. The switches detect the movement or entry of water in the system. Activation of these switches causes an initiation signal to be sent to an FACU that signals a fire alarm.

In addition to devices that signal sprinkler system activation, other devices may be used to monitor the status of the system using an FACU. For instance, the position of a control valve may be monitored so that a supervisory signal is sent whenever the control valve is turned to shut off the water to the sprinkler system. If this valve is turned off, water cannot flow through the sprinkler system, which means the system is inactive.

In some systems, water pressure from the municipal water supply may not be strong enough to push enough water to all parts of a building. In these cases, an engine- or electric motor-powered fire pump is required. When the fire pump runs, a supervisory signal is sent indicating that the fire pump is activated. If the pump runs to maintain system pressure and does not shut down within a reasonable time, a site visit may be required. When water is scarce or unavailable, an on-site water tank may be required. Supervisory signals may be generated when the temperature of the water drops to a level that is low enough to freeze or if the water level or pressure drops below a safe level. In a dry system, a high/low pressure supervisory switch signals a trouble alarm if the air

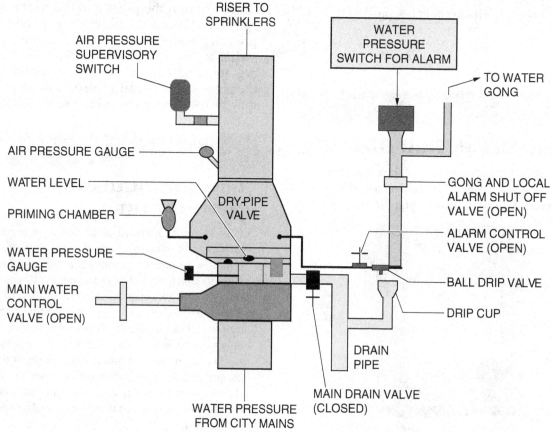

Figure 27 Dry sprinkler system.

33408-11_F27.EPS

Water Flow Detectors

This water flow detector is typical of those used with wet pipe sprinkler systems. Water flow switches are only used with wet pipe sprinkler systems. Water flow through the associated pipe deflects the detector's vane. This activates the internal switch contacts, initiating an alarm or auxiliary indication. Water flow in the pipe can be caused by the opening of one or more sprinklers because of a fire, the opening of a test valve, or a leaking or ruptured pipe.

33408-11_SA12.EPS

pressure exceeds or drops below a usable level. If fire pump power or fuel is monitored, lack of power, power phase reversal, or low fuel causes a supervisory signal.

5.7.4 Preaction Sprinkler Systems

Dry sprinkler systems have some disadvantages that can cause problems. The fire may be well along before there is enough heat to open the sprinkler. Once the sprinkler opens, the air in the pipes has to be purged before water can begin to flow. This purge time can allow the fire to grow. A preaction sprinkler system can overcome these problems. In a preaction system, heat, smoke, or flame detectors are used to detect the fire long before it builds to a point where it opens the sprinkler. When a fire is detected, a preaction valve opens to fill the dry pipes with water. Once the sprinklers open, water is there and ready to flow. Another advantage of the system is that the preaction valve prevents water flow if there is an accidental opening of a sprinkler. Types of preaction systems include non-interlock, single-interlock, and double-interlock.

A non-interlock system allows water to flow if a sprinkler is activated first or if a detector senses a fire. This system does not prevent water flow if a sprinkler is accidentally opened. A single-interlock system requires that a detector sense a fire and open the water valve before the sprinkler is activated. It provides some protection from accidental water discharges. A double-interlock system requires that the sprinkler must activate and the detector sense a fire before water can flow in

the system. If both conditions are not met, water does not flow. Preaction systems are used in areas where water damage from a damaged or defective sprinkler is not acceptable. Applications include museums, art galleries, computer rooms, and areas where paper archives are stored.

6.0.0 FIRE ALARM CONTROL UNITS

Different manufacturers produce fire alarm control units. *Figure 28* shows a typical intelligent, addressable control unit. It is a software-controlled unit capable of monitoring 1,980 individually identifiable and controllable detection/control points. It is field-programmable from the front panel keyboard, or remotely from a laptop computer.

The unit shown has a voice command center added for use in high-rise buildings. It provides for automatic evacuation messages, firefighter paging, and two-way communication to a central or supervising station through a telephone network. Although many control units may have unique features, all perform some specific basic functions. Control units detect problems through the sensor devices connected to them. They sound alerts or report these problems to a central or supervising location (*Figure 29*).

A control unit can allow the user to reprogram the system and, in some cases, to activate or deactivate the system or zones. The user may also change the sensitivity of detectors in certain zones of the system. On units that can be used for both intrusion and fire alarms, controls are provided to allow the user to enter or leave the monitored area with-

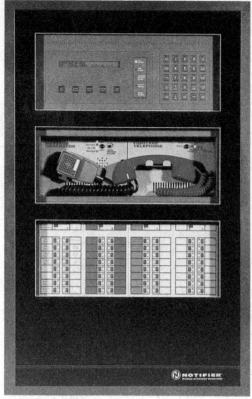

Figure 28 Intelligent, addressable FACU.

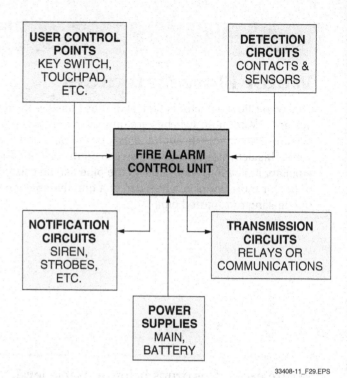

Figure 29 Fire alarm control unit inputs and outputs.

out setting off the intrusion detection system. In addition, controls allow some portions of the system, such as fire, panic, and holdup, to remain armed 24 hours a day, 365 days a year. Control units also provide the alarm user, responding authority, or inspector with a way to silence notification appliance circuits (NACs) or control other system features.

Control units are usually equipped with displays to communicate information, such as active zones or detectors, to an alarm user or monitor. These types of displays are usually included on the FACU, but can also be located remotely if required.

The control unit organizes all the components into a working and functional system. The control unit coordinates the actions the system takes in response to messages it gets from initiating de-

vices or, in some cases, notification devices that are connected to it. Depending on the inputs from the devices, it may activate the notification devices and may also transmit data to a remote location. In most cases, the control unit also conditions the power it receives from the building power system or from a backup power source so that it can be used by the fire alarm system.

6.1.0 User Control Points

User controls allow the alarm user to turn all or portions of the system on or off. User controls also allow the alarm user to monitor the system status, including the sensors or zones that are active, trouble reports, and other system parameters. The controls also allow the user to reset indicators of system events, such as alarms or indications of trouble with phone lines, equipment, or circuits. In today's systems, many devices can be used to control the system, including keypads, touch screens, telephones (including wireless phones), and computers.

6.1.1 Keypads

Two general types of keypads are in use today: alphanumeric (*Figure 30*) and LED. Either type can be mounted at the control or at a separate location. Both allow the entry of a numerical code to program various system functions. An alphanumeric keypad combines a keypad that is similar to a pushbutton telephone dial with an alphanumeric display that is capable of showing letters and numbers. LED keypads use a similar keypad, but display information by lighting small LEDs.

6.1.2 Touch Screens

Touch screens at the system allow the user to access multiple customized displays in graphic or menu formats. This enables rapid and easy interaction with the system. The displays provide alarm supervisory capability, and status and diagnostic features.

6.1.3 Telephone/Computer Control

Because most fire alarm systems are connected to a telephone network for alarm transmission purposes, and telephones are often in locations where user control devices have traditionally been located, some control unit manufacturers have incorporated ways for the telephone to function as a user interface. When telephones or cell phones are used as an interface for the system, feedback on system status and events is given audibly over the phone. Because the system is connected to the telephone network, a system with this feature can be monitored and controlled from anywhere a landline or cell phone call can be made. Most new systems can also be connected to a computer

33408-11_F30.EPS

Figure 30 Alphanumeric keypad and display.

either locally or via a telephone modem or a high-speed data line to allow control of events, receipt of information, and diagnostics using displays similar to those available at the system.

> **NOTE**
>
> *NFPA 72®* does not permit remote resetting of fire alarm systems.

6.2.0 FACU Initiating Device Circuits

An initiating circuit monitors the various types of initiating devices. A typical initiating circuit can monitor for three states: normal, trouble, and alarm. Initiating circuits can be used for monitoring fire devices, non-fire devices (supervisory), or devices for watch patrol or security. The signals from the initiating devices can be separately indicated at the premises and remotely monitored at a central or supervising station.

6.2.1 Initiating Circuit Zones

The term *zone* as it relates to modern fire alarm systems can have several definitions. Building codes restrict the location and size of a zone to enable emergency response personnel to quickly locate the source of an alarm. The traditional use of a zone in a conventional system was that each initiating circuit was a zone. Today, a single device or multiple initiating circuits or devices may occupy a zone. Generally, a zone may not cover more than one floor, exceed a certain number of square feet, or exceed a certain number of feet in length or width. In addition, some codes require different types of initiating devices to be on different initiating circuits within a physical area or zone. With modern addressable systems, each device is like a zone because each is displayed at the panel. It is common in addressable systems to group devices of several types into a zone. This is usually accomplished by programming the panel rather than by hardwiring the devices. Ultimately, the purpose of a zone is to provide the system monitor with information as to the location of a system alarm or problem. Local codes provide the specific guidelines.

6.2.2 Alarm Verification

To reduce nuisance alarms caused by temporary conditions only, *NFPA 72®* allows the use of alarm verification. To accommodate this, some FACUs allow for a delay in the activation of notification devices upon receiving an alarm signal from an initiating device. The code specifically restricts the use of this feature to situations where transient conditions or activities cause nuisance alarms. Several methods of verification are used. In conventional systems, verification can be of the reset-and-resample type. In this type of system, the panel resets the detector and waits a number of seconds, depending on programming, for the sensor to retransmit the alarm. If there is an actual fire, the sensor should detect it both times, and an alarm is then activated. In addressable systems, alarm verification can be of the wait-and-check type. In this case, the panel notes the initiation signal and then waits for the programmed time to see if the initiation signal remains constant before activating an alarm.

Another verification method for preaction alarms associated with deluge systems in large areas using multiple sensors is to wait until two or more sensors are activated before an alarm is initiated. In this method, known as cross zoning, a single detector activation may set a one minute or less pre-alarm condition warning signal at a manned monitoring location. If the pre-alarm signal is not cancelled within the allowed time, or if a second detector activates, a fire alarm is sounded. The pre-alarm signals are not required by the NFPA. In cross zoning, two detectors must occupy each space regardless of the size of the space, and each detector may cover only half of the normal detection area. Cross zoning, along with other methods of alarm verification, cannot be used on the same devices. Some local codes may prohibit alarm verification because of the time delay involved in reporting a fire.

When allowed or required, verification can be used in hotels, motels, and other institutions. It may also be used in monitored fire alarm systems for households to reduce unwanted or nuisance alarms.

A method similar to verification is used in some cases. This is called a positive alarm sequence feature. With this feature, alarm delay is usually adjustable for a period up to three minutes, but cannot exceed three minutes. This allows supervising or monitoring personnel to investigate the alarm. If a second detector activates during the investigation period, a fire alarm is immediately sounded.

6.2.3 FACU Zone or Address Identification

Effective system design and installation requires that zones or sensor addresses be labeled in a way that makes sense to all who use or respond to the system. In some cases, this may require two sets of labels—one for the alarm user and another for the police and fire authorities. A central or supervising station operator should be aware of both sets of labels, since he or she will be talking to both the police and the alarm user. Labeling for the alarm user is done using names familiar to the user (Johnny's room, kitchen, master bedroom, and so on). Labeling for the police and fire authorities is done from the perspective of looking at the building from the outside (first floor east, basement rear, second floor west, and so on).

On Site

Network Command Center

This PC-based network command center is used to display event information from local or wide area network devices in a text or graphic format. When a device initiates an alarm, the appropriate graphic floor plan is displayed along with operator instructions. Other capabilities of this command center include event history tracking and fire panel programming/control.

33408-11_SA13.EPS

6.3.0 Types of FACU Alarm Outputs

Most panels provide one or more of the following types of outputs:

- *Relay or dry contacts* – Relay contacts or other dry contacts are electrically isolated from the circuit controlling them, which provides some protection from external spikes and surges. Additional protection is sometimes needed. Always check the contact ratings to determine how much voltage and current the contacts can handle. Check with the manufacturer to determine if this provides adequate protection.
- *Built-in siren drivers* – Some household fire or burglary warning system units have a built-in siren driver. Built-in siren drivers use an amplifier circuit to drive a siren speaker. Impedance must be maintained within the manufacturer's specifications. Voltage drop caused by long wire runs composed of small-gauge conductors can greatly reduce the output level of the siren.
- *Voltage outputs* – Voltage outputs are not isolated from the control, and some protection may be required. Spikes and surges such as those generated by solenoid bells or the back EMF generated when a relay is de-energized can be a problem.
- *Open collector outputs* – Open collector outputs are outputs directly from a transistor and have limited current output. They are often used to drive a very low-current device or relay. Filtering may be required.

> **CAUTION**
>
> Do not overload these outputs, as the circuit board normally has to be returned to the manufacturer if an overload of even a very short duration occurs.

6.4.0 FACU Listings

It is mandatory that FACUs be Listed for the purpose for which they are being used. Common UL Listings for FACUs are *UL 864* (commercial fire alarm control units) and *UL 985* (household fire warning system units).

In combination fire/burglary panels, fire circuits must be on or active at all times, even if the burglary control is disarmed or turned off. Some codes or AHJs prohibit using combination intrusion and FACUs in commercial applications.

An FACU Listed for household use by UL may never be installed in any commercial application unless it also has the appropriate commercial approvals. The NFPA defines a household as a single- or two-family residential unit. This term applies to systems wholly within the confines of the unit. Except for monitoring, no initiation from, or notification to, locations outside the residence are permitted. Although an apartment building or condominium is considered commercial, a household system may be installed within an individual unit. However, any devices that are outside the confines of the individual unit such as manual pull boxes, must be connected to an FACU Listed as commercial.

> **NOTE**
>
> Devices with built-in verification cannot be combined in systems with FACU verification.

7.0.0 FACU PRIMARY AND SECONDARY POWER

Primary power for a fire alarm system and the FACU is normally provided by a utility. However, primary power for systems and panels can also be supplied from emergency uninterruptible backup primary power systems. To avoid service interruption, a fire alarm system may be connected on the line side of the electrical main service disconnect switch. The circuit must be protected with a circuit breaker or fuse no larger than 20A.

Secondary power for a fire alarm system can be provided by a battery or a battery with an approved backup generator as defined by the *NEC*®. Secondary power must be immediately supplied to the fire alarm system in the event of a primary power failure. As defined by *NFPA 72*®, standby time for operation of the system on secondary power must be no less than 24 hours for central, local, proprietary, voice communications, household, auxiliary, and remote systems. The battery standby time may be reduced if a properly configured battery-backup generator is available. During an alarm condition, the secondary power must operate the system under load for a specified minimum number of minutes after expiration of the standby period, depending on the type of system. Any batteries used for secondary power must be able to be recharged within 48 hours.

8.0.0 NOTIFICATION APPLIANCES

Notification to building occupants of the existence of a fire is the most important life safety function of a fire alarm system. The two primary types of notification are audible and visible. Concerns resulting from the Americans with Disabilities Act (ADA) have prompted the introduction of tactile (sense of touch) types of notification as well.

8.1.0 Visual Notification Devices

Strobes are high-intensity flashing lights. They can be separate devices or mounted on or near the audible device. Strobe lights are more effective in residential, industrial, and office areas where they don't compete with other bright objects. Because ADA requirements dictate clear or white xenon strobe lights, most interior visual notification devices are furnished with clear strobe lights (*Figure 31*). Strobe devices can be wall-mounted or ceiling-mounted and are usually combined with an audible notification device. When more than two strobes are visible from any location, the strobes must be synchronized to avoid random flashing, which can be disorienting and may cause seizures. Strobe units are rated in candelas (cd).

8.2.0 Audible Notification Devices

Audible alarm devices make noise. They include sirens, bells, or horns. In some cases, low-current chimes or buzzers are also used. Audible signals used for fire alarms in a facility that contains other sound-producing devices must produce

CEILING-MOUNTED

WALL-MOUNTED

33408-11_F31.EPS

Figure 31 Ceiling-mounted and wall-mounted strobe devices.

a unique sound pattern so that a fire alarm can be recognized. If more than one audible signal is used in a facility, they must be synchronized to maintain any sound pattern. *NFPA 72®* specifies that a standard evacuation signal known as Temporal 3 is required for those systems in which evacuation from the building or relocation is required, including household systems. This signal has three ½-second tones, a pause, and then a repeat of the pattern until the alarm is manually reset. Only a fire alarm may use this signal. However, a different evacuation signal may be used if approved by the AHJ.

Strobe lights are often combined with a bell (*Figure 32*). These types of solenoid-operated bells can also produce electrical noise, which interferes with controls unless proper filters are used. Bells draw relatively high current, typically 750 to 1,500 milliamps (mA). Another type of bell is a motor-driven bell. In this type of bell, a motor drives the clapper and produces louder sounds than a solenoid-operated bell. Electrical interference is eliminated and less current is required to operate the bell.

Self-contained sirens are combinations of speakers and sound equipment. They produce siren sounds and can be used for voice announcements. Self-contained siren packaging saves installation time. If more than one siren is required, they must be synchronized.

A horn with a strobe light is shown in *Figure 33*. A buzzer uses less power and consists of a vibrating membrane like a horn. Chimes are electronic devices that have a very low current draw. In supervised notification circuits, all devices are polarized, allowing current to flow in one direction only.

8.3.0 Voice Evacuation Systems

With a voice evacuation system (*Figure 34*), building occupants can be given instructions in an emergency. A voice evacuation system can be used with an FACU or as a stand-alone unit with

Figure 32 Bell with a strobe light.

Figure 33 Horn with a strobe light.

a built-in power supply and battery charger. Voice announcements (*Figure 35*) can be made to inform occupants about the problem or how to evacuate. In most cases, the voice announcements are pre-recorded and selected as required by the system. Live announcements can also be made from a microphone located at the FACU or at a remote panel. *Figure 36* shows several types of speakers used for

voice evacuation systems. Temporal 3 signaling is not used with zoned voice evacuation systems.

Codes require that voice systems have adequate clarity and audibility. Sound clarity is defined as freedom from distortion of all kinds by *IEC 60849, Sound Systems for Emergency Purposes*.

8.4.0 Signal Considerations

The types of code-authorized signals supplied to various notification devices from a typical fire control unit follow. Signal considerations for notification devices are also provided:

- *General signal* – General signals operate throughout the entire building. Evacuation signals require a distinctive signal. The code requires that a Temporal 3 signal pattern be used that is in accordance with *ANSI/ASA S3.41, American National Standard Audible Emergency Evacuation Signal,* and *ISO 8201, NFPA 101®,* and *NFPA 72®*. Temporal 3 signals consist of three short ½-second tones with ½-second pauses between the tones. This is followed by a 1-second silent period, and then the process is repeated.
- *Attendant signal* – Attendant signaling is used where assisted evacuation is required due to such factors as age, disability, and restraint. Such systems are commonly used in conjunction with coded chimes throughout the area or a coded voice message to advise people of

Figure 34 Voice evacuation system.

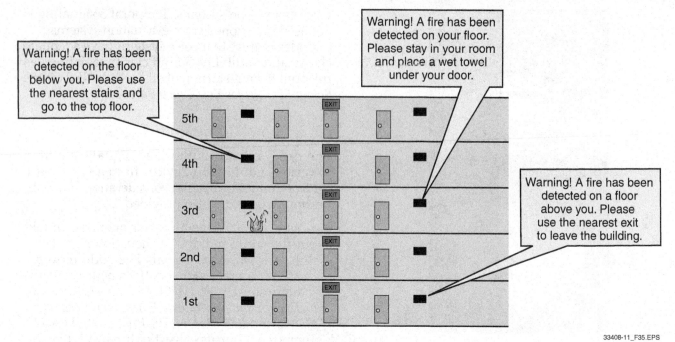

Figure 35 Voice evacuation messages.

33408-11_F35.EPS

Figure 36 Examples of speakers.

33408-11_F36.EPS

the location of the alarm source. A coded message might be announced such as, "Mr. Green, please report to the third floor nurses' station." No general alarm needs to be sounded, and no indicating appliances need to be activated throughout the area. The attendant signal feature requires the approval of the AHJ.

- *Pre-signal* – A pre-signal is an attendant signal with the addition of human action to activate a general signal. It is also used when the control delays the general alarm by more than one minute. The pre-signal feature requires the approval of the AHJ. A signal to remote locations must activate upon an initial alarm signal.
- *Positive alarm sequence* – When an alarm is initiated, the general alarm is not activated for

15 seconds. If a staff person acknowledges the alarm within the 15-second window, the general alarm is delayed for three minutes so that the staff can investigate. Failure to manually acknowledge the initial attendant signal automatically causes a general alarm. Failure to abort the acknowledged signal within 180 seconds also automatically causes a general alarm. If another automatic detector is activated during the delay window, the system immediately causes a general alarm. An activation of a manual station automatically causes a general alarm. This technique prevents nuisance general alarms. However, trained personnel are an integral part of these systems. The positive alarm sequence feature requires the approval of the AHJ.

- *Voice evacuation* – Voice evacuation may be either live or prerecorded and automatically or manually initiated. Voice evacuation systems are a permitted form of general alarm and are required in certain occupancies, as specified in the applicable chapter of *NFPA 101®*. Voice evacuation systems are often zoned so that only the floors threatened by the fire or smoke are immediately evacuated. Zoned evacuation is used where total evacuation is physically impractical. Temporal 3 signaling is not used with zoned evacuation. Voice evacuation systems must maintain their ability to communicate even when one or more zones are disabled due to fire damage. Buildings with voice evacuation usually include a two-way communication system (fireman's phone) for emergency communications. High-rise buildings are required to have a voice evacuation system.
- *Alarm sound levels* – Where a general alarm is required throughout the premises, audible signals must be clearly heard above maximum ambient noise under normal occupancy conditions. Public area audible alarms should be 75dBA at 10' and a maximum of 110dBA at minimum hearing distance. Public area audible alarms must be at least 15dBA above ambient sound level, or a maximum sound level that lasts over 60 seconds measured at 5' above finished floor (AFF) level. Audible alarms to alert persons responsible for implementing emergency plans (guards, monitors, supervisory personnel, or others) must be between 45dBA and 110dBA at minimum hearing distance. If an average sound level greater than 105dBA exists, the use of a visible signal is required. Typical average ambient sound levels from *NFPA 72®* are given in *Table 5*.

On Site

Digital Voice Messages

Digital voice messages used with various voice evacuation systems are stored in nonvolatile read-only memory integrated circuit chips. These stored messages are used in applications where clear and defined messages of evacuation are required. One major fire alarm manufacturer of voice evacuation systems has more than 100 pre-stored evacuation messages available.

Closed doors reduce the dB levels below those required to wake children or the hearing-impaired. Air conditioners, room humidifiers, and similar equipment may cause the noise levels to increase (55dBA typical). Sirens, horns, bells, and other indicating devices (appliances) may fail or be disabled by a fire. To reduce the effect of these disabled devices, multiple fire sounders should be considered. This may also provide ample dB levels to wake sleeping occupants throughout the structure (75dBA at the pillow is commonly accepted as sufficient to wake a sleeping person).

As specified in *NFPA 72®*, residential alarms installed in sleeping areas must have a sound level of at least 15dB above the average ambient sound level or 5dB above the maximum sound level with a duration of 60 seconds, or a sound level of at least 75 dBA, whichever is greater. These values are measured at the pillow level in the area being served. Sound intensity doubles with each 3dB gain and is reduced by one-half with each 3dB loss. Doubling the distance to the sound source causes a 6dB loss. Some other loss considerations are given in *Table 6*.

- *Coded versus non-coded signals* – A coded signal is a signal that is pulsed in a prescribed code for each round of transmission. For example, four pulses would indicate an alarm on the fourth floor. Temporal 3 is not a coded signal

---

Table 5 Typical Average Ambient Sound Levels

Area	Sound Level (dBA)	Area	Sound Level (dBA)
Mechanical rooms	85	Educational occupancies	45
Industrial occupancies	80	Underground structures	40
Busy urban thoroughfares	70	Windowless structures	40
Urban thoroughfares	55	Mercantile occupancies	40
Institutional occupancies	50	Places of assembly	55
Vehicles and vessels	50	Residential occupancies	35
Business occupancies	45	Storage	30

33408-11_T05.EPS

Table 6 Typical Sound Loss at 1,000Hz

Area	Loss (dBA)
Stud wall	41
Open doorway	4
Typical interior door	11
Typical fire-rated door	20
Typical gasketed door	24

33408-11_T06.EPS

and is only intended to be a distinct, general fire alarm signal. A non-coded signal is a signal that is energized continuously by the control. It may pulse, but the pulsing is not be designed to indicate any code or message.

• *Visual appliance signals* – Notification signals for occupants to evacuate must be made by audible and visible signals in accordance with *NFPA 72®* and *ANSI A117.1, Accessible and Useable Buildings and Facilities*. However, there may or may not be exceptions to this rule. Under existing codes, only audible signals are required where one of the following conditions exist:

  - No hearing-impaired occupant is ever present under normal operation
  - In hotels and apartments where special rooms are made available to the hearing-impaired
  - Where the AHJ approves alternatives to visual signals (ADA regulations may or may not allow these exceptions)

9.0.0 COMMUNICATIONS AND MONITORING

Communication is a means of sending information to people who are too far away to directly see or hear a fire alarm system's notification devices. It is the transmission and reception of information from one location, point, person, or piece of equipment to another. Understanding the information that a fire alarm system sends makes it easier to determine what is happening at the alarm site. Knowing how that information gets from the alarm site to the monitoring site is helpful if a problem occurs somewhere in between.

9.1.0 Monitoring Options

There are several options for monitoring the signals of an alarm system. They include the following:

• *Supervising station* – A location where operators monitor receiving equipment for incoming fire alarm system signals. The supervising station

may be a part of the same company that sold and installed the fire alarm system. It is also common for the installing company to contract with another company to do the monitoring on its behalf.

• *Proprietary* – A facility similar to a central station except that the notification devices are located in a constantly staffed room maintained by the property owner for safety operations. The staff may respond to alarms, alert local fire departments when alarms are activated, or both.

• *Certified central station* – Monitoring facilities that are built and operated according to a standard and are inspected by a listing agency to verify compliance. Several organizations, including UL and FM Global, publish criteria and List central stations that conform to those criteria.

9.2.0 Communications Methods

Unless a fire alarm system is strictly local in nature, it must possess some means of communicating alarms to monitoring personnel or other individuals or organizations outside the protected area. There are a number of different methods for communicating this information (*Figure 37*), including the following:

• Internet
• Multiplex
• Fiber optic
• Long-range radio or satellite
• Digital
• Cellular/digital wireless backup

9.2.1 Internet

A fire alarm system may make use of a facility's Internet connection to communicate information through a computer to a computer monitor. One of the benefits of this method of communication is that monitoring personnel can check the status of the system from any computer that has Internet access.

9.2.2 Multiplex Systems

Multiplexing is the transmission of multiple pieces of information over the same communications channel. Thousands of signals can be sent over a single transmission medium. Transmission media include radio waves, metal wires, or fiber optics. The three types of multiplex systems are frequency division multiplexing (FDM), time division multiplexing (TDM), and wavelength division multiplexing (WDM). These three systems are described in the *Wireless Communication* module.

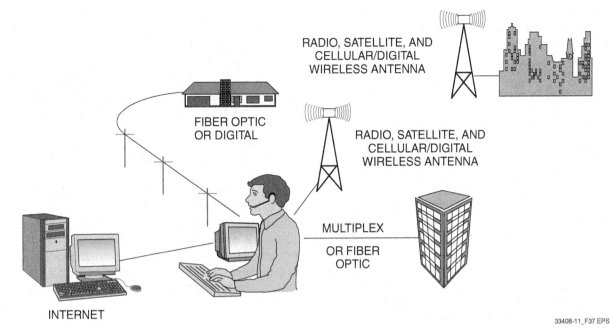

RADIO, SATELLITE, AND CELLULAR/DIGITAL WIRELESS ANTENNA

FIBER OPTIC OR DIGITAL

RADIO, SATELLITE, AND CELLULAR/DIGITAL WIRELESS ANTENNA

MULTIPLEX

OR FIBER OPTIC

INTERNET

33408-11_F37.EPS

Figure 37 Typical communications methods.

Polling can be done with multiplexing. In TDM, polling is used to control the communication between the alarm sites and the monitoring site. Each alarm site is polled or told when it can send its information at regular intervals. Computers can poll each site many times each second, and the alarm sites are set up to retain their information until it is their turn. The individual channels of FDM also can be polled at regular intervals. If a particular alarm site fails to respond to the poll, the monitoring site may be set up to indicate this as a communication failure. Monitoring equipment can be set up not only to indicate if the signal is present, but also to indicate if that signal is weak.

9.2.3 Fiber Optic Transmission

Glass fiber optic cables transmit large quantities of information over long distances. They are immune to most types of interference and do not conduct electricity. Fiber optic cables do not corrode, are unaffected by most chemicals, do not radiate electrical energy, and can handle a large bandwidth of signals. Fiber optics cannot be used for direct-wire or direct-connect systems. For more information, refer to the *Fiber Optics* module.

9.2.4 Long-Range Radio or Satellite Systems

Long-range radio or satellite systems allow equipment in the protected area to communicate signals to the monitoring site through the air. The electrical signal from the system's control panel is combined with a carrier signal. An altered or modulated signal produces radio

waves at the antenna of the system's transmitter. A radio receiver at the monitoring site, tuned to the frequency of the carrier wave, recreates the original transmitted information. Signals may be sent directly to the monitoring site, or relayed through a repeater or a satellite system. *NFPA 72*® defines this type of system as a digital alarm radio system (DARS). It contains a digital alarm radio transmitter (DART) and digital alarm communicator receiver (DACR). If the DARS is a backup communication system, a fire alarm system's digital alarm communicator must try to transmit over a telephone network a number of times in addition to transmitting via the DARS. For more information on radio systems, refer to the *Wireless Communication* module.

9.2.5 Digital Communication

Digital alarm communicator transmitters (DACTs) and digital alarm communicator receivers (DACRs) use standard telephone lines or wireless telephone service to send and receive data. Costs are low with this method because existing voice lines may be used, eliminating the need to purchase additional communication lines. Standard voice-grade telephone lines are also easier to repair than other communication links. For additional information on other digital communication protocols and systems, refer to the *Buses and Networks* and *Wireless Communication* modules.

The DACT is connected to a standard, voice-grade telephone line through a special connecting device called the RJ-31X (*Figure 38*). The RJ-31X is a Listed modular telephone jack into which a cord

---

from the DACT is plugged. The RJ-31X separates the telephone company's equipment from the fire alarm system equipment and is approved by the Federal Communications Commission (FCC).

Although using standard telephone lines has several advantages, problems may arise if the customer and the fire alarm system both need the phone at the same time. The DACT can control or seize the line whenever it needs to send an alarm signal. If the customer is using the telephone when the alarm system needs to send a signal, the DACT disconnects the customer until the signal has been sent. Once the signal is sent, the customer's phones are reconnected.

The typical sequence that occurs when the DACT for an alarm system is activated is shown in *Figure 39*. When an alarm is to be sent, the digital communicator energizes the seizure relay. The activated relay disconnects the house phones and connects the communicator to the telephone network. After the signal is sent and the DACT detects a signal called the kiss-off tone, it de-energizes the seizure relay. This disconnects the DACT and reconnects the house phones.

When an intrusion and fire alarm system are incorporated into a total system, it is important to remember that in the event of both an intrusion alert and a fire alarm, the fire alarm signal supersedes the intrusion alert signal, even though the intrusion may occur first.

The RJ-31X is a modular connection and, like a standard telephone cord, it can unplug easily. If the cord is unplugged from the jack, the digital communicator is disconnected from the telephone network and a local trouble alarm or indication occurs. However, even if the alarm system is disconnected from the telephone network, the local notification appliances are activated in the event of a fire.

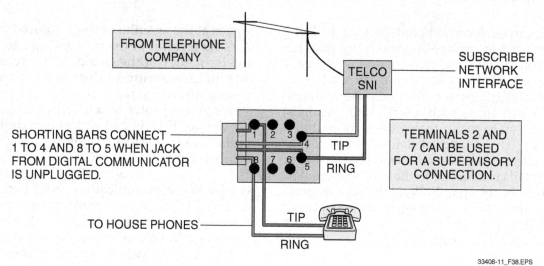

Figure 38 RJ-31X connection device.

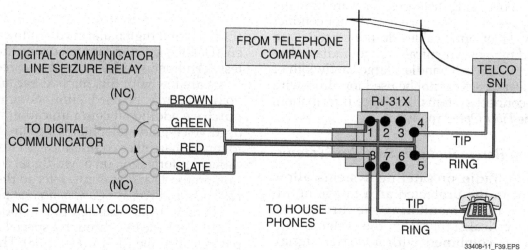

Figure 39 Line seizure.

The NFPA uses the following terms to refer to digital communications:

- *Digital alarm communicator receiver (DACR)* – This is a system component that accepts and displays signals from DACTs sent over the public switched telephone network.
- Digital alarm communicator system (DACS) – This is a system in which signals are transmitted from a DACT (located in the secured area) through the public switched telephone network to a DACR.
- *DACT* – This is a device that sends signals over the public switched telephone network to a DACR.

The following conditions have been established by *NFPA 72*® with regard to digital communicators:

- They can be used as a remote supervising station fire alarm system when acceptable to the AHJ.
- Only loop start and not ground start lines can be used.
- The communicator must have line-seizure capability.
- A failure-to-communicate signal must be shown if 10 attempts are made without getting through.
- They must connect to two separate phone lines at the protected premises. *Exception*: The secondary line may be a radio system or an Internet system, among others (this does not apply to household systems).
- Failure of either phone line must be annunciated at the premises within four minutes of the failure.
- If standard voice-grade long distance telephone service, including wide area telephone service (WATS) is used, the second telephone number shall be provided by a different long distance provider, where available.
- Each communicator shall initiate a test call to the central station at least once every 24 hours (this does not apply to household systems).

9.2.6 Cellular/Digital Wireless Backup

Some fire alarm systems that rely on phone line for communications have a cellular or digital wireless backup system that uses wireless phone technology to restore communications in the event of a disruption in normal telephone line service (*Figure 40*).

10.0.0 GENERAL INSTALLATION GUIDELINES

This section contains general installation information applicable to all types of fire alarm systems. For specific information, always refer to the manufacturer's instructions, the building drawings, and all applicable codes.

10.1.0 General Wiring Requirements

NEC Article 760 specifies the wiring methods and special cables required for fire protective signaling systems. The following special cable types are used in protective signaling systems; however, equivalent CM or MP cable may be substituted:

- Power-limited fire alarm (FPL) cable
- Power-limited fire alarm riser (FPLR) cable
- Power-limited fire alarm plenum (FPLP) cable
- Nonpower-limited fire alarm (NPLF) circuit cable
- Nonpower-limited fire alarm riser (NPLFR) circuit riser cable
- Nonpower-limited fire alarm plenum (NPLFP) circuit cable

In addition to *NEC Article 760*, the following *NEC*® articles cover other items of concern for fire alarm system installations. Some of these are briefly covered later in this module.

- *NEC Section 110.11, Deteriorating Agents*
- *NEC Section 300.6, Protection Against Corrosion and Deterioration*
- *NEC Section 300.21, Spread of Fire or Products of Combustion*

On Site

Telephone Line Problems

One method of determining if the fire alarm system has caused a problem with the telephone lines is to unplug the cord to the RJ-31X. This disconnects the fire alarm system and restores the connections of the telephone line. If the customer's telephones function properly when the cord is unplugged, the fire alarm system may be causing the problem. If the problem with the telephone line persists with the cord to the RJ-31X unplugged, the source of the problem is not the alarm system.

 33408-12 Fire Alarm Systems

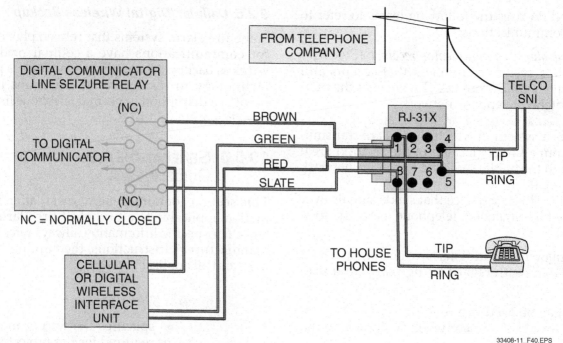

Figure 40 Cellular or digital wireless backup system.

33408-11_F40.EPS

- *NEC Section 300.22, Wiring in Ducts Not Used for Air Handling, Fabricated Ducts for Environmental Air, and Other Spaces for Environmental Air (Plenums)*
- *NEC Articles 500 through 516 and 517, Part IV, Hazardous (Classified) Locations, Classes I, II, and III, Divisions 1 and 2*
- *NEC Article 725, Class 1, 2, and 3 Remote-Control, Signaling, and Power-Limited Circuits*
- *NEC Article 760, Fire Alarm Systems*
- *NEC Article 770, Optical Fiber Cables and Raceways*
- *NEC Article 800, Communications Circuits*
- *NEC Article 810, Radio and Television Equipment*

Some AHJs may specify requirements that modify or add to the *NEC®*. It is essential that a person or firm engaged in fire alarm work be thoroughly familiar with the *NEC®* requirements, as well as any local requirements for fire alarm systems.

10.2.0 Workmanship

Fire alarm circuits must be installed in a neat and workmanlike manner. Cables must be supported by the building structure in such a manner that the cables are not damaged by normal building use. One way to determine accepted industry practice is to refer to nationally recognized standards, such as *ANSI/EIA/TIA 568, Commercial Building Telecommunications Cabling Standard; ANSI/EIA/TIA 569, Commercial Building Standard for Telecommunications Pathways and Spaces;* and

ANSI/EIA/TIA 570, Residential and Light Commercial Telecommunications Cabling Standard.

10.3.0 Access to Equipment

Access to equipment must not be blocked by wires and cables or suspended ceiling panels that prevent access to controls and removal of equipment panels.

10.4.0 Fire Alarm Circuit Identification

Fire alarm circuits must be identified at the control and at all junctions as fire alarm circuits. Junction boxes must be clearly marked as fire junction boxes to prevent confusion with commercial light and power. Check the marking requirements before starting work. The following are examples of some AHJ-acceptable markings:

- Red painted cover
- The words Fire Alarm on the cover
- Red painted box
- The word Fire on the cover
- Red stripe on the cover

10.5.0 Power-Limited Circuits in Raceways

Power-limited fire alarm (PLFA) circuits must not be run in the same cable, raceway, or conduit as high-voltage circuits. Examples of high-voltage circuits are electric light, power, and nonpower-limited fire (NPLF) circuits. When FPL cables must be run in the same junction box, they must be run in accordance with the *NEC®*, including maintaining a ¼" spacing from Class 1, power, and lighting circuits.

10.6.0 Mounting of Detectors

Observe the following precautions when mounting detectors:

- Circuit conductors are not supports and must not be used to support the detector.
- Plastic masonry anchors should not be used for mounting detectors to gypsum drywall, plaster, or drop ceilings.
- Toggle or winged expansion anchors should be the minimum used for gypsum drywall.
- The best choice for mounting is an electrical box. Most equipment is designed to be fastened to a standard electrical box.

All fire alarm devices should be mounted to the appropriate electrical box as specified by the manufacturer. If not mounted to an electrical box, fire alarm devices must be mounted by other means as specified by the manufacturer. Always read and follow the manufacturer's installation instructions to meet the requirements of the UL Listing.

10.7.0 Outdoor Wiring

Fire alarm circuits extending beyond one building are governed by the *NEC®*. The *NEC®* sets the standards on the size of cable and methods of fastening required for cabling. Some manufacturers prohibit any aerial wiring. The *NEC®* also specifies clearance requirements for cable from the ground. Overhead spans of open conductors and open multi-conductor cables of not over 600V, nominal, must be at least 10' (3.05m) above finished grade, sidewalks, or from any platform or projection from which they might be reached, where the supply conductors are limited to 150V to ground and accessible to pedestrians only. Additional requirements apply for areas with vehicle traffic. In addition, the *NEC®* states that fire alarm wiring can be attached to the building, but it must be kept not less than 3' (914mm) from the sides, top, bottom, and front of windows that are designed to be opened; doors; porches; balconies; ladders; stairs; fire escapes; and similar locations. However, conductors running above the top level of a window can be less than 3' (914mm) from the top of the window.

10.8.0 Fire Stopping

Electrical equipment and cables must not be installed in a way that might help the spread of fire. The integrity of all fire-rated walls, floors, partitions, and ceilings must be maintained. An approved sealant or sealing device must be used to fill all penetrations. Any wall that extends from the floor to the roof or from floor-to-floor should be considered a firewall. In addition, raceways and cables that go from one room to another through a fire barrier must be sealed with fire-stopping.

10.9.0 Wiring in Air Handling Spaces

Wiring in air handling spaces requires the use of approved wiring methods, including the following:

- Special plenum-rated cable (FPLP)
- Flexible metal conduit (sometimes called BX™)
- Electrical metallic tubing (EMT)
- Intermediate metallic conduit (IMC)
- Rigid metallic conduit (hard wall or Schedule 80 conduit)

Standard cable tie straps are not permissible in plenums and other air handling spaces. Ties must be plenum rated. Bare solid copper wire used in short sections as tie wraps may be permitted by most AHJs. Fire alarm equipment is permitted to be installed in ducts and plenums only to sense the air. All splices and equipment must be contained in approved fire-resistant and low smoke-producing boxes.

10.10.0 Wiring in Hazardous Locations

The *NEC®* includes requirements for wiring in hazardous locations. Particular attention to *NEC Articles 500 through 505* is required to prevent improper installation of fire alarm systems in Class I, II, and III environments. Class I environments consist of areas where flammable liquids, gases, or vapors occur or are processed. Class II environments consist of areas where various combustible dust hazards occur. Class III environments consist of areas where various combustible fiber hazards occur. Death can result if fire alarm systems are improperly installed in these environments. In addition to the requirements of *NEC Articles 500 through 505*, some specific areas that are considered hazardous and that have additional specific requirements are as follows:

- *NEC Article 511*, *Commercial Garages, Repair and Storage*
- *NEC Article 513*, *Aircraft Hangars*

- *NEC Article 514, Motor Fuel Dispensing Facilities*
- *NEC Article 515, Bulk Storage Plants*
- *NEC Article 516, Spray Application, Dipping, and Coating Processes*
- *NEC Article 517, Health Care Facilities*
- *NEC Article 518, Assembly Occupancies*
- *NEC Article 520, Theaters, Audience Areas of Motion Picture and Television Studios, Performance Areas, and Similar Locations*
- *NEC Article 545, Manufactured Buildings*
- *NEC Article 547, Agricultural Buildings*

10.11.0 Wet or Corrosive Environments

A minimum of ¼" spacing must be maintained between an exterior wall and the equipment or conduit. Conduits and cables must be Listed for the environment in which they are used. Outdoor-rated boxes must be used outdoors and in wet or corrosive locations.

10.12.0 Wiring Protection

Wiring that may be damaged by exposure or by being buried underground requires protection. That protection is described in the following sections.

10.12.1 Exposed Wiring

Protection for exposed wiring may be provided by conduit or as part of the building construction. Where wiring is installed exposed, cables must be adequately supported and installed in such a way that maximum protection against physical damage is afforded by the building construction. This includes baseboards, door frames, and ledges. Protection must also be provided where cable is located within 7' (2.13m) of the floor. Cables must be securely fastened in an approved manner at intervals of not more than 18" (457mm), except where fished through enclosed walls, ceilings, or floors.

10.12.2 Underground Wiring

Only conduit that will not be damaged by corrosion can be used for direct burial. Plain steel EMT must not be used for direct burial. Use PVC, rigid (galvanized), or direct-burial (UG) cable (*Figure 41*).

10.13.0 Fire Pumps

The powered control wiring between the start relay and the fire alarm is governed by *NEC Article 760* and must be supervised. The wiring between the start relay and the fire pump controller is governed by *NEC Article 695*. Any short or open of these wires must not prevent any other method of starting the fire pump, although it may cause the fire pump to start and continuously run (fail safe). This should be tested by removing a load-side wire from the start relay. The fire pump must either start and run or be able to be started by all other provided methods.

10.14.0 Remote Control Signaling Circuits

Building control circuits (*Figure 42*) are normally governed by *NEC Article 725*. However, circuit wiring that is both powered and controlled by the fire alarm system is governed by *NEC Article 760*. A common residential problem occurs when using a system-type fire alarm that is a combination intrusion and fire alarm. Many believe that the keypad is only an intrusion system device. If the keypad is also used to control the fire alarm, it is a fire alarm device, and the cable used to connect it to the control unit must comply with *NEC Article 760*. Also, any intrusion system devices that are controlled and powered by the combination fire alarm system must be wired in accordance with *NEC Article 760* (fire-rated or permissible substitute cable).

10.15.0 Cables Running Floor to Floor

Riser cable is required when wiring runs from floor to floor and is not enclosed in conduit. The cable must be labeled as passing a test to prevent fire from spreading from floor to floor. An example of riser cable is FPLR. This requirement does not apply to one- and two- family residential dwellings.

10.16.0 Cables Running in Raceways

All cables in a raceway must have insulation rated for the highest voltage used in the raceway. Power-limited wiring may be installed in raceways or conduit, exposed on the surface of a ceiling or wall, or fished in concealed spaces. Cable splices or terminations must be made in Listed fittings, boxes, enclosures, fire alarm devices, or utilization equipment. All wiring must enter boxes through approved fittings and be protected against physical damage.

10.17.0 Cable Spacing

Power-limited fire alarm circuit conductors must be separated at least 2" (50.8mm) from any electric light, power, Class 1, or nonpower-limited fire alarm circuit conductors. This is to prevent damage to the power-limited fire alarm circuits from induced currents caused by the electric light, power, Class 1, or nonpower-limited fire alarm circuits.

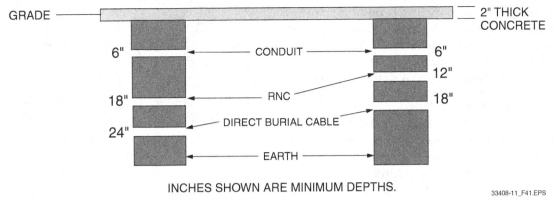

Figure 41 Minimum underground wiring burial depths under concrete.

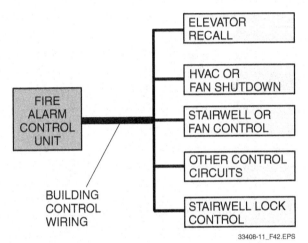

Figure 42 Building control circuits.

10.18.0 Elevator Shafts

Wiring in elevator shafts must directly relate to the elevator and be installed in rigid metallic conduit, rigid nonmetallic conduit, EMT, IMC, or up to 6' of flexible conduit.

10.19.0 Terminal Wiring Methods

The wiring for initiating device Class B circuits using end-of-line (EOL) terminations must be done so that removing the device causes a trouble signal (*Figure 43*).

10.20.0 Conventional Initiation Device Circuits

There are three styles of Class B and two styles of Class A conventional initiation device circuits Listed in *NFPA 72®*. Various local codes address what circuit types are required. *NFPA 72®* describes how they are to operate. A brief explanation of some of the circuits follows:

- *Class B, Style A* – FACUs using this style are no longer made in the United States, but a few sys-

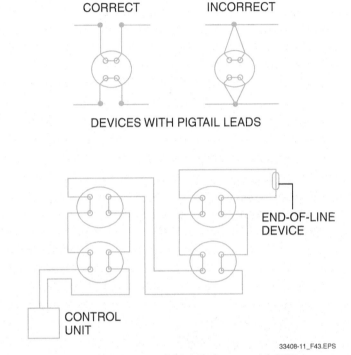

Figure 43 Correct wiring of devices for use with EOL terminations.

tems remain in operation. Single (wire) opens and ground faults are the only types of trouble that can be indicated, and the system does not receive an alarm in a ground fault condition. An alarm is initiated with a wire-to-wire short.
- *Class B, Style B* – (See *Figure 44*.) The FACU is required to receive an alarm from any device up to a break with a single open. An alarm is initiated with a wire-to-wire short. A trouble signal is generated for a circuit ground or open using an EOL device that is usually a resistor. An alarm can also be received with a single ground fault on the system.
- *Class B, Style C* – (See *Figure 45*.) This style, while used in the United States, is more common in Europe. An open circuit, ground, or wire-to-wire short causes a trouble indication.

Devices or detectors in this type of circuit often require a resistor in series with the contacts in order for the panel to detect an alarm condition. The panel receives an alarm signal with a single ground fault on the system. The current-limiting resistor is normally lower in resistance than the end-of-line resistor.

- *Class A, Style D* – (See *Figure 46*.) In this type of circuit, an open or ground causes a trouble signal. Shorting across the initiation loop causes an alarm. Activation of any initiation device results in an alarm, even when a single break or open exists anywhere in the circuit, because of the back loop circuit. The loop is returned to a special condition circuit, so there is no end of line, and therefore no EOL device.

- *Class A, Style E* – (See *Figure 47*.) This style is an enhanced version of Style D. An open circuit, ground, or wire-to-wire short is a trouble condition. All devices require another device (normally a resistor) in series with the contacts to generate an alarm. Activation of any of the initiating devices results in an alarm even if the initiation circuit has a single break or the system has a single ground.

The style of circuit can affect the maximum quantity of each type of device permitted on each circuit and the maximum quantity of circuits allowed for a fire alarm control unit/communicator.

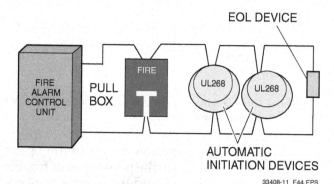

Figure 44 Class B, Style B initiation circuit.

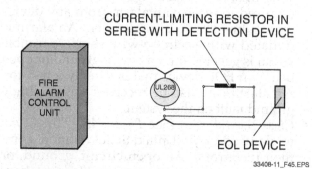

Figure 45 Class B, Style C initiation circuit.

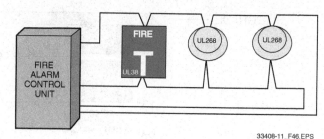

Figure 46 Class A, Style D initiation circuit.

Figure 47 Class A, Style E initiation device.

10.21.0 Signaling Line Circuits (SLCs)

The two classes of signaling line circuits (SLCs) are Class B and Class A. Hybrid circuits, composed of both Class A and B, are also used.

- *Class B circuits* – Class B SLCs for an addressable system essentially require that two conductors reach each device on the circuit by any means, as long as the wire type and physical installation rules are followed. It does not require wiring to pass in and out of each device in a series arrangement. With a circuit capable of 100 devices, it is permissible to go in 100 different directions from the FACU. Supervision occurs because the control unit polls and receives information from each device. The route taken is not important. A break in the wiring results in the loss of communication with one or more devices.

- *Class A circuits* – Class A SLCs for an addressable system require that the conductors loop into and out of each device. At the last device, the signaling line circuit is returned to the control unit by a different route. The control unit normally communicates with the devices via the outbound circuit but has the ability to communicate to the back side of a break through the return circuit. The control unit detects the fact that a complete loop no longer exists and shifts the unit into Class A mode. All devices remain in operation. Some systems even have the ability to identify which conductor has broken and how many devices are on each side of the break.

A hybrid system may consist of a Class A main trunk with Class B spur circuits in each area. A good example would be a Class A circuit leaving the control panel, entering a junction box on each floor of a multistory building, and then returning to the control panel by a different route. The signaling line circuits on the floors are wired as Class B from the junction box. This provides good system reliability and keeps costs down.

10.22.0 Notification Appliance Circuits

The following classes and styles of circuits are used for notification circuits (*Figure 48*):

- *Class B, Style W* – Class B, Style W is a two-wire circuit with an end-of-line device. Devices operate up to the location of a fault. A ground may disable the circuit.
- *Class B, Style X* – Class B, Style X is a four-wire circuit. It has alarm capability with a single open, but not during a ground fault.
- *Class B, Style Y* – Class B, Style Y is a two-wire circuit with an end-of-line device. Devices on this style of circuit operate up to the location of a fault. Ground faults are indicated differently than other circuit troubles.
- *Class A, Style Z* – Class A, Style Z is a four-wire circuit. All devices should operate with a single ground or open on the circuit. Ground faults are indicated differently than other circuit troubles.

Style X is similar to Style Z, except that during a ground fault, Style X does not operate and the panel is not able to tell what type of trouble exists. Style W is similar to Style Y, except that during a ground fault, Style W does not operate and the panel is not able to tell what type of trouble exists. Only Style Z operates all devices with a single open or a single ground fault. Only Style Z is a Class A notification appliance circuit (NAC).

All styles of circuits indicate a trouble alarm at the premises with a single open and/or a single ground fault. Styles Y and Z have alarm capability with a single ground fault. Styles X and Z have alarm capability with a single open. In Class B circuits, the panel monitors whether or not the wire is intact by using an EOL device (wire supervision). Electrically, the EOL device must be at the end of the indicating circuit; however, Style X is an exception. Examples of EOL devices are resistors to limit current, diodes for polarity, and capacitors for filtering.

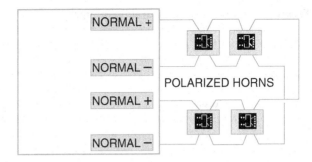

CLASS B, STYLE X OR CLASS A, STYLE Z

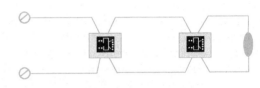

NOTIFICATION APPLIANCES

CLASS B, STYLES W & Y

33408-11_F48.EPS

Figure 48 Notification appliance circuits.

10.23.0 Primary Power Requirements

If more power is needed than can be supplied by one 20A circuit, additional circuits may be used, but none may exceed 20A. To help prevent system damage or nuisance alarms caused by electrical surges or spikes, surge protection devices should be installed in the primary power circuits unless the fire alarm equipment has self-contained surge protection. In addition, some jurisdictions may require breaker locks and/or other means of identifying circuit breakers for FACUs.

10.24.0 Secondary Power Requirements

An approved generator supply or backup batteries must be used to supply the secondary power of a fire alarm system. The secondary power system must, upon loss of primary power, immediately keep the fire alarm functioning for at least as long as indicated in *Table 7*.

Figure 49 provides an example of two forms designed to help calculate the proper size battery for an alarm system. *Figure 50* shows the forms filled out using the devices for a typical fire alarm system. In the Amps per Device column of the first form, the normal (standby mode) and alarm (or action) current draw shown for each device must

Table 7 Secondary Power Duration Requirements

NFPA Standard	Maximum Normal Load	Maximum Alarm Load
Central station	24 hours	See local system
Local system	24 hours	5 minutes
Auxiliary systems	24 hours	5 minutes
Remote stations	24 hours	5 minutes
Proprietary systems	24 hours	5 minutes
Household system	24 hours	4 minutes
Emergency voice alarm communication systems	24 hours	15 minutes maximum load 2 hours emergency operation

33408-11_T07.EPS

be obtained from the device label or the device manufacturer's data sheet. The current draw shown in the Total Amps column is the product of the number of devices multiplied by the value in the Amps per Device column. For this example, the system is an auxiliary or remote station system and the totals calculated in the first form are entered and calculated in the appropriate column of the second form. The result is the proper battery capacity (in amp/hr) for whatever battery voltage the system requires (12V, 24V, etc.). Calculations for an emergency voice alarm communication system could be done in a similar manner using different forms and the power/current draw of the various items of communications equipment.

10.25.0 Grounding

Fire alarm control units and enclosures must be properly grounded to a building unified ground, using a minimum of 14-gauge copper wire and a Listed type of connection. A unified ground can consist of one or more of the following, bonded together by appropriately-sized copper wire:

- Metal underground water pipe (10' or more)
- Metal building frame
- A ground ring
- Driven or concrete-encased ground rod (a made ground using ½" diameter rod at least 8' long)

Proper grounding of the panel and enclosure allows protection in case of a ground fault or accidental contact with higher voltage wires. It also drains leakage current and static charges to ground. The grounded panel and enclosure also acts to shield any internal electronic circuits from disruption by external electromagnetic interference (EMI).

On Site

Low Power Radio Smoke Detection System

The components that form a low power radio smoke detection system are shown here. This type of system can be used in situations where the building design or installation costs make hard wiring of the detectors impractical or too expensive. The heart of this system is the translation unit, called a gateway. It can communicate with up to four remote receiver units. The receivers can monitor radio frequency signals from up to 80 wireless smoke detectors. Each receiver unit transmits the status of the related wireless detectors via communication wiring to the translator. The translator then communicates this status to an intelligent FACU via a signaling line circuit loop.

33408-11_SA14.EPS

Fire Alarm System Standby Battery Calculation Form

Based on NFPA 72

Job Name:

Job Address:

Account#		Phone#		Contact	
Device	Number of This Type of Device	Amps per Device		Total Amps	
		Normal	Alarm	Normal	Alarm
Totals					

33408-11_F49A.EPS

Figure 49 Battery calculation form (1 of 2).

Fire Alarm System Standby Battery Calculation Form   Page 2

Based on NFPA 72

Job Name:

Job Address:

Account#		Phone#		Contact	
System Type Pick the System That Applies	Central Station	Local & Proprietary	Auxiliary & Remote Station	Household	
Total Normal Amps					
Time	**24 Hrs**	**24 Hrs**	**4 Hrs**	**4 Hrs**	
= Total Normal Amps					
Total Alarm Amps	**N/A**				
Time	**N/A**	0.08333*	0.08333*	0.06666*	
= Total Alarm Amps					
Total Normal Amps + Total Alarm Amps **= System Demand**					
+ 20% Contingency					
Minimum Proper Battery					

*0.08333 Hrs = 5 Minutes and 0.06666 Hrs = 4 Minutes

Figure 49 Battery calculation form (2 of 2).

11.0.0 TOTAL PREMISES FIRE ALARM SYSTEM INSTALLATION GUIDELINES

This section covers the requirements for the proper installation, testing, and certification of fire alarm systems and related systems for totally protected premises.

11.1.0 Manual Fire Alarm Box (Pull Box) Installation

The following guidelines apply to the installation of a manual fire alarm box (pull box):

- Manual pull boxes must be UL 38-Listed or the equivalent. Multi-purpose keypads cannot be used as fire alarm manual pull boxes unless UL 38-Listed for that purpose.
- Manual pull boxes must be installed in the natural path of escape, near each required exit from an area, in occupancies that require manual initiation. Ideally, they should be located near the doorknob edge of the exit door. In any case, they must be no more than 5' from the exit (*Figure 51*). In most cases, manual pull boxes are installed with the actuators at the heights of 42" to 54" to conform to both ADA and NFPA requirements.

Most new pull boxes are supplied with Grade II Braille on them for the visually impaired.

- The force required to operate manual pull boxes must be no more than 5 foot-pounds.
- A manual pull box must be within 200' of horizontal travel on the same floor from any part of the building (*Figure 52*). If the distance is exceeded, additional pull boxes must be installed on the floor.

11.2.0 Flame Detector Installation

When installing UV or IR flame detectors, the manufacturer's instructions and *NFPA 72*® should be consulted. The following should be observed when installing these detectors:

- UV flame detectors
  - Response is based on the distance from the fire, angle of view, and fire size. While some units have a 180-degree field of view, sensitivity drops substantially with angles of more than 45 to 50 degrees. Normally, the field of view is limited to less than 90 degrees (*Figure 53*).
  - If used outdoors, the unit must be Listed for outdoor use.

Fire Alarm System Standby Battery Calculation Form — Page 1

Based on NFPA 72

Job Name: *NBFAA Fire Alarm Installation Class Example*

Job Address: *This City*
This City, USA

Account# *001*   **Phone#** *(301) 907-3202*   **Contact:** *NTS*

Device	Number of This Type of Device	Amps per Device		Total Amps	
		Normal	Alarm	Normal	Alarm
Control Panel	1	0.350	0.400	0.350	0.400
Keypad	1	0.0100	0.100	0.100	0.100
Smoke Detectors	12	0.000100	0.100	0.0012	0.0111
Heat Detectors	10	0.0	0.0	0.0	0.0
Pull Stations	4	0.0	0.0	0.0	0.0
Strobe/Horn Combinations	8		0.085		0.085
Beam Detector	2	0.200	0.230	0.400	0.430
Auxiliary Relays	4	0	0.010		0.040
Totals				0.8512	1.0661

Fire Alarm System Standby Battery Calculation Form — Page 2

Based on NFPA 72

Job Name: *NBFAA Fire Alarm Installation Class Example*

Job Address: *This City*
This City, USA

Account# *001*   **Phone#** *(301) 907-3202*   **Contact** *NTS*

System Type Pick the System That Applies	Central Station	Local & Proprietary	Auxiliary & Remote Station	Household
Total Normal Amps			0.8512	
Time	24 Hrs	24 Hrs	24 Hrs	24 Hrs
= Total Normal Amps			20.4288	
Total Alarm Amps	N/A		1.0661	
Time	N/A	0.08333*	0.08333*	0.06666*
= Total Alarm Amps			0.089	
Total Normal Amps + Total Alarm Amps = System Demand			20.517	
+ 20% Contingency			4.103	
Minimum Proper Battery			25 amp/hrs	

*.08333 Hrs = 5 Minutes and .06666 Hrs = 4 Minutes

33408-11_F50.EPS

Figure 50 Sample battery calculations.

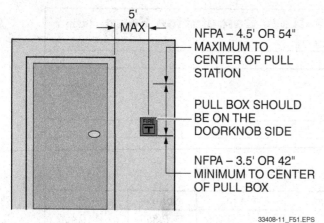

Figure 51 Pull box location and mounting height.

NFPA – 4.5' OR 54"
MAXIMUM TO
CENTER OF PULL
STATION

PULL BOX SHOULD
BE ON THE
DOORKNOB SIDE

NFPA – 3.5' OR 42"
MINIMUM TO CENTER
OF PULL BOX

33408-11_F51.EPS

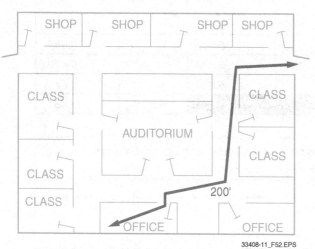

Figure 52 Maximum horizontal pull box distance from an exit.

33408-11_F52.EPS

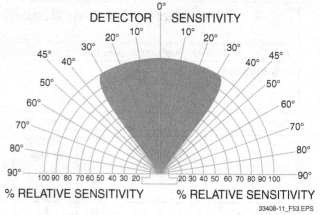

Figure 53 UV flame detector field of view.

DETECTOR SENSITIVITY

% RELATIVE SENSITIVITY   % RELATIVE SENSITIVITY

33408-11_F53.EPS

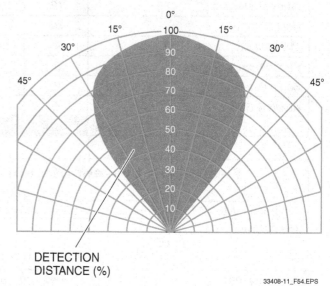

DETECTION DISTANCE (%)

Figure 54 IR flame detector field of view.

33408-11_F54.EPS

- A UV detector is considered solar blind. But in order to prevent unwanted or nuisance alarms, it should never be aimed near or directly at the sun.
- Units must be able to detect the light frequency of the fuel source.
- Units must never be aimed into areas where electric arc welding or cutting may be performed.

• IR flame detectors
  - Response varies depending on the angle of view. At 45 degrees, the sensitivity drops to 60 percent of the 0-degree sensitivity (*Figure 54*).
  - IR detectors cannot be used to detect alcohol, liquefied natural gas, hydrogen, or magnesium fires.

- IR detectors work best in low light level installations. High light levels desensitize the units. Discriminating units can tolerate up to 10 footcandles (fc) of ambient light. Non-discriminating units can tolerate up to two footcandles of ambient light.
- Units must never be used outdoors.
- Units must be able to detect the light frequency of the fuel source.

11.3.0 Smoke Chamber, Smoke Spread, and Stratification Phenomena

The following sections cover what defines a smoke chamber, smoke spread phenomena, and stratification phenomena.

11.3.1 Smoke Chamber Definition

Before automatic smoke detectors can be installed, the area of coverage known as the smoke chamber must be defined. The smoke chamber is the continuous, smoke-resistant perimeter boundary of a room, space, or area to be protected by one or more automatic smoke detectors between the upper surface of the floor and the lower surface of the ceiling. For the purposes of determining the area to be protected by smoke detectors, the smoke chamber is not the same as a smoke-tight compartment. It should be noted that some rooms that may have a raised floor and a false ceiling actually have three smoke chambers: the chamber beneath a raised floor, the chamber between the raised floor and the visible ceiling above, and the chamber between the room's visible ceiling and the floor above (or the lower portion of the roof). Few cases require detection in all these areas. However, some computer equipment rooms with a great deal of electrical power and communications cable under the raised floor may require detection in the chamber under the floor in addition to the chamber above the raised floor. In any case, the amount of detection required or specified is based on codes and/or the fire detection goals of the fire alarm system.

The simplest example of a smoke chamber would be a room with the door closed. However, if an intervening closed door exists between this and adjoining (communicating) rooms, the line denoting the barrier becomes less clear. The determining factor would be the depth of the wall section above the open archway or doorway, based on the following criteria:

- An archway or doorway that extends down from the ceiling greatly delays smoke travel and may be considered a boundary or a barrier, based on how far it extends below the ceiling (*Figure 55*). A low wall that does not fully extend to the ceiling can be treated as a barrier.
- Open grids above doors or walls that allow free flow of air and smoke are not considered barriers. To be considered an open grid, the opening must meet all of the requirements defined and demonstrated in *Figure 56*.
- Smoke doors, which are kept open by hold-and-release devices and meet all of the standards applicable thereto, may also be considered boundaries of a smoke chamber.

11.3.2 Smoke Spread and Stratification Phenomena

In a fire, smoke and heat rise in a plume toward the ceiling because they are lighter than the more dense, surrounding cooler air. In an area with a relatively low, flat, smooth ceiling, the smoke and heat quickly spread across the entire ceiling, triggering smoke or heat detectors. When the ceiling is irregular, smoke and heated air tend to collect, perhaps stratifying near a peak or collecting in the bays of a beamed or joist ceiling.

In the case of a beamed or joist ceiling, the smoke and heated air fills the nearest bays and begins to overflow to adjacent bays (*Figure 57*). The process continues until smoke and heat reach either a smoke or heat detector in sufficient quantity to cause activation. Because each bay must fill before it overflows, a substantial amount of time may pass before a detector placed at its maximum listed spacing activates. To reduce this time, the spacing for the detectors is reduced. If the beams or joists extend down too low, a detector would be placed in each bay.

> **NOTE**
>
> Ceiling beams and door arches form pockets that trap smoke and prevent its spread across a ceiling or between rooms. The placement of spot-type smoke detectors in different types of ceilings is covered in the 2010 edition of *NFPA 72®*, Section 17.7.3.2.

When smoke must rise a long distance, it tends to cool off and become denser. As its density becomes equal to that of the air around it, it stratifies, or stops rising (*Figure 58*). As a fire grows, additional heat is added, and the smoke eventually rises to the ceiling. However, if a great deal of time is lost, the fire gets much larger, and a large quantity of toxic gases is present at the floor level. Due to this delay, alternating detectors are lowered at least 3' (*Figure 59*). The science of stratification is very complex. It requires a fire protection engineer's evaluation to determine if stratification is a factor and, if so, to determine how much to lower the detectors to compensate for this condition.

Some conditions that cause stratification include the following:

- Uninsulated roofs that are heated by the sun, creating a heated air thermal block
- Roofs that are cooled by low outside temperatures, cooling the gases before they reach a detector
- HVAC systems that produce a hot layer of ceiling air
- Ambient air that is at the same temperature as the fire gases and smoke

There are no clear, set rules regarding which environments are susceptible to stratification. The factors and variables involved are beyond the

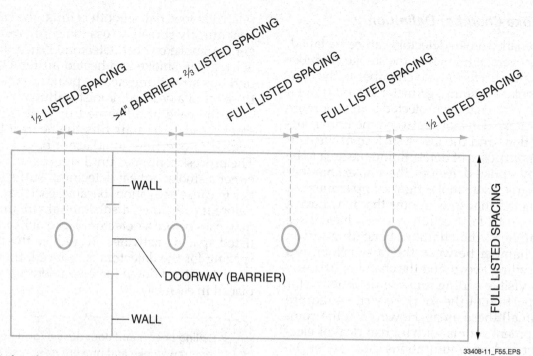

Figure 55 Reduced spacing required for a barrier.

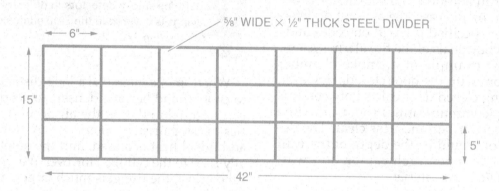

GRID OPENINGS MUST BE AT LEAST ¼" IN THE LEAST DIMENSION.
OPENINGS ARE 5⅜" (6" MINUS ⅝") × 4⅜" (5" MINUS ⅝").
REQUIREMENT MET.

THE THICKNESS OF THE MATERIAL DOES NOT EXCEED THE LEAST DIMENSION.
OPENINGS ARE 5⅜" (6" MINUS ⅝") × 4⅜" (5" MINUS ⅝"). THICKNESS IS ½".
REQUIREMENT MET.

THE OPENINGS CONSTITUTE AT LEAST 70% OF THE AREA OF THE PERFORATED MATERIAL.
OPENINGS ARE 5⅜" (6" MINUS ⅝") × 4⅜" (5" MINUS ⅝")
5⅜" × 4⅜" = 23½ SQ IN

21 OPENINGS × 23½ SQ IN   =   TOTAL OPENING = 493½ SQ IN
15" × 42" = 630 SQ IN   630 SQ IN × 0.70 = 441 SQ IN

REQUIREMENT MET.

Figure 56 Detailed grid definition.

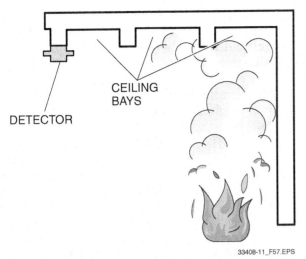

Figure 57 Smoke spread across a beam or joist ceiling.

Figure 59 Smoke stratification countermeasure.

Figure 58 Smoke stratification.

scope of this course. Note that only alternating detectors are suspended for stratification. Some fires can result in stratification caused by superheated air and gases reaching the ceiling prior to the smoke, causing a barrier that holds the smoke below the detectors. Smoke detectors with integral heat detectors should be effective in such cases, without the need to suspend alternating detectors below the ceiling.

Stratification can also occur in high-rise buildings because of the stack effect principle. This occurs when smoke rises until it reaches a neutral pressure level, where it begins to stratify (*Figure 60*). Stack effect can move smoke from the source to distant parts of the building. Penetrations left unsealed can be a major contributor in such smoke migration. Stack effect factors include the following:

- Building height
- Air tightness
- Air leakage between floors
- Interior/exterior temperature differential

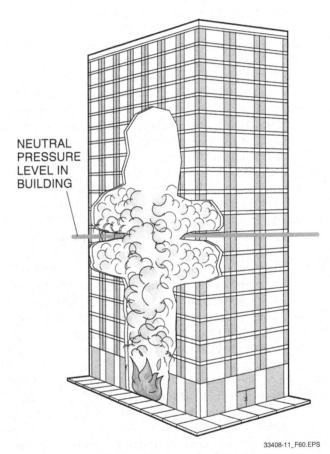

Figure 60 Stack effect in a high-rise building.

- Vertical openings
- Wind force and direction

Some factors that influence stack effect are variable, such as the weather conditions and which interior and exterior doors might be open. For this reason, the neutral pressure level may not be the same in a given building at different times. One thing is very clear concerning stratification—penetrations made in smoke barriers, particularly those in vertical openings such as shafts, must be sealed.

11.4.0 General Precautions for Detector Installation

The following are general precautions for detector installation:

- *Recessed mounting* – Detectors must not be recess-mounted unless specially Listed for recessed mounting.
- *Air diffusers* – Air movement can have a number of undesirable effects on detectors. The introduction of air from outside the fire area can dilute the smoke, which delays activation of the detector. The air movement may create a barrier, which delays or prevents smoke from reaching the detector chamber. Smoke detectors are Listed for a specific range of air velocity. This means that they are expected to function with air (and smoke) moving through the detection chamber at any speed within the Listed range. Detectors may not function properly if air movement is above or below the Listed velocity. Air movement can be measured with an instrument called a velocimeter. Unless the airflow exceeds the Listed velocity, spot-type smoke detectors must be a minimum of 3' from air diffusers.
- *Problem locations and sources of unwanted or nuisance alarms* – Detectors, especially smoke detectors, can register nuisance alarms as a result of the following:
  - *Electrical interference:* Keep detector locations away from fluorescent lights and radio transmitters, including cellular phones. Electrical noise or radio frequencies radiated from these devices can cause nuisance alarms.
  - *Heating equipment:* High temperatures, dust accumulation, improper exhaust, and incomplete combustion from heating equipment cause problems for detectors.
  - *Engine exhaust:* Exhaust from engine-powered forklifts, vehicles, and generators is a potential problem for ionization-type smoke detectors.
  - *Solvent and chemical fumes:* Cleaning solvents and adhesives are a problem for ionization detectors.
  - *Other gases and fumes:* Fumes from machining operations, paint spraying, industrial or battery gases, curing ovens/dryers, cooking equipment, sawing/drilling/grinding operations, and welding/cutting operations cause problems for smoke detectors.

- *High temperatures:* Avoid very hot or cold environments for smoke detector locations. Temperatures below 32°F (0°C) can cause nuisance alarms, and temperatures above 120°F (48.8°C) can prevent proper smoke detector operation. Extreme temperatures affect beam, ionization, and photoelectric detectors.
- *Dampness or humidity:* Smoke detectors must be located in areas where the humidity is less than 93 percent. In ionization detectors, dampness and high humidity can cause tiny water droplets to condense inside the sensing chamber, making it overly sensitive and causing unwanted or nuisance alarms. In photoelectric detectors, humidity can cause light refraction and loss of current flow, either of which can lead to nuisance alarms. Common sources of moisture to avoid include slop sinks, steam tables, showers, water spray operations, humidifiers, and live steam sources.
- *Lightning:* Nearby lightning can cause electrical damage to a fire alarm system. It may also induce electrical noise or spikes in the alarm system wiring or detectors, resulting in nuisance alarms. Surge arrestors installed in the system's primary power supply to protect the system, in conjunction with alarm verification, can reduce the chance of system damage and unwanted alarms.
- *Dusty or dirty environments:* Dust and dirt accumulated on a smoke detector's sensing chamber make the detector overly sensitive. Avoid areas where fumigants, fog, dust, or mist-producing materials are consistently used.
- *Outdoor locations:* Dust, air currents, and humidity typically affect outdoor structures, including sheds, barns, stables, and other open structures. This makes outdoor structures unsuitable for smoke detectors.
- *Insect-infested areas:* Insects in a smoke detector can cause an unwanted alarm. Good bug screens on a detector can prevent most adult insects from entering the detector. However, newly hatched insects may still be able to enter. An insecticide strip next to the detector may help solve the problem, but it may also cause unwanted alarms because of fumes. Check with the manufacturer for the use of an approved strip. Ionization detectors are less prone to unwanted alarms from insects.

– *Construction:* Smoke detectors must not be installed until after construction cleanup unless required by the AHJ for protection during construction. Detectors that are installed prior to construction cleanup must be cleaned or replaced. Prior to 1993, the code permitted covering smoke detectors to protect them from dirt, dust, or paint mist. However, the covers did not work very well and resulted in clogged, oversensitive, and damaged detectors.

As an alternative to the traditional prescriptive designs for detection equipment installation explained in this module, *NFPA 72*® allows performance-based fire alarm system design. Because of the mathematical complexity of such a design, it must be implemented only by qualified fire protection engineers. This involves basing detector location and spacing on objectives set for the system, size and rate of fire growth to be detected, ceiling heights, ambient temperatures, and response characteristics of the detectors. The design method will use strategically placed plume-dependent heat detectors/sensors and smoke detectors/sensors, including radiant energy detectors/sensors. For further information, refer to *NFPA 72*®, *Annex B*.

On Site

Troubleshooting

Troubleshooting older systems requires that you trace and check each detector in order to isolate the source of an alarm. The fire alarm panels of newer systems, such as the one shown here, have fault isolation messages and LCD display panels to help pinpoint which detector has alarmed or has a problem.

33408-11_SA15.EPS

11.5.0 Spot Detector Installations on Flat, Smooth Ceilings

A smooth ceiling is defined as having a surface uninterrupted by continuous projections such as solid joists, beams, or ducts that project more than 4" below the ceiling. A level ceiling is defined as one with a slope equal to or less than 1.5" per foot (1' to 8').

11.5.1 Conventional Installation Method

The following applies to spot detector installation on flat, smooth ceilings:

- When a smoke detector manufacturer's specifications do not specify a particular spacing, a 30' spacing guide may be applied.
- The distance between heat detectors must not exceed their Listed spacing.
- There must be detectors located within one-half the Listed spacing measured at right angles from all sidewalls.

> **NOTE**
> This section assumes detectors Listed for 30' spacing mounted on a smooth and flat ceiling of less than 10' in height, where 15' is one-half the Listed spacing. Spacing is reduced as indicated in later sections when the ceiling height exceeds 10', or the ceiling is not smooth and flat as defined by the code.

In the following example, the detector locations for a simple room that is 30' × 60' long will be determined. The locations are determined by the intersection of columns and rows marked on a sketch of the ceiling:

- The first column is located by a line that is parallel to the end wall and not more than one-half the Listed spacing from the end wall (*Figure 61*).

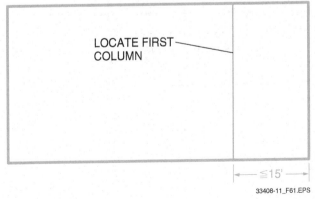

33408-11_F61.EPS

Figure 61 Locating the first column.

- The first row is located by a line parallel to the sidewall that is not more than one-half the Listed spacing from that sidewall (*Figure 62*).
- The first detector is located at the intersection of the row and column lines.
- The second column is located by a line that is parallel to the opposite end wall and not more than one-half the Listed spacing from the end wall.
- The second detector is located at the intersection of the row and second column lines, provided that the distance between the first and second detectors does not exceed the Listed spacing (*Figure 63*). The only time that the full Listed spacing of any detector is used is when measuring from one detector to the next.

11.5.2 The 0.7 Rule

The 0.7 rule states that all points on the ceiling must have a detector within 0.7 times the detector's Listed spacing. This rule, also called the rule of point 7, applies to spot-type, automatic heat and smoke detectors. According to the code, regardless of the method used to determine automatic detector placement, the 0.7 rule is the overriding requirement. In one sense, this method is essentially the same as the conventional placement method. By locating a detector at the intersection

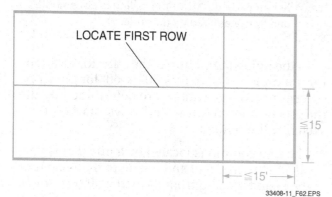

Figure 62 Locating the first row.

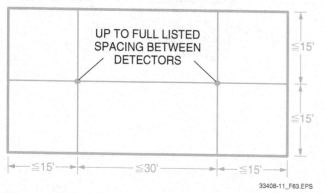

Figure 63 First and second detector locations.

of right angle lines which are one-half the Listed spacing from each adjoining wall, the maximum distance to the adjacent corner will be 0.7 times the Listed 30' spacing, or 21' (*Figure 64*).

For the following example, assume that a set of plans shows a room that is 30' × 30' (900 ft², which is the maximum square-foot coverage of a detector Listed for 30' spacing):

- Using the conventional method where one-half the Listed spacing is required from each wall, the detector would be placed at precisely the center of the room.
- Using the 0.7 rule, draw a circle on the plan with a compass set to the scale of the plan (*Figure 65*) for a radius of 21' (0.7 times 30'), as shown in *Figure 66*. The resulting circle shows that the room is code-compliant, because no point on the ceiling is more than 0.7 times the Listed spacing from the detector.

When planning detector placement on paper, it is common to use the maximum allowable spacing provided by the 0.7 rule circle radius. In the 30' × 30' room shown in *Figure 67*, an obstruction exists that prevents placement of the detector at the precise center of the room. The obstruction may be a light fixture, HVAC diffuser, or similar feature. When verifying compliance with the 0.7 rule circle, it is found that the room is no longer in compliance, because areas of the ceiling exist that are more than 0.7 times the 30' Listed spacing from the detector. By using approved architectural drawings and reviewing the ceiling plans, most conflicts can be determined in the design and planning stage to prevent costly last-minute change orders. Unfortunately, most architects and designers leave these conflicts to be worked out in the field. Regardless of how careful a planner may be, some conflicts are bound to occur. Therefore, it is important that both the installer and the inspector be familiar with the code requirements.

In order to protect this room in compliance with the code, a second detector must be added on the opposite side of the obstruction to eliminate the unprotected areas. The room could also be protected by sidewall-mounted detectors. In this case, each detector covers one-half of its Listed spacing out from the detector toward the center of the room, as shown in *Figure 68*. The same 0.7 rule can be applied to sidewall-mounted detectors as well as ceiling-mounted detectors. However, in sidewall mounting, the full coverage of the detector can never be used, which increases the number of detectors required.

To illustrate the need for verification of coverage using the 0.7 rule, assume that the top and bottom sidewalls of the room are lengthened by

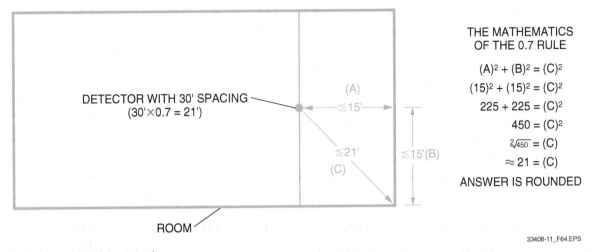

DETECTOR WITH 30' SPACING
(30'×0.7 = 21')

(A)
≤15'

≤21'
(C)

≤15'(B)

ROOM

THE MATHEMATICS
OF THE 0.7 RULE

$$(A)^2 + (B)^2 = (C)^2$$
$$(15)^2 + (15)^2 = (C)^2$$
$$225 + 225 = (C)^2$$
$$450 = (C)^2$$
$$\sqrt[2]{450} = (C)$$
$$\approx 21 = (C)$$

ANSWER IS ROUNDED

33408-11_F64.EPS

Figure 64 Explanation of the 0.7 rule.

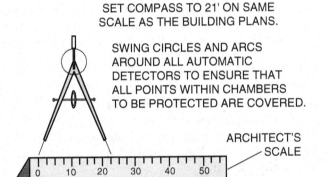

SET COMPASS TO 21' ON SAME
SCALE AS THE BUILDING PLANS.

SWING CIRCLES AND ARCS
AROUND ALL AUTOMATIC
DETECTORS TO ENSURE THAT
ALL POINTS WITHIN CHAMBERS
TO BE PROTECTED ARE COVERED.

ARCHITECT'S
SCALE

0   10   20   30   40   50

33408-11_F65.EPS

Figure 65 Setting a compass to scale for the 0.7 rule circle radius.

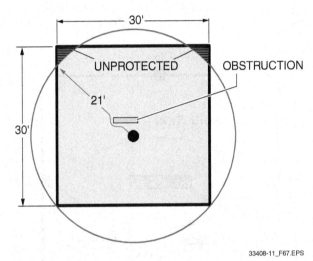

30'

UNPROTECTED   OBSTRUCTION

21'

30'

33408-11_F67.EPS

Figure 67 Room with a ceiling obstruction.

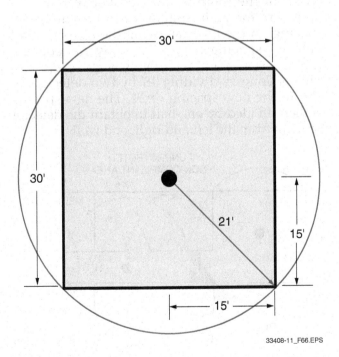

30'

30'

21'

15'

15'

33408-11_F66.EPS

Figure 66 Confirmation of code compliance using a 0.7 rule circle.

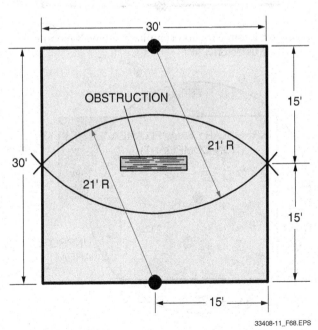

30'

OBSTRUCTION

15'

21' R

30'

21' R

15'

15'

33408-11_F68.EPS

Figure 68 Room with sidewall-mounted detectors.

5', resulting in a room that is 30' × 35' (*Figure 69*). When verification of compliance is attempted using the 0.7 rule circle arcs, it is found that two unprotected areas exist. This results in non-compliance for the mounting of sidewall detectors in this room. In this case, only ceiling-mounted detectors provide compliant coverage.

The 0.7 rule method is very useful for irregularly shaped areas. An example of an irregular area is shown in *Figure 70*. In this example, no location in the room can achieve compliance using only one detector.

11.5.3 Planning and Review Techniques

In preparing or reviewing building plans for the placement of automatic detectors, there are usually many irregularly shaped areas, including corridors and L-shaped rooms. Most rooms or areas

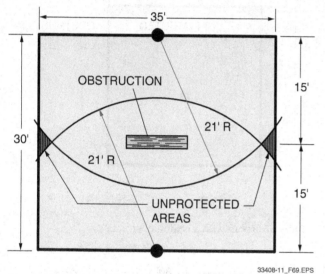

Figure 69 Room with non-compliant sidewall-mounted detectors.

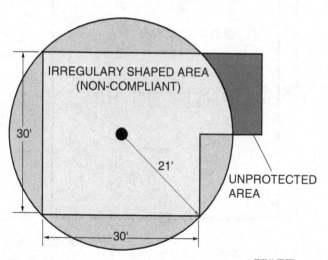

Figure 70 Non-compliant irregularly shaped area.

to be protected are not be dimensioned such that the maximum Listed spacing of the required number of detectors can be used. By drawing 0.7 rule circles and arcs throughout each area on the building plans to be protected by each detector, it can be easily determined if compliance has been met with the planned detector locations (*Figure 71*). If areas of non-compliance are noted, detector locations can be replanned to provide the proper coverage. As shown in *Figure 71*, the lobby is compliant, but the corridor detectors must be shifted toward the lobby to eliminate the unprotected areas.

11.5.4 Planning Techniques for Average Sized-Rooms

For average-sized rooms, it is common practice to begin planning coverage using the conventional method. The first detector is located from one of the corners by measuring one-half the Listed spacing from the two adjacent sidewalls. The second detector is then located along the same row, 30' from the first detector (*Figure 72*). This usually results in the detectors being off-center in the room, because the dimensions of most rooms are not even multiples of the detector spacing. While the coverage is compliant, it does not have the best location for the detectors, and it is not aesthetically pleasing.

Rows and columns of detectors should be evenly distributed to cover the room or area. This can be accomplished by dividing the length and width of the room by the number of required detectors for each dimension to obtain a new spacing. In the room showing the redistributed detector spacing (*Figure 73*), only the spacing across the length of the room has to be adjusted. This is done by dividing 48' by two detectors to obtain the new spacing of 24'. The new spacing is then divided by one-half to obtain the detector spacing from the left and right end walls.

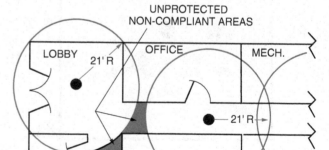

Figure 71 Examples of non-compliant areas in a commercial facility.

11.5.5 Planning Techniques for Large Areas

A form of the conventional method can be used to determine detector coverage for large areas that are rectangular or square or that can be broken down into two or more large rectangular or square sections. In this modified conventional method, the number of detectors required is easily determined by subdividing the length and width of the area or each section of the area, using a square representing the Listed detector spacing (coverage). This is accomplished by dividing both the length and width of the area, or each section of the area, by the Listed spacing for the desired detector, and rounding the results up to the next whole detector for each dimension. Then, the numbers of detectors for the length and width are multiplied together to yield the total number of detectors required for the area or each section of the area.

For example, assume that a set of plans shows a room that is 40' wide by 70' in length, and the desired detector has a Listed spacing of 30':

$$40' \div 30' = 1.33 \text{ (rounded up)}$$
$$= 2 \text{ detectors to cover the width of the room}$$
$$70' \div 30' = 2.33 \text{ (rounded up)}$$
$$= 3 \text{ detectors to cover the length of the room}$$

Hence, the total number of detectors required is:

$$2 \times 3 = 6 \text{ detectors for the room}$$

Because the coverage of the six detectors would actually be larger than the room area if the detectors were spaced at full coverage (such as six 30' squares with the detectors at the centers), the spacing for the detectors can be adjusted for appearance by dividing both the width and length of the room by the applicable number of detectors:

$$40' \div 2 \text{ detectors}$$
$$= 20' \text{ detector spacing for the room width}$$
$$70' \div 3 \text{ detectors} = 23.33',$$
$$\text{or } 23'\text{-}4'', \text{ detector spacing for the room length}$$

Spacing from the walls would be one-half of the adjusted detector spacing:

$$\text{Room width wall spacing} = 20 \div 2 = 10'$$
$$\text{Room length wall spacing} = 23'\text{-}4'' \div 2 = 11'\text{-}8''$$

Figure 74 shows the adjusted spacing for the six required detectors. If desired, coverage could be verified by drawing scaled 0.7 rule circles around each detector location using the full Listed spacing for the detector.

11.5.6 Planning Techniques for Long, Narrow Spaces

Detector coverage for long, narrow spaces such as corridors is generally not determined using the conventional method. The following example illustrates the advantage of using the 0.7 rule method of determining the required number of detectors.

Assume that a set of plans shows a corridor that is 10' wide and 120' long, and the desired detector has a Listed spacing of 30'. In the conventional method, detectors would be placed at one-half their Listed spacing from the end walls, and at their full Listed spacing between detectors. This would result in the need for four detectors. While the use of four detectors is more than compliant, it is not cost effective. By drawing 0.7 rule circles on the center line of the corridor so that they just overlay all areas of the corridor (*Figure 75*), it is evident that only three detectors (with a spacing of 40') are required for compliant coverage.

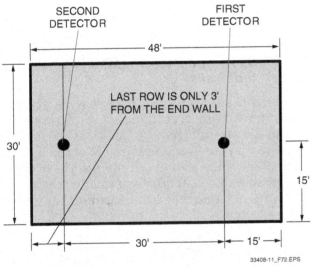

Figure 72 Initial detector layout in a typical room.

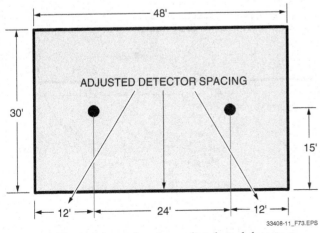

Figure 73 Typical room showing redistributed detectors.

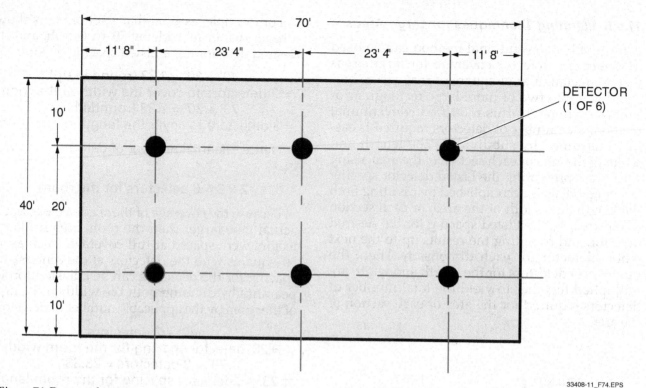

Figure 74 Detector locations with spacing adjusted for a 40' × 70' room.

How can a detector Listed for 30' spacing be used at 40' spacing? The answer is that since the required coverage area is narrow, nearly the entire length (20') of the 21' radius can be used for coverage. This is because the 0.7 rule states that all points on the ceiling of the area to be protected must be within 0.7 times the Listed spacing from the detector. As shown by the 0.7 rule circles, the corridor is compliant with the use of three instead of four detectors.

11.5.7 Spot-Type Detector Installations on Sidewalls

Automatic spot-type detectors may be mounted on sidewalls in accordance with the requirements shown in *Figure 76*. Detectors must not be mounted in the 4" triangular area at the intersection of the sidewall and ceiling. Tests have shown that smoke and heat do not normally migrate into this area at a rate that is sufficient for early warning detection.

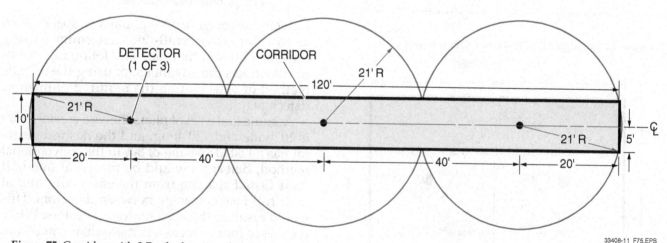

Figure 75 Corridor with 0.7 rule detector placement.

11.6.0 Photoelectric Beam Smoke Detector Installations on Flat, Smooth Ceilings

Photoelectric beam smoke detectors have two Listings—one for width of coverage and one for the minimum/maximum length of coverage (beam length), as shown in *Figure 77*. The beam length is Listed by the manufacturer and, usually, so is the width coverage spacing (S). When a manufacturer does not List the width coverage, the 60' guideline specified in *NFPA 72®* must be used as the width coverage. The distance of the beam from the ceiling should normally be between 4" and

12". However, NFPA allows a greater distance to compensate for stratification. Always ensure the beam does not cross any expansion joint or other point of slippage that could eventually cause misalignment of the beam. In large areas, parallel beam detectors may be installed, separated by no more than their Listed spacing (S).

11.7.0 Spot Detector Installations on Irregular Ceilings

Any ceiling that is not flat and smooth is considered irregular. For heat detectors, ceilings of any description that are over 10' above floor level are considered irregular. Heat detectors located on any ceiling over 10' above finished floor (AFF) level must have their spacing reduced below their Listed spacing in accordance with *NFPA 72®*, *Table 17.6.3.5.1*. This is because hot air is diluted by the surrounding cooler air as it rises, which reduces its temperature. Smoke detector spacing is not adjusted for high ceilings.

Irregular ceilings include sloped, solid joist, and beam ceilings:

- *Sloped ceilings* – A sloped ceiling is defined as a ceiling having a slope of more than 1.5" per foot (1' to 8'). Any smooth ceiling with a slope equal to or less than 1.5" per foot is considered a flat ceiling. Sloped ceilings are usually shed or peaked types.
  - *Shed:* A shed ceiling is defined as having the high point at one side with the slope extending toward the opposite side.
  - *Peaked:* A peaked ceiling is defined as sloping in two directions from its highest point. Peaked ceilings include domed or curved ceilings.

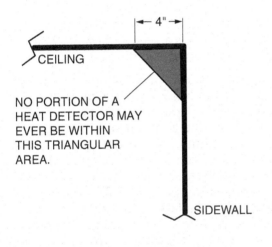

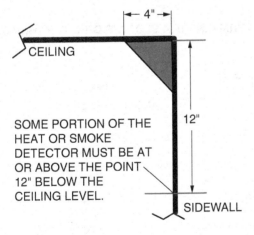

33408-11_F76.EPS

Figure 76 Sidewall mounting requirements for spot-type detectors.

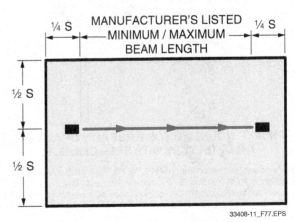

33408-11_F77.EPS

Figure 77 Maximum straight-line single-beam smoke detector coverage (ceiling view).

- *Solid joist or beam ceilings* – Solid joist or beam ceilings are defined as those with joists or beams spaced less than 3' center-to-center with a solid member extending down from the ceiling for a specified distance. For heat detectors, the solid member must extend more than 4" down from the ceiling. For smoke detectors, the solid member must extend down more than 8".

11.7.1 Shed Ceiling Detector (or Alarm) Installation

Shed (sloped) ceilings having a rise greater than 1.5"/1" run (1'/8') must have the first row of detectors (heat or smoke) located on the ceiling within 3' of the high side of the ceiling (measured horizontally from the sidewall), but not closer than 4" from the adjoining wall surface (*Figure 78*).

Heat detectors must have their spacing reduced in areas with high ceilings (over 10'). Smoke detector spacing is not adjusted due to the ceiling height.

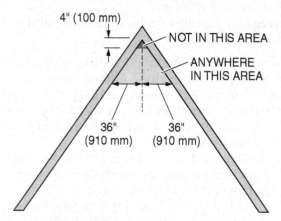

PROPER MOUNTING FOR FIRST ROW OF ALARMS OR DETECTORS WITH PEAKED CEILINGS

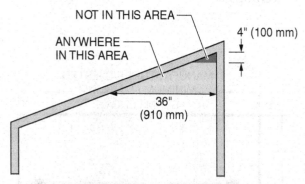

PROPER MOUNTING FOR FIRST ROW OF ALARMS OR DETECTORS WITH SHED CEILINGS

33408-11 F78.EPS

Figure 78 Shed and peaked ceiling alarm/detector mounting.

For a roof slope of less than 30 degrees, determined as shown in *Figure 79*, all heat detector spacing must be reduced based on the height at the peak. For a roof slope of greater than 30 degrees, the average ceiling height must be used for all heat detectors other than those located in the peak. The average ceiling height can be determined by adding the high sidewall and low sidewall heights together and dividing by two. In the case of slopes greater than 30 degrees, the spacing to the second row is measured from the first row and not the sidewall for shed ceilings. See *NFPA 72®* for heat detector spacing reduction based on peak or average ceiling heights.

Once you have determined that the ceiling you are working with is a shed ceiling, use the following guidelines for determining detector placement:

- Place the first row of detectors within 3' of the high sidewall (measured horizontally).
- Use the Listed spacing (adjusted for the height of the heat detectors on ceilings over 10') for each additional row of detectors (measured from the detector location, not the sidewall).
- For heat detectors on ceilings over 10' high, adjust the spacing per *NFPA 72®*. Smoke detector spacing is not adjusted for ceiling height.
- If the slope is less than 30 degrees, all heat detector spacing is based on the peak height. If

DETERMINING IF A SLOPE IS GREATER THAN 30°

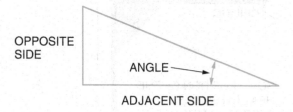

If the opposite side divided by the adjacent side is > 0.5774, the angle is > 30°.

The tangent of the smallest angle of a right triangle equals the opposite side divided by the adjacent side.

The sum of all the angles of a triangle equals 180°.

The tangent of a 30° angle is 0.5774.

33408-11_F79.EPS

Figure 79 Determining degree of slope.

the slope is more than 30 degrees, peak heat detectors are based on the peak height, and all other detectors are based on the average height.

- Additional columns of detectors that run at right angles to the slope of the ceiling do not need to be within 3' of the end walls.

To calculate the average ceiling height, use one of the following methods:

Method 1:

Step 1 Subtract the height of the low ceiling from the height of the high ceiling.

Step 2 Divide the result by 2.

Step 3 Subtract the result from the height of the high ceiling.

Method 2:

Step 1 Add the height of the low ceiling to the height of the high ceiling.

Step 2 Divide the result by 2.

11.7.2 Peaked Ceiling Detector (or Alarm) Installation

A ceiling must be shaped as defined by the code in order to be considered a peaked ceiling. It must slope in more than one direction from its highest point. Because domed and curved ceilings do not have clean 90-degree lines to clearly show that the ceiling slopes in more than one direction, they should be viewed from an imaginary vertical center line. From this perspective, it can be seen if the ceiling slopes in more than one direction.

- The first row of detectors (heat or smoke) on a peaked ceiling shall be located within 3' of the peak, measured horizontally from the center line of the peak, but not less than 4" down from the center of the peak (*Figure 78*). The detectors should be located alternately on either side of the peak.
- Regardless of where the detector is located in the 3' space on either side of the peak, the measurement for the next row of heat detectors is taken from the peak (measured horizontally) not from the heat detector located near the peak. This is different from the way detector location is determined for shed ceilings.
- Heat detectors must have their spacing reduced in areas with ceilings over 10'. Smoke detector spacing is not adjusted due to the ceiling height.
- As with sloped ceilings, use of the peak height for just the row of heat detectors at the peak, or for all the ceiling heat detectors, depends on whether or not the ceiling is sloped more than

30 degrees. Use the peak height to adjust heat detector spacing of all the room's heat detectors when the slope of the ceiling is less than 30 degrees. When the slope is greater than 30 degrees, use the peak height for only the heat detectors located within 3' of the peak. Use the average height for all other heat detectors in the room.

Smoke detectors should be located on a peaked ceiling using the following guidelines:

- A row of detectors is within 3' of the peak vertical center line on either side, or alternated from side to side.
- The next row of detectors is within the Listed spacing of the peak measured horizontally from the peak vertical center line. Do not measure from the detectors within 3' of the peak. It does not matter where within 3' of the peak the first row is; the next row on each side of the peak is installed within the Listed spacing of the detector from the peak.
- Additional rows of detectors are installed using the full Listed spacing of the detector.
- The sidewall must be within one-half the Listed spacing of the last row of detectors.
- Columns of detectors are installed in the depth dimension of the diagrams using the full Listed spacing (the same as on a smooth, flat ceiling).
- There is no need to reduce smoke detector spacing for ceilings over 10'.

11.7.3 Solid Joist Ceiling Detector Installation

Solid joists are defined for heat detectors as being solid members that are spaced less than 3' center-to-center and that extend down from the ceiling more than 4". Solid joists are defined for smoke detectors as being solid members that are spaced less than 3' center-to-center and that extend down from the ceiling more than 8".

Heat detector spacing at right angles to the solid joist is reduced by 50 percent. In the direction running parallel to the joists, standard spacing principles are applied. If the ceiling height exceeds 10', spacing is adjusted for the high ceiling in addition to the solid joist as defined in *NFPA 72*®.

Smoke detector spacing at right angles to the solid joists is reduced by one-half the Listed spacing for joists 1' or less in depth and ceilings 12' or less in height. In the direction running parallel to the joists, standard spacing principles are applied. If the ceiling height is over 12' or the depth of the joist exceeds 1', spot-type detectors must be located on the ceiling in every pocket. Additional reductions for sloped joist ceilings also apply as defined in *NFPA 72*®.

11.7.4 *Beamed Ceiling Detector Installation*

Beamed ceilings consist of solid structural or solid nonstructural members projecting down from the ceiling surface more than 4" and spaced more than 3' apart center-to-center as defined by *NFPA 72®*.

- The following guidelines apply to heat detector installation:
  - If the beams project more than 4" below the ceiling, the spacing of spot-type heat detectors at right angles to the direction of beam travel must not be more than two-thirds the smooth ceiling spacing.
  - If the beams project more than 12" below the ceiling and are more than 8' on center, each bay formed by the beams must be treated as a separate area.
  - Reductions of heat detector spacing in accordance with *NFPA 72®* are also required for ceilings over 10' in height.
- The following guidelines apply to smoke detector installation:
  - For smoke detectors, if beams are 4" to 12" in depth with an AFF level of 12' or less, the spacing of spot-type detectors in the direction perpendicular to the beams must be reduced 50 percent, and the detectors may be mounted on the bottoms of the beams.
  - If beams are greater than 12" in depth with an AFF level of more than 12', each beam bay must contain its own detectors. There are additional rules for sloped beam ceilings as defined in *NFPA 72®*.
  - The spacing of projected light beam detectors that run perpendicular to the ceiling beams need not be reduced. However, if the projected light beams are run parallel to the ceiling beams, the spacing must be reduced per *NFPA 72®*.
  - For smoke detectors, other than projected light beam detectors, located in high air movement areas, the spacing must comply with *NFPA 72, Table 17.7.6.3.3.1* and *Figure 17.7.6.3.3.1*.

11.8.0 Notification Appliance Installation

Notification appliances include the following types:

- Audible appliances, such as bells, horns, chimes, speakers, sirens, and mini-sounders
- Visual appliances, such as strobe lights
- Tactile (sense of touch) appliances, such as bed shakers and special sprinkling devices

The appropriate appliances for the occupancy must be determined. All appliances must be Listed for the purposes for which they are used. For example, *UL 1480* speakers are used for fire protective signaling systems, and *UL 1638* visual signaling appliances are used for private mode emergency and general utility signaling. While *NFPA 72®* recognizes tactile devices, it does not specify installation requirements. If an occupant requires this type of notification, use equipment Listed for the purpose and follow the manufacturer's instructions. As always, consult the local AHJ.

11.8.1 Notification Device Installation

The following guidelines apply to the installation of notification appliances:

- Ensure that notification appliances are wired using the applicable circuit style (Class A or B).
- In Class B circuits, the panel supervises the wire using an end-of-line device. Electrically, the EOL device must be at the end of the indicating circuit. Examples of EOL devices are resistors to limit current, diodes for polarity, and capacitors for filtering.
- It is extremely important that any polarized appliances be installed correctly. The polarization of the leads or terminals of these devices are marked on the device or are noted in the manufacturer's installation data. If the leads of a polarized notification appliance are reversed, the panel does not detect the problem, and the appliance does not activate. Because it is very easy to accidentally wire an appliance backwards, testing every appliance is extremely important. The general alarm should be activated, and every notification appliance checked for proper operation.
- Sidewall-mounted audible notification appliances must be mounted at least 90" AFF or at least 6" below the finished ceiling. Ceiling-mounted and recessed appliances are permitted.
- Visible notification appliances must be mounted at the minimum heights of 80" to 96" AFF (*NFPA 72®*) or, for ADA requirements, either 80" AFF or 6" below the ceiling. In any case, the device must be within 16' of the pillow in a sleeping area. Combination audible/visible ap-

pliances must follow the requirements for visible appliances. Non-coded visible appliances should be installed in all areas where required by local codes, *NFPA 101®*, or by the local AHJ. Consult ADA regulations for the required illumination levels in sleeping areas.

- *Table 8* provides spacing requirements for visual notification appliance spacing in corridors less than 20' wide. For corridors and rooms greater than 20', refer to *Tables 9* and *10*. In corridor applications, visible appliances must be rated at not less than 15 candelas (cd). Per *NFPA 72®*, visual appliances must be located no more than 15' from the end of the corridor with a separation of no more than 100' between appliances. Where there is an interruption of the concentrated viewing path, such as a fire door or elevation change, the area is to be considered as a separate corridor.

- The light source color for visual appliances must be clear or nominal white and must not exceed 1,000cd (*NFPA 72®*). In addition, special considerations apply when more than one visual appliance is installed in a room or corridor. *NFPA 72®* specifies that the separation between appliances must not exceed 100'. Visible notification appliances must be installed in accordance with *Tables 8* and *9*, using one of the following:
  - A single visible notification appliance
  - Two visible notification appliances located on opposite walls
  - More than two appliances for rooms 80' × 80' or larger (must be spaced a minimum of 55' from each other)

11.9.0 FACU Installation Guidelines

The guidelines for installing an FACU are as follows:

- When not located in an area that is continuously occupied, all fire alarm control equipment must be protected by a smoke detector (*Figure 80*). If the smoke detector is not designed to work properly in that environment, a heat detector must be used. It is not necessary to protect the entire space or room.

Table 8 Visual Notification Appliances Required for Corridors Not Exceeding 20'

Corridor Length (in ft)	Minimum Number of 15cd Appliances Required
0–30	1
31–130	2
131–230	3
231–330	4
331–430	5
431–530	6

33408-11_T08.EPS

Table 9 Room Spacing for Wall-Mounted Visual Notification Appliances

Maximum Room Size (in ft)	Minimum Required Light Output in Candelas (cd)		
	One Light per Room	Two Lights per Room*	Four Lights per Room**
20 × 20	15	Not allowable	Not allowable
30 × 30	30	15	Not allowable
40 × 40	60	30	Not allowable
50 × 50	95	60	Not allowable
60 × 60	135	95	Not allowable
70 × 70	185	95	Not allowable
80 × 80	240	135	60
90 × 90	305	185	95
100 × 100	375	240	95
110 × 110	455	240	135
120 × 120	540	305	135
130 × 130	635	375	185

* Locate on opposite walls
** One light per wall

33408-11_T09.EPS

Table 10 Room Spacing for Ceiling-Mounted Visual Notification Appliances

Maximum Room Size (in ft)	Maximum Ceiling Height (in ft)*	Minimum Required Light Output for One Light (cd)**
20 × 20	10	15
30 × 30	10	30
40 × 40	10	60
50 × 50	10	95
20 × 20	20	30
30 × 30	20	45
40 × 40	20	80
50 × 50	20	95
20 × 20	30	55
30 × 30	30	75
40 × 40	30	115
50 × 50	30	150

* Where ceiling heights exceed 30', visible signaling appliances must be suspended at or below 30' or wall mounted in accordance with *NFPA 72*.
** This table is based on locating the visible signaling appliance at the center of the room. Where it is not located at the center of the room, the effective intensity (cd) must be determined by doubling the distance from the appliance to the farthest wall to obtain the maximum room size.

33408-11_T10.EPS

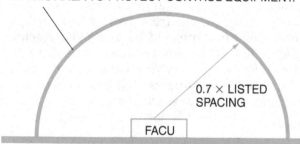

SMOKE DETECTOR REQUIRED ANYWHERE IN THIS AREA TO PROTECT CONTROL EQUIPMENT.

0.7 × LISTED SPACING

FACU

A smoke detector protecting control equipment must meet *NFPA 72*® spacing and placement standards, but the entire space (room) containing the FACP need not be protected.

33408-11_F80.EPS

Figure 80 Protection of an FACU.

- Detector spacing must be adjusted if the ceiling over the control equipment is irregular in one or more respects. This is considered protection against a specific hazard under *NFPA 72*® and does not require the entire chamber (room) containing the control equipment to be protected under the 0.7 rule.
- A means of silencing audible notification appliances from an FACU must be protected against unauthorized use. Most FACUs are located inside locked metal cabinets as shipped from the manufacturer. If the silencing switch is key-actuated or is locked within the cabinet, this provision should be considered satisfied. If the silencing means is via a keypad in a public area, key activation of the keypad or a user code is required. If the silencing means is within a room that is restricted to authorized use only, no additional measures should be required. The NFPA codes do not clearly define nor specify unauthorized or authorized use.
- FACU connections to the primary light and power circuits must be on a dedicated branch circuit with overcurrent protection rated at 20A or less. Any connections to the primary power circuit on the premises connected after the circuit breaker box must be directly related to the fire alarm system. No other use is permitted. This requirement does not necessitate a direct tap into the power circuit ahead of the distribution panel, although connecting ahead of the main disconnect is acceptable with Listed service equipment.
- The circuit breaker must be clearly identified as a fire alarm circuit control.

11.10.0 Trouble Signal Device Installation

Many fire alarm systems are self-diagnostic and can register trouble alarms by means of trouble signal indicators. In order to realize the benefits of the self-diagnostic ability, one must be made aware of any problem that develops in the system. For example, in the event of a loss of the primary power circuit, the system would switch to secondary power and send a trouble signal. If that trouble signal went unnoticed for 24 hours or more, the secondary power could fail, no protection would exist, and no one would be aware of this condition if the signal indicators were located in an unoccupied area.

Typically, trouble signal indicators are single gang wall plates featuring an audible and visual signal, and sometimes a means of silencing the audio trouble signal. Remote annunciators generally include such features and duplicate all or many of the other indicators that might be present at the FACU. If the FACU is located in an area where the trouble signal is likely to be heard, no remote indicators should be required. Therefore, unless the FACU is located in a manned area:

- Trouble signal devices or remote indicators must be located in a manned area where they are likely to be heard and seen, as determined by the AHJ.
- It would be reasonable for an AHJ to grant a written equivalency for a local trouble signal where a system included 24-hour off-premises monitoring service. This is because the monitoring station is a location where the signal is likely to be heard or seen.

12.0.0 FIRE ALARM-RELATED SYSTEMS AND INSTALLATION GUIDELINES

This section discusses various fire alarm-related systems, as well as the installation guidelines for each system.

12.1.0 Ancillary Control Relay Installation Guidelines

Ancillary functions, called protected premises fire safety functions, include such controls as elevator capture (recall), elevator shaft pressurization, HVAC system shutdown, stairwell pressurization, smoke management systems, emergency lighting, door unlocking, door hold-open device control, and building music system shutoff. For example, sound systems are commonly powered down by the fire alarm system so that the evacuation signal may be heard. In the normal state (*Figure 81*), an energized relay completes the power circuit to the device being controlled. Relays that are not located within 3' of the controlled device must be supervised.

> **NOTE**
> The circuit from the relay to the background music system is a remote control signaling circuit (see *NEC Article 725*). If this circuit is both powered by and controlled by the fire alarm, the circuit is a fire alarm circuit (see *NEC Article 760*).

12.2.0 Duct Smoke Detectors

Duct smoke detectors are not simply conventional detectors applied to HVAC systems. Duct smoke detectors are normally Listed for a slightly higher velocity of air movement and are tested under *UL 268A*. The primary function of a duct smoke detector is to turn off the HVAC system. This prevents the blower from spreading smoke throughout the building and stops the system from providing a forced supply of oxygen to the fire. Duct smoke detectors are not intended for early warning and notification. Relevant fire alarm-related provisions pertaining to HVAC systems can be found in *NFPA 90A* and *NFPA 90B*.

NFPA 90A covers when and where fire alarm-related provisions pertaining to HVAC systems apply, which includes each of the following types of buildings:

- HVAC systems serving spaces over 25,000 cubic feet in volume
- Buildings of Type III, IV, or V construction that are over three stories in height (see *NFPA 220*)
- Buildings that are not covered by other applicable standards
- Buildings that serve occupants or processes not covered by other applicable standards

NFPA 90B also covers when and where fire alarm-related provisions pertaining to HVAC systems apply, which includes the following:

- HVAC systems that service one- or two- family dwellings
- HVAC systems that service spaces not exceeding 25,000 cubic feet in volume

> **NOTE**
> No duct smoke detector requirements are found under *NFPA 90B*; however, any codes adopted by an AHJ must be checked for specific requirements. Duct smoke detector requirements may vary based on local or IBC-based codes.

12.2.1 Duct Detector Location

When determining the location of duct detectors prior to installation, use the following guidelines:

- In HVAC units over 2,000 cubic feet per minute (cfm), duct detectors must be installed on the supply side.
- Duct detectors must be located downstream of any air filters and upstream of any branch connection in the air supply.
- Duct detectors should be located upstream of any in-duct heating element.

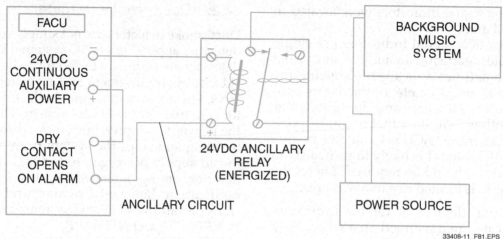

Figure 81 Music system control in the normal state.

33408-11_F81.EPS

- Duct detectors must be installed at each story prior to the connection to a common return and prior to any recirculation of fresh air inlet in the air return of systems over 15,000 cfm serving more than one story.
- Return air system smoke detectors are not required when the entire space served by the HVAC system is protected by a system of automatic smoke detectors and when the HVAC is shut down upon activation of any of the smoke detectors.

Refer to *NFPA 72®*, *Sections 17.7.4.2.1* and *17.7.4.2.2* for further information on the placement of duct detectors.

12.2.2 Conversion Approximations

The approximations given in *Table 11* are useful when the protected-premises personnel do not know the cfm rating of an air-handling unit, but do know either the tonnage or British thermal unit (Btu) rating. Additionally, the cfm rating may not always appear on air handling unit (AHU) nameplates or in building specifications. In this case, check the building mechanical plans.

Table 11 Conversions

Capacity Rating	CFM
1 ton	400
12,000 Btus	400
5 tons	2,000
60,000 Btus	2,000
37.5 tons	15,000
450,000 Btus	15,000

33408-11_T11.EPS

12.2.3 More Than One AHU Serving an Area

When more than one AHU is used to supply air to a common space, and the return air is drawn from this common space, the total capacity of all units must be used in determining the size of the HVAC system (*Figure 82*). This formal interpretation makes clear the fact that interconnected AHUs should be viewed as a system, as opposed to treating each AHU individually. When multiple AHUs serve a common space, the physical location of each AHU, relative to others interconnected to the same space, is irrelevant to the application of the formal interpretation.

The common space being served by multiple AHUs need not be contiguous (connected) for the formal interpretation to apply. In *Figure 83*, AHUs #3 and #4 serve common space and must be added together for consideration (1,100 + 1,100 = 2,200). Because 2,200 is greater than 2,000, duct smoke detectors are required on both. The same is true for AHU #1 and #2.

Duct smoke detectors may not be used as substitutes where open area detectors are required. This is because the HVAC unit may not be running when a fire occurs. Even if the fan is always on, the HVAC is not a Listed fire alarm device.

Duct smoke detectors must automatically stop their respective fans upon detecting smoke. It is also acceptable for the FACU to stop the fan(s) upon activation of the duct detector. However, fans that are part of an engineered smoke control or management system are an exception and are not always shut down in all cases when smoke is detected.

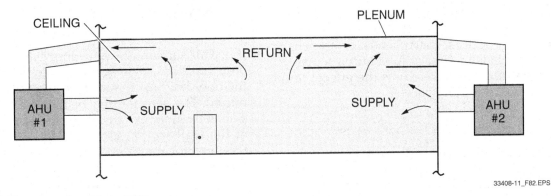

Figure 82 Non-ducted multiple AHU system.

33408-11_F82.EPS

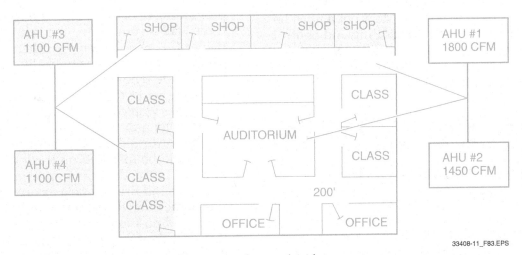

33408-11_F83.EPS

Figure 83 Example in which all AHUs require detectors on the supply side.

12.2.4 *Duct Smoke Detector Installation and Connections*

When a fire alarm system is installed in a building, all duct smoke detectors in that building must be connected to the fire alarm system as either initiating devices or as supervisory devices. The preference is as supervisory devices. Always check with the AHJ.

The code does not require the installation of a building fire alarm system. It does require that the duct smoke detectors be connected to the building fire alarm, if one exists. Duct smoke detectors, properly Listed and installed, accomplish their intended function when connected as initiating or supervisory devices.

When the building is not equipped with a fire alarm system, visual and audible alarm and trouble signal indicators (*Figure 84*) must be installed in a normally occupied area.

Most AHUs do not come with factory-installed duct detectors. If they are installed in a building with a fire alarm system, duct detectors must be installed and connected to the fire alarm system. In addition, the detectors may require standby power. Consult with the AHJ concerning how

33408-11_F84.EPS

Figure 84 Remote duct indicator.

many duct detectors can be placed in a zone. If it cannot be determined which detector activated after the power failure, the alarm indicator on the detector may be considered an additional function by the local AHJ and must be accessible.

Duct detectors for fresh air or return air ducts should be located six to ten duct widths from any openings, deflectors, sharp bends, or branch connections. This is necessary to obtain a representative air sample and to reduce the effects of stratification and dead air space.

12.3.0 Door Hold-Open Releasing Service

A door in a path or means of egress that is required or designed to be kept normally open is permitted to be automatically closed when special requirements are met as indicated in local codes, *NFPA 101*, or as specified by the AHJ. Door hold releases used to hold the door open must be actuated by the protective signaling and control system. In addition, if a smoke detector used for door release serving a stair enclosure is activated, all doors serving that stair enclosure must be re-

leased. Door hold release devices are not required to have secondary power per *NFPA 72*®.

- If a wall section above a door is 0" to 24" on both sides of the door, one smoke detector for the door hold-open and releasing service is required. The detector may be mounted on either side of the door at a distance equal to the depth of the wall section above the door. However, the detector may never be less than 12" from the wall section above the door, nor more than 5' away from the wall section above the door. The smoke detector must be centered on the doorway. Refer to *Figure 85* for the location and positioning of detectors for multiple doors.

- If the wall section above one side of the door is higher (24" to 60") than the other side, the single detector required is typically placed on the higher side. The theory is that smoke will rise to the higher level. When stratification or other conditions would delay the activation of the detector beyond the typical time that the detector on the lower side would activate, the detector should be installed on the lower side. The purpose of the doors is to prevent the spread of smoke and fire. If smoke is present on either side of the door, the doors should be released to the closed position immediately.

- If the depth of the wall section on both sides of the door is 24" to 60", detectors are required on both sides of the door. Location and spacing for both must be the same as for a single detector.

- When the depth of a wall section above a door is 60" or more, additional detectors may be required as indicated by an engineering evaluation.

- Where a detector is specifically Listed for doorframe mounting, or where a Listed combination or integral detector-door closer assembly is used, only one detector is required when it is installed in the manner recommended by the manufacturer. Where there are multiple doorways and Listed doorframe-mounted detectors, or where Listed combination or integral-detector door closer assemblies are used, there must be one detector for each single or double doorway.

- All detection devices used for door hold release service, whether integral or stand-alone, must be monitored for integrity in accordance with *NFPA 72*® when connected to the fire alarm system serving the protected premises.

12.4.0 Elevator Recall and Shutdown

Elevator recall is intended to route an elevator car to a non-fire floor, open its doors, and then put

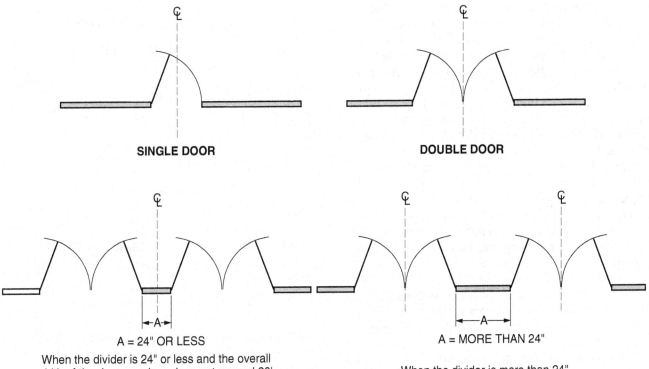

SINGLE DOOR

DOUBLE DOOR

A = 24" OR LESS

When the divider is 24" or less and the overall
width of the door openings does not exceed 20'.

A = MORE THAN 24"

When the divider is more than 24".

TWO DOUBLE DOORS

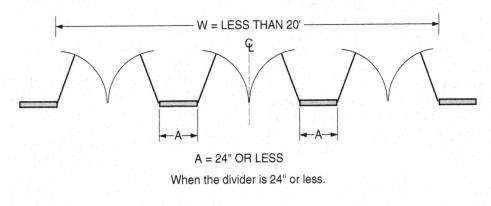

W = LESS THAN 20'

A = 24" OR LESS

When the divider is 24" or less.

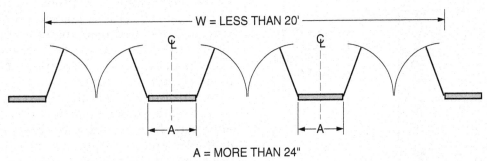

W = LESS THAN 20'

A = MORE THAN 24"

When the total width is greater than 20' or when the divider is over 24".

THREE DOUBLE DOORS

33408-11_F85.EPS

Figure 85 Detector locations for various door center line positions.

the car out of normal service until reset (Phase 1 recall). It is usually assumed that the recall floor is the level of primary exit discharge or a grade level exit. In the event that the primary level of exit discharge is the floor where the fire has been detected, the system should route the car to a pre-determined alternate floor, open its doors, and go out of normal service (Phase 1 recall to an alternate level). The operation of the elevators must be in accordance with *ANSI A17.1*.

Phase 1 recall (*Figure 86*) is the method of re-calling the elevator to the floor that provides the highest probability of safe evacuation (as determined by the AHJ) in the event of a fire.

Phase 1 recall to an alternate level (*Figure 87*) is the method of recalling the elevator to the floor that provides the next highest probability of safe evacuation (as determined by the AHJ) in the event of a fire. This type of recall is typically activated by the smoke detector in the lobby where the elevator would normally report during a Phase 1 recall.

Hoistway and elevator machine room smoke detectors must report an alarm condition to a building fire alarm control unit. However, notification devices may not be required to be activated if the control panel is in a constantly attended location. In facilities without a building fire alarm system, these smoke detectors must be connected to a dedicated fire alarm system control unit that must be designated as an Elevator Recall Control and Supervisory Panel.

Elevator recall must be initiated only by elevator lobby, hoistway, and machine room smoke detectors. Activation of manual pull boxes, heat detectors, duct detectors, and any smoke detec-

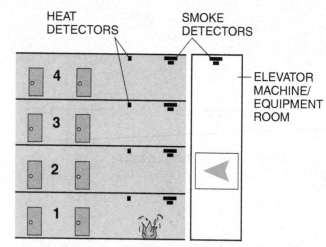

PHASE 1 RECALL TO AN ALTERNATE LEVEL:
Activation of the designated level detector sends all cars to the alternate floor.

33408-11_F87.EPS

Figure 87 Phase 1 recall to an alternate level.

tor not mentioned previously must not initiate elevator recall. In many systems, it would be inappropriate for the fire alarm control unit to initiate elevator recall. An exception is any system where the control function is selectable or programmable and the configuration limits the recall function to the specified detector activation only.

Caution should be used when using two-wire smoke detectors to recall elevators. Each elevator lobby, elevator hoistway, and elevator machine room smoke detector must be capable of initiating elevator recall when all other devices on the same IDC have been manually or automatically placed in the alarm condition.

Unless the area encompassing the elevator lobby is a part of a chamber being protected by smoke detectors, such as the continuation of a corridor, the smoke detectors serving the elevator lobby may be applied to protect against a specific hazard per *NFPA 72®*. The elevator lobby, hoistway, or associated machine room detectors may also be used to activate emergency control functions as permitted by local codes, *NFPA 101®*, or as specified by the AHJ.

Where ambient conditions prohibit installation of automatic smoke detection, other appropriate automatic fire detection must be permitted. For example, a heat detector may protect an elevator door that opens to a high-dust or low-temperature area. Always refer to local codes.

When sprinklers are installed in an elevator hoistway or machine room, elevator shutdown must be accomplished with a shunt-trip control circuit that removes all power from the elevator and elevator hoist. The shunt-trip circuit must be independent of all other elevator controls and

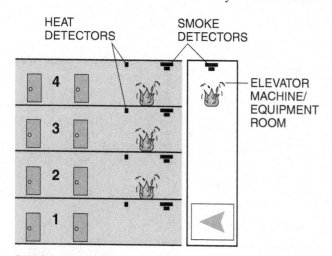

PHASE 1 RECALL:
Activation of any of these detectors sends all cars to the designated level.

33408-11_F86.EPS

Figure 86 Phase 1 recall.

must be activated before, or in conjunction with, activation of the sprinklers. Heat detectors, a water-flow switch, or a water pressure switch shall be used to activate the shunt-trip circuit. Smoke detectors cannot be used. If heat detectors are used, they must be located within 2' of each sprinkler. They must activate before the sprinklers so that sufficient time delay can be implemented to allow recall of the elevator from the top floor before the shunt-trip circuit is activated. If water flow or pressure switches are used, no time delay is permitted and the shunt-trip circuit must be activated immediately. The heat detectors and pressure or flow switches, along with the power for the shunt-trip circuit, must be supervised. Refer to *NFPA 72®, Section 21.4* for additional requirements and details.

12.5.0 Special Door Locking Arrangements

No lock, padlock, hasp, bar, chain, or other device intended to prevent free egress may be used at any door on which panic hardware or fire exit hardware is required (refer to local codes, *NFPA 101®*, or AHJ requirements). Note that these requirements apply only to exiting from the building. There is no restriction on locking doors against unrestricted entry from the exterior.

12.5.1 Delayed-Egress Locking Devices

Delayed-egress exiting is permissible for security reasons. A typical example might be a rear or remote side door in a large store, where shoplifting or employee pilferage is a potential problem (refer to local code requirements).

Approved, delayed-release locking devices are permitted if the following conditions are met:

- The applicable occupancy chapter of *NFPA 101®*, local codes, or the AHJ must allow it.
- Only low or ordinary hazard areas exist.
- The building is equipped with a supervised automatic detection system and/or supervised automatic sprinkler system, depending on local code.
- Doors automatically unlock when any two smoke detectors, a heat detector, or the sprinkler systems operate.
- Doors automatically unlock when power loss occurs to the locking mechanism.
- Doors automatically unlock within 15 seconds when force is applied to the release device for not more than three seconds (the AHJ may extend this 15 to 30 seconds). An audible alarm must sound at the doors.

- A sign with directions about the delay is provided at the doors.
- Emergency lighting must be provided at the doors.
- Doors may only relock by manual means.
- Not more than one delayed-locking arrangement in the path of egress to the nearest exit from any point in the building is permitted.
- Doors must unlock when primary power to the fire alarm system is lost.

If the locking system has backup power, it is acceptable for the doors to remain locked if the backup power is lost, provided the primary power is not lost. However, the code states that all emergency exits connected in accordance with *NFPA 72®* must unlock upon loss of the primary power to the fire alarm system serving the protected premises. The secondary power supply must not be used to maintain these doors in the locked condition.

The intent is to provide a delay to free egress, which deters unauthorized use, while allowing for emergency exiting within a safe time or upon primary power failure to the fire alarm system. The delay and alarm are intended to allow responsible personnel, such as security officials, to respond to and investigate unauthorized attempts to exit.

12.5.2 Stair Enclosure Doors

Upon activation of the fire alarm, stair enclosure doors must unlock to permit egress. They must also unlock to permit reentry into the building floors. This provision applies to buildings four stories and taller. The purpose of this requirement is to provide an escape from fire and smoke entering a lower level of the stair enclosure and blocking safe egress from the stair enclosure. See any local codes, *NFPA 101®*, or any AHJ requirements for additional details and exceptions.

12.6.0 Suppression System Fire Alarm Initiation

The following section discusses both dry and wet chemical extinguishing systems.

12.6.1 Dry Chemical Extinguishing Systems

Dry chemical extinguishing systems must be connected to the building fire alarm system if a fire alarm system exists in the structure. The extinguishing system must be connected as an initiating device. The standard (*NFPA 17*) does not require the installation of a building fire alarm system if one was not required elsewhere. Also see *NFPA 12, Standard on Carbon Dioxide Extin-*

guishing Systems, NFPA 12A, Standard on Halon 1301 Fire Extinguishing Systems, and *NFPA 2001, Standard on Clean Agent Fire Extinguishing Systems.*

12.6.2 Wet Chemical Extinguishing Systems

Wet chemical extinguishing systems must be connected to the building fire alarm system if a fire alarm system exists in the structure. As with dry chemical extinguishing systems, the standard (*NFPA 17A*) does not require the installation of a building fire alarm system. Also, see *NFPA 16, Standard for the Installation of Foam-Water Sprinkler and Foam-Water Spray Systems.*

12.7.0 Supervision of Suppression Systems

Each of the occupancy chapters of *NFPA 101®*, local codes, or AHJ requirements specifies the extinguishing requirements for that occupancy. The code specifies either an approved automatic sprinkler or an approved supervised automatic sprinkler system if a sprinkler system is required.

12.7.1 Supervised Automatic Sprinkler

Where a supervised automatic sprinkler is required, various sections of *NFPA 101®*, local codes, or AHJ requirements are applicable. These code sections usually begin with the phrase "Where required by another section of this code." Use of the word *supervised* in the extinguishing requirement section of each occupancy chapter is the method used to implement these two provisions. Sprinkler systems do not automatically require electronic supervision. Supervised automatic sprinkler systems require a distinct supervisory signal. This signal indicates a condition that would impair the proper operation of the sprinkler system, at a location constantly attended by qualified staff or at an approved remote monitoring facility. Water flow alarms from a supervised automatic sprinkler system must be transmitted to an approved monitoring station. When the supervised automatic sprinkler supervisory signal terminates on the protected premises in areas that are constantly attended by trained staff, the supervisory signal is not required to be transmitted to a monitoring facility. In such cases, only alarm and trouble signals need to be transmitted.

The following are some of the sprinkler elements that are required to be supervised where applicable:

- Water supply control valves
- Fire pump power
- Fire pump running
- Water tank levels
- Water tank temperatures
- Tank pressures
- Air pressure of dry-pipe systems

A sprinkler flow alarm must be initiated within 90 seconds of the flow of water equal to or greater than the flow from the sprinkler head, or from the smallest orifice (opening) size in the system. In actual field verification activities, the 90 seconds is measured from the time water begins to flow into the inspector's test drain, and not from the time the inspector's test valve is opened. The smallest orifice is the size of the opening at the smallest sprinkler head.

12.7.2 Sprinkler Systems and Manual Pull Boxes

Sprinkler systems, as well as other automatic systems that initiate a fire alarm system by a water flow or pressure switch, must include at least one manual box that is located where required by the AHJ. Manual boxes required elsewhere in the codes or standards can be considered to meet this requirement. Some occupancy chapters of *NFPA 101®*, local codes, or AHJ requirements allow the sprinkler flow switch to substitute for manual boxes at all the required exits. If such an option is used, this provision requires that at least one manual box be installed where acceptable to the AHJ.

12.7.3 Sprinkler System Shutoff Valves and Tamper Switches

Each shutoff valve must be supervised by a tamper switch. A distinctive signal must sound when the valve is moved from a fully open (off-normal) position. The supervisory signal must be initiated by the time the wheel reaches the end of the second pass.

Water flow and supervisory devices, in addition to their circuits, must be installed such that no unauthorized person may tamper with them, open them, remove them, or disconnect them without initiating a signal. Publicly accessible junction boxes must have tamper-resistant screws or tamper-alarm switches. Most water flow or pressure switch and valve tamper switch housings come from the manufacturer with tamper-resistant screws (hex or allen head), and they may be configured to signal when the housing cover is removed. If the device, circuit, or junction box requiring protection is in an area that is not accessible to unauthorized persons, no additional protective measures should be required. Simply sealing, locking, or removing the handle from a valve is not sufficient to meet the supervision requirement.

12.7.4 Tamper Switches versus Initiating Circuits

Water flow devices that are alarm-initiating devices cannot be connected on the same initiating circuit as valve supervisory devices. This is commonly done in violation of the code by connecting the valve tamper switch in series with the initiating circuit's EOL device, resulting in a trouble signal when activated. This method of wiring does not provide for a distinctive visual or audible signal. This statement is true for most conventional systems. It should be noted that at least one known addressable system has a Listed module capable of distinctly separating the two types of signals. These devices must be wired on the same circuit to meet the standard. Addressable devices are not connected to initiating circuits. They are connected to signaling line circuits.

12.7.5 Supervisory versus Trouble Signals

A supervisory signal must be visually and audibly distinctive from alarm signals and trouble signals. It must be possible to tell the difference between a fire alarm signal, a valve being off-normal (closed), and an open (broken wire) in the circuit.

12.7.6 Suppression Systems in High-Rise Buildings

Where a high-rise building is protected throughout by an approved, supervised automatic sprinkler system, valve supervision and water flow devices must be provided on each floor. In such buildings, a fire command center (central control system) is also required (refer to local code requirements).

13.0.0 HOUSEHOLD FIRE ALARM INSTALLATION GUIDELINES

This section contains the minimum requirements for the installation of fire warning equipment for use in family living quarters. The term *minimum requirements* cannot be stressed enough. The purpose of household fire alarms is life safety, not property protection. In many cases, it is desirable to provide a higher degree of fire detection and notification using superior equipment rated for commercial use, including system-type detection, additional detector and notification devices, and remote monitoring. These additional precautions may be used to provide a higher level of life safety, in consideration of who may be using the

---

On Site

Supervisory Switches

This supervisory switch is typical of those used with sprinkler systems to provide a tamper indication of valve movement. The switch is mounted on a valve with the actuator arm normally resting against the movable target indicator assembly. If the normal position of the valve is altered, the valve stem moves, forcing the actuator arm to operate the switch. The switch activates between the first and second revolutions of the control valve wheel.

33408-11_SA17.EPS

---

structure and all the risks and liabilities that may exist when designing a system. Some considerations that should be noted are the presence of children, the elderly, and handicapped persons, as well as other factors, such as construction material, emergency services response time, building age, and budget.

Today, most system-type installations in residential occupancies are accomplished using wireless devices. If a home has a hearing-impaired resident, visual notification devices must be installed. If a home has a mobility-impaired individual, windows may not be a viable means of escape. In this case, additional detectors are required to provide an earlier warning and more time to exit the building.

13.1.0 Smoke Detectors

System-type smoke detectors (*UL 268*) may be used anywhere single station alarms or interconnect station detectors (*UL 217*) are required. A system-type detector must be connected to a Listed/compatible control panel. A system-type detector, connected by wire or wireless means, provides a superior level of life safety compared to single station smoke alarms (*Figure 88*). A fire alarm system provides backup power for 24 hours or more and automatically maintains the battery charge. In addressable fire alarm or combination alarm panels, each detector reports to the monitoring station and provides the fire department with the structure location and the location of the activated detector. In addition, system-type detectors can also provide for the verified alarm function to reduce unwanted alarms.

13.1.1 Smoke Detector Locations

Smoke detectors must be in certain locations in order to provide optimal protection. The following list and figures show this:

> **NOTE**
>
> In the following figures, the backward S enclosed in a circle is not an error. It represents smoke that is rising and is the symbol for a smoke detector. Sub-symbols near the circle, such as P (photoelectric) or I (ionization), represent the type of smoke detector.

- A smoke detector must be located outside of each sleeping area and in the immediate vicinity of the bedrooms (*Figure 89*).
- At least one detector must be located on each floor of the family living unit, including the basement (not above the basement ceiling level) and excluding unfinished attics (*Figure 90*).

PHOTOELECTRIC SMOKE ALARM

IONIZATION SMOKE ALARM

33408-11_F88.EPS

Figure 88 Typical residential smoke alarms.

- In new construction, a smoke detector must be located in every bedroom in addition to those areas indicated above (*Figure 91*).

> **NOTE**
>
> When smoke travel is not delayed by a door, the lower level smoke detector may be eliminated. In split-level homes, the two levels that are split require only a smoke detector in the upper level if no door exists between the two levels. If a door does exist, even if never closed, a smoke detector is required on each level. Some AHJs interpret this to mean that if a doorway exists, both levels require a smoke detector. There are two reasons that this may be appropriate. First, a door may be installed later anywhere a doorframe exists. Second, if the doorway has a build-down from the ceiling, smoke travel to the smoke detector will be delayed. If there is any doubt about whether a detector is needed, consult the local AHJ.

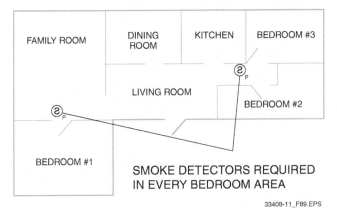

Figure 89 Existing structure bedroom smoke detector requirements.

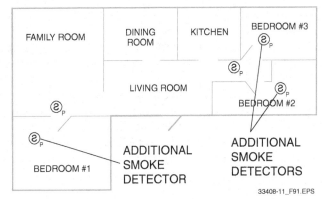

Figure 91 New construction bedroom smoke detector requirements.

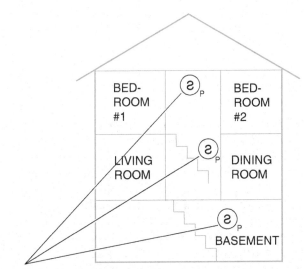

SMOKE DETECTORS ARE REQUIRED ON EACH LEVEL

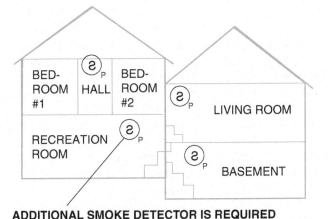

ADDITIONAL SMOKE DETECTOR IS REQUIRED
ONLY IF THERE IS A DOOR BETWEEN LEVELS

Figure 90 Multilevel structure smoke detector requirements.

• Smoke detectors should be located throughout the basement, existing bedrooms, dining room, utility room, and all hallways, except where a smoke detector is likely to generate unwanted alarms.
• Smoke detectors must not be installed in kitchens, attics, furnace rooms, or garages unless specifically Listed for the application.
• Smoke detectors must not be installed within 3' of a door to a kitchen or to a bathroom containing a tub or shower unless specifically Listed for the application.
• Smoke detectors must not be installed within 3' of a forced air system register unless specifically Listed for the application.

13.1.2 Smoke Detector Precautions/ Recommendations

Caution should be used when placing a smoke detector in a hallway outside a bathroom. Steam could activate the detector when the bathroom door is opened after a hot shower.

Keep detectors 3' or more away from electrical noise-producing devices, such as fluorescent lights.

Alarm verification is strongly recommended when the system is monitored for emergency services notification or alarm. Reducing unwanted alarms saves lives. However, local codes must be checked to determine that alarm verification is not prohibited.

13.2.0 Household Heat Detectors

Heat detectors should be installed in kitchens, unfinished attics, garages, furnace rooms, and utility rooms. Heat detectors used in household systems must be rated for 50' spacing. The rated temperature of the detector must be at least 20°F above the highest normal anticipated ambient temperature of the space, and as close to this temperature as possible.

If household heat detectors are mounted to open-joist ceilings, they must be mounted on the bottoms of joists with the horizontal spacing reduced by one-half in the direction perpendicular to the joist. Spacing is not reduced in the direction parallel to the joist.

13.3.0 Household Audibility Considerations

The activation of any smoke detector in the structure must cause an alarm to sound such that it can be heard throughout the structure. The activation of one detector does not have to activate a sounder in all detectors (in system-type detectors). However, if one detector is activated, the alarm must be clearly audible in all bedroom areas over background noise levels with all intervening doors closed (equivalent distribution of alarm signal). Because a typical window air conditioner or humidifier produces approximately 55dB, a sound level of at least 70dB would be needed in the bedroom area (not at the indicating device). Interconnect-type smoke detectors require wiring such that if any one detector is activated, all detectors sound an alarm.

> **NOTE**
> When existing interconnected smoke detectors are already installed, any detector that is added must be clearly heard throughout the structure when activated. This includes clear audibility through closed doors and over normal background noise, such as room air conditioners. Some multi-station detectors have interconnecting Listed relay modules available for connecting the detectors to existing systems.

When referring to notification signaling in household systems, the term *household* usually refers to single-family dwellings. *NFPA 101®* addresses living units, suites of rooms, and similar areas within hotels, apartments, dorms, and other buildings. This code requires that the alarms must sound only within an individual living unit, suite of rooms, or similar area and must not actuate the building fire alarm system. However, remote notification is permitted. Because of the number of unwanted alarms caused by household activities, it is not desirable to evacuate an entire building because of burnt toast or other incidental events. This only causes the residents to assume that an alarm is false, and the value of the life safety system is lost. By following the code, if an actual fire were to occur, the household smoke detector within the living unit would sound. The occupant would activate a manual pull box in the hallway on the way out, thereby notifying the rest of the building occupants.

13.4.0 Extra Sounders for Greater Life Safety

Sirens, horns, and bells may fail or be disabled by a fire. Multiple fire sounders provide redundancy and, if properly placed, provide ample sound levels to wake all sleeping occupants. Remember that the code is a minimum requirement and a higher level of life safety is always acceptable. Because the circuitry of household notification appliances is not required to be supervised, regular testing is very important.

13.5.0 Household Visible Notification Appliances

Visible notification appliances used in rooms where a hearing-impaired person sleeps must have a minimum rating of 177cd, if mounted on the ceiling, for a maximum room size of 14' × 16'. In larger rooms, the visible notification device must be within 16' of the pillow. When the visible notification appliance in a sleeping room is mounted more than 24" below the ceiling, a minimum rating of 110cd is permitted. Visible notification appliances in other areas must have a minimum rating of 15cd.

13.6.0 Primary Power

Make sure that the power source for a fire alarm is not turned on and off by a light switch. While a lighting circuit is a good choice to supply a household fire alarm, power supplied to the fire alarm must be taken from the circuit before any switch. The reason a lighting circuit is recommended is that the occupants are not very likely to leave this circuit disabled if a fuse blows or a circuit breaker trips. Do not use a circuit that is protected by a ground fault circuit interrupter (GFCI) device.

For electrically powered detectors, an AC primary power source must be used in all new construction. This does not mean the detector itself must be AC powered. The system's primary power must be provided by utility power, except as otherwise indicated by the code.

If an external voltage stepdown transformer is used as the fire alarm primary power supply, the transformer mounting screw is part of the UL Listing of the transformer. It must be used to hold the transformer in place.

13.7.0 Standby Power Requirements

Batteries must be capable of operating a residential fire alarm system for 24 hours without the primary power source. After 24 hours they must provide enough power to sound the alarm for no less than four minutes.

The sound level of all sounding devices must be maintained at the required level during the four-minute period. The system must be capable of performing with all equipment that may be drawing power from the secondary power supply during normal operation. In the case of combination panels, this would include all motion detectors, control relays, notification devices, and any other equipment that receives power from the same power supply.

A common misunderstanding of this requirement relates to a combination panel that has separate termination points for fire alarm power and burglar alarm power. This is the same power supply; it has only one transformer, one charging circuit, and one battery system. In short, the burglar alarm draws power from the fire alarm transformer and the fire alarm battery. The burglar alarm's consumption of power shortens the time that the fire alarm is able to operate on secondary power.

13.8.0 Combination Systems

Combination systems are permitted, provided the equipment Listed for such use. The system must be wired, programmed, operated, and maintained per the manufacturer's instructions for combination use.

13.9.0 Monitoring/Supervising Station Systems

Any communication method described in *NFPA 72®* is permitted for transmission of signals from household fire warning equipment to a supervising station. Monitoring of a household system is not required. However, if the system is monitored, all of the provisions of the NFPA apply as appropriate with the following exceptions:

- Only one telephone line is required for one- and two-family homes. If only one line is used, an external RJ-31X interface device must be used to connect the system to the telephone line, unless the FACU contains the interface internally. In either case, all telephones in the home must

be connected to the telephone line after the RJ-31X interface device. The RJ-31X device allows the FACU to terminate any use of the telephone line and capture the telephone line to report a fire, even if the telephone line is in use at the time.
- Each DACT may only be required to be programmed to call a single DACR.
- Each DACT serving a one- or two-family home must transmit a test signal to its associated receiver at least monthly.

13.10.0 Supervising Station Verification of Signals

On receipt of an alarm signal from household fire warning equipment, the supervising station must immediately (within 90 seconds) retransmit the alarm to the public fire communications center. An exception in the code allows the supervising station to contact the home for verification of an alarm condition. Where acceptable assurance is provided within 90 seconds that the fire service is not needed, retransmission of an alarm to the public fire communications center is not required. If a competent person does not indicate that the fire service is not needed within 90 seconds, the public fire communications center must be immediately notified. If the answering party does not sound old enough or awake enough to understand the risk, importance of immediate action, and what to do, or if the answering party sounds under the influence of drugs or alcohol, the public fire communications center must be notified.

13.11.0 User Instructions

The user must be provided with an instruction manual containing the following:

- Typical layouts
- Printed operating and testing instructions
- Printed information on establishing an emergency escape plan
- Printed information on repair, replacement, and warranty service

Since the system user often misplaces this information and may not recall ever receiving the manuals, a written receipt should be obtained for these documents. This information is normally shipped with every detector and control panel and must be provided to the user.

13.12.0 Wiring Methods

Residential fire alarm system wiring and wiring installation must be in accordance with *NEC Ar-*

ticle 760. With combination fire/intrusion alarm systems, all circuits controlled and powered by the fire alarm system must also be wired in accordance with *NEC Article 760*.

13.13.0 Residential Testing Requirements

The homeowner must test all smoke detectors at least once per month. In addition, every home fire alarm system must be tested by a qualified service technician at least every three years. All testing must be in accordance with the manufacturer's instructions.

14.0.0 INSPECTION, TESTING, COMMISSIONING, AND MAINTENANCE

This section contains information on the inspection, testing, commissioning, and maintenance of life safety systems.

Testing is the only way to determine that the system is providing the designed level of life safety. Improper wiring, damage to the system, defective equipment, and damaged equipment often cause the fire alarm panel to generate a trouble signal, but not always. Testing in accordance with the manufacturer's written testing procedure is required to determine that the equipment is likely to function correctly in the event of an emergency. *NFPA 72®* requires inspection, testing, and maintenance programs to:

- Satisfy the requirements of *NFPA 72®*
- Conform to the equipment manufacturer's recommendations
- Verify proper operation of the fire alarm system

NFPA 72® requires that inspection, testing, and maintenance of fire alarm systems be performed by qualified people. In addition to the qualifications indicated by this section, those persons should be aware of the responsibility undertaken to ensure life safety equipment is working properly. This includes both the moral and legal responsibilities of these duties.

14.1.0 Before Testing

Prior to any testing, it is extremely important to notify everyone who may be affected. This includes the monitoring station staff, building occupants, building engineers, owners, and anyone else who may become aware of the alarm and think it is an actual emergency instead of a test. Do not forget that people may continue to enter the building after an announcement that a test is being conducted. Be sure to post a notice at all entrances.

Review the results of activating a fire alarm system and the impact on other systems. Fire alarm systems often control HVAC systems, suppression systems, building security systems, building power systems, or other systems. Take appropriate action to prevent unnecessary operation of auxiliary systems. For example, unlocking all the doors may create a security risk unless extra security measures are taken prior to activating the fire alarm system. Discharging of suppression systems may be costly and possibly dangerous.

14.2.0 Precautions for Occupied Buildings

It is possible for a real fire to occur while a scheduled test is being conducted. Always be sure of the signal that is expected to result from the device being tested. If a signal from another device or zone is received, take immediate action in accordance with the policy of the AHJ. If the AHJ does not have a policy, your company should have a policy that provides for the highest level of life safety reasonably possible during a test. If you are uncertain what action you should take, evacuate the building and immediately notify emergency services. Someone familiar with the policy should remain at the control panel any time the supervising station has been instructed to ignore alarm signals from the system.

14.3.0 Definitions

The following terms are necessary for understanding the testing, maintenance, and commissioning of a system:

- *Test* – A test is an act intended to verify the functional operability of all or part of a device, circuit, or system.
- *Inspection* – An inspection is a visual evaluation to verify or confirm that something appears as desired, is in the proper location, appears to be attached properly, has a listing label attached, and so on.
- *Maintenance* – Maintenance is an act or acts intended to prevent failure and to keep equipment, devices, and systems in proper working order.
- *Acceptance* – Acceptance is when the AHJ has found the system acceptable.
- *Test methodology* – Test methodology is the procedure used to perform a test.
- *Test frequency* – Test frequency is the time interval at which a test is performed.

14.4.0 General Requirements

A 100-percent acceptance test must be performed in the presence of the AHJ following an installation or any alterations. Inspection, testing, and maintenance programs must satisfy code requirements and the equipment manufacturer's instructions.

Before connecting any devices to wiring, tests should be conducted for stray voltage, ground faults, short circuits, and loop resistance. Any voltage found on a pair of wires may be from an induced source. This may result in unwanted alarms and shorten the life of the system. Fire alarm wires should not be run parallel to power, lighting, or Class 1 circuits. A minimum spacing of 2" is required. Short circuits result in an alarm signal or an overload when connected to the control panel.

Ground faults must be cleared prior to connecting to the control panel. If more than one ground fault occurs on a system, it may destroy part of the control panel. Ground faults are very serious problems and must be corrected immediately.

The loop resistance should be recorded for future troubleshooting. Poor connections and corrosion may cause intermittent problems. These problems can often be found if the circuit resistance prior to the problem is known so that it may be compared.

14.5.0 Central Stations (Certificated Systems)

The central station must have competent people available to inspect, test, maintain, and repair the system. Repairs must start within four hours of discovering a problem and continue until the repair is completed. The central station must maintain spare parts for protected premises and monitoring center equipment.

14.6.0 All Systems

All systems must be under the supervision of a qualified person. The owner must provide for proper maintenance of the system. The system equipment must be kept in normal condition for operation. A record of tests and operations of the system must be kept for two years unless off-premise monitoring is provided. If off-premise monitoring is provided, the records must be kept for one year. Original installation completion testing records must be maintained for the life of the system.

14.7.0 Testing Methodology

Testing must be done in accordance with the manufacturer's instructions and *NFPA 72*®. *Table 12* is a summary of the required testing showing the testing cycle. A sample inspection and testing form is provided in *Appendix A*. A general description of some of the testing procedures is as follows:

- *Open and grounds* – A single open and a ground fault must be created when testing each device. A trouble signal must be indicated at the remote annunciators and monitoring station as applicable.
- *Manual boxes* – Actuate annually per manufacturer's instructions.
- *Non-restorable spot-type heat detectors* – Perform mechanical and electrical testing only. Two percent of the heat detectors must be sent to a testing lab at 15 years and every 5 years thereafter. If a failure occurs, additional detectors must be removed and tested to determine either a general problem or a localized problem.
- *Restorable spot-type heat detectors* – To test these detectors, expose them to a heat source such as a hair dryer or shielded heat lamp for one minute.

> **NOTE**
> In the 2010 edition of *NFPA 72*, *Table 14.3.1* lists visual inspection schedules. *NFPA 72*, *Table 14.4.2.2* describes test methods and *NFPA 72*, *Table 14.4.5* lists testing schedules. Visual inspection schedules and testing schedules may be different for the same component or device.

> **NOTE**
> In cool or cold locations, a rate-of-rise detector may be tested by placing both warm hands on the detector for one minute. This prevents the overheating of the fixed temperature element by accident. Test annually.

- *Smoke detectors (functional test)* – Test these detectors in place with a Listed test aerosol, punk, or other substance. Shorting across the circuit terminals is not a functional test. Follow the manufacturer's instructions. Some manufacturers do not recommend aerosols. Test annually.
- *Smoke detectors (calibrated sensitivity test)* – Any method Listed for universal use or for the specific make and model detector can be used. Some models can be tested with a magnet or by pressing a test switch. In other models, this

type of test just shorts across the termination points. Consult the manufacturer's instructions. Test during the first year of installation and then every other year.

- *Electric detectors thermal link* – Operate these devices electronically. Test semi-annually.
- *Other detectors (flame, spark, gas, etc.)* – Test radiant energy fire detectors semi-annually per the manufacturer's instructions.
- *Sprinkler water flow switch* – Allow water to flow through the inspector's test valve. An alarm must be activated within 90 seconds of water discharge. Know where the water is going before testing. Test semi-annually.
- *Extinguishing systems* – Do not discharge the agent. Test per the extinguishing system manufacturer's instructions in consultation with the system installer.
- *Sprinkler valves* – Fully open, then operate the valve to close it. The valve tamper should be detected within two revolutions or when the stem moves one-fifth of its travel distance. Test semi-annually.

> **CAUTION**
>
> Make sure that you have the proper certification to perform sprinkler system pressure tests; otherwise, you may accidentally activate the sprinkler system.

- *Pressure sources* – Bleed 10 psi off the normal pressure. A low air pressure signal should be activated. Restore the pressure to normal. Increase the pressure by 10 psi. A high air pressure signal should be activated. The high and low air pressure signals may be one signal (not distinctive from one another), but that signal must be activated under both conditions. Test semi-annually.
- *Pressure tank level* – Drain 3" of water from the normal level. A low water signal should be activated. Restore the water level to normal. Increase the water level. A high water signal should be activated. Test semi-annually.
- *Water storage tank levels* – Drain the water to 12" below normal or, if accessible, move the float on the water level switch. A low water signal should be activated. Test semi-annually.
- *Remote notification appliances* – Verify that all applicable FACU status changes are indicated at the remote notification appliance. Test annually.
- *Notification appliances (audibility)* – Take A-weighted measurements at 5' above the floor in all areas that can be occupied. The sound level must be 15dBA above the ambient sound.

Verify the voice clarity if applicable. The clarity of a voice evacuation system is extremely important. If the signal is not understood, the building occupants may not evacuate, may panic, or may not understand any special instructions. The purpose of a voice evacuation system is to increase the level of life safety by providing clear emergency signaling, providing special messages, and reducing panic. Test annually.

- *Notification appliances (functional)* – Initiate a general alarm. Verify that every audible, visual, and voice appliance is activated. Test every firefighter's two-way telephone. Test annually.
- *Control panels* – Illuminate all lamps and LEDs, remove the fuses, and verify that trouble signals occur. Test the primary power by interrupting the secondary power and verify occurrence of trouble signals. Induce the maximum load for five minutes while the secondary power is disconnected. Test the secondary power by interrupting the primary power. Verify the trouble signal is activated. Measure the standby current per the manufacturer's instructions and calculate adequacy for required standby time. Induce the maximum load for five minutes. Test quarterly in unmonitored systems and annually in monitored systems.
- *Control panel trouble silence* – Induce a trouble signal and verify audible and visual indications. Silence the audible indication and verify that visual indications remain on. Induce a supervisory signal and verify the audible and visual indications. If a common sounder is used, it should sound. Silence and correct all troubles. If the silence switch is a maintained switch, verify that the trouble signal sounds until the switch is restored to normal. Test quarterly in unmonitored systems and annually in monitored systems.
- *Control panel zone and city disconnect* – Actuate each of these separately and verify a trouble signal is activated. Test quarterly.
- *Control panel alarm silence* – Induce an alarm, then actuate a trouble signal and verify that it is indicated. Induce additional alarms on another circuit and verify that the general alarm signal is reactivated. Reset all alarm and trouble signals (reset panel). Actuate and verify a trouble signal when no alarm condition is present. Test quarterly in unmonitored systems and annually in monitored systems.
- *Control panel supervisory silence* – Induce a supervisory signal and verify that both visual and audible indications occur. Check that the signal is either visually or audibly distinct from the trouble signal. Silence, but do not restore,

the supervisory signal. Activate another supervisory circuit and verify indications. Test quarterly in unmonitored systems and annually in monitored systems.

- *Control panel signaling circuits* – Induce an alarm, trouble, and supervisory condition and verify that each is properly received at the monitoring center. Test every condition monitored. Test quarterly.
- *Batteries* – Test batteries per the manufacturer's instructions and as indicated in *NFPA 72®, Table 14.4.5*.

14.8.0 After Testing

After testing is completed, be sure to notify everyone involved. This should include the monitoring station, the building owners or their representatives, and the building occupants. Do not forget to remove any signage indicating the test at entrances. Failing to notify everyone may result in serious injury, property damage, and loss of life. The monitoring station may disregard an actual signal. Occupants may not evacuate the building. People may continue to enter the building. Building staff may not take emergency action.

Restore all systems to their normal states to ensure life safety. Verify that all indicating lights are in their normal condition. Remove all debris from the fire control room.

Complete a written record of your test and provide the customer with a copy. The record should include the name of the technician who conducted the test, a notice of any problems or deficiencies, and any recommendations. When required, notify the AHJ of any problems that result from the testing. When problems have been corrected and tested, a certificate of completion should be filled out. *Appendix B* contains a typical certificate of completion that may be modified and used if desired.

Place a copy of the final inspection and testing report along with the certificate of completion in the premise fire alarm records, and file a copy in your office in case the premise copy is lost.

15.0.0 COMMISSIONING

As previously described in the *System Commissioning and User Training* module, system commissioning is a process by which a formal and organized approach is taken to obtaining, verifying, and documenting the installation and perfor-

mance of a particular system or systems. The goal of commissioning is to make sure that a system operates as intended and at optimum efficiency. Commissioning is normally required for newly installed or retrofitted commercial and industrial systems. Because each building and its systems are different, the specific elements of the commissioning process must be tailored to fit each situation. However, the objectives for performing system commissioning are the same:

- Verify that the system design meets the functional requirements of the owner.
- Verify that all systems are properly installed in accordance with the design and specifications.
- Verify that all systems and components meet required local, state, and other codes.
- Verify and document the proper operation of all equipment, systems, and software.
- Verify that all documentation for the system is accurate and complete.
- Train building operator and maintenance personnel to efficiently operate and maintain the installed equipment and systems.

NFPA 3, Recommended Practice on Commissioning and Integrated Testing of Fire Protection and Life Safety Systems, formalizes the commissioning process. The objective of *NFPA 3* is to establish "the minimum requirements for procedures, methods and documentation for commissioning and the integrated testing of active and passive fire protection and life safety systems." Its purpose is to "provide a reasonable degree of assurance that fire protection and life safety systems function in accordance with the Basis of Design (BOD) and the Owner's Project Requirements (OPR)." In order to accomplish the stated mission, *NFPA 3* defines a step-by-step commissioning process. It also identifies the participants in the commissioning process, along with their required skills and qualifications, and describes the functions of each participant. Although *NFPA 3* is a recommended practice rather than a code, it is reasonable to expect that it will be adopted by owners and AHJs.

16.0.0 TROUBLESHOOTING

The troubleshooting approach to any fire alarm system is basically the same. Regardless of the situation or equipment, some basic steps can be followed to isolate problems:

Table 12 NFPA-Specified Fire Alarm Testing (1 of 9)

Data from NFPA 72®-2002, Table 10.4.3

Component	Initial/ Reacceptance	Monthly	Quarterly	Semiannually	Annually	Table 10.4.2.2 Reference
1. Control Equipment — Building Systems Connected to Supervising Station						1, 7, 16, 17
(a) Functions	X	—	—	—	X	—
(b) Fuses	X	—	—	—	X	—
(c) Interfaced equipment	X	—	—	—	X	—
(d) Lamps and LEDs	X	—	—	—	X	—
(e) Primary (main) power supply	X	—	—	—	X	—
(f) Transponders	X	—	—	—	X	—

Item 1 in Table 10.4.3 refers to systems connected to a supervising station that receives all three signals: alarm, trouble, and supervisory. In unmonitored systems, signals are not transmitted to a supervising station for appropriate action to be taken. Therefore, quarterly inspection of these items to ensure reliability is necessary. See item 2 below for quarterly testing for unmonitored systems.

Component	Initial/ Reacceptance	Monthly	Quarterly	Semiannually	Annually	Table 10.4.2.2 Reference
2. Control Equipment — Building Systems Not Connected to a Supervising Station	—	—	—	—	—	1
(a) Functions	X	—	X	—	—	—
(b) Fuses	X	—	X	—	—	—
(c) Interfaced equipment	X	—	X	—	—	—
(d) Lamps and LEDs	X	—	X	—	—	—
(e) Primary (main) power supply	X	—	X	—	—	—
(f) Transponders	X	—	X	—	—	—
3. Engine-Driven Generator — Central Station Facilities and Fire Alarm Systems	X	X	—	—	—	—
4. Engine-Driven Generator — Public Fire Alarm Reporting Systems	X (weekly)	—	—	—	—	—
5. Batteries — Central Station Facilities						
(a) Lead-acid type	—	—	—	—	—	6b
1. Charger test (replace battery as needed.)	X	—	—	—	X	—
2. Discharge test (30 minutes)	X	X	—	—	—	—
3. Load voltage test	X	X	—	—	—	—
4. Specific gravity	X	—	—	X	—	—
(b) Nickel-cadmium type	—	—	—	—	—	6c
1. Charger test (replace battery as needed.)	X	—	X	—	—	—
2. Discharge test (30 minutes)	X	—	—	—	X	—
3. Load voltage test	X	—	—	—	X	—
(c) Sealed lead-acid type	X	X	—	—	—	6d
1. Charger test (replace battery within 5 years after manufacture or more frequently as needed.)	—	X	X	—	—	—

Section 5(c) was modified to require sealed lead-acid batteries be replaced within 5 years of the date of manufacture or more frequently as needed. In the past, sealed lead-acid batteries were required to be replaced after 4 years with no mention of whether this was the date of manufacture or the date of installation. (Also see Table 10.4.3, items 6(d)1 and 7(c)1.)

33408-11_T12A.EPS

Table 12 NFPA-Specified Fire Alarm Testing (2 of 9)

Component	Initial/ Reacceptance	Monthly	Quarterly	Semiannually	Annually	Table 10.4.2.2 Reference
2. Discharge test (30 minutes)	X	X	—	—	—	—
3. Load voltage test	X	X	—	—	—	—
6. Batteries — Fire Alarm Systems						
(a) Lead-acid type	—	—	—	—	—	6b
1. Charger test (replace battery as needed.)	X	—	—	—	X	—
2. Discharge test (30 minutes)	X	—	—	X	—	—
3. Load voltage test	X	—	—	X	—	—
4. Specific gravity	X	—	—	X	—	—
(b) Nickel-cadmium type	—	—	—	—	—	6c
1. Charger test (replace battery as needed.)	X	—	—	—	X	—
2. Discharge test (30 minutes)	X	—	—	—	X	—
3. Load voltage test	X	—	—	X	—	—
(c) Primary type (dry cell)	—	—	—	—	—	6a
1. Load voltage test	X	X	—	—	—	—
(d) Sealed lead-acid type	—	—	—	—	—	6d
1. Charger test (replace battery within 5 years after manufacture or more frequently as needed.)	X	—	—	—	X	—
2. Discharge test (30 minutes)	X	—	—	—	X	—
3. Load voltage test	X	—	—	X	—	—
7. Batteries — Public Fire Alarm Reporting Systems Voltage tests in accordance with Table 10.4.2.2, items 7(1)–(6)	X (daily)	—	—	—	—	—
(a) Lead-acid type	—	—	—	—	—	6b
1. Charger test (replace battery as needed.)	X	—	—	—	X	—
2. Discharge test (2 hours)	X	—	X	—	—	—
3. Load voltage test	X	—	X	—	—	—
4. Specific gravity	X	—	—	X	—	—
(b) Nickel-cadmium type	—	—	—	—	—	6c
1. Charger test (replace battery as needed.)	X	—	—	—	X	—
2. Discharge test (2 hours)	X	—	—	—	X	—
3. Load voltage test	X	—	X	—	—	—
(c) Sealed lead-acid type	—	—	—	—	—	6d
1. Charger test (replace battery within 5 years after manufacture or more frequently as needed.)	X	—	—	—	X	—
2. Discharge test (2 hours)	X	—	—	—	X	—
3. Load voltage test	X	—	X	—	—	—
8. Fiber-Optic Cable Power	X	—	—	—	X	12b
9. Control Unit Trouble Signals	X	—	—	—	X	9
10. Conductors — Metallic	X	—	—	—	—	11
11. Conductors — Nonmetallic	X	—	—	—	—	12

33408-11_T12B.EPS

Table 12 NFPA-Specified Fire Alarm Testing (3 of 9)

Component	Initial/ Reacceptance	Monthly	Quarterly	Semiannually	Annually	Table 10.4.2.2 Reference
12. Emergency Voice/Alarm Communications Equipment	X	—	—	—	X	18
13. Retransmission Equipment (The requirements of 10.4.7 shall apply.)	X	—	—	—	—	—
14. Remote Annunciators	X	—	—	—	X	10
15. Initiating Devices	—	—	—	—	—	13
(a) Duct detectors	X	—	—	—	X	—
(b) Electromechanical releasing device	X	—	—	—	X	—
(c) Fire extinguishing system(s) or suppression system(s) switches	X	—	—	—	X	—
(d) Fire–gas and other detectors	X	—	—	—	X	—
(e) Heat detectors (The requirements of 10.4.3.4 shall apply.)	X	—	—	—	X	—
(f) Fire alarm boxes	X	—	—	—	X	—
(g) Radiant energy fire detectors	X	—	—	X	—	—
(h) System smoke detectors — functional	X	—	—	—	X	—
(i) Smoke detectors — sensitivity (The requirements of 10.4.3.2 shall apply.)	—	—	—	—	—	—
(j) Single- and multiple-station smoke alarms (The requirements for monthly testing in accordance with 10.4.4 shall also apply.)	X	—	—	—	X	—
(k) Single- and multiple-station heat alarms	X	—	—	—	X	—
(l) Supervisory signal devices (except valve tamper switches)	X	—	X	—	—	—
(m) Waterflow devices	X	—	—	X	—	—
(n) Valve tamper switches	X	—	—	X	—	—
16. Guard's Tour Equipment	X	—	—	—	X	—
17. Interface Equipment	X	—	—	—	X	19
18. Special Hazard Equipment	X	—	—	—	X	15
19. Alarm Notification Appliances	—	—	—	—	—	14
(a) Audible devices	X	—	—	—	X	—
(b) Audible textual notification appliances	X	—	—	—	X	—
(c) Visible devices	X	—	—	—	X	—
20. Off-Premises Transmission Equipment	X	—	X	—	—	—
21. Supervising Station Fire Alarm Systems — Transmitters	—	—	—	—	—	16
(a) DACT	X	—	—	—	X	—
(b) DART	X	—	—	—	X	—
(c) McCulloh	X	—	—	—	X	—
(d) RAT	X	—	—	—	X	—
22. Special Procedures	X	—	—	—	X	21

33408-11_T12C.EPS

Table 12 NFPA-Specified Fire Alarm Testing (4 of 9)

Component	Initial/ Reacceptance	Monthly	Quarterly	Semiannually	Annually	Table 10.4.2.2 Reference
3. Supervising Station Fire Alarm Systems	—	—	—	—	—	17
— Receivers						
(a) DACR	X	X	—	—	—	—
(b) DARR	X	X	—	—	—	—
(c) McCulloh systems	X	X	—	—	—	—
(d) Two-way RF multiplex	X	X	—	—	—	—
(e) RASSR	X	X	—	—	—	—
(f) RARSR	X	X	—	—	—	—
(g) Private microwave	X	X	—	—	—	—

33408-11_T12D.EPS

Table 12 NFPA-Specified Fire Alarm Testing (5 of 9)

Data from NFPA 72®-2010, Table 14.4.5

Component	Initial/ Reacceptance	Monthly	Quarterly	Semiannually	Annually	Table 14.4.2.2 Reference
1. Control equipment — building systems connected to supervising station						1, 7, 18, 19
(a) Functions	X	—	—	—	X	—
(b) Fuses	X	—	—	—	X	—
(c) Interfaced equipment	X	—	—	—	X	—
(d) Lamps and LEDs	X	—	—	—	X	—
(e) Primary (main) power supply	X	—	—	—	X	—
(f) Transponders	X	—	—	—	X	—
2. Control equipment — building systems not connected to a supervising station						1
(a) Functions	X	—	X	—	—	—
(b) Fuses	X	—	X	—	—	—
(c) Interfaced equipment	X	—	X	—	—	—
(d) Lamps and LEDs	X	—	X	—	—	—
(e) Primary (main) power supply	X	—	X	—	—	—
(f) Transponders	X	—	X	—	—	—
3. Engine-driven generator — central station facilities and fire alarm systems	X	X	—	—	—	—
4. Engine-driven generator — public emergency alarm reporting systems	X (weekly)	—	—	—	—	—
5. Batteries — central station facilities						
(a) Lead-acid type						6b
(1) Charger test (Replace battery as needed.)	X	—	—	—	X	—
(2) Discharge test (30 minutes)	X	X	—	—	—	—
(3) Load voltage test	X	X	—	—	—	—
(4) Specific gravity	X	—	—	X	—	—
(b) Nickel-cadmium type						6c
(1) Charger test (Replace battery as needed.)	X	—	X	—	—	—
(2) Discharge test (30 minutes)	X	—	—	—	X	—
(3) Load voltage test	X	—	—	—	X	—
(c) Sealed lead-acid type						6d
(1) Charger test (Replace battery within 5 years after manufacture or more frequently as needed.)	X	X	X	—	—	—
(2) Discharge test (30 minutes)	X	X	—	—	—	—
(3) Load voltage test	X	X	—	—	—	—
6. Batteries — fire alarm systems						
(a) Lead-acid type						6b
(1) Charger test (Replace battery as needed.)	X	—	—	—	X	—
(2) Discharge test (30 minutes)	X	—	—	X	—	—
(3) Load voltage test	X	—	—	X	—	—
(4) Specific gravity	X	—	—	X	—	—

(continues)

33408-11_T12F.EPS

Table 12 NFPA-Specified Fire Alarm Testing (6 of 9)

Component	Initial/ Reacceptance	Monthly	Quarterly	Semiannually	Annually	Table 14.4.2.2 Reference
6. Batteries — fire alarm systems (continued)						
(b) Nickel-cadmium type						6c
(1) Charger test (Replace battery as needed.)	X	—	—	—	X	—
(2) Discharge test (30 minutes)	X	—	—	—	X	—
(3) Load voltage test	X	—	—	X	—	—
(c) Primary type (dry cell)						6a
(1) Age test	X	X	—	—	—	—
(d) Sealed lead-acid type						6d
(1) Charger test (Replace battery within 5 years after manufacture or more frequently as needed.)	X	—	—	—	X	—
(2) Discharge test (30 minutes)	X	—	—	—	X	—
(3) Load voltage test	X	—	—	X	—	—
7. Power supply — public emergency alarm reporting systems						
(a) Lead-acid type batteries						6b
(1) Charger test (Replace battery as needed.)	X	—	—	—	X	—
(2) Discharge test (2 hours)	X	—	X	—	—	—
(3) Load voltage test	X	—	X	—	—	—
(4) Specific gravity	X	—	—	X	—	—
(b) Nickel-cadmium type batteries						6c
(1) Charger test (Replace battery as needed.)	X	—	—	—	X	—
(2) Discharge test (2 hours)	X	—	—	—	X	—
(3) Load voltage test	X	—	X	—	—	—
(c) Sealed lead-acid type batteries						6d
(1) Charger test (Replace battery within 5 years after manufacture or more frequently if needed	X	—	—	—	X	—
(2) Discharge test (2 hours)	X	—	—	—	X	—
(3) Load voltage test	X	—	X	—	—	—
(d) Wired system — voltage tests	X (daily)	—	—	—	—	7d
8. Fiber-optic cable power	X	—	—	—	X	13b
9. Control unit trouble signals	X	—	—	—	X	10
10. Conductors — metallic	X	—	—	—	—	12
11. Conductors — nonmetallic	X	—	—	—	—	13
12. In-building fire emergency voice/alarm communications equipment	X	—	—	—	X	20

33408-11_T12F.EPS

Table 12 NFPA-Specified Fire Alarm Testing (7 of 9)

	Component	Initial/ Reacceptance	Monthly	Quarterly	Semiannually	Annually	Table 14.4.2.2 Reference
13.	Retransmission Equipment (The requirements of 14.4.10 shall apply.)	X	—	—	—	—	—
14.	Remote Annunciators	X	—	—	—	X	11
15.	Initiating Devices*						14
	(a) Duct detectors	X	—	—	—	X	—
	(b) Electromechanical releasing device	X	—	—	—	X	—
	(c) Fire extinguishing system(s) or suppression system(s) switches	X	—	—	—	X	—
	(d) Fire–gas and other detectors	X	—	—	—	X	—
	(e) Heat detectors (The requirements of 14.4.5.5 shall apply.)	X	—	—	—	X	—
	(f) Manual fire alarm boxes	X	—	—	—	X	—
	(g) Radiant energy fire detectors	X	—	—	X	—	—
	(h) System smoke detectors — functional test	X	—	—	—	X	—
	(i) Smoke detectors — sensitivity testing in other than one- and two-family dwellings (The requirements of 14.4.5.3 shall apply.)	—	—	—	—	—	—
	(j) Single- and multiple-station smoke alarms (The requirements for monthly testing in accordance with 14.4.6 shall also apply.)	X	—	—	—	X	—
	(k) Single- and multiple-station heat alarms	X	—	—	—	X	—
	(l) Supervisory signal devices						
	(1) Valve supervisory switches	X	—	—	X	—	—
	(2) Pressure supervisory indicating devices	X	—	X	—	—	—
	(3) Water level supervisory indicating devices	X	—	X	—	—	—
	(4) Water temperature supervisory indicating devices	X	—	X	—	—	—
	(5) Room temperature supervisory indicating devices	X	—	X	—	—	—
	(6) Other suppression system supervisory initiating devices	X	—	X	—	—	—
	(7) Other supervisory initiating devices	X	—	—	—	X	—
	(m) Waterflow devices	X	—	—	X	—	—
16.	Guard's tour equipment	X	—	—	—	X	24
17.	Combination systems						
	(a) Fire extinguisher electronic monitoring device/systems	X	—	—	—	X	21a
	(b) Carbon monoxide detectors/systems	X	—	—	—	X	

(continues)

33408-11_T12G.EPS

Table 12 NFPA-Specified Fire Alarm Testing (8 of 9)

Component	Initial/ Reacceptance	Monthly	Quarterly	Semiannually	Annually	Table 14.4.2.2 Reference
18. Interface equipment and emergency control functions	X	—	—	—	X	22, 23
19. Special hazard equipment	X	—	—	—	X	17
20. Alarm notification appliances						15
(a) Audible devices	X	—	—	—	X	—
(b) Audible textual notification appliances	X	—	—	—	X	—
(c) Visible devices	X	—	—	—	X	—
21. Exit marking notification appliances	X	—	—	—	X	16
22. Supervising station alarm systems — transmitters	X	—	—	—	X	18
23. Special procedures	X	—	—	—	X	25
24. Supervising station alarm systems — receivers						19
(a) DACR	X	X	—	—	—	—
(b) DARR	X	X	—	—	—	—
(c) McCulloh systems	X	X	—	—	—	—
(d) Two-way RF multiplex	X	X	—	—	—	—
(e) RASSR	X	X	—	—	—	—
(f) RASSR	X	X	—	—	—	—
(g) Private microwave	X	X	—	—	—	—
25. Public emergency alarm reporting system transmission equipment						—
(a) Publicly accessible alarm box	X	—	—	X	—	8a
(b) Auxiliary box	X	—	—	X	—	8b
(c) Master box	—	—	—	—	—	8c
(1) Manual operation	X	—	—	X	—	—
(2) Auxiliary operation	X	—	—	—	X	—
26. Mass notification system — protected premise, supervised						27
(a) Control unit functions and no diagnostic failures are indicated	X	—	—	—	X	27
(b) Audible/visible functional test	X	—	—	—	X	27
(c) Secondary Power	X	—	—	—	X	27
(d) Verify content of prerecorded messages	X	—	—	—	X	27
(e) Verify activation of correct prerecorded messages	X	—	—	—	X	27
(f) Verify activation of correct prerecorded message based on a targeted area	X	—	—	—	X	27
(g) Verify control unit security mechanism is functional	X	—	—	—	X	27
27. Mass notification system — protected premise, nonsupervised systems installed prior to adoption of this Code						27

33408-11_T12H.EPS

Table 12 NFPA-Specified Fire Alarm Testing (9 of 9)

Component	Initial/ Reacceptance	Monthly	Quarterly	Semiannually	Annually	Table 14.4.2.2 Reference
(a) Control unit functions and no diagnostic failures are indicated	X		—	X	—	27
(b) Audible/visible functional test	X		—	X	—	27
(c) Secondary power	X		—	X	—	27
(d) Verify content of prerecorded messages	X		—	X	—	27
(e) Verify activation of correct prerecorded message based on a selected event	X		—	X	—	27
(f) Verify activation of correct prerecorded message based on a targeted area	X		—	X	—	27
(g) Verify control unit security mechanism is functional	X		—	X	—	27
28. Mass notification system — wide-area (UFC 4-021-01)						27
(a) Control unit functions and no diagnostic failures are indicated	X	—	—	—	X	27
(b) Control unit reset	X	—	—	—	X	27
(c) Control unit security	X	—	—	—	X	27
(d) Audible/visible functional test	X	—	—	—	X	27
(e) Software backup	X	—	—	—	X	27
(f) Secondary power test	X	—	—	—	X	27
(g) Antenna	X	—	—	—	X	27
(h) Transceivers	X	—	—	—	X	27
(i) Verify content of prerecorded messages	X	—	—	—	X	27
(j) Verify activation of correct prerecorded message based on a selected event	X	—	—	—	X	27
(k) Verify activation of correct prerecorded message base on a targeted area	X	—	—	—	X	27
(l) Verify control unit security mechanism is functional	X	—	—	—	X	27

*See A.14.4.5.

33408-11_T12I.EPS

Step 1 Know the equipment. For easy reference, keep specification sheets and instructions for equipment. Become familiar with the features of the equipment.

Step 2 Determine the symptoms. Try to make the system perform or fail to perform as it did when the problem was discovered.

Step 3 List and write down possible causes of the problem.

Step 4 Check the system in an orderly manner. Plan activities so that problem areas are not overlooked, in order to eliminate wasted time.

Step 5 Correct the problem. Once the problem has been located, repair it. If it is a component that cannot be repaired, replace it.

Step 6 Test the system. After the initial problem has been corrected, thoroughly check all the functions and features of the system to make sure other problems are not present.

16.1.0 Alarm System Troubleshooting Guidelines

A systematic troubleshooting approach is the most effective way to isolate a malfunction. The charts in *Figures 92, 93*, and *94* are examples of systematic troubleshooting.

The following guidelines provide information on resolving potential problems for specific conditions:

• *Sensors* – Always check sensor power, connections, environment, and settings. Lack of detection can be caused by a loose connection, obstacles in the area of the sensor, or a faulty unit. If unwanted alarms occur, recheck the in-

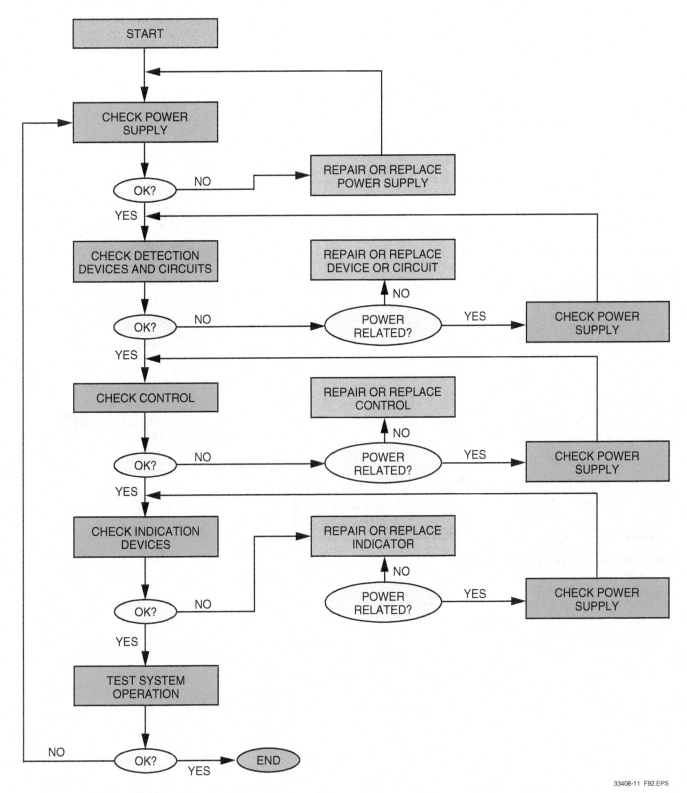

33408-11 F92.EPS

Figure 92 System troubleshooting.

stallation for changing conditions. If none are present, cover or seal the sensor to confirm that the alarm originated with the sensor. If alarms still occur, wiring problems, power problems or electromagnetic interference (EMI) could be the cause. If the alarm stops when the sensor is covered or sealed, the environment monitored by the sensor is the source of the problem. Replacing the unit should be the last resort after performing these checks.

- *Open circuit problems* – Opens cause trouble signals and account for the largest percentage of faults. Major causes of open circuits are loose connections, wire breaks, staple cuts, bad splices, cold-solder joints, wire fatigue, and defective wire.
- *Short circuit or ground fault problems* – Shorts or ground faults on a circuit cause alarms or trouble signals. Events occurring past a short are not seen by the system. Some of the com-

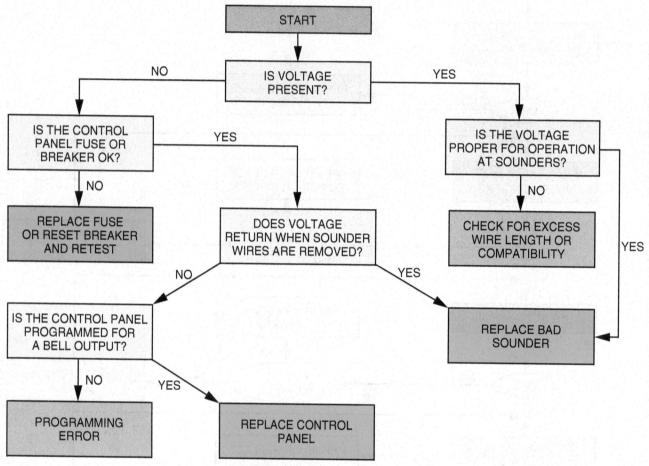

Figure 93 Alarm output troubleshooting.

33408-11_F93.EPS

mon causes of shorts or ground faults are staple cuts, sharp edge cuts, improper splices, moisture, and cold-flow of wiring insulation. Cold-flow of wiring insulation can occur in a short period of time, or over a long period, depending on the amount of tension or pressure that is applied to the wire insulation. Cold-flow is caused when wiring is under tension and routed around a corner of a conductive object, or when the wire is under external pressure. The pressure of the wire on the corner of the object eventually causes the wire insulation to move away from beneath the contact point of the wire, until the wire touches the conductive object. If a pair of wires, primarily a twisted pair, is under external pressure from c like furniture, the pressure can cause th. sulation to move from between the wires u; the wires touch and cause a short. To eliminat staple cuts, use other types of fasteners or leave wires under the staples loose. To eliminate insulation cold flow or edge cuts, do not pull wires tight around corners.

- *Improper splices* – Splices should properly insulate conductors from each other and from ground. Make sure soldered splices are properly taped. Make sure mechanical splices are properly crimped.
- *RJ-31X problems* – See *Figure 95* to troubleshoot various conditions of the RJ-31X interface.

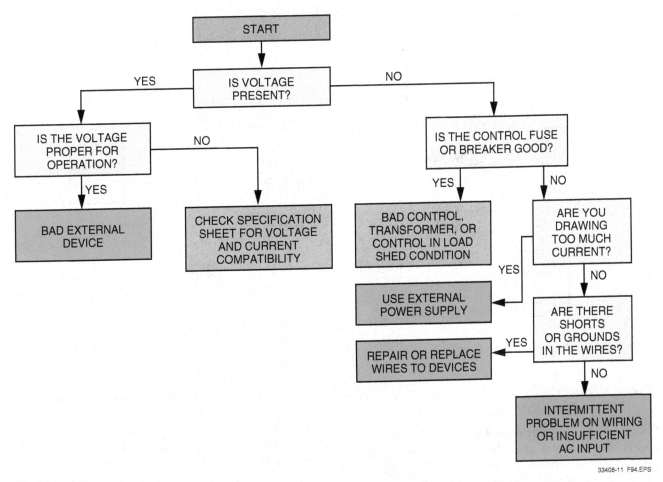

33408-11 F94.EPS

Figure 94 Auxiliary power troubleshooting.

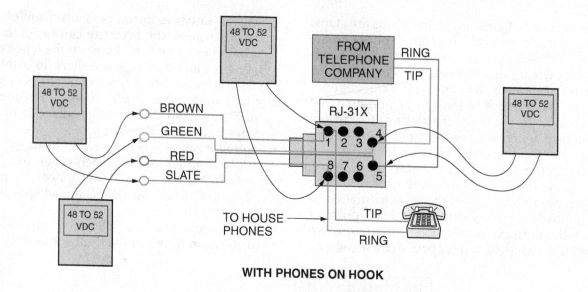

WITH PHONES ON HOOK

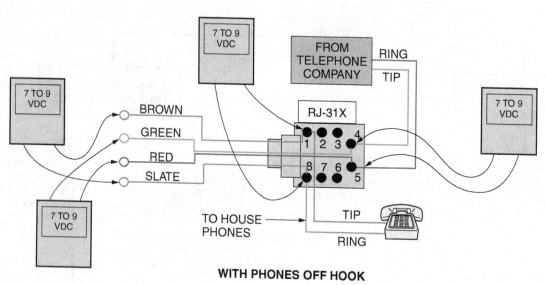

WITH PHONES OFF HOOK

Figure 95 RJ-31X troubleshooting. (1 of 2)

33408-11_F95A.EPS

16.2.0 Addressable System Troubleshooting Guidelines

There are a wide variety of designs, equipment, and configurations of addressable fire alarm systems. Some manufacturers of addressable systems provide troubleshooting data in their service literature, but others do not. Most of the general guidelines for troubleshooting discussed earlier also apply to troubleshooting addressable systems. In the event general troubleshooting methods fail to find the problem and troubleshooting information is not provided by the manufacturer, the best thing to do is to contact the manufacturer and ask for technical assistance. This will prevent a needless loss of time.

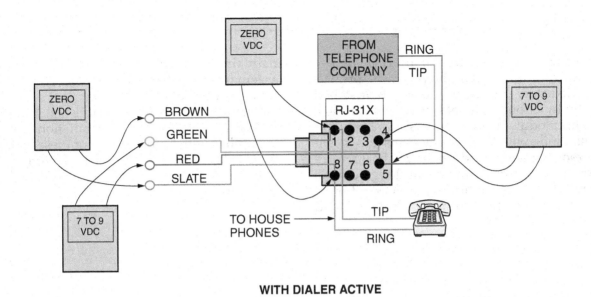

WITH DIALER ACTIVE

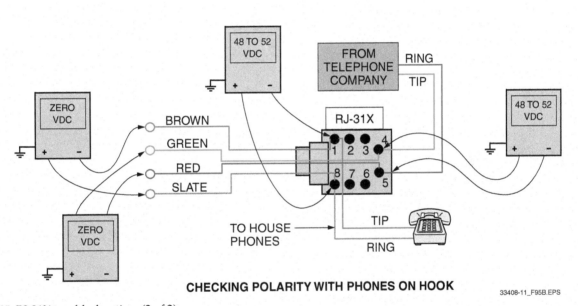

CHECKING POLARITY WITH PHONES ON HOOK

33408-11_F95B.EPS

Figure 95 RJ-31X troubleshooting. (2 of 2)

Summary

A fire alarm system is a combination of components designed to detect and report fires, primarily for life safety purposes. A properly installed and fully functioning fire alarm system can save lives. This module covered the basics of fire alarm systems, along with their components and installation requirements. Additionally, a basic introduction to some fire alarm system design criteria was provided. A fire alarm system does not function properly if it is not designed, installed, and maintained according to specifications. No two system installations are exactly alike. To help regulate the different applications that exist, codes and standards have been established that dictate the specifics of design and installation. You must be familiar with requirements of these codes and standards, as well as the basic design, components, installation, testing, and maintenance of typical fire alarm systems.

1. The specific requirements for wiring and equipment installation for fire protective signaling systems are covered in _____.
   a. *NFPA 70*®
   b. *NFPA 72*®
   c. *NFPA 101*®
   d. *NFPA 1*

2. The code established to help fire authorities continually develop safeguards against fire hazards is _____.
   a. *NFPA 1*
   b. *NFPA 72*®
   c. *NFPA 70*®
   d. *NFPA 101*®

3. Which class of circuits are run in non-redundant paths to increase reliability?
   a. A
   b. B
   c. C
   d. D

4. A defined area within the boundaries of a fire alarm system is known as a(n) _____.
   a. section
   b. signal destination
   c. alarm unit area
   d. zone

5. A fire alarm system with zones that allow multiple signals from several sources to be sent and received over a single communication line is known as a(n) _____.
   a. combination system
   b. hardwired system
   c. zoned system
   d. multiplex system

6. Which of the following is a type of spot detector?
   a. Projected beam
   b. Addressable
   c. Thermoelectric
   d. Aspirating

7. An automatic detector that draws air from the protected area back to the detector is called a(n) _____.
   a. line detector
   b. spot detector
   c. air sampling detector
   d. addressable detector

8. Which of the following is a type of fixed-temperature heat detector?
   a. Fusible link
   b. Rate-of-rise
   c. Rate compensation
   d. Ionization

9. *NFPA 72*® prohibits a duct detector from being used as a supervisory device.
   a. True
   b. False

10. The CO_2 spike phenomenon for hydrocarbon-based fires is used in _____.
    a. ultraviolet detectors
    b. infrared detectors
    c. fire-gas sensors
    d. laser flame sensors

11. Which type of pull box is restricted for use in special applications?
    a. Single-action
    b. Glass-break
    c. Double-action
    d. Key-operated

12. A verification method used by addressable systems is _____.
    a. positive alarm sequence
    b. wait-and-check
    c. reset-and-resample
    d. cross-zoning

13. Fire warning system units for commercial use must be listed in accordance with _____.
    a. *UL 780*
    b. *UL 864*
    c. *UL 985*
    d. *UL 995*

14. Primary power circuits supplying fire alarm systems must be protected by circuit breakers not larger than _____.

    a. 15A
    b. 20A
    c. 25A
    d. 30A

15. The *NFPA 72®* Temporal 3 sound pattern for a fire alarm signal is _____.

    a. three short (½ sec.) tones, three long (1 sec.) tones, then a repeat
    b. three long (1 sec.) tones, a pause, then a repeat
    c. three short (½ sec.) tones, a pause, then a repeat
    d. alternating single short (½ sec.) tones and long (1 sec.) tones

16. At 10 feet, public area audible notification devices must have a minimum rating of _____.

    a. 65 dBa
    b. 75 dBa
    c. 85 dBa
    d. 95 dBa

17. Which of the following code-authorized signals does not require the approval of the AHJ?

    a. Attendant signal
    b. Pre-signal
    c. General signal
    d. Positive alarm sequence

18. Which of the following conduits or cable cannot be buried?

    a. PVC
    b. Galvanized rigid
    c. Plain steel EMT
    d. UG cable

19. Commercial fire alarm cables that run from floor to floor must be _____.

    a. general-purpose type
    b. water-resistant
    c. rated for riser use
    d. run in PVC conduit

20. Power-limited fire alarm circuit conductors must be separated from nonpower-limited fire alarm circuit conductors by at least _____.

    a. 2"
    b. 6"
    c. 12"
    d. 24"

21. How many conductors does a Class B SLC require to reach each device on a circuit?

    a. Two
    b. Four
    c. Six
    d. Eight

22. An SLC that consists of a Class A main trunk with a Class B spur circuit is known as a _____.

    a. dual SLC
    b. combined SLC
    c. twin SLC
    d. hybrid SLC

23. For smoke chamber definition purposes, an archway that extends down from the ceiling is considered a(n) _____.

    a. obstruction
    b. boundary
    c. open grid
    d. beam pocket

24. The primary function of a duct smoke detector is to _____.

    a. sound an alarm for duct smoke
    b. shut down an associated HVAC system
    c. detect external open-area fires
    d. disable a music system for evacuation

25. In a fire alarm system, a visual or audible supervisory signal must be _____.

    a. different from a visual alarm signal
    b. the same as an alarm signal
    c. different from a trouble signal
    d. different from the alarm and trouble signals

Trade Terms Introduced in This Module

Addressable device: A fire alarm system component with discrete identification that can have its status individually identified or that is used to individually control other functions.

Air sampling (aspirating) detector: A detector consisting of piping or tubing distribution from the detector unit to the area or areas to be protected. An air pump draws air from the protected area back to the detector through the air sampling ports and piping or tubing. At the detector, the air is analyzed for fire products.

Alarm signal: A signal indicating an emergency requiring immediate action, such as an alarm for fire from a manual station, water flow device, or automatic fire alarm system.

Alarm verification: A feature of a fire control panel that allows for a delay in the activation of alarms upon receiving an initiating signal from one of its circuits. Alarm verification is commonly used in hotels, motels, and institutions.

Authority having jurisdiction (AHJ): The authority having jurisdiction is the organization, office, or individual responsible for approving equipment, installations, or procedures in a particular locality.

<aside>
NOTE

The NFPA does not approve, inspect, or certify any installations, procedures, equipment, or materials, nor does it approve or evaluate testing laboratories. In determining the acceptability of installations, procedures, equipment, or materials, the authority having jurisdiction may base acceptance on compliance with NFPA, or other appropriate standards. In the absence of such standards, said authority may require evidence of proper installation, procedure, or use. The authority having jurisdiction may also refer to the Listings or labeling practices of an organization concerned with product evaluations that is in a position to determine compliance with appropriate standards for the current production of listed items.
</aside>

Automatic fire alarm system: A system in which all or some of the circuits are actuated by automatic devices, such as fire detectors, smoke detectors, heat detectors, and flame detectors.

Backflow preventer: A device used in piping system to prevent a fluid from flowing backward. They are required in wet sprinkler systems that are filled with antifreeze to prevent the antifreeze from contaminating the potable water system.

Candela (cd): The International System (SI) unit of luminous intensity. It is roughly equal to 12.57 lumens.

Ceiling surfaces: Ceiling surfaces referred to in conjunction with the locations of initiating devices are as follows:

- **Beam construction:** Ceilings having solid structural or solid nonstructural members projecting down from the ceiling surface more than 4" (100mm) and spaced more than 3' (0.9m) center to center.

- **Girders:** Girders support beams or joists and run at right angles to the beams or joists. When the tops of girders are within 4" (100mm) of the ceiling, they are a factor in determining the number of detectors and are to be considered as beams. When the top of the girder is more than 4" (100mm) from the ceiling, it is not a factor in detector location.

Certification: A systematic program using randomly selected follow-up inspections of the certified system installed under the program, which allows the listing organization to verify that a fire alarm system complies with all the requirements of the *NFPA 72®* code. A system installed under such a program is identified by the issuance of a certificate and is designated as a certificated system.

Class A circuit: Class A refers to an arrangement of supervised initiating devices, signaling line circuits, or notification appliance circuits that prevents a single open or ground on the installation wiring of these circuits from causing loss of the system's intended function.

Class B circuit: Class B refers to an arrangement of initiating devices, signaling lines, or notification appliance circuits that does not prevent a single open or ground on the installation wiring of these circuits from causing loss of the system's intended function.

Digital alarm communicator receiver (DACR): A system component that will accept and display signals from digital alarm communicator transmitters (DACTs) sent over public switched telephone networks.

Digital alarm communicator system (DACS): A system in which signals are transmitted from a digital alarm communicator transmitter (DACT) located at the protected premises through the public switched telephone network to a digital alarm communicator receiver (DACR).

Digital alarm communicator transmitter (DACT): A system component at the protected premises to which initiating devices or groups of devices are connected. The DACT will seize the connected telephone line, dial a preselected number to connect to a DACR, and transmit signals indicating a status change of the initiating device.

End-of-line (EOL) device: A device used to terminate a supervised circuit. An EOL is normally a resistor or a diode placed at the end of a two-wire circuit to maintain supervision.

Fault: An open, ground, or short condition on any line(s) extending from a control unit, which could prevent normal operation.

Fire alarm control unit (FACU): A device with the control circuits necessary to furnish power to a fire alarm system, receive signals from alarm initiating devices (and transmit them to audible alarm indicating appliances and accessory equipment), and electrically supervise the system installation wiring and primary (main) power. The control unit can be contained in one or more cabinets in adjacent or remote locations.

Flame detector: A device that detects the infrared, ultraviolet, or visible radiation produced by a fire. Some devices are also capable of detecting the flicker rate (frequency) of the flame.

General alarm: A term usually applied to the simultaneous operation of all the audible alarm signals on a system, to indicate the need for evacuation of a building.

Ground fault: A condition in which the resistance between a conductor and ground reaches an unacceptably low level.

Heat detector: A device that detects an abnormally high temperature or rate-of-temperature rise.

Heat tracing: The practice of placing thermostatically controlled electric heat tapes around or along wet sprinkler system pipes to prevent them from freezing. Heat tracing is prohibited by some codes.

Indicating device: Any audible or visible signal employed to indicate a fire, supervisory, or trouble condition. Examples of audible signal appliances are bells, horns, sirens, electronic horns, buzzers, and chimes. A visible indicator consists of an incandescent lamp, strobe lamp, mechanical target or flag, meter deflection, or the equivalent. Also called a notification device (appliance).

Initiating device: A manually or automatically operated device, the normal intended operation of which results in a fire alarm or supervisory signal indication from the control unit. Examples of alarm signal initiating devices are thermostats, manual boxes (stations), smoke detectors, and water flow devices. Examples of supervisory signal initiating devices are water level indicators, sprinkler system valve-position switches, pressure supervisory switches, and water temperature switches.

Initiating device circuit (IDC): A circuit to which automatic or manual signal initiating devices, such as fire alarm manual boxes (pull boxes), heat and smoke detectors, and water flow alarm devices are connected.

Light scattering: The action of light being reflected or refracted off particles of combustion, for detection in a modern day photoelectric smoke detector. This is called the Tyndall effect.

Listed: Equipment or materials included in a list published by an organization acceptable to the authority having jurisdiction that is concerned with product evaluation and whose listing states either that the equipment or material meets appropriate standards or that it has been tested and found suitable for use in a specified manner.

> **NOTE**
> The means for identifying listed equipment may vary for each organization concerned with product evaluation (UL [Underwriters Laboratories], FM Global, IRI [Industrial Risk Insurers], and so on). Some organizations do not recognize equipment Listed unless it is also labeled. The authority having jurisdiction should use the system employed by the Listed organization to identify a Listed product.

Multiplexing: A signaling method that uses wire path, cable carrier, radio, fiber optics, or a combination of these techniques, and characterized by the simultaneous or sequential (or both simultaneous and sequential) transmission and reception of multiple signals in a communication channel, including means of positively identifying each signal.

Non-coded signal: A signal from any indicating appliance that is continuously energized.

Obscuration: A reduction in atmospheric transparency caused by smoke, usually expressed in percent per foot.

Path (pathway): Any conductor, optical fiber, radio carrier, or other means for transmitting fire alarm system information between two or more locations.

Photoelectric smoke detector: A detector employing the photoelectric principle of operation, using either the obscuration effect or the light-scattering effect for detecting smoke in its chamber.

Positive alarm sequence: An alarm verification feature that provides up to a three-minute delay that allows supervising or monitoring personnel time to investigate the alarm.

Preaction sprinkler system: A type of dry sprinkler that fills the pipes with water before the sprinkler head opens. It allows water to spray on a fire faster because the air in the pipes does not have to be purged after the sprinkler head opens.

Projected beam smoke detector: A type of photoelectric light-obscuration smoke detector in which the beam spans the protected area.

Rate compensation detector: A device that responds when the temperature of the air surrounding the device reaches a predetermined level, regardless of the rate-of-temperature rise.

Rate-of-rise detector: A device that responds when the temperature rises at a rate exceeding a predetermined value.

Remote supervising station fire alarm system: A system installed in accordance with the applicable code to transmit alarm, supervisory, and trouble signals from one or more protected premises to a remote location where appropriate action is taken.

Signaling line circuits (SLCs): A circuit or path between any combination of circuit interfaces, control units, or transmitters, over which multiple system input signals or output signals (or both input signals and output signals) are carried.

Smoke detector: A device that detects visible or invisible particles of combustion.

Spot-type detector: A device in which the detecting element is concentrated at a particular location. Typical examples are bimetallic detectors, fusible alloy detectors, certain pneumatic rate-of-rise detectors, certain smoke detectors, and thermoelectric detectors.

Stratification: The phenomenon in which the upward movement of smoke and gases ceases due to a loss of buoyancy.

Supervising station: A facility that receives signals and at which personnel are in attendance at all times to respond to these signals.

Supervisory signal: A signal indicating the need for action in connection with the supervision of guard tours, the fire suppression systems or equipment, or the maintenance features of related systems.

System unit: The active subassemblies at the central station used for signal receiving, processing, display, or recording of status change signals. The failure of one of these subassemblies causes the loss of a number of alarm signals by that unit.

Thermal lag: The time required to add heat to or remove heat from a substance before it reaches a set temperature.

Trouble signal: A signal initiated by the fire alarm system or device that indicates a fault in a monitored circuit or component.

Visible notification appliance: A notification appliance that alerts by the sense of sight.

Zone: A defined area within the protected premises. A zone can define an area from which a signal can be received, an area to which a signal can be sent, or an area in which a form of control can be executed.

SAMPLE INSPECTION AND TESTING FORM

INSPECTION AND TESTING FORM

DATE: __Nov 5, 2002__

TIME: __2:30 pm__

SERVICE ORGANIZATION

Name: __ABC Fire Service Co__

Address: __123 Main Steet, Anytown FL__

Representative: __Joe Alarm__

License No.: __F12345__

Telephone: __(555) 456 – 4321__

PROPERTY NAME (USER)

Name: __Buy American, Inc.__

Address: __1776 Freedom Lane, N. Anytown FL__

Owner Contact: __George Washington__

Telephone: __(555) 456 – 7123__

MONITORING ENTITY

Contact: __Sydney Forrester, Monitoring Co, Inc.__

Telephone: __(555) 457 – 1007__

Monitoring Account Ref. No.: __A103__

APPROVING AGENCY

Contact: __Olin Firestop, Anytown Fire Chief__

Telephone: __(555) 456 – 1234__

TYPE TRANSMISSION
- ☐ McCulloh
- ☐ Multiplex
- ☑ Digital
- ☐ Reverse Priority
- ☐ RF
- ☐ Other (Specify) _____

SERVICE
- ☐ Weekly
- ☐ Monthly
- ☐ Quarterly
- ☐ Semiannually
- ☐ Annually
- ☑ Other (Specify) __Initial Acceptance Inspection & Test__

Control Unit Manufacturer: __Alarm MFG Co.__

Model No.: __F301__

Circuit Styles: __4 & Y__

Number of Circuits: __3 & 4__

Software Rev.: __F301   V 1.01__

Last Date System Had Any Service Performed: __—__

Last Date that Any Software or Configuration Was Revised: __Aug. 29, 2002   (Factory)__

ALARM-INITIATING DEVICES AND CIRCUIT INFORMATION

Quantity	Circuit Style	
22	4	Manual Fire Alarm Boxes
0		Ion Detectors
42		Photo Detectors
16		Duct Detectors
8		Heat Detectors
10		Waterflow Switches
17		Supervisory Switches
0		Other (Specify): _____

Alarm verification feature is disabled __✔__ enabled _____ .

(NFPA Inspection and Testing, 1 of 4)

33408-11 A01.EPS

ALARM NOTIFICATION APPLIANCES AND CIRCUIT INFORMATION

Quantity	Circuit Style	
0		Bells
32	Y	Horns
0		Chimes
38	Y	Strobes
0		Speakers
		Other (Specify): _____

No. of alarm notification appliance circuits: _____ 4 _____
Are circuits monitored for integrity? ☑ Yes   ☐ No

SUPERVISORY SIGNAL-INITIATING DEVICES AND CIRCUIT INFORMATION

Quantity	Circuit Style	
0		Building Temp.
0		Site Water Temp.
0		Site Water Level
1	4	Fire Pump Power
1	4	Fire Pump Running
1	4	Fire Pump Auto Position
1	4	Fire Pump or Pump Controller Trouble
0		Fire Pump Running
0		Generator In Auto Position
0		Generator or Controller Trouble
0		Switch Transfer
0		Generator Engine Running
1	4	Other: *Electric FP phase reversal*
12	4	*valve position*

SIGNALING LINE CIRCUITS

Quantity and style of signaling line circuits connected to system (*see NFPA 72, Table 6.6.1*):
Quantity _____ 3 _____   Style(s) _____ 4 _____

SYSTEM POWER SUPPLIES

(a) Primary (Main):   Nominal Voltage _____ 120 VAC _____   Amps _____ 8.5 (FACP) _____
    Overcurrent Protection: Type _____ Circuit breaker _____   Amps _____ 15 _____
    Location (of Primary Supply Panelboard): *Electrical Room 103 Panel EP2*
    Disconnecting Means Location: *Breaker 26 Panel EP2*
(b) Secondary (Standby):
    _____ ✓ _____ Storage Battery: Amp-Hr. Rating _____ 10 A-H _____
    Calculated capacity to operate system, in hours: _____ 5.6 A-H _____ 24 _____ — _____ 60
    _____Engine-driven generator dedicated to fire alarm system:
    Location of fuel storage: _____ — _____

TYPE BATTERY

☐ Dry Cell
☐ Nickel-Cadmium
☑ Sealed Lead-Acid
☐ Lead-Acid
☐ Other (Specify):
(c) Emergency or standby system used as a backup to primary power supply, instead of using a secondary power supply:
_____ Emergency system described in NFPA 70, Article 700
_____ Legally required standby described in NFPA 70, Article 701
_____ Optional standby system described in NFPA 70, Article 702, which also meets the performance
    requirements of Article 700 or 701.

(NFPA Inspection and Testing, 2 of 4)

33408-11_A02.EPS

PRIOR TO ANY TESTING

NOTIFICATIONS ARE MADE

	Yes	No	Who	Time
Monitoring Entity	☑	☐	*Operator 115*	*8:15 AM*
Building Occupants	☑	☐		
Building Management	☑	☐		
Other (Specify)	☐	☐		
AHJ Notified of Any Impairments	☐	☑		

SYSTEM TESTS AND INSPECTIONS

TYPE	Visual	Functional	Comments
Control Unit	☑	☑	
Interface Equipment	☑	☑	
Lamps/LEDS	☑	☐	
Fuses	☑	☐	
Primary Power Supply	☑	☐	
Trouble Signals	☑	☑	
Disconnect Switches	☑	☐	
Ground-Fault Monitoring	☑	☐	*Tested OK*

SECONDARY POWER

TYPE	Visual	Functional	Comments
Battery Condition	☑		
Load Voltage		☑	*26.2 VDC*
Discharge Test		☑	*simulated*
Charger Test		☐	
Specific Gravity		☐	
TRANSIENT SUPPRESSORS	☑		
REMOTE ANNUNCIATORS	☑	☑	

NOTIFICATION APPLIANCES

	Visual	Functional	Comments
Audible	☑	☑	*Low in Long Rm 2*
Visible	☑	☑	
Speakers	☐	☐	
Voice Clarity		☐	

INITIATING AND SUPERVISORY DEVICE TESTS AND INSPECTIONS

Loc. & S/N	Device Type	Visual Check	Functional Test	Factory Setting	Measured Setting	Pass	Fail
1st FL E	*man box*	☑	☑			☑	☐
Corr NR 103	*SD*	☑	☑	*3.2*	*3.1*	☑	☐
Corr NR 105	*SD*	☑	☑	*3.2*	*3.2*	☑	☐
Corr NR 107	*SD*	☑	☑	*3.2*	*3.1*	☑	☐
E stairs	*WF*	☑	☑		*62 sec*	☑	☐
E stairs	*Tamper*	☑	☑			☑	☐

Comments: *Additional inspection and test sheets attached*

(NFPA Inspection and Testing, 3 of 4)

33408-11_A03.EPS

EMERGENCY COMMUNICATIONS EQUIPMENT	Visual	Functional	Comments
Phone Set	☐	☐	
Phone Jacks	☐	☐	
Off-Hook Indicator	☐	☐	
Amplifier(s)	☐	☐	
Tone Generator(s)	☐	☐	
Call-in Signal	☐	☐	
System Performance	☐	☐	

INTERFACE EQUIPMENT	Visual	Device Operation	Simulated Operation
(Specify) AHU #1 shotdown	☐	☑	
(Specify) Elevator recall	☐	☑	☐
(Specify) Elevator Fire Hat	☐	☑	☐

SPECIAL HAZARD SYSTEMS

	Visual	Device Operation	Simulated Operation
(Specify) Kitchen Hood System	☑	☐	☑
(Specify)	☐	☐	☐
(Specify)	☐	☐	☐

Special Procedures: _____

_____

Comments: _____

_____

SUPERVISING STATION MONITORING	Yes	No	Time	Comments
Alarm Signal	☑	☐	9:45 AM	OK
Alarm Restoration	☑	☐		OK
Trouble Signal	☑	☐		OK
Supervisory Signal	☑	☐		OK
Supervisory Restoration	☑	☐		OK

NOTIFICATIONS THAT TESTING IS COMPLETE	Yes	No	Who	Time
Building Management	☑	☐	Bldg manager	2:15
Monitoring Agency	☑	☐	Operator 97	2:15
Building Occupants	☑	☐		
Other (Specify)	☐	☐		

The following did not operate correctly: Audibility low in conference Room 2. Corrected by Installer and retested OK

System restored to normal operation: Date: Nov 5, 2002   Time: 2:15

THIS TESTING WAS PERFORMED IN ACCORDANCE WITH APPLICABLE NFPA STANDARDS.

Name of Inspector: Tom Technician   Date: Nov 5, 2002   Time: 2:30

Signature: Tom Technician

Name of Owner or Representative: George Washington

Date: Nov 5, 2002   Time: 2:35

Signature: George Washington

(NFPA Inspection and Testing, 4 of 4)

33408-11_A04.EPS

CERTIFICATE OF COMPLETION

FIRE ALARM SYSTEM
RECORD OF COMPLETION

Name of protected property: _Buy American, Inc._

Address: _1776 Freedom Lane, North Anytown, FL_

Representative of protected property (name/phone): _George Washington (555) 456-7123_

Authority having jurisdiction: _Olin Firestop, Fire Chief, Anytown FL_

Address/telephone number: _1302 Center Street (555) 456-1234_

	Organization name/phone	_Representative name/phone_
Installer	_Alarm System Installations, Inc._	_John Q. Smith (555) 458-3001_
Supplier	_Protection Products Supply Co._	_Charles A. Bundant (555) 811-1244_
Service organization	_ABC Fire Service Co._	_Joe Alarm (555) 456-4321_

Location of record (as-built) drawings: _Buy American, Inc - Maintenance Shop_

Location of operation and maintenance manuals: _Buy American, Inc - Maintenance Shop_

Location of test reports: _Buy American, Inc. - Maintenance Shop_

A contract for test and inspection in accordance with NFPA standard(s)

Contract No(s): _CC-00987_   Effective date: _Oct 15, 2002_   Expiration date: _Oct 15, 2005_

System Software

(a) Operating system (executive) software revision level(s): _System F Bios V2.3_

(b) Site-specific software revision date: _F 301 V1.01_

(c) Revision completed by: _Abel Programmer (Factory) Alarm MFG Co._

    (name)   (firm)

1. Type(s) of System or Service

✓ _NFPA 72_, Chapter 6 — Local

If alarm is transmitted to location(s) off premises, list where received: _Monitoring Company, Inc._

✓ _NFPA 72_, Chapter 8 — Remote Station

Telephone numbers of the organization receiving alarm:

Alarm: _Monitoring Company, Inc._

Supervisory: _Monitoring Company, Inc._

Trouble: _Monitoring Company, Inc._

If alarms are retransmitted to public fire service communications centers or others, indicate location and telephone numbers of the organization receiving alarm: _Anytown Fire Department_
1302 Center Street, Anytown FL   (555) 456-1212

Indicate how alarm is retransmitted: _Dedicated Circuit_

_____ _NFPA 72_, Chapter 8 — Proprietary

Telephone numbers of the organization receiving alarm:

Alarm: _____

Supervisory: _____

Trouble: _____

If alarms are retransmitted to public fire service communications centers or others, indicate location and telephone numbers of the organization receiving alarm: _____

Indicate how alarm is retransmitted: _____

_____ _NFPA 72_, Chapter 8 — Central Station

Prime contractor: _____

Central station location: _____

(NFPA 72, 1 of 4)

33408-11_A05.EPS

Means of transmission of signals from the protected premises to the central station:

_____ McCulloh _____ Multiplex _____ One-way radio

_____ Digital alarm communicator _____ Two-way radio _____ Others

Means of transmission of alarms to the public fire service communications center:

(a) _____

(b) _____

System location: _____

_____ NFPA 72, Chapter 9 — Auxillary

Indicate type of connection: _____ Local energy _____ Shunt _____ Parallel telephone

Location of telephone number for receipt of signals: _____

2. Record of System Installation

(Fill out after installation is complete and wiring is checked for opens, shorts, ground faults, and improper branching, but prior to conducting operational acceptance tests.)

This system has been installed in accordance with the NFPA standards as shown below, was inspected by

Henry C. Installer on _Nov. 1, 2002_, includes the devices shown in 5 and 6, and has been in service since _Nov. 1, 2002_

✓ NFPA 72, Chapters ⓵ 2 3 4 5 6 7 ⑧ 9 ⑩ 11 (circle all that apply)

✓ NFPA 70, *National Electrical Code*, Article 760

✓ Manufacturer's instructions

_____ Other (specify): _____

Signed: _Henry C. Installer_ Date: _Nov. 1, 2002_

Organization: _Alarm System Installations, Inc._

3. Record of System Operation

Documentation in accordance with Inspection Testing Form, Figure 10.6.2.3, is attached _Yes_.

All operational features and functions of this system were tested by _Tom Technician_ date _Nov. 5, 2002_

and found to be operating properly in accordance with the requirements of:

✓ NFPA 72, Chapters ⓵ 2 3 4 5 6 7 ⑧ 9 ⑩ 11 (circle all that apply)

_____ NFPA 70, *National Electrical Code*, Article 760

✓ Manufacturer's instructions

_____ Other (specify): _____

Signed: _Tom Technician_ Date: _Nov. 5, 2002_

Organization: _ABC Fire Service Co._

4. Signaling Line Circuits

Quantity and class of signaling line circuits connected to system (*see NFPA 72, Table 6.6.1*):

Quantity: _3_ Style: _4_ Class: _B_

(*NFPA 72, 2 of 4*)

33408-11_A06.EPS

5. Alarm-Initiating Devices and Circuits

Quantity and class of initiating device circuits *(see NFPA 72, Table 6.5)*:

Quantity: _____ Style: _____ Class: _____

MANUAL

(a) Manual stations   Noncoded _____   Transmitters _____   Coded _____   Addressable __22__

(b) Combination manual fire alarm and guard's tour coded stations _____

AUTOMATIC

Coverage: Complete _____   Partial ___✓_____

   Selective _____   Nonrequired _____

(a) Smoke detectors _____   Ion _____   Photo __42__   Addressable ___✓_____

(b) Duct detectors _____   Ion _____   Photo __16__   Addressable ___✓_____

(c) Heat detectors _____   FT __8__   RR _____   FT/RR _____   RC _____   Addressable ___✓_____

(d) Sprinkler waterflow indicators: Transmitters _____   Noncoded _____   Coded _____   Addressable __10__

(e) The alarm verification feature is disabled ___✓___ or enabled _____ , changed from _____ seconds to _____ seconds.

(f) Other (list): _____

_____

6. Supervisory Signal-Initiating Devices and Circuits (use blanks to indicate quantity of devices)

GUARD'S TOUR

(a)_____ Coded stations

(b)_____ Noncoded stations

(c) _____ Compulsory guard's tour system comprised of _____ transmitter stations and intermediate stations

Note: Combination devices are recorded under 5(b), Manual, and 6(a), Guard's Tour.

SPRINKLER SYSTEM

Check if provided

(a) __✓__ Valve supervisory switches

(b)_____ Building temperature points

(c) _____ Site water temperature points

(d)_____ Site water supply level points

Electric fire pump:

(e) __✓__ Fire pump power

(f) __✓__ Fire pump running

(g) __✓__ Phase reversal

Engine-driven fire pump:

(h)_____ Selector in auto position

(i) _____ Engine or control panel trouble

(j) _____ Fire pump running

ENGINE-DRIVEN GENERATOR:

(a)_____ Selector in auto position

(b)_____ Control panel trouble

(c) _____ Transfer switches

(d)_____ Engine running

Other supervisory function(s) (specify): ___*Electric fire pump "Auto position"*_____

*Electric fire pump "Controller trouble"*_____

(NFPA 72, 3 of 4)

33408-11_A07.EPS

7. Annunciator(s)

Number: __1__   Type: __Graphic__   Location: __Main entrance__

8. Alarm Notification Appliances and Circuits

NFPA 72, Chapter 6 — Emergency Voice/Alarm Service

Quantity of voice/alarm channels: ___—___   Single: _____   Multiple: _____

Quantity of speakers installed: _____   Quantity of speaker zones: _____

Quantity of telephones or telephone jacks included in system: _____

Quantity and the class of notification appliance circuits connected to system *(see NFPA 72, Table 6.7)*:

Quantity: __4__   Style: __Y__   Class: __B__

Types and quantities of notification appliances installed:

(a) Bells _____   With Visible _____
(b) Speakers _____   With Visible _____
(c) Horns __32__   With Visible __32__
(d) Chimes _____   With Visible _____
(e) Other: _____   With Visible _____
(f) Visible appliances without audible: __6__

9. System Power Supplies

(a) Fire Alarm Control Panel: Nominal voltage: __120 VAC__   Current rating: __8.5__

   Overcurrent protection: Type: __Circuit Breaker__   Current rating: __15__

   Location: __Electrical Room 103 Panel EP2__

(b) Secondary (standby):

   Storage battery: __✓__   Amp-hour rating: __10 A-H__

   Calculated capacity to drive system, in hours: __5.6 A-H (24 hours)__

   Engine-driven generator dedicated to fire alarm system: ___—___

   Location of fuel storage: _____

(c) Emergency system used as backup to primary power supply: ___—___

   Emergency system described in NFPA 70, Article 700: _____

10. Comments

Frequency of routine tests and inspections, if other than in accordance with the referenced NFPA standard(s):

_____

System deviations from the referenced NFPA standard(s) are: __Audible level in Conference Room 2__
__Found low – corrected by installer Nov 5, 2002__

Henry C. Installer	*Installation Supervisor*	*Nov. 5, 2002*
(signed) for installation contractor/supplier	(title) (date)	
Tom Technician	*Lead Test Technician*	*Nov. 5, 2002*
(signed) for alarm service company	(title) (date)	
(signed) for central station	(title) (date)	

Upon completion of the system(s) satisfactory test(s) witnessed (if required by the authority having jurisdiction):

Olin Firestop	*Fire Chief, Anytown*	*Nov. 5, 2002*
(signed) representative of the authority having jurisdiction	(title) (date)	

(NFPA 72, 4 of 4)

33408-11_A08.EPS

Additional Resources

This module presents thorough resources for task training. The following resource material is suggested for further study.

A Designer's Guide to Fire Alarm Systems. Robert M. Gagnon, Ronald H. Kirby. Quincy, MA: National Fire Protection Association.

Electrical Installations in Hazardous Locations. Peter J. Schram, Robert Bendetti, Mark W. Earley. Sudbury, MA: Jones and Bartlett Learning.

Fire Alarm Signaling Systems. Richard W. Bukowski, Wayne D. Moore. Quincy, MA: National Fire Protection Association.

NFPA Pocket Guide to Fire Alarm and Signaling System Installation. Merton W. Bunker, Jr., Richard J. Roux. Quincy, MA: National Fire Protection Association.

Figure Credits

NCCER CURRICULA — USER UPDATE

NCCER makes every effort to keep its textbooks up-to-date and free of technical errors. We appreciate your help in this process. If you find an error, a typographical mistake, or an inaccuracy in NCCER's curricula, please fill out this form (or a photocopy), or complete the online form at **www.nccer.org/olf**. Be sure to include the exact module ID number, page number, a detailed description, and your recommended correction. Your input will be brought to the attention of the Authoring Team. Thank you for your assistance.

Instructors – If you have an idea for improving this textbook, or have found that additional materials were necessary to teach this module effectively, please let us know so that we may present your suggestions to the Authoring Team.

NCCER Product Development and Revision

13614 Progress Blvd., Alachua, FL 32615

Email: curriculum@nccer.org
Online: www.nccer.org/olf

❏ Trainee Guide   ❏ AIG   ❏ Exam   ❏ PowerPoints   Other _____

Craft / Level: _____   Copyright Date: _____

Module ID Number / Title: _____

Section Number(s): _____

Description: _____

_____

_____

_____

_____

Recommended Correction: _____

_____

_____

_____

_____

Your Name: _____

Address: _____

_____

Email: _____   Phone: _____

33409-12

Overview of Nurse Call and Signaling Systems

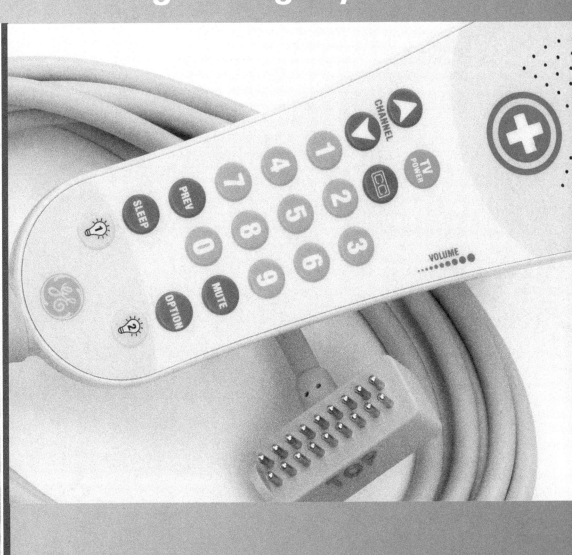

Module Nine

Trainees with successful module completions may be eligible for credentialing through NCCER's National Registry. To learn more, go to **www.nccer.org** or contact us at **1.888.622.3720**. Our website has information on the latest product releases and training, as well as online versions of our *Cornerstone* newsletter and Pearson's product catalog.

Your feedback is welcome. You may email your comments to **curriculum@nccer.org,** send general comments and inquiries to **info@nccer.org**, or fill in the User Update form at the back of this module.

Objectives

When you have completed this module, you will be able to do the following:

1. Explain key terms related to nurse call systems.
2. Identify the codes and standards that apply to the installation and operation of nurse call systems.
3. Describe the different types of nurse call systems and explain their differences.
4. Explain the limitations on connections between nurse call systems and other systems.
5. State the general installation guidelines that apply to nurse call systems.
6. Install and connect nurse call system components.

Performance Task

Under the supervision of your instructor, you should be able to do the following:

1. Install and connect nurse call system components.

Required Trainee Materials

1. *ANSI/UL 1069, Hospital Signaling and Nurse Call Equipment*
2. *NFPA 70®, National Electrical Code®*
3. *NEMA Installation Guide for Nurse Call Systems*

Industry Recognized Credentials

If you're training through an NCCER-accredited sponsor you may be eligible for credentials from NC-CER's Registry. The ID number for this module is 33409-12. Note that this module may have been used in other NCCER curricula and may apply to other level completions. Contact NCCER's Registry at 888.622.3720 or go to nccer.org for more information.

Note: *NFPA 70®, National Electrical Code®* and *NEC®* are registered trademarks of the National Fire Protection Association, Inc., Quincy, MA 02269. All *National Electrical Code®* and *NEC®* references in this module refer to the 2011 edition of the *National Electrical Code®*.

Contents

Topics to be presented in this module include:

Figures and Tables

1.0.0 INTRODUCTION

Patients in hospitals and other nursing facilities need the ability to summon help from the medical staff. Years ago, this was simply a switch that would turn on a light over the door of the patient's room. Today, the light over the door still exists, but it is likely to be a multi-colored, multi-layered signaling device. There are also other methods of reaching out to the nursing staff through hardwired and wireless devices. Today's systems go beyond patient-staff communications; they also provide peer-to-peer communication and enable supervisory personnel to quickly locate critical equipment and to locate and communicate with staff members. *Figure 1* provides an overview of a basic nurse call system.

When the patient presses the call button, the light above the door to the patient's room lights. The bed station shown in *Figure 1* is only one of several devices used for signaling; bath stations and code stations are among the other types of signaling stations. The call signal is sent to the nurse master station, which is usually located at the nurses' desk. Depending on the type and sophistication of the system, the master station can take on a variety of forms. At its most basic, the station is simply an annunciator panel containing one light for each patient room. With this type of system, a staff member has to go to the

patient's room to find out what kind of assistance is needed.

A more common type of master station has an LCD display that indicates the source and type of alarm and has the ability to display text messages. This type of unit has a telephone handset that allows the person at the station to communicate with patients and staff members. With such a system, the person at the master station can determine the nature and priority of the call before dispatching assistance. Except for the most basic visual systems, most nurse call systems in use today have the ability to interface with optional paging and cell phone systems.

The most advanced nurse call systems are driven by computers or dedicated central equipment linked through a data network. This type of system will have a full array of integrated features, including cell phone and pager systems, access to the facility's telephone and PA systems, and the ability to track personnel and equipment. In addition, a computer interface is provided to allow management to obtain records of patient calls and responses. *Figure 2* shows the various options that are available with such a system.

This module provides an overview of nurse call systems. Unlike the mix-and-match systems you see in video, audio, and RF systems, nurse call systems are largely proprietary. That is, with

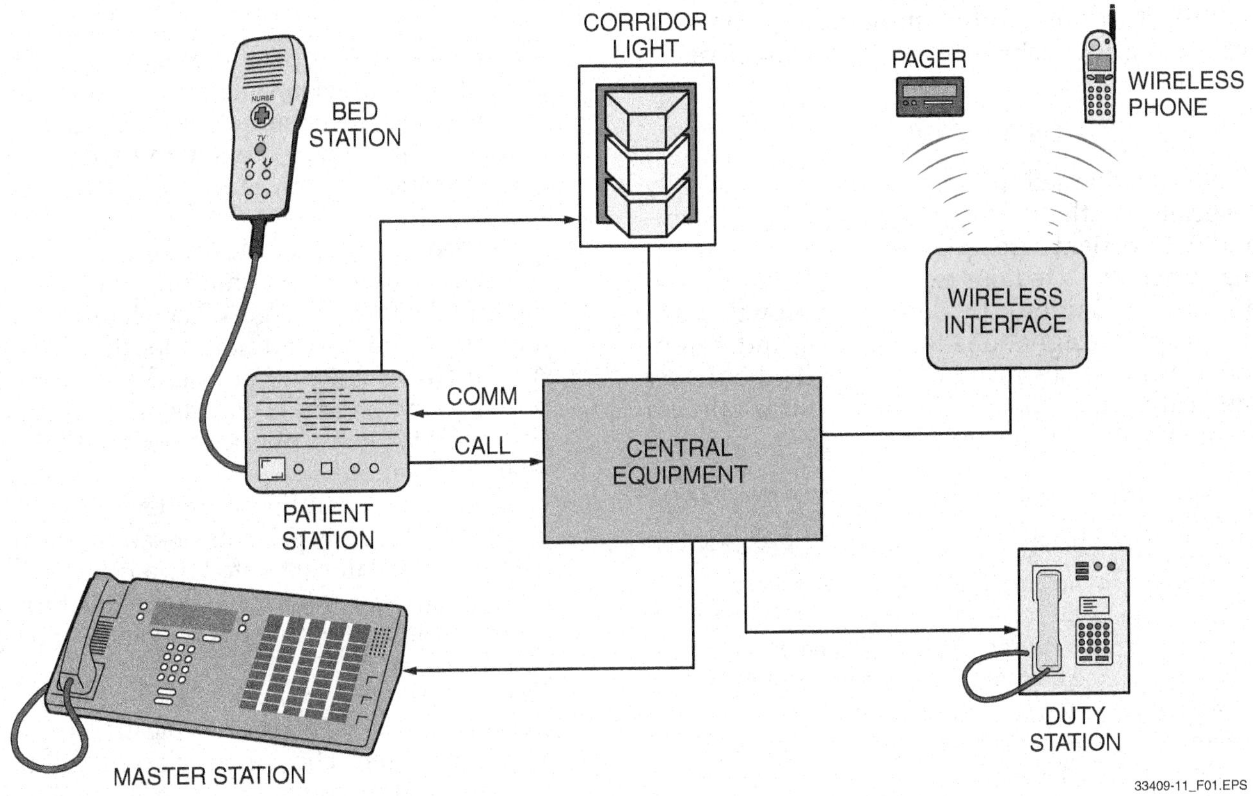

Figure 1 Basic nurse call system.

33409-11_F01.EPS

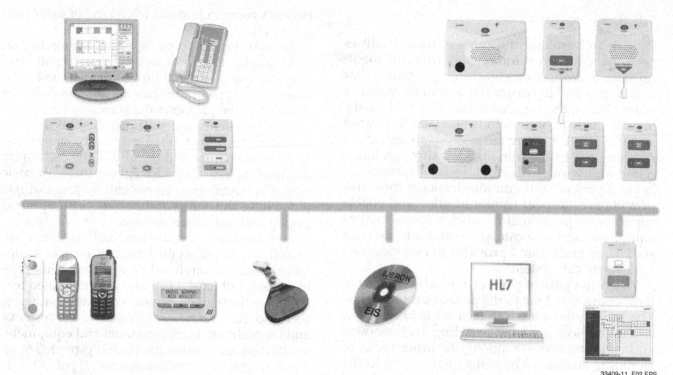

Figure 2 Nurse call system options.

33409-11_F02.EPS

the exception of auxiliary devices such as cell phones and pagers, all the system components will come from the same manufacturer and will be part of an integrated system package. While the objectives of the systems are the same, each manufacturer uses different hardware and software designs to achieve the system's objectives.

2.0.0 IMPORTANT TERMS

The terms defined in this section are specifically associated with nurse call and signaling systems and with patient care facilities in general. While it may seem like a tedious exercise, you should read all these definitions. By doing so, you will learn the proper designations for facilities and equipment. You will also gain some insight about how medical facilities operate and how nurse call and signaling systems are used.

> **NOTE**
> These definitions are derived from the 2007 *NEMA Installation Guide for Nurse Call Systems*. Refer to this document for a more complete list of terms.

- *Alarm* – A signal, typically generated automatically, that annunciates an abnormal condition. Examples are the opening of an emergency-exit door, an unauthorized entry to narcotics storage, or a smoke alarm. Also see *code call* in this section.
- *Annunciator* – An electrically operated visual-indicating appliance containing one or more identified lamps or targets. Each lamp or target indicates a circuit condition, a location, or both.
- *Annunciator panel* – A device that signals the presence and locations of calls or alarms in the system. It typically provides both audible and visual indications.
- *Bath station* – A call-initiating device located in a bath area to allow patients or staff to summon help. It is typically activated by a pull cord.
- *Battery backup* – A battery source used to provide power to the nurse call system during a power outage.
- *Call cord* – A cord with a switch at one end and a connector at the other. It typically plugs into a patient station. The patient places a call to the nursing staff by momentarily pulling the cord

or pressing the switch. A call is automatically placed when the cord is removed from the station receptacle.

- *Central equipment* – The components needed to process and distribute signals among nurse and patient stations and other peripheral devices. Typical components are low-voltage power supplies, logic and control circuits, and a terminal.
- *Code call* – A distinctive audible and visual signal representing a life-threatening situation that requires immediate action. Typical codes are Code Blue, which indicates that a patient requires immediate resuscitation, and Code Pink, which indicates that a baby has been removed from the nursery. Additional codes may be used by some hospitals.
- *Corridor lamp* – A visual annunciator, mounted on the wall or ceiling outside a patient room, that indicates calling activities and the presence of staff members. It may have a single bulb or several bulbs that are used to light color-coded segments to indicate the types of calls and staff members. It is also referred to as a dome light.
- *Day/night transfer* – A feature that allows the staff in one nursing unit to temporarily answer calls from patients in a nearby nursing unit, in addition to their normally assigned calls.
- *Duty station* – A station that uses tones and lamps to annunciate calls by their type or priority. It is normally installed in a location where nurses tend to be when they are not at the nurse control station or in patient rooms, such as the nurses' lounge.
- *General care area* – Patient bedrooms, examining rooms, treatment rooms, clinics, and similar areas in which the patient come into contact with ordinary appliances, such as a nurse call system or electrical devices.
- *Geriatric call cord* – A call cord designed for patients who cannot easily manipulate a pushbutton. It normally has a pressure-pad switch rather than a pushbutton switch.
- *Headwall* – An assembly, usually prefabricated, set into the wall at the head of a patient's bed. It provides connections for and helps integrate a variety of systems, including medical gas, vacuum, electrical power, lighting, communication, and emergency signaling.
- *Intensive care unit (ICU)* – An area where patients are subjected to invasive procedures and connected to line-operated, electro-medical devices. Examples are special care units, coronary care units, angiography laboratories, cardiac catheterization laboratories, delivery rooms, and operating rooms. It may also be called a critical care unit (CCU).

- *Nurse call system* – A system of components that provides audible and visual communication between patients and hospital personnel. It must be listed for the intended use and interconnected according to the manufacturer's instructions.
- *Nurse control station* – A component, intended to be located at the nurses' station, that provides audible tones and visual annunciation of incoming calls. Typically, it also provides audio communication between the nurse and the patient. Many other features are optionally available. It is also called a nurse master station.
- *Nursing unit* – An area of the facility that includes rooms for patients and the specific group of personnel dedicated to those beds.
- *Parallel call* – A system that provides for one or more patient stations to simultaneously annunciate at more than one nurse control station.

Code Blue

When a Code Blue alarm is sounded, trained staff come running with a crash cart containing special equipment and medication designed to revive a patient whose heart has stopped beating.

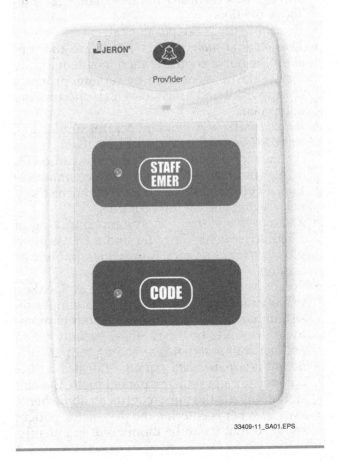

33409-11_SA01.EPS

- *Patient bed location* – The location of an in-patient sleeping bed or the bed or procedure table used in a critical-care area.
- *Patient care areas* – Any portion of a health care facility devoted to examining or treating patients.
- *Patient-controlled lighting* – Room and reading lights that can be regulated from the patient's bed.
- *Patient monitoring* – The use of a nurse control station to monitor the sounds from one or more patient rooms.
- *Patient station* – A device located on the wall behind the patient bed that allows patients or staff to summon help. It is typically activated by a call cord or pillow speaker. It normally has a call-assurance lamp, which lights when a call is placed, and a reset switch for canceling a call. Common options include an intercom speaker/ microphone and entertainment circuits for television control.
- *Patient vicinity* – In an area where patients are normally cared for, the patient vicinity is the space with surfaces likely to be contacted by the patient or an attendant who can simultaneously touch the patient. In a typical patient room, this encloses a space not less than 6' (1.83 m) beyond the perimeter of the bed in its normal location and extends vertically not less than 7' (2.29 m) above the floor.
- *Pendant control* – See *pillow speaker*.
- *Physiological monitoring* – A feature that enables the nurse control station to automatically signal alarms generated by remote, properly equipped, listed physiological monitoring equipment.
- *Pillow speaker* – A pendant control similar to a call cord but with additional features such as a speaker for personal-entertainment audio (TV or programmed audio) and nurse communication; a volume control; an entertainment-channel selector; and lighting controls.
- *Radio paging system* – A system consisting of a radio-frequency encoder and a transmitter, plus portable pagers, sometimes known as beepers.
- *Secure psychiatric ceiling speaker* – A speaker, usually mounted in the ceiling with tamper-resistant hardware, that allows the staff to monitor the sounds in the room, especially when another staff member is attending the patient.
- *Secure psychiatric hall station* – Located in the hall outside a psychiatric patient room, the hall station signals the nurse control station when a staff member is entering the psychiatric patient room so that it can be monitored. Frequently,

such a call also enables (arms) the psychiatric emergency switches in the room.
- *Secure psychiatric emergency switch* – A tamper-resistant switch mounted inside a secure psychiatric patient room. When enabled, it can be pressed by a staff member to signal an emergency. This switch cannot be reset from inside the patient room.
- *Shower station* – A pull cord-activated and waterproof station that can be installed inside a shower stall and used by a patient to summon help. A unit is considered a shower station if it is listed under *ANSI/UL 1069* as a shower station as a result of passing the appropriate water resistance testing.
- *Staff emergency station* – A station that places staff emergency calls.
- *Staff register station* – A station that signals the type of staff, such as an RN or LPN, present in a room.
- *Staff station* – An intercom station typically used by the staff to place calls to the nurse control station. It is usually located in staff areas but can also be located in areas used by ambulatory patients.
- *Swing room* – A room wherein the patient station can selectively annunciate at a different nurse control station, typically in response to a change in the facility's patient load.
- *Toilet station* – A station in a toilet area that allows a patient to signal for help.
- *Visual nurse call system* – A nurse call system that annunciates calls by visually identifying the location of origin and sounding a tone, but which does not provide audio communication.
- *Wet area* – Patient-care areas that are commonly wet while patients are present. These include standing fluids on the floor or the drenching of the work area where the fluids would likely come in contact with the patient or the staff. Areas that are routinely made wet by housekeeping procedures and incidental spills are not considered wet.
- *Zone lamp* – This is similar to a corridor lamp, but it visually annunciates calls from a group of rooms in a particular area. Normally, a zone lamp is located at corridor intersections, where the observer cannot see the room corridor lamps.

3.0.0 CODES AND STANDARDS

All equipment and facilities associated with health care are governed by codes and standards. The primary governing organizations and the roles they play in standardizing health care are described here. Any person who designs, installs, or services nurse call systems must be familiar with the requirements of these codes and standards.

3.1.0 National Fire Protection Association (NFPA)

The NFPA publishes two important codes that govern the installation of nurse call systems:

- *NFPA 70®, National Electrical Code® (NEC®)* – *NEC Article 517* establishes the electrical wiring standards for all types of health care facilities; *NEC Article 725* establishes wiring standards for power-limited circuits.
- *NFPA 99, Standard for Health Care Facilities* – *NFPA 99* establishes criteria for performance, maintenance, testing, and safe practices in order to minimize the risks associated with explosion, fire, and electricity in health care facilities.

3.2.0 Underwriters Laboratory (UL)

The UL establishes performance standards for all types of electrical and electronic equipment. UL *Standard 1069, Hospital Signaling and Nurse Call Equipment* is the industry standard for manufacturing and testing nurse call systems.

3.3.0 Joint Commission on Accreditation of Health Care Organizations (JCAHO)

Known simply as the Joint Commission, the JCAHO evaluates and accredits its member health care organizations in the United States, including hospitals, nursing homes, assisted living facilities, mental health facilities and organizations, and laboratories. A main component of its accreditation is periodic evaluation of facilities to determine that they meet standards established for patient care. Most states require Joint Commission accreditation as a condition of licensing of health care facilities.

3.4.0 National Electrical Manufacturers Association (NEMA)

NEMA publishes manufacturing standards for electrical equipment. The *NEMA Installation Guide for Nurse Call Systems* provides guidelines for the proper installation of these systems.

4.0.0 TYPES OF NURSE CALL SYSTEMS

There are several types of nurse call systems. The type selected depends on the level of care. For example, the system needed for an acute care facility such as a hospital will be much more sophisticated than a system needed for an assisted living facility. The three main categories of nurse call systems are as follows:

- Visual systems
- Audiovisual systems
- Microprocessor-controlled systems

Both hard-wired and wireless systems are defined by these categories. *Table 1* shows the differences in equipment and options among the three main types. Note that some items required for audiovisual systems are optional for visual systems. System selection is based on the level of care required and, to some extent, the facility's budget.

4.1.0 Visual Systems

Visual systems annunciate calls by visually identifying the source of the call and sounding a tone. They are basically push-a-button, light-a-light systems. *Figure 3* shows the annunciator station for a visual nurse call system. They provide patient-to-staff and staff-to-staff annunciations. Devices used to initiate calls can include patient stations, emergency stations, and a variety of other types of call devices and stations. *Figure 4* shows a basic call button that is used with such a system. The system may be able to distinguish different types of calls, such as emergency calls, bath calls, and routine calls. Typically, a patient call will light the light over the door of the room. At the same time, a tone will sound and the room/bed number will light up at the annunciator station. A visual nurse call system might be found in a nursing home, assisted living facility, emergency department, outpatient clinic, or critical-care unit where two-way communication between patient and staff is not a high priority.

Duty stations, such as the one shown in *Figure 5*, are placed in locations where nurses may be when they are away from the nurses' station. Such locations might include the linen closet, the nurses' lounge, or the floor kitchen. Duty stations sound a tone that alerts staff to the existence of a call. Some duty stations are also able to indicate call priority.

Emergency stations such as the one shown in *Figure 6* are placed in locations that may require immediate action. Help can be summoned by simply pulling on the cord.

Table 1 Types of Nurse Call Systems

	Required Equipment	Optional Equipment	Other Systems
Visual Systems			
Wired Systems	Central equipment including power supplies Nurse control station or annunciator panel Patient call stations (pull cord or push button) Dome light over door Bath stations (patient rooms; community bath and showers)		Duty stations Zone lights Door contacts Code call device Other alarm points
Wireless Systems	Central equipment including power supplies Nurse control station or annunciator panel Pocket pagers Patient call stations (pull cord or push button) Dome light over door Bath stations (patient rooms; community bath and showers)		Door contacts Other alarm points
Audiovisual Systems			
Wired Systems	Central equipment including power supplies Nurse control station or annunciator panel Patient call stations (pull cord or push button) Dome light over door Bath stations (patient rooms; community bath and showers)	Duty stations Staff stations Staff registration Zone lights Door contacts Code call device Other alarm points (pumps, bed exit) Music entertainment Light control TV control	**Optional Systems** Overhead paging systems Pocket paging systems
Microprocessor-Confined Systems	Computer and central equipment incl. power supplies Nurse control station and/or annunciator panel Patient call stations (pull cord or push button) Dome light over door Bath stations (patient rooms; community bath and showers)	Duty stations Staff stations Staff registration Zone lights Door contacts Code call device Other alarm points (pumps, bed exit) Music entertainment Light control TV control TV display of calls including messaging Management call reports	**Integrated Systems** Overhead paging systems Pocket paging systems Cell phone systems Door monitoring systems Smoke detection systems Infant protection systems Dementia alarm systems Man down systems Personnel tracking systems Equipment tracking systems

Figure 3 Example of an annunciator for a visual nurse call system.

Figure 4 Example of a call initiation device.

33409-11_F05.EPS

Figure 5 Example of a duty station.

4.2.0 Audiovisual Systems

The main difference between audiovisual and visual systems is that audiovisual systems not only light call buttons and sound tones, they also provide voice communication between the nurse control stations and the patient care areas (*Figure 7*). The staff member receiving the call is able to communicate with the calling patient or staff member. One of the benefits of the two-way system is that the responding staff member can answer questions or dispatch the required staff member without going to the room. Another benefit is that necessary equipment or materials can be assembled before the staff member goes to the room.

4.3.0 Microprocessor-Based Audiovisual Systems

Microprocessor-based systems have the same kinds of equipment as visual and audiovisual systems, but have a much more sophisticated array of features (*Figure 8*). These features include the

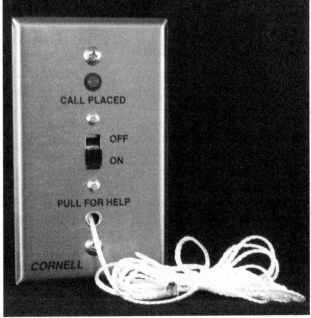

33409-11_F06.EPS

Figure 6 Example of an emergency pull-cord station.

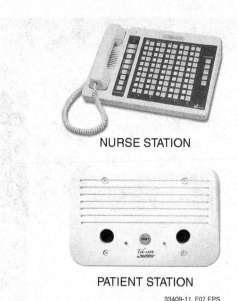

NURSE STATION

PATIENT STATION

33409-11_F07.EPS

Figure 7 Master console and patient station for an audiovisual nurse call system.

ability to activate pagers and cell phones in order to contact staff members. Specialized software programs available with these systems allow management to collect and analyze information, such as number and locations of calls and staff response times. Much of this software is designed to run on a Microsoft Windows® or Linux platform. The fact that they are microprocessor-controlled allows the user greater flexibility in customizing the system to meet their particular needs. This system is able to handle incoming signals from a variety of compatible devices.

5.0.0 CALL MANAGEMENT

Some facilities use a centralized call management approach in which all patient calls are processed at a call center. Other facilities use a decentralized approach in which a master control console is located at the central staff station for each floor, wing, or designated treatment area.

In a centralized system, an operator makes the initial communication with the patient or staff member and determines what assistance is needed. In many systems, the operator has only to touch a display screen (*Figure 9*) or click on a room location to obtain information about the patient that will help in handling the call and dispatching aid. The operator usually has a variety of options for contacting the required staff member, including the following:

- Paging the staff member over the loudspeaker system
- Sending a location or text message (*Figure 10*) to a wireless phone or pager
- Sending a signal to the staff member's normal duty station or auxiliary duty station

A key feature of some nurse call systems is a staff and equipment locator. This system uses small transmitters that are worn by staff members (*Figure 11*). Similar devices can be attached to key pieces of equipment so they can be quickly located in an emergency. Staff members log onto the system when they arrive on duty. From then on, the system can locate them when needed. IR

On Site

Special Options

Many facilities have special needs. For example, a pediatric ward may have a special system that allows the nursery staff to communicate with visitors. A psychiatric hospital needs special alerting and security systems to provide staff protection. Various types of facilities need alarm systems for exit doors. Many computer-based systems include these features. This photo shows a high-risk environment station that can be used in psychiatric hospitals. Patient tracking systems that use infrared and RFID technology are also used in such environments.

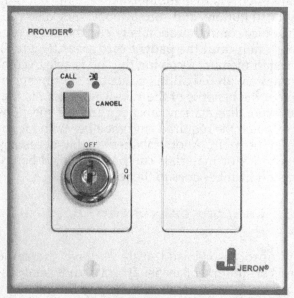

33409-11_SA02.EPS

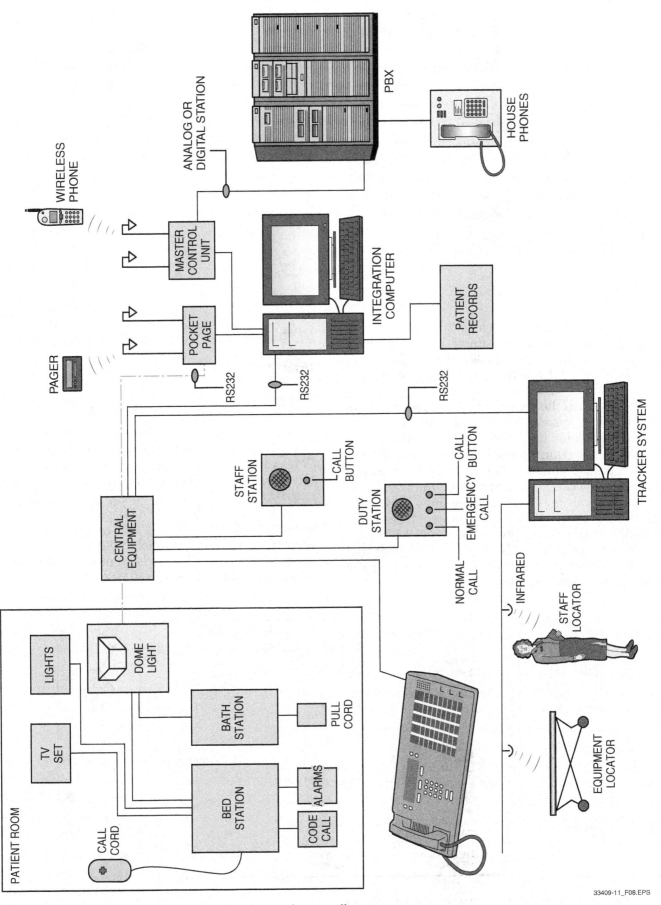

Figure 8 Diagram of a microprocessor-based audiovisual nurse call system.

33409-11_F08.EPS

Figure 9 Touch screen display.

33409-11_F09.EPS

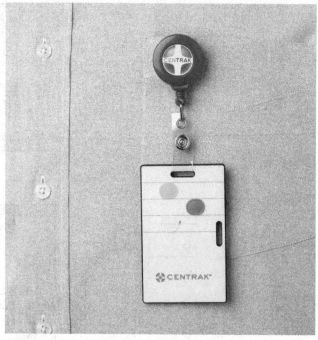

33409-11_F11.EPS

Figure 11 Staff locator transmitter badge.

or ultrasonic receivers such as the one shown in *Figure 12* are installed in patient rooms and other locations. These receivers detect an individual's transmitter and tell the control console where the person is. When a staff member responds to a patient call, the receiver in the patient's room notifies the system, cancels the call, and turns off the corridor light. As an option, the system may be programmed to turn on a multi-level dome light (*Figure 13*) to indicate the level of staff present. If a call goes unanswered for a period of time, the call is returned to the master console to be reassigned. When operators are looking for equipment, they simply locate the last place an entry was recorded. They can then contact that room or dispatch a staff member to see if the equipment is still in use.

6.0.0 SKILLED NURSING AND ASSISTED LIVING FACILITIES

A skilled nursing facility, or nursing home, provides care for people who need constant medical attention, but are not suffering from acute illnesses or injuries that would require placement in an acute care facility like a hospital. People in a nursing home are often not able to get around on their own and may need help with the most basic of human needs. A basic visual nurse call system is generally used in such facilities.

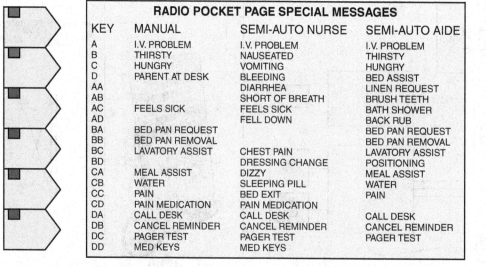
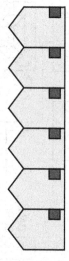

RADIO POCKET PAGE SPECIAL MESSAGES

KEY	MANUAL	SEMI-AUTO NURSE	SEMI-AUTO AIDE
A	I.V. PROBLEM	I.V. PROBLEM	I.V. PROBLEM
B	THIRSTY	NAUSEATED	THIRSTY
C	HUNGRY	VOMITING	HUNGRY
D	PARENT AT DESK	BLEEDING	BED ASSIST
AA		DIARRHEA	LINEN REQUEST
AB		SHORT OF BREATH	BRUSH TEETH
AC	FEELS SICK	FEELS SICK	BATH SHOWER
AD		FELL DOWN	BACK RUB
BA	BED PAN REQUEST		BED PAN REQUEST
BB	BED PAN REMOVAL		BED PAN REMOVAL
BC	LAVATORY ASSIST	CHEST PAIN	LAVATORY ASSIST
BD		DRESSING CHANGE	POSITIONING
CA	MEAL ASSIST	DIZZY	MEAL ASSIST
CB	WATER	SLEEPING PILL	WATER
CC	PAIN	BED EXIT	PAIN
CD	PAIN MEDICATION	PAIN MEDICATION	
DA	CALL DESK	CALL DESK	CALL DESK
DB	CANCEL REMINDER	CANCEL REMINDER	CANCEL REMINDER
DC	PAGER TEST	PAGER TEST	PAGER TEST
DD	MED KEYS	MED KEYS	

33409-11_F10.EPS

Figure 10 Example of a wireless phone or pager text message.

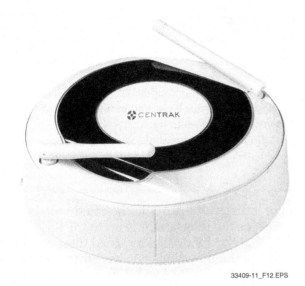

Figure 12 Staff locator receiver.

Figure 13 Multi-level dome light.

Assisted living facilities are designed to accommodate people, usually the elderly, who are able to take care of themselves to varying degrees, but need help with some daily living needs, as well as supervision and limited support from medical staff. Residents of such facilities may suffer from chronic, but not disabling, physical or mental disorders. In many cases, however, they are elderly retired persons who are relatively healthy and simply want to live in a supervised environment where medical care and living assistance are available if they need it.

The nurse call systems used in these facilities differ from those used in acute care facilities such as hospitals. For example, doors, elevators, and stairways may be monitored to keep mentally impaired residents from wandering away,

which is a common problem in facilities designed to supervise such residents. The residents will have emergency call stations, including bath call stations in their rooms or apartments, in order to summon help in case of an emergency. They may also wear an emergency call pendant (*Figure 14*) to summon help.

Because many of their patients are bedridden or confined to a chair, skilled nursing facilities may use devices designed for fall monitoring. A pad on the bed or chair alerts the staff if a person leaves a bed or chair while unattended. Incontinence detectors are also used in these facilities.

7.0.0 SYSTEM INTERFACES

Microprocessor-controlled nurse call systems are able to interface with a variety of devices, including monitoring devices such as smoke detectors and door monitors. It is important to note, however, that any device connected to a nurse call system must be listed for that system. That is one reason why components of a nurse call system are usually purchased from the same manufacturer. Any component that is not listed for the specific system must provide an electronically isolated interface that will prevent a failure in the component from affecting the nurse call system.

7.1.0 Telephone Equipment

Nurse call systems may be connected to a wireless phone system in the hospital via an integration computer. The wireless phone systems receive their dial tone from the in-house PBX,

Figure 14 Emergency call pendant.

Assisted Living

According to the Assisted Living Federation of America, as of 2011 there were more than 36,000 assisted living communities serving more than one million residents. Assisted living differs from independent living in that assisted living residents typically receive their meals from the facility, have access to health care, and have 24-hour emergency call systems. Assisted living facilities also have staff to provide personal assistance to residents. Independent living communities, on the other hand, are communal living arrangements designed for senior citizens who are able to care for themselves.

thus allowing the user of the wireless phone to receive incoming calls via the PBX, make outgoing calls via the PBX, and make station-to-station calls with any wireless or desktop phone in the facility. However, if a nurse call system is connected to central office telephone lines, the nurse call system must have FCC approval. The nurse call integration package allows the user to receive calls from the nurse call systems, identifying the room number, bed calling, and type of call being placed. The wireless phone may also receive calls from integrated systems, identifying the origin of the call and type of call being made.

Several companies make in-building wireless telephones. There is no charge for air time with these phones, as there is for standard cell phones. These special phones have LCD displays and can be used to display text messages. When the patient initiates a call, the corridor lamp automatically lights. At the same time, the call appears at the nurses' station and is forwarded to the assigned caregiver's wireless phone. Because these phones are not direct-connected to the nurse call system, they are not subject to the same restrictions as hard-wired devices.

7.2.0 Entertainment Equipment

In many hospitals, the patient bed control is much more than a nurse call button. Many pillow speakers are designed to control the room television and radio, as well as environmental features such as draperies and lighting. The more modern of these pillow speakers uses a five-wire configuration for audio, channel up, channel down, on/off, and common (*Figure 15*). Older models use a three-wire arrangement.

Any television set connected to the nurse call system must be approved for hospital use and must be compatible with the nurse call pillow speaker. The nurse call system manufacturer must specify that only a listed hospital-grade television receiver can be connected to the system. The

manufacturer will provide the rated voltage and current for the interface. It is up to the installer to verify that the TV does not exceed these ratings.

Programmed audio is received through an unused channel of the MATV or SMATV system. The signal is carried on 25V program lines to an electrical or mechanical stepper switch on the patient's pillow speaker.

7.3.0 Paging Systems

Many nurse call systems can be connected to pagers and the building's public address system. The connection to the paging system gives the staff the ability to page someone from the nurse call station through the building's overhead speaker system.

A Novel Approach

In one Wisconsin hospital, the telemetry technician is designated to act as the call center monitor. The telemetry technician is stationed in a monitoring room and monitors the vital signs of at-risk patients for any problems. By having the telemetry technician evaluate and respond to patient calls, the nursing staff is no longer tied to the floor station and can spend more time with patients.

This centralized approach has an additional benefit: it allows the telemetry technician to immediately dispatch the appropriate staff member in the event the telemetry indicated an irregularity in vital signs. In responding to a patient call, the operator of this system is able to communicate directly with the patient, turn on the corridor light, determine the level and type of care needed, and select from a list of text messages to be sent to the selected staff member's pager.

Source: Rauland-Borg Corporation

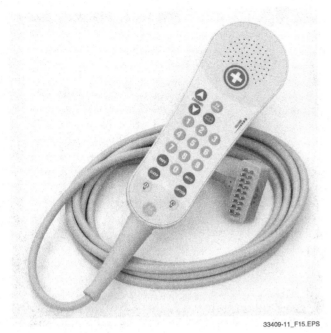

Figure 15 Pillow speaker controls.

33409-11_F15.EPS

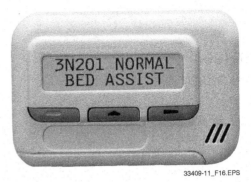

33409-11_F16.EPS

Figure 16 Pager.

Any paging device or PA system connected to a nurse call system must be listed for that system, or must have an electronically isolated interface. For pagers (*Figure 16*), the nurse call manufacturer's literature must identify pager models that will work with the system and include specific instructions for installing and configuring the pagers. If the pager is not listed for use with the system, it may only be used as a secondary annunciator.

7.4.0 Fire Alarm Systems

The smoke detectors in patient rooms must be connected as determined by the AHJ. They may be connected to the fire alarm system. A nurse call system can be used as an annunciator for a smoke alarms, but may not be used as the primary annunciator unless the nurse call system is listed as a fire alarm system. The smoke detector's normally-open, dry contacts may be used to operate an isolated lamp in the nurse call corridor lamp.

On microprocessor-based nurse call systems, the smoke alarm may be connected to a dedicated port on the patient station. However, it must generate a unique tone that will be recognized by staff members as a smoke detector activation call.

On Site

Visibility

Many nurse call master stations have touch screens that allow staff to quickly access information. This screen shows pending calls from patients and indicates that the responsible staff member for each call has been notified. The lower left portion of the screen shows that the staff members will be reminded if the call signal hasn't been turned off within a set time. At the bottom right of the screen, the staff locator system is showing the location of selected staff members.

33409-11_SA03.EPS

Pillow Speaker

This unique pillow speaker by WestCall has extra call buttons so the patient can communicate specific needs, such as the need for pain medication. This type of feature allows the provider to come prepared, thus saving steps.

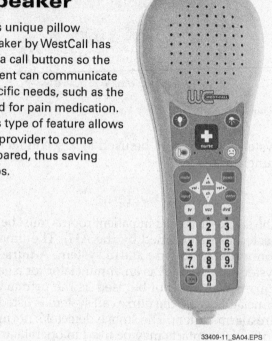

33409-11_SA04.EPS

Figure 17 Door alarm.

33409-11_F17.EPS

7.5.0 Security Systems

The nurse call system may be used to monitor the status of doors, elevators, narcotics cabinets, and other areas that require protection (*Figure 17*). Door and elevator monitoring is required in areas such as psychiatric wards and assisted living facilities where residents are apt to wander. These calls have a high priority and must remain in the system until canceled at the point of origin. In addition to access control, psychiatric wards and hospitals have special emergency call stations for staff to summon help, such as the emergency station shown in *Figure 18*.

The narcotics cabinet monitor also is a high-priority call that must be specifically canceled at the point of origin. A bypass key switch (*Figure 19*) may be provided so that those with approved access may open the cabinet without placing an alarm call. Like all circuits that interface with the nurse call system, these must either be tested and listed as part of the nurse call system or meet the isolation requirements of *UL-1069*.

7.6.0 Auxiliary Alarm Devices

Nurse call systems can interface with many types of alarm devices, such as intravenous (IV) drip monitors, bed exit alarms, incontinence alarms,

Figure 18 Emergency call station.

33409-11_F18.EPS

pulse monitors, and pump alarms. Such equipment must have an auxiliary alarm output plug designed to fit into the auxiliary connector of the call device. Alarms can be transmitted to central stations, wireless phones, and pagers in the same way as other alarms.

7.7.0 Computers and Printers

Many microprocessor-based systems are designed to interface directly with a printer. These printers can be used for a variety of purposes:

• Recording system activity for liability purposes

Figure 19 Bypass key switch.

33409-11_F19.EPS

- Obtaining statistical data on the numbers, types, and duration of call responses
- Obtaining system maintenance and diagnostic information
- Recording system configuration settings

A printer connected directly to a nurse call system must be UL Listed for connection to that particular system, or the nurse call system interface must be UL Listed for connection to ancillary equipment.

Computers are often connected to nurse call systems and are used to support many functions:

- Setting up and revising system configuration selections and priority levels
- Accessing nurse call system reporting and management software
- Entering patient information
- Accessing staff records and status information

8.0.0 INSTALLATION PRACTICES

This section covers common practices that apply to the installation of nurse call systems. However, note that the manufacturer of a given system is required by *UL Standard 1069, Section 45* to provide detailed instructions for installing, wiring, and programming their system. The manufacturer's manual will be the primary source of information and procedures related to the nurse call system, and it should be carefully studied by installers before attempting any installation work.

8.1.0 Electrical Power Requirements

NEC Article 517 requires an alternate power source, consisting of one or more generator sets or battery systems (where permitted) to be installed in health care facilities. Each patient bed location must be supplied by at least two branch circuits, one or more from the emergency system and one or more from the normal electrical system. Emergency system receptacles must be identified and must also indicate the panelboard and circuit number supplying them. All receptacles used in patient bed locations must be listed as hospital grade.

8.1.1 Electrical Systems

Hospitals must have two separate systems capable of supplying a limited amount of lighting and power for life safety and effective hospital operation during the time the normal electrical service is interrupted. These two systems are the emergency system and the equipment system.

In general, the emergency system is limited to circuits essential to life safety and critical patient care. These are designated the life safety branch and the critical branch. The equipment system must supply the major electrical equipment necessary for patient care and basic hospital operation.

The number of transfer switches to be used must be based upon reliability, design, and load considerations. Each branch of the essential electrical system must be served by one or more transfer switches, as shown in *Figure 20*. One transfer switch must be permitted to serve one or more branches or systems in a facility with a maximum demand on the essential electrical system of 150kVA.

The life safety branch and critical branch of the emergency system must be kept entirely independent of all other wiring and equipment and must not enter the same raceways, boxes, or cabinets with each other or other wiring, except as follows:

- In transfer switches
- In exit or emergency lighting fixtures supplied from two sources
- In a common junction box attached to exit or emergency lighting fixtures supplied from two sources

The wiring of the equipment system is permitted to occupy the same raceways, boxes, or cabinets as other circuits that are not part of the emergency system. The essential electrical system must have adequate capacity to meet the demand for the operation of all functions and equipment to be served by each system and branch. Further-

---

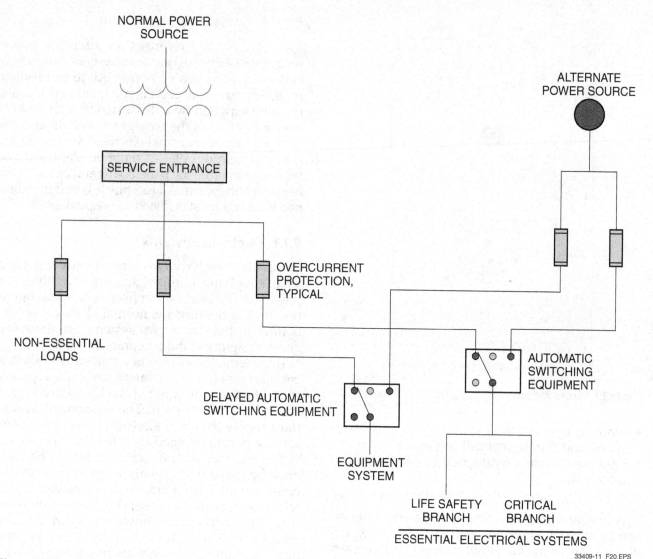

NORMAL POWER
SOURCE

ALTERNATE
POWER SOURCE

SERVICE ENTRANCE

OVERCURRENT
PROTECTION,
TYPICAL

NON-ESSENTIAL
LOADS

DELAYED AUTOMATIC
SWITCHING EQUIPMENT

AUTOMATIC
SWITCHING
EQUIPMENT

EQUIPMENT
SYSTEM

LIFE SAFETY
BRANCH

CRITICAL
BRANCH

ESSENTIAL ELECTRICAL SYSTEMS

33409-11_F20.EPS

Figure 20 Electrical system hookup.

more, the emergency system must be divided into two mandatory branches: the life safety branch and the critical branch, as described in *NEC Sections 517.32* and *517.33*, respectively. The life safety branch of the emergency system must supply power for the following:

- Lighting required for corridors, passageways, stairways, and landings at all exit doors
- Exit signs
- Fire and other alarms
- Communication systems
- Generator set location
- Elevators
- Automatic doors

The critical branch of the emergency system must supply power for task illumination, fixed equipment, selected receptacles, and special power circuits serving the following areas and functions related to patient care:

- Anesthetizing locations
- Isolated power systems in special environments
- Patient care areas
- Nurse call system
- Blood, bone, and tissue banks
- Telephone equipment room and closets
- Task illumination for specified areas

NEC Section 700.7(A) also requires labels, signs, or some other permanent marking to be used on all enclosures containing emergency circuits to readily identify them as components of an emergency system. See *Figure 21*.

8.1.2 Backup Power

Manufacturers of nurse call systems include backup power sources for their systems. Backup power can be supplied by a local battery-powered UPS that is installed between the normal power

Hazardous Locations

NEC Article 517, Section 517.60(A) places restrictions on electrical equipment used in hazardous locations. Areas where flammable anesthetizing gases are used and stored are classified as Class 1, Division I hazards. Any electrical equipment that operates at more than 10 volts must be specifically classified for use in those locations. *NFPA-99, Annex E2* provides further information on this subject.

source and the nurse call system. The UPS provides power during the switchover to emergency power. The UPS is typically designed to provide power for at least 30 minutes. The nurse call system must provide a means for periodically testing the backup power source.

8.2.0 Installation Guidelines

The *NEMA Installation Guide for Nurse Call Systems* lists the requirements for installing a system. The key installation requirements are summarized in this section.

8.2.1 Electrical Guidelines

The following electrical guidelines should be used for the installation of nurse call system components:

- The central equipment must have a dedicated 120V, 60Hz, 20A power line.

CONTAIN EMERGENCY CIRCUITS

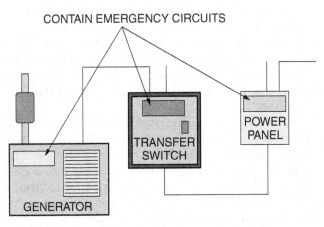

Labels, signs, or some other permanent marking must be used on all enclosures containing emergency circuits to readily identify them as components of an emergency system. *NEC Section 700.7(A)*

33409-11_F21.EPS

Figure 21 Summary of *NEC®* requirements for emergency circuit identification.

- Proper grounding of the system is necessary. The ground wire should be 10 AWG or larger and as short as possible.
- A surge suppressor and line conditioner should be provided at each AC input.
- A UPS is recommended. It should, at minimum, provide power for as long as it takes to get the facility's emergency power system on line.
- In order to minimize cable lengths, central equipment should be installed in a location that is central with respect to the nurse's console and system zones.
- Back boxes must be listed and must be compatible with the device and selected conduit diameter.
- Each back box must be grounded in accordance with the *NEC®*.

8.2.2 Location of Equipment

The following guidelines for height above the floor should be used for installation of nurse call system components:

- *Central equipment* – 4.5' (54") above finished floor (AFF)
- *Patient stations* – 4.5' (54") AFF
- *Code blue stations* – 3.5' (42") AFF
- *Emergency stations* – 3.5' (42") AFF
- *Dome lights* – 7.5' (90") AFF

> **NOTE**
> As required by the Americans with Disabilities ACT (ADA), any device mounted less then 84" (7') from the floor cannot protrude more than 4" from the wall.

Nurse master stations should be placed where they can be easily seen and accessed by the staff. Patient call stations should be placed within reach of the patient.

8.2.3 Electrical Safety Considerations

Steps must be taken to protect patients and staff from electrical harm. These steps include the following:

- Provide surge protection for lighting and entertainment control relays to damp voltage transients caused by the relay coils.
- Protect patients and staff from ground faults and leakage currents, and make sure the system grounds are isolated from earth ground and other AC-powered devices.

- Make sure components intended for use in oxygen-rich environments are UL-listed for that use. Non-electrical devices are normally required in an oxygen-rich environment.
- Make sure that only waterproof call devices are used in bath and shower areas.

8.2.4 System Wiring and Cabling

It is important to plan the wiring before you start running signal wire. Here are some things to consider when planning the wiring runs:

- The nurse call system must be on a dedicated AC circuit with a dedicated circuit breaker. All control equipment should be marked with the location and number of the circuit breaker that feeds it.
- At any location where both power and signal wiring exist, they must be separated by a barrier or other method that will prevent leakage current from being imposed on signal lines.
- Lay the cabling out in a logical pattern. Check the manufacturer's instructions for the maximum number of stations on a single cable run. Try to follow the floor layout so that troubleshooting is intuitive. For example, if the floor is T-shaped, a run for each leg of the T might be the best approach. If there are many stations on each leg, six runs—one for each side of each leg—might be the best approach.
- Refer to the manufacturer's instructions to determine the correct wire gauge. The manual will provide a table or formula for determining wire gauge based on projected current draw related to the number of possible calls on a wiring run, the number of stations, and the number of corridor lamps.
- Do not exceed the limits for the cross-sectional area of cables in a conduit, as specified by the *NEC*©.
- Be careful when pulling signal wire. The small conductors and light insulation will not bear as much strain as power wiring. It is a good idea to use powdered soapstone or other anti-friction agent when pulling the signal wiring.
- Do not use electrical tape and wire nuts for making signal wire splices. The *NEC*© recommends butt splices or equivalent crimp-style connectors for splicing signal wire. Check the manufacturer's instructions for other options.
- The use of metallic conduit is recommended for cable runs. If the cabling is run in trays instead of conduit, the nurse call system cables should be partitioned off from other cabling. The wire must be listed for cable tray.

A point-to-point wiring diagram showing all field connections, including wire color and terminations, must be prepared and either attached to the main control unit or included with the as-built drawings. This diagram must show the following information:

- How every system component is interconnected
- The physical location of each component
- Interconnections at all junction boxes and splice points
- The physical location of junction boxes and splice points
- The conduit routing and distances
- The location of the central equipment

8.3.0 Programming

Microprocessor-based systems will have a variety of customizable features, depending on the application and sophistication level of the system. Each health care facility will have its own protocols for annunciating and responding to calls. The installer must set up the system for those protocols at the time of installation. A typical way to plan the programming process is to meet with the staff and use programming checklists to allow them to select the options they want to use. *Figure 22* shows an example of one such checklist.

Programming choices depend a lot on hardware. Some programming options are available only if a particular hardware option has been purchased. Examples are pagers and staff locating systems.

One programming requirement that is common to all systems is entering the physical (architectural) location of each system device. This allows the system to identify the source of a message and to send messages to specific locations. For example, if a patient calls for assistance, the system needs to be able to tell the operator the room number or other location from which the call came. If the staff member needed to answer the call is in the nurses' lounge, the system needs to be able to route the message to the duty station in that specific location.

SYSTEM OPTIONS WORKSHEET

1. ACKNOWLEDGEMENT TONE Circle Choice: **ON** **OFF**	Turn ON for short pre-announce tone at both master and audio bus station to alert attendant and patient or staff that Nurse Master has made audio connection (call acknowledgement/origination) or group monitor/pager. Principal use is to alert patients when Nurse Master is monitoring.
2. PRESENCE STATIONS Circle Choice: **ON** **OFF**	Turn ON if Presence Stations are installed. Enables call forward operation; activates staff locator for nurse/aide needed; automatically programs group O for staff only page.
3. DUTY STATIONS Circle Choice: **ON** **OFF**	ON allows programming of installed Duty Stations and Electrical Supervision Stations to duty areas.
4. ZONE LIGHTS Circle Choice: **ON** **OFF**	ON allows programming of installed Zone Lights to zone areas.
5. NIGHT TRANSFER Circle Choice: **ON** **OFF**	ON allows one Nurse Master to take another Nurse Master's calls.
6. GROUP PAGE Circle Choice: **ON** **OFF**	ON allows groups of up to 10 rooms to be programmed for page.
7. GROUP MONITOR Circle Choice: **ON** **OFF**	ON allows groups of up to 10 rooms to be programmed for monitor.
8. P.A. PAGE Circle Choice: **ON** **OFF**	ON allows programming of all P.A. Page Stations.
9. CALL FORWARD Circle Choice: **ON** **OFF**	ON allows manual or automatic forwarding of call alert tones to Patient Stations (staff follower).
10. RADIO POCKET PAGE Circle Choice: **ON** **OFF**	ON allows programming of radio pocket page addresses for beds.
11. SWING ROOMS Circle Choice: **ON** **OFF**	Allows designated rooms to be swung from one master to another.
12. LOCKING CALLS ONLY Circle Choice: **ON** **OFF**	Causes all routine bed calls to be displayed as personal attention calls. These calls must be cancelled at the calling station.
13. TONE SILENCE OFF Circle Choice: **ON** **OFF**	Removes ability to silence the common audible tones at the Nurse Master Station.
14. EXPANDED ZONE AREA Circle Choice: **ON** **OFF**	ON allows for up to 128 zone areas (two members). Use Expanded Zone Area Worksheets, Sheets 7 and 8. OFF allows for up to 18 zone areas (four members). Use Sheet 6.
15. NUMERIC BEDS Circle Choice: **ON** **OFF**	ON allows for numeric-only bed designators. OFF allows for aplha-only bed designators.

33409-11_F22.EPS

Figure 22 System options worksheet.

9.0.0 System Checkout/ Commissioning

The system manufacturer's installation manual will contain procedures for checking out a system when the installation and programming are finished. The test parameters will be specific to the system and its options but, at a minimum, will include the following:

- Use the system diagnostics to verify that there are no failures present in the system. Clear any failures before further testing is done.
- At each room station, activate each call level and verify that the proper annunciation is displayed at the master station and any designated duty or staff stations. Also verify that the associated corridor lights and other display devices react properly to the call by displaying the proper color, flash rate, and/or tone rate.
- Verify that bed, shower, and bath stations provide the correct annunciation at the master station and corridor light.
- As applicable, verify the TV, radio, lighting, and curtain controls at each bed station.
- Place a call from each code station and verify that the correct message is displayed at applicable master stations and code display units.
- Respond to a call using the handset and non-handset modes. The message should be audible to the person on the receiving end and at a comfortable audio level.

- Verify that the person receiving the message can respond to the caller.
- Call each room and duty station from the master station and verify that the call is received.
- If pagers are used, initiate a call from the master station to the pager and verify that the message reaches the pager.
- Select messages at each priority level and verify the correct message reaches the selected pager.
- If audio paging is used, verify that pages reach the selected area(s).
- If the system is connected to fire or security alarms (smoke detectors and door alarms, for example), verify that the correct annunciation is received when the alarm is activated.
- Verify system memory integrity: power the system down, wait 10 to 15 minutes, then power it up again. Place a call from each zone and verify that the correct location number shows up at the master station.

> **NOTE**
>
> Training of facility personnel in the correct operation of the system is an important component of the installation process. This subject is covered in detail in *System Commissioning and User Training*.

SUMMARY

The basic function of a nurse call system is to allow a patient to obtain assistance at the push of a button. Modern nurse call systems go far beyond the concept of the old push-a-button, light-a-light concept. Through the use of microprocessor control, today's systems can locate staff and critical equipment, contact personnel through wireless phones and pagers, and allow management to keep track of patient calls and staff activity.

There are different levels of systems, which are selected to meet the needs and budget of the facility. Visual-only systems, for example, are effective in nursing homes, intensive care facilities, and emergency rooms, where two-way communication with the patient is not a requirement.

All components of a nurse call system must be electronically compatible. For that reason, a nurse call system is made up of components and software from the same manufacturer. The system is designed by the manufacturer to be acquired as a complete, integrated package. Nurse call system manufacturers are required by *UL 1069* to provide a complete technical manual for each system, including description, operation, installation, point-to-point wiring diagrams, troubleshooting, and programming. This manual serves as the guide for installing, checking out, and maintaining any nurse call system.

1. An annunciator panel is a device that _____.
   a. automatically generates an alarm in the event that an emergency door is opened
   b. is used by patients to summon help
   c. signals the presence and location of calls and alarms
   d. uses pull cords to sound an alarm

2. A staff register station is used to _____.
   a. signal the type of staff present in a room
   b. record when staff members arrive for work
   c. send a page to a staff member
   d. allow patients to locate nurses

3. Electrical wiring standards for health care facilities are governed by _____.
   a. NEC Article 90.7
   b. NEC Article 517
   c. NEC Article 770
   d. NEC Article 800

4. Visual nurse call systems are least likely to be found in a(n) _____.
   a. hospital patient care area
   b. assisted living facility
   c. outpatient clinic
   d. nursing home

5. An audiovisual nurse call system identifies the source of a call by _____.
   a. lighting a light and sounding a tone only
   b. lighting a light, sounding a tone, and providing voice communication
   c. providing voice communication and remote camera monitoring
   d. providing remote camera monitoring and lighting a call button

6. Emergency call pendants are typically worn by _____.
   a. the nursing staff
   b. nursing home residents
   c. ICU attendants
   d. hospital physicians

7. If a nurse call system is connected to central office telephone lines, the nurse call system must be _____.
   a. FCC approved
   b. NFPA approved
   c. wireless
   d. connected to the in-house PBX

8. Which of the following is true regarding system interfaces?
   a. A nurse call system can never be used as the primary annunciator for a fire alarm system.
   b. Pocket pagers that are not listed for use with the nurse call system can still be used as secondary annunciators.
   c. Any TV set can be connected to a pillow speaker as long as it is made in the United States.
   d. Wireless telephones are subject to the same restrictions as hard-wired equipment.

9. A hospital nurse call system must be powered through the hospital's _____.
   a. life safety branch
   b. critical branch
   c. equipment system
   d. non-essential loads system

10. Which of the following must be supplied by the life safety branch of a hospital's emergency system?
    a. Anesthetizing locations
    b. Automatic doors
    c. Patient care areas
    d. Isolated power systems

Figure Credits

GE Healthcare – Clinical Systems, Module opener, Figures 11, 12, 14, and 15

Jeron Electronic Systems, Inc., Figures 2, 7 (top photo), 16, 18, 22, SA01, and SA02

Cornell Communications, Inc., Figures 3–6, 17, and 19

Tektone Sound & Signal, Figures 7 (bottom photo) and 9

Dukane Communication Systems, Figure 10 and SA03

Rauland-Borg Corporation, Figure 13

West-Com Nurse Call Systems, Inc., SA04

NCCER CURRICULA — USER UPDATE

NCCER makes every effort to keep its textbooks up-to-date and free of technical errors. We appreciate your help in this process. If you find an error, a typographical mistake, or an inaccuracy in NCCER's curricula, please fill out this form (or a photocopy), or complete the online form at **www.nccer.org/olf**. Be sure to include the exact module ID number, page number, a detailed description, and your recommended correction. Your input will be brought to the attention of the Authoring Team. Thank you for your assistance.

Instructors – If you have an idea for improving this textbook, or have found that additional materials were necessary to teach this module effectively, please let us know so that we may present your suggestions to the Authoring Team.

NCCER Product Development and Revision
13614 Progress Blvd., Alachua, FL 32615

Email: curriculum@nccer.org
Online: www.nccer.org/olf

❏ Trainee Guide   ❏ AIG   ❏ Exam   ❏ PowerPoints   Other _____

Craft / Level: _____   Copyright Date: _____

Module ID Number / Title: _____

Section Number(s): _____

Description: _____

_____

_____

_____

_____

Recommended Correction: _____

_____

_____

_____

_____

Your Name: _____

Address: _____

Email: _____   Phone: _____

33410-12

CCTV Systems

Module Ten

Trainees with successful module completions may be eligible for credentialing through NCCER's National Registry. To learn more, go to **www.nccer.org** or contact us at **1.888.622.3720**. Our website has information on the latest product releases and training, as well as online versions of our *Cornerstone* newsletter and Pearson's product catalog.

Your feedback is welcome. You may email your comments to **curriculum@nccer.org,** send general comments and inquiries to **info@nccer.org**, or fill in the User Update form at the back of this module.

Objectives

When you have completed this module, you will be able to do the following:

1. Describe typical uses and configurations of CCTV systems.
2. Describe the operation of CCTV systems.
3. Identify and describe the components of a CCTV system, including:
   - Cameras
   - Lenses
   - Amplifiers
   - Recorders
   - Switchers/multiplexers
4. Describe how light affects camera selection.
5. Define installation and test requirements for CCTV systems.
6. Select the correct lens for a given CCTV application.
7. Select the correct equipment for a CCTV installation.
8. Troubleshoot a CCTV system.

Performance Tasks

Under the supervision of your instructor, you should be able to do the following:

1. Select the correct lens for a given CCTV application.
2. Select the correct equipment for a CCTV installation.
3. Troubleshoot a CCTV system.

Trade Terms

Aperture
Automatic gain control (AGC)
Automatic light control (ALC)
Backlight compensation (BLC)
Charge-coupled device (CCD)
Chrominance
Depth of field
Digital signal processor (DSP)
Field of view (FOV)
Fixed focal length (FFL)
Focal length (FL)
Footcandle
F-stop
Gain
Genlock
Ground loop
Illuminance

Infrared (IR)
Iris
Lens
Lumen
Luminance
Lux
Motion-JPEG (M-JPEG)
Multiplexer
Pan
Pan-tilt-zoom (PTZ)
Pixel
Quad
Redundant array of inexpensive disks (RAID)
Subcarrier
Tilt
Waveform monitor
Zoom

Industry Recognized Credentials

If you're training through an NCCER-accredited sponsor you may be eligible for credentials from NCCER's Registry. The ID number for this module is 33410-12. Note that this module may have been used in other NCCER curricula and may apply to other level completions. Contact NCCER's Registry at 888.622.3720 or go to nccer.org for more information.

Contents

Topics to be presented in this module include:

Figures and Tables

Figures and Tables (*continued*)

1.0.0 INTRODUCTION

Closed-circuit television (CCTV) is an off-air television system in which video signals are not publicly distributed. In a CCTV system, cameras are connected to video monitors covering a limited area, such as hallways and stairways in an office building, a college campus, or a store. The primary use of CCTV is in surveillance systems. Other uses for CCTV include the following:

- Traffic flow and pedestrian safety
- Intrusion detection
- Hospital and critical-care ward monitoring
- Public safety
- Theft and loss management
- Property management
- Remote monitoring at a factory, construction site, or mine
- Fire prevention and monitoring
- Hazardous-site monitoring
- Bank ATMs
- Machine vision and quality control
- Covert monitoring
- Combinations of the above
- Satellite TV

This module introduces the layout and key components of a CCTV system. Several CCTV configurations are covered to show how CCTV technology is used for small, medium, and large installations. The requirements for installing, testing, and troubleshooting CCTV components and systems are also covered.

2.0.0 CCTV SYSTEM OVERVIEW

This section examines a CCTV system by looking at CCTV setups. Starting with the most basic configuration, it gradually builds to a more complex CCTV system. Each of the elements of the system is explained as it is built.

2.1.0 A Typical CCTV System

A typical CCTV system consists of one or more cameras connected to a video monitor. *Figure 1* shows a single camera connected to a monitor. Both the monitor and video source require power. A camera is usually powered by either a 12VDC or 24VAC power source.

2.2.0 Multiple Cameras with a Switcher

Most buildings require many surveillance cameras. Typically, a switcher is placed between the cameras and the monitor (*Figure 2*). This allows

the operator to select which video source (camera) to observe.

Switchers can be manual or sequencing. A manual switcher requires the operator to select a specific video source by pressing a button on the switch faceplate. A sequencing switcher changes from one video source to another at timed intervals, usually every five to ten seconds. The image switches from one camera to the next, in sequence. If the operator wants to observe a particular video source, a manual override is provided at the switch.

When using a sequencing switcher, there may be a sizable delay before an individual video source is displayed. With a 10-second delay, and four cameras, it is a full 30 seconds before a given source is redisplayed. In the world of theft and loss prevention, that could be long enough for a shoplifter to complete the task of hiding merchandise. Even if the time interval were adjusted to 5 seconds, there would still be a 15-second delay before a given camera was displayed. This may be too long to catch a thief in the act.

2.3.0 Viewing Multiple Cameras with a Splitter

A splitter is a device that allows the transmission of a single signal based on the synchronization of two signals. In the example, the switch is replaced with a splitter. In *Figure 3*, two video signals are genlock synched. Genlock means that the signals are in phase using a master signal to synchronize. The output signal is a composite of the two input signals. Therefore, the output of both cameras is displayed together on a single video or television monitor. Some switches allow the operator to select the width of an image by adjusting controls on the switch. For example, the frame can be set wider for a parking garage and narrower for the entrance to the garage. Some switches also allow for a vertical or horizontal alignment of the two frames (side-by-side versus top-and-bottom alignment).

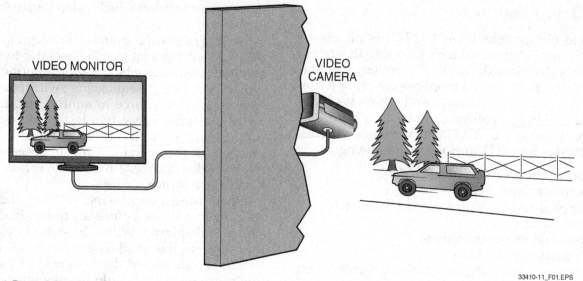

Figure 1 Basic CCTV system.

33410-11_F01.EPS

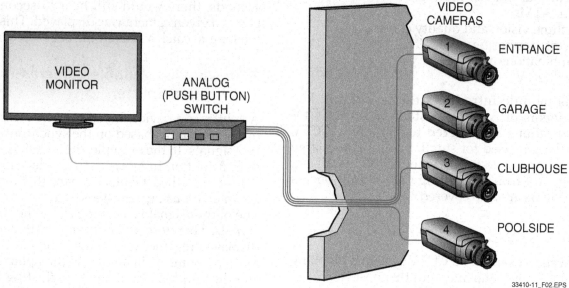

Figure 2 Basic CCTV system with four cameras.

33410-11_F02.EPS

2.4.0 Viewing Multiple Cameras with a Multiplexer

In the example, the splitter can be replaced with a four-way multiplexer (*Figure 4*), sometimes called a quad. The multiplexer combines the output of four cameras into a single composite video signal. This combined signal is displayed on the monitor. The resulting video image contains a four-up display, with each quadrant of the image devoted to a single camera. This permits the operator to view activity at all cameras at one time. A switch on the multiplexer allows the operator to select a specific camera for full screen viewing.

Multiplexers are available in increments of four, such as 4, 8, 12, or 16. The 4-way or 16-way multiplexers are the most common. There are also nine-channel multiplexers appearing in the marketplace.

2.5.0 Using a Video Recorder to Archive Video

Figure 5 adds a video recorder to the system. The operator may need to save the video image for successful prosecution of a shoplifter. An engineer may need a record of an event at a hazardous work site. The video may simply be needed for off-line review by management or other interested parties. Digital video recorders are used on newer systems, although VHS or other cassette recorders exist on some legacy systems. More information about video recorders is presented later in this module.

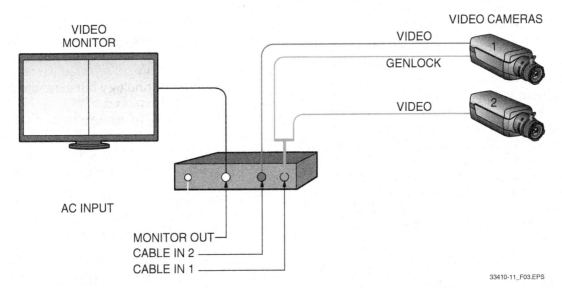

Figure 3 Basic CCTV system with a screen splitter.

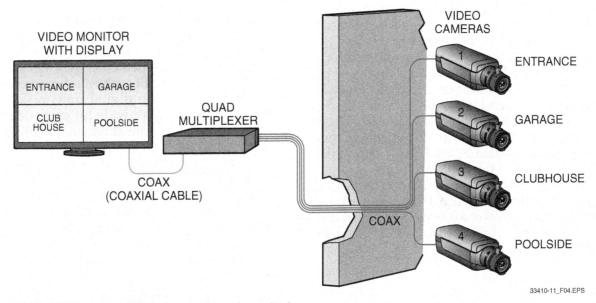

Figure 4 Basic CCTV system with four cameras and a multiplexer.

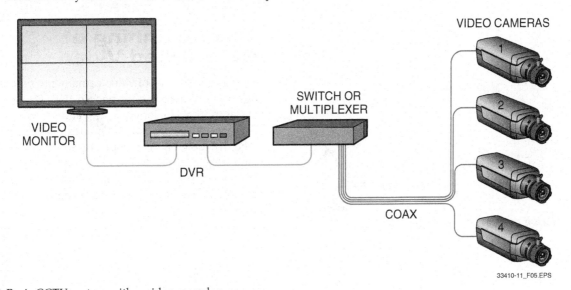

Figure 5 Basic CCTV system with a video recorder.

3.0.0 CCTV TECHNOLOGY

Although analog cameras and playback devices are still found in legacy systems, digital equipment has replaced analog equipment in newer installations, as it has in all types of video work. Most notable is the replacement of tape record/playback devices with digital video recorders (DVRs) and the replacement of cathode ray tube (CRT) monitors with plasma and LCD monitors. Manufacturers have developed equipment, such as video encoders and decoders, that allow older analog systems to be expanded or upgraded without sacrificing the investment in analog cameras and equipment.

3.1.0 Digital vs. Analog

At one time in the past, all television and video technologies were analog. Capture, distribution, and viewing were all accomplished using analog signals. An analog signal is a sine wave measured in two dimensions, amplitude and frequency.

A digital waveform is non-sinusoidal. It is a square wave with discrete states. It is produced by a discontinuous electrical signal, and is binary. If voltage is present, it is interpreted as a one. If voltage is not present, it is interpreted as a zero. These zeros and ones, called bits, are grouped together into sets of eight, called bytes. *Figure 6* shows the differences between the two waveforms.

Analog signals vary in amplitude and frequency. A composite video signal is a 1-volt peak-to-peak waveform that consists of a 0.7-volt peak signal and a 0.3-volt sync pulse. In the video portion, zero represents black and 0.7 represents white.

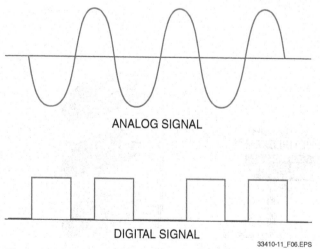

ANALOG SIGNAL

DIGITAL SIGNAL

33410-11_F06.EPS

Figure 6 Analog and digital waveforms.

In a previous section, it was shown that certain factors could distort a sine wave. These factors include line conditions, EMI, and RFI. Solid-state devices, such as charge-coupled devices (CCDs) that use digital technology but generate an analog signal, were also explored.

It is important to understand that modern video electronics uses both digital and analog signaling. When an image is captured from a CCD camera, it is a binary image. It consists of numerical values for each pixel. The camera uses a digital-to-analog converter and transforms these values into an analog signal. This signal is amplified and transmitted to the monitor.

Analog signals are subject to distortion and other effects that result in a loss of information. Each device in a CCTV system can act on the signal; therefore, each successive generation can amplify information loss.

Digital signals, on the other hand, are less sensitive to information loss. However, they do suffer from losses because of errors introduced into the signal and from dropped bits. They are less susceptible to EMI and RFI problems, however. What enhances or improves digital transmission the most is the ability to provide for error detection and correction. The rate at which bits are transmitted, called the bit rate, can also be controlled.

In a traditional CCTV system, cameras are wired to monitors, either directly or indirectly. In complex CCTV systems, the number of cameras is far greater than the number of monitors. The matrix switcher provides a means for selecting a given camera. Once selected, images from the camera can be viewed, recorded, or printed.

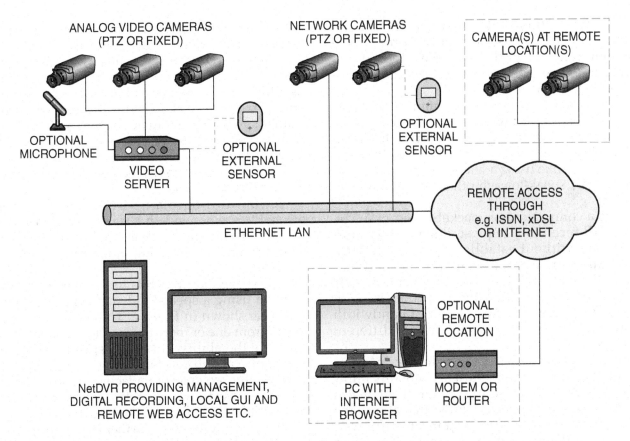

ANALOG VIDEO CAMERAS
(PTZ OR FIXED)

NETWORK CAMERAS
(PTZ OR FIXED)

CAMERA(S) AT REMOTE
LOCATION(S)

OPTIONAL
MICROPHONE

VIDEO
SERVER

OPTIONAL
EXTERNAL
SENSOR

OPTIONAL
EXTERNAL
SENSOR

REMOTE ACCESS
THROUGH
e.g. ISDN, xDSL
OR INTERNET

ETHERNET LAN

OPTIONAL
REMOTE
LOCATION

NetDVR PROVIDING MANAGEMENT,
DIGITAL RECORDING, LOCAL GUI AND
REMOTE WEB ACCESS ETC.

PC WITH
INTERNET
BROWSER

MODEM OR
ROUTER

SAMPLE CONFIGURATION ILLUSTRATING VIDEO AND AUDIO RECORDING WITH REMOTE ACCESS

33410-11_F07.EPS

Figure 7 Overview of Internet-based surveillance.

In a digital CCTV system, the situation is similar, but the camera is only one of many devices connected to the network. There are also printers, modems, routers, and other devices. *Figure 7* provides an overview of an Internet-based surveillance system. Modern digital networks allow a device to be selected using its Internet protocol (IP) address. This four-part address identifies a particular device on a specific network segment. Addresses are assigned in several ways. One method is called a static IP address. It is a permanent assignment, and does not change. Another way to get an IP address is through dynamic allocation. In this case, the IP address changes periodically. Cameras are usually provided with static IP addresses.

In network terminology, the camera serves up the broadcast. The device that requests a view of a camera is a client. With the coffee cam, a special program had to be written for both ends of the process. This required a server that took the camera output and sent it across the network, and a client that received it. This is referred to as client-server.

The following is a summary of the benefits derived from the introduction of digital technology to the CCTV world:

- A digital system can be configured to record for long periods. An analog system is limited by the length of a videotape.
- Picture quality is better with a digital system.
- A DVR can record all cameras at once.
- Digitally stored data can be searched by date and time.
- Authorized personnel can have remote access to video.
- The quality of digitally stored video remains the same when it is copied, unlike videotape, which loses quality in copying.
- It is possible to watch recorded video and record live video at the same time with the same camera.

3.2.0 Review of Internet Protocols

There are two types of client-server relationships when working with the Internet. They both relate to how data is transmitted between servers and clients. The Internet uses one of two protocols to communicate, TCP or IP. Collectively they are referred to as transmission control protocol/Internet protocol (TCP/IP).

TCP is a single, two-way communication protocol. It acts as a direct wire, or pipe. Once communication between the client and server has been established, there is a direct connection between the two. Just like using a telephone, the client and server only speak to each other over a TCP connection.

IP is a connectionless, packet-based protocol. When IP receives a request, it starts transmitting data without establishing a connection. The data is broken down into packets and sent out with an address targeting the client. Packets do not necessarily follow the same transmission path. Depending on the amount of bandwidth available, they may be routed across different networks.

IP packets have a header, which is an information field, attached to them. This contains information such as the IP address of the client. It also contains sequencing information that allows the packets to be re-assembled in a meaningful way. Otherwise, the data would be scrambled together and more than likely, meaningless. This sequencing information is used to put all the packets into the correct order to create a single, coherent message.

TCP/IP forms the underlying communication protocol for the Internet. There are additional, specialized protocols built on top of them, allowing different types of communication to take place. For example, File transfer protocol (FTP) uses TCP to send a file from a disk. Hypertext transfer protocol (http) uses IP to distribute web pages. The web page is assembled at the client from individual packets.

The Internet is a collection of loosely interconnected networks. TCP/IP is the fundamental underlying communication protocol and runs on each of the smaller, local networks. Each of these smaller networks is referred to as a LAN, or local area network. LANs are connected to each other forming a WAN, or wide area network. The Internet represents the sum total of all LANs and WANs connected together. They all have one thing in common: the use of TCP/IP.

3.3.0 Client-Server CCTV for the Internet Age

The key to using digital technology for CCTV involves the video server and the video client, usually a web browser. There are two approaches to implementing video servers and their clients.

The first approach is to connect a network camera to the local network. Keep in mind that there is a difference between a standard video camera and a true network camera. A standard video camera generates an analog signal and connects via coaxial cable to a CCTV system. A network camera generates digital images and uses TCP/IP to transfer them. It connects directly to a LAN using an Ethernet cable.

Technically, once a network camera is connected to the network, it becomes a video server. It only broadcasts its images when it receives a client request. An analog video camera always transmits its images.

The second approach is to connect an analog camera using a special video attachment device like that shown in *Figure 8*. These devices accept input from one or more analog video cameras and provide the ability to select a specific video input. This capitalizes on existing analog cameras, without the expense of replacing them. All of the necessary electronics for network communication are built into the device. Some models can act as a switcher or multiplexer, further eliminating costs.

Network video clients can be supported either by a standard web browser (*Figure 9*) or by custom software applications. Some web browsers may require a special add-in program, called a driver or plug-in. Custom video browsers may be created and sold by the manufacturer of the camera.

3.4.0 Recording and Retrieving Network Video

One of the key differences between traditional and digital CCTV systems is the recording and

33410-11_F08.EPS

Figure 8 Network attachment device for analog cameras.

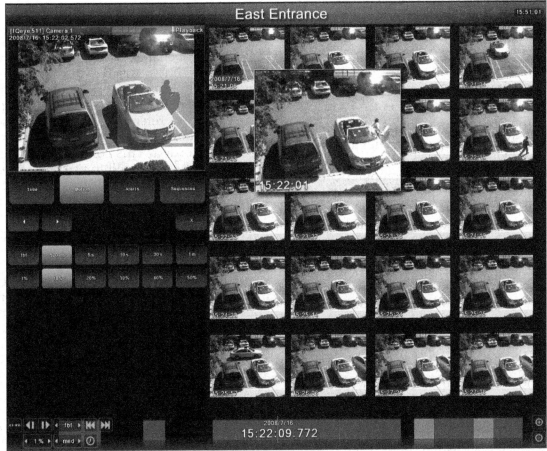

Figure 9 A web browser configured for network camera access.

playback of video. In traditional CCTV systems, video is directed to a VCR. For larger installations, it is common to find banks of VCRs managed by a VCR controller.

With network video, it is common to use a special DVR called a network DVR, as shown in *Figure 10*. A network DVR accepts a digital video stream and writes it directly to a computer hard-disk drive. There is no need to convert it first. It may be compressed using MPEG or motion-JPEG (M-JPEG) technology. It is stored in time-lapse or real-time modes, and can jump directly to the output of a single camera. This is accomplished using time or date information stored with the video.

When capturing video using a network-based CCTV system, video is already in a digital form. There is no need to convert it. It can be written directly to a file or database on a computer hard drive. This is usually done by a specialized computer with multiple hard drives and multiple fast processors. Its purpose is to provide compression and database support for the storage and retrieval of digital video.

The network DVR can be placed anywhere on the network, and may even be distributed. It is a

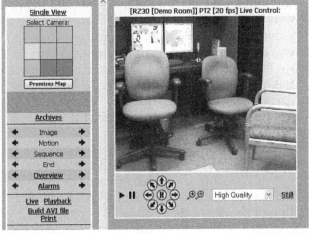

Figure 10 Web browser configured for PTZ control and monitoring of a network camera.

common practice to store the video remotely on a redundant array of inexpensive disks (RAID) device. RAID is a digital storage technology that writes data across several drives. This ensures data integrity in the case of a disk failure.

3.5.0 Factors that Affect Digital CCTV

There are three factors to consider for total digital CCTV solutions:

- Bandwidth
- Video throughput
- Video compression

Bandwidth and video throughput are related. Bandwidth is the total transmission capability of the network. Network video solutions use the same network infrastructure as computers, routers, and printers. Therefore, network bandwidth is shared among all users and devices. At different times of the day, the amount of data being transmitted across a network varies. From 8:00 AM to 9:00 AM, there is a tremendous pull on many corporate networks, as users log in and retrieve email. Many effects like this can reduce the amount of bandwidth available for video transmission. When you factor in the bandwidth requirements for video, particularly with multiple cameras, there can be a tremendous additional load applied to network resources.

Digital video throughput is determined by resolution, frame rate, and color data. Digital video resolution directly relates to the size and color of the images. Assuming a monochrome image and an 800 × 600 pixel video frame, each frame contains 480,000 bits, one bit for each pixel. This translates roughly into 60,000 bytes, or approximately 56K of data. Multiplying this by the frame rate of 30 fps yields 1.6 megabytes of data every second.

When color is added, the quantity of data increases tremendously. For minimum color, the increase is a factor of eight. For maximum color, it increases by a factor of 24. There are 8 bits of data per pixel for a 256-color image. For true color, there are 24 bits of data for each pixel (1.6 million colors).

One way to compensate for these increases is to reduce the frame rate. Rather than sending 30 frames per second, the video stream sends 15 or 20 frames per second. The overall result is a reduction of the data rate by 35 to 50 percent.

Video compression, on the other hand, serves to minimize the network costs of size, color, and frame rate. There are several techniques for compressing video. One technique, M-JPEG, compresses each video frame into a single JPEG image. The resulting image data is reduced by 60 percent to 70 percent. Therefore, the total video stream is reduced from 1.6 megabytes per second to about half a megabyte. JPEG compression has inherent problems with image loss. The resulting

video may not be useful for identifying details in the scene.

Another compression technique, MPEG2, uses the concept of change over time to compress images. The assumption is that most content in any scene does not change, or changes slowly over time. Using MPEG2, an initial frame is captured, and serves as a reference frame. As the scene is monitored, only those elements that change are actually recorded. This continues until enough of the scene has changed to warrant capturing a new reference frame. This significantly reduces the data transmission requirements.

3.6.0 Monitoring Video in a Digital CCTV System

As mentioned earlier, analog systems use video monitors to provide surveillance data to the operator. Monitors have to be near the control devices for switching, multiplexing, and recording. For large CCTV installations requiring multiple monitoring stations, expensive matrix switches and consoles are used. In the world of a digital CCTV installation, monitoring requirements are much less restrictive.

All elements of a digital system are distributed. That is, devices such as recorders, video printers, and monitoring stations can be located anywhere. There is no need for them to be connected in any specific sequence. Cabling and connection requirements are much less restrictive than in an analog system.

In an analog system, the number of monitors that can view a video stream at any one time is limited. With a digital system, any attached computer can access the video stream. Because a standard web browser is all that is typically needed to view the video, it is available to all security personnel and management. It can even be made available to outside users, such as fire personnel and police, through an Internet link.

On Site

Selecting the Correct Frame Rate

It is best to choose the lowest frame rate that provides the required video-quality level. The reason is simple: the higher the frame rate, the higher the bit rate, and the more bandwidth required. A frame rate of 3 fps is okay for general surveillance, and 7 fps picks up fine movement. A frame rate of 15 fps is equivalent to full motion.

Though the web browser provides the standard video client, factors, such as workstation CPU speed, display resolution, and network traffic, can still affect video quality. Some vendors configure high-end workstations to support digital CCTV applications. These workstations are typically desktop computers with expanded features. They support multiple high-resolution LCD displays and use fast graphics processors that have a lot of video memory (VRAM). They may also use multiple microprocessors to improve system performance. They provide for local recording and retrieval via high-speed disk drives. High-resolution video printing can be supported locally or remotely.

High-end workstations can reduce network load significantly. They can provide sophisticated software-supported post-processing of video and graphics images.

3.7.0 Network and User Authentication

For obvious reasons, not just anyone should be permitted to view surveillance video. In a small office or installation, the job of monitoring may be shared among several individuals. It may not be limited to a security guard. For example, a receptionist may monitor doors and hallways. During breaks and mealtime, other personnel may manage the observation tasks from their own workstation. In a larger organization or installation, monitoring tasks may be assigned strictly to security personnel.

As mentioned earlier, there may be supervisory and other personnel who need to respond to specific alarm situations. With digital video, this is much easier to do. There is no need to run special cable to these remote viewing locations. However, there is a need for controls to prevent unauthorized personnel from using CCTV functions.

The most common means is to use a hierarchical authorization scheme similar to network administration activities. In the data network, users are assigned privilege levels that determine their ability to access data or use devices such as printers. It has already been seen that with some console systems, passwords can be assigned to video operators. The same is true in the network CCTV environment.

In extreme situations, authentication relies on more than just password protection. Biometric inputs, such as thumbprints, voiceprints, and even retinal patterns, can be used to authorize users.

3.8.0 Encryption and Decryption

Administrative controls such as authentication provide a reasonable degree of system security. However, passwords are not always a reliable means for security. Users are historically lax in managing their private passwords. They often share them with others, write them on sticky notes, or base them on easy to guess personal data, such as a birthday or favorite color. Encryption adds an additional level of security to any system.

Encryption is the process of coding information so that only a person who has the encryption key can view (decrypt) it. Encryption has been used for centuries by soldiers, diplomats, and spies. In the CCTV world, the introduction of networked systems and wireless cameras has made it essential to be able to encrypt video in order to protect it from unauthorized access. There are both hardware and software methods available for video encryption. The selection of an encryption solution depends to a great extent on the level of security needed.

When it is necessary to use encryption in a CCTV system, someone who has expertise in the subject should be involved. There are quite a few encryption systems available, and each has its advantages and drawbacks. For example, some encryption algorithms are fast, but offer low security. Others offer high security, but may be too slow for real-time video streaming.

On Site

The Value of Encryption

In 2008, the US Army realized that Iraqi insurgents were capturing unencrypted live video feeds from Raven drones that were being used to locate and target the insurgents. The insurgents were able to hack into the video feeds using a $26 Russian software program called Skygrabber. This allowed them to thwart the Army's efforts at location. Soon after this discovery, the Army initiated a crash program to encrypt the video feeds and provide troops in the field with decryption capability.

Source: *The Army Times*, December 2009

4.0.0 CCTV SYSTEM COMPONENTS

There are a number of components attached to a CCTV system. Several have been identified already, such as a switch, a splitter, a multiplexer, and a video recorder. The nature and use of individual components vary as the applications for CCTV change, and as the size of the installation increases. Typically, larger systems are more elaborate because they monitor a greater range of activities. CCTV applications like critical-care ward monitoring can be extremely complex, even for small hospital ICUs. In examining a large facility CCTV layout, it is necessary to know about and understand several other components.

The module addresses the following components and how they improve, support, and interact in a CCTV installation:

- Cameras
- Camera lenses
- Camera mounts and housings
- Date and time generators
- Controllers
- Alarm interface units
- Motion detectors
- Signal and distribution amplifiers
- CCTV keyboards
- Recorder controllers

Each of these devices plays a critical role in the overall success of a CCTV system. Each may have unique installation requirements and prerequisites.

4.1.0 Cameras

The camera is the eyes of the system. The camera uses CCD technology to capture moving images. Selection of a particular camera is determined by the intended purpose of the system and its placement in the environment. Different types of cameras are available for different uses. The wireless camera (*Figure 11*) became popular because of the introduction of networked CCTV systems.

The CCD camera uses solid-state electronics to create a camera on a chip (*Figure 12*). The CCD chip is a rectangular array of imaging elements called pixels, or photosites. A photosite is often called a detector. The CCD detector converts incoming photons of light into electrons through a process known as the photoelectric effect. These freed electrons are captured and stored in the pixels until the chip is read out. The more photons that strike a pixel, the higher the numerical value recorded for that pixel. The light gathering area of a typical CCD is only a few millimeters on each side, but contains 400,000 or more light-sensitive photosites.

33410-11_F11.EPS

Figure 11 Wireless CCTV camera.

33410-11_F12.EPS

Figure 12 CCD camera on a chip.

A photonic input is converted to an electronic output by the CCD. Electrons are then transferred in a bucket brigade fashion to an output amplifier. The output amplifier converts the charge to a voltage output signal. An analog processing chain further amplifies this signal.

When working with CCD cameras, there are both shutter and shutterless models. A shuttered CCD camera allows the input aperture to close during the reading (transfer) of pixels. This, in essence, freezes the image so it is not affected by incoming photons during the reading process. Shutterless CCD cameras use a frame-transfer approach. The pixel values are first transferred to a frame buffer and then read out.

CCD cameras are available in either a monochrome or color format. Although monochrome cameras historically have had the finer resolution, the resolution of color CCDs today is comparable to that of a monochrome CCD.

While CCD cameras are the state of the art for all video applications, it is important to understand that there are two technologies involved. Many CCD cameras generate an analog signal. Though the CCD imaging array is digital, the output is converted to a standard National Television System Committee (NTSC) video stream. This is in contrast to what is known as an IP (Internet protocol) camera, where the digital output is transmitted across a standard computer network. The most common sizes of CCD chips are ¼", ½", and 1". The larger the chip, the more light it lets in. Thus, the size of the chip relates directly to the field of view.

When comparing monochrome and color cameras based on resolution, there are a few interesting facts. Monochrome images are not black and white. They are actually a single color and the absence of that color. Color cameras on the other hand capture the variances between hues of different colors of light. They use filters, lenses, or in some cases multiple video tubes or CCD chips. If a color image is displayed on a monochrome monitor, it is displayed in black and white. Black and white images contain more information than monochrome images. In black and white images, color hue and saturation differences are translated into shades of gray.

4.1.1 Network Cameras

A network, or IP, camera has an on-board video processor and web server, which allows the user to view video from any location that has Internet access. Network cameras are solid-state CCD devices. Just like their analog counterparts, they can be equipped with PTZ mechanisms, enclosed in special housings, and mounted indoors or out. The operator has the ability to change focal length (FL) and alter the field of view (FOV). Auto iris and motor-driven iris support are available. To control the various elements of the camera, the manufacturer or integrator provides plug-in or custom software.

4.1.2 Low-Light, Day-Night and Infrared Cameras

Some CCTV cameras need to be able to perform surveillance in low-light conditions up to and including full darkness. The infrared (IR) or thermal imaging camera, also called a night-vision camera, is designed to operate in full darkness. It is able to capture images in the IR spectrum, which is invisible to the eye. Therefore, it does not require any ambient light in order to do its job. It weakness is that, while it is able to detect an image in total darkness, it is not useful for recognition and identification because it simply produces a ghost-like image.

Other cameras are specifically designed to operate in conditions where there is very limited ambient light. There are a number of technologies used to accomplish this, and they vary from one manufacturer to another. Some of these cameras use IR filters that are automatically switched on and off depending on light conditions. Some use light amplification techniques, while others in-

On Site

Early CCTV Cameras – The Vacuum Tube

Vacuum tube cameras were historically used in CCTV installations and other video applications. Though these cameras have generally been replaced by smaller solid-state devices, they can still be found in surveillance applications. They are sometimes called a video tube or vidicon tube because of the materials used to make them. A vidicon is a vacuum tube containing a photosensitive area or target. It also has an electron gun that scans the signal from the target area.

The vacuum tube is surrounded by coils to create a magnetic field. The tube contains an image receptor that is made of a photoconductive material. The light from the image is focused through a glass plate onto the image receptor or anode. Electrons are generated relative to the intensity of light and are attracted to the anode. After striking the anode surface, the electrons are removed.

After exposure and removal of the photoelectrons, a residual image is stored in the photoconductor. This image is actually a distribution of positive charges. The stored image is then read out by scanning with an electron beam. The electron beam is emitted from the cathode located at the opposite end of the camera tube. The video signal is then amplified and transmitted across a semiconductor to a monitor.

crease the light sensitivity of the camera by using extremely sensitive CCD chips. Still others simply convert from color to black and white in low-light conditions. Camera selection depends on the light conditions in which the camera has to operate, as well as the required image resolution. It is often a tradeoff between the cost of providing illumination during all hours and the cost of cameras able to see in low-light or no-light conditions.

4.2.0 The Camera Lens

The camera lens must be considered for the selection and installation of any camera. The camera position, the observing location, and the type of monitoring determine which lens should be used. This section identifies the key attributes of lenses in general, as well as the features used to select the best lens for a given application.

> **NOTE**
> When the camera and lens are bought as a unit, far fewer field adjustments are required.

A lens is an optical device that collects reflected light from a scene. It focuses the light on the camera image sensor to construct an image of the scene. This image is then scanned by the camera's image sensor. The lens determines the actual area viewed by the camera.

Lenses are more than just a piece of glass. They have several other attributes that are critical to the effective operation of an imaging device. These are as follows:

- Size
- Mount
- Focal length (FL)
- Field of view (FOV)
- Iris
- F-stop
- Depth of field

4.2.1 Lens Size

The size of a lens is partially determined by the type of camera. A CCD camera used in TV broadcasting has a different size lens than a camera used for surveillance. Physically, these lenses are different sizes and have different mounting requirements. They operate the same, however. Likewise, a lens for a small CCD camera is smaller than one used with a full-sized CCD camera. The lens size simply refers to the physical dimensions of the lens housing and mount, which is sometimes called the form-factor. When considering lens size, the focal length (FL) and the field of view are also important.

4.2.2 The Lens Mount

The lens mount connects a lens to a camera. There are several standard mounts. They are not mechanically or optically interchangeable for the most part.

The most common mount is the C-mount. It is a 1"-diameter hole with 32 threads per inch. The lens has a corresponding thread that is screwed into the mount. The rear-mounting surface of the lens and the camera image sensor are separated by 0.69" (17.526mm). As a standard, the C-mount accommodates all ¼", ⅔", and 1" cameras.

The CS-mount was devised for CCD cameras and other smaller formats. The CS-mount accommodates ¼", ⅓", and ½" cameras. The CS-mount matches the C-mount in diameter and thread requirements. However, it changes the distance between the image sensor and the lens rear-mounting surface to 0.492" (12.5mm).

There is some degree of interoperability between C- and CS-mount lenses. A C-mount lens can be attached to a CS-mount housing, provided a 5mm spacer is used. The spacer allows the correct positioning of the lens at the proper distance from the image sensor. The opposite is not true, however. A CS-mount lens cannot be mounted on a C-mount camera. There is no way to compensate for the 5mm difference.

The two fairly standard lens mounts now in use are a 10mm or a 12mm diameter mount, both with a 0.5mm pitch. Other lens mounts are typically non-standard. The growth of CCD cameras has resulted in metric threads, different diameters, and different pitches. Pitch is the distance between the lens rear mount and the image sensor. Typically, the camera manufacturer provides the lenses for these specialty mounts.

4.2.3 The Focal Length of a Lens

The focal length (FL) of a lens is determined by the focal center of the lens and the principal convergent focal point. The shorter the focal length, the wider the field of view is going to be. This means that the image as focused from the center of the lens is projected onto the image sensor. The focal length is then the distance required for the projection to be an accurate reproduction of the image (*Figure 13*). FL is measured in millimeters (mm) and can be either fixed or variable. Some cameras have a fixed focal length (FFL), while others can be varied. A lens with a variable focal length, called a varifocal lens, allows for manually zooming in and out. Variable focal length lenses that can be adjusted electrically are called zoom lenses. The focal length of a lens is stated in

millimeters (mm). FFL lenses range from 3.6 to 16 mm, while zoom lenses have a focal length of up to 70 mm.

An FFL provides for a given level of magnification and therefore a fixed field of view. Manufacturers and resellers publish charts used to select the most appropriate focal length for a given application. It is important to match the focal length based on the form factor for the camera (¼", ⅓", ½", or 1"). As the focal length increases, the field of view becomes narrower and provides better definition for distant objects.

4.2.4 The Field of View Provided by a Lens

The field of view (FOV) is best thought of as the maximum angle of view. When thinking about FOV, it is important to keep magnification (M) in mind. If an image has the same magnification level as perceived by the human eye, it is referred to as M=1. The camera-imaging sensor and the lens create an image equal to that of the human eye. As the magnification increases, the height, width, and diameter of the image increase. In other words, as magnifica-

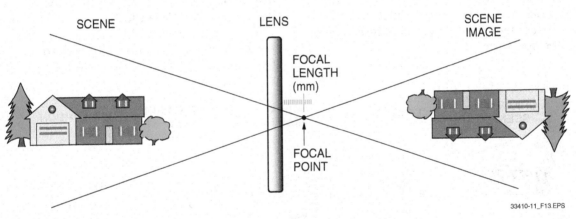

Figure 13 Lens focal length.

Camera Resolution Facts

Video resolution is a measure of the ability of a video camera to reproduce fine detail. Resolution is measured along both the horizontal and vertical planes. The National Television System Committee (NTSC) standard specifies that a full frame of video is composed of 525 horizontal lines. The vertical plane, however, depends on the receiving electronics of the monitor. It is typically limited to about 70 percent of the horizontal, or for NTSC, approximately 340 vertical lines.

The higher the camera resolution, the sharper the image produced is going to be. Standard broadcast TV systems can produce a picture resolution equal to about 300 lines of horizontal resolution. Cable TV, DVD, and satellite TV, as viewed on a home receiver, can reach 400 lines of horizontal resolution.

When considering resolution, it is important to know two things. First, it is measured at the monitor. Second, horizontal resolution actually measures the number of vertical lines. These measurements are made from a distance equal to the diagonal width of the screen.

tion increases, the FOV decreases. As the FOV increases, magnification decreases and exposes more of the image (*Figure 14*).

The FOV and the focal length are important attributes of a lens and camera pair. Close attention should be paid to both FL and FOV when selecting a lens.

4.2.5 *The Iris of a Camera Lens*

The iris adjusts the amount of incoming light that passes through a lens. It is used to adjust the aperture of the lens to allow more or less light to reach the imaging element. The iris is an integral part of a lens assembly and may be called an iris diaphragm. An iris may be manual or automatic, depending on the application and the specific camera-lens pair selected.

An automatic iris opens or closes itself when the amount of light moves beyond a certain threshold. The light can increase or decrease. The threshold is often determined by the camera, but newer lenses, called drive lenses, include the electronics directly within the lens housing. A manual iris is controlled by an operator. Manual control may not result in the best picture contrast. An auto iris on the other hand, provides reasonably good contrast. An auto iris uses the camera signal level to determine the width of the aperture.

Solid-state cameras that use electronic shuttering do not require an iris adjustment. For cameras with automatic gain control (AGC), a fixed-iris lens can be used. AGC is a camera feature that adjusts the incoming light much like an automatic iris. Cameras with AGC simply increase the video signal under low light conditions. This is in contrast to automatic light control (ALC), a form of backlight control. ALC makes an image brighter to help discern detail in darker portions of the image.

4.2.6 *F-Stop*

The F-stop, or F-number, defines the light-gathering capability of the lens. Light-gathering capability is directly related to the quality of an image. The lower the F-number, the more light it passes. An F-1.4 lens passes more light than an F-4 lens, for example. F-stops range from F-1.4 to F-22.

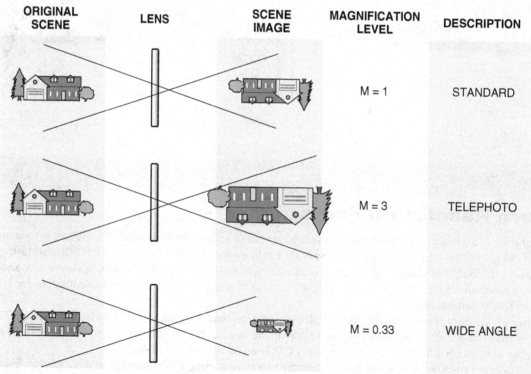

ORIGINAL SCENE	LENS	SCENE IMAGE	MAGNIFICATION LEVEL	DESCRIPTION
			M = 1	STANDARD
			M = 3	TELEPHOTO
			M = 0.33	WIDE ANGLE

33410-11_F14.EPS

Figure 14 Lens field of view and magnification versus wide angle.

4.2.7 Depth of Field

Depth of field is the distance in front of and beyond the object of interest that remains in focus. Depth of field is inversely proportional to the focal length and the F-stop. A wide-angle (short) lens and narrow aperture would yield a long depth of field.

4.3.0 Camera Mounts and Housings

Cameras are designed to operate within an established range of temperatures, humidity levels, and even light levels. For cameras to work and serve their purpose, it is important to provide protection from the elements. There are many types of camera mounts and housings. These provide varying degrees of protection from exposure to sunlight, rain, dust, and even vandals.

Camera mounts can be light-duty mounts intended for wall, ceiling, or T-rail installations. Other mounts are heavy duty and are designed to support housings.

Some housings include camera mounts within the housing. They may also include power supplies. They offer features such as thermostats, blowers, heaters, and defrosters. There are even models with window wipers similar to those found on car windshields. Some housings require an external camera mount, which must be acquired and installed before the housing. There are five primary types of housings:

- Indoor
- Outdoor
- High-security
- Specialty
- Dome

> NOTE
>
> Some cameras come equipped with housings.

On Site

High-Speed Shutters

High-speed shutters are used in the analysis of sports activities. They are also used in machine vision applications on the manufacturing assembly line. When viewing products that are moving along a conveyor belt, a high-speed shutter is invaluable for identifying flaws and defects. A high-speed shutter requires more light than lower speed shutters.

4.3.1 Indoor Housings

Indoor cameras can be mounted in a housing or not. It depends on the facility, the environment, and the goals of the system. Typically, though, many indoor cameras are secured in some kind of housing. Some types of indoor housings include the following:

- Fixed housings that are mounted to ceilings and walls in such a way as to be semi-permanent. They do not provide for camera mobility.
- Corner housings that are mounted high in a corner. When equipped with a camera and a wide-angle lens, they allow for good coverage.
- Wall or pedestal-mounted housings that are attached to a wall using a bracket. It is possible to install electrical-mechanical pan-tilt-zoom (PTZ) devices between the housing and the mount. These devices allow the position and angle of the camera to be changed remotely.
- Ceiling housings are mounted flush with the ceiling, partially contained within the ceiling, or below the ceiling. They can be of a wedge, flat, or flush mount design.

Figure 15 shows two indoor camera pedestal housings.

4.3.2 Outdoor Housings

Cameras mounted outdoors are subject to extreme environmental effects, including rain, snow, and extremes of heat or cold. Outdoor housings are typically of heavy-duty construction as compared to indoor housings. They may include additional features, such as heaters, blowers, or defrosters. The voltages must match for all components. Like indoor housings, there are several shapes and mounting location types. *Figure 16* shows examples of outdoor housings. At the top is a standard

33410-11_F15.EPS

Figure 15 Indoor camera housings.

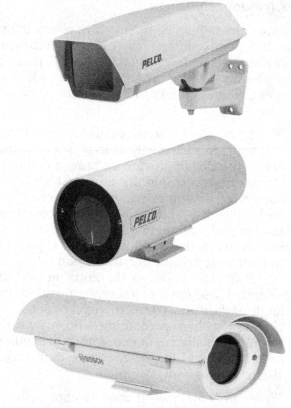

Figure 16 Examples of outdoor camera housings.

Figure 17 An outdoor camera with a window wiper.

outdoor housing, which has a sun hood, that is good for use in a garage, entry way, or other application. The camera in the middle is designed to be rainproof and dustproof. The bottom housing has a hinged side opening for mounting and servicing the camera within. *Figure 17* shows the front of an outdoor housing equipped with a window wiper.

4.3.3 High-Security Housings

Cameras mounted in high-security areas, such as prisons, bank vaults, elevators, or garages, are often armored. Such housings can be bulletproof, and most have tamper switches. Some even have speaker grills for two-way communication between the camera site and the operator. Like indoor housings, they can come in different shapes and mounting location types. A corner-mount housing is shown in *Figure 18*.

4.3.4 Specialty Housings

There are a number of camera housings designed for special circumstances, as shown in *Figure 19*. They include dustproof housings, explosion-proof housings (top), and pressurized housings (bottom). A pressurized housing is sealed and then

Figure 18 High-security camera housing.

pumped full of nitrogen, an inert gas. Pressurized housings can be bought with the camera already mounted within them and pressurized at the factory. Otherwise, they need to be pressurized in the field.

4.3.5 Dome Housings

A dome housing is a type of specialty housing consisting of a transparent glass or plastic bowl (*Figure 20*). It is typically mounted either partially embedded in the ceiling, or on a pedestal. The glass is usually dark or reverse mirrored to prevent the observant from seeing the camera inside. Domes can be mounted indoors or outdoors. Dome housings are available with many of the special features previously discussed.

Figure 19 Examples of specialty camera housings.

33410-11_F19.EPS

Figure 20 Example of dome housing.

33410-11_F20.EPS

Dome mounts are popular for surveillance applications for two reasons. First, they often support multiple cameras. Second, they may contain either fast turntables for scanning, or PTZ platforms. Dome housings are used in department stores, casinos, and public terminals. Dome housings are often acquired as a complete solution with one or more cameras, a PTZ or turntable platform, and the housing. Domes with darkened glass affect the F-stop of the camera. Clear domes should be used for low-light surveillance applications.

4.4.0 Date and Time Generators

A date and time generator superimposes date and/or time characters onto a video signal. The signal is then passed on to be displayed on a television or video monitor. Date and time generators can be mounted in racks, or can stand alone. They

are video through-looped, which means they do not alter the signal except to overlay date and time data.

Figure 21 shows a typical configuration of a date and time generator in a simple CCTV system. In the real world, the output would probably be routed through a switcher, a splitter, or more likely a multiplexer, before being displayed on a monitor. The date-time generator is less likely to be used in newer installations, as date and time functions are usually provided by DVRs.

4.5.0 Controllers

A controller allows the operator to manipulate remote cameras in a CCTV system. Operations provided by a CCTV video controller include camera pan, tilt, and zoom. Cameras must be placed on a special motor or servo-driven mount to support pan and tilt operations (see *Figure 22*). Using a special video keyboard, the operator can select the desired camera and send control signals to it

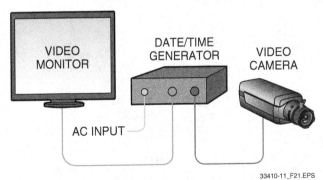

33410-11_F21.EPS

Figure 21 Date and time generator in a basic CCTV system.

(*Figure 23*). Actions the operator can perform include the following:

- Move the camera to focus on a new location
- Adjust for changes in light levels
- Adjust the viewing angle
- Adjust the FOV
- Swing the camera across an arc

Individual controllers often support hundreds of cameras. Multiple controllers can be connected in such a way as to support hundreds, if not thousands, of cameras. When working with multiple controllers, one is designated as the master controller and the others are daisy-chained to it. Each individual controller is addressable by the operator, and every camera attached to a given controller is accessible.

4.5.1 Elements of Control

A camera controller is actually two components, a transmitter and a receiver, as shown in *Figure 24*. The transmitter is connected to a special keyboard, and the receiver connects to the camera.

33410-11_F23.EPS

Figure 23 A camera PTZ control unit and keyboards.

Camera controllers are most often serial communication devices. They use RS232 and other serial protocols to transmit control messages and codes to camera receivers. Some receivers can be connected to several cameras. The receiver can be built directly into an assembly that includes the camera and the pan/tilt/zoom platform. They are all enclosed in a single housing, such as the dome-type PTZ camera shown in *Figure 25*.

Panning and tilting operate in both the vertical and horizontal planes. Once a camera has been repositioned, it may need to be returned to its original position. Many PTZ mechanisms and controllers allow prearranged parameters, called presets, to be entered into them. The device can be programmed to return to a specific state or to conduct a tour. Parameters for field of view, magnification, and home location can also be preset.

There are camera features other than pan, tilt, and zoom that can be controlled in this fashion. An example is the lens iris. Remember that the iris serves to adjust the amount of light entering the lens. An iris may need adjusting as light levels change over the course of a day. Iris and other lens actions, such as zooming, are usually controlled with a small motor. When the iris needs to be adjusted, a signal triggers the motor to adjust the lens. Many lenses and controllers support preset iris and zoom parameters.

4.5.2 CCTV Communication Networks

A communication link is needed to connect the controller and the camera. This connection allows the controller to send a command signal to the transmitter. The transmitter in turn communicates with the receiver, which provides the command interface to the electronics at the PTZ mechanism, the lens, or the camera. There are several standard means for connecting equipment to support this communication process:

33410-11_F22.EPS

Figure 22 A pan-tilt platform and PTZ camera.

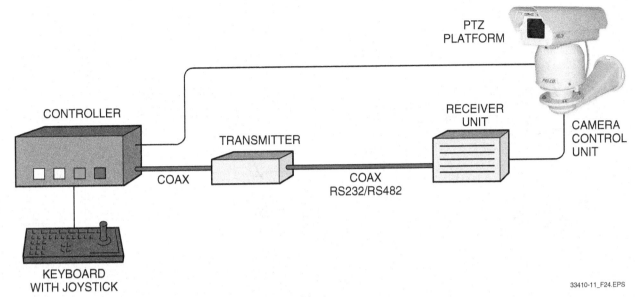

Figure 24 Transmitter and receiver portion of a PTZ control loop.

Figure 25 Dome-type PTZ camera assembly.

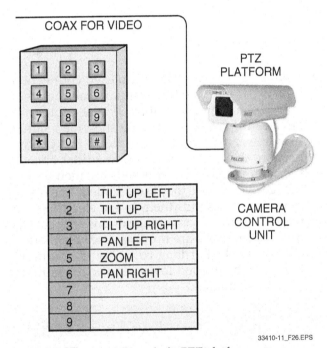

1	TILT UP LEFT
2	TILT UP
3	TILT UP RIGHT
4	PAN LEFT
5	ZOOM
6	PAN RIGHT
7	
8	
9	

Figure 26 Touchpad control of a PTZ platform.

- *Telephone interface* – The telephone interface uses existing touch-tone protocols. An operator uses a keypad similar that on a telephone. The transmitter generates a tone for each key pressed, which is then sent to the receiver. The receiver converts the tone into a command signal to operate the features of the camera (*Figure 26*). The advantage of this approach is that existing unshielded twisted pair cabling (telephone wires) can be used, thereby minimizing the cost of cabling. The downside is that the telephone touchpad interface provides only a minimum of control functions.
- *Direct-wired connection* – To communicate with the remote camera, a special multi-wire cable is

used to connect the controller to the transmitter, the transmitter to the receiver, and the receiver to the camera control mechanism. Each wire carries signals for a specific camera operation, as shown in the example in *Figure 27*. This cabling approach is often proprietary and may require controllers and PTZ units from the same manufacturer. For installations requiring long cable runs, this approach can get quite expensive. It is still necessary to connect the camera to monitoring equipment with a separate coaxial cable.

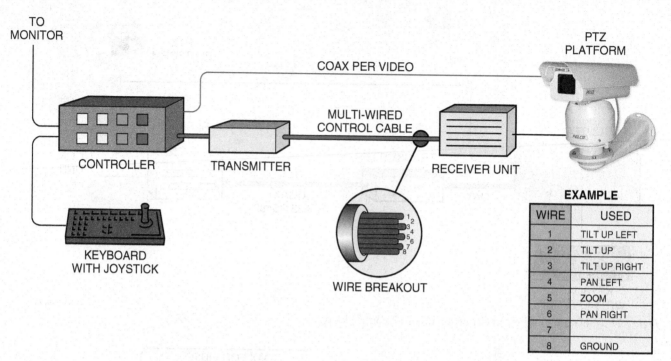

Figure 27 Controller cabling with wire breakout.

WIRE	USED
1	TILT UP LEFT
2	TILT UP
3	TILT UP RIGHT
4	PAN LEFT
5	ZOOM
6	PAN RIGHT
7	
8	GROUND

EXAMPLE

33410-11_F27.EPS

• *Multiplexed system* – A multiplexed connection uses existing video transmission cabling, usually coaxial or UTP cables. These cables are already in place for carrying the video signal from the camera to the monitor. In this configuration, the controller and transmitter encode the commands. The commands are then sent to the receiver, where they are decoded. The transmitter in turn sends them to the electronics of the PTZ unit. The advantage of this approach is that cabling costs are minimal because there is a single cable. For long runs, however, amplifiers may be needed to contend with resulting signal attenuation.

• *Wireless* – There are three types of wireless communication: infrared (IR), radio frequency (RF), and true microwave. Infrared uses light as a transmission media. Commands are encoded at the transmitter and pulsed from a special light-emitting diode (LED) to a receiver unit. An IR signal can be carried up to 700' or so, but the transmitter and receiver must be in the line of sight. A camera around a corner, or with an obstruction between the transmitter and the receiver, cannot receive the signal.

A second wireless option is to use radio frequencies with microwaves. In this case, an RF signal is modulated across a microwave car-

---

On Site

Understanding Pan, Tilt, and Zoom

In most CCTV applications, the camera location is fixed. The camera is mounted in the corner of a room, in the ceiling, or on a pole, wall, or pedestal. Because the camera location cannot be changed, there needs to be a way to adjust the field of view.

• *Panning* means to move the camera from side to side. The camera pivots on its mount, moving from left to right, but the horizontal image plane remains the same. The camera sweeps the scene up to a 360-degree panorama. This is sometimes called scanning.
• *Tilting* refers to moving the camera up and down in a vertical plane. By tilting the camera, one can see aspects of the scene that are above or below the fixed image plane.
• *Zooming* means adjusting the magnification of a target. The lens must be adjusted to focus in close on something, making it larger. Zooming out (pulling back) shows more of the scene, and makes the target smaller.

When pan and tilt are combined, the camera can be repositioned in an arc, aiming it at any given location on the scene. The addition of zooming, provides significant control over the field of view of a scene. Collectively, pan, tilt, and zoom compensate for the limitations resulting from a fixed camera location.

---

rier frequency. This is the same technology used for a household portable telephone. Microwave communication of this sort operates in the 900MHz range (portable telephone), or in the 2.4GHz range (Wi-fi), and newer 5.4GHz ranges. *Figure 28* shows a transmitter and receiver for RF transmission of video signals.

Many other modern devices use the 2.4GHz frequency. This includes microwave ovens, cellular phones, and wireless data networks, such as wireless LANs. Keep in mind that a 900MHz portable phone uses the standard wired telephone network to provide a connection. 900MHz equipment has a limited range, usually 700' or less, between the base unit and the handset; a cellular phone uses radio towers and has a range of several miles.

A wireless data network uses a transceiver, which is a device that both transmits and receives on the same carrier band, and has a range of three miles or less. What is important to note here is they all use the same band of frequencies and do not require licensing.

The problem with many wireless camera solutions is they have a limited set of frequencies for broadcasting. This means that only two to four cameras can be connected to these systems. The security of the broadcast is more important. Because the frequencies for these units are published and known, it is easy for someone to intercept the signals.

Microwave communication is a viable option for connecting locations across an extended campus or industrial facility. For large installations, microwaves are often used to communicate between buildings or locations. *Figure 29* shows a microwave antenna commonly used for this purpose. However, these types of installations use parts of the radio spectrum that are regulated, so they require FCC licensing.

- *CCTV over IP networks* – A traditional CCTV network uses coaxial cable to connect video cameras to video recorders and monitors. An IP camera contains an onboard processor and a web server, so it can deliver digital video as packetized video streams directly to a network. In an IP network (*Figure 30*), video data is carried on Ethernet or fiber optic cable to a central computer, where it can be viewed and stored.

33410-11_F28.EPS

Figure 28 Transmitter and receiver for video distribution using RF.

33410-11_F29.EPS

Figure 29 Microwave antenna for FCC-licensed transmission.

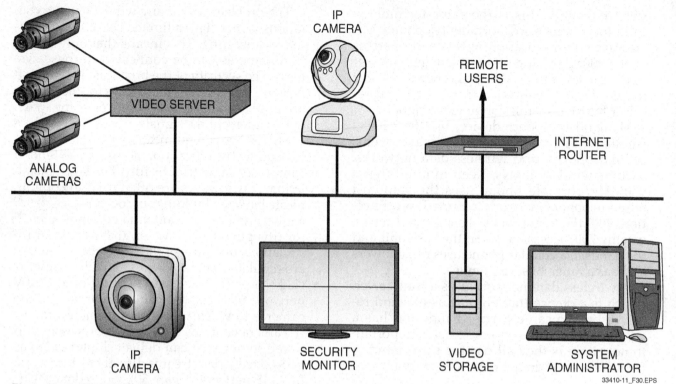

Figure 30 An IP CCTV network.

33410-11_F30.EPS

If the network uses digital (CCD) cameras, the camera can be connected directly to the network via the local network switch. For analog cameras, the cameras are connected to a video server that converts the analog data to digital form before it is sent to the network switch. The video server, like the IP camera, contains an onboard processor and web server software, which makes each analog camera addressable. This technology allows an existing system to be converted to an IP network without sacrificing the existing investment in analog video cameras.

Both live and stored video can be viewed by authorized persons on a LAN PC or laptop. In addition, remote viewers with the proper authorization can view the video through a web browser via the Internet.

4.6.0 Alarm Interface Units

An alarm interface unit (*Figure 31*) is a special device that receives a signal generated from a viewing location. For example, a tamper switch on a secure camera housing may generate a signal if the housing is vandalized. This alerts the operator that there is an incident requiring observation. This is true with other similar types of alarm signals. A motion detector may trigger an alarm based on movement at a camera location such as an entranceway or garage.

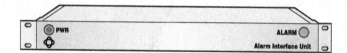

33410-11_F31.EPS

Figure 31 Alarm interface unit.

Alarms can be generated when devices sense a malfunction in their operation, such as when a PTZ unit freezes because the servomechanism is stuck. A modern video camera can contain circuitry that constantly tests and monitors its own performance. If a camera is failing, it may be smart enough to notify the operator that it is in need of service or repair.

When an alarm interface unit receives an alert signal, it identifies which zone and/or camera is affected. It then notifies the operator for manual intervention, and may automatically switch to the appropriate zone or camera. These alerts then trigger a number of preprogrammed responses, including turning on video recorders or moving cameras to pre-arranged positions.

4.7.0 Motion Detection

A motion detection device is sensitive to movement. Motion detection is used extensively in intrusion detection systems, including burglar alarm and lighting systems. A motion detector generates an alert, which is forwarded to a control unit. In the case of a CCTV installation, when motion is detected, a signal is sent to the alarm unit. This in turn activates some specific protocol, such as switching the monitoring station to the appropriate camera or notifying the operator of the need for observation.

There are different types of motion detection devices. Some are stand-alone units that sense motion based on air pressure and other environmental factors. With these devices, a signal is generated and sent to the CCTV alarm unit. Once a signal is received at the alarm unit, a camera view is activated.

Modern cameras can have motion detectors built into them. They may also use technology known as video motion detection. In this case, the camera uses a sophisticated microchip called a digital signal processor (DSP) to determine when a scene has changed. Many digital video recorders with built-in video motion detection operate in a similar manner; that is, they detect motion by examining changes in individual pixels.

In order to use video motion detection outdoors, or in extremely low-light environments, it may be necessary to use both an infrared-sensitive camera and an external infrared light source. Another option is to use readily available passive infrared (PIR) sensors.

4.8.0 CCTV Keyboards

An alarm may require a response from the operator. For example, a door in a monitored entranceway has been opened, or a motion detector has indicated movement in a secure area. A controlled access door may have been open, or a fire alarm has been activated. Regardless of the cause, the operator may need to view some aspect of a monitored area.

The CCTV keyboard (*Figure 32*) is designed to support the remote operation of cameras. These keyboards can range from a simple touch-tone keypad, such as those found on telephones, all the way to sophisticated devices with camera selection features, joystick control of camera movement, and the ability to select pan, tilt, or scan.

33410-11_F32.EPS

Figure 32 CCTV keyboard with joystick controller.

4.9.0 Recorder-Controllers

Earlier in this module, a recorder was shown inserted between a switcher or multiplexer, and the video monitor. There are several issues regarding the use of recorders in general that need to be examined, including the equipment to which recorders are attached.

First, there can be unwanted side effects from the type of inline equipment used to distribute the video. For example, when using a splitter, a recorder receives only the image being viewed. If the operator has manually switched to one input, the other is not recorded.

The same could be true of switchers and multiplexers; however, this problem has been resolved to a certain extent. To understand the nature of the problem, CCTV video recording must be examined in more detail, along with how multiplexers and other devices behave relative to recording.

4.9.1 Digital Video Recorders

The CCTV industry has historically used VCRs for its recording needs, but new recording methods have emerged to support longer recording times and multiple cameras. Over the past several years, the digital video recorder (DVR) has begun to replace traditional VCRs in the CCTV industry. The DVR uses a computer hard-disk drive to store recorded video. There are actually two types of DVR. One, for general use, is similar to analog recorders. It takes a composite video source, digitizes and optionally compresses it, and then stores it on digital media. *Figure 33* shows a typical DVR of this type. A network DVR, on the other hand, receives a digital video stream from network cameras connected to a LAN and stores it directly to digital media.

DVRs provide a number of controls in addition to those provided by a traditional VCR. For example, a DVR provides the ability to search stored video based on time, date, or even camera num-

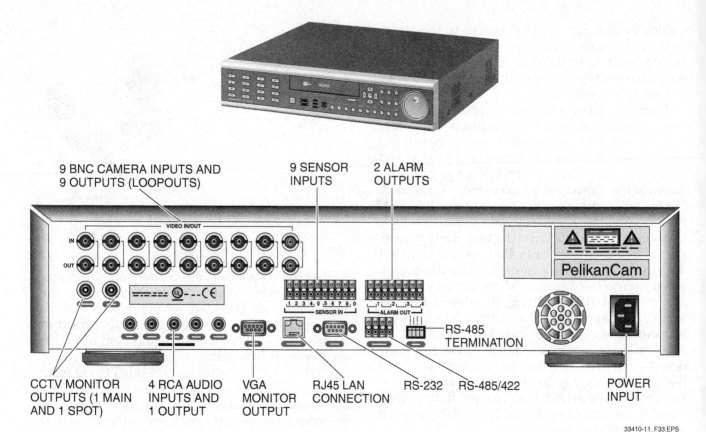

9 BNC CAMERA INPUTS AND
9 OUTPUTS (LOOPOUTS)

9 SENSOR
INPUTS

2 ALARM
OUTPUTS

PelikanCam

CCTV MONITOR
OUTPUTS (1 MAIN
AND 1 SPOT)

4 RCA AUDIO
INPUTS AND
1 OUTPUT

VGA
MONITOR
OUTPUT

RJ45 LAN
CONNECTION

RS-232

RS-485/422

RS-485
TERMINATION

POWER
INPUT

33410-11_F33.EPS

Figure 33 Digital video recorder for analog CCTV system.

ber. Unlike a VCR, a DVR can incorporate motion detection. They also provide some form of video compression, usually based on either MPEG or M-JPEG standards.

4.9.2 Video Recording in a CCTV System

Recording video in a CCTV system with a single camera is straightforward. Simply connect the camera to the recorder and the recorder to the monitor. However, with two or more cameras it becomes more complicated. A recorder can be provided for each camera, but this approach can get quite expensive. A monitor and recorder would be needed for each camera.

The problem starts with switched or multiplexed cameras. Typical questions that need to be answered include:

- How are multiple cameras recorded on a single tape?
- How is video retrieved from a specific camera?
- What happens to the video from other cameras while reviewing the video recorded for one camera?

To solve this problem, engineers have introduced a time-based recording method called time-lapse recording. Time-lapse recording takes the input from each camera and allots a specific amount of time for recording that camera. All inputs are split across the timing interval, and stored on a single tape.

When using an automatic switcher, the recording issues are the same as the viewing issues. While viewing one camera, the output of the other cameras is not available. Modern switchers allow the operator to program the recording time interval. This means that even while viewing one camera, all cameras are still being recorded. Therefore, time-lapse recording significantly improves the amount of video data available for offline review.

Another engineering solution is the use of real-time recorders. In reality, real-time means 30 frames per second for full motion. However, in the security field, generally only 20 frames per second are captured for 24-hour recording. Like a time-lapse recorder, these 20 frames are split among all the cameras and spread out across the one-second timing interval. Another advantage associated with real-time recorders is they can be set to record for extended periods, such as 12, 24, or 72 hours, on a single cassette. This eliminates the need to change tapes as often, which is a useful feature for unattended recording over a weekend or holiday period. T160 cassette tapes are used for extended recording periods greater than 48 hours, while T120 cassettes can be used for shorter recording periods.

4.9.3 Multiplexers and Video Recording

A multiplexer is a device that takes the inputs from multiple video sources and combines them into a single composite video signal. When this signal is displayed on a video monitor, it shows a segmented view of all the inputs. In the case of a quad unit, four images are displayed at the same time, each confined to a specific quadrant on the screen. For a 16- or 32-camera multiplexer, the screen may be divided into larger segments of four to eight cameras. Several screens are then displayed in sequence to expose all of the cameras. The operator can usually decide on the number of cameras per screen. For some applications, 16 cameras or more are displayed, but generally only on a larger monitor.

Multiplexers are available in three styles: simplex, duplex, and triplex. A simplex multiplexer allows either viewing, recording, or playback at any given time. A duplex unit permits any two of the three functions to be preformed at the same time, and a triplex unit can support all three functions simultaneously.

In the event of an alert or an alarm, a multiplexer switches to the camera that covers the area of the alarm. If there are four cameras being monitored, three of them are no longer displayed. However, just like the switcher, all cameras are still recorded. The operator can set a recording interval.

> NOTE
>
> A typical DVR used for CCTV has a built-in multiplexer. So a stand-alone unit is unnecessary.

4.9.4 Using a Controller in a CCTV System

A controller allows the selective control of multiple video recorders. A typical controller provides several operations to control the actions of a recording device. Most devices provide eight types of control, all of which are generally supported by a controller:

- Play
- Stop
- Fast forward
- Reverse
- Pause
- Record
- Search
- Eject

4.10.0 Video Monitors

The video monitor provides a means for viewing the outputs of cameras and video recorders. It is the last device in the system. Video monitors are not necessarily televisions. They usually lack the electronic tuner needed to receive television broadcasts. Originally, the monitor was a device that operated pretty much like a tube camera, only in reverse.

4.10.1 LCD Monitors

More than anything, the movement towards digital CCTV systems drives the use of liquid crystal displays (LCDs), as shown in *Figure 34*. A digital CCTV system looks and behaves more like a computer system.

An LCD is a special low-power display commonly used in electronic instruments and portable computers. LCD technology uses the characteristic of certain materials which, when heated, create an almost plastic state. The molecules are solid but flow as a liquid, hence the term *liquid crystal*.

Unlike CRTs, an LCD forms an image by blocking light to an area that is energized with a small electrical charge. Many LCDs use reflective light, which makes them difficult to see. They actually have a mirror on the back to reflect incoming light. Backlit models have a light source that is projected onto this mirrored surface.

On Site

Understanding Signal Sync

Video synchronization means that signals from two or more devices are in phase with each other. For example, two video cameras can be monitoring the same location, but from different angles. Each camera generates a video signal, and each signal has common characteristics. If they are not in sync, and you switch between them, there is a delay before one image replaces the other. There may also be significant distortions, rolling lines, and even blackness between images.

These effects show up on the monitor, but also affect the recording and retrieval of the images from videotape. Without synchronized signals, the video recorder also records these glitches when switching between cameras. This results in empty and partial frames being stored on tape.

Figure 34 LCD monitor for CCTV system.

Newer LCDs are created using miniature transistors laid out in a matrix on a thin film. These are called thin film transistors (TFTs). Collectively, they are called active-matrix TFT displays. There are also capacitors associated with each of these tiny transistors. Each transistor is addressable and is called a pixel. To turn on a particular pixel, a specific row is switched on. An electrical charge is then sent down a specific column. Since all of the other rows in that column are off, only the capacitor at the specific pixel receives the electrical charge. The capacitor holds the charge until the next refresh cycle. This is the origin of the word active in the phrase active-matrix.

LCD technology found rapid acceptance in the CCTV industry. This acceptance is partially due to its lower cost. Many manufacturers have stopped producing CRTs entirely in favor of LCDs. As shown in *Figure 35*, LCDs are available in a rackmount configuration.

4.10.2 LED Displays

The current generation of LCD monitors uses LEDs to provide backlighting for the screen. The use of LEDs allows for better energy efficiency, lower heat production, and a slimmer monitor.

Figure 35 Rack-mounted LCD monitors.

In contrast to the LED-backlighting used on LCD displays, an LED display uses LEDs as the display medium. LCD and plasma displays are limited in size, whereas the LED display size is virtually unlimited. LED displays are used for the giant viewing screens found in sports stadiums and arenas. The largest LED display, which is located in the nation of Dubai, is 33 stories high. Another application of LED display technology is the organic LED (OLED) display. It uses organic material placed between glass plates and can be used to produce super-thin screens. With this technology comes the potential for flexible TV screens.

4.10.3 CRT Monitors

Today, CRT monitors (*Figure 36*) are fairly common in legacy systems. But, like the TVs in our homes, these monitors are being replaced with flat-panel LCD monitors. The main component of a CRT monitor is the cathode ray tube (CRT). The cathode generates an electron stream or beam. This beam is manipulated both horizontally and vertically by a magnetic field. This causes the electron beam to bend and scan across a phosphor-coated wire screen. As the beam strikes the screen, it excites the phosphor molecules, causing them to glow for a short period.

In a video monitor, the beam actually makes two passes across the screen, scanning from left to right and top to bottom. Based on NTSC

On Site

Composite Video Input

When acquiring an LCD for use in CCTV applications, make sure it has a composite video input. Many commercial, off the shelf LCD monitors do not. If there is not a composite video input, the LCD needs a scan converter or a scaler. This device converts the scanning output of a video stream to make it conform to the input requirements of computer displays (VGA, SVGA, XGA, and others).

standards, a video monitor displays 525 lines. As the beam scans, it paints every other line on the first pass, and then fills in the missing lines on the next. This is called interlacing. Most CRT video monitors produce an interlaced image. CRT computer monitors, on the other hand, may use a CRT, but they are non-interlaced. They scan from left to right and top to bottom, painting each line in succession.

CRT computer monitors often have greater resolution, but experience a side effect called flicker, which makes them unsuitable for use as CCTV monitors. Flicker is a noticeable flashing of the picture based on the refresh rate, or the amount of time that the phosphor coating takes to lose its luminance. A video monitor refreshes the screen 60 times per second. A computer display may have to refresh 75 times per second or more to retain an image.

5.0.0 SIGNAL DISTRIBUTION

A CCTV installation has many of the same requirements as a CATV, SATV, or MATV system regarding signal propagation. The transmission of CCTV video uses a single composite video signal. On the other hand, the signal can be transmitted through the air using RF technology. RF also allows multiple video signals to be carried across a single transmission medium, such as cable, microwave, or infrared.

Figure 36 CRT monitors in a rack-mount configuration.

Attenuation results anytime a signal propagates across a conductor or semiconductor. This means the signal fades over distance and may no longer be sufficient to drive a monitor or receiver. The effect shows up with RG-59 and RG-6 coaxial cable, as well as unshielded twisted pair, and even custom cabling options. Each cable introduces its own attenuation characteristics onto the signal.

5.1.0 CCTV Signal Amplifiers

One way for CCTV installations to manage signal attenuation is to use an amplifier (Figure 37). The amplifier would normally be inserted between the video source and the monitoring device. CCTV applications use an unbalanced amplifier. Unbalanced means that the peak voltage should be the same at both ends. In the case of coaxial cable, the voltage is typically 1-volt peak to peak, represented in the literature as 1Vp-p. The two types of coaxial cable used for CCTV applications are RG-59 and RG-6.

Note also that all video equipment, including CCTV, has a characteristic impedance of 75 ohms. Cameras use the impedance to synchronize signaling. This is important when connecting monitors and other inline equipment to the output of a camera. For example, connecting two monitors to a single camera using a T-connector will not work properly. Assuming that both monitors are terminating monitors, they both have only 75-ohm input jacks. If both of them are simply connected to the same camera, negative effects, such as signal fading, alternating contrast, and even lost video frames, are experienced. Because both cameras present 75 ohms of impedance, signaling is affected.

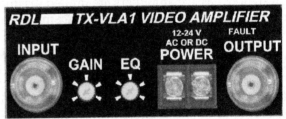

Figure 37 A video signal amplifier.

The correct way to connect two monitors to the same video source is to use a video loop-through. Most terminating monitors have a switch that toggles between 75 ohms and 0 ohms. This switch must be turned to the off position (0 ohms) at the first monitor, and on at the second (75 ohms). The output of the camera is connected to the first monitor and then looped through to the second. The second monitor provides the proper 75-ohm termination. Some monitors are auto-terminating. They sense whether a signal is being looped or not.

Two potential problems occur when connecting to multiple devices. One is caused by resistance, the other by signal loss. To solve signal loss problems, an unbalanced video amplifier is placed between the camera and the monitoring unit. This improves the signal by increasing the amplitude of the signal coupled with voltage. For example, the signal might be kicked-up to 5Vp-p. By the time it travels the length of the cable, it is received at the correct 1Vp-p as required. The length of the cable and the availability of power can determine where to place an amplifier. The amplifier must have power to operate, as it is considered an active device. Active amplifiers are used when the signal must be sent across an exceptionally long run of coaxial cable.

There are passive amplifiers, but they are used in noncurrent-carrying configurations. For example, a passive amplifier is used to carry video signals across unshielded twisted pair cabling. A transmitter at the camera side takes a composite video signal and converts it into an analog signal. It is then sent down the wire. At the other end, a receiver translates the analog signal back into a composite. It then resends it across coaxial cable to the monitoring station. Passive amplifiers are limited to runs of less than 700 feet. *Figure 38* shows transmitters and receivers used with twisted pair cabling.

One of the problems of using an amplifier is that it amplifies all aspects of the signal. This includes any noise introduced by the sending equipment. In addition, RFI is picked up through the shielding of the cable itself. Every wire, including coaxial cable, acts as an antenna. It can pick up unwanted RF, which is injected into the conductor core of the cable. The amplifier power supply can also inject magnetic field and RFI. When selecting an amplifier, it is important to select one that cancels unwanted noise from the signal.

5.2.0 Distribution Amplifiers for CCTV

There are conditions that require multiple monitors to review the output from a single camera. This is true regardless of the number of cameras

33410-11_F38.EPS

Figure 38 Twisted pair distribution transmitters with receivers.

attached to the system. A distribution amplifier (*Figure 39*) is a special post-equalizing amplifier. It is used to connect more than one monitor to a camera output.

As previously discussed, video loop-through means that the input signal is not affected by the equipment through which it is passing. Because the signal is not affected by the loop-through, it can be forwarded onto other equipment that is 75-ohm terminated.

To connect two monitors to the same video source, short of using a switch, splitter, or multiplexer, the cable needs to be through-looped at the first monitor. The signal is run out of the first monitor to a 75-ohm termination at the second monitor. Most monitors provide this type of support. Other inline devices already discussed, such as the multiplexer and the date/time generator, also provide this support.

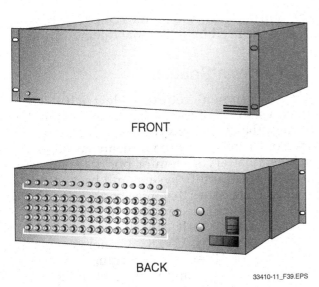

FRONT

BACK

33410-11_F39.EPS

Figure 39 Master distribution amplifier.

5.3.0 Signal-to-Noise Ratio

The human information processing circuit (eye, ear, brain) is fairly forgiving. There is some minimal threshold required for perception of a signal. The signal can still be picked out from all the noise in the environment. However, as the noise increases, it becomes more difficult to separate the information contained within the signal. The same is true of video transmission systems. Even though the image on the monitor may degrade, useful information can still be obtained.

Signal distortion is measured based on noise in dB. It is expressed as the signal-to-noise ratio (SNR). In general, the higher the signal-to-noise ratio, the better the picture is going to be. If there is a low signal-to-noise ratio, the picture can appear grainy or snowy, and sparkles of color may be noticeable. Equipment is not able to synchronize with extremely noisy signals.

To ensure that noise is not introduced from electrical and power sources, avoid running electrical cabling in tandem with signal cabling. Power should be run through conduit sufficient for the environmental conditions. Signal cable shielding is necessary. A coaxial cable with a solid copper conductor and braided copper wire shield is used. Select a conductor for power runs with a sufficient diameter to isolate voltage-induced distortions and interference.

5.4.0 The Importance of Impedance Matching

In order to transfer maximum power, especially where long cable runs are involved, it is important to match the impedance between the source and the load. The cable must have the same characteristic impedance as the equipment. Impedance mismatches can cause a signal to be reflected into the cable and result in signal attenuation and distortion. Some older British Naval Connectors (BNCs) have a 50-ohm impedance. If these connectors are used with 75-ohm cables and equipment, signal attenuation and distortion are likely to occur.

As discussed in the *Cable Selection* module, UTP data cable is sometimes used for long cable runs of the type commonly found in CCTV systems. Coaxial cable has a 75-ohm characteristic impedance, while data cable has a 100-ohm impedance. In order to match the impedance between the two, balun transformers are used to make the connections between the coax cable and the data cables.

5.5.0 Ground Loops

Signal problems often originate in the coaxial cable used to carry signal from a camera to a monitoring station. The subject of attenuation effects and how to compensate with amplifiers, video equalizers, and the separation of power cabling from signal cabling was previously discussed. Another problem with video signals is called a ground loop. This is where the ground potential between two different power sources is different. In the case of coaxial cable, the copper braided shielding provides the return path for the signal.

Cameras and monitors have independent power supplies. They may also be on different distribution paths because they are connected to the power grid at different locations. A difference in ground potential causes the coax shielding to carry the voltage differential to the closest ground. This can cause significant distortion to appear at the video monitor, VCR, or switch. This distortion is sometimes called a hum bar. There is a special isolation transformer called a humbucker that can be used, though usually on long cable runs. For shorter runs, the best option is an inline ground loop corrector, as shown in *Figure 40*, or a differential ground loop corrector.

It is important to keep ground potential equal across all equipment used. Wireless connections can alleviate these types of problems. Fiber optic cable can also eliminate them. However, where cameras are connected via coax or data cable, chassis grounds must be equivalent across the system. In addition, rack housings provide proper ground potential.

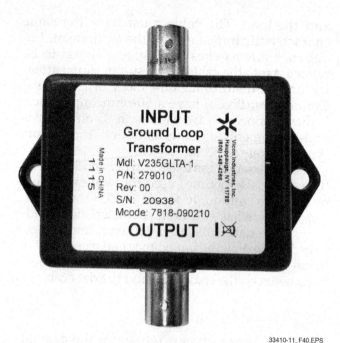

Figure 40 An in-line ground loop corrector.

33410-11_F40.EPS

5.6.0 The Advantages of Fiber Optic Cable in CCTV Applications

Because there is no metal used in its construction, the use of fiber optic cabling can eliminate the ground loop problem. It is therefore ideal for inter-building cabling runs. Fiber optic cabling has a number of other advantages as well. Its attenuation is typically less than a tenth that of co-axial cable, so it is ideal for long cabling runs. Its low attenuation level exists over a wide range of frequencies as well.

Fiber optic cabling is not susceptible to EMI and does not radiate signals that could be intercepted. Also, fiber optic cable is tiny in comparison to coax, so it can be used in very tight spaces where existing pathways do not support large bundles of cabling. An existing fiber optic link can be expanded to support more cameras and control channels through the use of multiplexing techniques. Thus, the system can be expanded without the need for additional cabling runs.

6.0.0 POWER SOURCES

CCTV equipment requires electricity. There are two sources of electricity. It can be supplied through the electrical grid from an energy provider, or it can be acquired from an on-site source, such as a battery system or portable generator. Both electrical utility system and backup power options create issues that must be accommodated.

CCTV equipment requires AC power, either 120VAC or 24VAC. It can also be powered in some instances by 12VDC. Cameras typically use 12VDC or 24VAC. Control equipment such as monitors, switchers, and recorders use 120VAC.

There are two ways power can be supplied to a camera using 120VAC. Usually, cameras are plugged into a parallel circuit. It is important to know the collective power consumption of all equipment in the circuit. This helps to ensure that that the proper gauge wire is used to minimize voltage drops. This improves the operation of the device.

Another way to connect equipment to the electrical distribution system is a radial method, better known as a star configuration. This means a direct circuit is wired from the electrical source to each device. This is usually done with equipment having large current requirements, such as a motor or compressor.

Even with radial attachment for large equipment, the high current draw of a compressor starting up can cause voltages to drop. This causes distortion and other signal problems. It also directly affects the operation of equipment.

Other factors that also influence the availability of current and proper voltages include the following:

- Power surges
- Voltage spikes and dips

- Brownouts and blackouts
- Electrical noise

All of these factors point to the need for line conditioning and filtering. The best way to provide a clean, consistent electrical source is by using a UPS (uninterrupted power supply). One type of UPS is a continuous-duty UPS; the other type is a true backup power supply.

If the power is off, as with a blackout, an alternate electrical source is needed. A UPS provides power within 10 milliseconds of power dropping. This is fast enough to keep most equipment running uninterrupted. With a basic backup power supply, the battery is kept charged, but the electrical current provided to equipment comes directly from the electrical source.

A continuous-duty UPS charges its battery constantly, but uses the battery to provide electric current. This allows the UPS to compensate for spikes, surges, and brownouts. The equipment continues operating with proper voltages. In the event of a blackout, it continues to provide uninterrupted power.

7.0.0 LIGHTING AND ILLUMINATION

Like the human eye, a video camera requires light. Without sufficient light, neither the camera nor the eye can receive enough information to form an image. This topic explores the role of light in video imaging. It identifies various approaches to working within lighting constraints. It also identifies specific solutions to working with lighting extremes.

7.1.0 Working with Light Conditions

During any 24-hour period, light conditions outdoors change from extremely bright to extremely dark. Indoor lighting conditions can also change as lights are selectively turned off or on, depending on the time of day, whether there are people in a work area, and other factors. Indoor lighting can be affected by light entering through windows. Light levels can change significantly by opening or closing a door. Outdoor lighting conditions can also change during the switch from natural to artificial light sources.

A camera receives light through its lens. Other factors, such as environmental and atmospheric conditions, can affect the amount of light received by the lens of a camera. For example, high levels of dust or liquid spray in a scene can reduce the amount of light and severely limit the field of view. Atmospheric conditions, such heat and

humidity, can cause lighting levels to fluctuate moment-to-moment, distorting incoming images.

In many cases, the lens iris can compensate for varying degrees of light. However, as lighting conditions reach extremes of light and dark, additional light may be needed to receive a useful video image. To understand the role light plays in video imaging, it is necessary to first explore the differences between natural and artificial light.

7.1.1 Natural Light

Natural light is light received from the sun. The human eye is sensitive to solar radiation with wavelengths from 400 to 750 nanometers (nm), and most sensitive to light at 555nm. The wavelength of the light determines the color. The longer wavelengths are at the red end of the spectrum, while shorter wavelengths are towards the blue. Light in the center is green. *Figure 41* shows this sensitivity to light. A camera, like the human eye, must receive light in this range in order to create a video image.

7.1.2 Artificial Light

Artificial light is produced by a number of processes, including heat, chemical reaction, and photoelectrical activity. Incandescent and fluorescent lights are the most popular for indoor lighting. Halogen, sodium, and metal-arc lamps (like those used in streetlights) are used outdoors, as are mercury and aluminum oxide lamps. Artificial lights, depending on the materials with which they are constructed, generate light at various wavelengths, and therefore different colors. For example, a sodium lamp generates light at a wavelength around 560nm, which appears yellow. A fluorescent lamp has a phosphor coating, and this coating determines the color, and therefore the wavelength of the light. Fluorescent lamps can have phosphor coatings that produce a full spectrum light, a white-blue light, or the more popular cool white, as well as many other colors.

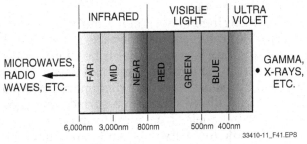

Figure 41 The visible light spectrum.

7.2.0 Measuring Light

Light is measured in photometric units. The basic measure of the amount of light produced by a light source is the lumen. A lumen is defined as the total amount of light, at one candela intensity, that falls on a one square foot section of the interior surface of a sphere. If the sphere has a one-foot radius, the total surface area is 12.7 square feet. The total number of lumens is therefore also 12.7.

To understand lumens, another measurement called candlepower (cp), or more simply, a candle, is used. Originally, candlepower was determined as being the intensity of light emitted by a standard candle, much like a plumber's candle, at a distance of one foot. It was expressed as a footcandle.

The concept of candle (candlepower) was later replaced with candela, which is based on the temperature of platinum where it converts from a liquid into a solid state. A candela is a metric value and provides a much more precise and consistent measurement.

7.3.0 Measuring Reflectivity

When working with CCTV and video cameras, the reflectivity, or illuminance, of objects and surfaces must be considered. When a light source strikes an object, the amount of light reflected by the object determines what is seen by either the camera or the eye.

Different materials reflect light to varying degrees. For example, snow reflects as much as 90 percent of the light that falls on it. Black velvet fabric reflects less than 20 percent of the light.

The challenge with measuring reflectivity is that materials absorb some wavelengths while reflecting others. This is important when selecting or using a camera. A black and white camera is immune to the spectral shifts, while color cameras are extremely sensitive to them.

As the object moves further away from a light source, the intensity of the light falling on a surface diminishes. Light measured on an object is inversely proportional to the distance the object is from the light source.

When engineers, designers, and manufacturers talk about reflectivity, they use a metric scale called lux. Lux is used to describe illumination, or the density of light falling on an object. Lux is defined by Webster's online dictionary as "equal to the direct illumination on a surface that is everywhere one meter from a uniform point source of one candle intensity or equal to one lumen per square meter." The term *illumination*, or *lux*, refers to the actual light available at a given distance.

Remember that lumens are used to measure how much light is produced by the light source.

Lux and footcandles are equivalent measures. That is, they both measure the same thing, but use different scales. A footcandle is 10.57 lux and a lux is 0.093 footcandles.

7.4.0 Light Sensitivity of Cameras

Most CCD cameras are sensitive to light that ranges from bright sunlight (10,000 footcandles, or 107,500 lux) to twilight (1 footcandle, or 10.7 lux). This means a camera can operate in relatively high or low light levels. As the light diminishes, though, the sensitivity of the camera decreases. As darkness falls, they may become useless for surveillance tasks without some kind of artificial light to enhance their performance. In such cases, either a low light level camera or an infrared camera is needed. Lux is the metric normally used in camera manufacturers' specifications and marketing material to define the light sensitivity of cameras.

7.5.0 Infrared Lighting and Cameras

Infrared light (IR) is light below the threshold of human vision. The infrared spectrum is actually quite broad, with wavelengths from around 800nm to 6,000nm. In fact the IR spectrum is often separated into two categories, near infrared and far infrared. In the near IR ranges, light has a wavelength that ranges from 800nm to 3,000nm.

Infrared sensitive cameras use a region in the near infrared spectrum, which is as close to visible light as possible with wavelengths from 800nm to 900nm. Light in this range behaves as visible light and can illuminate a scene for surveillance, but still be undetectable to the human eye. It can, however, be detected by instrumentation designed for visible light. There are two types of infrared sensitive cameras. Either they require an external IR light source or they amplify existing light.

For typical CCTV applications, cameras that work with both visible light and zero-to-no light are preferred. These cameras, sometimes called night-vision, or zero-lux cameras, are far more useful. They work by first converting the incoming light to an electrical signal. This signal is then amplified, and converted back into light as shown in *Figure 42*.

When available light is converted for use by a CCD camera, it is converted into green light. Remember that green is right in the middle of the visible spectrum, and that the human eye is most sensitive to green light. By converting to green,

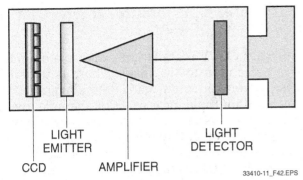

LIGHT
EMITTER

LIGHT
DETECTOR

CCD   AMPLIFIER

33410-11_F42.EPS

Figure 42 Simplified light amplification circuit for a night-vision camera.

the range of amplification is greater. Red and blue light both have narrow wavelengths in comparison. For this reason, images from low light cameras are usually green, though some cameras may switch from color to black and white to capture images during dawn and dusk hours.

One problem with available-light cameras is that they may generate visual noise, or visual artifacts, as part of the conversion process. In addition, because they amplify available light, the light from sources such as moonlight or streetlights may actually be too bright to capture. Fortunately, most cameras have automatic gain control, which allows the camera to compensate by adjusting its sensitivity. The images they provide are more useful, but details in shadow areas can be blacked out as a result.

For cameras that require an external IR light source, an array of LEDs is generally used to produce infrared light in the range of 900nm to 1,100nm. Zero-lux cameras contain a CCD for imaging, and some cameras are designed with an infrared LED array built into the camera housing. The challenge is to produce enough IR light to illuminate a scene. Cameras with built-in infrared LEDs typically have a short illumination distance, sometimes only 10' to 20'. To increase the distance, an external infrared illuminator is required.

External infrared illuminators come in two varieties—the infrared LED array already mentioned, and special, high-intensity lamps with a visible light filter. Lamps with filters are less than perfect, however. They can still emit visible light, though in the neighborhood of 0.01 lux. LED arrays are most likely to be used today.

7.6.0 Illumination and Beam Angles

When selecting illumination-enhancing light sources, it is necessary to know the beam angle and the intensity of the light. Beam angle de-

scribes the shape of the light source output. An incandescent light puts out light equally in all directions, or 360 degrees. This provides a uniform lighting condition, but one whose intensity diminishes noticeably as it moves further away from the source.

Technically, light travels forever. When it strikes a surface it either is reflected, or passes through it. As the concentration of light diminishes over space and time, it appears to be less bright. To support illumination requirements, it is important to select a light source with an intensity that provides enough illumination at distances determined by the application.

The two primary light beam shapes are the spotlight and the flood light, as shown in *Figure 43*. A spot light concentrates its light output into a single beam and projects it in a narrow angle. Spotlights have a beam of approximately 15 to 20 degrees. This allows them to project their light a much longer distance. A typical flood light has a beam angle of 40 to 60 degrees. These lamps project light across a shorter distance, but provide consistent light intensity across a wider area.

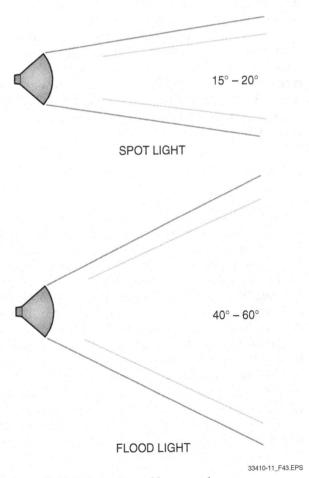

15° – 20°

SPOT LIGHT

40° – 60°

FLOOD LIGHT

33410-11_F43.EPS

Figure 43 Light intensity and beam angle.

7.7.0 Managing Backlighting

Backlighting can be an unwanted effect in CCTV. Backlighting means that details in the video image are blacked-out because of a bright light source behind the object being viewed.

To solve backlighting problems, many cameras provide either a backlight compensation (BLC) or ALC. BLC provides camera circuitry to achieve the optimum balance between the iris and gain settings. By using average light metering across a scene, it can calculate the best exposure and adjust the settings as needed. An ALC allows the camera to ignore bright light sources and therefore focus on darker areas to extract more detail.

> **NOTE**
>
> Backlight compensation must be done when the camera and lens are not matched.

Another solution for backlighting problems is to minimize the effects by altering the angle of either the camera or the lighting sources. Determining optimal angles may require testing, but can offer reasonable improvements in image quality when combined with BLC or ALC.

A third solution is to use multiple cameras. With each camera aimed at the target from a different angle, the effects of backlighting can be minimized by switching to a different camera, depending on the time of day. Matrix switchers allow cameras to be selected based on criteria such as time of day.

8.0.0 MEDIUM TO LARGE CCTV SYSTEMS

Some suppliers think of a small system as consisting of eight cameras or less. Others define a small system as having fewer than 16 cameras. However, with more than eight cameras, the complexity of monitoring increases significantly. Many devices like multiplexers and switchers support only four or eight cameras and require multiple devices to scale up beyond that limit. As the size of the CCTV installation grows, the ratio of monitors to cameras decreases.

This topic examines the equipment found in larger installations, such as manufacturing plants, gaming casinos, or hospitality environments such as a multi-floor hotel. In these environments, complexity is managed by complex control systems and flexible camera setups.

8.1.0 Control Systems for Large CCTV Installations

In larger CCTV installations, multiple operators share the monitoring tasks. There is usually a central control room, with secondary monitoring from multiple locations. For example, two or more operators may be in the control room, with a supervisor or shift manager in an office. For corporate environments or multi-building campus applications, there may be a control room for each building manned by one or more operators, and a central facility with access to monitoring equipment in all locations.

A large installation includes all of the basic equipment discussed earlier, but it is mounted in racks. In addition, the monitoring station is typically expanded into a console layout.

8.1.1 CCTV Rack Systems

A CCTV rack system consists of standard EIA rack housings as shown in *Figure 44*. These are 19" wide, with a standard, repeating three-hole mounting configuration. The advantage of rack units is they can support a host of environmental requirements. For example, they may have covers on the sides and top, with latching cabinet doors. They may also provide heating and cooling support, as well as UL/CSA-approved power blocks.

Most CCTV devices are available for either stand-alone use, or for integration with other components. For example, a VCR controller can be mounted in a rack. All the recorders it addresses can then be placed in adjacent racks. If there are 30, 40, or more DVRs, they must be organized and accessible, and all the cabling can be neatly routed. This improves the serviceability and maintenance of the component as well.

Devices are either mounted directly to the rack, or placed on shelves, on sliders, or in special housings. The equipment is specifically designed for insertion into the racks. In addition, rack mounting allows cable routing panels to be used for easy cable and wire routing. Routing panels are often attached at the rear of the rack. Patch panels are usually mounted to the front.

Power is provided to devices in the rack via approved electrical strips and power cords. These usually have a single, accessible emergency power off (EPO) switch. An EPO switch allows power to be removed from the entire rack in case of an emergency. Power can be removed from all devices at one time, rather than unplugging several devices to service them.

Figure 44 Standard equipment rack.

8.1.2 CCTV Console Systems

A CCTV console is where monitoring takes place. Physically, a console consists of monitors, keyboards, and communication devices. Telephone and radio transceivers are typically included at the monitoring station. A simple console can sit on a table. A complex console must often be custom built, or use a vendor-provided console frame (*Figure 45*). These in turn provide for custom configuration of the equipment and devices available to the operator.

Because of the nature of a large CCTV installation, specialized CCTV system components are used. There are a number of devices and reasons for using them.

8.1.3 Matrix Switchers

A matrix switcher is a special CCTV control device used to switch from one video signal to the other. It is not the same thing as a multiplexer. It allows an operator to manage medium to large CCTV installations and supports use by more than one operator. A matrix switcher for a medium-sized installation has the following:

- Multiple video inputs and outputs: for example, 48 inputs, 8 outputs
- Viewing and control of individual cameras, multiplexers, and recorders

Monitors are available for rack mounting as well. They can be placed on shelf units. Some monitors are packaged in groups of two or three 5" or 9" monitors. The monitors are then mounted directly, using pre-drilled mounting holes in the rack.

Accessories for racks include cooling units, doors, and covers, as well as environmental protection, such as fireproofing, dust-proofing, and even seismic (earthquake and tremor) protection. Wheels and casters allow racks to be moved easily.

Because a CCTV system is part of an overall security function within an organization or community, control and monitoring equipment need to be secure from tampering. Racks can provide additional security for the equipment used in the CCTV system.

Figure 45 A console rack system for a medium-sized CCTV installation.

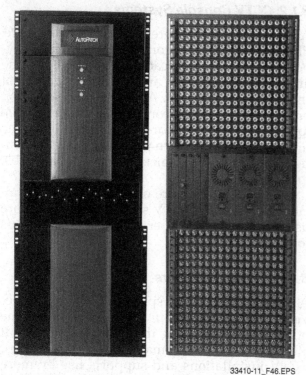

33410-11_F46.EPS

Figure 46 Matrix switcher for a medium-sized CCTV installation.

- Logical camera selection
- Terminating or looping video inputs
- ASCII and serial communication ports
- Password control and user authentication
- Multiple keyboard support
- Macro programming and event timers

A single matrix switcher makes it easy to manage multiple devices that control a CCTV system (*Figure 46*). These devices include cameras, recorders, and multiplexers. Multiple monitors can be connected to a matrix switcher. They can be independently controlled and have direct video output to them. For example, banks of 9" monochrome monitors can each display output from different quad multiplexers. When an alarm is triggered, a single 12" monitor can be provided with video from the appropriate camera.

ASCII communication capability allows alarm information to be forwarded to a data communication network. For example, when an alert is received, a command is sent to the data network. An email can then be created and forwarded to appropriate personnel. An alert could be forwarded to the police or fire department as needed.

Password control is needed when multiple users have access to the console. For example, the operator can have access to all cameras and recorders. A supervisor may have additional access to set passwords and may establish access protocols for individual users.

Macros are used to designate predetermined activities and alarm responses. When an alert is received, a macro can be activated to perform several key tasks. For example:

- The display is switched to a given camera.
- The camera is repositioned to a given preset.
- A recorder is switched from time-lapse to real-time.
- A command is sent to the data network.

8.1.4 Matrix Switchers in Large CCTV Installations

Matrix switchers used in medium installations are usually self-contained. That is, all of the features are enclosed in a single assembly. Video inputs and outputs are located on the rear of the device. Keyboard controller interfaces and communication connection ports are also at the rear of the unit.

Matrix switchers for large installations are modular. The control unit, also referred to as the CPU, is a separate device. It interfaces with a switching unit, which is highly expandable. Often, the CPU is expandable but in a more limited fashion.

A switching unit is actually a printed circuit card cage and is referred to as a bay. To expand the system's coverage, simply insert an additional circuit board. It is possible to have multiple switching bays attached to a single CPU controller. A typical switching bay can support up to 256 cameras. Switching bays will often support output to monitors by providing 4, 8, or 16 BNC jacks.

The CPU controller is microprocessor-controlled and communicates with external devices. It can respond to commands from external computers programmed to communicate with it. It has outputs that connect to expansion bays, such as switching units, PTZ controllers, and keyboards. Important features may include a VGA output for screen management tools provided by multiplexers and other equipment.

The CCTV keyboard provides the means for using the CPU to control the numerous inputs and outputs. Typical features supported via the keyboard include the following:

- Controlling receivers
- Camera and monitor switching
- Accessing and activating multiplexer screen functions
- Defining zones and presets
- Arming or disarming alarms

9.0.0 TESTING CCTV SYSTEM VIDEO

CCTV systems, large or small, are complex video signaling systems. They are subject to many of the same problems found in other video applications. Cable, satellite, and television recording and broadcast applications all require calibration and video signal conditioning. The same is true for CCTV. This topic examines key aspects of a CCTV video signal. It also identifies the testing requirements for maintaining quality video signals. It identifies test and calibration equipment commonly used.

9.1.0 Video Equipment Calibration

A standard video signal is always a monochrome signal. For applications using color, the color information is overlaid on the monochrome elements to reproduce color at the monitor, recorder, or printer. Analog video signals carry two types of color information: chrominance and luminance.

A luminance signal represents a series of voltage levels and determines the brightness of an image. The brighter an image element is, the lower the amplitude of the signal is. The chrominance signal is a sine wave. It contains information representing the properties of hue and saturation for video color.

Devices, such as cameras, video recorders, and monitors, may need periodic adjustment or calibration. This allows for correct display of chrominance and luminance signals. A waveform monitor (*Figure 47*) is used in conjunction with a color bar generator (*Figure 48*) for this purpose. A waveform monitor is a special oscilloscope. It is used to evaluate video signals. A waveform monitor displays amplitude information for the signal (*Figure 49*). A color bar generator creates an ideal color signal, allowing for ease of measurement. Color bar information is often recorded on a videotape leader.

Another test and calibration device is the vectorscope (*Figure 50*). It displays information about the chrominance, or color portion, of the video signal. The vectorscope allows presentation of color to be adjusted by synchronizing on a subcarrier of the video signal. The subcarrier information is sometimes called the burst or color burst. This subcarrier is used to provide a synchronizing signal for representing variations in hue and color saturation.

A vectorscope is used to test three key video attributes: chrominance phase, gain, and white balance. Chrominance should be in phase with the color burst. If it is out of phase, color hues are in-

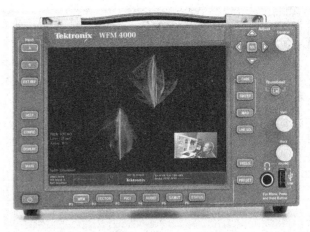

33410-11_F47.EPS

Figure 47 Waveform analyzer.

33410-11_F48.EPS

Figure 48 Handheld color bar generator.

correct and cause distortions, such as green faces or blue foliage.

Gain is a measure of amplitude. This measurement can be used to verify that the amplitude of the signal is within acceptable thresholds. White balancing is the process of calibrating the red, blue, and green channels. If they are out of balance, whites in the scene are affected, resulting in excessive or diminished brightness.

There are two important things to consider when testing and calibrating equipment. First, because it is a video signal it is necessary to en-

---

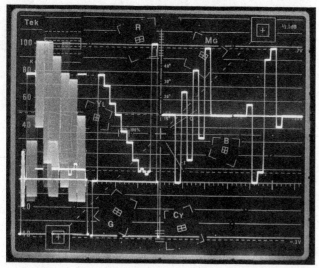

Figure 49 Active waveform analyzer display.

33410-11_F49.EPS

33410-11_F50.EPS

Figure 50 A vectorscope.

sure that each piece of equipment does not generate noise or distortion. The calibrating equipment (vectorscope or waveform monitor) does not affect the signal at all. The recorder, monitor, and camera must be adjusted in order to affect the signal. The idea is to start with the first piece of equipment in the circuit and work to the end of the circuit. All adjustments to the signal have to be made at the generating device.

Calibration is the process of adjusting a measurement to a known standard and is applied at two levels. Some equipment can be calibrated using built-in, internal calibration features. Many pieces of equipment have controls that allow for operator adjustment. Serious calibration requires expert technical and electronics skills. These types of calibrations should only be performed by qualified service technicians in a lab. They should never be performed in the field.

9.2.0 Signal Synchronization

Video signals are produced independently by cameras and recorders in a CCTV system. Switching between two cameras with uncoordinated signals can cause distortions. There may be a momentary tearing and rolling between the displaying of images. The screen may go black or suffer from horizontal roll.

Synchronization is the process of providing a single master signal to every device. Many cameras and recorders have an input specifically for this signal. Each device in the system can then coordinate the phase of their signal to the master signal. This is sometimes called a genlock because a master synch generator is used.

Both the vectorscope and waveform monitor are used in this process. Again, each piece of equipment, starting with the first in line, should be calibrated.

WARNING!

Do not remove equipment covers to perform calibration of equipment in the field. Any calibration beyond the external adjustments (knobs, dials, pots, etc.) provided to the user should only be performed by qualified technical personnel in a properly equipped lab environment. Many electronic devices have significant shock hazards, even when unplugged, and require proper grounding techniques and knowledge.

Did You Know?

Waveform Monitor

A waveform monitor, paired with a color bar generator can be used to determine if a cable is good or bad.

SUMMARY

A CCTV system uses cameras, television monitors, recorders and other equipment to provide surveillance support for a building, campus, or other environment. A system can consist of as little as a single camera and monitor. A large system can consist of hundreds of cameras and a matrix switcher for managing the review of individual video streams at a select number of monitors. Video recorders are used to capture surveillance video for playback, storage, and long-term archiving.

Digital cameras, long in use within analog CCTV installations, can now be connected to an organization's data network, allowing video to be distributed to the desktop and to be viewed anywhere in the world by any authorized person who has access to the Internet. Network video provides some challenges, however, as data compression and encryption support are applied to video streams, in some cases reducing the quality or throughput of the video image. Bandwidth therefore becomes an important consideration in the planning and use of networked CCTV systems.

Video is video, whether analog or digital, and there are important technical aspects that the technician must know in order to support a CCTV installation. Cabling, equipment calibration, test instruments, and a broad understanding of broadcast standards are needed to successfully set up and validate a quality installation.

This module has introduced you to the technology, applications, and terminology of CCTV, and has provided you with insight into the world of CCTV that will help you achieve success. Learning is a lifelong endeavor and you are encouraged to continue the process started here.

Review Questions

1. In a CCTV system, a video switcher _____.
   a. changes voltages on a coaxial cable
   b. is sometimes called a quad
   c. changes the monitor from one video source to another
   d. prevents video from being reduced

2. The CCD in a CCD camera refers to _____.
   a. coupled camera definition
   b. charge-coupled device
   c. coupled charging device
   d. charge-changing device

3. An image captured from a CCD camera is a(n) _____.
   a. analog image
   b. sinusoidal image
   c. wave image
   d. binary image

4. A network camera is connected to a LAN using a(n) _____.
   a. coaxial cable
   b. Ethernet cable
   c. twisted pair cable
   d. UTP cable

5. Which of the following is a correct statement regarding C and CS lens mounts?
   a. A C-mount lens can be adapted for use with a CS-mount housing.
   b. A CS-mount lens can be adapted for use with a C-mount housing.
   c. The two types of mounts are completely interchangeable.
   d. There is no compatibility between the two types.

6. The iris of a camera is used to change the _____.
   a. field of view
   b. white balance
   c. focal length
   d. aperture

7. A PTZ device is used to _____.
   a. increase the amount of light available to the lens
   b. protect the camera from extremes of heat or cold
   c. remotely adjust the position of the camera
   d. conceal the camera partially inside the ceiling

8. The device that allows remote cameras in a CCTV system to be manipulated is a(n) _____.
   a. modulator
   b. controller
   c. receiver
   d. amplifier

9. A communication link needed to control a camera is established with a _____.
   a. transmitter and keyboard
   b. keyboard and controller
   c. controller and receiver
   d. transmitter and receiver

10. In the security field, a real-time recorder captures video at the rate of _____.
    a. 10 frames per second
    b. 15 frames per second
    c. 20 frames per second
    d. 30 frames per second

11. A CRT computer monitor should not be used in a CCTV system due to _____.
    a. high resolution
    b. aspect ratio
    c. flicker
    d. poor resolution

12. Two monitors connected to a video source using a T-connector will result in _____.
    a. signal fading
    b. a good system
    c. a split screen
    d. a video loop-through

13. A video loop-through is used _____.

    a. when recording two cameras at once
    b. when connecting two monitors to the same source
    c. to provide instant replay
    d. to provide continuous playback

14. A distribution amplifier is used to _____.

    a. connect two or more cameras to a monitor
    b. connect two or more monitors to a camera output
    c. equalize the focal length of the camera
    d. expand the field of view of the camera

15. A balun transformer is used to _____.

    a. filter out noise
    b. match impedance between coax and data cable
    c. step down high-amplitude signals
    d. isolate components from noise sources

16. CCTV equipment is designed to run on 24VDC.

    a. True
    b. False

17. What part of a video camera is adjusted to allow more or less light into the camera?

    a. Matrix switcher
    b. Focal length
    c. Zoom
    d. Iris

18. A sodium lamp generating light at a wavelength around 560nm appears _____.

    a. cool white
    b. full spectrum
    c. white-blue
    d. yellow

19. A type of camera that works with both visible and zero-to-no light is a(n) _____.

    a. night lux camera
    b. zero vision camera
    c. infrared camera
    d. green light camera

20. The device needed when multiple users have access to a console is a(n) _____.

    a. multiplexer
    b. encryption program
    c. matrix switcher
    d. ASCII communication capability

Trade Terms Introduced in This Module

Aperture: An adjustable opening in a lens that limits the amount of light entering a camera. It operates much like the iris of the human eye. The size of an aperture is controlled by an iris adjustment on the camera lens and is measured in F-stops. Smaller F-stop numbers indicate a larger opening in the lens.

Automatic gain control (AGC): Automatic gain control is an electronic circuit that uses signal feedback to regulate a certain voltage level, ensuring that it falls within predetermined margins.

Automatic light control (ALC): Electronic circuitry in an automatic-iris lens that causes darker images to be displayed brighter in order to discern detail in the darker areas.

Backlight compensation (BLC): Provides camera circuitry to achieve the optimum balance between the iris and gain settings. By using average light metering across a scene, it can calculate the best exposure and adjust the settings as needed.

Charge-coupled device (CCD): A large scale integrated device made out of silicon. A CCD contains thousands of small light-sensitive diodes arranged in a two-dimensional matrix.

Chrominance: The color information in a video signal, representing the properties of hue and saturation for color video.

Depth of field: The distance in front of and beyond the object of interest that will remain in focus.

Digital signal processor (DSP): A special type of microprocessor. Digital signal processors used in cameras can fine tune or adjust the aspects of an image or a video stream before delivering it to a monitor for viewing.

Field of view (FOV): The width, height, or diameter of a scene to be monitored, determined by the lens focal length, the sensor size, and the lens-to-subject distance. FOV can also be considered the maximum angle of view that can be seen through a lens or optical assembly. FOV is usually described in degrees for a vertical, horizontal, or circular dimension.

Fixed focal length (FFL): A lens with a predetermined focal length. An FFL lens usually has an iris control and focus ring.

Focal length (FL): The distance between the optical center of a lens and the principal convergent focal point.

Footcandle: The US unit of measure for illumination from a uniform light source of 1 square foot, placed 1' away. (See *lux.*) When used in calculations, footcandle is represented as fc. A footcandle is equal to about 10 lux.

F-stop: The F-stop is a ratio of the focal length of a lens to the diameter of its iris. The F-stop determines how much light is allowed into the lens (how long the shutter is open). The lower the F-stop number, the faster the shutter. Also referred to as the F-number.

Gain: In CCD imaging, gain refers to the magnitude of amplification a given system will produce. Gain is reported in terms of electrons/ADU (analog-to-digital unit). A gain of 8 means that the camera system digitizes the CCD signal so that each ADU corresponds to 8 photoelectrons.

Genlock: The process of locking both the synch and burst of one signal to the synch and burst of another signal making the two signals synchronous. This allows the receiver or decoder to reconstruct the picture, including luminance, chrominance, and timing synchronization pulses from the transmitted signal.

Ground loop: The result of multiple groundpaths. In video it causes significant distortion of the video signal.

Illuminance: Relates to the reflectivity of light off an object.

Infrared (IR): Radiation that lies outside the visible spectrum at its red end.

Iris: The part of a camera lens that regulates the amount of light entering the camera.

Lens: The part of a camera that collects the light from a scene and forms an image of the scene on the light-sensitive camera sensor.

Lumen: The total amount of light, at one candela intensity, that falls on a one-square-foot section of the interior surface of a sphere.

Luminance: The brightness of an image.

Lux: The metric unit of measure for illumination from a uniform light source of 1 square meter, placed 1 meter away. Some common measurements of illumination include full sunlight: 100,000 lux; full moonlight: 0.01 lux; clear starlight: 0.001 lux, overcast sky at night: 0.0001 lux. When used in calculations, lux is represented as a lowercase L (l). See *footcandle*.

Motion-JPEG (M-JPEG): A way of capturing and compressing full motion video images by a computer for transmission across a network. MJPEG is a standard for motion video derived by the Joint Photographic Experts Group.

Multiplexer: A device for combining two or more signals into a single, composite signal.

Pan: To swing a camera from left to right, or vice versa, across a scene.

Pan-tilt-zoom (PTZ): The ability to remotely control the movement and positioning of a camera. See *pan*, *tilt*, and *zoom*.

Pixel: A picture element. On a computer display, pixels are individual dots of light. On a CCD chip, they are individual photosites.

Quad: An industry term for a four-way multiplexer.

Redundant array of inexpensive disks (RAID): RAID is both an architecture and applied technology for using multiple computer hard drives to store and retrieve data.

Subcarrier: A sine wave which is imposed on the luminance portion of a video signal and modulated to carry color information. Subcarrier is also used to form burst. The frequency of the subcarrier is 3.58MHz in NTSC and PAL-M and 4.43MHz in PAL.

Tilt: To move a camera up and down across a scene, with the camera position remaining stationary.

Waveform monitor: An instrument used to measure luminance or picture brightness.

Zoom: To move the field of view forward or backward, with the camera position remaining stationary.

Additional Resources

This module presents thorough resources for task training. The following resource material is suggested for further study.

www.video-surveillance-guide.com contains articles and links related to video surveillance.

Figure Credits

Bosch Security Systems, Inc., Module opener, Figures 15 (bottom photo), 16 (bottom photo), and 18

On-net Surveillance Systems, Inc., Figures 7 and 10

Axis Communications, Figure 8

OnSSI, Figure 9

Uniden America Corporation, Figure 11

Courtesy NASA/JPL-Caltech, Figure 12

Pelco, Figures 15 (top photo), 16 (top and middle photo), 17, 19, 20, 22–27 (photos), and 32

Premier Wireless, Inc., Figure 28

Radio Waves, Inc., Figure 29

PelikanCam, Figure 33

Unix CCTV Corporation, Figure 34

Marshall Electronics, Inc., Figure 35

Radio Design Labs, Figure 37

Nitek, Figure 38

Vicon Industries, Inc., Figure 40

Middle Atlantic Productions, Inc., Figures 44 and 45

AutoPatch division of XN Technologies, Inc., Figure 46

Tektronix, Inc., Figures 47–50

NCCER CURRICULA — USER UPDATE

NCCER makes every effort to keep its textbooks up-to-date and free of technical errors. We appreciate your help in this process. If you find an error, a typographical mistake, or an inaccuracy in NCCER's curricula, please fill out this form (or a photocopy), or complete the online form at **www.nccer.org/olf**. Be sure to include the exact module ID number, page number, a detailed description, and your recommended correction. Your input will be brought to the attention of the Authoring Team. Thank you for your assistance.

Instructors – If you have an idea for improving this textbook, or have found that additional materials were necessary to teach this module effectively, please let us know so that we may present your suggestions to the Authoring Team.

NCCER Product Development and Revision
13614 Progress Blvd., Alachua, FL 32615

Email: curriculum@nccer.org
Online: www.nccer.org/olf

❏ Trainee Guide   ❏ AIG   ❏ Exam   ❏ PowerPoints   Other _____

Craft / Level: _____   Copyright Date: _____

Module ID Number / Title: _____

Section Number(s): _____

Description: _____

_____

_____

_____

Recommended Correction: _____

_____

_____

_____

_____

Your Name: _____

Address: _____

_____

Email: _____   Phone: _____

33411-12

Access Control Systems

Module Eleven

V.1 3/12

Objectives

When you have completed this module, you will be able to do the following:

1. Explain the application and operation of access control systems.
2. Identify and explain the uses of the following types of entry equipment:
   - Entry barriers
   - Locking devices, fail safe, and fail secure
   - Entry/exit readers
3. Explain the types of controller topologies.
4. Describe general installation guidelines for entry control system equipment.
5. Install a reader for an entry control system.
6. Install an access control system.

Performance Tasks

Under the supervision of your instructor, you should be able to do the following:

1. Select components for an access control system.
2. Install an access control system.
3. Troubleshoot an access control system.

Trade Terms

Anti-passback
Biometric readers
Coded credentials
Contraband
Controller
Covertly
Hammer-puller

Mantrap
Overtly
Piggybacking
Protective distribution system (PDS)
Reader
Tailgating

Industry Recognized Credentials

If you're training through an NCCER-accredited sponsor you may be eligible for credentials from NCCER's Registry. The ID number for this module is 33411-12. Note that this module may have been used in other NCCER curricula and may apply to other level completions. Contact NCCER's Registry at 888.622.3720 or go to nccer.org for more information.

Contents

Topics to be presented in this module include:

Figures and Tables

Figures and Tables (*continued*)

1.0.0 INTRODUCTION

Access control is a general term used to describe various systems that enable some authority to control entry and access to areas and resources in a facility. Entry and access control systems can range from simple locking devices for doors and gates to complex scanners and readers that are part of a highly sophisticated security system.

The security industry sometimes uses the terms *entry control system* and *access control system* interchangeably, because both entry control equipment and access control equipment and software are usually supplied as a package by manufacturers. But because of the different technical considerations involved, some security experts treat entry control systems and their equipment separately from any associated access control system software, data records, and coded credentials (cards). In this module, entry control systems are defined as the physical equipment used to allow access to a restricted area or facility. This equipment includes entry control barriers, readers, controllers, locking devices, and accessories. Access control systems are defined as the central computers, application software, data records, communications, and coded credentials used to authorize passage through entry control systems.

This module describes typical entry control systems and their associated access control systems, as used in commercial and industrial applications. Included are brief examinations of the relative advantages and disadvantages of different entry control equipment and access control methods. General installation and testing guidelines for entry control systems are also provided.

1.1.0 Important Terms

There are a number of unique terms used in access control work. The following is a list of these terms and their definitions.

- *Door forced open* – A signal indicating that the door was opened without a valid card or request-to-exit (REX) device. This requires a status switch and some type of REX device.
- *Door held open* – A signal indicating that the door was held open longer than it should have been. This is used to prevent someone from using a card to enter and then leaving the door open for others. This requires a status switch on the door.
- *Egress* – A means of exiting.
- *Egress (free)* – A system in which there is no security to prevent someone from leaving a secure area. The system either detects someone approaching an exit and unlocks the door, or has a release button or bar that allows people to leave.
- *Egress (controlled)* – A system that has a controller that requires a valid coded credential to leave the area. By law, access control systems have to be set up to allow people to exit if the system fails or power goes out.
- *Fail-secure* – A locking device that is locked when no power is applied. These are not normally used for exit doors since they cannot be unlocked during a power outage.
- *Non-fail safe* – Same as fail-secure.
- *Personal identification number (PIN)* – A personal identification number assigned to a user. It is used either by itself or in conjunction with a card.
- *Request to exit (REX)* – A device that shunts a door contact and/or releases a locked door.
- *RTE-fail safe* – A locking device that is unlocked when no power is applied.
- *Shunt time* – A period of time after a door is released, during which the status switch is automatically shunted to allow a person to enter/exit. If the time is exceeded, a door-held-open signal occurs.
- *Status switch* – A magnetic contact mounted on the controlled door. It is used to detect a door held or door forced condition.

2.0.0 ENTRY AND ACCESS CONTROL SYSTEMS

Entry control and access control systems are an integral part of a physical protection system. Besides being a visible deterrent, physical protection systems criteria are summarized in the following three functions, with an emphasis on entry control systems:

- *Intrusion detection* – The intrusion detection function includes sensing of attempts to penetrate a facility externally and internally, either overtly or covertly. Entry and access control systems are subsystems of intrusion detection. Entry control systems allow entry by authorized personnel and material, while detecting and preventing covert entry by unauthorized personnel or material. Access control systems validate authorized entry by the use of coded credentials in conjunction with software and/or managed databases containing entry parameters. In some cases, the equipment used for access control, such as CCTV viewing and recording, intercom, alarm reporting, and record keeping of events, is also used for intrusion detection and fire alarm functions. Some entry control systems are totally manual and use guard forces and photo identification badges, stored video image comparisons, or badge exchange methods for access control purposes. These manual methods are not addressed in this module.

- *Delay* – The delay function of a physical protection system slows down the unauthorized, overt penetration of a facility externally and internally by the use of guards, fixed barriers, locked barriers, and activated delays. Entry control systems usually make use of guards and/or some form of physical equipment, including barriers at entry points to permit or prevent passage. If the entry point is not staffed, the entry control equipment should be physically strong enough to allow response forces time to intercept the unauthorized individual(s) or material during or shortly after penetration.
- *Response* – The response function deals with the interruption of an unauthorized penetration by response forces arriving at or near the point of penetration. The response time includes the time to notify the response forces, the time for their deployment, and the time to intercept and interrupt a penetration.

Many modern entry and access control systems are capable of providing functions such as remote authorization and monitoring, record keeping, and text and email notifications. In such a system, input devices can include card readers, scanners, and radio modules that function much like a garage door opener to allow remote access. Local controllers and system computers provide record storage, decision-making functions, and text and email notifications. Output devices, such as locking mechanisms, annunciators, and printers, carry out commands from the controllers and computers.

2.1.0 Non-Staffed Entry Control System

A large variety of entry control systems are available. They range from simple home-type electric lock systems to large, complex systems used for industrial or government facility security. *Figure 1* shows a partial diagram of a typical non-staffed entry control system used in an industrial setting. In this particular example, a controller is shown interfacing via a LAN or WAN to a central computer for updating data records that are stored in the controller and used for entry/exit authorization. In addition, actual entry/exit requests (who, where, and at what time) are transmitted from the controller back to the central computer for access control recording and, depending on the system configuration, anti-passback tracking. Anti-pass-

back tracking is used to deter a credential from being used a number of times within a defined period to admit unauthorized individuals.

Each controller on the network can support multiple types of readers for door, gate, or turnstile entry/exit points. The readers may scan some form of coded credential used by the individual to request authorization to enter/exit the facility or to move from one area to another within the facility. Coded credentials can be cards that are swiped through or inserted into a reader, or passed by a proximity reader within a certain distance of the reader. In high-security applications, biometric readers are normally used to relate an individual to a prerecorded physical characteristic. The most common biometric scans are for hand/finger geometry, fingerprints, eye patterns (retinal or iris), and, more recently, facial recognition. In addition to coded credential or biometric authorization, some facilities may also require that a personal identification number (PIN) be entered by the individual on the keypad of a reader as an added security precaution. In many cases, the number of times the PIN can be entered is limited before it is locked out or an alarm is sounded.

As shown in *Figure 1*, some entry points may also have both an entry and exit reader. These are used to track the movements of individuals into, out of, and within the facility. In other cases, a touch bar, panic button, or break-glass device that initiates an alarm and a short time-delay release can be used on fire exits to release an exterior door lock if approved by the AHJ. The alarm and time delay release allows guard forces to monitor the exit with CCTV or physically respond to the area of the exit. Contact switches mounted at doors or gates may be used to provide a positive signal that the entry point is closed and latched.

Power in the form of 12VDC or 24VDC for operation of the system components may be supplied by the controller and/or separate power supplies. In this example, power for the readers is supplied from the controller; however, because of higher current requirements, each locking device has either a small, separate power supply or a central power supply. The power supplies are normally de-energized in the event of a fire alarm, sprinkler alarm, or loss of power to the building. There is typically hardware on exterior exit doors that enable free egress during an emergency, as well. A relay module, switched by a low-current signal from the reader after authorization is granted, controls the power applied to the locking device. This reduces the amount of power required from the controller and the number of wires to and from the controllers and readers. In other ap-

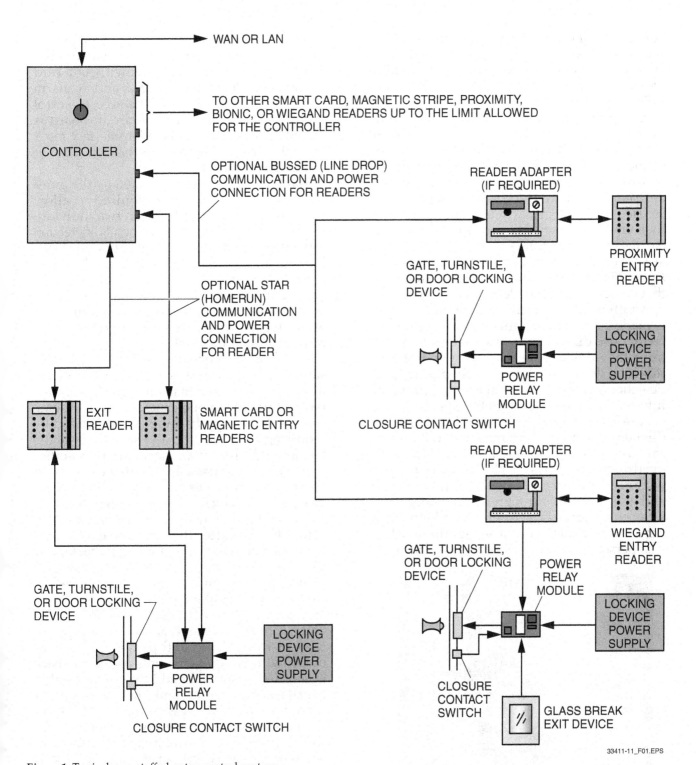

Figure 1 Typical non-staffed entry control system.

plications, manufacturers may use a variety of schemes for wiring the system components.

In addition to personnel verification from a reader, some entry control systems may also use metal, passive or active explosive, or other types of material detectors to provide contraband detection clearance before entry or exit authorizations are granted. Various types of contraband detection systems exist or are under development.

2.2.0 Non-Staffed Entry Control System Considerations

Non-staffed entry control systems have some inherent problems that must be considered when determining the risk and speed of penetration of a secure area:

- *Entry barrier strength* – As mentioned earlier, the strength of the entry barriers determines how easily and quickly an overt penetration can occur. The advantages and disadvantages of various types of doors, gates, and turnstiles are covered later in this module.

- *Tailgating* – Tailgating and piggybacking are terms applied to an unauthorized entry that closely follows an authorized entry through an entry barrier without detection. This can be vehicular traffic or pedestrian traffic, depending on the type of entry barrier. In the case of pedestrian entry barriers, tailgating sensors can be used to signal an alarm to guard forces and then close the barrier or activate a secondary barrier in a device called a mantrap. Mantraps are covered in later sections of this module. *Figure 2* is one type of multiple-beam IR sensor that is used to detect pedestrian tailgating.

Another type of tailgate detection system is made by Kouba Systems. These tailgate detectors use directional sensor arrays to determine the number of entries and the direction in which those entries are traveling. A tailgate detection activates a local alarm and provides information to the access control system.

A tailgate with a door control system uses directional sensor arrays mounted at the vertical faces of the doorframe in conjunction with overhead sensor arrays. This design provides enhanced monitoring, pedestrian counting, and access control through secured doorways. The sensor array can differentiate between a person and a cart. It can also determine when two persons are walking side by side through

Figure 2 Anti-tailgating unit.

33411-11_F02.EPS

a doorway and generate an alarm. One person is allowed to pass though a doorway for each valid card presented. A person without a card would generate a tailgate alarm, and an alarm relay contact would notify the access control system. The tailgate control circuit board is mounted next to the card reader interface module, either in an equipment box near the door or in the equipment room.

Many tailgating units are rated by the minimum separation distance required to allow the unit to distinguish between two individuals. For example, a unit rated for a ¼" detection capability means that the unit can detect two individuals separated by as little as ¼" as they pass through the unit. These units use beam pattern algorithms to distinguish the difference between two individuals or an individual with luggage, purse, briefcase, a wheelchair, cane, or similar material.

In the case of vehicular traffic, vehicle sensors buried in the ground or IR beams can signal an alarm if more than one vehicle at a time passes through an open entry barrier. In some cases, vehicle piggybacking is discouraged by requiring a second stop and scan to close the barrier and prevent an alarm after an authorized vehicle has passed through the barrier. In other cases, two gates are used, separated by a space for the vehicle. To prevent piggybacking, the second gate may require a second authorization to open after the first gate closes, or it may open automatically if no piggybacking is detected. Pedestrian piggybacking through a vehicle entry point can only be monitored by CCTV or a physical guard.

- *Multiple use of a coded credential* – In non-staffed entry control systems that do not employ entry/exit tracking for anti-passback purposes or CCTV monitoring, multiple use or passback of a coded credential and PIN can be used to overtly defeat the system and allow one or more individuals access to the facility. Even if a non-staffed entry control system uses the safeguards listed above, it is still vulnerable to single-person covert penetrations if a coded credential and PIN are forged, or if they are forcibly or voluntarily surrendered to an unauthorized individual. Only biometric readers can prevent a covert penetration by an unauthorized individual using a captured or forged coded credential and PIN.
- *Restricted area control* – In some applications, multiple readers are used on both sides of an entry point to enforce anti-passback procedures that discourage individuals from sneaking into certain areas of a facility. The readers on both

sides must be used in a certain sequence for entry or exit. If not, an alarm is sounded and guard forces are dispatched. In more sophisticated systems, the coded credentials used by individuals are uniquely programmed so that only individuals authorized to be in a certain restricted area are allowed entry.

2.3.0 Access Control Systems

As with entry control systems, there are many types of access control systems available. Typical access control systems include a computer system running an access control software application that includes data recording, event logging/tracking functions, and text and email notification capabilities.

Coded credentials that are compatible with the data requirements of the access control software are also part of the access control system. With some software applications, the access control system can also perform the coding or programming of certain types of credentials. *Figure 3* is a diagram of a typical scalable industrial access control system that can be used in large or small facilities. It can also be used to integrate third party digital surveillance video systems, fire alarm systems, and other intrusion systems. It can display and record the activities of these third party systems. It should be noted, however, that if an access control system is integrated with a fire alarm system, it needs to be UL Listed for life safety. Otherwise, the system needs to tie through a fire alarm relay.

In the application shown in the figure, separate locations (regions) with their own monitoring stations, servers, and entry system controllers connected by a LAN Ethernet are tied together via a WAN to remotely located central monitoring and card administration functions. All credentials are entered and updated for use at all controller locations at the central card administration point. As shown in the figure, one or more locations may have video surveillance camera interfaces as well as a badging client station. The badging client sta-

tion (*Figure 4*), if supplied, can be used to create facility photo identification (ID) badges with or without certain types of credential coding. It can also be used to enter photo IDs for tracking and monitoring purposes in the access control system database.

Access control systems are primarily used to provide authorization data to the entry system controllers, so that only individuals with the proper clearance are admitted to the installation locations. As explained earlier, it is very desirable that the access control system perform detailed event logging (tracking) of all entry and exit activities, including who, where, and when (*Figure 5*). This is so that guard forces can monitor any unusual activity. In some systems, multiple entry and/or exit attempts by the same individual within a defined period are automatically flagged and alarmed for monitoring personnel. Systems may also use door contacts to monitor doors that are left ajar. If a door-closed contact is not activated within a defined period after the door is opened, an alarm is sounded.

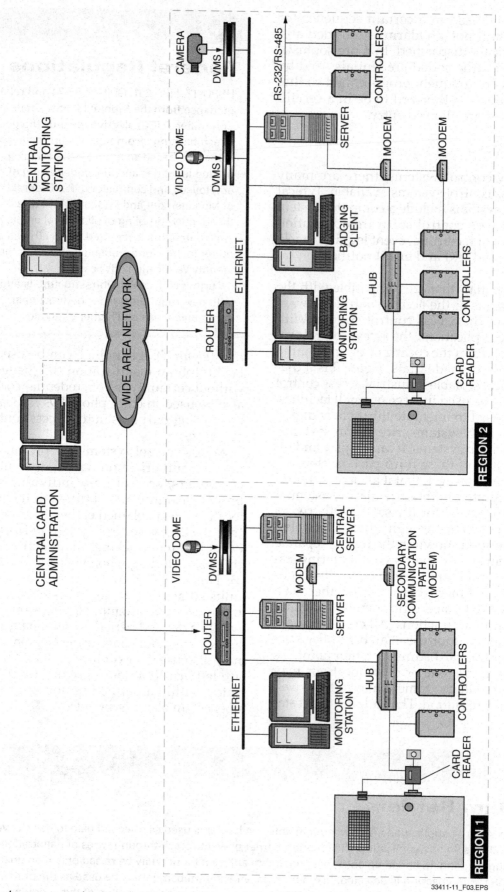

Figure 3 Typical access control system.

33411-11_F03.EPS

NCCER – *Electronic Systems Technician Level Four* 33411-12

Figure 4 Badging client station.

33411-11_F04.EPS

2.4.0 Coded Credentials

A coded credential refers to a device or card that contains coded information about a facility and an individual. The coded information is read from the device or card by a reader and transmitted to a controller where it is used to authorize passage of the individual into a secure facility or area.

A wide range of coded credentials is available commercially. They are sometimes called key cards. The most common cards used for access control systems include magnetic stripe, Wiegand wire, and smart cards. Also available are proximity cards or devices, as well as mixed-technology cards that use two or more types of coding on the same card. In most access control applications, these technologies are incorporated into photo ID badges.

The protocol for the data that is encoded onto these cards is typically 26-bit, 32-bit, or 64-bit. The 26-bit Wiegand data format is considered to be the standard worldwide.

2.4.1 Magnetic Stripe Cards

Magnetic stripe cards (*Figure 6*) are the most common cards in general use. Most credit cards are this type. The cards have a black stripe of thin magnetic oxide on the back of the card. This magnetically sensitive oxide is subject to erasure by stray magnetic fields and can be damaged due to rubbing during normal handling and swiping. With frequent use, the cards tend to wear out within a few years. These cards can be rather easily duplicated unless a proprietary coding system is used during their preparation. Unfortunately,

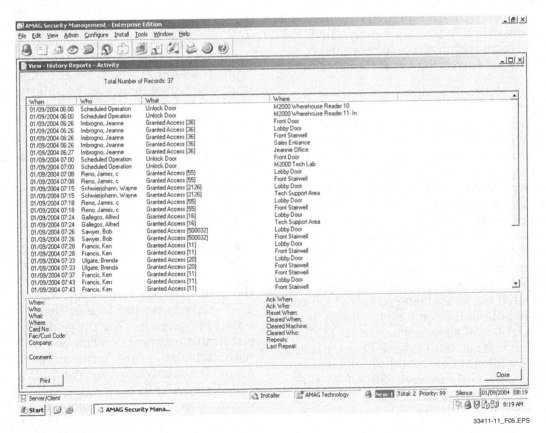

33411-11_F05.EPS

Figure 5 Typical event log.

Figure 6 Typical magnetic stripe card.

33411-11_F06.EPS

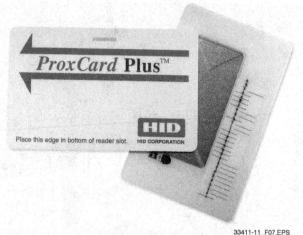

33411-11_F07.EPS

Figure 7 Wiegand wire card.

proprietary coding may limit the use of the cards with other subsystems or equipment.

The versions used for security purposes are usually combined with a photo ID badge and used with a PIN. This combination provides a reasonable level of security in general applications. In some cases, the cards may be mixed-technology cards that contain Wiegand wire, optical, proximity, or smart card technology as well as a magnetic stripe so they can be used in more than one type of reader. On security cards, the magnetic stripe is normally made of a magnetic oxide with a high degree of resistance to erasure. This is referred to as a high coercive force rating. In new installations, it is important to start a rotating scheduled card replacement within a year to avoid the problem of all the cards wearing out at the same time.

2.4.2 Wiegand Wire Cards

Wiegand wire cards have two rows of embedded magnetic wires (*Figure 7*) made of a proprietary metal alloy that has special magnetic properties. The two data wires are called DATA 0 and DATA 1. The pattern of the wires in the rows forms the code. The DATA 0 line carries the 0 bits of binary code, while the DATA 1 line carries the 1 bits.

When the card is exposed to a fluctuating magnetic interrogation field within a Wiegand reader, each wire in the two rows produces a very precise electrical pulse that is read by two reading heads in the reader. If no data is being sent, DATA 0 and DATA 1 are both at the same high voltage. If a 0 is sent, DATA 0 is at a low voltage while DATA 1 is at the high voltage. If a 1 is sent, DATA 1 is at a low voltage while DATA 0 is at the high voltage.

Wiegand wire cards are very difficult to duplicate and are considered very durable and secure. Unfortunately, the cards can only be produced at the factory and delivery times may be a consideration when using this technology.

2.4.3 Proximity Cards and Devices

Proximity cards and devices (*Figure 8*) can be read by readers that are out of sight just by passing the card or device past the area of the reader. The cards and devices are available as either passive or active types. Passive cards are powered by an RF interrogation field generated by the reader. Once powered, they transmit their coded information with an RF signal at a different frequency. *Figure 9* shows the embedded pickup and transmit coil with an RF chip in a passive proximity card. Active cards are self-powered by a battery. Some transmit all the time and others transmit only when they detect an RF interrogation field from a reader. The cards are also classified by their operating frequency range and read/write capability. Some types are read-only cards that have a code assigned by the manufacturer. Others have a read/write capability that allows reprogramming by the security system manager.

2.4.4 Smart Cards

Smart card technology is relatively new in the United States. Conventional smart cards (*Figure 10*) have an integrated circuit embedded within the card that is activated through five to eight gold contacts on the face of the card when inserted into a reader. The integrated circuit can be just simple memory or it can be a microprocessor with nonvolatile read/write memory and a program. The microprocessor can perform a proper coded response to a changing coded inquiry using complex encryption functions when the card is inserted in a reader. This makes microprocessor-based cards extremely difficult to counterfeit.

STANDARD LIGHTWEIGHT
PROXIMITY CARD

PROXIMITY-KEY
RING TAG

33411-11_F08.EPS

Figure 8 Proximity card and key ring tag.

33411-11_F09.EPS

Figure 9 Passive proximity card.

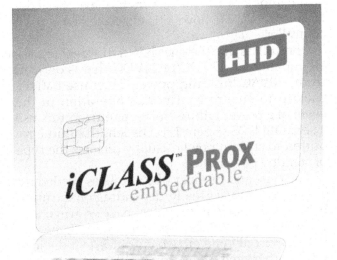

33411-11_F10.EPS

Figure 10 Example of a smart card.

Long-Range Proximity Devices

Larger programmable proximity devices with a stronger battery-powered transmitter are fastened to vehicle windshields to authorize the automatic opening of powered security gates. These devices are activated by high-powered readers located near the gate. They are similar to the type used by various toll road authorities.

33411-11_SA01.EPS

This card is also capable of storing a major part of the bionic template information that is used by a bionic scanner to authorize a cardholder. This eliminates the delay and communications problems caused by large downloads from the entry system controllers or from a remote central computer.

One example of a smart card is a Common Access Card (CAC). A CAC is a US Department of Defense smart card issued to military personnel and civilian employees. The card serves as an identification card and can be used for authentication to access Department of Defense facilities and equipment.

Recent advances in smart card technology have led to powerful contactless cards. Contactless cards differ from contact cards in how data is transferred to the reader. With a contact card, the contacts on the face of the card must touch corresponding contacts in the reader to enable data transfer. With a contactless card, the data transfer occurs through a 13.56MHz radio frequency. The card simply needs to be close to the reader.

One example of contactless smart cards is HID® iCLASS® cards. These cards are fast, versatile, and very secure, which makes them suitable for access control applications. Often, iCLASS® cards include proprietary chips called MIFARE®. Combining iCLASS® and MIFARE® technologies makes the cards multifunctional. The versatility of iCLASS®/MIFARE® cards allows users to manage multiple applications with a single card.

Smart cards cost much more than other credentials. In high-security systems, this must be weighed against their data storage capability and resistance to counterfeiting. Smart card technology is also being incorporated in battery-powered proximity devices for increased security using these devices. Numerous communication methods exist for proximity applications, but the *ISO/IEC 14443* international standard applies to contactless integrated circuit cards and proximity cards used for identification.

2.5.0 Protective Distribution Systems

If unencrypted classified information is transmitted across a wireline or fiber optic telecommunication system, a protective distribution system (PDS) is required. A PDS is designed to deter and detect attempts to access the system in order to protect national security information. The safeguards that are used to deter access must include electrical, electromagnetic, and physical defenses. Guidelines for the design, installation, and maintenance of protective distribution systems are provided by the *National Security Telecommunications and Information Systems Security Instruction (NSTISSI) No. 7003*.

Two basic categories of protective distribution systems are identified in the *NSTISSI No. 7003*: hardened distribution systems and simple distribution systems. Hardened distribution systems require significant physical security protection. Simple distribution systems are less vulnerable and require less protection than hardened systems.

Hardened distribution systems are further divided into three subcategories: hardened carrier PDS, alarmed carrier PDS, and continuously viewed carrier PDS.

In a hardened carrier PDS:

- Data cables must be installed in a carrier made of electrical metallic tubing, ferrous conduit or pipe, or rigid sheet steel ducting that uses elbows, couplings, nipples, and connectors that are made of the same material.
- All connectors must be permanently sealed using an appropriate method such as welding, compression, epoxying, fusion, and so on. If pull boxes are used, the covers should be sealed to the boxes to prevent access to the interior.

- If a hardened carrier is buried, the carrier must be encased in concrete for protection.
- In an alarmed carrier PDS, the carrier uses electronic monitoring to detect intrusions on the carrier. The carrier is alarmed so that any acoustic vibrations that suggest an intrusion can be detected.
- In a continuously viewed carrier PDS, the carrier should be under 24-hour continuous observation. Any intrusion attempt on the carrier should be investigated by the appropriate personnel within 15 minutes of its detection.

Simple distribution systems use less physical security protection than hardened systems. They use what is called a Simple Carrier System (SCS) and require the following criteria:

- Data cables should be installed in a carrier.
- The carrier can be constructed of any material, such as wood, PVC, or electrical metallic tubing (EMT).
- All joints and access points should be secured.
- The carrier must be inspected according to the requirements specified in *NSTISSI No. 7003*.

3.0.0 CONTROLLERS AND POWER SUPPLIES

In many ways, controllers (or control panels) are the brains of modern access control systems. These microprocessor-based devices use information from cards or other components to make the decisions necessary to manage access to secure areas. Some systems use separate control panels to manage groups of access control devices. Other systems have intelligent readers that make the control decisions locally. And some systems use host computers and/or servers to make these decisions. *Figure 11* shows one example of a typical controller.

Controllers like the one shown in *Figure 11* usually operate on 12VDC or 24VDC that is derived from 110VAC building power. They use battery backup to ensure continuous operation in the event of a power failure. Some manufacturers use a separate UPS (*Figure 12*). The length of time required to sustain power usually dictates the type of standby power used.

The type and number of controllers needed for a system can vary greatly, depending on the number of doors, elevators, gates, barrier arms, and other entrance/exit devices being used. Some systems have only one or two entrance/exit devices, while others may have hundreds of devices. In modern control systems, the number of entrance/exit devices being controlled can be virtually un-

Figure 11 Typical complex controller.

Figure 12 Uninterruptible standby power supply.

limited. Some of the controlled devices may be at remote locations ranging from several miles away to several states away, and occasionally even to other continents. Some systems cannot be expanded beyond one controller. Others are virtually unlimited.

The architecture, or topology, of a control system is based on the input and output devices, as well as the type of control that is needed. There are numerous access control system topologies in use. One common topology is a hub and spoke, or ring, topology (*Figure 13*). In this configuration, a control panel functions as the hub for the system, and individual readers function as the spokes. The control panel performs the control function for the system and is connected to a host computer that functions as a database for record keeping. The readers communicate with the control panel via data links, in this case RS-485

connections. Other data links include Ethernet and RS-482. Manufacturers incorporate different microprocessors, expandable memory chips, embedded operating systems, and secure communication methods.

Control system topologies can be very complex. One of the primary factors that affects control system topology is the sophistication of the reader or readers being used. In some system layouts like the one in *Figure 14*, groups of readers are connected to sub-controllers which, in turn, are connected to a main controller. The main controller is connected to a central host computer.

The trend in access control systems is to place the decision-making capability at each individual access point. This is accomplished by using intelligent readers and IP readers. Intelligent readers are capable of making all the control decisions necessary to control the access point. IP readers are a type of intelligent reader that can communicate directly with a host computer over a network or the Internet without the need for separate control panels. *Figure 15* shows the layout of a system that uses a host computer and several IP readers. In some applications, the number of readers, inputs, and outputs can be expanded using different module interfaces. *Figure 16* shows an example of an application in which control modules support multiple inputs and outputs.

4.0.0 ENTRY/EXIT READERS

A great number of different readers exist. They range from plain keypads to various card readers to biometric readers. Readers are classified according to their capabilities. Four basic categories of readers are:

- *Basic (non-intelligent) readers* – Basic readers simply read the information encoded in a credential and forward that information to a control panel. This type of reader is relatively inexpensive and has widespread use.
- *Semi-intelligent readers* – Semi-intelligent readers can read the information encoded in a credential and use the decision-making output from a control panel to activate whatever locking mechanism is in place at the access point. For instance, a semi-intelligent reader can lock and unlock a door, but it does not have the ability to make decisions about access. The decision-making is left up to the control panel.
- *Intelligent readers* – Intelligent readers are able to read the information from a credential, make decisions about access, and activate the applicable locking mechanism at the access point. These readers essentially provide stand-alone

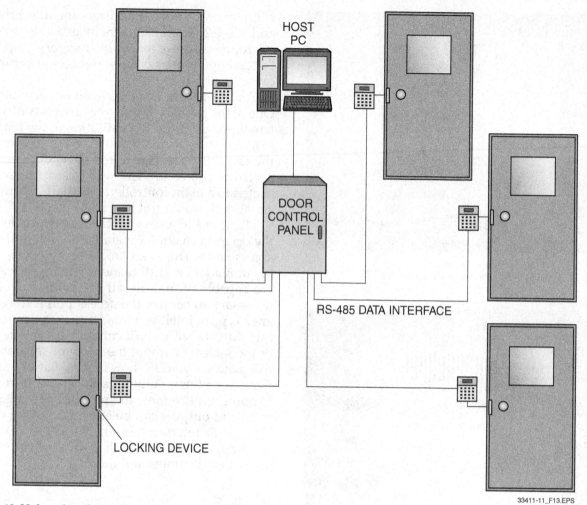

HOST
PC

DOOR
CONTROL
PANEL

RS-485 DATA INTERFACE

LOCKING DEVICE

33411-11_F13.EPS

Figure 13 Hub and spoke (star) topology.

access control, although intelligent readers are often connected to a controller or computer for reporting and programming.

- *IP readers* – IP readers are a type of intelligent reader that is capable of communicating directly with a host computer without the need for a separate control panel. These readers communicate over a private network or the Internet. Like other intelligent readers, IP readers provide stand-alone access control at the access point. The host computer provides the reporting and programming functions.

Some common types readers are described in the following sections.

4.1.0 Swipe, Insert, and Proximity Readers

A few typical swipe, insert, and proximity readers are shown in *Figure 17*. These devices accommodate the magnetic stripe, Wiegand wire, and proximity credential technologies described earlier. Some readers accommodate mixed-technol-

ogy credentials and some have keypads so that a PIN can be entered as an extra security precaution. Some readers use 20mA current to extend the range of the reader.

- *Swipe readers* – Swipe readers require the user to swipe a card through a slot on the reader. The reader reads the magnetic stripe on the card and either forwards that information to a controller that makes the access decision about opening or locking the access point or, in the case of a stand-alone reader, makes the access decision locally.
- *Insert readers* – Insert readers are similar to swipe readers. The main difference is the user inserts the card into the reader versus swiping it through a reader slot. The reader reads the magnetic stripe on the card and either forwards that information to a controller or makes the access decision locally to open or lock the access point.
- *Proximity readers* – Proximity readers are able to read the information on a card without the card being physically inserted or swiped in the reader. Instead, a proximity card is waved or

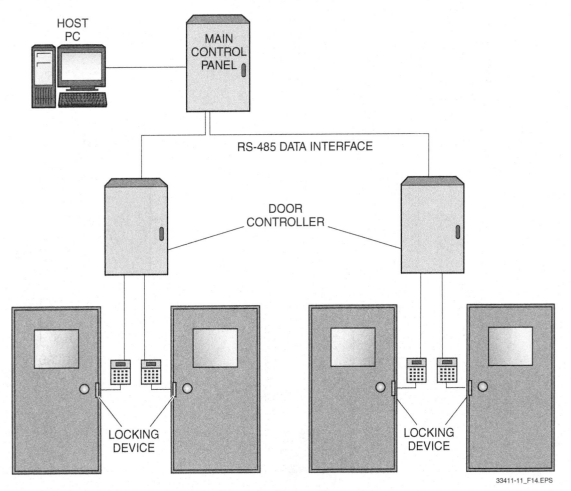

Figure 14 System with host computer, main controller, and sub-controllers.

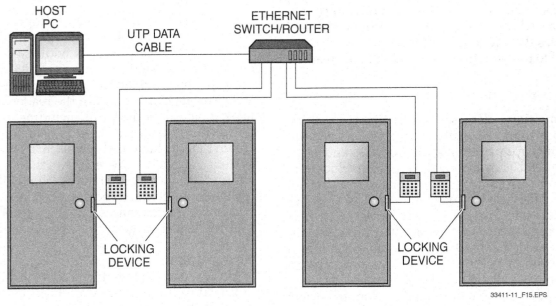

Figure 15 System with host computer and IP readers.

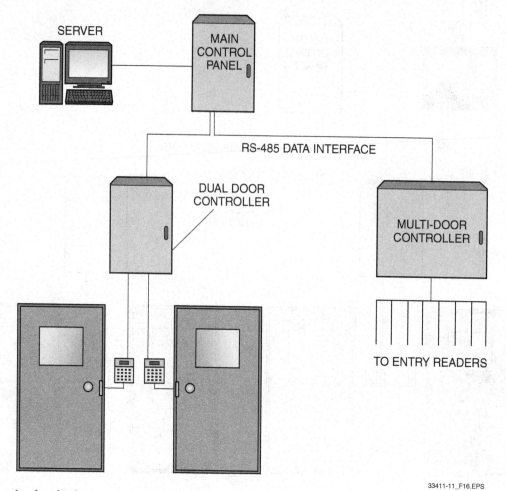

SERVER

MAIN CONTROL PANEL

RS-485 DATA INTERFACE

DUAL DOOR CONTROLLER

MULTI-DOOR CONTROLLER

TO ENTRY READERS

33411-11_F16.EPS

Figure 16 Example of multi-door control module application.

passed near the reader. A proximity reader generates an electrical field. When a proximity card is passed through the electrical field, the field excites the circuit in the card which enables the card to transmit the card number to the reader.

4.2.0 Biometric Readers

A biometric reader uses one of five recognition methods: hand geometry, finger geometry, fingerprints, eye patterns (retinal or iris), and facial recognition. They are used for personnel identity verification. To reduce data-record search time for the large files used to verify an individual's identity, a PIN is usually used to select a specific file. In other cases, a smart card that has a stored encrypted file may be used. If desired, and communication time is fast enough, all of these units can operate in a recognition mode without a PIN or smart card data storage. In recognition mode, the data from the reader or scanner is compared against all stored files at a central location to identify an individual.

Biometric readers provide a higher level of security than some systems because they eliminate the need for cards, which can be lost or stolen.

4.2.1 Hand/Finger Geometry Readers

The technology used in hand/finger geometry readers is based on the three-dimensional shape of the hand. Such things as widths and lengths of fingers and hand thickness are unique to individuals. These are the features that are measured. A solid-state camera within the reader (*Figure 18*) takes a picture of the hand, including a side view for thickness, using infrared light and a reflective platen that the hand rests on. From this image, the system creates a feature vector that is matched against a stored template by the system controller. If the feature vector and template match within certain tolerances, passage authorization is granted. For passage authorization, the template file is usually selected via a PIN or stored on a smart card.

Hand/finger geometry readers typically are not used in high-security applications. They are more likely to be used by businesses and schools to measure attendance or keep time-tracking information.

WIEGAND
CARD READERS

SMART CARD
READER

MAGNETIC STRIPE
READERS

PROXIMITY READER

MAGNETIC STRIPE INSERT READER

33411-11_F17.EPS

Figure 17 Swipe, insert, and proximity readers.

33411-11_F18.EPS

Figure 18 Hand/finger geometry reader.

4.2.2 Fingerprint Readers

Fingerprint readers are appropriate for many security applications because everyone has a unique fingerprint. By obtaining an image of a candidate's fingerprint and then comparing the ridges and valleys of that image to pre-scanned images on file, a fingerprint reader is able to grant or withhold access to a secure area.

Fingerprint readers (*Figure 19*) use one of a number of imaging techniques to obtain a fingerprint template, including the following:

- *Optical systems* – This technique uses a solid-state camera and prism to capture a digital image of the fingerprint using visible light. These systems are similar to a digital camera.
- *Ultrasonic sensors* – This technique involves the use of ultrasonic sensors that generate very high frequency sound waves which reflect off the finger and are detected by the sensor. The reflected waves are measured and used to create a visual image of the fingerprint.

- *Capacitive sensors* – These sensors use the difference in capacitance between the ridges and valleys of a fingerprint to create an image of the fingerprint.
- *Thermal sensors* – These sensors measure the difference in temperature between the ridges and valleys in a fingerprint and convert those measurements into an electrical signal that represents the print.

All of these methods have been commercially developed. Like hand/geometry readers, data files for comparison are selected by use of a PIN or stored on a smart card.

4.2.3 Retinal Scanners

A retinal scanner scans the pattern of blood vessels on the retina at the back of the eye. The user looks through a viewer with one eye and stares at a target. The scan is done in a circular path around the center of vision using a very low-intensity non-laser light from infrared light-emitting diodes. The intensity of the reflected light determines the location of the retinal blood vessels. The initial reference file is made using a number of rapid scans that are algorithmically combined. For passage authorizations, only one scan is made to compare against the reference file selected by a PIN or stored on a smart card.

4.2.4 Iris Scanners

Instead of scanning the retina, the device shown in *Figure 20* scans the iris structure of an eye. The colored portion of the eye that admits light to the eye is the iris. A video camera in the scanner images the structure of the iris for comparison against a stored image. An advantage of this system is that the eye can be imaged at a distance

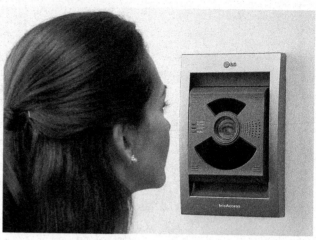

33411-11_F20.EPS

Figure 20 Iris scanner.

of about 10 inches using normal external light. No facial contact with the scanner is required. Older models had some false readings caused by glare from glasses, but newer models have compensated for this by using glare detection and reduction features. Blind individuals, or those who have iris damage, extremely dilated eyes, or very dark irises normally cannot be identified using iris scanners. For passage authorizations, a reference file is usually selected by a PIN or stored on a smart card.

4.2.5 Facial Scanners

Facial scanners are relatively new and, in some applications, controversial. They capture facial images using a high-resolution video camera or an infrared imager. Certain distinguishing features, such as bone structure and distances between the eyes, nose, and mouth, are captured and compared by a very sophisticated program against a stored reference data file. If a match is made within certain tolerances, passage authorization is granted.

The advantages of a facial scanning system are that the imaging is done on the fly. No special effort by the individual requesting authorization is required. However, for rapid authorization, a reference data file is usually selected by entry of a PIN or from data stored on a smart card.

In certain security applications, such as airport access, where thousands of candidates are matched against an enormous database, problems with accurate recognition have been reported. But for typical commercial and industrial applications, facial scanners have been proven to be accurate. In fact, these systems are approaching the point that accurate facial recognition can be accomplished even if the individual is disguised.

33411-11_F19.EPS

Figure 19 Fingerprint reader.

5.0.0 LOCKING DEVICES AND ACCESSORIES

The following sections cover various locking devices and accessories used for entry doors and gates. The electrically operated devices covered are all 12V AC or DC or 24V AC or DC units. Whenever installing electric locking devices, always check with the local authority that has jurisdiction over how to properly handle emergency exit situations.

5.1.0 Electric Strikes

Figure 21 shows electric strikes, along with a local power module. Electric strikes are very common and provide a remote release of a locked door without requiring the retraction of a latch bolt. They are available as fail-safe or fail-secure. Fail-secure means that on loss of power, the strike remains locked. Fail-safe means that on loss of power, the strike opens. Always check the local jurisdiction before installing electric strikes. With the possible exception of prisons or mental institutions, most local jurisdictions require a locking device on an exterior exit door to fail-safe upon a fire alarm, sprinkler alarm, or loss of AC power. For interior fire doors or stairway doors, most codes prohibit the use of fail-safe strikes to prevent the spread of fire and smoke. In these applications, electric locks or latches must be used to allow the doors to remain latched when they are electrically unlocked.

The devices are available with internal switches that monitor the latch bolt position and the strike lip. On most models, the strike lip, sometimes called a keeper or gate, closes shortly after the door is opened. When the door is closed, the beveled spring-loaded latch bolt rides over the lip and falls into the pocket. Most of these devices can be mounted on either side of the doorframe for a right-hand or left-hand opening door. With these devices, a portion of the outer edge of the doorframe must be removed to accommodate the back box that allows the strike lip to swing open. This can weaken the doorframe, allowing easier penetration.

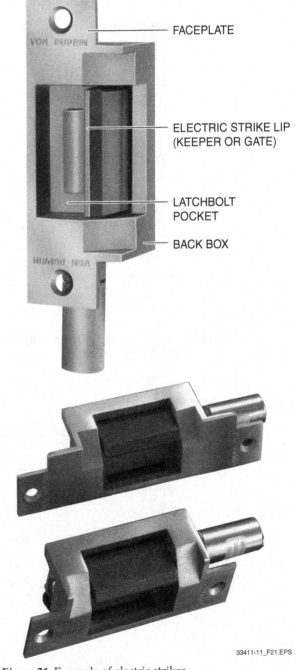

Figure 21 Example of electric strikes.

5.2.0 Electric Bolt Locks

Electric bolt locks (*Figure 22*) are an alternative to electric strikes or magnetic locks because the bolting device that locks a door or gate is mounted on the top and/or sides of the doorframe. The door itself has no latch. Multiple electric bolt locks can be used on a door to provide very strong security. The locks are available as fail-secure or fail-safe units. In most cases, neither can be used on exterior exit doors because of code restrictions. Some

of these devices are designed to fit in narrow doorframes and do not require removal of a portion of the doorframe edge. Others are designed for surface mount or for use as sliding or swinging gate locks.

5.3.0 Electric Locksets (Latches)

The primary use for electric locksets (*Figure 23*), also called electric latches, is in stairway fire doors on each floor of a building. Building codes

EXTRA HEAVY-DUTY COMMERCIAL/INDUSTRIAL
GRADE ELECTRIC BOLT LOCK

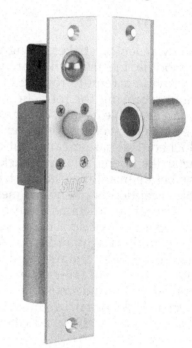

SPACESAVER CONCEALED
NARROW ELECTRIC BOLT LOCK

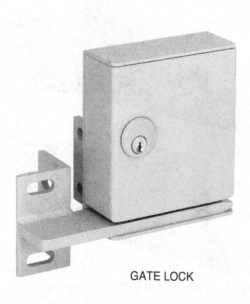

GATE LOCK

CONCEALED DIRECT THROW DESIGN
ELECTRIC BOLT LOCK

SURFACE MOUNT
DOOR LOCK

33411-11_F22.EPS

Figure 22 Electric bolt locks.

generally require that stairway fire doors not be locked on the stair side unless they may be remotely unlocked without unlatching. Electric locksets provide the required locking, unlocking, and latching features. While providing controlled access and remote unlocking capability, the doors stay latched even when unlocked to maintain fire door integrity. Because these locksets are mounted in the door, power transfer hinges like those shown in *Figure 23* are required. The hinges are available with two-, four-, and ten-wire conductors that are protected on the inner face of the hinge.

5.4.0 Electromagnetic Locks

Electromagnetic locks (*Figure 24*), which are often referred to by a specific manufacturer's name (Maglock®), are fail-safe and can be used on interior doors and exterior exit doors. They have no moving parts and are not subject to wear. Electromagnetic locks are available as directhold and shearhold (concealed) styles. The directhold styles are graded for use by ANSI as follows:

- *Grade 1* – 1,650 pounds direct holding force, medium security
- *Grade 2* – 1,200 pounds direct holding force, light security
- *Grade 3* – 650 pounds direct holding force, door holding only

> **NOTE**
>
> When installing magnetic locks, check national, state, and local codes related to interfaces with life safety systems.

There are some electromagnetic locks with 2,000 pounds or more of direct holding force. These locks stay joined even when the door they secure is destroyed. Shear types have holding forces of 2,700 pounds, but they are rated as only grade 1 because of the 90-degree pulling angle. Most electromagnetic locks have integral door position switches to indicate that the door is locked and secure. The shear locks have relock delay timers activated by the position switch so that the door is at rest before the lock reactivates.

5.5.0 Exit Devices

Some type of request to exit (REX) device is generally used to control and monitor egress from a secure area. This section describes the most common methods of egress control.

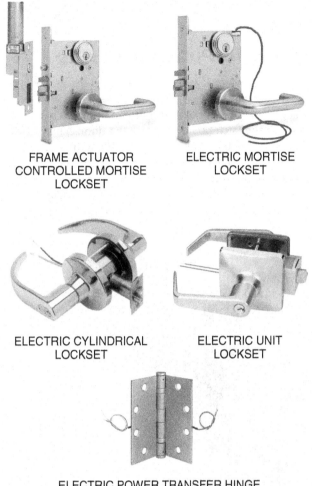

FRAME ACTUATOR CONTROLLED MORTISE LOCKSET

ELECTRIC MORTISE LOCKSET

ELECTRIC CYLINDRICAL LOCKSET

ELECTRIC UNIT LOCKSET

ELECTRIC POWER TRANSFER HINGE

33411-11_F23.EPS

Figure 23 Examples of electric locksets.

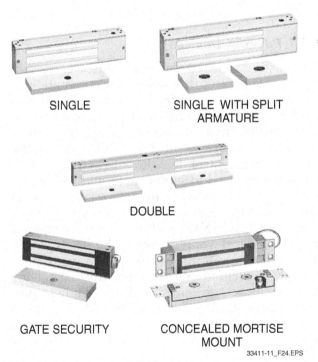

SINGLE

SINGLE WITH SPLIT ARMATURE

DOUBLE

GATE SECURITY

CONCEALED MORTISE MOUNT

33411-11_F24.EPS

Figure 24 Electromagnetic locks.

Electric Plunger Strike

This type of electric strike is activated in the same manner as electric lip strikes. However, instead of a lip, the device uses a motor-driven plunger to depress the latch bolt and unlock the door. When de-energized, the plunger retracts. When the door closes, the beveled latch bolt rides over the strike plate and falls into the latch pocket. The advantage of this device is that it does not require removing part of the outer frame edge to accommodate the strike.

33411-11_SA02.EPS

5.5.1 PIR-REX Devices

The passive infrared (PIR) exit sensor is used to detect motion and is designed for automatic unlocking of doors for uninhibited exit. PIR devices emit a pattern of IR radiation that is roughly the width of the door. Anyone entering the pattern disrupts it and causes a signal to be sent to the door release control. REX PIRs (when activated) are typically used to supply an input to an access control panel to either shunt a door contact or release a locking device. They typically have form C contacts (N.O./N.C.), and a timing circuit that is adjustable for the application. REX PIRs typically are used in high-traffic exit areas, or areas of public egress such as multi-tenant buildings. They have been used on standard doors simply to provide a door shunt for egress. These devices need to be used selectively, as anyone approaching an exit door can activate it. They are not generally used on double glass exterior doors, as someone can slip an item through the crack between the doors and activate them. A stairwell door is a good example of an application for shunting a door contact.

5.5.2 Exit Switches

Exit switches (*Figure 25*), sometimes called RTE or REX switches, are available in a variety of mechanical and electrical configurations. The most common applications are as bypass, momentary or delayed unlock, and alarm shunt switches. They are connected to the RTE or REX inputs of an access control system or to the electric lock directly. The mini-timer shown is used in the delayed unlock applications and fits into two-gang exit switch assemblies, inside narrow doorframes or in a remote junction box.

5.5.3 Delayed Exit Alert Locks

Delayed exit alert locks (*Figure 26*), sometimes called RTE or REX locks, legally delay an exit through exterior exit doors. When an exit is attempted, an alarm is sounded and a signal is sent to guards for CCTV or physical monitoring purposes. After 15 seconds, the exit door is unlocked, permitting an exit. A signal from a fire alarm system can also release the lock, allowing unrestricted exits during an emergency.

5.5.4 Exit Door Accessories

In addition to accessories covered in previous sections, other door accessories are available, as shown in *Figure 27*.

- *Break-glass exit device* – This device should be used on all emergency exit doors equipped with a fail-safe locking system. Breaking the glass on the device immediately unlocks the door to allow emergency egress. It also sounds an alarm. The alarm and the glass discourage nuisance alarms, but provide a redundant means for emergency door release if required.
- *Emergency exit button* – This is a manual device designed to cut power. It will cause a door-forced alarm.
- *Concealed mortise-mount power transfer loop* – This device can be used to conceal wiring between a hollow door and hollow doorframe to prevent tampering. It provides a way to run electric wiring from the frame to doors that have electric locking mechanisms and exit devices.

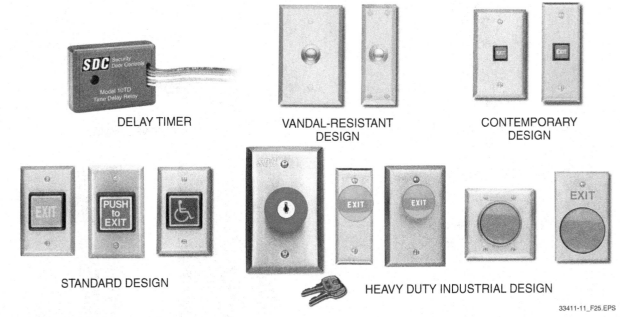

Figure 25 Examples of exit switches.

DELAY TIMER

VANDAL-RESISTANT
DESIGN

CONTEMPORARY
DESIGN

STANDARD DESIGN

HEAVY DUTY INDUSTRIAL DESIGN

33411-11_F25.EPS

Figure 26 Delayed exit alert locks.

GRN – SECURE
YEL – ALARM
RED – UNLOCKED
OFF – BYPASS

AUDIBLE
SHUNT

33411-11_F26.EPS

• *Door monitoring contacts* – These contacts are used to monitor whether a door is closed or left ajar. If the door is intended to stay closed, the monitoring contacts will detect if the door is ajar and sound an alarm after a predetermined period of time.

5.5.5 Touch Bars and Handles

In place of exit switches or readers, touch sense bars and handles (*Figure 28*) and switch bars are commonly used to turn off electromagnetic locks. The touch sense bars and handles are capacitive touch-sensitive switches and have no moving parts. The switch bars have mechanical switches. Some touch sense or switch bars have electronic timers that delay de-energizing an emergency exit for a set amount of time. Armored cables are used to connect the bars or handles to the hinge side of the doorframe, where the wiring is routed to a controller or to the electromagnetic lock. These touch bars/handles are part of a group of exit devices that are sometimes called request-to-exit (RTE or REX) devices.

5.6.0 Cable Supervision

Some access control systems allow cables to door contact switches, exit switches, and other contact devices to be monitored for short and open circuits by adding terminating resistors to the end of the cable, as near to the device as possible. The

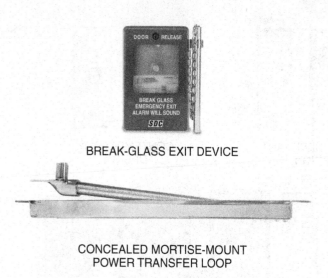

BREAK-GLASS EXIT DEVICE

CONCEALED MORTISE-MOUNT
POWER TRANSFER LOOP

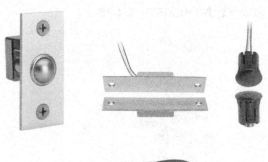

DOOR MONITORING CONTACTS

33411-11_F27.EPS

Figure 27 Exit door accessories.

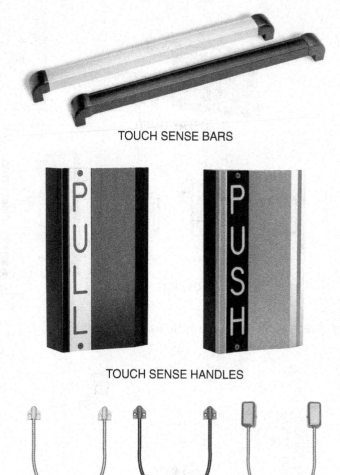

TOUCH SENSE BARS

TOUCH SENSE HANDLES

ARMORED CABLES

33411-11_F28.EPS

Figure 28 Typical touch sense bars, handles, and armored cables.

resistor configuration determines whether three-state or four-state supervision is in use. For four-state supervision, both short circuits and open circuits on the cable can be detected. If three-state supervision is used and the contact is normally open, only an open circuit on the cable can be detected. If the contact is normally closed, only a short circuit can be detected. The term *two-state supervision* means that the cable is not monitored for faults.

Figure 29 shows how to set up two-, three-, and four-state supervision. The level of supervision can be different for each contact, and must be configured using the access control software.

5.7.0 Door Status Devices

There are several types of devices that can be used to provide a door status indication—that is, whether a door is open or closed. Two of the simpler types of door monitoring contacts, magnetic door contacts and electromechanical ball switches, are shown in *Figure 27*. These devices are typically connected to an access control system so that the door status can be monitored remotely.

Door status information is sometimes included as part of an electric locking device. For instance, some delayed egress magnetic locks provide door status to the control system for remote monitoring.

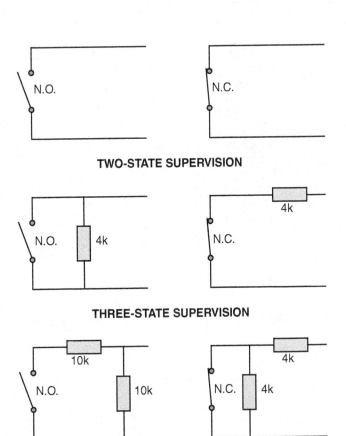

TWO-STATE SUPERVISION

THREE-STATE SUPERVISION

FOUR-STATE SUPERVISION

N.O. = NORMALLY OPEN CONTACT
N.C. = NORMALLY CLOSED CONTACT

33411-11_F29.EPS

Figure 29 Cable supervision states.

6.0.0 ENTRY CONTROL BARRIERS

Entry control barriers basically consist of various types of gates, turnstiles, doors, mantraps, and the locking devices used to secure them. As explained earlier, barriers are required to delay overt entry into a secured area. Given enough time, tools and materials, any barrier can be breached. The degree of security and the amount of delay required dictate the strength of the barriers employed. Entry control barriers must be at least as strong as the rest of the exterior or interior barrier materials.

While gates, doors, and similar barriers typically function as the access points to a secure area, the fencing itself can also be part of an access control system. For instance, fencing can be equipped with sensing devices that detect disturbances such as sounds, motions, or vibrations on or near the fence. IntelliFIBER™, a product made by Senstar Corporation®, is an example of fencing that offers perimeter protection. It uses optical fiber and sensors to detect disturbances. A controller then interprets those disturbances as either intrusion attempts or normal ambient conditions and signals the system's central alarm monitoring system (CAMS). This type of system can provide perimeter protection for miles of fencing. *Figure 30* shows chain link fencing that has a fiber optic sensing cable installed to detect disturbances.

6.1.0 Gates

Entry and exit through a fenced or, in some cases, walled area is normally associated with a gate of some type. For the purposes of this module, gates are defined as perimeter vehicular entry and exit barriers. These types of gates are powered by direct electric motor or hydraulic drives employing hydraulic cylinders, gear tracks, friction wheels, or gear chains to move the gate. The drive system is activated when a vehicle is granted authorized passage via a coded credential.

6.1.1 Swing Gates

Powered swing gates are usually installed so that for an entry road or lane the gate swings inward, away from an entering vehicle. For an exit road or lane, the gate swings outward for an exiting vehicle. The entry gate shown in *Figure 31* is 7' high. It is powered by a hydraulic unit in the swing post (*Figure 32*) that raises the gate 6" to disengage a locking pin in a hole in the pavement at the center. Once the pin is disengaged, the power unit swings the gate open 90 degrees and gradually raises it another 6" to clear any incline or curbing.

6.1.2 Sliding Gates

Figure 33 shows a conventional sliding gate used for entry control at a secure industrial facility. This particular installation uses a 6'-high chain link fence gate, topped with strands of barbed wire. A conventional electric motor drive and gear chain are used to open the gate. Note the insert-type entry card reader and the security guard phone at the left side between the two bollards. This type of barrier prevents casual penetration by law-abiding vehicles and pedestrians. However, it requires fence sensing or constant CCTV or physical monitoring by guard forces because it can be penetrated, overtly or covertly, rather easily. To help increase penetration time, these types of gates should be equipped with an electric locking device.

6.1.3 Anti-Ram Sliding Gates

Powered anti-ram sliding gates (*Figure 34*) are used in high-security installations. The gate

Figure 31 Swing gate.

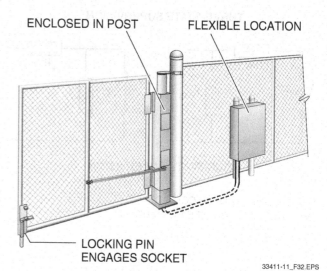

ENCLOSED IN POST   FLEXIBLE LOCATION

LOCKING PIN
ENGAGES SOCKET

33411-11_F32.EPS

Figure 32 Swing gate hydraulic unit and operator.

33411-11_F30.EPS

Figure 30 Fencing with fiber optic sensing cable.

33411-11_F33.EPS

Figure 33 Conventional sliding gate.

Bank Automatic Teller Machine Access

A common application of an electromagnetic lock is for after-hours access to a bank automatic teller machine. This is a low-security access system designed to keep individuals without a credit card or bankcard out of an area housing the ATM. It comprises an exterior swipe or insertion-type magnetic stripe reader and the electromagnetic lock and touch sense bar on the secure side of the door.

MAGNETIC STRIPE INSERT READER

EXTERIOR ENTRANCE TO ATM

ELECTROMAGNETIC LOCK

TOUCH SENSE BAR

ATM LOBBY EXIT

33411-11_SA03.EPS

shown in the figure is 12' high with an opening of 12'. Scaling the gate is difficult without using ladders and ropes. Penetrating it with a vehicle is extremely difficult. It is certified to meet a Department of State (DOS) anti-ram vehicle barrier specification rating of K12/L3. The K number represents the speed and the L number represents the maximum penetration distance for a 15,000-pound vehicle as follows:

K12 = 50 mph   L3 = 3' or less
K8 = 40 mph   L2 = 3' to 20'
K4 = 30 mph   L1 = 20' to 50'

6.1.4 Drop-Bar Gates

Figure 35 shows a conventional drop-bar gate used in a low-level security installation for separate entry and exit lanes. These gates offer no barrier to pedestrian entry. They are only for law-abiding vehicle entry and exit. For liability reasons, the bar is designed to break away without causing

injury to the driver in the event of an accidental or deliberate vehicle entry or exit. For actual security, they must be monitored continuously by sensors or CCTV, or physically by guard forces. Again, note the entry reader and guard phone on the second entry bollard on the right hand entry lane. The exit lane drop bar is activated by an in-ground vehicle sensor.

6.1.5 Anti-Ram Cable-Beam Gates

Anti-ram cable-beam gates (*Figure 36*) look very much like conventional drop-bar gates; however, the one shown is certified for a DOS rating of K4/L2 and is powered by a hydraulic lift motor. The difference between a conventional drop bar and this drop bar is that this bar is actually a cable beam made of many high-strength cable strands within a fiberglass bar or tube. The cable strands at the pivot end are indirectly secured to a large concrete anchor. The cable strands at the raised end are directly captured by a large concrete

SLIDING GATE

CRASH TEST

33411-11_F34.EPS

Figure 34 Anti-ram sliding gate.

33411-11_F35.EPS

Figure 35 Conventional drop-bar gates.

anchor when the gate is closed. While this gate presents no barrier to pedestrians, it can cut into a vehicle attempting to crash through the barrier at its rated speed. If the vehicle is a car, severe injury or decapitation of the driver or passenger may occur.

33411-11_F36.EPS

Figure 36 Anti-ram cable-beam gate.

6.1.6 Pop-Up Barriers

Anti-ram wedge barriers or retractable bollards are sometimes called pop-up barriers and, depending on their anti-ram rating, are used in medium to very high security applications. These barriers (*Figure 37*) are hydraulic powered with two seconds or less activation times. As explained earlier, these types of barriers are often used as secondary entry barriers beyond a primary gate, especially drop-bar gates. They may also be used as wrong-way barriers on a facility exit lane. In either application, depending on security risks, they may be normally raised at all times and lowered as required; or they may be normally lowered and raised upon detection of an unauthorized entry. Like drop-bar gates, they offer no barrier to pedestrian penetration and require constant sensor, CCTV, or physical monitoring; however, they are very resistant to vehicle penetrations. The bollards shown are rated K12/L2 and the wedge barrier shown is rated K12/L3. This particular wedge barrier is extremely strong and exceeds its rating by a large margin. In a documented crash-through attempt by a 15,000-pound vehicle traveling in excess of 75 mph, the vehicle failed to penetrate or damage the barrier, which continued to operate normally. The cabin and engine compartment of the crash vehicle were totally crushed.

6.2.0 Turnstiles and Rotary Security Doors

Turnstiles are usually employed as pedestrian perimeter entry and exit control barriers for a secure facility. *Figure 38* is an example of a common, full-height, open-bar, small-diameter, three-quadrant (120-degree sector) rotary turnstile used as an entry/exit barrier for the security of a facility. These types of turnstiles are usually used in perimeter fences. The three-quadrant version is comfortable

Friction Drive Slide Gates

Slide gates like those shown in this figure use a drive rail and friction drive to open and close the gate. This type of drive reduces the high wear of the metal-to-metal contact inherent with conventional chain drive or gear drive systems.

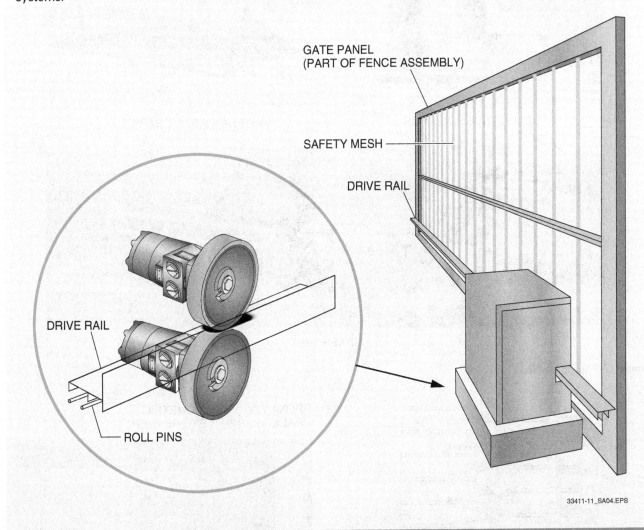

GATE PANEL (PART OF FENCE ASSEMBLY)

SAFETY MESH

DRIVE RAIL

DRIVE RAIL

ROLL PINS

33411-11_SA04.EPS

to use, but because of the size of the quadrants, it is subject to tailgating problems. A small-diameter four-quadrant (90-degree sector) turnstile virtually eliminates tailgating because of the small size of the quadrants.

A very high security mantrap version of a small-diameter four-quadrant turnstile is shown in *Figure 39*. In this particular application, which is internal to a facility, the turnstile operates in two stages. Authorization is required to pass through each stage. In this example, a proximity card is passed by the entry reader on the outside wall to unlock and rotate the turnstile 90 degrees to the left to enter an enclosed quadrant space. Once in the enclosed space, a biometric eye reader must be used to unlock and rotate the turnstile another 90 degrees to enter the restricted area. Turnstiles can be used in combinations of two or more to accommodate the maximum traffic for a facility. The units shown are physically strong enough to cause considerable delay to forced penetration. Because they can be used in the reverse direction of entry, the same turnstile can be used for authorized exit or unrestricted exit.

Another type of full-height turnstile is a locking, one-way revolving security door with an entry reader located in a pedestal to the right of the door. Depending on the authorization devices used, this is a low to high security device, and its doors are equipped with thick impact-resistant

DECORATIVE BOLLARDS

ANTI-RAM WEDGE BARRIER

CRASH TEST

33411-11_F37.EPS

Figure 37 Pop-up barriers.

33411-11_F38.EPS

Figure 38 Common full-height three-quadrant turnstile.

PROXIMITY READER   BIOMETRIC EYE READER

33411-11_F39.EPS

Figure 39 Very high security mantrap turnstile.

polycarbonate plastic lights. These types of revolving security doors are used for aesthetically pleasing entrances to residential apartment or commercial business buildings. Because of the size of the 90-degree quadrants, tailgating can be a problem with these units if an authorized person is not vigilant or unless two-stage entry and CCTV are used. Like turnstiles, rotary security doors can be used in combinations of two or more

Turnstiles with Detectors

Turnstiles are sometimes coupled with various types of detectors or scanners in conjunction with, or in place of, coded or biometric entry readers. The unit shown is coupled with a metal detector. Other detectors could include explosive detectors, biological agent detectors, chemical detectors, or full body scanners.

33411-11_SA05.EPS

to accommodate the maximum entry traffic for the building. Separate revolving doors with exit devices must be used for outgoing traffic.

6.3.0 Mantraps

In addition to the two-stage turnstiles or rotary doors already mentioned, another high-security type of mantrap is shown in *Figure 40*. In this example, it is a room or vestibule equipped with two doors and magnetic locks (B). Both doors are locked at all times. Entry and exit readers or exit devices (C) are included for both doors. A keylock switch (D) activates the controller (A). This type of mantrap can use two different types of readers for increased two-stage security during entry or exit if desired. The interior can also be equipped with metal detectors or other types of detectors and interior sensors. CCTV can be used to determine that only one person is in the room at a time. A number of configurations of this type of mantrap can be used. Four doors and two physically separated lanes can be used to allow differ-

Apartment Building Main Entrance Mantraps

In some areas of the country, double-lane mantraps have been used on the main entrances of apartment buildings in high crime areas for entrance and exit security of residents. For primary and secondary entrance, residents use entry readers and/or a PIN. Visitors are authorized for primary entrance via audio. Then, CCTV is used for secondary entrance authorization. In some cases, metal detection is also included on the entrance lane. While such mantraps can be expensive, especially with a metal detector, they have greatly reduced crime within the apartment buildings where they have been installed.

ent exit and entry security methods. For instance, the entry side could require a full two-stage entry authorization with metal detection, while the exit side could only require a momentary exit request device, such as a touch bar or pushbutton switch for both doors. In this case, the exit doors could be interlocked so that the outer door exit device is not active until the inner door has closed. This prevents unauthorized individuals from overtly penetrating the mantrap on the exit side. These exit locks could be disabled by a signal from the building's fire alarm system in the event of a fire; however, exit side locking is subject to fire and building codes and must strictly follow the codes of the AHJ.

6.4.0 Doors

Doors can be one or more of several types, including the following:

- *Industrial/commercial doors* – These doors can be selected based on the frequency of use and impact probability, as shown in *Table 1*. The door level referred to in *Table 1* is established by the Steel Door Institute (SDI) in *ANSI/SDI A250.8-2003 (R2008)*. The level and model criteria established by SDI may also used to select a door, as shown in *Table 2*. Only level 3 or 4 industrial/commercial doors should be used for security.
- *Security doors* – These doors are graded in accordance with *ANSI/NAAMM/HMMA 863-04*, as shown in *Table 3*. The term *ANSI* stands for American National Standards Institute; *NAAMM* is the acronym for National Association of Architectural Metal Manufacturers; and *HMMA* stands for Hollow Metal Manufactur-

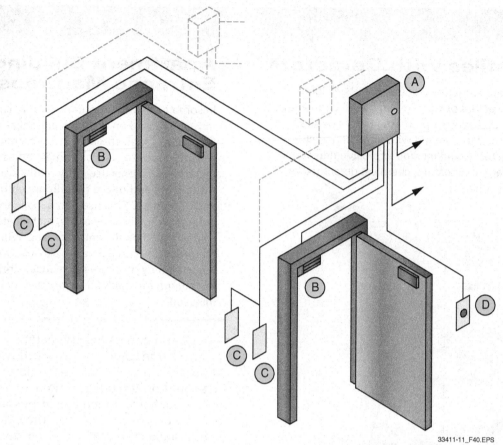

Figure 40 Single-lane mantrap.

33411-11_F40.EPS

High-Speed Security Portal

Recognizing the need for a security portal that would allow a high volume of foot traffic, Carl Kellem, an engineer and security consultant, designed and developed this free-flow, high-speed security portal that performs the function of a mantrap. In operation, its doors can be normally open, permitting an unrestricted passage through the unit until an unauthorized entry is detected. Used in conjunction with an entry reader and/or a detector at the entrance, this device can accommodate the passage of 50 or more individuals a minute. Multi-beam sensors make the unit capable of ¼" detection for tailgating. In the event of tailgating or an unauthorized entrance, an alarm is sounded, the bi-fold doors at the center of the unit quickly close, and a side port opens to allow discharge of the unauthorized individual into a holding area. All door edges are sensored to prevent them from injuring a person. Once the doors strike something, they instantly stop and back off slightly to release pressure. Algorithms used for tailgating detection can distinguish wheelchairs, canes, luggage, briefcases, and purses to reduce false alarms. The unit can also permit exiting of individuals in the reverse direction. Multiple units can be used in a straight or staggered line configuration to allow large flows of individuals into a walled-off restricted area, such as a factory, commercial office building, or airport terminal concourse.

33411-11_SA07.EPS

BUILDING TYPE AND OPENING LOCATION	FREQUENCY OF USE	IMPACT PROBABILITY	DOOR LEVEL				DOOR THICKNESS	SPECIAL CONDITIONS
OFFICES - PROFESSIONAL BUILDINGS			1	2	3	4		
Entrance	High	High				■	1-3/4"	Two way vision
Individual Office	Low	Low	■				1-3/4"	Privacy & Acoustics
Toilet Rooms	High	High		■			1-3/4"	
Stairwells	Moderate	Moderate		■	■		1-3/4"	Two way vision
Equipment Rooms	Moderate	High		■	■	■	1-3/4"	
Boiler Room	Moderate	High		■	■	■	1-3/4"	Fire Label
Storm Shelters	Low	High				■	1-3/4"	Impact Threat (Tornados)
SCHOOLS - UNIVERSITIES - LIBRARIES								
Entrance & Exits	High	High			■	■	1-3/4"	Two way vision
Classrooms	Moderate	Moderate		■			1-3/4"	Two way vision & Acoustics
Toilet Rooms	High	High		■	■	■	1-3/4"	
Gymnasiums	Moderate	High		■	■	■	1-3/4"	Impact
Cafeteria	Moderate	Moderate to High		■	■	■	1-3/4"	Two way vision
Stairwells	High	High		■	■	■	1-3/4"	Two way vision
Storm Shelters	Low	High				■	1-3/4"	Impact Threat (Tornados)
Safe Rooms	Low	High				■	2"	Aggression
HOSPITALS - NURSING HOMES - CONVALESCENT HOMES								
Entrance	High	High			■	■	1-3/4"	Two way vision
Patient Rooms	Moderate	Moderate		■			1-3/4"	Privacy & Acoustics
Stairwells	Moderate	Moderate		■	■		1-3/4"	Two way vision
Operating Rooms	High	High		■	■		1-3/4"	Sanitation
Examination Rooms	High	High			■	■	1-3/4"	Sanitation
Recreation Rooms	Moderate	Moderate		■			1-3/4"	Acoustics
Kitchen	Moderate	Moderate		■	■		1-3/4"	Two way vision
Cafeteria	High	High			■	■	1-3/4"	Two way vision
Storm Shelters	Low	High				■	1-3/4"	Impact Threat (Tornados)
APARTMENTS - DORMITORIES - CONDOMINIUMS								
Building Entrance	High	High			■	■	1-3/4"	Two way vision
Room Entrance	Moderate	Moderate	■	■			1-3/4"	Privacy & Acoustics
Bedrooms	Low	Low	■				1-3/4"	
Bathrooms	Low	Low	■				1-3/4"	
Closets	Low	Low	■				1-3/4"	
Stairwells	Moderate	High		■			1-3/4"	Two way vision
Storm Shelters	Low	High				■	1-3/4"	Impact Threat (Tornados)
HOTELS - MOTELS - TOURIST CABINS								
Building Entrance	High	Moderate	■	■			1-3/4"	Two way vision
Room Entrance	Moderate	Moderate	■				1-3/4"	Privacy, Acoustics, Insulation
Bathrooms	Low	Low	■				1-3/4"	
Closets	Low	Moderate	■				1-3/4"	
Stairwells	Moderate	High		■			1-3/4"	Two way vision
Storage & Utility	Moderate	High		■	■		1-3/4"	
Offices	Low	Low	■				1-3/4"	Privacy
COMMUNITY SHELTERS - TORNADO SHELTERS								
Entrance & Exits	High	High			■	■	1-3/4"	Impact Threat (Tornados)
Toilet Rooms	High	High		■	■	■	1-3/4"	
Storage & Utility	Moderate	High		■	■		1-3/4"	

33411-11_T01A.EPS

BUILDING TYPE AND OPENING LOCATION	FREQUENCY OF USE	IMPACT PROBABILITY	DOOR LEVEL 1	2	3	4	DOOR THICKNESS	SPECIAL CONDITIONS
SUPERMARKETS								
Entrance	High	High			■	■	1-3/4"	Two way vision
Service Doors	High	High		■	■		1-3/4"	Cart Impact, Two way Vision
Toilet Rooms	Moderate	Moderate		■			1-3/4"	
Pharmacy	Moderate	Moderate		■	■		1-3/4"	Security
Individual Offices	Low	Low	■				1-3/4"	
Equipment Rooms	Moderate	High		■	■	■	1-3/4"	
Rear Exits	Moderate	Moderate		■	■			
THEATERS - AUDITORIUMS - RECREATION BUILDINGS								
Entrance & Exits	High	High			■	■	1-3/4"	Two way vision
Theater Entrance	High	Moderate		■	■		1-3/4"	Two way vision & Acoustics
Toilet Rooms	High	High		■	■	■	1-3/4"	
Employee Areas	Moderate	Moderate		■	■		1-3/4"	
Storage	Low	Moderate to High		■	■	■	1-3/4"	
Stairwells	High	Moderate		■	■	■	1-3/4"	Two way vision
Individual Offices	Low	Low	■				1-3/4"	
Emergency Exits	Moderate	Moderate		■	■		1-3/4"	
Holding Cells	Moderate	High			■	■	2"	Two way vision
PUBLIC BUILDINGS - COURT HOUSES - UTILITY BUILDINGS								
Entrance	High	High			■	■	1-3/4"	Two way vision
Public Offices	High	High			■		1-3/4"	Two way vision
Private Offices	Moderate	Moderate		■	■		1-3/4"	Privacy & Acoustics
Stairwells	High	High		■	■		1-3/4"	Fire Label
Cafeteria	High	High			■	■	1-3/4"	Sanitation
Toilet Rooms	High	High		■			1-3/4"	Acoustics
Janitor - Storage	Low	Moderate		■	■		1-3/4"	
Holding Cells	Moderate	High			■	■	2"	Two way vision
Community Shelters	Low	High				■	1-3/4"	Impact Threat (Tornados)
STORES - RESTAURANTS - NIGHT CLUBS - BARBER								
Entrance	High	High			■	■	1-3/4"	Two way vision
Individual Office	Low	Moderate	■	■	■		1-3/4"	Privacy & Acoustics
Toilet Rooms	High	High		■			1-3/4"	
Stairwells	Moderate	Moderate		■	■		1-3/4"	Two way vision
Equipment Rooms	Moderate	High		■	■	■	1-3/4"	
Boiler Room	Moderate	High		■	■	■	1-3/4"	
Storm Shelters	Low	High				■	1-3/4"	Impact Threat (Tornados)
INDUSTRIAL PRODUCTION & PROCESSING - FACTORIES								
Entrance & Exits	High	High			■	■	1-3/4"	Two way vision
Offices	Moderate	Moderate	■				1-3/4"	Acoustics, Privacy
Office / Warehouse	High	High		■	■		1-3/4"	
Display Areas	High	Moderate		■	■	■	1-3/4"	
Service Entrance	High	High		■			1-3/4"	Insulation, Security
Chemical Storage	Low	Low					1-3/4"	

33411-11_T01B.EPS

ers Association. Security door grades are in the reverse order of the levels for commercial doors. The most secure is grade 1 and the least secure is grade 4. A grade 4 security door is equivalent to a level 4 maximum duty commercial door.

- *Bullet-resistant doors* – Bullet-resistant doors with or without lights are rated in eight levels or grades. The levels or grades are in accordance with *Underwriters Laboratory (UL) Specification 752*, as shown in *Table 4*.

Table 2 SDI 100-98 Model and Level Criteria

DOOR STYLE		DOOR FACE Min. Steel Thickness			DOOR STYLE		DOOR FACE Min. Steel Thickness		
		Gauge	Inches	mm			Gauge	Inches	mm
SDI 100-98 LEVEL 1 - LIGHT COMMERCIAL					**SDI 100-98 LEVEL 2 - HEAVY DUTY COMMERCIAL**				
Model 1	Full Flush	20	0.032	0.8	Model 1	Full Flush	18	0.042	1.0
Model 2	Seamless	20	0.032	0.8	Model 2	Seamless	18	0.042	1.0
SDI 100-98 LEVEL 3 - EXTRA HEAVY DUTY COMMERCIAL					**SDI 100-98 LEVEL 4 - MAXIMUM DUTY COMMERCIAL**				
Model 1	Full Flush	16	0.053	1.3	Model 1	Full Flush	14	0.067	1.6
Model 2	Seamless	16	0.053	1.3	Model 2	Seamless	14	0.067	1.6
Model 3	Stile & Rail	16	0.053	1.3	SDI (STEEL DOOR INSTITUTE)				

33411-11_T02.EPS

Table 3 Security Door Grades

DOOR STYLE	DOOR FACE Min. Steel Thickness		
	Gauge	Inches	mm

ANSI (AMERICAN NATIONAL STANDARDS INSTITUTE)
NAAMM (NATIONAL ASSOCIATION of ARCHITECTURAL METAL MANUFACTURERS)
HMMA (HOLLOW METAL MANUFACTURERS ASSOCIATION)

ANSI / NAAMM / HMMA 861-00 COMMERCIAL HOLLOW METAL PRODUCTS				
		18	0.042	1.0
		16	0.053	1.3
		16	0.053	1.3
		14	0.067	1.6

ANSI / NAAMM / HMMA 863-98 COMMERCIAL SECURITY PRODUCTS				
Grade 1		12	0.093	2.5	Detention (Prison) Grade Doors and Frames
Grade 2		12	0.093	2.5	Detention (Prison) Grade Doors and Frames
Grade 3		14	0.067	1.6	
Grade 4		14	0.067	1.6	

33411-11_T03.EPS

- *Blast-resistant doors* – These doors are rated for use in four classes of protection, as shown in *Table 5*. They are also rated for the following blast overpressures:
  - VLRB: Very low-range blast (blast overpressures up to 1.0 psi)
  - LRB: Low-range blast (overpressures up to 3.0 psi)
  - MRB: Medium-range blast (overpressures up to 20.0 psi)
  - HRB: High-range blast (overpressures above 20.0 psi)

Table 4 Bullet-Resistant Door Levels or Grades

UL LEVEL	UL752 WEAPONS CRITERIA	SHOTS
1	9 mm Parabellum (Luger) Pistol	3
2	0.357 Magnum Pistol	3
3	0.44 Magnum Pistol	3
4	0.30-06 Springfield Rifle	1
5	7.62 mm NATO (0.308 Winchester)	1
6	9 mm Parabellum (SMG loading)	5
7	5.56 mm U.S. Army (0.223 Remington)	5
8	7.62 mm NATO (0.308 Winchester)	5

33411-11_T04.EPS

Table 5 Blast Door Classes of Protection

Class A Protection	Provides life safety; protects personnel from fragments, falling portions of a structure or equipment; attenuates blast pressures and structural motion to a level consistent with safety requirement.
Class B Protection	Protects equipment and supplies from fragment impact, blast pressures, and structural motions; protects against uncontrolled releases of hazardous materials including toxic chemicals, radioactive materials, and similar items.
Class C Protection	Protects against communication of detonation by fragments and high blast pressures.
Class D Protection	Protects against mass detonations of explosives as a result of sympathetic detonations produced by communication of detonation between two adjoining areas.

33411-11_T05.EPS

Figure 41 shows an HRB blast-resistant door. Blast-resistant doors should not be confused with pressure-rated doors. Blast-resistant doors are designed to withstand very high, short-duration overpressures. Pressure-rated doors are designed to withstand constant pressures built up over a longer period.

Most exterior doors in commercial and industrial applications open outward due to fire regulations. Even if they are rated for security, bullet resistance, or blast resistance, their hinges and locking mechanisms are vulnerable to penetration using hand tools, cutting torches, or cutting tools. Penetrations of a door with a cutting torch or tool can occur in less than three minutes. The hinges, even if the pins are non-removable, can be defeated in about one minute using hand tools. In less than a minute, most exposed locking mechanisms can be picked by a locksmith or defeated using a large hammer-puller. Inside panic bars on an exit door can be activated through a drilled hole below the bar using a wire or bent rod. A number of methods can be used on outward-swing exterior doors to extend the penetration time if only hand tools are used. These methods are shown in *Figure 42* and are explained as follows:

- No external mechanical lock hardware should be used; only an external entry reader and electric lock should be used if needed.
- If an electric strike or bolt is used, a ⅜" thick hardened-steel plate should be welded to the

33411-11_F41.EPS

Figure 41 HRB blast-resistant door.

door so that the plate extends to cover the area where the electric bolt or strike is located.
- If a panic bar is located on the interior side of the door, a flanged ⅛" or ³⁄₁₆" thick hardened-steel plate should be mounted under the panic bar to prevent drilling and using wires or bent rods to activate the panic bar.
- Mount a ⅛" Z-strip on the interior of the door on the hinge side. This prevents removal of that side of the door from the doorframe if the hinges are defeated.
- If possible, fill the frame with concrete.

7.0.0 INSTALLATION GUIDELINES

Although entry control systems are normally installed and wired with security systems, they are sometimes added later or may be stand-alone systems. The following sections offer some general guidelines for their installation.

7.1.0 Installation Tips

The best advice for installing any system is to follow the manufacturer's installation instructions. This cannot be overemphasized. Many problems have been caused by not following the directions.

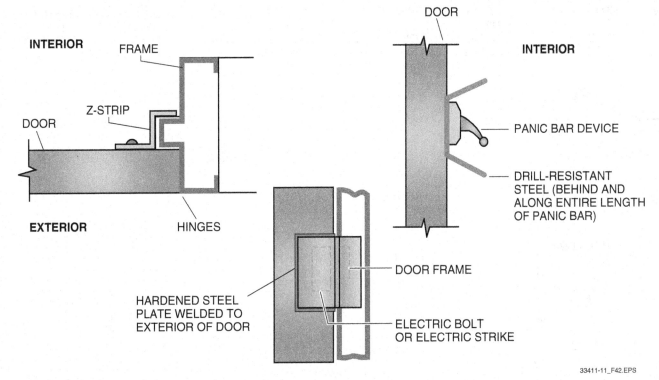

Figure 42 Door modifications to increase security.

Even though an electronic systems technician has installed the same or similar systems many times, no two installations are exactly the same, and some recent changes may have been made to the device.

Wiring should be concealed so that it is not accessible from inside or outside the protected area. This precludes tampering or attempts to defeat the system.

Another important consideration prior to installing an entry control system is to identify exactly how each door swings. The direction in which a door opens is called the hand or swing of the door. *Figure 43* illustrates one way to identify door swing.

Another method of identifying door swing is known as the hinge location method. This method, shown in *Figure 44*, involves mentally placing yourself in the floor plan of the building at the doorway in question so that your back is against the hinge locations.

7.2.0 Installation Procedures

Specific procedures for an installation start with cable selection and equipment enclosure installation. Following equipment enclosure installation, the readers, door position detectors, request-to-exit devices, and door actuating devices are installed.

The cable selection is based upon the desired use of the cable. For example, Belden 8761 shielded #22 AWG cable is acceptable for use with door switches and request-to-exit devices up to 500', but West Penn 3280 shielded #18 AWG is needed for reader connections for distances up to 500'. Belden 8723 shielded #22 AWG cable can be used up to 4,000' for RS-485 communication duties.

The controllers need to be securely mounted within the protected area. Remove the appropriate knockouts from the controller cabinets for wire routing (the AC supply and signal wires must occupy separate knockouts). If a power supply is not going to be installed, connect AC power to the terminal block, and connect the green grounding wire to the green screw inside the cabinet. Place the battery assembly (if provided) at the bottom of the cabinet, but do not connect it until the circuit boards are installed.

Following controller installation, the included or optional power supply must be installed in accordance with the manufacturer's instructions.

If communication with a command center is desired, the proper communication links must be installed.

Following the installation of the controller, the readers, keypads, electric door strikes, or similar devices can be installed. This section covers the installation for a proximity reader, a swipe reader, and an electric door strike.

LEFT-HAND SWING:
HINGES AT LEFT.
DOOR OPENS INWARD.
HANDED LOCK = LH

LEFT-HAND REVERSE:
HINGES AT LEFT.
DOOR OPENS OUTWARD.
HANDED LOCK = LHR

RIGHT-HAND SWING:
HINGES AT RIGHT.
DOOR OPENS INWARD.
HANDED LOCK = RH

RIGHT-HAND REVERSE:
HINGES AT RIGHT.
DOOR OPENS OUTWARD.
HANDED LOCK = RHR

OUTSIDE OF DOOR   OUTSIDE OF DOOR   OUTSIDE OF DOOR   OUTSIDE OF DOOR

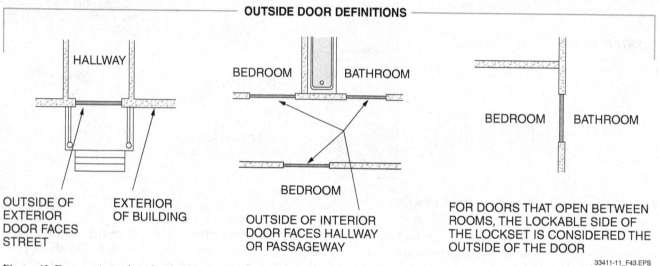

OUTSIDE DOOR DEFINITIONS

HALLWAY

OUTSIDE OF EXTERIOR DOOR FACES STREET

EXTERIOR OF BUILDING

BEDROOM   BATHROOM

BEDROOM

OUTSIDE OF INTERIOR DOOR FACES HALLWAY OR PASSAGEWAY

BEDROOM   BATHROOM

FOR DOORS THAT OPEN BETWEEN ROOMS, THE LOCKABLE SIDE OF THE LOCKSET IS CONSIDERED THE OUTSIDE OF THE DOOR

33411-11_F43.EPS

Figure 43 Door swing when facing the outside of a door.

Mounting restrictions are normally associated with proximity readers, as they can be susceptible to RFI or their signal can be attenuated by metal.

The installation of a proximity reader consists of mounting the reader, verifying any internal switch and/or jumper settings, connecting the cable to the host, and performing operational tests. The installation should be performed as follows:

Step 1 Determine the appropriate mounting position for the reader.

Step 2 Route the interface cable and/or power supply from the reader to the host.

Step 3 Install the cable fitting to the rear of the reader and connect the wiring according to the wiring list.

Step 4 Connect the tamper switch (if provided) to the host.

Step 5 Set any dipswitches and/or jumpers in accordance with the instructions.

Step 6 Attach the reader to the mounting surface.

Step 7 Test the operation of the reader.

A swipe reader can be installed indoors or outdoors, with the slot horizontal or vertical. In cold climates, an outdoor reader should always be mounted with the slot horizontal to prevent ice accumulation. The reader should be mounted 42" to 54" above the ground, using either a single-gang or double-gang handy box, a flat surface with or without an adapter plate, or a structural doorframe.

When mounting swipe readers, consider the angle of the reader card slot in relation to the natural angle of the wrist when holding the card. This usually means the card slot is horizontal with the card face down and the swipe direction toward the door; or vertical with the card face away from the door with a downward swipe direction.

The installation of a swipe reader consists of pulling the cable, mounting the gang box, connecting the cable, setting any dipswitches or jumpers, and performing operational tests. It should be performed in the following order:

Step 1 Determine the appropriate mounting location for the reader.

Step 2 Mount any adapter plate or reader base to a suitable flat surface or handy box.

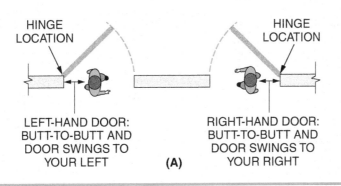

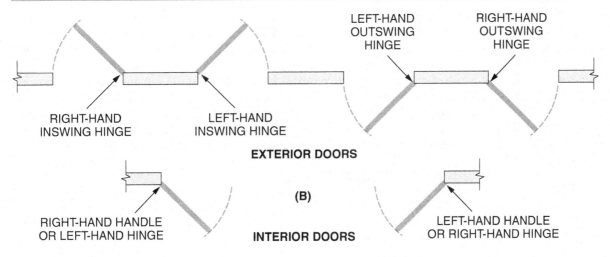

Figure 44 Hinge location method of determining door swing.

Step 3 Pull an appropriate length and gauge of multi-conductor cable between the controller and the reader.

Step 4 Connect the wires to the reader cable assembly using crimp or wire nut connections. Do not plug the cable assembly into the reader yet.

Step 5 Connect the cable to the controller using the pin-out description in the manufacturer's instructions.

Step 6 Set any dipswitches or jumpers in accordance with the manufacturer's instructions.

Step 7 Plug the cable assembly into the reader and secure the reader to the baseplate.

Step 8 Perform operational tests.

Once the readers are installed, operational tests of the entire system can be performed. Operational tests include valid and invalid card reads, tamper switch and alarm settings, and operational verification of supervised devices.

A door position sensor is a magnetic switch that senses whether a door is open or closed, and sends this condition to the master controller. The door sensing circuit can be wired in either a supervised or non-supervised condition.

The request-to-exit device provides a means of activating an electronically locked door to permit a person to exit without using an access card. On some systems where a person may turn a knob to exit, the request-to-exit device notifies the access system that a door is about to be opened for exit purposes. The device can be wired in either a supervised or unsupervised condition.

Installing an electric door strike consists of preparing the door for the strike, running the wiring to the strike location, installing the strike, connecting the wiring, and testing the installation.

The installation should be performed in the following order:

Step 1 If the electric strike is replacing an existing strike, remove the old strike and use the new strike as a template to enlarge the existing holes (if necessary).

Step 2 Run the electric wiring that connects the strike with the power supply in the access control system.

Step 3 Mount the electric strike onto the door and test the fit by opening and closing the door several times to ensure proper clearances.

Step 4 Connect the wiring between the electric strike and the control system to power up the strike.

Step 5 Perform operational tests.

SUMMARY

This module provides an overview of entry and access control systems. Entry control and access control are sometimes used interchangeably, but they have specific meanings to security professionals. Entry control includes the security barriers, readers, and controllers, while access control includes the computers, software, and credentials used to control the entry devices. Together, entry and access control systems are designed to control access to buildings and facilities and to alert security personnel when access control systems have been breached.

Readers are used to authorize access, including swipe, insert, and proximity readers; biometric readers; hand and finger geometry readers; fingerprint readers; retina and iris scanners; and facial scanners. Unauthorized access is prevented through a variety of electrical and electromagnetic locking mechanisms for doors and gates. On a larger scale, primarily for use at building and facility entrances, are various types of control barriers. These include swing, sliding, and drop bar gates; turnstiles and security doors; and mantraps.

1. The basic function of an access control system is to validate authorized entry by the use of coded credentials in conjunction with _____.
   a. software and/or managed databases containing entry parameters
   b. Wiegand principles that identify who, what, when, and where
   c. fail-safe mechanisms that must take priority over security
   d. at least one anti-passback biometric reading

2. Of the following physical protection system functions, covert penetration of an entry control system is most likely to be restricted by _____.
   a. system response
   b. tailgating
   c. controller delay
   d. intrusion detection

3. Anti-passback is typically implemented for entry control systems by using _____.
   a. special barriers
   b. tracking
   c. coded credentials
   d. IR sensors

4. High-security entry control systems typically use _____.
   a. PIN pads
   b. swiped/inserted credentials
   c. biometric readers
   d. proximity credentials

5. In a non-staffed entry control system, tailgating of individuals is most likely to be detected by _____.
   a. buried sensors near the access point
   b. multiple IR beams and pattern algorithms
   c. hub and spoke turnstiles
   d. security staff at the access point

6. How easily and quickly an overt penetration can occur is determined by the _____.
   a. strength of the entry barrier(s)
   b. amount of host computer memory
   c. number of sub-controllers
   d. scalability of the control system

7. A type of coded card that can be rather easily duplicated is a _____.
   a. proximity card
   b. Wiegand wire card
   c. smart card
   d. magnetic stripe card

8. In a hub and spoke topology, the spokes are represented by _____.
   a. host computers
   b. individual readers
   c. network cables
   d. control panels

9. To ensure continuous operation during a power outage, battery backups and uninterruptible power supplies are typically provided for _____.
   a. controllers
   b. card readers
   c. Wiegand sensors
   d. biometric relays

10. The trend in modern access control systems is to place the decision making capability at _____.
    a. a single remote host computer
    b. numerous sub-controllers
    c. each individual access point
    d. a centralized control panel

11. A type of intelligent reader that is capable of communicating directly with a host computer without the need for a separate control panel is a(n) _____.
    a. basic (non-intelligent) reader
    b. semi-independent reader
    c. IP reader
    d. sub-controller reader

12. A type of reader that generates an electrical field that excites the circuit in a card is a(n) _____.
    a. swipe reader
    b. proximity reader
    c. insert reader
    d. magnetic stripe reader

13. A fingerprint reader that generates very high frequency sound waves and then measures the waves that are reflected by a finger uses _____.
    a. ultrasonic sensors
    b. optical sensors
    c. thermal sensors
    d. capacitance sensors

14. For recognition purposes, a retinal scanner scans the _____.
    a. overall shape of the eye
    b. colored portion of the eye
    c. pupil detail of the eye
    d. pattern of blood vessels at the back of the eye

15. A person who is blind or has extremely dilated pupils normally cannot be identified using a(n) _____.
    a. capacitive scanner
    b. iris scanner
    c. retinal scanner
    d. facial recognition scanner

16. An electric strike that is fail-safe means that if power is lost, the strike _____.
    a. remains locked
    b. switches to backup power and locks
    c. opens
    d. cycles between locked and open every 15 seconds

17. A type of fail-safe door lock that has no moving parts and is available in directhold or shearhold styles is called a(n) _____.
    a. electromagnetic lock
    b. PIR exit sensor
    c. electromechanical ball switch
    d. delayed exit alert lock

18. One type of device that can be used to provide a door status indication is a(n) _____.
    a. RTE switch
    b. electric strike
    c. REX lock
    d. magnetic door contact

19. The hand of a door can be described as _____.
    a. an area of one square foot surrounding the knob
    b. the direction in which the door opens
    c. the opposite of the door's swing
    d. the thickness of the door at the hinges

20. During the installation of access control system equipment, the wiring that is used to connect components should always be _____.
    a. concealed so that it is not accessible from inside or outside the protected area
    b. larger than what the manufacturer recommends to ensure superior performance
    c. run at least twice to provide redundancy for system failures
    d. easily accessible from inside the protected area to reduce maintenance time

21. When it comes to installing access control equipment, a good rule is to _____.
    a. trust your instincts and remember the value of common sense
    b. take the advice of security experts and then increase what they recommend
    c. anticipate needs by putting yourself in an intruder's position
    d. always follow the equipment manufacturer's guidelines

22. Radio frequency interference (RFI) and signal attenuation issues often lead to mounting restrictions for _____.
    a. proximity readers
    b. door position detectors
    c. request-to-exit devices
    d. sub-controllers

23. During the installation of a card reader, operational tests should be performed _____.
    a. prior to installing any of the components
    b. after each step of the procedure
    c. once the installation is complete
    d. only if security staff deems them to be necessary

24. To prevent ice accumulation in cold climates, a swipe reader should always be mounted so that _____.

    a. heat from the network wiring can be put to use

    b. the swipe slot is horizontal

    c. a proximity reader can be used for back-up

    d. the swipe slot is vertical

25. When a card reader is being installed in an entry control system, it should be mounted so that it is above the ground or floor by _____.

    a. 36" to 46"

    b. 40" to 50"

    c. 46" to 52"

    d. 42" to 54"

Trade Terms Introduced in This Module

Anti-passback: A term applied to various methods used to deter a coded credential from being used a number of times within a defined period to admit unauthorized individuals into a secure facility or area.

Biometric readers: See *reader*.

Coded credentials: A device or card containing coded information for an individual and a facility. The coded information is read from the device or card by a reader and used to authorize passage of the individual into a secure facility or area.

Contraband: Unlawful or prohibited materials or goods.

Controller: An electronic device that interfaces with a number of readers and locking/activation equipment for barriers at entry/exit points. The device may contain a microprocessor, memory, and communications to remote computers for the purpose of authorizing the passage of individuals through the barriers at the entry/exit points.

Covertly: Done in a disguised or hidden manner.

Hammer-puller: This device, sometimes called a slide hammer, consists of a heavyweight hammer that slides on a shaft. One end of the shaft has a hardened threaded screw that is forced into the lock cylinder. The hammer is repeatedly slammed against a stop on the opposite end of the shaft and the lock cylinder is eventually broken and pulled out of the lock mechanism.

Mantrap: A room or device that provides a secondary barrier that does not open until the primary barrier is closed. In some applications, a second reader must be used to open the second barrier.

Overtly: Done in a visible, undisguised (unhidden) manner.

Piggybacking: An unauthorized entry closely following an authorized entry through an entry barrier. This can be vehicular or pedestrian traffic depending on the type of entry barrier.

Protective distribution system (PDS): A system that uses some type of protective carrier (such as metal conduit) to deter and detect attempts to access wireline or fiber optic telecommunication cabling that is used for transmitting unencrypted classified data in order to protect national security information.

Reader: A device that is used to scan a coded credential or a physical characteristic of an individual. Readers that scan physical characteristics are called biometric readers.

Tailgating: See *piggybacking*.

Additional Resources

This module presents thorough resources for task training. The following reference material is suggested for further study.

The Design and Evaluation of Physical Protection Systems, 2001. M.L. Garcia. Burlington, MA: Butterworth Heinemann.

Security, ID Systems, and Locks: The Book on Electronic Access Control, 1997. J. Konicek/K. Little. Burlington, MA: Butterworth-Heinemann.

Figure Credits

Boon Edam, Inc., Module opener, Figures 38, 39, SA05, and SA06 (left photo)

PathMinder Optical Turnstiles, Figure 2 and SA06 (right photo), SA07

Software House, Figure 3, 4, 11, 12, and 17 (magnetic stripe, smart card, and Wiegand card readers)

AMAG Technology, Inc., Figure 5

Keri Systems, Inc., Figures 6 and 8

HID Corporation, Figures 7, 9, 10, and 17 (proximity reader)

Topaz Publications, Inc., Figures 17 (magnetic stripe insert reader), 33, 35, SA01, and SA03

Recognition Systems, Inc., Figure 18

Identix Incorporated, Figure 19

Iridian Technologies, Inc., Figure 20

IR Safety and Security, Figure 21

SDC Security Door Controls, Figures 22–28 (armored cables) and 40

Securitron Magnalok Corp. An ASSA ABLOY Group Company, Figure 28 (touch sense bar and handles) and SA02

Senstar Corporation, Figure 30

HySecurity, Figure 31, 32, and SA04

B&B ARMR Corporation, Figure 34

Delta Scientific, Inc., Figures 36 and 37

Ceco Door Products/YSG Door Security Consultants, Tables 1–3

Temet, Figure 41

NCCER CURRICULA — USER UPDATE

NCCER makes every effort to keep its textbooks up-to-date and free of technical errors. We appreciate your help in this process. If you find an error, a typographical mistake, or an inaccuracy in NCCER's curricula, please fill out this form (or a photocopy), or complete the online form at **www.nccer.org/olf**. Be sure to include the exact module ID number, page number, a detailed description, and your recommended correction. Your input will be brought to the attention of the Authoring Team. Thank you for your assistance.

Instructors – If you have an idea for improving this textbook, or have found that additional materials were necessary to teach this module effectively, please let us know so that we may present your suggestions to the Authoring Team.

NCCER Product Development and Revision
13614 Progress Blvd., Alachua, FL 32615

Email: curriculum@nccer.org
Online: www.nccer.org/olf

❏ Trainee Guide   ❏ AIG   ❏ Exam   ❏ PowerPoints   Other _____

Craft / Level: _____   Copyright Date: _____

Module ID Number / Title: _____

Section Number(s): _____

Description: _____

_____

_____

_____

_____

Recommended Correction: _____

_____

_____

_____

_____

Your Name: _____

Address: _____

_____

Email: _____   Phone: _____

Glossary

24-hour zone: A zone that remains active 24 hours a day without interruption.

3D TV: A video image in which three dimensions can be perceived.

Access control: Any means of monitoring and restricting human traffic through doors, gates, and elevators.

Acknowledgement byte (ACK): A signal sent back to a transmitter indicating that parity bits match. See *non-acknowledgement signal*.

Acoustical: Audio properties of a device, space, or material.

Active infrared (photoelectric beam) sensor: A sensor that transmits a beam of infrared light to a photocell. Interruption of the beam activates the sensor.

Active sensor: A sensor that continuously emits energy, such as low-power microwave or photoelectric beams, into a protected area.

Addressable device: A fire alarm system component with discrete identification that can have its status individually identified or that is used to individually control other functions.

Advanced Television Systems Committee (ATSC): The committee that is responsible for DTV standards.

Air sampling (aspirating) detector: A detector consisting of piping or tubing distribution from the detector unit to the area or areas to be protected. An air pump draws air from the protected area back to the detector through the air sampling ports and piping or tubing. At the detector, the air is analyzed for fire products.

Alarm signal: A signal indicating an emergency requiring immediate action, such as an alarm for fire from a manual station, water flow device, or automatic fire alarm system.

Alarm verification: A feature of a fire control panel that allows for a delay in the activation of alarms upon receiving an initiating signal from one of its circuits. Alarm verification is commonly used in hotels, motels, and institutions.

Alarm verification: A feature of a security system control panel that allows for a delay in the activation of alarms after receiving an alarm signal from one of its detection circuits.

Alternate mark inversion (AMI): A method for adding timing signals to the bit streams of transmitted T-1 signals.

Americans with Disabilities Act Accessibility Guidelines (ADAAG): Building construction recommendations developed as the result of an act of Congress intended to ensure civil rights for physically challenged people.

Amplifier: A device that increases the power of an audio signal.

Amplitude modulated (AM): A process by which the amplitude (size or strength) of a carrier signal is varied in accordance with the information being sent.

Amplitude: The strength of an audio signal, measured in decibels.

Anti-passback: A term applied to various methods used to deter a coded credential from being used a number of times within a defined period to admit unauthorized individuals into a secure facility or area.

Aperture: An adjustable opening in a lens that limits the amount of light entering a camera. It operates much like the iris of the human eye. The size of an aperture is controlled by an iris adjustment on the camera lens and is measured in F-stops. Smaller F-stop numbers indicate a larger opening in the lens.

Armed: An activated alarm system.

Aspect ratio: The ratio between the horizontal and vertical axes on a display or monitor.

Attack: An attempt to burglarize or vandalize or an attempt to defeat a security system.

Audio detection system: A system that detects the sounds or vibrations caused by attempted forceful entry into a protected structure.

Audio discriminator: A sound detection and evaluation device capable of discriminating between different types of sounds, such as the difference between a passing truck and breaking glass.

Authority having jurisdiction (AHJ): The authority having jurisdiction is the organization, office, or individual responsible for approving equipment, installations, or procedures in a particular locality.

Automatic fire alarm system: A system in which all or some of the circuits are actuated by automatic devices, such as fire detectors, smoke detectors, heat detectors, and flame detectors.

Automatic gain control (AGC): A type of circuit that maintains the output signal level of an amplifier or receiver constant within a range, regardless of variations in the signal strength of the applied input signal; an electronic circuit that uses signal feedback to regulate a certain voltage level, ensuring that it falls within predetermined margins.

Automatic light control (ALC): Electronic circuitry in an automatic-iris lens that causes darker images to be displayed brighter in order to discern detail in the darker areas.

Azimuth: Angular position in a horizontal axis, usually related to true North.

Backflow preventer: A device used in piping system to prevent a fluid from flowing backward. They are required in wet sprinkler systems that are filled with antifreeze to prevent the antifreeze from contaminating the potable water system.

Backlight compensation (BLC): Provides camera circuitry to achieve the optimum balance between the iris and gain settings. By using average light metering across a scene, it can calculate the best exposure and adjust the settings as needed.

Bandpass filter: A filter which cuts off frequencies outside of an established range.

Bandwidth: The range of frequencies, expressed in MHz, over which the amplitude of a signal remains constant.

Bandwidth: The range of usable frequencies processed by an electronic circuit or piece of equipment.

Baseband: The frequency band occupied by a signal used to modulate a carrier. For example, baseband audio frequencies range from about 15 to 20kHz and baseband video frequencies from about 30kHz to 4MHz.

Basic rate interface (BRI): The first level of service in an integrated services digital network (ISDN).

Biometric readers: See reader.

Bit rate: The number of ones and zeros contained in one second.

Blanking interval: The time between the end of one horizontal scanning line and the beginning of the next. Blanking occurs when a monitor's electron beam is positioned to start a new line or a new field.

Blu-ray: An optical disk storage format that allows playback of high-definition TV.

Bollard: An upright wood, metal, or concrete post used as a vehicle barrier.

Broadband: A term used to describe radio frequency (RF) systems or equipment that process a relatively broad range of frequencies.

Broadcast television systems committee (BTSC) stereo: A television sound system that allows simultaneous transmissions of a 15Hz to 15kHz stereo audio signal along with a completely second audio program (SAP) signal, all within the 6MHz television channel bandwidth.

Built-in siren driver: A driver that uses a transistor amplifier circuit to drive a siren speaker.

Burglar alarm screen: A window screen with a fine gauge wire that is woven through the fabric and connected to a control device.

Buried line intrusion sensor: See seismic sensor.

Bus: Signal lines in an audio mixer.

Candela (cd): The International System (SI) unit of luminous intensity. It is roughly equal to 12.57 lumens.

Carrier-to-noise (C/N) ratio: Ratio of amplitude of a carrier to the noise power that is present in that portion of spectrum occupied by the carrier, expressed in decibels. The more negative the number, the better.

Ceiling surfaces: Ceiling surfaces referred to in conjunction with the locations of initiating devices are as follows:

- **Beam construction:** Ceilings having solid structural or solid nonstructural members projecting down from the ceiling surface more than 4" (100mm) and spaced more than 3' (0.9m) center to center.

- **Girders:** Girders support beams or joists and run at right angles to the beams or joists. When the tops of girders are within 4" (100mm) of the ceiling, they are a factor in determining the number of detectors and are to be considered as beams. When the top of the girder is more than 4" (100mm) from the ceiling, it is not a factor in detector location.

Central Station system: A system or group of systems in which the alarm or supervisory signaling devices are received, recorded, maintained, and supervised from an approved central station.

Certification: A systematic program using randomly selected follow-up inspections of the certified system installed under the program, which allows the listing organization to verify that a fire alarm system complies with all the requirements of the *NFPA 72®* code. A system installed under such a program is identified by the issuance of a certificate and is designated as a certificated system.

Channel bank: A termination point for a T-1 that performs the analog to digital conversion for analog communications.

Channel detection: Narrow detection in an area where an intruder is expected to cross.

Charge-coupled device (CCD): A large scale integrated device made out of silicon. A CCD contains thousands of small light-sensitive diodes arranged in a two-dimensional matrix.

Chrominance: The color information in a television image. Chrominance has two properties relating to color: hue and saturation.

Chrominance: The color information in a video signal, representing the properties of hue and saturation for color video.

Circuit response time: The minimum length of time that an open condition or short must last before the circuit responds.

Circumvent: To defeat an alarm system by avoiding its detection devices, such as by jumping over a pressure mat, entering through a hole in an unprotected wall (rather than entering through a protected door), or keeping outside the range of an ultrasonic motion detector.

Class A circuit: Class A refers to an arrangement of supervised initiating devices, signaling line circuits, or notification appliance circuits that prevents a single open or ground on the installation wiring of these circuits from causing loss of the system's intended function.

Class B circuit: Class B refers to an arrangement of initiating devices, signaling lines, or notification appliance circuits that does not prevent a single open or ground on the installation wiring of these circuits from causing loss of the system's intended function.

Clipping: An unwanted effect that occurs when the peaks of an audio signal are cut off by an amplifier or other device.

CMYK color: Cyan, magenta, yellow, and black are the four colors that make up the subtractive color space for printed materials.

Coded credentials: A device or card containing coded information for an individual and a facility. The coded information is read from the device or card by a reader and used to authorize passage of the individual into a secure facility or area.

Cold-flow: A condition in which pressure on a wire causes the insulation to flow away from the conductor, leaving it exposed.

Color space: A geometric representation of colors in space, usually of three dimensions.

Combination system: A local protective signaling system for security alarm supervisory, or guard's tour service, whose components may be used in whole or in part with a non-security signaling system, such as a fire alarm system, paging system, musical program system, or a process-monitoring service system, without degradation of or hazard to the protective signaling system.

Combined technology sensor: A sensor that uses two technologies, usually one passive and one active, to prevent false alarms and verify that an intrusion is caused by human motion or presence.

Commission Internationale de l'éclairage (CIE): A set of color matching functions and coordinate systems.

Component video: Video in the form of three separate signals, all of which are required to specify the color picture and sound. The signals are composed of luminance information, chrominance information, and sync.

Composite video: A video signal that contains all of the information needed to reproduce a color picture. Luminance and chrominance information is used to carry brightness and color. Composite video also contains horizontal, vertical, and color synchronizing information.

Compression: The forward movement of a speaker's cone.

Compressor: A device that reduces or levels the dynamic range of a signal, making loud parts softer and soft parts louder. A compressor is a limiter with a ratio of 4:1.

Contraband: A term meaning unlawful or prohibited materials or goods.

Control unit: The means through which security personnel operate a security system.

Controlled zone: Zones that turn on and off through a control point.

Controller: An electronic device that interfaces with a number of readers and locking/activation equipment for barriers at entry/exit points. The device may contain a microprocessor, memory, and communications to remote computers for the purpose of authorizing the passage of individuals through the barriers at the entry/exit points.

Convergence: The point where different technologies such as telephony, computers, security, audio, and video merge together.

Covertly: Done in a disguised or hidden manner.

Crossover: A circuit found in loudspeaker systems that directs certain frequencies to specific speakers.

Decibel (dB): A logarithmic measure of signal power. Decibels indicate the ratio of power output to power input, expressed as follows: $dB = 10 \log_{10}(P1/P2)$.

Decibel (dB): One-tenth of a bel. A decibel is a logarithmic measurement of electrical, acoustical, or power ratios.

Decibel-microvolt (dBμV): This is the abbreviation for decibels referenced to 1μV across a 75-ohm impedance.

Decibel-millivolt (dBmV): This is the abbreviation for decibels referenced to 1mV across a 75-ohm impedance.

Decibel-milliwatt (dBm): This is the abbreviation for decibels referenced to 1mW across a 50-ohm impedance.

Delay zone: A detection circuit on which a delay has been applied that does not apply to all the parts of a security system.

Depth of field: The distance in front of and beyond the object of interest that will remain in focus.

Detection device: A sensor used to initiate a signal.

Diaphragm: A thin membrane that vibrates, either in response to sound waves to produce electric signals, or the reverse, in response to electric signals to produce sound waves.

Digital alarm communicator receiver (DACR): A system component that accepts and displays signals from a digital alarm communicator transmitter (DACT) that are sent over public switched telephone networks.

Digital alarm communicator receiver (DACR): A system component that will accept and display signals from digital alarm communicator transmitters (DACTs) sent over public switched telephone networks.

Digital alarm communicator system (DACS): A system in which signals are transmitted through the public switched telephone network from a digital alarm communicator transmitter (DACT) located at the protected premises to a digital alarm communicator receiver (DACR).

Digital alarm communicator system (DACS): A system in which signals are transmitted from a digital alarm communicator transmitter (DACT) located at the protected premises through the public switched telephone network to a digital alarm communicator receiver (DACR).

Digital alarm communicator transmitter (DACT): A system component in the secured area to which initiating devices or groups of devices are connected. The DACT seizes the connected telephone line, dials a preset number to connect to a DACR, and transmits signals that indicate a status change of the initiating device.

Digital alarm communicator transmitter (DACT): A system component at the protected premises to which initiating devices or groups of devices are connected. The DACT will seize the connected telephone line, dial a preselected number to connect to a DACR, and transmit signals indicating a status change of the initiating device.

Digital communicator: A device that uses standard telephone lines or wireless telephone service to send and receive data.

Digital signal processor (DSP): A special type of microprocessor. Digital signal processors used in cameras can fine tune or adjust the aspects of an image or a video stream before delivering it to a monitor for viewing.

Digital subscriber line (DSL): A subscriber line that is able to handle both voice and data over the same line.

Digital visual interface (DVI): A digital computer video transmission standard developed by the digital display working group (DDWG) consortium. The DVI standard involves two sub-types: a digital-only version (DVI-D) and an integrated analog and digital version (DVI-I).

Distortion: Undesired changes in the waveform of a signal so that a spurious element is added. All distortion is undesirable.

Downlink: The transmission of signals sent from a communications satellite to a ground receiving station.

Driver: Another name for a loudspeaker.

Drops: In CATV/SMATV/MATV systems, the cables that connect individual ports to the feeder/trunk cable.

Dynamic host configuration protocol (DHCP): A protocol that allows centralized management and automation of the assignment of IP addresses on a network.

Electric field sensor: See *proximity sensor*.

End-of-line (EOL) device: A device used to terminate a supervised circuit. An EOL is normally a resistor or a diode placed at the end of a two-wire circuit to maintain supervision.

End-of-line (EOL): A device used to terminate a supervised circuit; normally a resistor or a diode placed at the end of a two-wire circuit to maintain supervision.

Entry delay: A delay that allows a security system user to enter the secured area and disarm the system before an alarm is triggered.

Equalization: To modify the frequency range of an audio signal.

Exit delay: A delay that allows a security system user to arm the system and exit the secured area before an alarm is triggered.

Expander: A device that enhances or extends the dynamic range. It is used to push down the noise floor, eliminating unwanted effects of hiss and other noise.

F-stop: The F-stop is a ratio of the focal length of a lens to the diameter of its iris. The F-stop determines how much light is allowed into the lens (how long the shutter is open). The lower the F-stop number, the faster the shutter. Also referred to as the F-number.

Fader: A control that allows a variable attenuation of the audio signal and, therefore, its waveform.

False alarm control team (FACT): An association of security system users, security personnel, and responding authorities, whose purpose and responsibility is to reduce false alarms.

Fault: An open, ground, or short condition on any line(s) extending from a control unit, which could prevent normal operation.

Feedback: A loud screeching noise that results when sound from an audio system is reintroduced into the system, forming a loop.

Feeders: In CATV systems, the cables that run in front of homes, from which individual homes are connected by drops. Also see *drops*.

Field of view (FOV): The width, height, or diameter of a scene to be monitored, determined by the lens focal length, the sensor size, and the lens-to-subject distance. FOV can also be considered the maximum angle of view that can be seen through a lens or optical assembly. FOV is usually described in degrees for a vertical, horizontal, or circular dimension.

Fire alarm control unit (FACU): A device with the control circuits necessary to furnish power to a fire alarm system, receive signals from alarm initiating devices (and transmit them to audible alarm indicating appliances and accessory equipment), and electrically supervise the system installation wiring and primary (main) power. The control unit can be contained in one or more cabinets in adjacent or remote locations.

Fixed focal length (FFL): A lens with a predetermined focal length. A FFL lens usually has an iris control and focus ring.

Flame detector: A device that detects the infrared, ultraviolet, or visible radiation produced by a fire. Some devices are also capable of detecting the flicker rate (frequency) of the flame.

Flicker: A distortion on a video display due to interlacing at slower refresh rates.

Focal length (FL): The distance between the optical center of a lens and the principal convergent focal point.

Footcandle: The US unit of measure for illumination from a uniform light source of 1 square foot, placed 1' away. (See *lux*.) When used in calculations, footcandle is represented as fc. A footcandle is equal to about 10 lux.

Foreign exchange service (FX): A telephone service that provides local exchange switch service to a business office that may be located hundreds or thousands of miles from the main business office.

Frame rate: The number of video frames per second. NTSC video consists of 30 frames per second. ATSC video consists of 60 frames per second but shows 30 frames twice each in succession.

Framing bit: A bit added to a T-1 bit stream that identifies channel one.

Frequency modulated (FM): A process by which the instantaneous frequency of a carrier signal is varied by an amount proportionate to the amplitude of a modulating wave.

Frequency response: The ratio between the input power level and the output (sound pressure).

Frequency: Defines how often a periodic wave occurs. Frequency is measured in hertz, or the number of times per second. In audio, frequency relates to the pitch of the sound.

Gain: An increase in power or signal level provided by a device. Each device adds its gain to the overall strength of the signal.

Gain: In CCD imaging, gain refers to the magnitude of amplification a given system will produce. Gain is reported in terms of electrons/ADU (analog-to-digital unit). A gain of 8 means that the camera system digitizes the CCD signal so that each ADU corresponds to 8 photoelectrons.

Gamma: The nonlinear relationship between the electrical voltage of the signal and the intensity of a video image.

Gate: A circuit that acts like a switch, either on or off, allowing frequencies above a certain threshold to pass.

General alarm: A term usually applied to the simultaneous operation of all the audible alarm signals on a system, to indicate the need for evacuation of a building.

Genlock: The process of locking both the synch and burst of one signal to the synch and burst of another signal making the two signals synchronous. This allows the receiver or decoder to reconstruct the picture, including luminance, chrominance, and timing synchronization pulses from the transmitted signal.

Geostationary orbit: A satellite orbit that matches Earth's rotation, causing the satellite to remain over the same point on Earth (22,300 miles above the surface).

Glare: A condition in which the central office senses an off-hook condition when the PBX is still in an on-hook state. Glare is caused by a false ground in long subscriber loops.

Glass-break detector: A device used to detect the frequencies generated by breaking glass.

Ground fault: A condition in which the resistance between a conductor and ground reaches an unacceptably low level.

Ground loop: The result of multiple ground-paths. In video it causes significant distortion of the video signal.

Haas effect: A 40-millisecond delay built into human hearing. Delayed sounds are integrated by the sound equipment if they reach the ears within 40 milliseconds of the first sound.

Hammer-puller: This device, sometimes called a slide hammer, consists of a heavyweight hammer that slides on a shaft. One end of the shaft has a hardened threaded screw that is forced into the lock cylinder. The hammer is repeatedly slammed against a stop on the opposite end of the shaft and the lock cylinder is eventually broken and pulled out of the lock mechanism.

Hardwired intrusion system: An intrusion system that uses conventional sensors and conventional notification devices connected together by wires.

Harmonic distortion: Harmonics of an input signal that are added to an output signal.

Harmonics: Similar to an overtone, harmonics are integral multiples of the frequency of a fundamental tone.

Headend: In CATV/SMATV/MATV systems, the equipment that creates and amplifies the signals to drive the distribution network. This is the starting point in all networks.

Headroom: The difference between the nominal operating level of a device and the maximum level that the device can pass without distortion.

Heat detector: A device that detects an abnormally high temperature or rate-of-temperature rise.

Heat tracing: The practice of placing thermostatically controlled electric heat tapes around or along wet sprinkler system pipes to prevent them from freezing. Heat tracing is prohibited by some codes.

Heterodyning: A process by which two frequencies are mixed together in order to produce two other frequencies that are equal to the sum and difference of the first two.

High definition multimedia interface (HDMI): A digital HDTV interface that uses a single cable with 19-pin connectors to carry both video and audio.

High definition TV (HDTV): High definition television.

High-level data link control (HDLC): An algorithm that provides a numbering sequence while data is being transmitted.

Holdup alarm (HUA): An alarm system that uses a device whereby signal transmission is initiated by the action of a bank employee or other person without alerting the intruder. Money clips and cash drawer alarms are examples of HUAs.

Home and away: Programming features that allow parts of a security system, specifically perimeter sensors, to remain active when the secured area is occupied, whereby false alarms within the perimeter are prevented.

Horizontal blanking interval: During this brief moment, the electron beam in a CRT repositions itself at the beginning of the next horizontal line.

Hot-swappable: A hard disk drive system in which additional drives can be added by simply sliding a new drive into the rack. The new drives are automatically added to the system for storage.

Hue: The color that appears on the display screen or monitor.

Hybrid fiber-coaxial (HFC): A common configuration in a CATV system in which the trunk cables and feeder cables are fiber optic and the subscriber drop cables are coaxial.

Illuminance: Relates to the reflectivity of light off an object.

Indicating device: Any audible or visible signal employed to indicate a fire, supervisory, or trouble condition. Examples of audible signal appliances are bells, horns, sirens, electronic horns, buzzers, and chimes. A visible indicator consists of an incandescent lamp, strobe lamp, mechanical target or flag, meter deflection, or the equivalent. Also called a notification device (appliance).

Infrared (IR): Radiation that lies outside the visible spectrum at its red end.

Initiating device circuit (IDC): A circuit to which automatic or manual signal initiating devices, such as fire alarm manual boxes (pull boxes), heat and smoke detectors, and water flow alarm devices are connected.

Initiating device: A manually or automatically operated device, the normal intended operation of which results in a fire alarm or supervisory signal indication from the control unit. Examples of alarm signal initiating devices are thermostats, manual boxes (stations), smoke detectors, and water flow devices. Examples of supervisory signal initiating devices are water level indicators, sprinkler system valve-position switches, pressure supervisory switches, and water temperature switches.

Insertion loss: The signal loss encountered between the input and output ports of a passive device. Insertion loss is expressed in decibels. The lower the insertion loss, the better.

Instant zone: A detection circuit that immediately triggers an alarm when activated.

Integrated services digital network (ISDN): A total digital service that can handle voice, video, and data at a customer's location.

Interior follower zone: A zone that establishes how an entry delay is applied. This zone is temporarily ignored by the alarm system during entry/exit delay periods.

Interior: The area inside the perimeter.

Interlaced video: An image is recreated on a CRT by first tracing every odd-numbered line and then tracing every even-numbered line. The electron beam in the CRT makes two complete, consecutive passes every second in order to recreate a single field of video.

Interlacing: A process in which every other line on a monitor is refreshed on every pass.

Intrusion detection system (IDS): A detection and alarm system for signaling the entry or attempted entry of an object or person into the area or volume protected by the system.

Inverse square law: A decrease of 6dB for each doubling of the distance from a sound source.

Iris: The part of a camera lens that regulates the amount of light entering the camera.

Isolation: The signal loss encountered between adjacent and non-adjacent ports of a passive device. Isolation is expressed in decibels.

Key switch: A switch operated by the use of a mechanical key.

Key telephone system (KTS): A telephone system in which multiple telephones share multiple predetermined Telco central office phone lines to make or receive calls. Selection of an unused outside line to make a call, or to answer a ringing line, is made by the individual telephone users.

Kiss-off tone: A signal from a receiving device confirming that its transmission has been received.

KVM: Keyboard, video, mouse. A KVM switch allows high-performance workstations to share a common monitor, keyboard, and mouse.

Labeled: Equipment or materials to which a label, symbol, or other identifying mark of an organization acceptable to the authority having jurisdiction (AHJ) has been attached.

Lens: The part of a camera that collects the light from a scene and forms an image of the scene on the light-sensitive camera sensor.

Light scattering: The action of light being reflected or refracted off particles of combustion, for detection in a modern day photoelectric smoke detector. This is called the Tyndall effect.

Limiter: A device that limits the dynamic range of an audio waveform.

Line carrier system: A security system that uses existing electrical wiring in the secured area to transmit messages among the system components.

Listed: Equipment or materials included in a list published by an organization acceptable to the authority having jurisdiction that is concerned with product evaluation and whose listing states either that the equipment or material meets appropriate standards or that it has been tested and found suitable for use in a specified manner.

> **NOTE**
>
> The means for identifying listed equipment may vary for each organization concerned with product evaluation (UL [Underwriters Laboratories], FM Global, IRI [Industrial Risk Insurers], and so on). Some organizations do not recognize equipment Listed unless it is also labeled. The authority having jurisdiction should use the system employed by the Listed organization to identify a Listed product.

Lobe: A rounded energy projection. The lobe of a microphone is that portion of the microphone's three-dimensional sensitivity range where sound is picked up.

Local control unit (LCU): A piece of hardware that allows the central MMS computers to communicate with remote display equipment.

Local subscriber loop: The hard-wired lines that connect the subscriber to the telephone company central office.

Local system: An intrusion system designed to alert only the occupants of a secured area or only those people within range of an alarm bell, light, or voice warning in the secured area.

Logical address: An address assigned within a network that can be changed.

Logical link control (LLC): A sub-layer in Ethernet that deals with addressing and multiplexing when multi-access media is required.

Lumen: The total amount of light, at one candela intensity, that falls on a one-square-foot section of the interior surface of a sphere.

Luminance: The brightness of an image.

Luminance: The portion of a video waveform signal required for black and white images. Color systems use luminance as well, but it is obtained as the weighted sum of the red, blue, and green (RGB) signals.

Mantrap: A room or device that provides a secondary barrier that does not open until the primary barrier is closed. In some applications, a second reader must be used to open the second barrier.

Masking: Using one sound to inhibit the hearing of another sound.

Media access control (MAC): The hardware address of a device connected to the network. The equipment manufacturer assigns the address.

Media management system (MMS): A centralized system to store, retrieve, and distribute all types of electronic media.

Microwave detector: A sensor that transmits microwave energy and analyzes the reflection of that energy.

Mid-range: An audio driver designed to reproduce the middle range of audio frequencies.

Modulation: The process by which a baseband signal is added to or encoded onto a carrier wave.

Modulator: A device in which the modulation of a signal occurs.

Monitored intrusion system: An intrusion system connected to a central station or designed to alert property owners, police, or security personnel outside the secured area.

Monitoring site: The location where a security system is monitored, also known as the central station.

Motion (space) detector: A device that responds to changes in the environment caused by the movement of human beings.

Multiplexed: The simultaneous transmission of two or more signals over a common transmission line or other medium.

Multiplexing: A signaling method using wire path, cable carrier, radio, fiber optics, or a combination of these techniques and characterized by the simultaneous or sequential transmission (or both) and reception of multiple signals in a communication channel, including a means of positively identifying each signal.

Multiplexing: The mixing of multiple signals on or in a single communications path.

National Television System Committee (NTSC): This committee established both the 525-scanning-lines-per-frame/30-frames-per-second standard and the color television system currently used in the United States. It is also the common name of the NTSC-established color television system.

Noise figure: A measure of the level of thermal noise at the output of an amplifier, receiver, or system, expressed in decibels. It is the ratio of the measured output noise to the noise generated by an ideal noise standard. It is assumed to equal the thermally generated noise appearing across an ideal resistor.

Non-acknowledgement signal (NAK): A signal sent back to the transmitter indicating that parity bits do not match. See acknowledgement byte.

Non-coded signal: A signal from any indicating appliance that is continuously energized.

Notification device: An electrically or mechanically operated visible or audible signaling device. Examples of audible signals are bells, horns, sirens, electronic horns, buzzers, and chimes. A visible signal consists of an incandescent lamp, strobe lamp, mechanical target or flag, meter deflection, graphical display, or equivalent. Also known as an annunciation or indicating device.

Null modem cable: A cable required when terminal-to-terminal or modem-to-modem connections are required. It is used because handshaking signals will be incomplete.

Obscuration: A reduction in atmospheric transparency caused by smoke, usually expressed in percent per foot.

Octave: An interval between two frequencies, where the second frequency is twice the frequency of the first.

Off-premise extension (OPX): A telephone that is installed in a location that is remote from the business where the main telephone system equipment is located. A typical application is for a worker who has his/her office at home.

Open collector output: An output directly from a transistor that has limited current output.

Orbital slot: The arc of the orbit allocated to a single geostationary satellite. They are identified by their longitude.

Overtly: Done in a visible, undisguised (not hidden) manner.

Pair gain system: A technology that allows a large number of individual subscriber lines to be installed in a building without investing in new cables.

Panning: Adjusting sound between two speakers, as in a stereo mix.

Parity bit: Used to tell the number of odd or even ones in a data byte. It is used for error detection. If parity bits do not match, it indicates that information has been corrupted and cannot be used.

Passive sensor: A sensor that detects natural radiation or radiation disturbances but does not emit the radiation on which its operation depends.

Path (pathway): Any conductor, optical fiber, radio carrier, or other means for transmitting fire alarm system information between two or more locations.

Perimeter follower zone: A non-entry/exit zone, typically a perimeter zone located on an entry/exit path, which is treated as an entry/exit zone during an entry/exit time.

Perimeter: The outer limits of the secured area.

Period: For sine waves, the distance between two amplitude peaks.

Periodic: A repeating sine wave.

Phantom power: Using direct current electricity to power a condenser microphone over the same cable that connects the microphone to the equipment.

Phono-cartridge: A low-impedance transducer that converts the motion of a stylus on a phonograph record to voltage.

Photoelectric beam detector: A type of sensor in which an infrared light is transmitted to a photocell. A break in the beam activates the sensor.

Photoelectric detector: An active infrared sensor that transmits infrared light.

Photoelectric smoke detector: A detector employing the photoelectric principle of operation, using either the obscuration effect or the light-scattering effect for detecting smoke in its chamber.

Piezoelectric sensor: A device that uses piezoelectric crystals to generate electrical signals when the sensor is activated.

Piggybacking: An unauthorized entry closely following an authorized entry through an entry barrier. This can be vehicular or pedestrian traffic depending on the type of entry barrier.

Pink noise: A complete, harmonic waveform resulting from random noise. Pink noise has a constant amount of energy in all octaves.

Plain old telephone service (POTS): A basic telephone service involving analog telephones, telephone lines, and access to the public switched network.

Positive alarm sequence: An alarm verification feature that provides up to a three-minute delay that allows supervising or monitoring personnel time to investigate the alarm.

Preaction sprinkler system: A type of dry sprinkler that fills the pipes with water before the sprinkler head opens. It allows water to spray on a fire faster because the air in the pipes does not have to be purged after the sprinkler head opens.

Pressure mat: A thin rubber mat containing metal contact strips. When an intruder walks on the mat, the contacts close, which activates an alarm.

Primary rate interface (PRI): The next higher level of service in an integrated services digital network (ISDN).

Private branch exchange (PBX): Privately owned switching equipment that allows communication within a business and between businesses. It is connected to a common group of lines from one or more Telco central offices.

Private line automatic ringdown (PLAR): A system of two directly connected phones. If either phone is picked up, the other phone rings. Phones used in this system do not require a rotary dial or numbered keypad.

Progressive video: Results from the raster of a CRT, which traces each line of a video image, one at a time, from the top to bottom. Sometimes called non-interlaced video.

Projected beam smoke detector: A type of photoelectric light-obscuration smoke detector in which the beam spans the protected area.

Protected distribution system (PDS): Wiring and/or installation method that prevents or detects physical access to alarm system communication lines.

Protected premises: See *secured area*.

Protective distribution system (PDS): A system that uses some type of protective carrier (such as metal conduit) to deter and detect attempts to access wireline or fiber optic telecommunication cabling that is used for transmitting unencrypted classified data in order to protect national security information.

Proximity sensor: A device that detects a change in the capacitance of a metal object caused by the presence of an intruder.

Public switched telephone network (PSTN): The various telephone companies that are connected together to form an international telephone service and switching system.

Pulse code modulation (PCM): A modulation scheme used to convert analog voice signals to digital voice signals.

Pulse dialing: The type of dialing used with rotary dial telephones. The rotary dial breaks and reconnects a voltage on the ring line based on the number dialed. For example, dialing five breaks and reconnects the voltage five times.

Rarefaction: The backwards movement of a speaker's cone.

Rate compensation detector: A device that responds when the temperature of the air surrounding the device reaches a predetermined level, regardless of the rate-of-temperature rise.

Rate-of-rise detector: A device that responds when the temperature rises at a rate exceeding a predetermined value.

Reader: A device that is used to scan a coded credential or a physical characteristic of an individual. Readers that scan physical characteristics are called biometric readers.

Relational database: A software application designed to manage large amounts of information stored in separate tables.

Remote supervising station fire alarm system: A system installed in accordance with the applicable code to transmit alarm, supervisory, and trouble signals from one or more protected premises to a remote location where appropriate action is taken.

Repeater: Equipment that relays signals between supervising stations, subsidiary stations, and secured areas.

Resolution: The sharpness or crispness of a picture. For a display device, it is measured based on three factors: the number of scan lines used to create each frame of video, the number of picture elements (pixels) contained in each line, and the number of bits per color.

Resonance: The intensity and continued expression of a sound vibration.

Resonant dipole antenna: The most elementary type of antenna, consisting of two $\frac{1}{4}$ wavelength elements that are insulated from each other. Each element is connected by a transmission line to a receiver. It is resonant when receiving a signal that has a wavelength that precisely matches that of the antenna.

Reverb: Reflections of a sound wave that reach the ear with different delays. If the delay is too long, the sound is heard as an echo.

RGB sync-on-green: A form of component video where the three colors of the red, blue, green color space are carried on separate wires; however, synchronization signals are combined with the green color signals and carried on the green signal wire.

RGB: A color space consisting of the three primary colors, red, green, and blue. Each color is carried on a separate wire in a three-wire RGB cable.

RGBHV: A form of component video where each color of the red, blue, green color space is carried on a separate wire, with synchronization signals for vertical and horizontal control also transmitted on individual wires.

Ring generator: A pulsating +90VDC signal that causes a called party's telephone to ring.

Ring: In an analog subscriber line, it is the wire that connects to the ring of the telephone patch cord.

RJ-31X: An interface that connects telephone line signaling devices to the telephone line.

Roll-off: The portion of an audio waveform where high and low frequencies are attenuated.

S-video: A form of component video called separated video, commonly referred to as Y/C video. With S-video, the color information (C) for a video signal is separated from the brightness information (Y), and each is transmitted on a separate wire.

Saturation: The amount of gray in a color (as opposed to hue).

Secured area: The physical location protected by a security system.

Security system: A combination of components designed to detect and report possible threats to property or to human life.

Seismic sensor: A sensor that responds to vibrations transmitted through the ground.

Shock (vibration) detector: A device designed to detect the shock of attack before intrusion.

Signal-to-noise (S/N) ratio: The ratio of signal power to noise power in a specified bandwidth. It usually is expressed in decibels.

Signal-to-noise ratio (SNR): A measurement of the amount of noise present after the value of a signal has been removed. It is expressed as ratio and measured in decibels. In audio, it relates to the strength of the signal over the amount of noise present.

Signaling line circuits (SLCs): A circuit or path between any combination of circuit interfaces, control units, or transmitters, over which multiple system input signals or output signals (or both input signals and output signals) are carried.

Silent alarm: An alarm without an obvious local indication that the alarm has been sounded.

Simple mail transport protocol (SMTP): The protocol used to send email on the Internet.

Simple network management protocol (SNMP): A standard for gathering statistical data about network traffic and the behavior of network components.

Smoke detector: A device that detects visible or invisible particles of combustion.

Snake: A multi-cable microphone connector.

Sound damping: Removing energy from an audio signal.

Sound pressure level (SPL): A measurement of the amount of barometric change in the atmosphere due to a sound wave. SPL is expressed in decibels.

Spot detection: Detection focused on a particular object or on high-value areas such as safes, vaults, and storage areas.

Spot-type detector: A device in which the detecting element is concentrated at a particular location. Typical examples are bimetallic detectors, fusible alloy detectors, certain pneumatic rate-of-rise detectors, certain smoke detectors, and thermoelectric detectors.

Standard definition television (SDTV): Part of the DTV standards, SDTV is used to display video on a standard definition television (NTSC). It is also used to package up to five different NTSC television channels for transmission on a single 6MHz channel band.

Static address: An address assigned to a specific port or device within a network.

Stratification: The phenomenon in which the upward movement of smoke and gases ceases due to a loss of buoyancy.

Stress sensor: A device that reacts to the flexing of the material to which it is mounted.

Structural-attack piezoelectric sensor: A device that uses a piezoelectric crystal and is designed to detect the vibrations associated with forced entry into a secure area.

Supervising station: A facility that receives signals and at which personnel are in attendance at all times to respond to these signals.

Supervisory signal: A signal indicating the need for action in connection with the supervision of guard tours, the fire suppression systems or equipment, or the maintenance features of related systems.

Surface acoustic wave (SAW): The abbreviation for a type of solid-state device that processes signals via a technology that excites and detects minute acoustic waves that travel over the surface of the device substrate, much like earthquake waves travel over the crust of the earth. These devices are typically used for broad bandwidth signal delay, custom-designed filters, and complex signal generation and correlation applications.

Switched telephone network: An assembly of communications facilities and central office equipment, operated jointly by authorized service providers, that provides the general public with the ability to establish transmission channels via discrete dialing.

Synchronization (sync): Video signals are complex signals that are shared among different devices. Synchronization signals are timing signals that allow you to match the chrominance and luminance information into a coherent image on the screen.

System unit: The active subassemblies at the central station used for signal receiving, processing, display, or recording of status change signals. The failure of one of these subassemblies causes the loss of a number of alarm signals by that unit.

Systems controller: A configurable and programmable computer that acts as the hub of an integrated system.

Tailgating: See *piggybacking*.

Telco: A shortened name for a telephone company.

Telnet: A protocol for remote computing on a LAN as well as on the Internet.

Thermal lag: The time required to add heat to or remove heat from a substance before it reaches a set temperature.

Tie line: A line that connects two switches that have no time and charges applied.

Time division multiplexing (TDM): A multiplexing scheme in which time slots are assigned to different T-1 channels.

Tip: In an analog subscriber line, it is the wire that connects to the tip of the telephone patch cord.

Toll center: A telephone switch in which time and charges are assigned to a call.

Touch-Tone®: A dialing method that creates a series of tones when numbered buttons are touched on a keypad on the phone. Each number or key function has its own distinct tone.

Transducer: A device that converts energy from one form to another.

Transition minimized differential signaling (TMDS): The signaling method or protocol used for the transmission of DVI video signals.

Transmitter: A system component that provides an interface between signaling line circuits, initiating device circuits, or control units, and the transmission channel.

Transponder: A microwave repeater. In a satellite, it amplifies and downconverts received uplink signals, and then retransmits the signals as downlink signals back to Earth.

Trap detection: Narrow or wide detection for high-traffic areas or for the path an intruder is expected to take to get from one point to another within a secured area.

Trouble signal: A signal initiated by a security system or device that is indicative of a fault in a monitored circuit or component.

Trouble signal: A signal initiated by the fire alarm system or device that indicates a fault in a monitored circuit or component.

Trunk: A telephone line between the local exchange and the toll center.

Trunks: In CATV systems, the cables that carry the TV signals from the headend to major portions of the service area, where they connect to feeders. Also see *drops*.

Tweeter: An audio driver designed to reproduce the highest range of audio frequencies.

Uplink: The transmission of signals sent from a ground transmitting station to a communications satellite.

Vertical blanking interval: With a CRT, when the end of the first of two video fields is reached, the electron beam in the CRT is repositioned to the top of the display before tracing the next field. The amount of time needed to do this is called the vertical blanking interval.

Visible notification appliance: A notification appliance that alerts by the sense of sight.

Voice over Internet Protocol (VoIP): A technology that allows phone conversations to take place over the Internet.

Volumetric detection: Wide detection in a defined area.

Walk test light: A light that indicates a sensor is functioning in response to human motion.

Waveform: The component frequencies of a complex sound wave.

White noise: Random noise with equal amplitude at each frequency.

Wi-fi protected access (WPA): A newer encryption code for wireless Internet access that is more secure than older encryption codes.

Wired equivalent privacy (WEP): An early wireless Internet encryption code. It is not as secure as newer encryption codes.

Woofer: An audio driver designed to reproduce the lowest range of audio frequencies.

WORM: Write once, read many times. This method of data storage allows for permanent storage without data deterioration.

Y/C video: A term used to describe the separation of an NTSC video signal into its luminance (Y) and chrominance or color (C) signals. Y/C video is used in S-video component video distribution.

Zone: A defined area within the protected premises. A zone can define an area from which a signal can be received, an area to which a signal can be sent, or an area in which a form of control can be executed.

Index

R

Raceways, (Module 6): 28, (Module 7): 33, 36, (Module 8): 40–41, 42

Rack mounting, (Module 3): 15, 22, 23, (Module 10): 34–35

Radiators. *See* Speakers

Radio
  AM, (Module 3): 1, 17, 18, 52, (Module 5): 16, 17
  fire alarm system, (Module 8): 37, 46
  FM, (Module 3): 1, 52, (Module 5): 16, 17
  frequency allocation, (Module 3): 7
  Internet stations, (Module 4): 15
  long-range, (Module 7): 23
  media management system, (Module 4): 18

Radioactive material, (Module 8): 16, 19, (Module 11): 34

Radio frequency (RF), wireless applications
  access control card, (Module 11): 8, 9
  CCTV system, (Module 10): 20–21, 27
  control code file, (Module 6): 23
  media management system, (Module 4): 18
  network subsystem interface, (Module 6): 15
  paging system, (Module 9): 1, 9
  touch panel, (Module 6): 26

Radio frequency identification (RFID), (Module 9): 8

Radio system, digital alarm (DARS), (Module 8): 37

RADSL. *See* Rate adaptive DSL

RAID. *See* Redundant array of inexpensive disks

Randomizing, (Module 2): 10

Rarefaction, (Module 1): 1, 6, 14, 49

Rate adaptive DSL (RADSL), (Module 5): 20

Rauland Telecenter®, (Module 6): 50

RCA, (Module 1): 29, 32, (Module 2): 24, (Module 4): 19, (Module 6): 50, 51, (Module 10): 24

Readers and scanners, entry/exit
  air sample, for particulates, (Module 8): 18, 20
  definition, (Module 11): 42
  fingerprint, (Module 10): 9, (Module 11): 2, 15–16
  full body, (Module 11): 29
  hand/finger geometry, (Module 11): 2, 14, 15
  insert, (Module 11): 12, 15
  location in access control system, (Module 11): 3, 11, 13
  overview, (Module 11): 2
  proximity, (Module 11): 12, 14, 15, 28, 36
  retinal or iris, (Module 10): 9, (Module 11): 2, 16
  swipe, (Module 11): 12, 15, 36–37
  types, (Module 11): 11–12, 14–16
  voiceprint, (Module 10): 9
  Wiegand wire card, (Module 11): 8, 15

RealNetworks, (Module 4): 15

RealOne™, (Module 4): 15

Real-time analyzer (RTA), (Module 1): 31, 33

Receivers
  bounce-back active infrared sensor, (Module 7): 15
  CCTV controller, (Module 10): 18, 19, 21, 28
  digital, (Module 3): 8
  digital alarm communicator (DACR), (Module 7): 5, 24, (Module 8): 37, 39, 91, 102
  intrusion detection system, (Module 7): 8–9, 15, 24, 28, 42
  outdoor microwave sensor, (Module 7): 8–9, 42, 43
  portable color TV, (Module 3): 42
  radio smoke detection system, (Module 8): 46
  satellite, (Module 3): 17, 25
  staff locator, in nurse call system, (Module 9): 11
  symbols, (Module 3): 58
  troubleshooting, (Module 3): 46, 47

Recommended standards, OSI reference model. *See* RS *entries*

Record, database, (Module 4): 3

Rectifier, (Module 1): 19

Redundancy, (Module 6): 28–29, (Module 8): 78

Redundant array of inexpensive disks (RAID), (Module 4): 13–14, (Module 10): 7–8, 43

Reference frame, (Module 2): 8, (Module 10): 8

Reflectivity, (Module 10): 32

Regulations
  Americans with Disabilities Act, (Module 6): 46, (Module 8): 31, 36, 48, 64–65
  audio prohibited in CCTV, (Module 10): 30
  credentials of federal employees, (Module 11): 5
  Digital Television Transition and Public Safety Act, (Module 3): 1
  Disclosure of Non-public Personnel Information, (Module 4): 5
  Health Insurance Portability and Accountability Act, (Module 1): 25

Relay module, (Module 8): 78, (Module 11): 2, 3

Remote control
  building automation system, (Module 6): 30, 32
  duct smoke indicator, (Module 8): 69, 70
  fire alarm system, (Module 8): 29, 39, 42, 43, 82, 103
  infrared, for local control unit, (Module 4): 6
  integrated school system, (Module 6): 53
  intrusion detection system, (Module 7): 35, 36
  projector, (Module 2): 19, (Module 4): 10
  television, (Module 3): 3

REN. *See* Ringer equivalence number

Renkus Heinz, (Module 6): 51

Repeater, (Module 5): 8, (Module 7): 23, 59, (Module 8): 37

Request-to-exit device (REX; RTE), (Module 11): 1, 19–20, 21, 22, 37

Residences
  apartment buildings, (Module 3): 5, (Module 8): 31, 36, 78, (Module 11): 28–29, 31. *See also* Buildings, high-rise
  condominiums, (Module 8): 31, (Module 11): 31
  distributed audio system, (Module 1): 25–26
  dormitories, (Module 8): 78, (Module 11): 31
  fire alarm system, (Module 6): 29, (Module 8): 2, 31, 75–80
  household (single-family), (Module 8): 75–80
  intercom system, (Module 1): 22
  intrusion alarm system, (Module 7): 31
  network applications, (Module 6): 29–33
  security system, (Module 6): 29
  smart home system, (Module 6): 30–33

Resistance, electrical
  and audio systems, (Module 1): 19, 20, 26
  and conductor size, (Module 1): 27
  and heat detectors, (Module 8): 15
  loop resistance, (Module 7): 47, (Module 8): 81
  telephone system lines, (Module 5): 4, 21
  and wire length, (Module 1): 20. *See also* Attenuation

Resolution
  CCD camera, (Module 10): 11
  definition, (Module 2): 29
  overview, (Module 10): 13
  projection system, (Module 4): 9–10
  TV broadcast system, (Module 10): 13
  video signal, (Module 2): 7, 9, 13, 16–19, 29, (Module 10): 8

Response time
  access control system, (Module 11): 2
  circuit, (Module 7): 30, 57
  sprinkler system, (Module 8): 26, 27
  video display, (Module 2): 15

Restaurant, (Module 6): 1, (Module 8): 54, (Module 11): 32

Y

Z